Book companion website:

cme.eecs.umich.edu

Calculus for the Modern Engineer was piloted in Fall 2024 with this cover design by T. Renay Hutchings, Student Administrative Assistant Sr., Michigan Robotics.

Calculus for the Modern Engineer

Math and Programming at the Scale of Life

Jessy Grizzle

Robotics Department
The University of Michigan

Published in the United States of America by Michigan Publishing.
Manufactured in the United States of America.

ISBN 978-1-60785-917-8 (print)
ISBN 978-1-60785-918-5 (OA)
DOI: doi.org/10.3998/mpub.13076384

The Free ECE Textbook Initiative is sponsored by the Robotics Department, the ECE Department, and the College of Engineering of the University of Michigan.

MICHIGAN ENGINEERING
ROBOTICS
UNIVERSITY OF MICHIGAN

ECE ELECTRICAL & COMPUTER ENGINEERING
UNIVERSITY OF MICHIGAN

MICHIGAN ENGINEERING
UNIVERSITY OF MICHIGAN

Work together. Create smart machines. Serve society.

Robotics Department, University of Michigan

Preface

In the ever-evolving landscape of engineering education, calculus stands as a foundational pillar, yet its teaching methods have remained largely unchanged for decades. This textbook, a trailblazing effort in calculus education, is born out of a deep-seated need for reform. Since my tenure at the University of Michigan began in 1987, I have witnessed firsthand the challenges and limitations of the traditional calculus sequence in mathematics education, particularly within our College of Engineering (CoE).

Traditional calculus courses struggle to incorporate relevant, real-world examples. My colleagues in mathematics have been given an impossible job: present realistic examples in domains far from their fields of expertise. Students see right through this. The disconnect between mathematical theory and its real-world uses hinders the learning process and fails to fully engage students in the practical applications of calculus. A second core issue lies not in the examples used but in the selection of material and the overemphasis on lengthy, manual calculations that lack meaningful context.

This book represents a radical departure from conventional approaches, inspired by my experience developing and teaching the first-semester course ROB 101 *Computational Linear Algebra.* It aims to breathe life into mathematics through programming and realistic examples. In collaboration with my colleague Prof. Chad Jenkins, we reimagined the calculus curriculum, placing integration at the beginning of the learning process. This approach, starting with sums and progressing to integration, is not only more intuitive but also lends itself well to programming applications, such as calculating the change in position of mobile robots from velocity curves.

Following integration, the book delves into differentiation (both single-variable and multivariable) and its myriad applications, culminating in a comprehensive chapter that instructors can tailor to their needs. Traditional methods of computing integrals, such as antiderivatives and substitution methods, are then introduced and framed within the context of both manual calculations and modern symbolic tools.

The journey continues through improper integrals, ordinary differential equations (ODEs), Laplace Transforms, and Feedback Control, encompassing a breadth of topics rarely seen in a single calculus course. This ambitious curriculum is not just theoretical; it has been put to the test in a pilot program, reminiscent of the successful pilot of the year-one first-semester course, ROB 101, which included LU Factorization in its fifth week.

Prerequisites and Future Pathways

The prerequisites for this course are Linear Algebra and familiarity with the Julia programming language. Students who are proficient in either MATLAB or Python can pick up Julia quickly; the Lab Manual for Julia Instruction: Computational Linear Algebra by Prof. Jessy Grizzle is one potential resource.

A second consideration is which courses this new course can replace. At the University of Michigan at Ann Arbor's Robotics Department, students must earn 12 credits in mathematics, with *Calculus for the Modern Engineer* counting toward four of those credits. If, before taking *Calculus for the Modern Engineer*, students already have credit for one or more of Calc I (Differential Calculus of a Single Variable), Calc II (Integral Calculus of a Single Variable), or Calc IV (Ordinary Differential Equations and Laplace Transforms), those credits still count[1]. To enroll in any of those courses after completing *Calculus for the Modern Engineer*,

[1]For students who have already completed Calc I and Calc II, one might be concerned about redundancy. This has not been our experience. In particular, these students remain highly engaged through the programming assignments and major projects; see Teaching Reviews.

students must obtain advisor permission.

Students requiring additional courses to reach 12 total math credits are guided to select from a carefully curated list maintained by the advising office, including courses such as *Multivariable and Vector Calculus*, *Linear Algebra (with proofs)*, *Applied Modern Algebra*, *Principles of Analysis*, *Numerical Methods*, *Abstract Algebra*, *Theory of Algorithms*, and *Advanced Calculus I*, a proof-based treatment of calculus grounded in Real Analysis over $\mathbb{R}^n$.

In Robotics, mathematical knowledge is power—especially when paired with the ability to quickly translate it into working code. This curriculum complements traditional math courses, offering both depth and breadth to our students' mathematical toolkits.

Why Julia?

This textbook proudly uses Julia for its computational elements, embracing its open-source nature and cost-free availability as fundamental to our commitment to accessibility. Unlike the commonly used MATLAB, which comes with substantial licensing fees, Julia ensures that all students, regardless of their university's resources, can freely access state-of-the-art computational tools. This choice reflects our commitment to leveling the educational playing field, allowing every student to explore and excel in engineering without financial barriers.

A New Paradigm in Calculus Education

This textbook is not just a guide to calculus theory; it is a manual for applying calculus in the real world of engineering. It balances the need for theoretical understanding, computational skills, and practical application. While some manual calculations are included to bridge the gap with traditional calculus education, the emphasis is on programming-based solutions to complex, real-life engineering problems. This approach not only enhances understanding but also provides a clear, verifiable, and efficient method for problem-solving.

In summary, this textbook is more than just a collection of mathematical principles; it is a blueprint for a new way of teaching and learning calculus. It is designed not just to educate, but to inspire, opening doors to a world where calculus is not merely understood but applied with confidence and creativity.

Homework Problems and Supplemental Material

The textbook contains a large number of fully worked examples designed to illustrate both fundamental concepts and their application to engineering problems. These examples are intended not only to support learning but also to serve as a foundation for the creation of homework assignments.

A collection of sample homework problems—both written and programming-based—is available at ROB 201 Calculus for the Modern Engineer, Free Online Course, along with lecture video recordings and lecture notes. These materials were developed through the pilot offerings of the course and provide instructors with a starting point for building assignments aligned with the textbook's philosophy.

Modern large language models (LLMs) provide flexible tools for generating new homework problems. Several effective approaches include:

1. Providing your favorite LLM with existing homework problems and requesting variations of greater, equal, or lesser difficulty, with or without solutions.

2. Providing your favorite LLM with worked examples from the textbook and requesting related homework problems, again with control over difficulty and the inclusion of solutions.

3. Uploading to your favorite LLM a chapter of the textbook (e.g., as a PDF) and specifying the type, scope, and difficulty of desired homework problems, with or without solutions.

These approaches enable instructors to rapidly create customized assignments that match their course objectives while maintaining consistency with the material and style of the textbook.

Rigor

While the textbook covers extensive material, it does not do so at the expense of mathematical rigor. More than 90% of the fundamental results of Calculus are proved in the textbook. When a proof is short and edifying, it is fully integrated into the relevant chapter; when it would cause a significant break in the flow, it is relegated to an optional reading section at the end of the chapter.

The proofs are included to satisfy the curiosity of mathematically advanced undergraduate students and to provide clear explanations for engineering instructors who may not have a background in Real Analysis, where the fundamental principles of Calculus are typically developed. Each instructor must balance the blend of fundamentals and applications to meet the needs of their students.

Acknowledgment

This textbook is first and foremost dedicated to the students who were brave enough to enroll in the pilot offering of ROB 201, *Calculus for the Modern Engineer: Math at the Scale of Life*, in Fall 2024 and Winter-2025. Their willingness to engage with a new and untested approach to teaching calculus within a robotics curriculum made this work possible and shaped its early development in fundamental ways.

I also wish to express my deep gratitude to the undergraduate Instructional Assistants who worked with me from January 2024 through April 2026 to improve, refine, and extend the contents of this textbook. These students contributed directly to the creation of homework sets, projects, instructional materials, and successive drafts of the manuscript. Their careful feedback, sustained effort, and intellectual engagement were instrumental in shaping this work. The Instructional Assistants are: Kaylee Johnson (lead), David Anapolsky, Advaith (Adi) Balaji, Madeline (Maddy) Bezzina, Justin Boverhof, Ashrith Edukulla, Yulei Fu, Annie Li, Elaina Mann, Reina Mezher, Sydney Ragla, Sajjy Shah, Alvin Tuo, Kane Weng, Maxwell West, Meiyi Yang, and Elvis Zhang. I also gratefully acknowledge the Graduate Student Instructor, Maxwell Gonzalez, whose contributions supported the continued development and delivery of the course.

ROB 201 was conceived by Prof. Chad Jenkins and the author as part of a complete undergraduate curriculum in robotics. Chad and I also thank the former Associate Dean for Undergraduate Education, Prof. Joanna Mirecki Millunchick, for her encouragement and support in rethinking how mathematics is taught to engineers.

ROB 201 was piloted in both semesters of the 2024–25 Academic Year thanks to the strong support of Department Chair Dawn Tilbury. I thank Dawn for her unequivocal encouragement. She has consistently highlighted ROB 201, *Calculus for the Modern Engineer: Math at the Scale of Life*, as a shining example of the curricular innovation taking place in the Robotics Department at the University of Michigan College of Engineering.

Finally, I thank the publishing team of Richard Carnes and Prof. Fawwaz Ulaby for their excellent guidance, editorial expertise, and sustained support throughout the preparation of this textbook.

Jessy Grizzle
Ann Arbor, Michigan, USA
April 2026

Contents

Chapter 1

Precalculus: Notation, Functions, and Various Algebraic Facts

Learning Objectives

By the end of this chapter, the student should be able to:

- Recognize calculus as the science of approximations.
- Interpret mathematical notation and appreciate its efficacy in conveying complex mathematical ideas succinctly.
- Revisit and refresh knowledge on key mathematical concepts that are crucial for understanding calculus.
- Develop an understanding of the importance of algorithms in mathematical problem-solving and analysis.
- Cultivate a taste for careful and precise mathematical reasoning.
- Explore Euler's number, e, and learn how it arose from a simple everyday question.

Outcomes

Upon successful completion of this chapter, students will be able to:

- Recognize the utility and importance of mathematical notation in the precise expression of mathematical concepts.
- Observe the Approximation Principle at work through the study of numbers like π, $\sqrt{2}$, and e.
- Understand and apply the Bisection Algorithm as an example of the Approximation Principle in numerical methods, and apply rules for manipulating inequalities.
- Reaffirm understanding of fundamental concepts such as functions, domains, ranges, and inverses.
- Conduct a thorough examination of roots and powers and their properties.
- Review and consolidate knowledge of the key characteristics of exponential and logarithmic functions.
- Utilize Euler's Formula to simplify complex trigonometric expressions effectively.
- Revisit (or learn) the Binomial Theorem and its applications in algebraic expansions.
- Learn how to effectively apply shifting and scaling operations to functions for various analytical purposes.

1.1 Introduction

This chapter "reviews" a selection of topics from Algebra I & II, and a tiny bit of Trigonometry, as typically taught in American high schools. "Review" is in quotes because, for Algebra I & II, we make an attempt to state the results in more precise terms than a typical student would have seen them in high school. For Trigonometry, we provide an idea of how it relates to building mathematical models of robots, another topic not typically covered in high school.

1.1.1 Julia and LLM/GenAI Resources

Throughout the textbook, including this chapter, we will take advantage of *Computational Linear Algebra*, and especially the use of Julia. Associated resources include:

- **Computational Linear Algebra:**
  - Computational Linear Algebra Textbook: what today's undergraduate engineer needs to know about linear algebra to excel in courses, internships, and jobs.
  - Computational Linear Algebra Lab Manual: teaches Julia from scratch in the context of linear algebra.
  - Textbook companion website.
- **Julia Sites**
  - Julia Programming Language.
  - JuliaPackages.com (paraphrased) "Hello Julians, Big news from JuliaPackages.com. We've given the site a major refresh ... your perfect package is just a click away."
  - JuliaHub: Ask AI. Access a Julia-speaking LLM.
- Access point to ChatGPT and more. How does this help with calculus?
  - By pasting in a Julia code stub/block and error message, we can ask ChatGPT to fix it. Once ChatGPT has the code, we can give it feedback via any new error messages.
  - If we provide ChaptGPT a draft of our function and the names of key variables, we can ask it to complete the code for us. The results vary with our level of specificity. **You must ALWAYS have simple examples ready on which to test the code because ChatGPT can/does make mistakes.** With practice, you will become efficient at using ChatGPT as your coding buddy.
  - If you give ChatGPT explicit enough instructions, you can get a good start on a draft piece of code. "Hello, ChatGPT! I'm interested in implementing the Gram-Schmidt algorithm, but with a twist. I'd like the algorithm to handle linearly dependent vectors as well. Additionally, I'd like the output to include both normalized and unnormalized orthogonal vectors. Could you provide a detailed explanation and a Julia implementation of this enhanced Gram-Schmidt process?"
  - **You can also ask for background information on any topic, for example:** "Hello, ChatGPT! I'd like to explore the concepts of domains, codomains, and ranges of functions in more detail. Can you provide a comprehensive explanation of these concepts, highlighting their differences and significance? Additionally, I'd appreciate practical examples in the Julia programming language that demonstrate how to determine and work with these aspects of functions. This will help enhance my understanding of these foundational algebraic concepts" When this answer is too superficial for your needs, it is then up to you to request more information, such as "I'd like to delve deeper into the concepts of domains, codomains, and ranges of functions. Specifically, I'm interested in understanding the clear distinction between the range and the codomain of a function. Could you provide a detailed explanation, perhaps using a function where the range and codomain are different? Additionally, I'd appreciate a demonstration in Julia that visually represents this difference." From here on, each of you will have a unique experience with ChatGPT.

1.1.2 Got Calculus Dread?

In case you are dreading calculus because "you feel that you are bad at math" ...

- We recommend reading this string of reddit posts. They emphasize that math is like any other skill: it takes practice, and putting in the time to practice is easier in the context of examples that you find meaningful. In this book, we have taken a novel approach to revealing the secrets of calculus, emphasizing its real-world applications in engineering, particularly in the realm of robotics. While traditional calculus courses often delve deep into theory before allowing students to tackle "illustrative" problems, we believe in bringing mathematical concepts to life through computation and compelling case studies. This approach is evident in our use of small mobile robots as examples. However, it is worth noting that while this book offers a fresh perspective on calculus, it does not necessarily cover every traditional topic in depth, and some chapters may not be as rich in engineering examples as others. Nevertheless, we hope that the large number of solved problems for each topic will serve as a valuable review and resource for the student. It is essential to understand that our goal is to equip the student with practical skills for the modern era of Information, AI, Data Science, and Robotics. Yet, we also emphasize the importance of mastering traditional calculus techniques, as they remain a foundational requirement in many engineering courses.

- In The Hard Truth About Intelligence and Learning, The Math Sorcerer discusses intelligence, learning, not being smart enough, how talent may have gotten you through high school, but in college, work habits win the game. The comments on this video are more informative than most. Here is one example of many: *@samuelakwensivie1734. As a person who aced almost everything in high school, the university really taught me this lesson. I had never failed in my studies as I did in the University. It was a really humbling experience. It has made me appreciate the process of failure more and has helped me improve my work ethic in my quest to become a robotics engineer.* The Math Sorcerer also provides a list of resources.

- And then there is: Is Success Luck or Hard Work? by Derek Alexander Muller of Veritasium. Stick through the first minute of the video, and you'll be hooked until the end. His point: "In a competitive world, tiny advantages can make all the difference." Your author's hope: this course gives you a competitive advantage in achieving career success.

Learners often seek additional practice problems. Here are some sources:

- Essential Calculus Skills Practice Workbook by Chris McMullen. It includes solutions.

- Calculus Workbook. Lots of problems, mostly on the easier side. Solutions are not included, but your favorite LLM/ChatBot can solve them all.

- Exercises and Problems in calculus. Includes answers to odd-numbered exercises.

- Upload a portion of the textbook to your favorite LLM and ask it to generate exercises with solutions. Currently, if you try to upload the entire PDF of the textbook, you will overwhelm the system.

1.2 The Approximation Principle: The Essence of Calculus through the Lens of Irrational Numbers

In the following, we illustrate how to compute some basic numbers that we use "all the time", yet the numbers cannot be written as the ratio of two integers. In each case, we will provide a means for determining lower and upper bounds to an approximation of the number in the form

$$\boxed{x^{\rm low} \le x \le x^{\rm up},} \tag{1.1}$$

where x is the quantity we seek to estimate or approximate, $x^{\rm low}$ is a lower bound, and $x^{\rm up}$ is an upper bound. If we can make the lower and upper bounds "converge" to one another, i.e., become arbitrarily close to one another, we will then have an effective means of computing the quantity x with arbitrary accuracy. **This is the essence of calculus**.

When we write $x^{\text{low}} \leq x \leq x^{\text{up}}$, we are not giving an explicit approximation for x. This can be remedied by defining

$$\begin{aligned} x^{\text{est}} &:= \frac{x^{\text{up}} + x^{\text{low}}}{2}, \quad \text{and} \\ x^{\text{error}} &:= \frac{x^{\text{up}} - x^{\text{low}}}{2}, \end{aligned} \tag{1.2}$$

and we can then write

$$\boxed{x = x^{\text{est}} \pm x^{\text{error}},} \tag{1.3}$$

which we read as x is equal to x^{est} plus or minus x^{error}. The plus-minus term provides our range of "confidence" in the approximation of x. We are familiar with the $\pm$ sign from the quadratic formula. Here, it means something a bit different: it means the value of x can be anything within the **range given by the** $\pm$ **sign**. For example, $x = 1 \pm 0.2$ means that x can be anything from $1 - 0.2 = 0.8$ to $1 + 0.2 = 1.2$, i.e, any value in the closed interval $[0.8, 1.2]$.

1.2.1 Approximating π as Archimedes Did Around 200 BCE

We recall that π was **defined** by the ancient Greeks to be the ratio of the circumference of a circle to its diameter D,

$$\boxed{\pi := C/D.} \tag{1.4}$$

While this is wonderful as a definition, it does not specify the value of π. **Figure 1.1** illustrates how Archimedes approximated it. Measuring the diameter of a circle is straightforward. It is the circumference that is hard to compute. The ancients understood right triangles. Archimedes had the brilliant idea to inscribe[1] a set of equally sized isosceles triangles in a circle so that he could add up their outer edges as a **lower bound** on the circumference of the circle. He then circumscribed a set of equally sized isosceles triangles and added up their outer edges as an **upper bound** on the circumference. He understood that as he used more and more triangles, he would obtain better and better approximations of the circumference. Moreover, he could then normalize by the diameter of the circle to obtain lower and upper bounds on an approximation of π. The method is purely geometric, and around 200 BCE, **Fig. 1-1** constituted proof of its well-foundedness.

Using trigonometry and the half-angle formulas for sine and tangent, Archimedes could compute everything he needed! Indeed, given a circle with radius $\frac{1}{2}$ and diameter 1, and using the trigonometric values for $\frac{\pi}{6}$, the side length of the inscribed polygon (inner approximation) and the side length of the circumscribed polygon (outer approximation) could be calculated as follows:

- For $\frac{\theta}{2} = \frac{\pi}{6} = 30$ (degrees), the exact values of sine and tangent are: $\sin\left(\frac{\pi}{6}\right) = \frac{1}{2}$ and $\tan\left(\frac{\pi}{6}\right) = \frac{\sqrt{3}}{3}$, and hence
  - **Inner Approximation:** $d_i = r \cdot \sin\left(\frac{\theta}{2}\right) = \frac{1}{2} \cdot \frac{1}{2} \implies C_i = 12 \cdot \frac{1}{4} = 3.$
  - **Outer Approximation:** $d_0 = r \cdot \tan\left(\frac{\theta}{2}\right) = \frac{1}{2} \cdot \frac{\sqrt{3}}{3} \implies C_o = 12 \cdot \frac{\sqrt{3}}{6} = 2\sqrt{3} \approx 3.464.$

  Hence:
  $$\pi^{\text{est}} \approx \frac{3.464 + 3}{2} = 3.232 \pm 0.232.$$

- For $\frac{\theta}{2} = \frac{\pi}{12} = 15$ (degrees), $\frac{\pi}{12}$ can be represented as $\frac{\pi}{3} - \frac{\pi}{4}$. Archimedes could use the sum and difference formulas for sine and tangent to compute exact values, namely, $\sin\left(\frac{\pi}{12}\right) = \frac{\sqrt{6}-\sqrt{2}}{4}$ and $\tan\left(\frac{\pi}{12}\right) = 2 - \sqrt{3}$. Consequently,
  - **Inner Approximation:** $d_i = r \cdot \sin\left(\frac{\theta}{2}\right) = \frac{1}{2} \cdot \left(\frac{\sqrt{6}-\sqrt{2}}{4}\right) \implies C_i = 24 \cdot \frac{\sqrt{6}-\sqrt{2}}{8} \approx 3.106.$
  - **Outer Approximation:** $d_0 = r \cdot \tan\left(\frac{\theta}{2}\right) = \frac{1}{2} \cdot (2 - \sqrt{3}) \implies C_o = 24 \cdot \left(1 - \frac{\sqrt{3}}{2}\right) \approx 3.215.$

[1] Inscribe means to place inside. Circumscribe means to place outside.

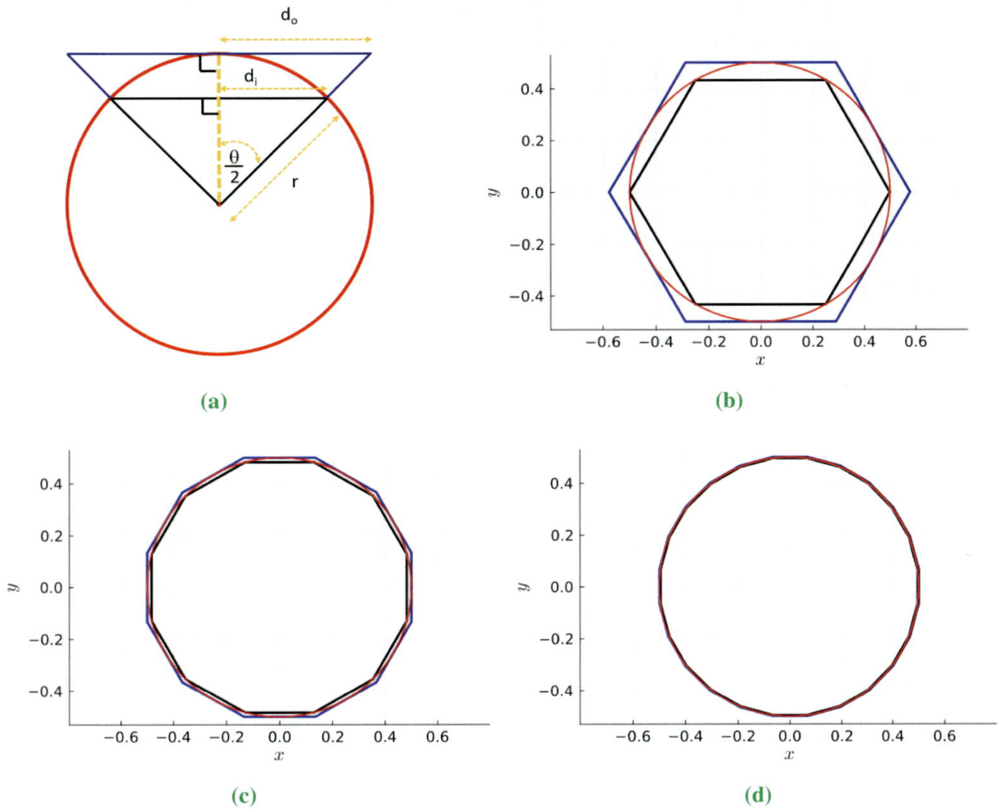

Figure 1.1: Archimedes' approach to estimating π by approximating a circle with a polygon. (a) The radius of the circle is r, $d_i = r\sin(\theta/2)$ is half the length of the inner triangle's base, and $d_o = r \cdot \tan(\theta/2)$ is half the length of the outer triangle's base. The total length of the sides of the inner polygon with n sides is then $2nd_i$, which provides a **lower bound** on the circle's circumference, while the total length of the sides of the outer polygon is $2nd_o$ and provides an **upper bound** on the circumference. As the number of sides of the polygon increases from six in (b), to 12 in (c), and 24 in (d), the lower and upper bounds become closer to one another and to the true circumference of the circle. **Archimedes understood that as the number of sides tends to infinity (aka an arbitrarily large number), the upper and lower bounds converge to one another, providing a means to compute the true circumference of the circle with arbitrary accuracy.** Not bad for 2,200 years ago!

Hence:

$$\pi^{\text{est}} \approx \frac{3.215 + 3.106}{2} = 3.161 \pm 0.0545.$$

Below is a Julia implementation of Archimedes' method, with illustrative results presented in Table 1.1. You are encouraged to play with the code. Using 1,000 triangles, we have estimated $\pi = 3.1415874858795636 \pm 7.75158829635636e\text{-}6$. That's not too bad! To be clear, this algorithm "converges" very slowly and is not used to compute π to billions of decimal places. To do that requires algorithms based on calculus! Nevertheless, Archimedes' Algorithm is a valid means to compute lower and upper bounds for π. A crucial point is that we know the accuracy of the approximation. For us engineers, that can be important.

```
function archimedes_pi(n, radius=0.5)
    # Circle with radius one half has a diameter of one
    diameter = 2*radius
    n = floor(Int,n) # force n to be an integer
    if n < 3
        n = 3
    end
    theta = 360/n
```

```
    #
    circumInnerApprox = radius*2*n*sind(theta/2)
    circumOuterApprox = radius*2*n*tand(theta/2)
    #
    piLowerBound = circumInnerApprox/diameter
    piUpperBound = circumOuterApprox/diameter
    println("$piLowerBound < pi < $piUpperBound for n = $n")
    piApprox = (piUpperBound + piLowerBound)/2
    piErrorBound = (piUpperBound - piLowerBound)/2
    return (piEst=piApprox, piErr=piErrorBound, piL=piLowerBound, piU=piUpperBound)
end

# Estimate pi using polygons with n sides
F = archimedes_pi(1000)
```

Output

```
3.141587485879563 <  pi  <  3.141602989056156 for n = 1000

(piEst = 3.1415952374678593, piErr = 7.75158829635636e-6,
piL = 3.141587485879563, piU = 3.141602989056156)
```

n	π^{low}	π^{est}	π^{up}	$\pm\pi^{\text{error}}$
6	3.0000	3.2320	3.4641	± 0.2320
24	3.1326	3.1461	3.1597	± 0.0135
128	3.1416	3.1416	3.1416	± 1.3e-5
512	3.1416	3.1416	3.1416	± 8.2e-7

Table 1.1: Archimedes estimated π around 200 BCE. By the time one reaches 24 sides on the polygon, the estimate is approximately 22/7. At 128 sides, the upper and lower bounds agree to four decimal places.

Here is a video on Archimedes' method entitled How Archimedes Trapped Pi by Dubious Insights. It is amazingly hard to prove that π is irrational. An animated proof can be found on Mathologer.

(Optional Read):

A Few Historical Remarks. The ancient Indian mathematician Madhava of Sangamagrama (c. 1340 – c. 1425) is credited with discovering a series approximation for π, much before the invention of calculus in Europe. His pioneering work laid the foundations for later developments in mathematical analysis. Here's a breakdown of his contributions for further exploration:

1. **Pi Approximation**: Madhava is known for deriving a series expansion for calculating the value of π, which is now known as the Madhava-Leibniz series. This series is given by:

$$\frac{\pi}{4} = 1 - \frac{1}{3} + \frac{1}{5} - \frac{1}{7} + \frac{1}{9} - \cdots$$

2. **Infinite Series**: Madhava developed various infinite series expressions for trigonometric functions like sine, cosine, and arctangent. **His work showcased a deep understanding of the concept of infinity**. He was the first to use infinite series approximations for a range of trigonometric functions, a precursor to the development of calculus.

3. **Legacy**: Madhava's discoveries laid the foundation for later mathematicians both within the Kerala School and eventually in Europe, where his series for π was rediscovered by Leibniz. The Madhava-Leibniz series is named in honor of both Madhava's and Leibniz's contributions to the development of calculus.

This last point, "rediscovery", is a constant battle in science, even today with Google Scholar at our fingertips: results that are already known are "rediscovered" and published as new. Leibniz (1646–1716) did his work

300 years after Madhava; he did not have Google Scholar, so we give him a pass! During the so-called "Early Modern Period" (15th–17th centuries), Europe and India had extensive trade relations. It seems unclear how much scientific exchange took place. Imagine where we might be today if revolutionary discoveries like Madhava's had made it to Europe during his lifetime.

1.2.2 Approximating $\sqrt{2}$ as the Greeks Only Dreamed of Doing

The ancient Greeks puzzled over the length of the hypotenuse of a right triangle when the other two sides had length one. We know from the Pythagorean Theorem that $a^2 + b^2 = c^2$ and hence $c = \sqrt{a^2 + b^2}$; well, Pythagoras was a Greek philosopher/mathematician, so the Greeks eventually figured out that the hypotenuse squared equaled two, but they hotly debated whether or not $\sqrt{2}$ could be expressed as the ratio of two counting numbers, in other words, whether $\sqrt{2}$ was rational or not. Eventually, Euclid, another Greek philosopher/-mathematician, formally proved, by a method we now call "Proof by Contradiction", that $\sqrt{2}$ is **irrational**, meaning, it is not a **rational** number. The proof is given in Prop. 1.72.

The proof aside, how does one compute $\sqrt{2}$? Since the square of $\sqrt{2}$ is 2, it follows that $\sqrt{2}$ is the (positive) solution to:

$$\boxed{x^2 - 2 = 0.} \tag{1.5}$$

In *Computational Linear Algebra*, we use two methods for computing the roots of an equation, $f(x) = 0$. The simplest one is the Bisection Algorithm, where we start with two values a and b that "bracket" the root (i.e., a is a lower bound for the root and b is an upper bound). We then compute the midpoint, $c := \frac{a+b}{2}$, and use it to update a or b depending on the sign of $f(a) \cdot f(c)$. The details are given in the Julia code below. A key point is that at **each step of the algorithm**, the bracketing values a and b provide **lower and upper bounds** on the root, respectively, so we actually know how accurately the root has been computed. In other words, we have an effective means for computing an approximation of $\sqrt{2}$ along with a bound on the accuracy of the estimate! (And yes, the Greeks had an algorithm for square roots, but they did not have Julia and the Bisection Algorithm.)

```julia
function bisection(f, a, b, tol)
    if f(a) * f(b) > 0
        error("f(a) and f(b) must have opposite signs")
        return NaN
    end
    kmax = 1e5 # max number of loops so that the algorithm terminates
    k=1
    c=NaN #define c out of the loop so we can access it
    while ( (b - a) / 2 > tol || abs(f(c))> tol ) && k < kmax
        c = (a + b) / 2
        #
        if f(a) * f(c) < 0
            b = c
        else
            a = c
        end
        k = k+1
    end
    root_ErrorBound = (b - a)/2
    return (est=(a + b) / 2, low=a, up=b, error=root_ErrorBound)
end

# Define the function whose root we want to find
f(x) = x^2 - 2

# Call the bisection method on this function
F = bisection(f, 1.0, 2.0, 1e-10)
```

```
@show F

println(" ")
println("The root is approximately ", F.est)
println("The square of the root is ", F.est^2)
```

Output

```
F = (est = 1.414213562355144, low = 1.414213562355144,
up = 1.4142135623842478, error = 1.4551915228366852e-11)

The root is approximately 1.414213562355144
The square of the root is 1.9999999999492266
```

From the above, we see that $\sqrt{2} = 1.414213562355144 \pm 1.4551915228366852e\text{-}11$. Moreover, in principle, we can determine its value to as many (finite) digits as we want.

n	$\sqrt{2}^{\text{low}}$	$\sqrt{2}^{\text{est}}$	$\sqrt{2}^{\text{up}}$	$\pm\sqrt{2}^{\text{error}}$
5	1.406250	1.421875	1.437500	± 0.01562500
10	1.414551	1.414063	1.415039	± 0.00048828
50	1.414214	1.414214	1.414214	± 4.4409e-16

Table 1.2: The ancient Greeks discovered the existence of irrational numbers, such as $\sqrt{2}$. Here, n is the number of iterations used in the Bisection Algorithm. To generate this table, you can change the "while loop" of the function algorithm into a "for loop" so that the number of iterations can be fixed.

Remark 1.1. *The ancient Greek mathematician Eudoxus of Cnidus (408-355 BCE) is sometimes credited with the first use of the Bisection Algorithm,[2] as part of his "method of exhaustion for calculating areas and volumes." However, its systematic development and formalization in the context of algorithmic and numerical methods took place much later with the advent of modern computing. It's important to note that while the method itself is ancient, the terminology "bisection method" or "algorithm" is much more recent, linked to the development of computer science and numerical methods in the 20th century. Because the Bisection Algorithm halves the distance between the bracketing values at each iteration, the lower and upper bounds will converge to one another, thereby providing an arbitrarily good approximation to* $\sqrt{2}$.

1.2.3 Defining and Approximating Euler's Constant, e

Wikipedia recounts how in 1683, Jacob Bernoulli discovered the constant $e \approx 2.718281828\ldots$ by asking a question regarding compound interest in a bank account. It is fun to see how math is discovered! Yes, facts in math are often discoveries. Bernoulli imagined he had a bank account that started with \$1.00 and paid 100 percent interest per year. If the interest were credited once at year's end, the account's value would be \$2.00.

Bernoulli asked: What if the interest were computed and credited more frequently during the year?

If the interest is credited twice a year, in the second half of the year, you will earn interest on the interest paid in the first half of the year in addition to the interest on your initial deposit! How does that work? Well, the interest rate for every 6 months will be 50%, so the initial \$1 is multiplied by 1.5 at the end of six months, and that amount is multiplied by 1.5 again at the end of the year. Hence, your \$1 gets multiplied by 1.5 twice, yielding $\$1.00 \times (1+\frac{1}{2})^2 = \$2.25$ at the end of the year. Compounding quarterly yields $\$1.00 \times (1+\frac{1}{4})^4 = \$2.44140625$, and compounding monthly yields $\$1.00 \times (1+\frac{1}{12})^{12} = \$2.613035....$ If there are n compounding intervals, the interest for each interval will be $100/n$ % and the value at the end of the year will be $\$1.00 \times (1+\frac{1}{n})^n$.

[2] Knorr, W. R. (1986), *The Ancient Tradition of Geometric Problems,* New York: Dover. This book includes a comprehensive review of the geometric techniques used by ancient Greek mathematicians, including Eudoxus.

Bernoulli surmised that the sequence of numbers, $(1+\frac{1}{n})^n$, approaches a limiting value with each larger value for n and, thus, more frequent, smaller compounding intervals. Compounding weekly ($n=52$) yields \$2.692596..., while compounding daily ($n=365$) yields \$2.714567... (approximately two cents more than when compounding weekly). The limit as n grows arbitrarily large is the number that came to be known as e. With continuous compounding, the account value would reach \$2.718281828... after one year.

$$\boxed{e :\approx \left(1+\frac{1}{n}\right)^n \quad \text{for } n \text{ very large,}} \tag{1.6}$$

with the approximation becoming better as n grows larger.

Remark 1.2. *You may wonder why the constant is denoted "e" and not "b" since Bernoulli discovered it. Well, life is not fair, as you well know. Euler*[3] *later found much more efficient algorithms for computing "Bernoulli's constant", and he denoted it by "e". Euler's notation stuck, and that was the end of the story.*

The proof that the upper and lower bounds converge to one another is more difficult in this case than in the previous two. Archimedes' method is "geometrically intuitive", the Bisection Algorithm is "mathematically intuitive", while "continuous compounding" only becomes "intuitively obvious" when you code it up, because ***"Coding is Believing!"*** *(Prof. Chad Jenkins, circa summer 2020.*

```
function bernoulli_e(n)
    n = floor(Int,n) # force n to be an integer
    if n < 2
        n = 2
    end
    e_LowerApprox = (1.0 + 1.0/n)^n
    e_UpperApprox = 1.0
    for k = 1:n
        e_UpperApprox = e_UpperApprox + 1.0/factorial(big(k))
    end
    e_UpperApprox = e_UpperApprox + 1.0/(n*factorial(big(n)))
    e_Approx = (e_UpperApprox + e_LowerApprox)/2.0
    e_ErrorBound = (e_UpperApprox - e_LowerApprox)/2.0
    e_Approx = Float64(e_Approx)
    e_ErrorBound =  Float64(e_ErrorBound)
    @show e_UpperApprox
    @show e_LowerApprox
    return (est=e_Approx, low=e_LowerApprox, up=e_UpperApprox, error=e_ErrorBound)
end

F = bernoulli_e(52*7*24) #Compounding hourly!
```

Output

```
e_UpperApprox = 2.718281828459045235360287471352662497757 2
e_LowerApprox = 2.7181262654676885

(est = 2.718204046963367, low = 2.7181262654676885,
up = 2.71828182845904523536, error = 7.778149567834747e-5)
```

Hence, $e \approx 2.7182 \pm 7.77815e\text{-}5$. More information on the "many faces" of e is given by Michael Penn on his YouTube channel. A proof that e is irrational is given by Numberphile, in song and verse by TheGermanFox, and in what BriTheMathGuy calls the "The Most Beautiful Proof". Intrigued enough to take a look?

[3]"Euler" rhymes with "spoiler".

n	e^{low}	e^{est}	e^{up}	$\pm e^{\text{error}}$
2	2.250000	2.500000	2.750000	± 0.250000
4	2.441406	2.580078	2.718750	± 0.138672
12	2.613035	2.665659	2.718283	± 0.052623
52	2.658970	2.688626	2.718282	± 0.029656
365	2.714567	2.716425	2.718282	±0.001857
10^5	2.718126	2.718204	2.718282	±7.7782e-5

Table 1.3: Bernoulli discovered the constant used in the natural logarithm, but Euler got to name it "e".

1.2.4 Future Icons in Mathematics

Because algebra was invented more than 2000 years ago and calculus was invented in the mid-17th century (aka mid 1600's), you may find it hard to relate to its creators. **In case you would like to learn about people who have done amazing mathematical work in this century, you may enjoy investigating a few of the Fields Medal[4] winners from 2022:**

- Hugo Duminil-Copin Lifted from the Comments: **"[making mistakes is just an important component of the creative process]; I wish this poster was up there in every school!"**
- Maryna Viazovska Lifted from the Comments: **"People like you fill my heart with hope for tomorrow. Good going, Professor. Hope the scars of war will be over soon and you will be able to return to Kyiv."**
- June Huh Lifted from the Comments:**" [Your] story is truly an inspiration to a struggling undergrad like myself, congratulations Dr. June Huh!**
- James Maynard Lifted from the Comments: **"Thank you so much for this video! It's really helpful for me as a secondary school teacher trying to get teenagers to see maths as an attractive career choice."**
- MigOroEdu provides a continuously updated source of videos featuring inspiring talent, many of whom did not follow a traditional path, some of whom were marked early on as "unlikely to succeed".

1.3 Important Notation

Important Notation:

Definition 1.3. *Symbols for Defining Other Symbols and Frequently Used Sets of Numbers:*

(a) $:=$ *means that whatever is on the left side of "$:=$" is,* ***by definition****, equal to what is on the right. You may be more familiar with "$\triangleq$", a triangle over the equals sign. As an example, the discriminant of the quadratic equation $ax^2 + bx + c = 0$ is defined to be $\Delta := b^2 - 4ac$, meaning that Δ is by definition equal to $b^2 - 4ac$ and the colon being on the left side is telling you that Δ is a new symbol assuming the value $b^2 - 4ac$ and that the symbols a, b, and c, which are on the side of the equals sign without the colon, are already known to you (in other words, they have been defined previously in the text, or are known to everyone who arrives at this level of mathematical training).*

(b) $=:$ *means that whatever is on the right side of "$=:$" is,* ***by definition****, equal to what is on the left. You cannot do this with the more familiar "$\triangleq$". We will not need this flexibility very often, but just in case!*

[4]The Fields Medal is a prestigious award given to two to four mathematicians under 40 years old at the International Congress of the International Mathematical Union (IMU), held every four years. Named in honor of the Canadian mathematician John Charles Fields, it is often likened to the Nobel Prize of Mathematics but has significant differences, such as age limits, frequency of award, and criteria.

(c) $\in$ *means "**element of**" as in "$x \in A$" stands for the entire phrase "the unknown x is an element of the set A."*

(d) $\notin$ *means "**not an element of**" as in "$x \notin A$" stands for the entire phrase "the unknown x is not an element of the set A."*

(e) $\mathbb{N} := \{1, 2, 3, \cdots\}$ *the set of **natural numbers or counting numbers**. Here, you are assumed to know the numbers 1, 2, 3, and so on from your previous mathematical training. What is being defined here is the notation $\mathbb{N}$ for the set of all counting numbers or the set of all natural numbers. Sometimes 0, the number zero, is included in the natural numbers, and sometimes not! Here, we are NOT including zero.*

(f) $\mathbb{Z} := \{\cdots, -3, -2, -1, 0, 1, 2, 3, \cdots\}$ *the set of all **integers or whole numbers**. Once again, you are assumed to know already the counting numbers, their negatives, and the number zero. What is being defined here is the notation $\mathbb{Z}$ for the set of all integers or whole numbers.*

(g) $\mathbb{Q} := \left\{ \frac{p}{q} \mid p \in \mathbb{Z}, q \in \mathbb{N}, \text{no common factors other than one (that is, reduce all fractions)} \right\}$ *the set of **rational numbers**. There is a lot going on here. First of all, because the denominator is a natural (counting) number, it cannot be zero. Secondly, in English, you would have to write "the set of rational numbers is defined to be the collection of all ratios of two integers p and q, where $q > 0$, and moreover, p and q do not have any common factors other than one." You are assumed to know from your past mathematical training that $3/12 = 1/4$, $-1/(-4) = 1/4$, and so on. Why does one insist on removing common factors? It is so that rational numbers are unique. If we do not remove common factors, then every rational number can be written in an infinite number of ways, such as $\frac{1}{2} = \frac{k}{2k}$ for all $k \in \mathbb{N}$. In practice, we don't worry about it too much. For example, when doing computations, a computer is just as happy with $3/12$ as it is with $1/4$.*

(h) $\mathbb{R}$ *denotes the set of **real numbers**. In this course, we assume that the real numbers are familiar to you and will not delve into their formal construction, which is commonly covered in a course on real analysis.*

(i) ∞ *denotes **infinity**, and stands for "a quantity that is boundless, endless, or larger than any natural [or real] number". We use $-\infty$ to denote a quantity that decreases without bound. Sometimes we use $+\infty$ instead of ∞ to emphasize that we are concerned with increasing without bound.*

(j) *For $a < b$ real numbers, the set $[a, b]$ is a **bounded closed interval**, (a, b) is a **bounded open interval**, while $[a, b)$, and $(a, b]$ are **bounded half-open intervals**. The sets $[a, \infty)$, $(-\infty, b]$, and (∞, ∞) are **unbounded closed intervals**, while (a, ∞), $(-\infty, b)$, and $(-\infty, \infty)$ are **unbounded open intervals**[a]. A single point and an empty set are not considered intervals. For definiteness, $[a, b] := \{x \in \mathbb{R} \mid a \leq x \leq b\}$, $[a, b) := \{x \in \mathbb{R} \mid a \leq x < b\}$, and $(-\infty, \infty) := \mathbb{R}$. The learner can fill in the other definitions. When a point a or b is included in an interval, it is called an **endpoint**. For definiteness, a and b are endpoints of $[a, b]$, while only b is an endpoint of $(a, b]$. Infinity can never be an endpoint.*

(k) ***Decimal expansions (also known as decimal representations) of real numbers**. Consider $[0, 1) = \{x \in \mathbb{R} \mid 0 \leq x < 1\}$ and note that the number 1.0 has been excluded from the set. You have seen that every $x \in [0, 1)$ has a decimal expansion of the form*

$$x = 0.d_1 d_2 d_3 \ldots = d_1 10^{-1} + d_2 10^{-2} + d_3 10^{-3} + \cdots =: \sum_{i=1}^{\infty} d_i 10^{-i}, \ d_i \in \{0, 1, 2, \ldots, 9\},$$

where, for now, we are not formally defining the sum to infinity except to say that it stands for

$$\sum_{i=1}^{N} d_i 10^{-i} := d_1 10^{-1} + d_2 10^{-2} + \cdots + d_N 10^{-N},$$

with N allowed to grow without bound.

*You may or may not recall that **for decimal representations to be unique**, you must exclude those that terminate in an infinite string of nines. Why? Consider, for example, $0.9999\ldots$ We agree that $1.0 - 0.9 = 0.1$, $1.0 - 0.99 = 0.01$, $1.0 - 0.999 = 0.001$, etc. Hence, $1.0 - 0.9999\ldots = 0.0$, and therefore, $0.9999\ldots = 1.0$. The same goes for $0.073999\ldots = 0.074$.*

(l) $\mathbf{i}$ *stands for the **imaginary number** whose square equals -1, that is, $\mathbf{i}^2 := -1$. If you prefer to write this as $\mathbf{i} := \sqrt{-1}$, that is fine for the purposes of the course.*

(m) $\mathbb{C} := \{\alpha + \mathbf{i}\beta \mid \alpha, \beta \in \mathbb{R}\}$ *the set of **complex numbers**. You have used them to solve quadratic equations $ax^2 + bx + c = 0$, when the discriminant is negative, that is, $b^2 - 4ac < 0$. That is all we will say for now.*

[a] Yes, $(-\infty, \infty)$ is both open and closed. This only becomes clear in a course on real analysis.

The list of notation could have been longer. Many math textbooks use $\forall$, which means "for all", "for every", or "for each" element, and $\exists$, which means "there exists", "for some", or "for at least one" element. In this textbook, we've chosen to spell them out. Hence, in this textbook, the logical statement, "$\forall\, a \in A$, $\exists\, b \in B$ such that $f(a) = b$" becomes "for all $a \in A$ there is at least one $b \in B$ such $f(a) = b$."

1.4 Algebraic Manipulation and Inequalities

We now present a quick review of key properties when manipulating inequalities. We need these properties to understand upper and lower bounds for calculations at the heart of calculus.

Proposition 1.4: Triangle and Reverse Triangle Inequalities, Plus a Few Other Handy Rules

For all real numbers x and y, the following relations hold:

(a) $|x + y| \le |x| + |y|$ (Triangle Inequality).

(b) $|x - y| \ge \big|\, |x| - |y| \,\big| = \text{abs}(|x| - |y|)$ (Reverse Triangle Inequality).

(c) $|x| \le y \iff -y \le x \le y$ (No Name).

(d) For $a > 0$, $x < y \iff a \cdot x < a \cdot y$.

(e) For $a < 0$, $x < y \iff a \cdot x > a \cdot y$.

(f) For $a > 0$, $a \cdot x < y \iff x < \frac{y}{a}$ and $x < a \cdot y \iff \frac{x}{a} < y$.

(g) For $a < 0$, $a \cdot x < y \iff x > \frac{y}{a}$ and $x < a \cdot y \iff \frac{x}{a} > y$.

Proof: We will prove the first two properties. The others are very standard, even though it is always good to review them!

(a) $|x + y| \le |x| + |y|$:

If one or more of x and y is zero, the result is immediate. If x and y are both nonzero and have the same sign, then $|x + y| = |x| + |y|$. If x and y are both nonzero and have opposite signs, then $|x + y| < |x| + |y|$.

(b) $|x - y| \ge \text{abs}(|x| - |y|)$:

By the triangle inequality:

$$|x| = |x - y + y| \le |x - y| + |y|$$

and

$$|y| = |y - x + x| \le |y - x| + |x|,$$

which yields

$$|x| - |y| \le |x - y|$$

and

$$|y| - |x| \le |y - x|.$$

Because $|y - x| = |x - y|$ and $|y| - |x| = -\left(|x| - |y|\right)$, we have

$$|x| - |y| \le |x - y|$$

and

$$-\left(|x| - |y|\right) \le |x - y|,$$

which proves inequality (b) once property (c) is established or accepted. ■

Remark 1.5. *How to recall all these properties and ensure you use them correctly? Check special cases:* $2 < 3 \implies \frac{1}{2} > \frac{1}{3}$ *and* $2 < 3 \implies -2 > -3$.

Example 1.6. *Show the following properties associated with inequalities.*

(a) Show that if $x > 0$, then $x < y \iff 1 < \frac{y}{x}$.

(b) Show that if $x < 0$, then $x < y \iff 1 > \frac{y}{x}$.

(c) Show that if $x > 0$, then $x < y \iff \frac{1}{x} > \frac{1}{y}$. Note: Both x and y are positive.

(d) Show that if $y < 0$, then $x < y \iff \frac{1}{x} > \frac{1}{y}$. Note: Both x and y are negative.

(e) Show that if $x < 0$ and $y > 0$, then $x < y \iff \frac{1}{x} < \frac{1}{y}$. Note: x and y have opposite signs.

(f) Show that for all $x > 0$, $x + \frac{1}{x} \ge 2$.

Solutions:

(a) **Claim:** If $x > 0$, then $x < y \iff 1 < \frac{y}{x}$.
Proof: Because $x > 0$, in Prop. 1.4-(d), we can take $a := \frac{1}{x} > 0$, and deduce that $x < y \iff \frac{x}{x} < \frac{y}{x}$, which proves the result in (a).

(b) **Claim:** If $x < 0$, then $x < y \iff 1 > \frac{y}{x}$.
Proof: Because $x < 0$, in Prop. 1.4-(e), we can take $a := \frac{1}{x} < 0$, and deduce that $x < y \iff \frac{x}{x} > \frac{y}{x}$, which proves the result in (b).

(c) **Claim:** If $x > 0$, then $x < y \iff \frac{1}{x} > \frac{1}{y}$.
Proof: We note that $0 < x < y$ and hence both x and y are positive. In Prop. 1.4-(d), we can take $a := \frac{1}{x \cdot y} > 0$, and deduce that $x < y \iff \frac{x}{x \cdot y} < \frac{y}{x \cdot y}$, which proves the result in (c).

(d) **Claim:** If $y < 0$, then $x < y \iff \frac{1}{x} > \frac{1}{y}$.
Proof: We note that $x < y < 0$ and hence both x and y are negative. In Prop. 1.4-(d), we can take $a := \frac{1}{x \cdot y} > 0$, and deduce that $x < y \iff \frac{x}{x \cdot y} < \frac{y}{x \cdot y}$, which proves the result in (d).

(e) **Claim:** If $x < 0$ and $y > 0$, then $x < y \iff \frac{1}{x} < \frac{1}{y}$.
Because x and y are nonzero and have opposite signs, in Prop. 1.4-(e), we can take $a := \frac{1}{x \cdot y} < 0$, and deduce that $x < y \iff \frac{x}{x \cdot y} > \frac{y}{x \cdot y}$, which proves the result in (e).

(f) **Claim:** If $x > 0$, then $x + \frac{1}{x} \geq 2$.
Proof: See Sum of positive number and its reciprocal by @MathVisualProofs.

■

Example 1.7. *Solve for all $x \in \mathbb{R}$ satisfying $12 \geq 6 - 3x \geq -3$. The problem comes from TC Math Academy.*

Solution: **Ans.** $-2 \leq x \leq 3$

$$\begin{aligned} 12 \geq 6 - 3x &\geq -3 \\ &\Updownarrow \\ 6 \geq -3x \geq -9 &\quad \text{(Subtracted 6 from all parts of the equation)} \\ &\Updownarrow \\ -2 \leq x \leq 3 &\quad \text{(Divide by } -3\text{, which flips the direction of the inequalities).} \end{aligned}$$

■

Example 1.8. *Solve the following problems dealing with inequalities, which we will encounter later in the study of "limits".*

(a) *For $\epsilon > 0$, find all $x > 0$ such that $\frac{1}{x} < \epsilon$.*

(b) *For $\epsilon > 0$, solve for all $\delta > 0$, if any values exist, such that $|\frac{1}{3} - \frac{1}{\delta+3}| < \epsilon$.*

(c) *For $\epsilon > 0$, solve for all $x > 0$, if any values exist, such that such that $\frac{1}{3} - \frac{1}{\frac{1}{x^2}+3} < \epsilon$.*

Solution:

(a) **Ans.** $\{x \in \mathbb{R} \mid x > 0, \frac{1}{x} < \epsilon\} = \{x \in \mathbb{R} \mid x > \frac{1}{\epsilon}\}$ by applying Example 1.6-(c); for positive quantities, when we take reciprocals, we have to flip the inequality.

(b) **Ans.** For $\epsilon > 0$, $\left\{\delta > 0 \,\middle|\, |\frac{1}{3} - \frac{1}{\delta+3}| < \epsilon\right\} = \left\{\delta > 0 \,\middle|\, \delta < \frac{9\epsilon}{1-3\epsilon}\right\}$. For this set to be nonempty, we need $\epsilon < \frac{1}{3}$ so that $1 - 3\epsilon > 0$.

We first note that for all $\delta > 0$, we have $\frac{1}{\delta+3} < \frac{1}{3}$ and hence $|\frac{1}{3} - \frac{1}{\delta+3}| = \frac{1}{3} - \frac{1}{\delta+3}$. Hence, we can delete the absolute value symbol, which simplifies the problem immensely. The next steps involve the use of careful, though tedious, algebraic manipulation:

$$0 < \frac{1}{3} - \frac{1}{\delta + 3} < \epsilon$$
$$\Updownarrow$$
$$0 < \frac{\delta + 3}{3(\delta + 3)} - \frac{3}{3(\delta + 3)} < \epsilon \quad \text{(put over common denominator)}$$
$$\Updownarrow$$
$$\frac{\delta}{\delta + 3} < 3\epsilon \quad \text{(algebra)}$$
$$\Updownarrow$$
$$\frac{\delta + 3}{\delta} > \frac{1}{3\epsilon} \quad \text{(positive reciprocals)}$$
$$\Updownarrow$$
$$1 + \frac{3}{\delta} > \frac{1}{3\epsilon} \quad \text{(simplify)}$$
$$\Updownarrow$$
$$\frac{3}{\delta} > \frac{1}{3\epsilon} - 1 \quad \text{(algebra)}$$
$$\Updownarrow$$
$$\frac{1}{\delta} > \frac{1 - 3\epsilon}{9\epsilon} \quad \text{(more algebra).}$$

Hence, if $\epsilon \geq \frac{1}{3}$, then there is no solution $\delta > 0$ because $1 - 3\epsilon < 0$. For $0 < \epsilon < \frac{1}{3}$, we have

$$0 < \delta < \frac{9\epsilon}{1 - 3\epsilon}$$

which satisfies $\frac{1}{3} - \frac{1}{\delta+3} < \epsilon$.

(c) **Ans.** This problem is nearly the same as the previous one, with δ replaced by $1/x^2$. Hence, we can reuse our previous work and obtain: For $\epsilon > 0$,

$$\left\{x > 0 \,\middle|\, \left|\frac{1}{3} - \frac{1}{\frac{1}{x^2} + 3}\right| < \epsilon\right\} = \left\{x > 0 \,\middle|\, \frac{1}{x^2} < \frac{9\epsilon}{1 - 3\epsilon}\right\}$$
$$= \left\{x > 0 \,\middle|\, x^2 > \frac{1 - 3\epsilon}{9\epsilon}\right\}$$
$$= \left\{x > 0 \,\middle|\, x > \sqrt{\frac{1 - 3\epsilon}{9\epsilon}}\right\},$$

where, for this set to be nonempty, we need $\epsilon < \frac{1}{3}$ so that $1 - 3\epsilon > 0$. ■

1.5 Functions, Domains, Ranges, Inverses, and Compositions

The topics of domains, ranges, and inverses become crucial when we compose functions and when we compute inverses of functions. To get us started, we will use the following types of functions, all defined for a real number x:

- **monomials** x^k, for k a counting number.
- **polynomials** $a_n x^n + a_{n-1} x^{n-1} + \cdots + a_1 x + a_0$, for n a counting number and a_i real numbers.
- **basic trigonometric functions** $\sin(x)$, $\cos(x)$, and $\tan(x)$ (where $\tan(x)$ is not defined at odd

multiples of $\frac{\pi}{2}$, because it "blows up" there!).

In addition, we will use **rational functions,** $\frac{n(x)}{d(x)}$, where $n(x)$, the numerator, is a polynomial and $d(x)$, the denominator, is a (nonzero) polynomial. Also, for values of x where $d(x) = 0$, a rational function is not well defined because we do not attach a meaning to the division of a real number[5] by zero. We will also be careful about the values of x when we define $\sqrt[n]{x}$ because, unless we are willing to work with complex numbers, the even roots are not defined for negative numbers. The values of x for which a function is defined are called its **domain** and the values that y can take on when $y = f(x)$ and x is in the domain are called the **range** of the function. Another relevant term is **codomain**, which is introduced in the following subsection.

1.5.1 Intuitive Notion of a Function and Related Concepts

What Is a Function: Consider a machine that takes an input, does something to it, and produces an output. In mathematics and computer science, such a machine is called a **function**.

- **Domain**: The **domain** of a function resembles a basket containing all possible *inputs* that can be fed into the machine. Every piece in this basket is something the function can "handle". If the machine makes soccer balls, it can likely "handle" rubber, leather, thread, and glue as inputs. If aluminum is fed into the machine, the machine would break. Similarly, the function arcsine can "handle" any number $y \in [-1, 1]$, but numbers larger than one in magnitude are the function's kryptonite and are not in its domain.
- **Codomain**: The **codomain** consists of every possible *outcome* that could ever come out of the machine, even if some outcomes do not actually occur with the available inputs.
- **Function Rule**: The function itself is a specific rule that defines how to transform an input from the domain into an output. This rule must apply for every input in the domain, and for each input, there is exactly one output.
- **Range**: The **range** is what actually comes out of the machine after it has been fed all the inputs in the domain. It is a collection of the actual outcomes, which might be less than all the possibilities that can exist in the codomain. For the soccer ball machine, we may have originally set the codomain as the set of all balls (not just soccer balls), but in the end, the machine produces only high-quality soccer balls for use in professional sports, which would be its range.

Conceptual Example: Consider a machine (function) designed to accept a number (input part) and then generate a number (output part) whose value is double that of the input part.

- **Domain** (input bin to the machine): Any subset of the real numbers, which we may define as $[0, 1 \times 10^{307}]$, which is close to the largest number that can be represented in double-precision floating point.
- **Codomain** (output bin to the machine): Can be taken as large as all real numbers (doubling any real number gives a real number), or as $[0, 1.8 \times 10^{308}]$, based on the largest number that can be represented in double-precision floating point. The domain, however, cannot be smaller than $[0, 2 \times 10^{307}]$ because the codomain (output bin) must be large enough to contain the outputs of the machine (function). The machine (function) is not obliged to completely fill the output bin (codomain).
- **Function Rule**: Multiply by 2.
- **Range**: $[0, 2 \times 10^{307}]$, the part of the output bin (codomain) that is actually filled with products of the machine (numbers produced by the function).

Summary:

- The **domain** is where the function is defined, the set of points on which it is allowed to act.

[5]To be clear, not even zero divided by zero is given a definition. Eventually, we will learn "to compute a limiting value for it" under specific assumptions.

- The **codomain** can include every conceivable output value, not just the ones that actually occur.

- The **range** is where the function actually goes, the true results you see when applying the function to all points in its domain.

> **Intuitive Notion of Continuity:** For now, we will informally consider a function $f : \mathbb{R} \to \mathbb{R}$ to be **continuous** if we can draw the graph of $y = f(x)$ on a sheet of paper without lifting our pencil (from the paper)!

Figure 1.2-(a) clearly passes this test while **Fig. 1.2-(b)** does not. **Figure 1-2-(c)** "seems" to pass the "without lifting your pencil" test, but the graph does not represent a function! The *vertical line test* is a quick way to check whether a graph represents a function. If any vertical line touches the graph at more than one point, then the graph is not a function. This is because a function should give only one output for each input." In **Fig. 1-2-(c)**, for $x = 0$, we have $f(x) = y$ for all $y \in [-1, 2]$, which makes it not a function. What about the part of the graph where the "non-function" is constant? Does that also make it not a function? No, it is fine for the single value of $y = -1.0$ to be associated with many values of x (constant functions are functions); it is the other way around that is a problem. Functions map points to points and not points to something bigger, such as the interval $[-1, 2]$.

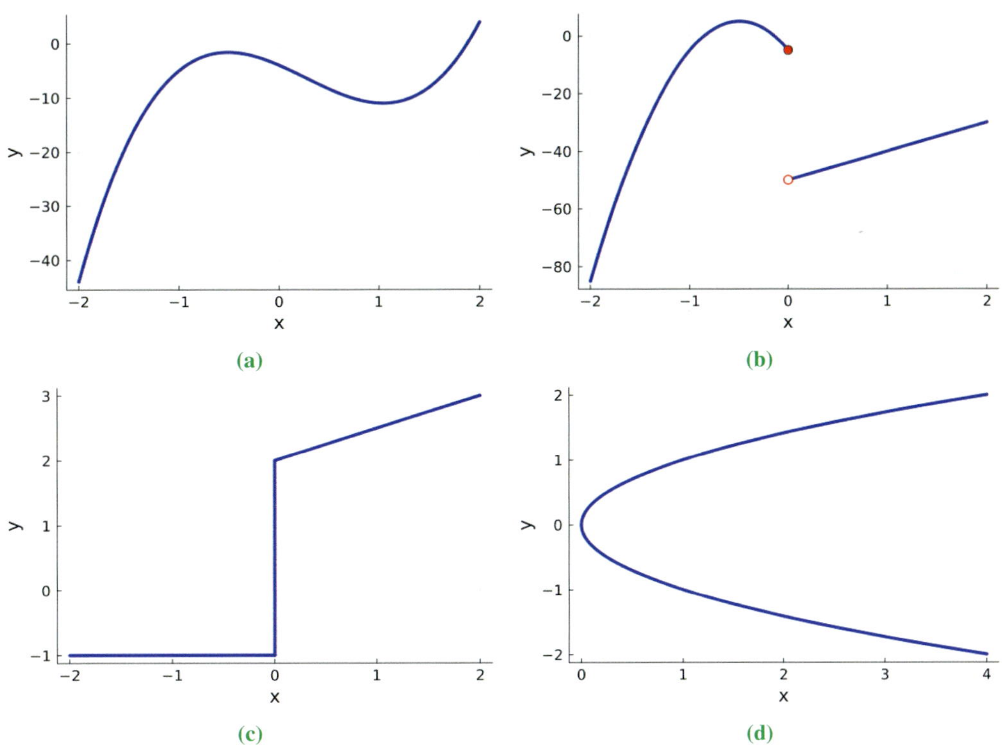

Figure 1.2: Examples of a continuous function, a discontinuous function, and two graphs that are not functions. Observe that in (b), at the jump, the value of the function has been indicated by a solid dot. Note that in (c), the point $x = 0$ is mapped to the interval $[-1, 2]$, while in (d), each $0 < x \leq 4$ is mapped to two points. To qualify as a function, each point in the domain can be mapped to only a single point in the range.

1.5.2 Formal Notions of Domain, Codomain, and Range of a Function

Defining Functions in a Careful Manner

What is a function?

(a) For A and B arbitrary sets, $f : A \to B$ denotes a **function from** A **to** B if, for all $a \in A$, $f(a) \in B$. Two attributes are important: (1) $f(a)$ is defined for every point $a \in A$, and (2) $f(a)$ is a single element of B, meaning it cannot contain two or more elements.

(b) A is called the function's **domain**. It is the set of values for which the function is defined or is being defined. The domain contains all of the possible "input values" of the function.

(c) B is called its **codomain** . It is the set in which the function takes its values; for all $x \in A$, we have $f(x) \in B$. We can think of the codomain as containing the POTENTIAL "output values" of the function.

(d) $C \subset B$ is the **range** of the function f if two conditions hold:

(a) For all $x \in A$, it is true that $f(x) \in C$ (C contains all of the values of f).

(b) For all $y \in C$, there exists $x \in A$ such that $f(x) = y$ (every element of C is an output of f).

The range contains the ACTUAL "output values" of the function because every point in C is an output of the function. Alternatively, the range is the set of points in B that are "landing points from A under the action of the function f". If f does not "land on" $b \in B$, then b is not in the range of f.

Note: The **range of $f : A \to B$ is the smallest possible codomain of the function**. In other symbols, the range is

$$f(A) := \{f(a) \mid a \in A\}.$$

Looking Ahead: The difference between range and codomain becomes important when we consider the inverse of a function. For example, $\arcsin(x)$, the inverse function of $\sin(x)$, is not defined for $|x| > 1$, unless one allows complex numbers. When considering the inverse of the sine function, we must be careful about how we define the function's domain and codomain. The same is true for x^2; its inverse function is not defined for $x < 0$. In other situations, there may be no need to specify the domain and codomain of a function.

Potentially helpful videos:

- How to Find the Domain of Any Function by Nancy Pi.
- Domain and Range – Get Ready to Understand! by TabletClass Math; easy to understand lesson on what the domain and range of a function represent.

Example 1.9. *We can write* $\sin : \mathbb{R} \to \mathbb{R}$, *and in this case, the sine function's domain and codomain have both been defined to be the set of real numbers,* $\mathbb{R}$. *What is the function's range?*

Solution: For all $x \in \mathbb{R}$,

$$-1 \le \sin(x) \le 1.$$

Moreover, from the actual definition of the sine function illustrated in Fig. 1.3, it follows that for every $y \in [-1, 1]$, there is a real number $x \in [-\frac{\pi}{2}, \frac{\pi}{2}]$ such that $\sin(x) = y$. In particular,

- $\sin(\frac{\pi}{2}) = 1$
- $\sin(-\frac{\pi}{2}) = -1$, and

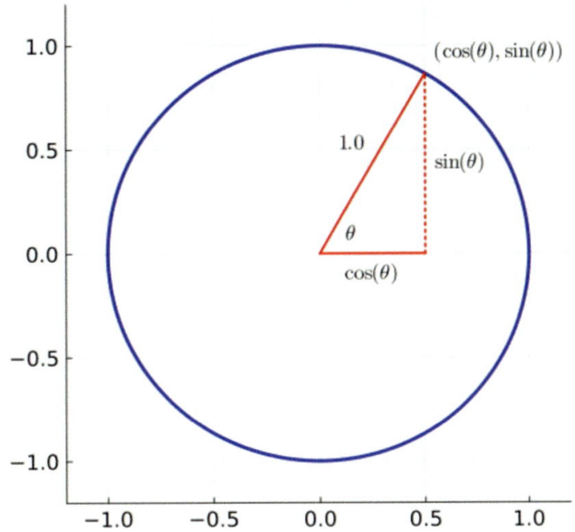

Figure 1.3: Definitions of $\sin(\theta)$ and $\cos(\theta)$ using a circle of radius 1.

- by selecting $-\frac{\pi}{2} < x < \frac{\pi}{2}$, all the values between -1 and 1 are obtained by $\sin(x)$.

The set $[-1, 1]$ is therefore the **range** of the function $\sin(x)$.

The set $[-1, 1]$ is the smallest subset of the codomain that contains all the values of $\sin(x)$. Hence, even though $[-2, 4] \subset \mathbb{R}$, and $\sin(x) \in [-2, 4]$ for all $x \in \mathbb{R}$, $[-2, 4]$ **is not the range of the sine function** because there does not exist any value of $x \in \mathbb{R}$ such that $\sin(x) = -2$. Moreover, $(0, 1]$ is **not the range** of the sine function because $\sin(-\frac{\pi}{2}) = -1$, and $-1 \notin (0, 1]$.

Conclusion: The range is a **Goldilocks notion:** it cannot be too big, and it cannot be too small; it cannot contain extra points on which the function f never lands, but it must be big enough to contain all of the values of f. ■

Example 1.10. *As another example, consider the function, x^2, which is defined for all real numbers. Hence, its domain can be specified as $\mathbb{R}$. Its values are real numbers, so we can write $(\bullet)^2 : \mathbb{R} \to \mathbb{R}$ (note that the "bullet point or dot" is just indicating a placeholder where you can plug in a value to be squared). We could also write*

$$(\bullet)^2 : \mathbb{R} \to [0, \infty).$$

Both are correct. In the second case, we have used the function's range as its codomain. ■

Fact 1.11: Range vs. Codomain

Why do we not always write functions as $f : A \to B$, where B has to be the **range** of the function?

The simple reason is that **we often do not know the range of a given function**. Consider $f : \mathbb{R} \to \mathbb{R}$ with

$$f(x) := \operatorname{atan}\left(\sin(x^3 + 4) - 2^{-1-x^2}\right), \tag{1.7}$$

where $\operatorname{atan}(x)$ is the inverse of the tangent function. What is the range of f? It would require a lot of work to find out the true answer, but a plot as in Fig. 1.4 can be very helpful. Often, it may not be necessary to know the range of a particular function for the intended application.

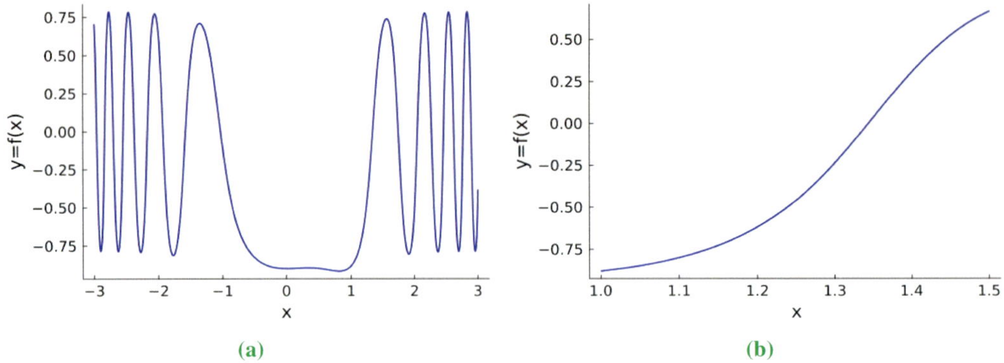

Figure 1.4: A graph of the function $f : A \to \mathbb{R}$ for $f(x) := \text{atan}(\sin(x^3 + 4) - 2^{-1-x^2})$ for two choices of its domain. In (a), the domain is $A = [-3.0, 3.0]$, while in (b), the domain is $A = [1.0, 1.5]$.

Example 1.12. *Determine a valid domain for the function* $\tan(x) := \frac{\sin(x)}{\cos(x)}$.

Solution: Let $A := \{x \in \mathbb{R} \mid x \neq \frac{\pi}{2} \mod \pi\}$, that is, $x \neq \frac{\pi}{2} + k\pi$, for all $k \in \mathbb{Z}$, or stated yet another way, x is not an odd multiple of $\frac{\pi}{2}$. Then, $\tan : A \to \mathbb{R}$ is a function, because $\cos(x) \neq 0$, for all $x \in A$. We have avoided division by zero. The real numbers, by the way, are both a valid codomain and range for the function $\tan(x)$.

■

1.5.3 Composing Functions

Function composition is a fundamental concept in mathematics and computer programming. It refers to the process of applying one function to the result of another. In essence, we are "chaining" functions together so that the output of one becomes the input of the next. In a "transformer neural network", such as the ones used in ChatGPT (2022), the chains are up to 96 units long.

The Composition of Two or More Functions

Definition 1.13. *Given a "first" function, $f : A \to B$, and a "second" function $g : C \to D$, their **composition** is denoted by $(g \circ f)(x)$ or simply $g(f(x))$. For the composition $g \circ f$ to be defined (which means to make sense),* ***the range of f must be contained within the domain of g.***

In simple words, every output of f must be a valid input for g. This is crucial. If f produces a single value that g cannot accept, then their composition is undefined. In Julia, an undefined composition will (eventually) generate an error, but in normal math, we are likely to generate a nonsensical outcome.

In other symbols, given functions $f : A \to B$ and $g : C \to D$,

(a) their composition is defined if $f(A) \subset C$, that is, the range of f is contained in C, the domain of g.

(b) The function composition is $g \circ f : A \to D$, meaning $g \circ f$ has domain A and codomain D.

Note: Suppose $f_i : A_i \to B_i$, for $1 \leq i \leq N$. Then, $f_N \circ f_{N-1} \circ \cdots \circ f_1 : A_1 \to B_N$ as long as $f(A_i) \subset A_{i+1}$ for $1 \leq i \leq N - 1$.

Example 1.14. *Consider two functions:*

- $f:(0,\infty)\to\mathbb{R}$ *by* $f(x)=\frac{1}{x}$, *and*
- $g:\mathbb{R}\to\mathbb{R}$ *by* $g(x)=x^2$.

(a) Does f composed with g make sense, and if it does, what is the resulting function?

(b) Does g composed with f make sense, and if it does, what is the resulting function?

Solution:

(a) The range of f is $(0,\infty)$. The function g is defined for all real numbers. Hence, their composition, $g\circ f$, does make sense.

The composition $g\circ f:(0,\infty)\to\mathbb{R}$ by $(g\circ f)(x)=g(f(x))=g\left(\frac{1}{x}\right)=\frac{1}{x^2}$.

(b) The range of g is $\mathbb{R}=(-\infty,\infty)$, which contains zero. The function f is undefined for $x=0$. Hence, their composition, $f\circ g$, does NOT make sense. The solution stops here because we do not need to compute something that is not defined!

Remark 1.15. ***Key Takeaways on Function Composition:***

- *Function composition allows us to create new functions by chaining existing ones.*
- *The order matters! In general, $g\circ f$ is not the same as $f\circ g$. In fact, one can be defined, and the other not.*
- *For the composition to be defined, the range of the INNERMOST function must be contained within the domain of the NEXT function. This way of thinking about it avoids the difficult task of saying which function is FIRST when writing $(g\circ f)(x)=g(f(x))$ and $(f\circ g)(x)=f(g(x))$.*
- *The best way to remember it is: we need g to accept $f(x)$ for $g(f(x))$ to make sense, and we need f to accept $g(x)$ for $f(g(x))$ to make sense. A function only accepts values in its own domain.*

Function composition is a powerful tool in mathematics and programming, enabling us to build complex operations from simpler ones.

1.5.4 Inverse Functions

Inverse of a Function

Definition 1.16. *Let A and B be arbitrary sets and $f:A\to B$ a function. A function $g:B\to A$ is the* ***inverse of*** *f if,*

- *for all $a\in A$, $g(f(a))=a$,* ***and***
- *for all $b\in B$, $f(g(b))=b$.*

Both properties must hold.

Because it can be shown that if an inverse function exists, it is necessarily unique (meaning there cannot be more than one of them), we use the notation $f^{-1}:B\to A$. We say that $f:A\to B$ is ***invertible*** *if it has an inverse function.* ***We note that $f:A\to B$ is the inverse of $f^{-1}:B\to A$, i.e., $(f^{-1})^{-1}=f$, as long as f^{-1} exists.***

The following example relates the computation of inverse functions to root finding (as highlighted earlier when we computed $\sqrt{2}$).

Example 1.17. *Consider the function* $f : [1.0, 1.5] \to \mathbb{R}$ *by* $f(x) := \text{atan}\left(\sin(x^3 + 4) - 2^{-1-x^2}\right)$, *with the plot given in* **Fig. 1.4-(b)**. *Compute* $f^{-1}(y)$ *for* $y = -0.5$.

Solution: We need to find x such that $f(x) = y$, which is equivalent to $f(x) - y = 0$. In other words, computing the inverse $f^{-1}(y)$ is the same as computing a root of $f(x) - y$. From **Fig. 1.4-(b)**, we see that $y = -0.5$ when $x \approx 1.25$; in other symbols, $f^{-1}(-0.5) \approx 1.25$. Below, we use the Bisection Algorithm to compute a more precise value than we can eyeball from the graph, and we obtain $f^{-1}(-0.5) = 1.2386 \pm 1.907e\text{-}6$.

```
f(x) = atan(sin(x^3 + 4) - 2.0^(-1 - x^2) )

# Call the bisection method on this function
y = -0.5
h(x) = f(x) - y

# for
xmin = 1
xmax = 1.5

# One line of code
F = bisection(h , xmin, xmax, 1e-5)
@show F

y_est = f(F.est)

# Print relevant information
using IJulia
latex_string = "The value of \$f^{-1}(y = 0.5) = $(F.est)\$"
display("text/latex", latex_string)

latex_string = "The quality of the inverse is \$\\left(f^{-1}(y = 0.5)\\right) = $(y_est)\$ is pretty good!"
display("text/latex", latex_string)
```

Output

```
F = (est = 1.2386131286621094, low = 1.2386093139648438,
up = 1.2386131286621094, error = 1.9073486328125e-6)
```

The value of $f^{-1}(y = 0.5) = 1.2386112213134766$. The quality of the inverse is very good:

$$f(1.2386112213134766)) = -0.5000045415576979.$$

Ideally, $f\left(f^{-1}(0.5)\right) = -0.5$

More Function Vocabulary

Definition 1.18. *Let* A *and* B *be arbitrary subsets and* $f : A \to B$ *a function.*

- f *is* ***one-to-one or 1:1*** *if for all* $a_1, a_2 \in A$, $a_1 \neq a_2 \implies f(a_1) \neq f(a_2)$. *We also call the function* ***injective***.
- f *is* ***onto*** *if for all* $b \in B$, *there is some* $a \in A$ *such that* $f(a) = b$. *We also call the function* ***surjective***.

Notes: When both A and B are subsets of the real numbers, we can use the **horizontal line test** to determine if a function is injective (i.e., 1:1). It goes like this: Let $A \subset \mathbb{R}$ and $B \subset \mathbb{R}$ be subsets of $\mathbb{R}$ and let $f : A \to B$ be a function represented by its graph. The function f is one-to-one (or injective) if and only if no horizontal line intersects the graph of f more than once. And while it is a perfectly valid test when A and B are subsets of $\mathbb{R}$, it fails in more general situations.

- A function $f : A \to B$ is **surjective or onto** if for each $b \in B$, the set $\{a \in A \mid f(a) = b\}$ (the set of all points in A that get mapped onto b) is nonempty. There could be multiple points $a \in A$ that get mapped onto the same $b \in B$, and the function would still be surjective, but there cannot exist a point $b \in B$ for which there are no points in A that get mapped onto it.

- **Said another way, the function's codomain is also its range.**

Example 1.19. *Determine if the following are surjective functions.*

(a) $f : \mathbb{R} \to \mathbb{R}$ *by* $f(x) = \cos(x)$.

(b) $f : \mathbb{R} \to [-1, 1]$ *by* $f(x) = \cos(x)$.

(c) $f : (-\pi, \pi] \to [-1, 1]$ *by* $f(x) = \cos(x)$.

(d) $f : [-\pi/2, 0] \to [-1, 1]$ *by* $f(x) = \cos(x)$.

(e) $f : (-\pi, \pi] \to [0, 1]$ *by* $f(x) = \cos(x)$.

Solution:

(a) $\cos : \mathbb{R} \to \mathbb{R}$ is not surjective (not onto) because there is no value $x \in \mathbb{R}$ such that $\cos(x) = 2$.

(b) $\cos : \mathbb{R} \to [-1, 1]$ is onto (surjective) because $\cos(-\pi/2) = -1$, $\cos(0) = 1$, and for all $-1 < y < 1$, there is an $-\pi/2 < x < 0$ such that $\cos(x) = y$.

(c) $\cos : (-\pi, \pi] \to [-1, 1]$ is onto (surjective) by the same reasoning as in (b).

(d) $\cos : [-\pi/2, 0] \to [-1, 1]$ is onto (surjective) by the same reasoning as item (b)

(e) $\cos : (-\pi, \pi] \to [0, 1]$ is not a function and hence it cannot be surjective (onto). Why is it not a function? Because $-1 = \cos(-\pi/2) \notin [0, 1]$. In other words, $[0, 1]$ is not a valid codomain for cosine.

■

Just as we have alternative ways to determine if a function is surjective (onto), we can do the same if the function is injective (1:1).

- A function $f : A \to B$ is **injective or 1:1** if for each $b \in B$, the set $\{a \in A \mid f(a) = b\}$ (the set of all points in A that get mapped onto b) is either empty or a singleton. There could be no points $a \in A$ that get mapped onto a given $b \in B$, but if a point gets mapped onto b, there cannot exist a second point distinct from the first point.

- **Stated another way, for an injective function, if there exists $a_1, a_2 \in A$ such that $f(a_1) = b$ and $f(a_2) = b$, then $a_1 = a_2$.**

Example 1.20. *Determine if the following functions are injective.*

(a) $f : \mathbb{R} \to \mathbb{R}$ *by* $f(x) = \cos(x)$.

(b) $f : [-\pi/2, \pi/2] \to [-1, 1]$ *by* $f(x) = \cos(x)$.

(c) $f : (-\pi, 0] \to [-1, 1]$ *by* $f(x) = \cos(x)$.

(d) $f : [-\pi/2, 0] \to [-1, 1]$ *by* $f(x) = \cos(x)$.

(e) $f : (-\pi, \pi] \to [0, 1]$ *by* $f(x) = \cos(x)$.

Solution:

(a) $\cos : \mathbb{R} \to \mathbb{R}$ is not injective (not 1:1) because $\cos(0) = 1$ and $\cos(2\pi) = 1$. To conclude that a function is **not injective**, all we need to do is to find a single pair of distinct points that get mapped to a common point. The fact that there are many other pairs of points that also get mapped to a common point is irrelevant to showing the definition of injectivity fails (aka, it is not satisfied).

(b) $\cos : [-\pi/2, \pi/2] \to [-1,1]$ is not injective (not 1:1) because $\cos(-\pi/2) = 0$ and $\cos(\pi/2) = 0$.

(c) $\cos : (-\pi, 0] \to [-1,1]$ is 1:1 (injective) because for all $x, y \in (-\pi, 0]$, $x < y \implies \cos(x) < \cos(y)$, and hence distinct values in the domain produce distinct values in the range.

(d) $\cos : [-\pi/2, 0] \to [-1,1]$ is injective (1:1) because we already showed that $\cos : (-\pi, 0] \to [-1,1]$ is injective and $[-\pi/2, 0] \subset (-\pi, 0]$. Alternatively, we can use the same reasoning as in (c).

(e) $\cos : (-\pi, \pi] \to [0,1]$ is not a function and hence it cannot be injective (1:1). Why is it not a function? Because $-1 = \cos(-\pi/2) \notin [0,1]$.

■

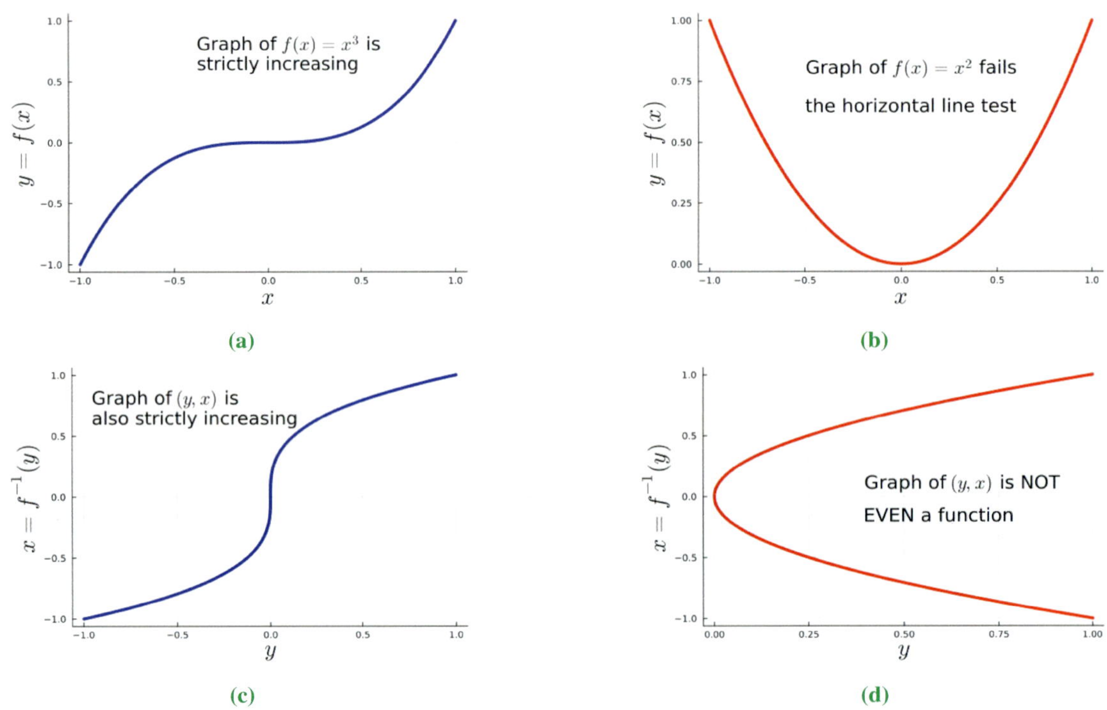

Figure 1.5: Comparison of "strictly monotonic" vs. "nonmonotonic" functions and how that affects the existence of an inverse. In (a), the graph of $f : [-1,1] \to \mathbb{R}$ by $f(x) = x^3$, and right below it in (c) is plotted (y, x), that is, x and y have been swapped. Because $f : [-1,1] \to \mathbb{R}$ by $f(x) = x^3$ is strictly increasing, it passes the horizontal line test and swapping the roles of x and y in the graph automatically produces the inverse of f. In (b), the graph of $f : [-1,1] \to \mathbb{R}$ by $f(x) = x^2$, and right below it in (d) is plotted (y, x) (again, x and y have been swapped). When $f : [-1,1] \to \mathbb{R}$ by $f(x) = x^2$ fails the horizontal line test, swapping the roles of x and y in the graph does NOT even produce a function, much less an inverse for f. The graph in (d) fails the vertical line test.

Proposition 1.21: Existence of Inverse Functions

A function $f : A \to B$ has an inverse if, and only if, it is both injective (1:1) and surjective (onto).

Note: Such functions have a name: they are called **bijective**.

1.5.5 Strictly Monotonic Functions and Relation to Existence of Inverse Functions

While it is mathematically appropriate to have necessary and sufficient conditions for the existence of an inverse, in most cases, checking the conditions for the existence of an inverse is "nearly impossible". There are a few easy cases; in that light, we introduce the following notions.

Strictly Monotonic Functions

Definition 1.22. *Let A and B be subsets of the real numbers and $f : A \to B$ a function.*

- *f is **strictly increasing** if for all $x, y \in A$, $y > x \implies f(y) > f(x)$.*
- *f is **strictly decreasing** if for all $x, y \in A$, $y > x \implies f(y) < f(x)$.*

*When the inequalities used above are **not strict**, we say that f is **increasing** or **decreasing**, respectively. Increasing and decreasing functions are also called **monotonically increasing or decreasing** to further emphasize that they never reverse direction; they can, however, remain "flat or constant" over a set of points.*

Remark 1.23. *Among the powerful results of calculus is an effective means to check if a function is (strictly) increasing or decreasing! For now, it is suggested that we graph the function and apply the horizontal line test.*

Proposition 1.24: Strict Montonicity Helps with Invertibility

Suppose that A and B are subsets of the real numbers.

(a) If $f : A \to B$ is strictly monotonic (increasing or decreasing), then it is injective (one-to-one).

(b) If $f : A \to B$ is strictly monotonic (increasing or decreasing) and surjective (onto), then it is invertible.

1.6 Trigonometric and Inverse Trigonometric Functions

Figure 1.6 shows an idealized diagram of a robot with three links that we will use to illustrate how Roboticists use trigonometry. In Robotics, we call such a diagram "a kinematic diagram of the robot" or a "kinematic chain representation of the robot". Because the word "kinematic" is new to this textbook, let us describe it.

Kinematics delves into the study of the motion of rigid bodies, focusing on the positions (not the speeds, which is dynamics) of points on a body as it moves through space, that is, $\mathbb{R}^2$ or $\mathbb{R}^3$. In the context of robotics, such as the robot with three links in Fig. 1.6, kinematics provides insights into the possible movements and configurations of the robot without considering the forces and physical properties influencing these movements (no masses, no inertias, no motor torques, which we will examine later). By emphasizing the geometric and algebraic aspects of motion, kinematics offers a mathematical framework to describe and analyze the geometry of motion, allowing Roboticists to achieve precise and coordinated movements in robotic systems.

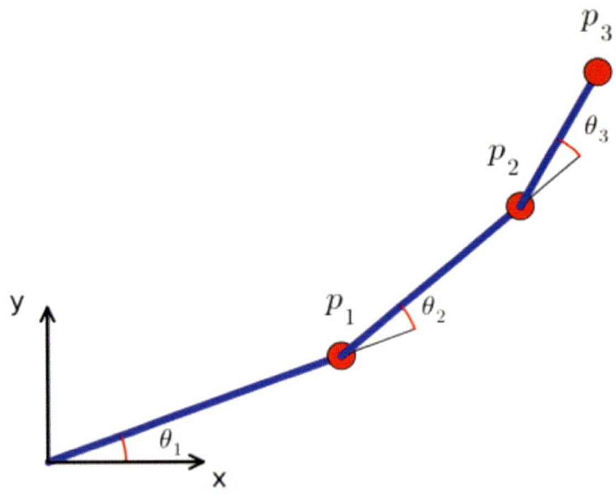

Figure 1.6: This is a robot kinematic chain, along with a Cartesian coordinate system, (x, y). The angle θ_1 is measured counterclockwise (aka CCW) from the x-axis, while θ_2 is measured CCW with respect to link-1 and θ_3 is measured CCW with respect to link-2. The angles are defined this way because we can measure all of the indicated angles with practical instruments. The term "chain" just means one body is attached to another to form a chain. But what about "kinematic"?

1.6.1 Trigonometric Functions

The main trigonometric functions we need are $\sin(x)$, $\cos(x)$, and $\tan(x)$. The others,

- $\sec(x) := \frac{1}{\cos(x)}$, called the **secant** of x,
- $\csc(x) := \frac{1}{\sin(x)}$, called the **cosecant** of x, and
- $\cot(x) := \frac{\cos(x)}{\sin(x)} = \frac{1}{\tan(x)}$, called the **cotangent** of x,

are reciprocals of the main trigonometric functions. They show up when we perform trigonometric substitutions in Section 7.3.6. The following video is highly recommended: Mathematical Visual Proofs: Trig Visualized as a review. An informative comment left by a viewer: "I've never seen this diagram. **Even a static version is quite informative, but the animation knocks it out of the park**."

We also need the inverse $\sin(x)$, $\cos(x)$, and $\tan(x)$ functions. Those are treated separately because they can be more complicated.

1.6.2 Trigonometric Identities

There are "innumerable" trigonometric identities. The ones we list here show up in the book in at least one location.

- $\cos^2(x) + \sin^2(x) = 1$.
- $\sec^2(x) = 1 + \tan^2(x)$.
- $\sin(2x) = 2\sin(x) \cdot \cos(x)$.
- $\cos(2x) = 2\cos^2(x) - 1$.
- $\cos^2(x) = \frac{1}{2} + \frac{1}{2}\cos(2x)$.
- $\sin^2(x) = \frac{1}{2} - \frac{1}{2}\cos(2x)$.

- $\cos(2x) = \cos^2(x) - \sin^2(x)$.

Even though several of the above identities are "redundant" in the sense that they can be easily derived from other listed identities, it is more efficient to have them handy for when we need them.

1.6.3 A 3-Link Robot Kinematic Chain in the Plane

Let us define the lengths of the three links shown in **Fig. 1.6** to be L_1, L_2, and L_3. Their lengths are 2, 1.5, and 1, respectively. Let us define p_0 to be the origin,

$$p_0 = \begin{bmatrix} 0.0 \\ 0.0 \end{bmatrix},$$

and we assume that the base of the first link is attached at the origin with a **revolute joint**. This means that its position is fixed at p_0, but the link can rotate in a full circle, that is, 2π radians, or $360°$. The angle in a revolute joint is measured with a joint encoder, which usually is shortened to simply **encoder**. We need to specify a zero point (origin) for the encoder and the positive direction. We will select the positive direction of the x-axis as the zero point ($0°$), and counterclockwise (CCW) as the positive direction. With this convention, the y-axis is at $\frac{\pi}{2}$ or $90°$.

Given these specifications, we can compute the Cartesian position of the end of link-1, labeled p_1 in **Fig. 1.6** using:

$$p_1 = p_0 + \begin{bmatrix} L_1 \cdot \cos(\theta_1) \\ L_1 \cdot \sin(\theta_1) \end{bmatrix}. \tag{1.8}$$

Because p_0 is the origin, we can leave it out of the equation, but including it allows us to mount the base of the robot at any point we wish and offers recursive structure for writing down the equations.

Continuing, we have

$$p_2 = p_1 + \begin{bmatrix} L_2 \cdot \cos(\theta_1 + \theta_2) \\ L_2 \cdot \sin(\theta_1 + \theta_2) \end{bmatrix}.$$

This looks just like (1.8), except that we have $(\theta_1 + \theta_2)$. Why the sum of the angles? As illustrated in **Fig. 1.7**-(b), the angle θ_2 is measured with an encoder installed at the connecting point of links 1 and 2 of the robot. The encoder measures the angle of rotation of link 2 with respect to link 1, or in the vocabulary of **Fig. 1.7**, the relative angle between the second and first link. We need the absolute angle[6] of the second link in order to calculate the position of its tip in the Cartesian coordinate system (x, y). That angle is the sum of the absolute angle of the first link and the angle of the second link relative to the first link, namely, $(\theta_1 + \theta_2)$.

Based on the above discussion, computing the Cartesian position of the end of the third link is straightforward:

$$p_3 = p_2 + \begin{bmatrix} L_3 \cdot \cos(\theta_1 + \theta_2 + \theta_3) \\ L_3 \cdot \sin(\theta_1 + \theta_2 + \theta_3) \end{bmatrix}. \tag{1.9}$$

We illustrate the process with the following computer program.

```
# functions for

function modelParameters()
    # Place the parameters in a single location so that
    # they can be updated at a moment's notice
    L1, L2, L3 = [2, 1.5, 1]
    return (L1=L1, L2=L2, L3=L3)
end

```

[6]While it is "theoretically possible" to measure the absolute angle of the second link directly, by using a laser, for example, the only practical way to do it is via an encoder attached at the joints between successive links.

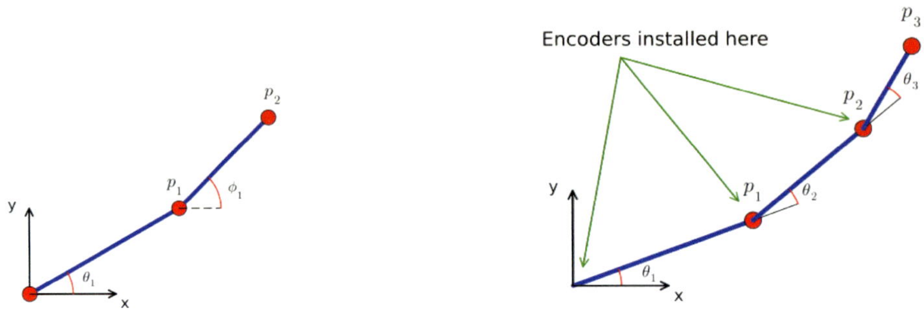

(a) (b)

Figure 1.7: **Absolute angles** are those referenced to an agreed-upon stationary coordinate system (called a world frame in mechanics and robotics). In (a), both θ and ϕ are absolute angles; each is measured with respect to a fixed horizontal axis. **Relative angles** are those referenced to a coordinate system that moves with respect to the agreed-upon world frame. Typically, the angle is measured *relative* to another point on the robot's body. In (b), θ_1 is the absolute angle of the first link, while θ_2 measures the relative angle of the second link with respect to the first. The absolute angle of the second link is $\theta_1 + \theta_2$, the sum of the absolute angle of the first link and the relative angle between the second link and the first link. Similarly, the absolute angle of the third link is $\theta_1 + \theta_2 + \theta_3$.

```
function linkPositions(th1, th2, th3)
    params = modelParameters()
    p0=[0.0; 0.0]
    p1 = p0 + [ params.L1 * cos(th1); params.L1 * sin(th1)]
    p2 = p1 + [ params.L2 * cos(th1+th2); params.L2 * sin(th1+th2)]
    p3 = p2 + [ params.L3 * cos(th1+th2+th3); params.L3 * sin(th1+th2+th3)]
    return (p0=p0, p1=p1, p2=p2, p3=p3)
end

th1, th2, th3 = (2*pi/360) * [20, 20, 20]
positions = linkPositions(th1, th2, th3)
```

Output

```
(p0 = [0.0, 0.0], p1 = [1.8793852415718169, 0.6840402866513374],
p2 = [3.028451906250284, 1.6482217011811464],
p3 = [3.528451906250284, 2.514247104965585])
```

Next, we exercise the above function by computing p_1, p_2, and p_3 for various angles θ_1, θ_2, and θ_3, and then plotting the links of the robot. Once we write the code, we can run it as many times as needed . We can also put in nominal angles, such as all angles equal to zero or all angles equal to $\frac{\pi}{2}$, and check that we obtain what we know has to be the correct configuration of the mechanism.

```
using Plots

# Global variable to keep track of the number of times the function has been
# called
global call_count = 0

function plot_points(positions; line_thickness=5, ball_size=10)
    global call_count
    p0 = positions.p0
```

```
    p1 = positions.p1
    p2 = positions.p2
    p3 = positions.p3

    # Colors for the lines for each call
    line_colors = [:blue, :green, :orange, :purple, :cyan, :magenta, :yellow, :black]

    # Increment the call count and determine the color for this call
    call_count += 1
    current_color = line_colors[mod(call_count, length(line_colors)) + 1]

    # Plot the lines
    plot!([p0[1], p1[1], p2[1], p3[1]], [p0[2], p1[2], p2[2], p3[2]],
        linewidth=line_thickness, color=current_color, label="Call $(call_count)")

    # Plot the balls
    scatter!([p0[1]], [p0[2]], color=:black, markersize=ball_size, label=nothing)
    scatter!([p1[1], p2[1], p3[1]], [p1[2], p2[2], p3[2]], color=:red,
        markersize=ball_size, label=nothing)

    display(current())
end

# Initialize the plot
#plot(; legend=:topright, aspect_ratio=:equal)
fig = plot(; legend=:topleft, aspect_ratio=:equal)

# Example usage
th1, th2, th3 = (2*pi/360) * [20, 20, 20]
positions = linkPositions(th1,th2,th3)
plot_points(positions)

# Call again with different angles to see the effect
th1, th2, th3 = (2*pi/360) * [-20, 20, 20]
positions = linkPositions(th1,th2,th3)
plot_points(positions)

# Call again with different angles to see the effect
th1, th2, th3 = (2*pi/360) * [-140, -90, -90]
positions = linkPositions(th1,th2,th3)
plot_points(positions)

# How to print a png for inclusion in a report
png(fig, "robotKinematicChainWithOurNewFunction")
```

Output: The image is shown in Fig. 1.8.

As promised, we give the general formula for the position of the end effector, that is, the end of link 3.

$$\begin{bmatrix} p_{3,x} \\ p_{3,y} \end{bmatrix} = \begin{bmatrix} L1 \cdot \cos(\theta_1) + L2 \cdot \cos(\theta_1 + \theta_2) + L3 \cdot \cos(\theta_1 + \theta_2 + \theta_3) \\ L1 \cdot \sin(\theta_1) + L2 \cdot \sin(\theta_1 + \theta_2) + L3 \cdot \sin(\theta_1 + \theta_2 + \theta_3) \end{bmatrix} \tag{1.10}$$

1.6.4 Notation for Inverse Trigonometric Functions

The trigonometric functions, sine, cosine, and tangent, have **inverse functions** if their domains and codomains are properly defined. The inverse functions help us find **angles in radians** when we know the trigonometric values. There are a few ways to denote the inverse functions, and we go through them below.

- The most popular way to write inverse functions is with an "arc" in front. For example:

  $\arcsin(x)$ for inverse sine,

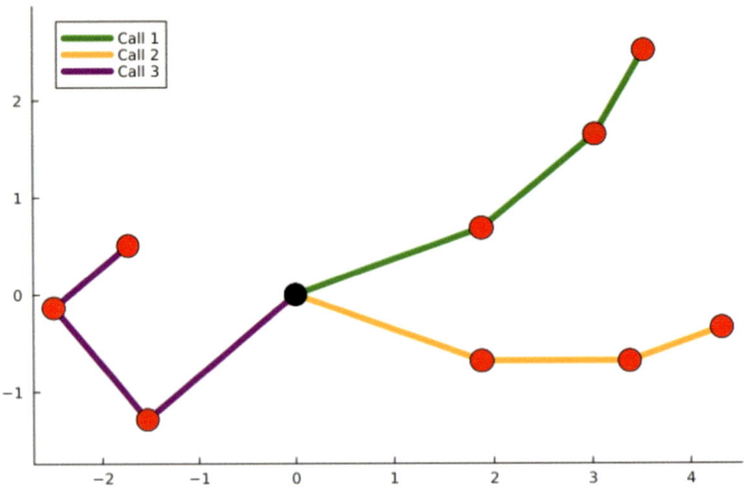

Figure 1.8: Image produced by the Julia functions, `linkPositions` and `plot_points`, on three consecutive calls, with Call 1 corresponding to $(\theta_1, \theta_2, \theta_2) = (20°, 20°, 20°)$, Call 2 corresponding to $(\theta_1, \theta_2, \theta_2) = (-20°, 20°, 20°)$, and Call 3 corresponding to $(\theta_1, \theta_2, \theta_2) = (-140°, -90°, -90°)$.

$\arccos(x)$ for inverse cosine, and

$\arctan(x)$ for inverse tangent.

- In Julia and many other programming languages, they are written a bit shorter, namely:

  `asin` for inverse sine,

  `acos` for inverse cosine, and

  `atan` for inverse tangent.

- Another way, which was introduced in the 1800s by John Herschel, is to write them with a "−1" as a power:

  $\sin^{-1}(x)$ for inverse sine,

  $\cos^{-1}(x)$ for inverse cosine, and

  $\tan^{-1}(x)$ for inverse tangent.

  The Julia notation is the most practical.

Remark 1.25. ***In Julia, the trigonometric functions*** sin, cos, *and* tan, *as well as their inverses,* ***operate in radians by default****. However, if you want to use degrees, Julia provides specific functions for that:* $\text{sind}(x)$, $\text{cosd}(x)$, *and* $\text{tand}(x)$ *to compute the sine, cosine, and tangent of* x*, where* x *is in degrees. For the inverse trigonometric functions in degrees, the functions* $\text{asind}(x)$, $\text{acosd}(x)$, *and* $\text{atand}(x)$ *compute the arcsine, arccosine, and arctangent of* x*, returning the result in degrees.* ■

1.6.5 Understanding Inverse Trigonometric Functions

In **Figs. 1.9-(a)**, **1.10-(a)**, and **1.11-(a)**, we show the three functions, namely $\sin(x)$, $\cos(x)$, and $\tan(x)$, as we would normally plot them. In each case, the functions fail the horizontal line test for being injective (1:1), and hence they do not have inverses. If you properly restrict their domains, as shown in the middle plots, then they do become injective (1:1) and have an inverse, as shown in the plot to the right, that is, **Figs. 1.9-(c)**, **1.10-(c)**, and **1.11-(c)**.

Consider the sine function. It is injective (1:1) and surjective (onto) for all of the following domains:

- $\sin : [-\frac{\pi}{2}, \frac{\pi}{2}] \to [-1, 1]$.

- $\sin : [\frac{\pi}{2}, \frac{3\pi}{2}] \to [-1,1]$.
- $\sin : [-\frac{3\pi}{2}, -\frac{\pi}{2}] \to [-1,1]$.
- $\sin : [-\frac{\pi}{2} + k\pi, -\frac{\pi}{2} + k\pi] \to [-1,1]$, $k \in \mathbb{Z}$.

For all of these domains, $\sin(x)$ satisfies the "horizontal line test", and hence, an inverse can be defined in all of these situations. The key to insuring that a function is injective is to define an appropriate domain.

The domain choices made in Table 1.4 and Figs. 1.9, 1.10, and 1.11 are called the **Principal Domains**. Paraphrasing, the principal domain of a trigonometric function is the largest interval near the origin (positive or symmetric about zero, if possible), where the function is one-to-one, strictly increasing or decreasing, and takes on all possible values.

We summarize this information in Table 1.4 and suggest the following videos:

- Evaluating Inverse Trigonometric Functions (arcsin, arccos, arctan) Using Unit Circle by Mario's Math Tutoring.
- Inverse Trigonometric Functions by Professor Dave Explains.

	$\sin(x)$	$\cos(x)$	$\tan(x)$	$\arcsin(x)$	$\arccos(x)$	$\arctan(x)$
Domain	$[-\frac{\pi}{2}, \frac{\pi}{2}]$	$[0, \pi]$	$(-\frac{\pi}{2}, \frac{\pi}{2})$	$[-1,1]$	$[-1,1]$	$(-\infty, \infty)$
Codomain	$\mathbb{R}$	$\mathbb{R}$	$\mathbb{R}$	$\mathbb{R}$	$\mathbb{R}$	$\mathbb{R}$
Range	$[-1,1]$	$[-1,1]$	$(-\infty, \infty)$	$[-\frac{\pi}{2}, \frac{\pi}{2}]$	$[0, \pi]$	$(-\frac{\pi}{2}, \frac{\pi}{2})$

Table 1.4: Domains, Codomains, and Ranges of Trigonometric and Inverse Trigonometric Functions

We illustrate the importance of principal domains with the following code in which we evaluate sin, cos, and tan for x in their principal domains and outside them. Then, we apply the inverse functions and see if we recover the original x values or not!

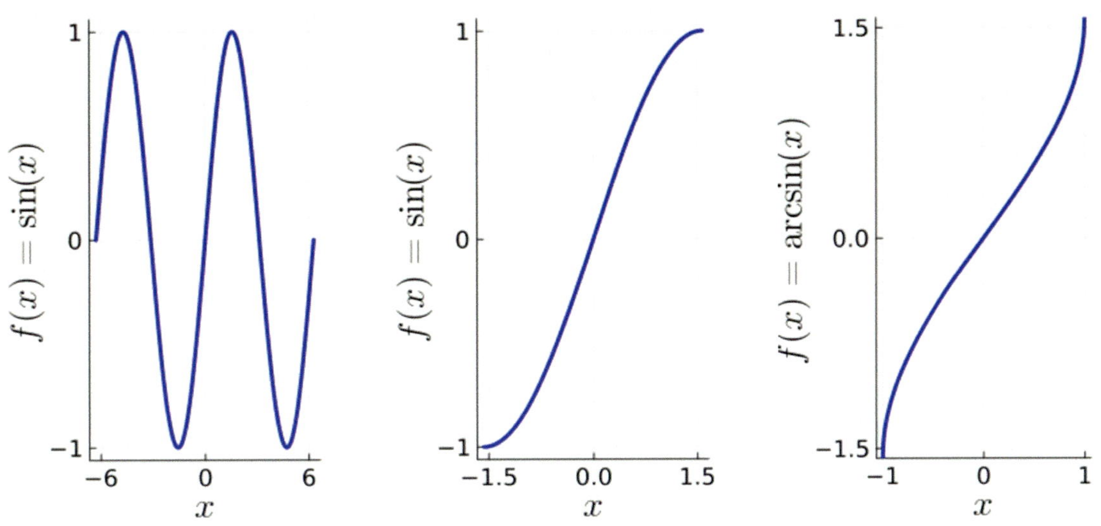

(a) (b) (c)

Figure 1.9: The sine function and its inverse. The sine function $\sin : \mathbb{R} \to [-1,1]$ in (a) is surjective (onto) $[-1,1]$ but does not have an inverse over its entire domain. You can see that it fails the "horizontal line test". However, when restricted to $[-\frac{\pi}{2}, \frac{\pi}{2}]$, as in (b), it becomes injective (1:1) and has an inverse, shown on the right in (c).

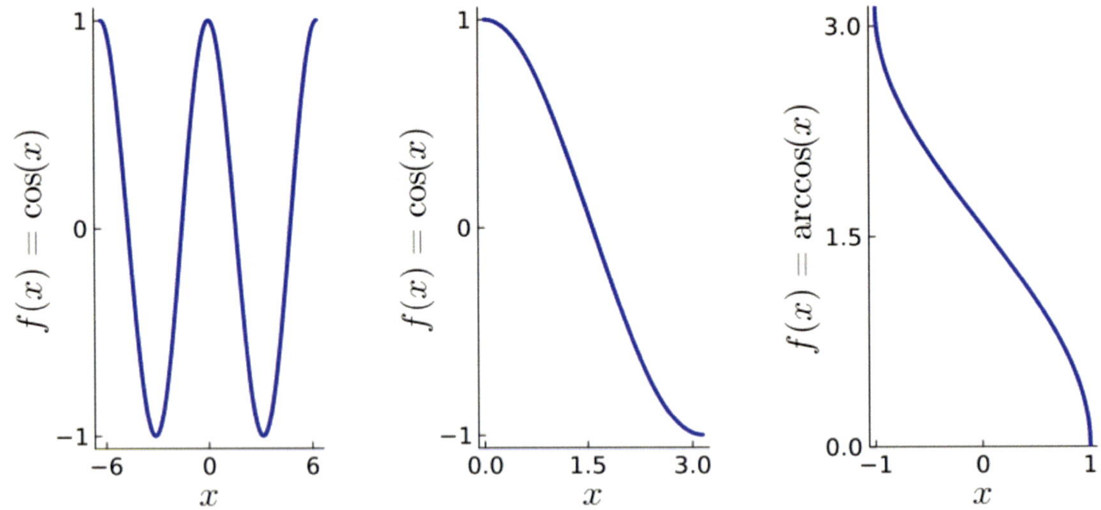

(a) (b) (c)

Figure 1.10: The cosine function and its inverse. The cosine function $\cos : \mathbb{R} \to [-1, 1]$ in (a) is surjective (onto) $[-1, 1]$ but does not have an inverse over its entire domain. You can see that it fails the "horizontal line test". However, when restricted to $[0, \pi]$, as in (b), it becomes injective (1:1) and has an inverse, shown on the right in (c).

```julia
# A few examples illustrating principal domains
# we take values of sin, cos, and tan within their principal
# domains and outside them. Then we apply the inverse functions
# and see if we recover the original values or not!

using Printf

x = pi*[-1.3, -1, -0.4, -0.25, 0, 0.25, 0.4, 1, 1.3]
function_names = ["sin(x)", "cos(x)", "tan(x)"]

for k = 1:length(x)
    xk = x[k]
    InData = [sin(xk) cos(xk) tan(xk)]
    OutData = [asin(sin(xk)) acos(cos(xk)) atan(tan(xk))]
    OriginalXk = [xk xk xk]
    FlagDomainProblem = (abs.(OutData - OriginalXk) .> 1e-8) # Using a small
                                                #threshold to account for floating-point errors

    println("For x = $(@sprintf("%.10f", xk)):")
    println("Trig function values:     ", InData)
    println("After applying inverse: ", OutData)
    println("Hoping for:                ", OriginalXk)
    for j = 1:length(FlagDomainProblem)
        if FlagDomainProblem[j]
            println("x = $xk  is outside the principal domain for $(function_names[j]) .")
        else
            println("x = $xk  is inside the principal domain for $(function_names[j]) .")
        end
    end

    println("------")
end
```

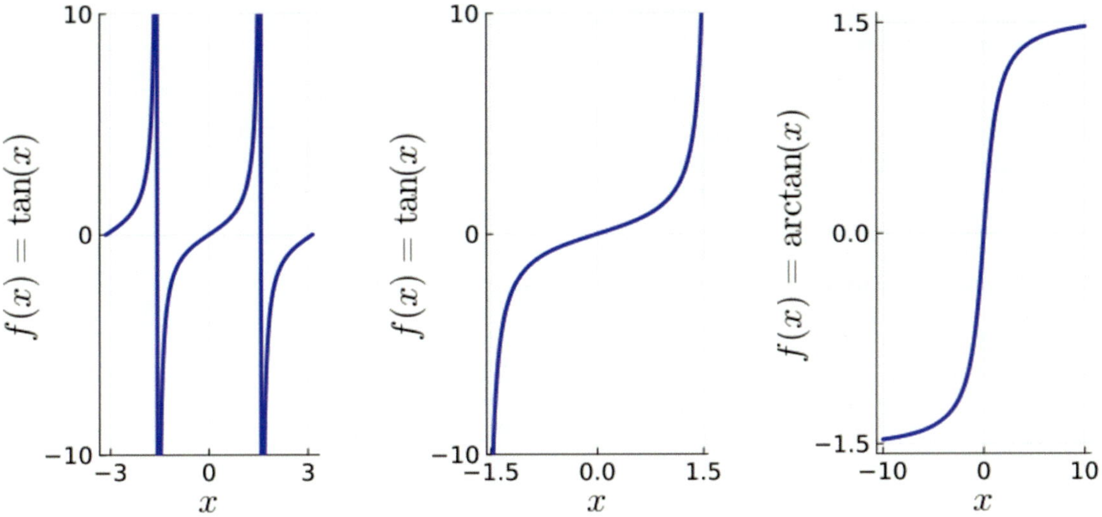

(a) (b) (c)

Figure 1.11: The tangent function and its inverse. The tangent function as shown in (a) does not have an inverse over $\mathbb{R}$, even after setting $\tan(x)$ to zero at odd multiples of $\frac{\pi}{2}$, where it blows up. However, when restricted to the open interval $(-\frac{\pi}{2}, \frac{\pi}{2})$, as shown in (b), it becomes injective (1:1) and has an inverse, shown on the right in (c).

Output (after truncating numbers to 4 decimal places)

```
For x = -4.0841:
Trig function values:    0.8090, -0.5878, -1.3764
After applying inverse: 0.9425, 2.1991, -0.9425
Hoping for:              -4.0841, -4.0841, -4.0841
x = -4.0841 is outside the principal domain for sin(x).
x = -4.0841 is outside the principal domain for cos(x).
x = -4.0841 is outside the principal domain for tan(x).
------
For x = -3.1416:
Trig function values:    -0.0000, -1.0000, 0.0000
After applying inverse: -0.0000, 3.1416, 0.0000
Hoping for:              -3.1416, -3.1416, -3.1416
x = -3.1416 is outside the principal domain for sin(x).
x = -3.1416 is outside the principal domain for cos(x).
x = -3.1416 is outside the principal domain for tan(x).
------
For x = -1.2566:
Trig function values:    -0.9511, 0.3090, -3.0777
After applying inverse: -1.2566, 1.2566, -1.2566
Hoping for:              -1.2566, -1.2566, -1.2566
x = -1.2566 is inside the principal domain for sin(x).
x = -1.2566 is outside the principal domain for cos(x).
x = -1.2566 is inside the principal domain for tan(x).
------
For x = -0.7854:
Trig function values:    -0.7071, 0.7071, -1.0000
After applying inverse: -0.7854, 0.7854, -0.7854
Hoping for:              -0.7854, -0.7854, -0.7854
x = -0.7854 is inside the principal domain for sin(x).
```

```
x = -0.7854 is outside the principal domain for cos(x).
x = -0.7854 is inside the principal domain for tan(x).
------
For x = 0.0000:
Trig function values:   0.0000, 1.0000, 0.0000
After applying inverse: 0.0000, 0.0000, 0.0000
Hoping for:             0.0000, 0.0000, 0.0000
x = 0.0000 is inside the principal domain for sin(x).
x = 0.0000 is inside the principal domain for cos(x).
x = 0.0000 is inside the principal domain for tan(x).
------
For x = 0.7854:
Trig function values:   0.7071, 0.7071, 1.0000
After applying inverse: 0.7854, 0.7854, 0.7854
Hoping for:             0.7854, 0.7854, 0.7854
x = 0.7854 is inside the principal domain for sin(x).
x = 0.7854 is inside the principal domain for cos(x).
x = 0.7854 is inside the principal domain for tan(x).
------
For x = 1.2566:
Trig function values:   0.9511, 0.3090, 3.0777
After applying inverse: 1.2566, 1.2566, 1.2566
Hoping for:             1.2566, 1.2566, 1.2566
x = 1.2566 is inside the principal domain for sin(x).
x = 1.2566 is inside the principal domain for cos(x).
x = 1.2566 is inside the principal domain for tan(x).
------
For x = 3.1416:
Trig function values:   0.0000, -1.0000, -0.0000
After applying inverse: 0.0000, 3.1416, -0.0000
Hoping for:             3.1416, 3.1416, 3.1416
x = 3.1416 is outside the principal domain for sin(x).
x = 3.1416 is inside the principal domain for cos(x).
x = 3.1416 is outside the principal domain for tan(x).
------
For x = 4.0841:
Trig function values:   -0.8090, -0.5878, 1.3764
After applying inverse: -0.9425, 2.1991, 0.9425
Hoping for:             4.0841, 4.0841, 4.0841
x = 4.0841 is outside the principal domain for sin(x).
x = 4.0841 is outside the principal domain for cos(x).
x = 4.0841 is outside the principal domain for tan(x).
------
```

1.6.6 Composing Trigonometric Functions and Inverse Trigonometric Functions

The video Precalculus Algebraic Expression Inverse Trig and Double Angle by Malmos Math shows how to use right triangles and basic trigonometric identities to derive algebraic formulas for expressions such as:

$$\boxed{\cos\left(2\,\mathrm{atan}\left(\frac{x}{a}\right)\right).} \tag{1.11}$$

Let $\theta := \mathrm{atan}\left(\frac{x}{a}\right) \iff \tan(\theta) = \frac{x}{a}$. Because tangent equals opposite over adjacent, θ is as represented in Fig. 1.12. From our handy list of trigonometric identities, we have $\cos(2\theta) = 2\cos^2(\theta) - 1$. From our right triangle, because cosine equals adjacent over hypotenuse, we have $\cos(\theta) = \frac{a}{\sqrt{a^2+x^2}}$. Now, we just have some algebra to do:

$$\begin{aligned}\cos\left(2\,\mathrm{atan}\left(\frac{x}{a}\right)\right) &= \cos(2\theta)\\ &= 2\cos^2(\theta) - 1\\ &= 2\left(\frac{a}{\sqrt{a^2+x^2}}\right)^2 - 1\\ &= \frac{2a^2}{a^2+x^2} - \frac{a^2+x^2}{a^2+x^2}\\ &= \frac{a^2-x^2}{a^2+x^2}.\end{aligned}$$

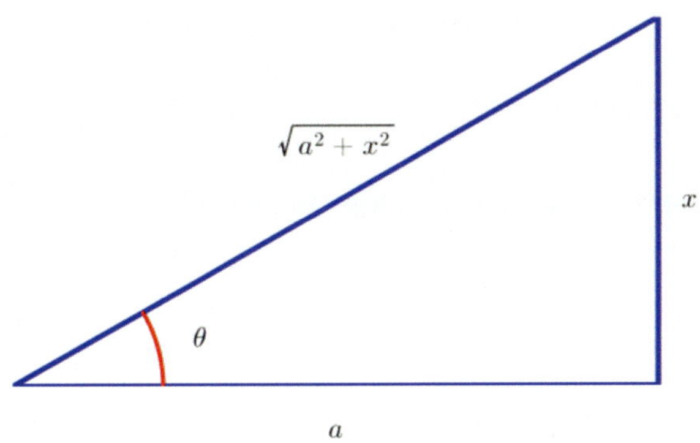

Figure 1.12: Right triangle with base a and height x.

1.7 Powers and Roots: Common, but Often Less Understood

Consider the function x^y, where y is not an integer. For example, $x^{3/2}$. What does it really mean? We want it to mean $\sqrt{x^3}$, but then we have to wrestle with the fact that, as long as we work within the real numbers, the expression only makes sense for $x \geq 0$. Moreover, when we do calculations on a computer, we want it to return a single value, even though we know that square roots can be negative. So, let us define once and for all what we mean by whole-number roots, $\sqrt[n]{x}$, $n \geq 1$, which are often denoted as $x^{1/n}$, and then rational powers, $x^{p/q}$, for integers p and q. Finally, we will consider the subject of x^y for a real number y.

1.7.1 nth Roots, aka Integer Roots

Defining Roots in a Careful Manner and the Notation $x^{1/n}$

Definition 1.26. *nth Roots*

(a) *For $n \geq 1$ an **odd integer** and x any real number, we define $\sqrt[n]{x}$ to be the unique real number y such that $y^n = x$. We note that x and y will have the same sign.*

(b) *For $n \geq 1$ an **even integer** and x a nonnegative real number, we define $\sqrt[n]{x}$ to be the unique nonnegative real number y such that $y^n = x$.*

(c) *For **all positive integers** n, we define $x^{1/n}$ to be the same as $\sqrt[n]{x}$. Hence, if n is odd, $x^{1/n}$ is* *defined for all real-valued numbers x, and if n is even, it is defined for only nonnegative x.* **The zeroth root of x is not defined.**

(d) *For* **all positive integers** *n and $x \neq 0$, we define $x^{-1/n}$ to be the same as $\frac{1}{x^{1/n}}$. Hence, if n is odd, $x^{-1/n}$ is defined for nonzero x, and if n is even, it is defined for only positive x.*

It is worth recalling that, for x any nonzero real number, we define $x^0 := 1$. This explains why a zeroth root is not defined (the zeroth root of 1 could be anything). **In calculus, it is important to leave 0^0 as undefined.** For more on zero to power zero, check out this link. There are plenty of sources that can lead one astray, such as Math is Fun.

Example 1.27. *Use the Bisection Algorithm to compute $\sqrt[7]{x}$ for $x = 3$ and $x = -5$. Compute the answer to a tolerance of 1e-5*

Solution:

The Bisection Algorithm implements the literal definition of a root: $\sqrt[n]{x}$ is the unique real number y such that $y^n = x$. Here, it is solving $y^7 - x = 0$.

```
function bisection(f, a, b, tol)
    if f(a) * f(b) > 0
        error("f(a) and f(b) must have opposite signs")
        return NaN
    end
    kmax = 1e5 # max number of loops so that the algorithm terminates
    k=1
    c=NaN #define c out of the loop so we can access it
    while ( (b - a) / 2 > tol || abs(f(c))> tol ) && k < kmax
        c = (a + b) / 2
        #
        if f(a) * f(c) < 0
            b = c
        else
            a = c
        end
        k = k+1
    end
    root_ErrorBound = (b - a)/2
    return (est=c, low=a, up=b, error=root_ErrorBound)
end
```

```
# Define the function whose root we want to find
x=3
f(y) = y^7 - x

# Call the bisection method on this function
F = bisection(f, 1.0, 2.0, 1e-5)
@show F

println(" ")
println("The 7th root is approximately ", F.est,)
println("The 7th power of the root is ", F.est^7)
```

Output

```
F = (est = 1.1699304580688477, low = 1.1699304580688477,
up = 1.169931411743164, error = 4.76837158203125e-7)

The 7th root is approximately 1.1699304580688477
The 7th power of the root is 2.9999936334014214
```

We call the function with a different set of bracketing points and $x = -5$.

```
# Define the function whose root we want to find
x=-5
f(y) = y^7 - x

# Call the bisection method on this function
F = bisection(f, -3.0, 0.0, 1e-5)
@show F

println(" ")
println("The 7th root is approximately ", F.est,)
println("The 7th power of the root is ", F.est^7)
```

Output

```
F = (est = -1.2584989070892334, low = -1.2584996223449707,
up = -1.2584989070892334, error = 3.5762786865234375e-7)

The 7th root is approximately -1.2584989070892334
The 7th power of the root is -4.999998788762909
```

Example 1.28. *Compute the following roots for $x = 3$ and $x = -5$. When defined, give the answers to 3 decimal points. You can use a calculator or the Julia function for x^y.*

(a) $\sqrt[3]{x}$

(b) $\sqrt[2]{x}$

(c) $\sqrt[0]{x}$

(d) $\sqrt[1]{x}$

(e) $\sqrt[5]{x}$

Solution: We use a built-in Julia function to perform the calculations.

(a) $\sqrt[3]{3} = 1.442$ and $\sqrt[3]{-5} = -1.710$. We use the command `cbrt(x)` when computing the cubed root of a negative number in Julia. Otherwise, Julia returns a domain error with $(-5)^{1/3}$. The reason is explained when we cover rational powers. A second method is shown below.

(b) $\sqrt[2]{3} = 1.732$ while $\sqrt[2]{-5}$ is undefined (when working with the real numbers). We note that per our previously defined notations, $\sqrt{x} := \sqrt[2]{x}$.

(c) $\sqrt[0]{3}$ and $\sqrt[0]{-5}$ are both undefined.

(d) $\sqrt[1]{3} = 3$ and $\sqrt[1]{-5} = -5$.

(e) $\sqrt[5]{3} = 1.246$ and $\sqrt[5]{-5} = -1.380$. In Julia, you can use the command `sign(x)*(abs(x)^(1/5))`, where

$$\text{sign}(x) = \begin{cases} +1 & x > 0 \\ -1 & x < 0 \\ \ \ 0 & x = 0 \end{cases}$$

By taking the absolute value, Julia has no trouble computing the root and you restore the correct sign with the sign function.[7]

```
function myRoot(x::Float64,n::Int)
    if n % 2 == 0
        println(n, " is even.")
        if x < 0
            println(n, "Root is not defined for n even and x negative.")
            root = NaN
        end
    else
        println(n, " is odd.")
    end
    root = sign(x)*( abs(x)^(1/n) )
    return root
end

myRoot(-5.0,5)
```

Output

```
-1.379729661461215
```

■

Proposition 1.29: Root of a Product Equals the Product of Roots

For all natural numbers $n \geq 1$ and real numbers $x > 0$ and $y > 0$,

$$\sqrt[n]{x \cdot y} = \sqrt[n]{x} \cdot \sqrt[n]{y}. \tag{1.12}$$

Proof: This proof, like many proofs in early math courses, is a "simple application" of the definitions. Let $z_1 := \sqrt[n]{x}$ and $z_2 := \sqrt[n]{y}$. Then, by the definition of an nth root, z_1 and z_2 must satisfy

$$(z_1)^n = x \text{ and } (z_2)^n = y.$$

Hence,

$$x \cdot y = (z_1)^n \cdot (z_2)^n = (z_1 \cdot z_2)^n$$

by standard rules for multiplication. Indeed,

$$\underbrace{z_1 \cdot z_1 \cdots z_1}_{\text{n times}} \cdot \underbrace{z_2 \cdot z_2 \cdots z_2}_{\text{n times}} = \underbrace{(z_1 \cdot z_2) \cdots (z_1 \cdot z_2)}_{\text{n times}} = (z_1 \cdot z_2)^n.$$

Because $(z_1 \cdot z_2)^n = x \cdot y$, we can conclude that $z_1 \cdot z_2 = \sqrt[n]{x \cdot y}$, thereby proving that $\sqrt[n]{x} \cdot \sqrt[n]{y} = \sqrt[n]{x \cdot y}$.

■

Remark 1.30. *What if we have $\sqrt[n]{x^m}$ for n and m positive integers? You can successively apply Prop. 1.29 to obtain:*

$$\sqrt[n]{x^m} = \sqrt[n]{x} \cdot \sqrt[n]{x^{m-1}} = \sqrt[n]{x} \cdot \sqrt[n]{x} \cdot \sqrt[n]{x^{m-2}} = \underbrace{\left(\sqrt[n]{x}\right) \cdot \left(\sqrt[n]{x}\right) \cdots \left(\sqrt[n]{x}\right)}_{m \text{ times}} = \left(\sqrt[n]{x}\right)^m.$$

For the first factorization of the nth root, we apply $x = x$ and $y = x^{m-1}$ to factor $\sqrt[n]{x^m}$ as $\sqrt[n]{x} \cdot \sqrt[n]{x^{m-1}}$. For the second factorization, we apply $x = x$ and $y = x^{m-2}$ to factor $\sqrt[n]{x^{m-1}}$ as $\sqrt[n]{x} \cdot \sqrt[n]{x^{m-2}}$; etc., until the last factorization, when we apply $x = x$ and $y = x$ to factor $\sqrt[n]{x^2}$ as $\sqrt[n]{x} \cdot \sqrt[n]{x}$. ■

[7] The sign function is sometimes called the signum function because "signum" is Latin for "sign".

Proposition 1.31: Nested Radicals (aka Nested Roots) Can Be Combined

For all natural numbers $n \geq 1$, $m \geq 1$ and real numbers $x > 0$,

$$\sqrt[n]{\sqrt[m]{x}} = \sqrt[n \cdot m]{x}.$$

In other symbols, $\left(x^{1/m}\right)^{1/n} = (x)^{\frac{1}{m \cdot n}}$.

Important: $\sqrt[n]{\sqrt[m]{x}} = \sqrt[n \cdot m]{x}$ **and NOT** $\sqrt[n+m]{x}$. This is analogous to $(x^n)^m = x^{n \cdot m}$ and **NOT** x^{n+m}, because the exponential of an exponential means to multiply the exponents, NOT to add them. See Prop. 1.40 for general "Power Towers".

Proof: Let $z_1 := \sqrt[m]{x}$ so that $(z_1)^m = x$. Next, set $z_2 = \sqrt[n]{z_1}$ so that $(z_2)^n = z_1$. Combining these two results, we have

$$z_1 = (z_2)^n \implies \underbrace{(z_1)^m}_{x} = ((z_2)^n)^m = (z_2)^{n \cdot m} \implies x = (z_2)^{n \cdot m}.$$

Hence,

$$x = (z_2)^{n \cdot m},$$

and therefore, $z_2 = \sqrt[n \cdot m]{x}$, that is, $\sqrt[n]{\sqrt[m]{x}} = \sqrt[n \cdot m]{x}$. ∎

In a similar fashion, for any finite number of nested (integer) roots of $x > 0$, we can combine them into a single (integer) root of x by multiplying the integers specifying the nested roots. A more general procedure is called "denesting of radicals".

Proposition 1.32: Roots and Powers are Inverse Operations

For all natural numbers $n \geq 1$ and real numbers $x > 0$,

$$\sqrt[n]{x^n} = x \text{ and } \left(\sqrt[n]{x}\right)^n = x.$$

In other symbols, $(x^n)^{1/n} = x$ and $\left(x^{1/n}\right)^n = x$. We can also write this as $(x^n)^{1/n} = x^{\frac{n}{n}} = x^1 = x$ and $\left(x^{1/n}\right)^n = x^{\frac{n}{n}} = x^1 = x$.

Proof: Let $z_1 := \sqrt[n]{x^n}$. Then, by the definition of the nth root, this holds if, and only if, $(z_1)^n = x^n$. Hence, $z_1 = x$ by standard properties of multiplication.[8] This proves the first statement.

Next, let $z_2 := \sqrt[n]{x}$. Then, once again by the definition of the nth root, this holds if, and only if, $(z_2)^n = x$. Hence, $\left(\sqrt[n]{x}\right)^n = x$. ∎

1.7.2 Rational Exponents or Powers

Now we seek to explore the meaning of $x^{p/q}$. A key observation is that we wish to obtain the same answer, whether we use $\sqrt[q]{x^p}$ or $\left(\sqrt[q]{x}\right)^p$ as the definition. Just as with roots, for negative values of x, we have to be careful.

Defining Rational Exponents (Powers) in a Careful Manner

Definition 1.33. *Rational Powers*

(a) *For $x \geq 0$ any nonnegative real number, and for p and q any positive integers, both $\sqrt[q]{x^p}$ and $\left(\sqrt[q]{x}\right)^p$ return the same answer.[a] Hence, we can define $x^{p/q}$ using either one. For definiteness,*

[8] Suppose that x and y are both positive real numbers. Then $x = y \iff x^n = y^n$ for all positive integers, n. Indeed, if $x = y$, then $x^n = y^n$ is immediate. Suppose, therefore, that $x \neq y$. Without loss of generality, we can assume $0 < x < y$. Then, $x \cdot x < x \cdot y < y \cdot y$, which proves that $x^2 < y^2$. Other powers are similar. Hence, $x^n = y^n$ and $x > 0, y > 0$ together imply that $x = y$.

we define $x^{p/q} := \sqrt[q]{x^p}$.

(b) *For* $x > 0$ *any positive real number, and for* p *and* q *any positive integers, we define* $x^{-p/q} := \frac{1}{x^{p/q}}$.

(c) *For* $x < 0$ *any negative real number, and for all* p *and* q *integers, we are leaving* $x^{p/q}$ *as* **undefined**. *The reason is indicated below: we do not want to deal with complex roots at this time.*

[a]This follows from Prop. 1.29

Remark 1.34. *When* p *and* q *are both* **odd integers**, *it is OK to define* $x^{p/q} := \text{sign}(x)\sqrt[q]{|x|^p}$. *For* $p = 1$ *and* $q = 3$, *this is consistent with how we defined the cube root of* -1. *However, how would we define* $x^{0.6}$ *for* $x < 0$? *We are leaving it as* **undefined** *when* $x < 0$ *because 0.6 could be interpreted as 3/5 or 6/10, which is problematic, as we show next.*

- $\sqrt[5]{(-1)^3} = \sqrt[5]{-1} = -1$ *(when all of the powers and roots are odd, it all works), whereas*
- $\sqrt[10]{(-1)^6} = \sqrt[10]{1} = +1$, *where we added the plus sign for emphasis.*

To avoid these inconsistencies, we recommend that we adopt $x^{p/q}$ *as being undefined for* $x < 0$.

Note: *Some programming languages, such as Julia, respond with a TYPE*[9] *error, when the user tries to compute* $(-1)^{0.6}$, *while others, such as MATLAB, return a complex number.*

■

Example 1.35. *Compute the following rational powers for* $x = 3$ *and* $x = -5$:

(a) $x^{3/2}$

(b) $x^{2/3}$

(c) $x^{-3/2}$

(d) $x^{-2/3}$

(e) $x^{8/12}$

(f) $x^{3/5}$

Solution:

(a)	$3^{3/2} = \sqrt{3^3} \approx 5.196$	$(-5)^{3/2} = \text{undefined}$
(b)	$3^{2/3} = \sqrt[3]{3^2} \approx 2.080$	$(-5)^{2/3} = \text{undefined}$
(c)	$3^{-3/2} = \frac{1}{3^{3/2}} \approx 0.192$	$(-5)^{-3/2} = \text{undefined}$
(d)	$3^{-2/3} = \frac{1}{3^{2/3}} \approx 0.481$	$(-5)^{-2/3} = \text{undefined}$
(e)	$3^{8/12} = \sqrt[12]{3^8} = \sqrt[3]{3^2}$	$(-5)^{8/12} = \text{undefined}$
(f)	$3^{3/5} = \sqrt[5]{3^3} \approx 1.933$	$(-5)^{3/5} = \text{undefined}$
(f′)	$3^{3/5} = \sqrt[5]{3^3} \approx 1.933$	$(-5)^{3/5} = -\sqrt[5]{5^3} \approx -2.627$

[9]The TYPE error in Julia arises when we use real numbers (Float64 and Int64) in the expression $(-1)^{0.6}$, and yet the answer is a complex number. We show later that if you use numbers that are of complex TYPE, Julia returns a complex answer, and there is no longer a TYPE error. This is a FEATURE and not a BUG! Julia warns you, while MATLAB assumes you know what you are doing.

Note: While the solution in (f'), $(-5)^{3/5} = -\sqrt[5]{5^3}$, is **not wrong, neither is it a recommended solution.** Why? Because, if instead, we wrote $0.6 = 6/10$ for $\frac{3}{5}$, which seems totally legit, we would then likely define $(-5)^{6/10}$ as $\sqrt[10]{(-5)^6}$, which gives a positive number due to $(-5)^6 = \left((-5)^2\right)^3 = 25^3 > 0$. When $0.6 = \frac{6}{10}$ is inconsistent with writing $0.6 = \frac{3}{5}$, we can have all sorts of confusion on hand. We have to be careful! ■

Julia has a command for x^y, but not one for $x^{p/q}$ with p and q called out separately. We can create our own function, of course. What happens if we use Julia's power command ^ to solve the above problem?

```
p = 3
q = 2
a = p/q
@show a

x = 3
y = -5
@show x^a
@show y^a
```

Output

```
a = 1.5
x ^ a = 5.196152422706632

DomainError with -5.0:
Exponentiation yielding a complex result requires a complex argument.
Replace x^y with (x+0im)^y, Complex(x)^y, or similar.
```

Note: `im` is the Julia symbol for $\mathbf{i}$, the imaginary number satisfying $(\mathbf{i})^2 = -1$. We learned in algebra how to use complex numbers when computing square roots of a negative number, but not necessarily how to use complex numbers to compute $\sqrt[3]{-1}$ or $\sqrt[8]{-1}$, for example, because, when we allow complex solutions, the nth root, $\sqrt[n]{-1}$, involves n numbers, with at most one of them being real.

Example 1.36. *Solve the following SAT problem for p/q:*

$$\frac{\sqrt{x^5}}{\sqrt[3]{x^4}} = x^{\frac{p}{q}}.$$

Solution:

$$\sqrt{x^5} = x^{5/2}$$
$$\sqrt[3]{x^4} = x^{4/3}$$
$$\Downarrow$$
$$\frac{\sqrt{x^5}}{\sqrt[3]{x^4}} = x^{5/2} \cdot x^{-4/3}$$
$$= x^{5/2-4/3}$$
$$= x^{7/6}.$$

Hence, $p/q = 7/6$. ■

Example 1.37. *Solve the following Math Olympiad problem for $K > 0$:*

$$\sqrt{K \cdot \sqrt{K \cdot \sqrt{K}}} = 3.$$

Solution:
Squaring both sides gives $K \cdot \sqrt{K \cdot \sqrt{K}} = 3^2$. Then, using our notation, $x^{\frac{1}{2}} = \sqrt{x}$, and the two facts $x^a \cdot x^b = x^{a+b}$ and $(x^a)^b = x^{a \cdot b}$, we obtain:

$$\begin{gathered} K \cdot \left(K \cdot K^{\frac{1}{2}}\right)^{\frac{1}{2}} = 3^2 \\ \Updownarrow \\ K \cdot \left(K^{\frac{3}{2}}\right)^{\frac{1}{2}} = 3^2 \\ \Updownarrow \\ K \cdot K^{\frac{3}{4}} = 3^2 \\ \Updownarrow \\ K^{\frac{7}{4}} = 3^2 \\ \Updownarrow \\ K^7 = 3^8 \\ \Updownarrow \\ K = 3^{\frac{8}{7}}. \end{gathered}$$

Hence, $K \approx 3.509792$.

An alternative solution is to deduce $\sqrt{K \cdot \sqrt{K \cdot \sqrt{K}}} = 3 \iff \left(K \cdot \left(K \cdot K^{\frac{1}{2}}\right)^{\frac{1}{2}}\right)^{\frac{1}{2}} = 3 \iff K^{\frac{7}{8}} = 3$, and then to conclude that $K = \left(K^{\frac{7}{8}}\right)^{\frac{8}{7}} = 3^{\frac{8}{7}}$. ■

1.7.3 Real Exponents

What is x^y when the exponent y is a real number? It is traditional to define x^y using the natural logarithm, as introduced in the next section. It can also be defined using "limits" of rational powers, which can be reduced to integer roots and integer powers. Let us consider **conceptually how to define $\mathbf{x}^{\pi}$**. If we approximate the real number π by a rational number $r = p/q$, then we can use that number in place of π to compute

$$x^{\pi} \approx x^{p/q} := \sqrt[q]{x^p}.$$

Then, while holding x constant, we take successively "better" rational approximations of π (which means, $|\pi - \frac{p}{q}|$ becomes closer to zero) and check that the result $x^{p/q}$ "tends" to a fixed value. The same process works for an arbitrary real number y in place of π.

Of course, when we form a rational approximation of a real number y, p/q will not necessarily be a ratio of odd numbers, so we must only define x^y for $x \geq 0$.

What is x^y when the exponent y is a real number?

Definition 1.38. *For $x \geq 0$, x^y is the "limiting" value obtained as we approximate y more and more closely by a rational number, such as a decimal expansion.*

Table 1.5 illustrates this process for 2^{π}; that is, $x = 2$ and $y = \pi$.

Example 1.39. *Compute the following powers using Julia.*

(a) $\left(\sqrt{2}\right)^{\pi}$

Approximation of π	Approximation of 2^π
3	8.00000
22/7	8.51396
3.14	8.81524
3.14159	8.82498
3.141592653589793	8.824977827076287

Table 1.5: Computation of 2^π for different rational approximations of π.

(b) $\left(\sqrt{2}\right)^{-\pi}$

(c) $e^{2/3}$

(d) π^e

(e) $(-3)^e$

Solution:

(a) $\left(\sqrt{2}\right)^{\pi} \approx 2.97069$

(b) $\left(\sqrt{2}\right)^{-\pi} \approx 0.336623$

(c) $e^{2/3} \approx 1.94773$

(d) $\pi^e \approx 22.4592$

(e) $(-3)^e$ is **undefined**.

```
x = [sqrt(2) sqrt(2) exp(1) π]
y = [π -π 2/3 exp(1)]
x.^y
```

Output

```
The 1×4 Matrix{Float64}:
 2.97069  0.336623  1.94773  22.4592
```

```
# Two ways to access Euler's constant in Julia
@show e # obtained by \euler followed by the Tab key
println("The value of Euler's constant is ",e) # leaves it as the symbol e
@show exp(1);
println("The value of Euler's constant is ", exp(1))
@show e^e # definitely uses the value of Euler's constant
@show exp(1)^exp(1);
```

Output

```
e = e
The value of Euler's constant is e
exp(1) = 2.718281828459045
The value of Euler's constant is 2.718281828459045
e ^ e = 15.154262241479262
exp(1) ^ exp(1) = 15.15426224147926W
```

```
(-3)^exp(1)
```

Output

```
DomainError with -3.0:
Exponentiation yielding a complex result requires a complex argument.
Replace x^y with (x+0im)^y, Complex(x)^y, or similar.
```

```
(-3 + 0*im)^exp(1)
```

Output

```
-12.546688358893388 + 15.334119260562304im
```

■

Proposition 1.40: Power Towers

For all $x > 0$, and all $y, z \in \mathbb{R}$

$$(x^y)^z = x^{y \cdot z}. \tag{1.13}$$

From Prop. 1.40, we have that $(x^y)^z = (x^z)^y$ because both are equal to $x^{y \cdot z}$. Moreover, we can use this fact to reduce power towers of any (finite) length to a single power, such as, for all $x > 0$:

$$\left(\left(\left(x^{y_1}\right)^{y_2}\right)^{y_3} \cdots\right)^{y_n} = x^{y_1 \cdot y_2 \cdot \cdots \cdot y_n}.$$

The proof of Prop. 1.40 is given in Section 1.19.

Proposition 1.41: Other Useful Properties of Real Exponents

Suppose that $x > 0$ is a positive real number. Then the following relations hold:

(a) For all $y \in \mathbb{R}$, $x^y > 0$.

(b) For all $y \in \mathbb{R}$, $x^{-y} = 1/x^y$.

(c) For all $y, z \in \mathbb{R}$, $x^{y+z} = x^y \cdot x^z$.

(d) For all $y > 0$, and $z \in \mathbb{R}$, $x^z \cdot y^z = (x \cdot y)^z$.

To prove these properties, we would replace the exponents with rational approximations and then apply properties of rational powers.

1.8 Exponentials and Logarithms

When working with x^y, which variable(s) are we changing: x, y, or both? Exponential functions remove this vagueness by fixing x to a constant value and then varying the exponent. The constant value is called a base, and following Descartes' notation[10], we should use a lowercase a, b, or c to denote it.

Exponential Function of Base a

Definition 1.42. *For a real number $a > 0$, the* ***exponential function of base a*** *is $\exp_a : \mathbb{R} \to (0, \infty)$ divided by $\exp_a(x) := a^x$.*

[10]In 1637, René Descartes "invented" the convention of representing unknowns in equations by x, y, and z, and known quantities by a, b, and c.

Remark 1.43. *We* **do not define the exponential function for bases** $a < 0$. *See the earlier discussion of rational powers,* $x^{p/q}$. *And because we will define a new "inverse operation" called the "logarithm",* **we also do not define the exponential function for** $a = 0$.

Remark 1.44. *The domain of* $\exp_a(x)$ *is all real numbers, and its codomain is all positive real numbers. In fact, its range is also all positive real numbers.*

Proposition 1.45: Properties of the Exponential Function

Suppose that $a > 0$ is a positive real number. Then the following relations hold:

(a) For all $x \in \mathbb{R}$, $\exp_a(x) > 0$.

(b) If $a > 1$, then $\exp_a(x)$ is strictly increasing; that is, for $y > x$, $\exp_a(y) > \exp_a(x)$.

(c) If $0 < a < 1$, then $\exp_a(x)$ is strictly decreasing; that is, for $y > x$, $\exp_a(y) < \exp_a(x)$.

(d) If $a = 1$, then $\exp_a(x) := 1^x = 1$ for all $x \in \mathbb{R}$.

(e) For all $x \in \mathbb{R}$, $\exp_a(-x) = 1/\exp_a(x)$.

(f) For all $x, y \in \mathbb{R}$, $\exp_a(x + y) = \exp_a(x) \cdot \exp_a(y)$.

(g) For all $x, y \in \mathbb{R}$, $(\exp_a(x))^y = \exp_a(x \cdot y)$.

Remark 1.46. *Proposition 1.45-(d) states that* $\exp_1(x)$ *is a constant function and is not very useful! Also, because the function is constant, it cannot have an inverse.*

Because $0 < a < 1 \iff 1 < \frac{1}{a} = a^{-1} < \infty$, *it follows that for all* $x \in \mathbb{R}$ *and* $0 < a < 1$, $\exp_a(x) = \exp_{\frac{1}{a}}(-x)$. *Hence, it is not really necessary to use exponentials for bases* $0 < a < 1$ *as they can be easily replaced with exponentials of a base greater than one. This motivates some textbooks to only define the logarithm for bases larger than one.*

Logarithm Base a

For real numbers $a > 0$, $a \neq 1$, and $x > 0$:

- $y \in \mathbb{R}$ is the **logarithm base** a of x if $a^y = x$; that is, $\exp_a(y) = x$.
- The notation is $y = log_a(x)$, and is read: y equals log base a of x.
- The **logarithm function of base** a is $\log_a : (0, \infty) \to \mathbb{R}$; that is, its domain includes positive real numbers and its range includes real numbers.

Note: For $a = 1$ and $x \neq 1$, there is no $y \in \mathbb{R}$ such that $a^y = x$. That is why $a = 1$ has to be excluded. From the definition $y = log_a(x) \iff a^y = x$ and our power rules, it follows that $\log_{\frac{1}{a}}(x) = -\log_a(x)$.

Remark 1.47. *To see an example worked out for logarithms to base* $\frac{1}{3}$, ***you may enjoy this video by the Math Sorcerer, aka Daniel Gabilondo****, a former math professor at Seminole College.*[a] *His YouTube channel is very down-to-earth. This video also talks about how math can be hard for anyone. Prof. Gabilondo traveled after high school and then restarted with math at age 24. Today, we would call that taking six gap years.*

[a]Here's a comment on Prof. Gabilondo from one of his students: "His class alone, his teaching style, turned everything I thought I knew about math on its head and I not only [under]stood the math, but why it was useful and how it was derived. I excelled, got an A, and have since gone on to self-study (through MIT open courses) linear algebra as well as complex analysis. [...] **I found that I'm super hard on myself and if I didn't understand something, I would just quit. But that doesn't help anything in your life.**"

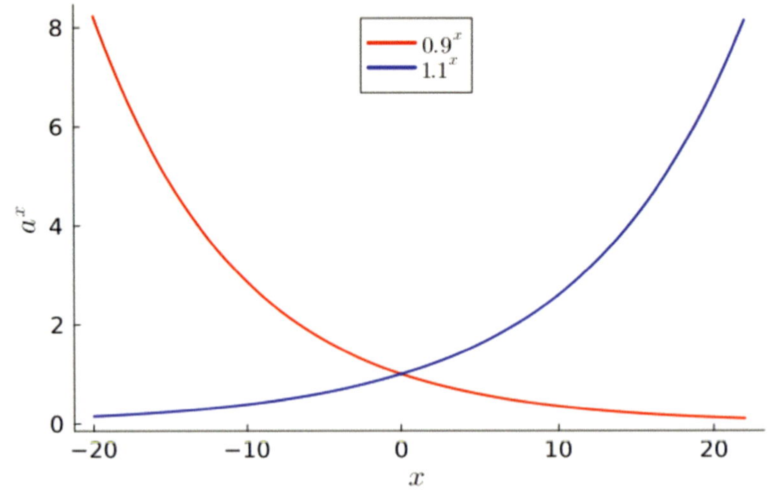

Figure 1.13: Shown are plots of $\exp_a(x) := a^x$ for $a = 0.9$ and $a = 1.1$. We observe the following trends: For $0 < a < 1$, a^x is strictly decreasing; in particular, $a^x \to \infty$ as $x \to -\infty$, and $a^x \to 0$ as $x \to \infty$. For $a > 1$, however, a^x is strictly increasing, and the trends are flipped, namely $a^x \to 0$ as $x \to -\infty$ and $a^x \to \infty$ as $x \to \infty$. The symbol $\to$ means "tends to" or "approaches".

Note how the definition of the logarithm follows a pattern similar to the definition of the nth root. They are both defined "indirectly",

$$\boxed{y = \sqrt[n]{x} \iff y^n = x \quad \text{and} \quad y = \log_a(x) \iff a^y = x;} \tag{1.14}$$

for roots, y is "downstairs" while for logarithms, y is "upstairs". Specifically, for y to be an nth root, we need to **raise** y **to power** n to recover the original number, whereas for y to be a logarithm base a, we need to **raise** a **to power** y to recover the original number.

Similar to the nth root, we can compute logarithms via the Bisection Algorithm.

Example 1.48. *For $x = 10$, compute its logarithm base 2, base e, base π, and base 10 using the Bisection Algorithm.*

Solution: Applying the Bisection Algorithm given in Section 1.7 yields the results in Table 1.6.

a (base)	x	$\log_a(x)$	Error
2.0	10.0	2.302585	$\pm 3.337860 \times 10^{-6}$
e	10.0	3.321927	$\pm 3.337860 \times 10^{-6}$
π	10.0	2.011467	$\pm 1.668930 \times 10^{-6}$
10.0	10.0	1.000000	$\pm 8.344650 \times 10^{-7}$

Table 1.6: Computation of $\log_a(x)$ using the Bisection Algorithm. Note that rounding 0.9999997615814209 to six decimal points gave the result 1.0 for $\log_{10}(10)$.

```
bases = [2.0 exp(1) π 10.0]
x = 10.0

for k = 1:length(bases)
    a = bases[k]
    # Define the function whose root we want to find
    f(y) = a^y - x

    # Call the bisection method on this function
```

```
    F = bisection(f, 0.5, 4.0, 1e-5)
    @show F


    println("The log base $a is approximately ", F.est,)
    println("The a to power loga(x) is approximately ", a^F.est)
    println(" ")
end
```

Output

```
F = (est = 3.321927070617676, low = 3.321927070617676,
up = 3.3219337463378906, error = 3.337860107421875e-6)
The log base 2.0 is approximately 3.321927070617676
The a to power loga(x) is approximately 9.999992900306067

F = (est = 2.302584648132324, low = 2.302584648132324,
up = 2.302591323852539, error = 3.337860107421875e-6)
The log base 2.718281828459045 is approximately 2.302584648132324
The a to power loga(x) is approximately 9.999995551383774

F = (est = 2.0114665031433105, low = 2.011463165283203,
up = 2.0114665031433105, error = 1.6689300537109375e-6)
The log base 3.141592653589793 is approximately 2.0114665031433105
The a to power loga(x) is approximately 10.000007275393529

F = (est = 0.9999997615814209, low = 0.9999997615814209,
up = 1.0000014305114746, error = 8.344650268554688e-7)
The log base 10.0 is approximately 0.9999997615814209
The a to power loga(x) is approximately 9.999994510210845
```

■

Proposition 1.49: Properties of Logarithms (aka Logarithm Rules)

For all bases $a > 0$, $a \neq 1$,

$$\begin{aligned} a^{\log_a(x)} &= x, \text{ for all } x > 0. \\ \log_a(a^x) &= x, \text{ for all } x \in \mathbb{R} \\ \log_a(a) &= 1 \\ \log_a(1) &= 0 \\ \log_a(x \cdot y) &= \log_a(x) + \log_a(y), \text{ for all } x > 0, y > 0 \\ \log_a\left(\frac{x}{y}\right) &= \log_a(x) - \log_a(y), \text{ for all } x > 0, y > 0 \\ \log_a(x^y) &= y \log_a(x), \text{ for all } x > 0, y \in \mathbb{R}. \end{aligned} \tag{1.15}$$

In addition,

$$\begin{aligned} a > 1 &\implies \log_a(x) \text{ is strictly increasing for all } x > 0 \\ 0 < a < 1 &\implies \log_a(x) \text{ is strictly decreasing for all } x > 0. \end{aligned}$$

Remark 1.50. *The first two lines of Prop. 1.49 can be written as:*

$$\begin{aligned} \exp_a(\log_a(x)) &= x, \textit{ for all } x > 0, \textit{ and} \\ \log_a(\exp_a(x)) &= x, \textit{ for all } x \in \mathbb{R}, \end{aligned} \tag{1.16}$$

which underlines that $\log_a(x)$ is the inverse function of $\exp_a(x)$, and vice versa. ■

Remark 1.51. *Suppose that $f : \mathbb{R} \to \mathbb{R}$ is a function and for some $y \in \mathbb{R}$, $f(y) > 0$. Then $a^{\log_a(f(y))} = f(y)$. This follows from "$a^{\log_a(x)} = x$" by setting $x = f(y)$. Similarly, by setting $x = f(y)$, we have $\log_a(a^{f(y)}) = f(y)$ for all $f(y)$.* ■

Common logarithms are $\log_{10}$ (log base 10), $\log_2$ (log base two), and $\log_e$ (log base e, Euler's constant, the irrational number 2.71828182846...). In Julia, the notation is:

- `log(x)` is used for $\log_e(x)$
- `log10(x)` is used for $\log_{10}(x)$
- `log2(x)` is used for $\log_2(x)$
- We can generate Euler's constant, e, by typing "\euler" (without the quotes) and then pressing the TAB key. We can also use "$e = \exp(1)$". The latter has the advantage of being more commonly understood.

Example 1.52. *Solve for x in the following problem from a Chinese Math Olympiad:*

$$9^x - 6^x = 4^x.$$

Solution: We offer two solutions, one analytical and the other numerical.

Analytical solution:

$$\begin{aligned} 9^x - 6^x - 4^x &= 0 \quad \text{(express numbers as products of primes)} \\ &\Updownarrow \\ 3^{2x} - 2^x \cdot 3^x - 2^{2x} &= 0 \quad \text{(divide both sides by } 3^{2x}) \\ &\Updownarrow \\ 1 - \left(\frac{2}{3}\right)^x - \left(\frac{2}{3}\right)^{2x} &= 0 \quad \text{(look for structure in the equation).} \end{aligned} \tag{1.17}$$

Letting $y = \left(\frac{2}{3}\right)^x$ yields a quadratic equation:

$$\begin{aligned} 1 - y - y^2 &= 0 \\ &\Updownarrow \\ y^2 + y - 1 &= 0 \\ &\Updownarrow \\ y &= \frac{-1 \pm \sqrt{5}}{2}. \end{aligned} \tag{1.18}$$

Because $y := \left(\frac{2}{3}\right)^x > 0$, we have:

$$\begin{aligned} \left(\frac{2}{3}\right)^x &= \frac{-1+\sqrt{5}}{2} \quad \text{(take log on both sides, can use any base)} \\ &\Updownarrow \\ x \log_{10}\left(\frac{2}{3}\right) &= \log_{10}\left(\frac{-1+\sqrt{5}}{2}\right) \quad \text{(logarithm rules bring } x \text{ down so that we can solve for it)} \\ &\Updownarrow \\ x &= \frac{\log_{10}\left(\frac{-1+\sqrt{5}}{2}\right)}{\log_{10}\left(\frac{2}{3}\right)} \quad \text{(we can stop here; it is a valid answer, or)} \\ &\approx 1.1868 \quad \text{(we perform the indicated operations, in Julia or on calculator, to get a decimal approximation)} \end{aligned} \tag{1.19}$$

Numerical solution using the Bisection Algorithm:

```
f(x) = 9.0^x - 6.0^x -4.0^x
F = bisection(f,1.0, 4.0, 1e-9)
```

Output

```
(est = 1.1868143903557211, low = 1.1868143896572292,
up = 1.1868143903557211, error = 3.4924596548080444e-10)
```

■

Example 1.53. *Solve the following exponential Diophantine equation:*

$$2^x + x^3 = 4.$$

Solution: We observe that $2^1 + 1^3 = 3 < 4$ and $2^2 + 2^3 = 12 > 4$. Hence, we can use $a = 1$ and $b = 2$ as bracketing points in the Bisection Algorithm. The result is $x = 1.1956916 \pm 2.384185e\text{-}7$. ■

Example 1.54. *Solve the following problem from Vijay Maths, which in turn comes from one of the Math Olympiads:*

(a) Find all real x satisfying

$$x^{16\cdot(\log_5(x))^3-68\cdot\log_5(x)} = 5^{-16}. \tag{1.20}$$

(b) Compute the product of all solutions to the above equation.

Hints: *Useful logarithm properties: for any $b > 0$ and $b \neq 1$:*

$$\boxed{\begin{aligned}&\log_b\left(x^{f(x)}\right) = f(x)\cdot\log_b(x)\\&\log_b\left(b^c\right) = c.\end{aligned}} \tag{1.21}$$

Solution: We first take the log base 5 of both sides of (1.20), yielding:

$$\left(16\cdot(\log_5(x))^3 - 68\cdot\log_5(x)\right)\cdot\log_5(x) = -16.$$

Next, we make the substitution $a = \log_5(x)$, which yields

$$\left(16\cdot a^3 - 68\cdot a\right)\cdot a = -16,$$

which, after dividing both sides by -4 and rearranging terms, gives

$$4a^4 - 17a^2 + 4 = 0.$$

This is a quadratic equation in a^2. Indeed, if we let $y = a^2$, we have

$$4y^2 - 17y + 4 = 0,$$

and the quadratic formula provides us with

$$y = \frac{17 \pm \sqrt{17^2 - 4^3}}{8} = \frac{17 \pm \sqrt{225}}{8} = \frac{17 \pm 15}{8} = 4 \text{ and } \frac{1}{4}.$$

Recalling that $y = a^2$ and that for $c > 0$, $a^2 = c \iff a = \pm\sqrt{c}$, gives us

$$a \in \left\{-2, 2, -\frac{1}{2}, \frac{1}{2}\right\}.$$

Then, because $a = \log_5(x) \iff x = 5^a$, the answer to part (a) is

$$\boxed{x \in \{5^{-2}, 5^2, 5^{-\frac{1}{2}}, 5^{\frac{1}{2}}\}.}$$

Finally, the answer to part (b) is the product of these solutions, namely:

$$\boxed{\text{Ans} = 5^{-2} \cdot 5^2 \cdot 5^{-\frac{1}{2}} \cdot 5^{\frac{1}{2}} = 5^{-2+2-\frac{1}{2}+\frac{1}{2}} = 5^0 = 1.}$$

While the roots of $x^{16\cdot(\log_5(x))^3 - 68\cdot\log_5(x)} - 5^{-16} = 0$ can be determined numerically via the Bisection Algorithm, it is challenging to find the correct bracketing points for each of the four solutions. In many engineering problems, a combination of analytical and numerical prowess is required. ■

1.9 One Logarithm to Rule Them All

In engineering practice, we mostly use the so-called **natural logarithm**, which is log base e, Euler's constant. In mathematics, the natural logarithm is denoted $\ln(x)$, where the "n" is for natural. The corresponding exponential function base e is written e^x or $\exp(x)$, without a subscript. The terminology "natural logarithm" comes from calculus because the logarithm of base e is the only logarithm with a "slope" equal to 1 at $x = 0$ and its exponential function e^x is the only one of the exponential functions, a^x, that is equal to its own derivative.

Log Base a Rules	Natural Log Rules
$\log_a(x \cdot y) = \log_a(x) + \log_a(y)$	$\ln(x \cdot y) = \ln(x) + \ln(y)$
$\log_a\left(\frac{x}{y}\right) = \log_a(x) - \log_a(y)$	$\ln\left(\frac{x}{y}\right) = \ln(x) - \ln(y)$
$\log_a(x^y) = y \cdot \log_a(x)$	$\ln(x^y) = y \cdot \ln(x)$
$\log_a(a^x) = x$	$\ln(e^x) = x$
$a^{\log_a(x)} = x$	$e^{\ln(x)} = x$

Table 1.7: Side-by-side comparison of logarithm rules and natural logarithm rules (they are identical).

The logarithm rules make it easy to prove the result on "power towers", Prop. 1.40. In other words, does the equation

$$(x^y)^z = x^{yz} \tag{1.22}$$

hold for $x > 0$? If either y or z is zero, then the result is immediate because $x^0 = 1$ on both sides. Hence, we assume both x and y are positive. Taking the natural logarithm of the left-hand side of (1.22) and applying the logarithm rules yields

$$\ln((x^y)^z) = z \cdot \ln(x^y) = z \cdot y \cdot \ln(x).$$

Next, taking the natural logarithm of the right side of (1.22) and applying the logarithm rules leads to

$$\ln(x^{y\cdot z}) = y \cdot z \cdot \log(x).$$

Hence, when y and z are nonzero,

$$\ln((x^y)^z) = \ln(x^{y\cdot z}).$$

Taking the exponential of both sides of the above equation gives:

$$(x^y)^z = x^{yz}.$$

Hence, we have proven that (1.22) is valid. ■

It is sometimes useful to know how to compute all logs from just the natural log. From $x = a^{\log_a(x)}$, taking the natural log of both sides yields:

$$\ln(x) = \log_a(x) \cdot \ln(a).$$

Rearranging gives the result in Prop. 1-11.

Proposition 1.55: All Logs from the Natural Log

The natural logarithm can be used to compute any logarithm by the **change of base formula:** for $a > 0$, $a \neq 1$, and $x > 0$,

$$\log_a(x) = \frac{\ln(x)}{\ln(a)}. \tag{1.23}$$

Example 1.56. *Compute $log_a(\pi)$ for $a \in \{1.1, 2.0, e, \pi, 10.0\}$ using (1.23).*

Solution:

x	a	$\log_a(x) = \ln(x)/\ln(a)$	$a^{\log_a(x)}$
3.14159	1.10000	12.0106	3.14159
3.14159	2.00000	1.6515	3.14159
3.14159	2.71828	1.1447	3.14159
3.14159	3.14159	1.0000	3.14159
3.14159	10.0000	0.4972	3.14159

Table 1.8: Computing logarithms to an arbitrary base using the natural logarithm and (1.23).

```
function myLog(x::Real,a::Real)
    if !(x>0)
        println("x must be positive")
        return NaN
    end
    if !(a>0)
        println("the base-a must be positive")
        return NaN
    end
    return log(x)/log(a)
end

a = [1.1, 2.0, exp(1), pi, 10.0]
x = pi
y = 0*a

for i = 1:length(a)
    y[i] = myLog(x, a[i])
end

[x*ones(length(a), 1) a y a.^y]
```

Output

```
5×4 Matrix{Float64}:
 3.14159   1.1       12.0106   3.14159
 3.14159   2.0        1.6515   3.14159
 3.14159   2.71828    1.14473  3.14159
 3.14159   3.14159    1.0      3.14159
 3.14159  10.0        0.49715  3.14159
```

■

1.10 Set Notation and How to Make Sense of It

In mathematics, precise notation allows us to express complex ideas clearly and concisely. Let us consider the following specific examples.

What Does This Notation Mean?

Consider the following set:

$$S_{\text{A}} := \{x \in (0,\infty) \mid |\sin(x)| = 1.0\}.$$

Step-by-step explanation:

1. **Domain restriction:** The symbol $x \in (0,\infty)$ means we are considering only positive real numbers. This is similar to specifying the domain of a function.
2. **The vertical bar "|":** This symbol means "such that." The set includes only those values of x that satisfy the condition after the bar, namely, $|\sin(x)| = 1.0$.
3. **Interpretation:** This means we are selecting all x in $(0,\infty)$ such that the absolute value of $\sin(x)$ is exactly 1.

Enumerating the Set

It is sometimes possible and useful to list all of the elements in a set to make things more concrete. To do this for S_{A}, recall the following key properties of $\sin(x)$:

- $\sin(x) = 1$ at $x = \frac{\pi}{2} + 2k\pi$ for $k \in \mathbb{Z}$.
- $\sin(x) = -1$ at $x = \frac{3\pi}{2} + 2k\pi$ for $k \in \mathbb{Z}$.

The absolute value, $|\sin(x)|$, is periodic with period π. Therefore,

$$|\sin(x)| = 1 \implies x = \frac{\pi}{2} + k\pi, \quad k = 0,1,2,\ldots.$$

Since we are restricted to $x > 0$, negative values of k are excluded. Hence, we conclude that:

$$S_{\text{A}} = \left\{\frac{\pi}{2} + k\pi \;\middle|\; k = 0,1,2,\ldots\right\}.$$

This step of **enumerating the elements of the set** is central to understanding **countability**—a property that will arise later.

Working with More Complex Sets

Let us build on the previous result with another example, inspired by the **sawtooth wave** function $f(x)$ shown in **Fig. 1.14** and defined as:

$$f(x) = \begin{cases} x, & 0 \le x < 1, \\ x \bmod 1, & x \ge 1. \end{cases}$$

The goal is to understand sets defined by conditions such as:

$$S_{\text{B}} := \{x \in (0,\infty) \mid f(x) = \frac{0.9}{i}, \quad i = 1,2,3,\ldots\}.$$

Breaking it down: For simplicity, let us first analyze the subsets of S_{B} based on the intervals of x. For example:

- Points in $(0,1]$ that satisfy $f(x) = \frac{0.9}{i}$ are:

$$S_{\text{B},1} = \left\{x \in (0,1] \text{ such that } x = \frac{0.9}{i}, \quad i = 1,2,3,\ldots\right\}.$$

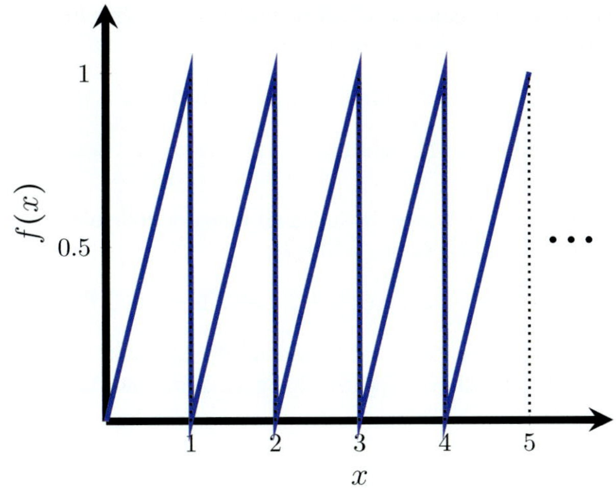

Figure 1.14: A sawtooth function with period one. It is plotted in the usual manner, with a vertical line connecting the points $(k,1)$ and $(k,0)$ for $k = 1,2,3,\ldots$, which, in theory, makes it not a function, but in reality, we ignore the vertical lines because they are "plotting artifacts" left to guide our eyes.

- Points in $(1,2]$ that satisfy $f(x) = \frac{0.9}{i}$ are:

$$S_{\mathrm{B},2} = \left\{x \in (1,2] \text{ such that } x = 1 + \frac{0.9}{i}, \quad i = 1,2,3,\ldots\right\}.$$

- Similarly, for $(j-1,j]$, we can write:

$$S_{\mathrm{B},j} = \left\{x \in (j-1,j] \text{ such that } x = (j-1) + \frac{0.9}{i}, \quad i = 1,2,3,\ldots\right\}.$$

Key Takeaway
Set-builder notation uses conditions to define a set, like restricting a domain or specifying a property.

Example 1.57. *Consider the following tasks:*

(a) *Write in set notation: The set of all positive integers divisible by 3 (same as being a multiple of 3).*

(b) *Enumerate the elements of the set* $T := \{x \in (0,1] \mid x = \frac{1}{i}, \quad i = 2,3,4,\ldots\}$.

(c) *Explain what is meant by* $\{n \in \mathbb{N} \mid n \bmod 3 = 0\}$ *and then explain how the set is different from* $\{n \in \mathbb{Z} \mid n \bmod 3 = 0\}$.

Solution:

(a) Let us define the set $S := \{x \in \mathbb{N} \mid x = 3k,\ k \in \mathbb{N}\}$. This means that S contains all positive integers that can be written as multiples of 3, where k is a positive integer. Enumerating a few elements of S, we have

$$S = \{3,6,9,12,\ldots\}.$$

(b) The condition $x = \frac{1}{i}$ means that x is the reciprocal of a positive integer $i \geq 2$. Substituting successive values of i, we get:

$$T = \left\{\frac{1}{2}, \frac{1}{3}, \frac{1}{4}, \frac{1}{5}, \ldots\right\}.$$

Observe that T is infinite, but we can systematically list all its elements, which helps us to understand the set.

(c) The set $\{n \in \mathbb{N} \mid n \mod 3 = 0\}$ consists of all natural numbers (positive integers) that are divisible by 3. In terms of set notation, this can be written as:

$$\{n \in \mathbb{N} \mid n \bmod 3 = 0\} = \{3, 6, 9, 12, \ldots\}.$$

This is the set of positive multiples of 3, starting from 3, since $\mathbb{N}$ includes only positive integers.

The set $\{n \in \mathbb{Z} \mid n \bmod 3 = 0\}$, however, includes all integers divisible by 3, zero as well as positive and negative integers. In terms of set notation, this is:

$$\{n \in \mathbb{Z} \mid n \bmod 3 = 0\} = \{\ldots, -9, -6, -3, 0, 3, 6, 9, \ldots\}.$$

The key difference is that the first set (with "domain" $\mathbb{N}$) only contains positive integers, while the second set (with "domain" $\mathbb{Z}$) includes negative integers, zero, and positive integers, representing all integer multiples of 3.

■

1.11 Euler's Formula

Euler's Formula

Definition 1.58. *Let* $\mathbf{i}$ *denote* $\sqrt{-1}$ *(that is,* $\mathbf{i}^2 = -1$*). Then for* $x \in \mathbb{R}$*,*

$$e^{\mathbf{i}x} := \cos(x) + \mathbf{i}\sin(x). \tag{1.24}$$

Let us note that $e^{-\mathbf{i}x} = e^{\mathbf{i}(-x)} := \cos(-x) + \mathbf{i}\sin(-x) = \cos(x) - \mathbf{i}\sin(x)$, because cosine is an even function (that is, $\cos(-x) = \cos(x)$) and sine is an odd function (that is, $\sin(-x) = -\sin(x)$). Hence,

$$e^{\mathbf{i}x} + e^{-\mathbf{i}x} = 2\cos(x),$$

yielding:

$$\boxed{\cos(x) = \frac{e^{\mathbf{i}x} + e^{-\mathbf{i}x}}{2}.} \tag{1.25}$$

Similarly,

$$e^{\mathbf{i}x} - e^{-\mathbf{i}x} = 2\mathbf{i}\sin(x),$$

yielding

$$\boxed{\sin(x) = \frac{e^{\mathbf{i}x} - e^{-\mathbf{i}x}}{2\mathbf{i}}.} \tag{1.26}$$

Example 1.59. *Derive a simpler formula for* $\cos^2(x) := (\cos(x))^2$.

Solution:

$$\begin{aligned}
\cos^2(x) &= \left(\frac{e^{\mathbf{i}x} + e^{-\mathbf{i}x}}{2}\right)^2 \\
&= \frac{e^{\mathbf{i}2x} + 2 + e^{-\mathbf{i}2x}}{4} \\
&= \frac{1}{2} + \frac{1}{2}\frac{e^{\mathbf{i}2x} + e^{-\mathbf{i}2x}}{2} \\
&= \frac{1}{2} + \frac{1}{2}\cos(2x).
\end{aligned} \tag{1.27}$$

The key step involves the use of the usual exponential rules, so that $e^{\mathbf{i}x} \cdot e^{\mathbf{i}x} = e^{\mathbf{i}2x}$ and $e^{\mathbf{i}x} \cdot e^{-\mathbf{i}x} = e^0 = 1$.

■

Example 1.60. *Use Euler's formula to derive a simpler expression for* $\sin^2(x)\cos(x)$.

Solution: We start by writing out $\sin(x)$ and $\cos(x)$ in terms of exponential functions and substituting these expressions into $\sin^2(x)\cos(x)$, yielding:

$$\begin{aligned}
\sin^2(x)\cos(x) &= \left(\frac{e^{\mathrm{i}x}-e^{-\mathrm{i}x}}{\mathrm{i}2}\right)^2\left(\frac{e^{\mathrm{i}x}+e^{-\mathrm{i}x}}{2}\right) \text{ (square the first term)}\\
&= \frac{(e^{\mathrm{i}2x}-2+e^{-\mathrm{i}2x})(e^{\mathrm{i}x}+e^{-\mathrm{i}x})}{(-4)(2)} \text{ (multiply out the terms)}\\
&= \frac{e^{\mathrm{i}3x}-2e^{\mathrm{i}x}+e^{-\mathrm{i}x}+e^{\mathrm{i}x}-2e^{-\mathrm{i}x}+e^{-\mathrm{i}3x}}{-8} \text{ (add like terms and group)}\\
&= \frac{\left(e^{\mathrm{i}x}+e^{-\mathrm{i}x}\right)-\left(e^{\mathrm{i}3x}+e^{-\mathrm{i}3x}\right)}{8}\\
&= \frac{1}{4}\frac{\left(e^{\mathrm{i}x}+e^{-\mathrm{i}x}\right)}{2}-\frac{1}{4}\frac{\left(e^{\mathrm{i}3x}+e^{-\mathrm{i}3x}\right)}{2}\\
&= \frac{1}{4}\cos(x)-\frac{1}{4}\cos(3x).
\end{aligned} \tag{1.28}$$

■

Fact 1.61: Consequence of Euler's Formula

Any (finite) product of powers of sines and cosines,

$$\Pi_{k=1}^{N}\sin^{n_k}(\omega_k x)\cdot\cos^{m_k}(\omega_k x), \tag{1.29}$$

with n_k and m_k being positive integers, can be expressed as a linear combination of sines and cosines to the first power plus a constant:

$$a_0+\sum_{i=1}^{J}a_i\sin(\nu_i x)+\sum_{i=1}^{K}b_i\cos(\nu_i x), \tag{1.30}$$

where the ν_i s are sums of $\pm\omega_k$, $1\le k\le N$.

Note: In (1.28), $N=1$, $\omega_1=1$, $\nu_1=\omega_1$ and $\nu_2=\omega_1+\omega_1+\omega_1=3\omega_1$.

Note: See Table 1.12 and Section 1.18.2 for a convenient set of power reductions for sine and cosine.

1.12 Hyperbolic Trigonometric Functions and Relation to Euler's Formula

The ordinary trigonometric functions arise by examining the unit circle. The **hyperbolic trigonometric functions**, on the other hand, arise from the study of the unit hyperbola. An excellent explanation is given in Hyperbolic functions and the unit hyperbola, from Khan Academy's Precalculus series. Aerospace, civil, and mechanical engineers encounter them daily, as explained in The applications of hyperbolic trig: Why do we even care about these things? by Zach Star. We introduce them here due to their parallels with Euler's formula in Section 1.58.

Hyperbolic Trigonometric Functions

Definition 1.62. *For all $x \in \mathbb{R}$,*

$$\cosh(x) := \frac{e^x + e^{-x}}{2}, \tag{1.31}$$

$$\sinh(x) := \frac{e^x - e^{-x}}{2}, \tag{1.32}$$

and

$$\tanh(x) := \frac{\sinh(x)}{\cosh(x)} \tag{1.33}$$

The functions are called hyperbolic cosine, hyperbolic sine, and hyperbolic tangent.

The hyperbolic trigonometric functions play a minor role in this book. We introduce them here to help explain Euler's Formula. We note that:

$$\cos(x) = \cosh(\mathbf{i}\,x) = \frac{e^{\mathbf{i}\,x} + e^{-\mathbf{i}\,x}}{2} \tag{1.34}$$

and

$$\sin(x) = \sinh(\mathbf{i}\,x) = \frac{e^{\mathbf{i}\,x} - e^{-\mathbf{i}\,x}}{2}, \tag{1.35}$$

even though the hyperbolic functions were discovered before Euler's formula was developed.

We would be remiss not to mention how the hyperbolic functions show up in architecture, in the form of the Catenary Arch, which is a mainstay of the beautiful buildings by Gaudi. Also, we know that $\sin^2(x) + \cos^2(x) = 1$. The hyperbolic version is:

$$\cosh^2(x) \underset{\uparrow}{-} \sinh^2(x) = 1.$$

The minus sign is not a typo!

Function	**Principal Value**	**Domain**
$\text{acosh}(x)$	$\text{acosh}(x) = \ln\left(x + \sqrt{x^2 - 1}\right)$	$x \geq 1$
$\text{asinh}(x)$	$\text{asinh}(x) = \ln\left(x + \sqrt{x^2 + 1}\right)$	All $x \in \mathbb{R}$

Table 1.9: Principal domains for $\text{acosh}(x)$ and $\text{asinh}(x)$.

To learn more:

- Hyperbolic Trig Identities by the The Organic Chemistry Tutor.
- Why hyperbolic functions are actually really nice by Dr. Trefor Bazett.

1.13 Summing Symbol and Change of Indices

The summing symbol $\Sigma_{i=1}^{n} x_i$ is shorthand for $x_1 + x_2 + \cdots + x_n$. In Julia, it can be implemented with a `for loop` or with the command `sum(x)`, if x is a vector with components x_i.

It is sometimes useful to perform a **change of index** in a sum. For example,

$$\boxed{\sum_{i=1}^{n}(i-1) = \sum_{j=0}^{n-1} j = \sum_{j=1}^{n-1} j} \tag{1.36}$$

by defining the change of index $j = i - 1$. Hence, when $i = 1$, the lower limit of j is $j = 0$, and when $i = n$, the upper limit is $j = (n-1)$. The simplification in the third sum is because $j = 0$ does not contribute to the sum, so we can remove it.

1.14 Binomial Theorem

The **Binomial Theorem** provides a method for expanding $(x+y)^n$, where $x+y$ and $(x+y)^n$ are called **binomials**. This theorem enables the expansion of the polynomial $(x+y)^n$ into a sum composed of terms of the form $x^{n-k} \cdot y^k$, where $0 \le k \le n$ is an integer. As an example, for $n = 4$, the expansion yields: $(x+y)^4 = x^4 + 4x^3y + 6x^2y^2 + 4xy^3 + y^4$. The coefficient in front of each term $x^{n-k}y^k$ is known as the **binomial coefficient**, often denoted as $\binom{n}{k}$.

The following video is helpful: Unveiling the Joy of Math: Inventing the Binomial Theorem by Quick QEDs.

1.14.1 Binomial Coefficients

In mathematics, binomial coefficients are positive integers that appear as coefficients in the Binomial Theorem. They are denoted by $\binom{n}{k}$, where n and k are integers such that $n \ge k \ge 0$. These coefficients represent the x^k term in the expansion of $(1+x)^n$. The binomial coefficients are defined by:

$$\boxed{\binom{n}{k} := \frac{n!}{k!(n-k)!}}, \tag{1.37}$$

where ! denotes factorial; for example, $5! = 5 \cdot 4 \cdot 3 \cdot 2 \cdot 1 = 120$. Equivalently, this formula can be written as

$$\binom{n}{k} = \frac{n \cdot (n-1) \cdots (n-k+1)}{k \cdot (k-1) \cdots 1} = \prod_{\ell=1}^{k} \frac{n-\ell+1}{\ell} = \prod_{\ell=0}^{k-1} \frac{n-\ell}{k-\ell}, \tag{1.38}$$

which is more efficient for doing the actual calculation.

Remark 1.63. *By convention,* $0! := 1$. *Hence* $\binom{n}{0} = 1$ *and* $\binom{n}{n} = 1$.

In the field of combinatorics, these coefficients denote the number of **distinct ways** that k elements can be selected from a set of n elements. Hence, $\binom{n}{k}$ is frequently pronounced as "n choose k". As an example, if we have a group of five people ($n = 5$) and we want to select a team of two people ($k = 2$) from this group, the number of different teams we can form is given by "5 choose 2" or $\binom{5}{2}$, which equals 10 according to (1.37). It is important to note that $\binom{n}{k}$ does not take into account the order of the elements chosen. If the order were important (for instance, if you were choosing a president and a vice president from a group), you would use a different formula (called a permutation) to determine the number of possible distinct combinations of a president and a vice president.

Proposition 1.64: Useful Lower Bound for Examining Euler's Number

For $n > 2k$,

$$\binom{n}{k} := \frac{n!}{k!(n-k)!} = \frac{n \cdot (n-1) \cdots (n-k+1)}{k!} > \frac{n^k}{2^k k!}. \tag{1.39}$$

Note: This is a **lower bound** because $\frac{n^k}{2^k k!} < \binom{n}{k} := \frac{n!}{k!(n-k)!}$. This simple bound will allow us in Section 2.5.3 to show that exponentials of the form a^x, with $a > 1$, **grow faster than any polynomial function** as x tends to infinity.

Proof: From $n > 2k$, we have $\frac{n}{2} > k$ and $-k > -\frac{n}{2}$. Hence, $(n-k+1) > (n - \frac{n}{2} + 1) > \frac{n}{2}$ and $n > (n-1) > \cdots > (n-k+1) > \frac{n}{2}$. Therefore,

$$\underbrace{n \cdot (n-1) \cdots (n-k+1)}_{k\text{-times}} > \frac{n^k}{2^k} \implies \frac{n \cdot (n-1) \cdots (n-k+1)}{k!} > \frac{n^k}{2^k k!}.$$

■

Proposition 1.65: Useful Upper Bounds for Examining Euler's number

For all $n \geq 1$ and $0 \leq k \leq n$,

$$\binom{n}{k} := \frac{n!}{k!(n-k)!} = \frac{n \cdot (n-1) \cdots (n-k+1)}{k!} \leq \frac{n^k}{k!}, \tag{1.40}$$

and for all $0 < k < n$, the inequality is strict, namely,

$$\binom{n}{k} := \frac{n!}{k!(n-k)!} = \frac{n \cdot (n-1) \cdots (n-k+1)}{k!} < \frac{n^k}{k!}. \tag{1.41}$$

Note: This even simpler bound will help us to approximate Euler's number.

Proof: Equation 1.41 holds because $(n-j) \leq n$ for all integers $j \geq 0$. Concerning the strict inequality, for $k = 1$, the result is immediate. For $2 \leq k < n$, we have

$$\underbrace{(n-k+1) < (n-k+2) < \cdots < (n-1) < n,}_{k \text{ terms}}$$

and thus $n \cdot (n-1) \cdots (n-k+1) < n^k$. Substituting this into (1.41) proves the result. ■

Proposition 1.66: A Second Useful Upper Bound for Examining Euler's Number

For all $n \geq 1$ and $0 \leq k \leq n$,

$$\binom{n}{k} \frac{1}{n^k} \leq \binom{n+1}{k} \frac{1}{(n+1)^k}, \tag{1.42}$$

and for $n \geq 2$ and $2 \leq k \leq n$, the inequality is strict, namely,

$$\binom{n}{k} \frac{1}{n^k} < \binom{n+1}{k} \frac{1}{(n+1)^k}. \tag{1.43}$$

Note: These bounds will help us show that $(1 + \frac{1}{n})^n < (1 + \frac{1}{n+1})^{n+1}$, a property that Bernoulli noted when he discovered e.

Proof: We now show the proof for (1.43). For all $n \geq 2$ and $2 \leq k \leq n$, we have

$$\begin{aligned} \binom{n}{k} \frac{1}{n^k} &= \frac{n \cdot (n-1) \cdot (n-2) \cdots (n-k+1)}{k!\, n^k} \\ \binom{n+1}{k} \frac{1}{(n+1)^k} &= \frac{(n+1) \cdot (n+1-1) \cdot (n+1-2) \cdots (n+1-k+1)}{k!\, (n+1)^k}. \end{aligned} \tag{1.44}$$

We can cancel the common term, $k!$, in the denominators. Then, (1.43) holds if, and only if:

$$\frac{n \cdot (n-1) \cdot (n-2) \cdots (n-k+1)}{n^k} < \frac{(n+1) \cdot (n+1-1) \cdot (n+1-2) \cdots (n+1-k+1)}{(n+1)^k}$$

$$\Updownarrow$$

$$\underbrace{\frac{n}{n} \cdot \frac{n-1}{n} \cdot \frac{n-2}{n} \cdots \frac{n-k+1}{n}}_{k \text{ terms}} < \underbrace{\frac{(n+1)}{(n+1)} \cdot \frac{(n+1)-1}{(n+1)} \cdot \frac{(n+1)-2}{(n+1)} \cdots \frac{(n+1)-k+1}{(n+1)}}_{k \text{ terms}}.$$

But this last inequality is true because, for all $n \ge 1$ and $m \ge 1$,

$$\frac{n-m}{n} = 1 - \frac{m}{n} < 1 - \frac{m}{(n+1)} = \frac{(n+1)-m}{(n+1)}.$$

■

Binomial	Expansion
$(x+y)^0$	1
$(x+y)^1$	$x+y$
$(x+y)^2$	$x^2+2xy+y^2$
$(x+y)^3$	$x^3+3x^2y+3xy^2+y^3$
$(x+y)^4$	$x^4+4x^3y+6x^2y^2+4xy^3+y^4$
$(x+y)^5$	$x^5+5x^4y+10x^3y^2+10x^2y^3+5xy^4+y^5$
$(x+y)^6$	$x^6+6x^5y+15x^4y^2+20x^3y^3+15x^2y^4+6xy^5+y^6$
$(x+y)^7$	$x^7+7x^6y+21x^5y^2+35x^4y^3+35x^3y^4+21x^2y^5+7xy^6+y^7$
$(x+y)^8$	$x^8+8x^7y+28x^6y^2+56x^5y^3+70x^4y^4+56x^3y^5+28x^2y^6+8xy^7+y^8$

Table 1.10: A list of binomial expansions. Note the symmetry in the coefficients relative to the middle of each formula. This is because $\binom{n}{k} = \binom{n}{n-k}$.

1.14.2 Expansion of $(x+y)^n$

Theorem 1.67: Binomial Theorem

For all integers $n \ge 1$,

$$\begin{aligned}(x+y)^n &= \sum_{k=0}^{n} \binom{n}{k} x^{n-k} y^k \\ &= \sum_{k=0}^{n} \binom{n}{k} x^k y^{n-k}.\end{aligned} \tag{1.45}$$

The above summation can also be written as:

$$(x+y)^n = \binom{n}{0} x^n y^0 + \binom{n}{1} x^{n-1} y^1 + \binom{n}{2} x^{n-2} y^2 + \cdots + \binom{n}{n-1} x^1 y^{n-1} + \binom{n}{n} x^0 y^n.$$

Table 1.10 provides explicit expansions for $0 \le n \le 8$. A proof by induction is given by Wikipedia, Khan Academy, Math is Fun, and many other sources.

Example 1.68. *Use the Binomial Theorem to determine the expansion of* $(1+x)^n$.

Solution:

$$
\begin{aligned}
(1+x)^n = (x+1)^n &= \sum_{k=0}^{n} \binom{n}{k} x^k \\
&= \binom{n}{0}x^0 + \binom{n}{1}x^1 + \binom{n}{2}x^2 + \binom{n}{3}x^3 + \cdots + \binom{n}{n-1}x^{n-1} + \binom{n}{n}x^n \\
&= 1 + nx + \frac{n(n-1)}{2}x^2 + \frac{n(n-1)(n-2)}{6}x^3 + \cdots + nx^{n-1} + x^n.
\end{aligned}
$$

■

1.15 Some Special Functions

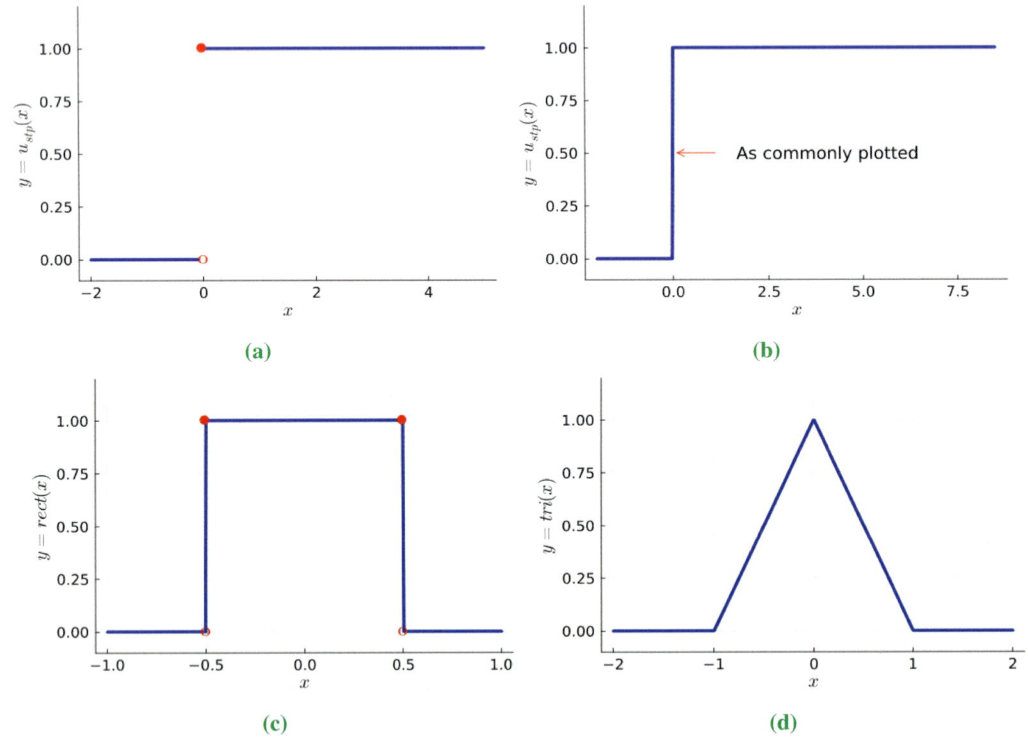

Figure 1.15: Some useful functions: (a) The unit step function. Here, the graph is done "by the book", defining clearly a value of the function at the point of discontinuity, the origin. (b) This is how the unit step function is normally plotted (graphed). While we must choose a SINGLE value for the function at the origin, it is usually not important what value we assign at the point of discontinuity. (c) The unit rectangle function is plotted in a "hybrid manner" so that its rectangular shape is evident. Normally, we do not care what value is assigned at each of the two points of discontinuity. (d) The unit triangle function is continuous everywhere (it has no jumps). Note that the base of the unit rectangle function has a width of one, while the base of the unit triangle function has a width of two. **This is done so that each function's "shape" encloses a "unit area", meaning the area is one.**

Figure 1.15 shows three functions, the Unit Step Function (aka Heaviside function, or Heaviside unit step function) in (a) and (b), the Unit Rectangle Function (aka rectangular function) in (c), and the Unit Triangle Function (aka triangular function, hat function, or tent function) in (d). The functions are defined as follows:

Unit Step:

$$
u_{\text{stp}}(x) := \begin{cases} 1 & x \geq 0, \\ 0 & x < 0. \end{cases} \tag{1.46}
$$

Unit Rectangle:

$$\operatorname{rect}(x) := \begin{cases} 1 & |x| \leq \frac{1}{2}\,, \\ 0 & \text{otherwise.} \end{cases} \tag{1.47}$$

The area under the function is equal to one because the base and height are both equal to one.

Unit Triangle:

$$\operatorname{tri}(x) := \begin{cases} 1 - |x| & |x| \leq 1, \\ 0 & \text{otherwise.} \end{cases} \tag{1.48}$$

The area under the function is also equal to one because the base of the triangle is equal to two, and the height is equal to one.

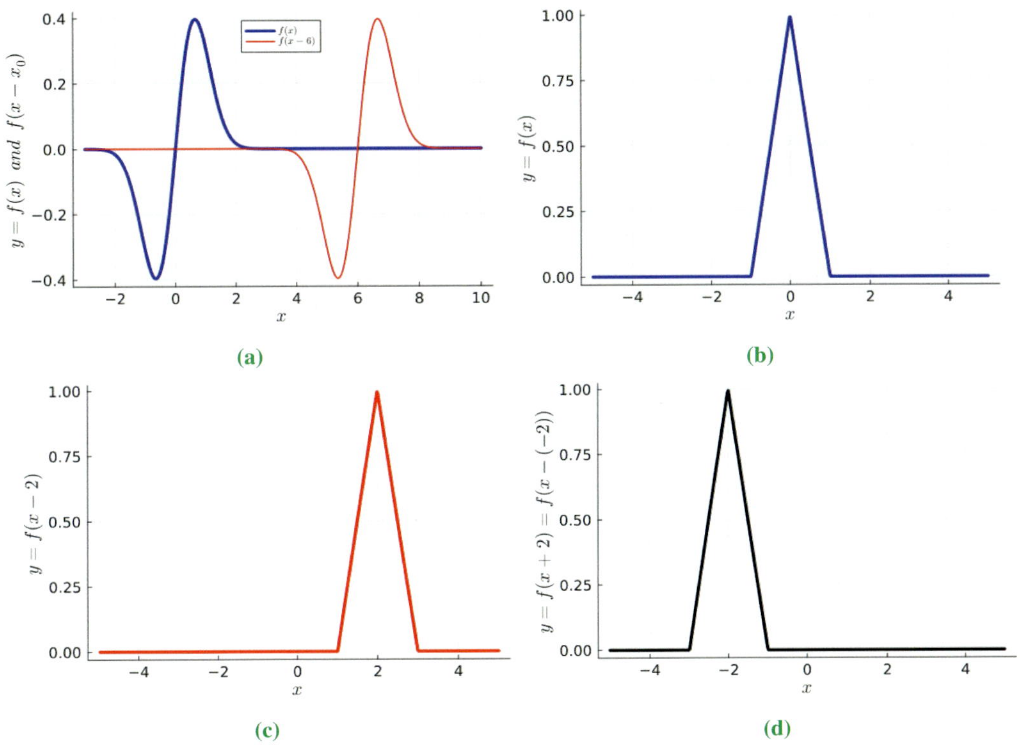

Figure 1.16: Shifted functions: (a) The nominal function is shown in blue and the shifted function is shown in red, The function is shifted by $x_0 = 6$. Because $x_0 > 0$, the function's graph is shifted to the right. The shape of the function is preserved. (b) A unit triangle function. (c) The unit triangle function has been shifted to the right by $x_0 = 2 > 0$. (d) The unit triangle function has been shifted to the left by two units because $x_0 = -2 < 0$. Note that the area under the triangles has not changed when shifted because their shape remains the same: the width of their bases are all the same, and their heights are the same. The same is true for rectangles.

1.16 Shifting and Scaling

Shifting and scaling are common ways to create new functions from existing functions. Let $f : \mathbb{R} \to \mathbb{R}$ be some function that we already have. Also, let x_0 and $c \neq 0$ be two real constants.

The shift of $f(x)$ by x_0 is denoted $f(x - x_0)$. As illustrated in Fig. 1.16, the shift operation applied to a function shifts its graph to the right by x_0 when $x_0 > 0$, and to the left by $|x_0|$ when $x_0 < 0$.

Example 1.69. *Show that with the exception of the two points of discontinuity of the rectangle function at* $x = \pm\frac{1}{2}$, *we have*

$$\text{rect}(x) = u_{\text{stp}}\left(x + \frac{1}{2}\right) - u_{\text{stp}}\left(x - \frac{1}{2}\right).$$

Solution: For $x < -\frac{1}{2}$, we have $x+\frac{1}{2} < 0$ and $x-\frac{1}{2} < 0$, and thus $u_{\text{stp}}(x+\frac{1}{2})-u_{\text{stp}}(x-\frac{1}{2}) = 0-0 = 0$, which is the value of $\text{rect}(x)$. Similarly, for $x > \frac{1}{2}$, we have $x+\frac{1}{2} > 0$ and $x-\frac{1}{2} > 0$, and thus $u_{\text{stp}}(x+\frac{1}{2})-u_{\text{stp}}(x-\frac{1}{2}) = 1-1 = 0$, which is also the value of $\text{rect}(x)$. Finally, we consider the interval $-\frac{1}{2} < x < \frac{1}{2}$. We have $x + \frac{1}{2} > 0$ and $x - \frac{1}{2} < 0$, and thus $u_{\text{stp}}(x + \frac{1}{2}) - u_{\text{stp}}(x - \frac{1}{2}) = 1 - 0 = 1$, which is the value of $\text{rect}(x)$.

Note: For $x = -\frac{1}{2}$, we have $u_{\text{stp}}(x+\frac{1}{2})-u_{\text{stp}}(x-\frac{1}{2}) = u_{\text{stp}}(0) - u_{\text{stp}}(-1) = 1-0 = 1$, which is the value of $\text{rect}(-\frac{1}{2})$, but for $x = \frac{1}{2}$, we have $u_{\text{stp}}(x + \frac{1}{2}) - u_{\text{stp}}(x - \frac{1}{2}) = u_{\text{stp}}(1) - u_{\text{stp}}(0) = 1 - 1 = 0$, which is NOT the value of $\text{rect}(\frac{1}{2})$. ■

The scale of $f(x)$ **by** $c \neq 0$ is denoted $f(c \cdot x)$. As illustrated in Fig. 1.17, the scale operation applied to a function "squeezes" its graph when $c > 1$ and expands its graph, when $0 < c < 1$. We are not defining here the scaling operation for negative values of c because it also "flips" the function about the y-axis.

Example 1.70. *Compute the area under the functions:*

(a) $\text{rect}(x)$

(b) $\text{rect}(3x)$

(c) $\text{rect}(0.2x)$

(d) $2 \cdot \text{tri}(x)$

(e) $2 \cdot \text{tri}(11x)$

(f) $2 \cdot \text{tri}(\frac{x}{3})$

Solutions:

(a) $\text{rect}(x)$ has a base of one and a height of one, and thus its area is also one; i.e., 1.0.

(b) $\text{rect}(3x)$ has a base of 1/3 and a height of 1, and thus its area is 1/3.

(c) $\text{rect}(0.2x)$ has a base of 5 and a height of 1, and thus its area is 5; i.e., 5.0.

(d) $2 \cdot \text{tri}(x)$ has a base of 2 and a height of 2, and thus its area is 2 (one half base times height); i.e., 2.0.

(e) $2 \cdot \text{tri}(11x)$ has a base of 2/11 and a height of 2, and thus its area is 2/11.

(f) $2 \cdot \text{tri}(\frac{x}{3})$ has a base of 6 and a height of 2, and thus its area is 6; i.e., 6.0.

■

1.17 A Peek at Feedback Control and a Partial Road Map to This Textbook

Why feedback control and why in a chapter on precalculus:

- Feedback control is a pervasive technology. At engineering colleges, feedback control courses can be found in approximately half of its engineering departments. Its widespread representation rivals physics and mathematics.

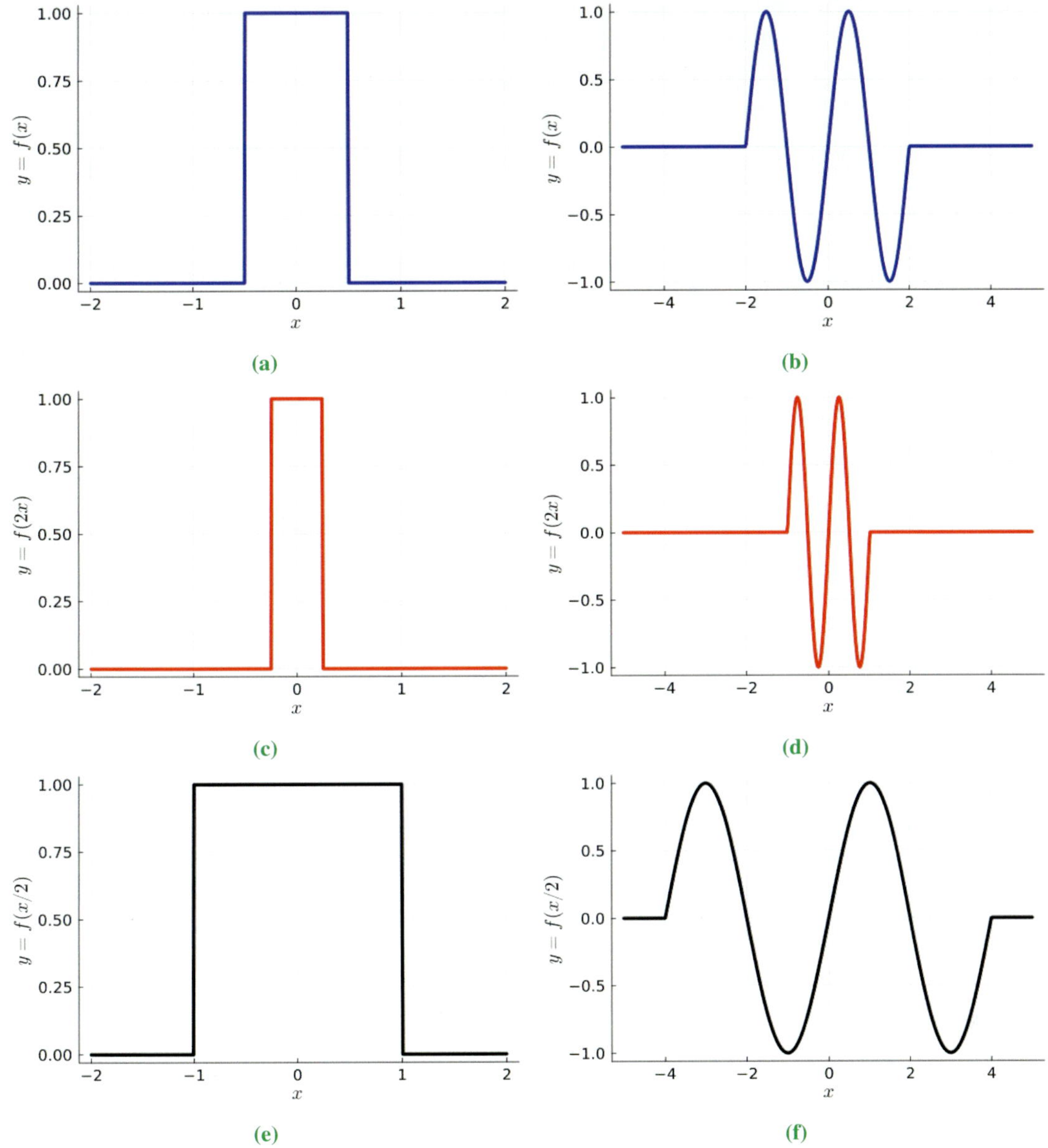

Figure 1.17: Scaled functions: (a) Unit rectangle function in blue. (c) Unit rectangle function scaled by $c = 2$ in red. The area of the rectangle has been decreased by a factor of two. (e) Unit rectangle function scaled by $c = \frac{1}{2}$ in black. The area of the rectangle has been increased by a factor of two. (b) Two periods of the function $\sin(x)$ in blue. (d) Two periods of the function $\sin(2x)$ in red. The frequency has been doubled. (f) Two periods of the function $\sin(\frac{x}{2})$. The frequency has been halved.

- The robot arms in Fig. 1.7 are not worth much if we cannot command them to reach an object at a given position. In many cases, that comes down to regulating each of the joints in the robot arm to a particular value, called a **setpoint**. Commanding robot joints to setpoints is the bread and butter of feedback control. Because some of the key ideas only involve basic algebra, Precalculus knowledge is sufficient to glean key insight into how a feedback controller works.

- The description of a feedback control system will also allow us to highlight where calculus is used in Robotics, partially offering a road map to this textbook.

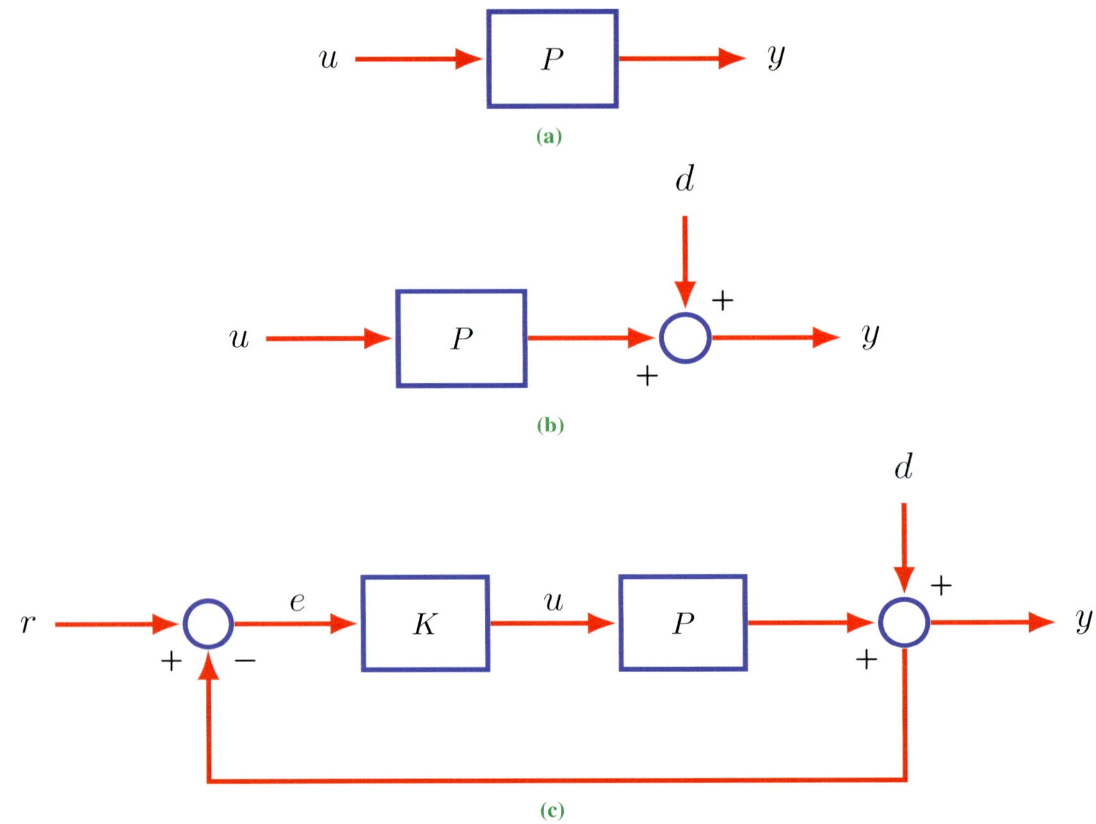

Figure 1.18: Feedback diagrams, where P stands for plant, which is the generic name for the system that is to be controlled (e.g., motor with a link connected to it), y is the measured output (e.g. joint angle), u is the input to the plant (e.g., motor torque), and d is a disturbance signal, also called a perturbation (e.g,, friction in the joint). (a) An open-loop system without a disturbance. (b) An open-loop system with a disturbance. (c) A **closed-loop system**, where K is the **controller gain**, r is the reference signal (what we want y to become), and e is the error term, the difference between the reference signal and the measured output. In (a) and (b), there is no loop, hence the term "open-loop system", while in (c), a loop is present, hence the term "closed-loop control".

Consider **Fig. 1.18-(b)**. The diagram represents the algebraic equation:

$$\boxed{y = P \cdot u + d,} \tag{1.49}$$

where

- u is the motor torque, d is a disturbance signal, and $P > 0$ is a positive constant representing the **gain** (aka multiplying factor) of the system to be controlled or regulated. Let us imagine that our plant is one of the links in **Fig. 1.6** attached to a DC motor. Our objective is to regulate the joint angle to a desired value, which we will denote by r because, in the control community, we call it a **reference signal**.
- u is the **input or control signal** to the plant, a signal we can manipulate to achieve our goal; let's imagine that u is the torque of the DC motor.
- y is the **measured output** of the system; let's imagine it is the output of the encoder at the joint angle.
- d is our nemesis, called a **disturbance signal** or **perturbation** in control parlance, which is universally present in real life; let's imagine that d is friction in the joint of the robot. In industrial robots, unfortunately, d can be very large. Yes, manufacturers try to minimize it, but alas, that is hard and expensive.

How to Read the Diagrams in Fig. 1.18: These diagrams represent linear equations. In (a), $y = P \cdot u$; the signal that comes into the box is multiplied by the system gain P. In (b), the circle with the plus signs is called a **summing junction**, aka **summer**; the output of the summing junction is equal to the sum of all the signals that enter it, with each input signal weighted by the sign nearest to it. Because in (b), they are all plus signs, $y = P \cdot u + d$. In (c), we have our first encounter with a feedback system, aka a closed-loop system. The first summer represents $e = r - y$, and the second summer represents $y = P \cdot u + d$, as in (b). These two equations are connected by $u = Ke$. We will solve these simultaneous equations shortly.

Objective: Regulate y to r, a desired angle, despite the fact that we are unable to measure the disturbance d.

Open-Loop Approach to the Problem

We know the plant model is $y = P \cdot u + d$, and we cannot measure d. So we propose

$$u = K \cdot r \tag{1.50}$$

and seek to select K so that $y \approx r$, namely its desired value. We write this as:

$$r \approx y = P \cdot u + d = P \cdot K \cdot r + d. \tag{1.51}$$

Because we know nothing about d, we assume it is as likely to be positive as negative, with a mean value of zero. This leads to

$$r \approx y \iff P \cdot K = 1 \iff K = \frac{1}{P}, \tag{1.52}$$

That is, the control gain is set to be equal to the reciprocal of the plant gain. Because $P \cdot K = 1$, our open-loop control system results in:

$$y = r + d. \tag{1.53}$$

There are two issues with this solution:

(a) the disturbance passes directly through to the output, causing error in the precision of the control system; and

(b) we may not know the plant gain exactly, and hence our control gain is really $K = \frac{1}{P^{\text{nom}}}$, where P^{om} is a nominal value for the plant gain, and the real output is then

$$y = \left(\frac{P}{P^{\text{nom}}}\right) \cdot r + d.$$

There must be a better way to handle the uncertainty in the model, namely the unmeasurable disturbance and any variations in the plant gain P, which leads us to the second solution.

Closed-Loop Approach to the Problem

The feedback diagram in **Fig. 1.18-(c)** yields:

$$\begin{aligned} y &= P \cdot u + d, \\ e &= r - y, \\ u &= K \cdot e, \end{aligned} \tag{1.54}$$

where K is a feedback gain. We get to adjust it so as to best meet our objective of making the output as close to the commanded reference value as possible, that is, $y \approx r$. Combining the bottom two equations of (1.54), we have

$$u = K \cdot e = K \cdot (r - y).$$

Plugging this into the top equation yields:

$$y = P \cdot K \cdot e = P \cdot K \cdot (r - y) + d,$$

which is a scalar linear equation. Solving for y gives:

$$(1 + P \cdot K) \cdot y = P \cdot K \cdot r + d,$$

and thus,

$$y = \frac{P \cdot K}{1 + P \cdot K} \cdot r + \frac{1}{1 + P \cdot K} \cdot d, \tag{1.55}$$

which is a linear equation relating the reference signal r and the disturbance signal d to the output y. **This is the most important equation in feedback control** because it tells us everything about how the free parameter K in the controller relates to the multiplying factor in front of the reference command, namely, $\frac{P \cdot K}{1+P \cdot K}$, and in front of the disturbance, namely $\frac{1}{1+P \cdot K}$.

In a "perfect world", we would have $\frac{P \cdot K}{1+P \cdot K} = 1$ and $\frac{1}{1+P \cdot K} = 0$, so that $y = r$. Perfection is hard to achieve, so we seek to make the following approximations:

$$\frac{P \cdot K}{1 + P \cdot K} \approx 1 \text{ and } \frac{1}{1 + P \cdot K} \approx 0$$

hold as closely as we can.

- Suppose that P has a nominal value of 2, and we select $K = 25$. Then, plugging in values gives us

  $$y = \frac{50}{51} \cdot r + \frac{1}{51} \cdot d \approx 0.98r + 0.0196d.$$

  The disturbance is greatly attenuated, and y is within 2% of its desired value, r.

- Suppose that P could vary between $1 \leq P \leq 3$, and we keep $K = 25$. Then we have:

  $$\begin{aligned} y &\approx 0.96r + 0.0385d \quad \text{when } P = 1, \\ y &\approx 0.99r + 0.0136d \quad \text{when } P = 3, \end{aligned}$$

  which is also great!

Note: Suppose we can select $K = 100$. How would the above numbers change?

Taking Stock

Closed-loop control is a game-changer. It allows one to "track" reference commands; that is, drive y to its desired value r, even in the face of:

- uncertainty in the mathematical model of the system being controlled, and
- nefarious disturbances trying to prevent us from achieving our tracking objective.

Where's the Calculus?

In the above analysis, we assumed that the system we wish to control can be represented by a constant gain, P. We know full well that "$F = ma$" has to enter into the picture when dealing with a robot. That is called **dynamics**, and that's where calculus first comes in.

- In Section 3.4, we will draw on our knowledge of physics to compute the displacement of an object from its speed of motion, using definite integration. Further applications of integration allow us to compute the mass of a robot link from its density and shape, and its "center of gravity", aka "center of mass", the point at which a single point mass can summarize the overall effect of gravity on the object.
- In Chapter 5, we will study derivatives and realize that velocity is the derivative of position with respect to time. In a similar manner, acceleration is the derivative of velocity with respect to time.
- Deriving dynamical models of robots via $F = ma$ is very tedious. Lagrange invented another method for creating dynamical models of collections of rigid bodies, such as our 3-link robot. It involves potential energy and kinetic energy. Section 6.4 explores Lagrange's Equations of Motion.

- In Chapter 9, we will learn about ordinary differential equations (ODEs for short) and how to solve them. We will also learn how to develop linear approximations of complicated nonlinear differential equations.
- In Chapter 10, we study the Laplace transform, a means to understand the stability of ODEs. The Laplace Transform is a kind of superpower: it transforms linear ODEs into algebraic equations! By examining the roots of the resulting algebraic equations, we can determine if small perturbations introduced into an ODE model of a robot eventually dissipate (this is what stability is all about) or cause the system to explode (this is the essence of instability).
- In Chapter 10, we will use all of this knowledge to design feedback control laws for linear ODEs, culminating in us stabilizing the BallBot. Potential followup courses include Control Systems Analysis and Design, Automatic Control, and Flight Dynamics and Control.

1.18 Binomial Theorem Meets Euler (Optional Read)

On the surface, the Binomial Theorem, which can be used to expand $(x + y)^n$, may seem to be limited to helping us with polynomials. However, Euler's number involves $\left(1 + \frac{1}{n}\right)^n$ and Euler's formula allows us to write

$$\cos^n(x) = \left(\frac{e^{\mathrm{i}x} + e^{-\mathrm{i}x}}{2}\right)^n$$

and

$$\sin^n(x) = \left(\frac{e^{\mathrm{i}x} - e^{-\mathrm{i}x}}{2\mathrm{i}}\right)^n.$$

Here, we apply the Binomial Theorem to understand these quantities.

1.18.1 Binomial Theorem Meets Euler's Constant

Applying the Binomial Theorem to the expression $\left(1 + \frac{1}{n}\right)^n$ yields another means to compute Euler's number, e. Indeed,

$$\boxed{\left(1 + \frac{1}{n}\right)^n = \sum_{k=0}^{n} \binom{n}{k} \frac{1}{n^k}.} \tag{1.56}$$

The kth term of this sum is:

$$\begin{aligned} \binom{n}{k}\frac{1}{n^k} &= \frac{n!}{k!(n-k)!} \cdot \frac{1}{n^k} \\ &= \frac{1}{k!} \cdot \frac{n \cdot (n-1) \cdot (n-2) \cdots (n-k+1)}{n^k} \\ &= \frac{1}{k!} \cdot \frac{n}{n} \cdot \frac{n-1}{n} \cdot \frac{n-2}{n} \cdots \frac{n-k+1}{n} \\ &= \frac{1}{k!} \cdot \left(1 - \frac{1}{n}\right) \cdot \left(1 - \frac{2}{n}\right) \cdots \left(1 - \frac{k+1}{n}\right), \end{aligned} \tag{1.57}$$

where we noted that $\frac{n}{n} = 1$, $\frac{n-1}{n} = 1 - \frac{1}{n}$, $\frac{n-2}{n} = 1 - \frac{2}{n}$, all the way to $\frac{n-k+1}{n} = 1 - \frac{k+1}{n}$.

We will now make some approximations in the "spirit of calculus" that will become more rigorous in the next chapter.

For k fixed and "n much larger than k", $1 - \frac{1}{n} \approx 1$, $1 - \frac{2}{n} \approx 1$, all the way to $1 - \frac{k+1}{n} \approx 1$. This leads to the next approximation, namely, the product of terms that are close to one is also close to one,

$$\left(1 - \frac{1}{n}\right) \cdot \left(1 - \frac{2}{n}\right) \cdots \left(1 - \frac{k+1}{n}\right) \approx 1.$$

Therefore, for k constant and "n large",

$$\binom{n}{k}\frac{1}{n^k} \approx \frac{1}{k!}.$$

This indicates that e can be approximated as

$$e \approx \left(1+\frac{1}{n}\right)^n = \sum_{k=0}^{n}\binom{n}{k}\frac{1}{n^k} \approx \sum_{k=0}^{n}\frac{1}{k!} = \frac{1}{0!}+\frac{1}{1!}+\frac{1}{2!}+\frac{1}{3!}+\cdots+\frac{1}{n!}, \tag{1.58}$$

with the approximation becoming more accurate as n gets larger and larger. The notion of approximations improving for large n will lead us to the concept of limits in Chapter 2.

Table 1.11 compares our two approximations for Euler's number. We see that the sum, $\sum_{k=0}^{n}\frac{1}{k!}$, "converges" to e much faster than the product, $\left(1+\frac{1}{n}\right)^n$. Euler's name is attached to the constant e because he proved rigorously,[11] among other things, that as n tends to infinity, the sum converges to e.

n	$(1+\frac{1}{n})^n$	$\sum_{k=0}^{n}\frac{1}{k!}$
5	2.48832000	2.71666667
10	2.59374246	2.71828180
10^2	2.70481383	2.71828183
10^3	2.71692393	2.71828183

Table 1.11: Comparison of the values of $e \approx 2.71828183\ldots$, the product $(1+\frac{1}{n})^n$, and the sum $\sum_{k=0}^{n}\frac{1}{k!}$ for various values of n. Values are shown to 8 significant digits. The sum seems to converge must faster, meaning that it provides a more practical way of computing e.

```julia
n=100

e_approxA = (1+1/n)^n

e_approxB = 0.0
for k = 0:n
    e_approxB = e_approxB + 1/factorial(big(k))
    # big(k) needed for 21! and beyond
end
[e_approxA; e_approxB; exp(1)]
```

Output

```
3-element Array{BigFloat,1}:
2.704813829421528481589120929129421710968017578125
2.718281828459045235360287471352662497757247093699959574966967627724076630353416
2.718281828459045090795598298427648842334747314453125
```

1.18.2 Binomial Theorem and Powers of Sine and Cosine

This section shows how powers of sine and cosine functions can be simplified through the Binomial Theorem. We first use Euler's formula to write:

$$\cos^n(\theta) = \left(\frac{e^{\mathrm{i}\theta}+e^{-\mathrm{i}\theta}}{2}\right)^n = \frac{1}{2^n}\cdot\left(e^{\mathrm{i}\theta}+e^{-\mathrm{i}\theta}\right)^n = \frac{1}{2^n}\cdot\left(x+x^{-1}\right)^n, \tag{1.59}$$

[11] Here, "rigorously" means that Euler filled in all the holes in the argument we sketched out. In mathematics, rigor refers to a level of thoroughness and consistency in reasoning that ensures the validity and correctness of results. It involves precise, clear assumptions, and sound logical techniques to establish the truth of mathematical statements. Rigor helps in avoiding errors, clarifying concepts, and ensuring that the conclusions reached are reliable and trustworthy. Through rigorous processes, mathematicians aim to build a solid foundation for the development of mathematical theories and the advancement of the field. Engineering is built upon these solid foundations.

for $x = e^{\mathrm{i}\theta}$. The binomial expansion of $\left(x + x^{-1}\right)^n$ for any integer n is given by the Binomial Theorem:

$$\left(x + x^{-1}\right)^n = \sum_{k=0}^{n} \binom{n}{k} x^k \cdot \left(x^{-1}\right)^{n-k} = \sum_{k=0}^{n} \binom{n}{k} x^k \cdot x^{k-n} = \sum_{k=0}^{n} \binom{n}{k} x^{2k-n}. \tag{1.60}$$

Case 1: n = odd.

Because the sum starts at zero and ends at n, there are $(n+1)$ terms in the expansion. When n is odd, $(n+1)$ is even, and we can pair the first term with the last term, the second term with the second-to-last term, and in general, the kth term from the start with the $(n-k)$th term from the end, to obtain:

$$\binom{n}{k} x^{2k-n} + \binom{n}{n-k} x^{n-2k}. \tag{1.61}$$

Since $\binom{n}{k} = \binom{n}{n-k}$, the expansion simplifies to:

$$\left(x + x^{-1}\right)^n = \sum_{k=0}^{\frac{n-1}{2}} \binom{n}{k} \left(x^{n-2k} + x^{-(n-2k)}\right). \tag{1.62}$$

Using (1.59) and substituting $x = e^{\mathrm{i}\theta}$, we have:

$$\begin{aligned} \cos^n(\theta) &= \frac{1}{2^n} \cdot \sum_{k=0}^{\frac{n-1}{2}} \binom{n}{k} \left(e^{\mathrm{i}(n-2k)\theta} + e^{-\mathrm{i}(n-2k)\theta}\right) \\ &= \frac{1}{2^{n-1}} \cdot \sum_{k=0}^{\frac{n-1}{2}} \binom{n}{k} \left(\frac{e^{\mathrm{i}(n-2k)\theta} + e^{-\mathrm{i}(n-2k)\theta}}{2}\right) \\ &= \frac{1}{2^{n-1}} \cdot \sum_{k=0}^{\frac{n-1}{2}} \binom{n}{k} \cos\left((n-2k)\theta\right). \end{aligned}$$

Case 2: n = even.

Similar steps yield the result for n even. The main difference is that when n is even, $(n+1)$ is odd. We can still perform the same pairing up of terms, except the middle term remains unpaired, giving:

$$\left(x + x^{-1}\right)^n = \sum_{k=0}^{\frac{n-2}{2}} \binom{n}{k} \left(x^{n-2k} + x^{-(n-2k)}\right) + \binom{n}{n/2}. \tag{1.63}$$

The result for $\sin^n(x)$ is developed in Section 1.19.

Proposition 1.71: Reducing Powers of Cosine and Sine

For all integers $n > 1$,

$$\cos^n(\theta) = \begin{cases} \dfrac{1}{2^{n-1}} \cdot \displaystyle\sum_{k=0}^{\frac{n-1}{2}} \binom{n}{k} \cos\left((n-2k)\theta\right) & n \text{ odd} \\ \dfrac{1}{2^n}\dbinom{n}{\frac{n}{2}} + \dfrac{1}{2^{n-1}} \cdot \displaystyle\sum_{k=0}^{\frac{n-2}{2}} \binom{n}{k} \cos\left((n-2k)\theta\right) & n \text{ even} \end{cases} \tag{1.64}$$

and

$$\sin^n(\theta) = \begin{cases} \dfrac{(-1)^{\frac{n-1}{2}}}{2^{n-1}} \cdot \displaystyle\sum_{k=0}^{\frac{n-1}{2}} \binom{n}{k} (-1)^k \sin\left((n-2k)\theta\right) & n \text{ odd} \\ \dfrac{1}{2^n}\dbinom{n}{\frac{n}{2}} + \dfrac{(-1)^{\frac{n}{2}}}{2^{n-1}} \cdot \displaystyle\sum_{k=0}^{\frac{n-2}{2}} \binom{n}{k} (-1)^k \cos\left((n-2k)\theta\right) & n \text{ even.} \end{cases} \tag{1.65}$$

Note: Powers two through seven are given in Table 1.12.

n	$\cos^n(\theta)$	$\sin^n(\theta)$
2	$\frac{1}{2}+\frac{1}{2}\cos(2\theta)$	$\frac{1}{2}-\frac{1}{2}\cos(2\theta)$
3	$\frac{3}{4}\cos(\theta)+\frac{1}{4}\cos(3\theta)$	$\frac{3}{4}\sin(\theta)-\frac{1}{4}\sin(3\theta)$
4	$\frac{3}{8}+\frac{1}{2}\cos(2\theta)+\frac{1}{8}\cos(4\theta)$	$\frac{3}{8}-\frac{1}{2}\cos(2\theta)+\frac{1}{8}\cos(4\theta)$
5	$\frac{5}{8}\cos(\theta)+\frac{5}{16}\cos(3\theta)+\frac{1}{16}\cos(5\theta)$	$\frac{5}{8}\sin(\theta)-\frac{5}{16}\sin(3\theta)+\frac{1}{16}\sin(5\theta)$
6	$\frac{5}{16}+\frac{15}{32}\cos(2\theta)+\frac{3}{16}\cos(4\theta)+\frac{1}{32}\cos(6\theta)$	$\frac{5}{16}-\frac{15}{32}\cos(2\theta)+\frac{3}{16}\cos(4\theta)-\frac{1}{32}\cos(6\theta)$
7	$\frac{35}{64}\cos(\theta)+\frac{21}{64}\cos(3\theta)+\frac{7}{64}\cos(5\theta)+\frac{1}{64}\cos(7\theta)$	$\frac{35}{64}\sin(\theta)-\frac{21}{64}\sin(3\theta)+\frac{7}{64}\sin(5\theta)-\frac{1}{64}\sin(7\theta)$

Table 1.12: Powers of sine and cosine.

```
using SpecialFunctions: binomial

function cos_power_n(θ, n)
    if n % 2 == 0  # n is even
        sum = binomial(n, n ÷ 2) / 2^n
        for k in 0:(n ÷ 2 - 1)
            sum += binomial(n, k) * cos((n - 2k) * θ) / 2^(n-1)
        end
        return sum
    else  # n is odd
        sum = 0.0
        for k in 0:((n - 1) ÷ 2)
            sum += binomial(n, k) * cos((n - 2k) * θ) / 2^(n-1)
        end
        return sum
    end
end

function sin_power_n(θ, n)
    if n % 2 == 0  # n is even
        sum = binomial(n, n ÷ 2) / 2^n
        for k in 0:((n - 2) ÷ 2)
            sum += (-1)^(n ÷ 2 + k) * binomial(n, k) * cos((n - 2k) * θ) / 2^(n-1)
        end
        return sum
    else  # n is odd
        sum = 0.0
        for k in 0:((n - 1) ÷ 2)
            sum += (-1)^((n - 1) ÷ 2 + k) * binomial(n, k) * sin((n - 2k) * θ) / 2^(n-1)
        end
        return sum
    end
end
```

Output

```
sin_power_n (generic function with 1 method)
```

```
using Symbolics
@variables θ
n=7
```

```
myAns = cos_power_n(θ, n); display(myAns)
myAns = sin_power_n(θ, n); display(myAns)
```

Output

$$\frac{21}{64}\cos(3\theta) + \frac{1}{64}\cos(7\theta) + \frac{35}{64}\cos(\theta) + \frac{7}{64}\cos(5\theta)$$

$$\frac{35}{64}\sin(\theta) + \frac{7}{64}\sin(5\theta) - \frac{21}{64}\sin(3\theta) - \frac{1}{64}\sin(7\theta)$$

1.19 Proofs Associated with the Chapter (Optional Read)

Proposition 1.72: Are There Any Irrational Numbers?

$\sqrt{2}$ is an irrational number.

Proof: Here is a proof by contradiction. We first define a logical statement such as:

$$P : \text{“}\sqrt{2} \text{ is an irrational number”}.$$

We then assume its logical negative, namely,

$$\neg P : \text{“}\sqrt{2} \text{ is a rational number”}.$$

We seek to show that assuming $\neg P$ is true leads to an “absurdity”; specifically, we seek to deduce from $\neg P$ a second logical statement R that is both true and false! In logic, statements that are both true and false are called **contradictions**. Moreover, *the only way to arrive at a contradiction by properly applying the rules of logic is to start with a false premise*. If a premise is false, then its negative is true, and vice versa.

The Logic of Euclid’s Proof by Contradiction

One assumes $\neg P$: “$\sqrt{2}$ is a rational number” is true (**T**) and deduces from that a contradiction. The only way out of the contradiction is for $\neg P$ to be false (**F**). Therefore, P : “$\sqrt{2}$ is an irrational number” is **T**.

Note: Not bad for 2,000 years ago!

For us, the logical statement R starts with a pair of integers “that do not have any common factors,” and yet, if their ratio is equal to $\sqrt{2}$, we will prove that “they both must be even,” and hence have two as a common factor, yielding a contradiction! We will then establish that our premise was false, and hence its negation is true.

In the proof, we will need the following fact about an integer m: **Fact:** m^2 is even if, and only if, m is even. In other words, if you square an integer and obtain an even number, you cannot have started with an odd[12] integer! Recall that by definition, an integer m is even if there exists an integer k such that $m = 2k$. Similarly, m is odd if there exists integer k such that $m = 2k + 1$.

If $\sqrt{2}$ is rational, then by the definition of the rational numbers, there must exist natural numbers m and n such that:

- m and n have no common factors,
- $n \neq 0$, and

[12] A counting number m is **odd** if there exists $k \in \mathbb{N}$ such that $m = (2k+1)$. We square the number and obtain $m^2 = (2k+1)^2 = 2(2k^2+k)+1$, which is an odd integer because $(2k^2+k)$ is an integer. Hence, if we square a number and obtain an even number, we cannot have started with an odd number; we must have started with an even number.

$$\sqrt{2} = \frac{m}{n}. \tag{1.66}$$

So far, we simply applied the definition of a rational number. Next, we square both sides of (1.66) to arrive at:

$$\left(2 = \frac{m^2}{n^2}\right) \implies \left(2n^2 = m^2\right) \implies \left(m^2 \text{ is even}\right),$$

because it is the product of 2 and the natural number n^2. From our fact about squares of integers, we deduce that m must be even, and hence there must exist an integer k such that $m = 2k$.

From $2n^2 = m^2$, we deduce that:

$$\left(2n^2 = (2k)^2\right) \implies \left(2n^2 = 4k^2\right) \implies \left(n^2 = 2k^2\right) \implies n^2 \text{ is even.}$$

Once again appealing to our fact about squares of integers, we deduce that n must be even, and hence there must exist an integer j such that $n = 2j$.

Because both m and n are even, they have 2 as a common factor, which contradicts the property of rational numbers that m and n have no common factors.

Because we arrived at this contradiction from the statement (premise) "$\sqrt{2}$ is rational", we deduce that our premise "$\sqrt{2}$ is rational" must be false. Hence, $\sqrt{2}$ is NOT RATIONAL, which means that $\sqrt{2}$ IS IRRATIONAL. ■.

Remark 1.73. *Estermann offers an alternative proof that is pretty cool. Here is its presentation by Michael Penn.*

Proposition 1.74: Bounds for Studying Euler's number

For $n \geq 1$, define $a_n := \left(1 + \frac{1}{n}\right)^n$. Then for all $n \geq 1$ and $m \geq 0$,

(a) $a_n \leq a_{n+1}$ (the terms are monotonically increasing),

(b) $a_n \leq \sum_{k=0}^{n} \frac{1}{k!}$, (we have an upper bound on the nth term) and, for all $m \geq 1$,

(c) $a_{n+m} < \sum_{k=0}^{n} \frac{1}{k!} + \frac{1}{n \cdot n!}$ (we have an upper bound for all terms).

(d) $a_m < 3$ (by taking $n = 3$, the above property gives $a_{3+m} < 3$ for all m; checking individually a_1, a_2, and a_3 gives the general result).

Proof: From the Binomial Theorem, for any $n \geq 1$,

$$a_n := \left(1 + \frac{1}{n}\right)^n = \sum_{k=0}^{n} \binom{n}{k} \left(\frac{1}{n}\right)^k = \sum_{k=0}^{n} \binom{n}{k} \cdot \frac{1}{n^k}. \tag{1.67}$$

Everything will follow from this equation.

(a) We consider the binomial expansions of a_n and a_{n+1}

$$a_n = \sum_{k=0}^{n} \binom{n}{k} \cdot \frac{1}{n^k}$$

and

$$a_{n+1} = \sum_{k=0}^{n} \binom{n+1}{k} \cdot \frac{1}{(n+1)^k} + \frac{1}{(n+1)^{n+1}}.$$

From Prop. 1.66, for all $0 \leq k \leq n$,

$$\binom{n}{k} \cdot \frac{1}{n^k} \leq \binom{n+1}{k} \cdot \frac{1}{(n+1)^k},$$

and thus $a_{n+1} - a_n \geq \frac{1}{(n+1)^{n+1}}$.

(b) For the upper bound, we can use Prop. 1.65, which states that the binomial coefficient $\binom{n}{k}$ is upper bounded by $n^k/k!$. This gives us

$$\left(1+\frac{1}{n}\right)^n = \sum_{k=0}^{n}\binom{n}{k}\cdot\frac{1}{n^k} \leq \sum_{k=0}^{n}\frac{n^k}{k!}\cdot\frac{1}{n^k} = \sum_{k=0}^{n}\frac{1}{k!}, \tag{1.68}$$

proving (b).

(c) Continuing from (b), for all $m \geq 1$,

$$\left(1+\frac{1}{n+m}\right)^{n+m} \leq \sum_{k=0}^{n}\frac{1}{k!} + \sum_{j=1}^{m}\frac{1}{(n+j)!}. \tag{1.69}$$

We note that

$$(n+j)! = \underbrace{(n+j)\cdot(n+j-1)\cdots(n+1)}_{\geq (n+1)^j}\cdot n! \geq (n+1)^j\cdot n!$$

Hence,

$$\left(1+\frac{1}{n+m}\right)^{n+m} \leq \sum_{k=0}^{n}\frac{1}{k!} + \frac{1}{n!}\sum_{j=1}^{m}\frac{1}{(n+1)^j}. \tag{1.70}$$

The term "$\sum_{j=1}^{m}\frac{1}{(n+1)^j}$" is nearly a geometric series, and we will show later that for all $m \geq 1$,

$$\sum_{j=1}^{m}\frac{1}{(n+1)^j} < \frac{1}{n}.$$

Hence, for all $n \geq 1$ and $m \geq 1$,

$$\left(1+\frac{1}{n+m}\right)^{n+m} < \sum_{k=0}^{n}\frac{1}{k!} + \frac{1}{n\cdot n!}. \tag{1.71}$$

■

Remark 1.75. *Because the right-hand side of (1.71) does not depend on m, it is valid for all $n \geq 1$, Euler's number is sandwiched between*

$$\left(1+\frac{1}{n}\right)^n \leq e \leq \sum_{k=0}^{n}\frac{1}{k!} + \frac{1}{n\cdot n!},$$

giving easily computable upper and lower bounds. Moreover, the upper and lower bounds approach one another as n becomes larger and larger. We do not have the tools at this point to prove this last statement, but it can be verified numerically. Later, we will be able to make sense of and write:

$$\boxed{\begin{aligned} &e := \lim_{n\to\infty}\left(1+\frac{1}{n}\right)^n \\ &\text{and} \\ &e = \sum_{k=0}^{\infty}\frac{1}{k!} := \lim_{n\to\infty}\sum_{k=0}^{n}\frac{1}{k!}. \end{aligned}} \tag{1.72}$$

Proof of Prop. 1.40 Power Towers That Avoid the Use of Logarithms

This kind of proof is called "First Principles" because we only use the basic properties of integer powers and integer roots that have been introduced in the chapter.

Let $x > 0$, and let $y \approx p/q$ and $z \approx r/s$ be rational approximations, where the denominators q and s are positive integers. We seek to show that:

$$\left(x^{p/q}\right)^{r/s} = x^{pr/qs}.$$

In other words, the exponents multiply. Indeed, by applying the definitions:

$$\begin{aligned}
\left(x^{p/q}\right)^{r/s} &= \sqrt[s]{\left(x^{p/q}\right)^{r}} && (r\text{th power and } s\text{th root of } x^{p/q}) \\
&= \sqrt[s]{\underbrace{\left(x^{p/q}\right)\cdots\left(x^{p/q}\right)}_{r \text{ times}}} && (\text{work out what it means to be the } r\text{th power}) \\
&= \sqrt[s]{\underbrace{\left(\sqrt[q]{x^p}\right)\cdots\left(\sqrt[q]{x^p}\right)}_{r \text{ times}}} && (\text{substitute for } x^{p/q} \text{ and group terms}).
\end{aligned} \tag{1.73}$$

By Prop. 1.29, a product of roots equals the root of the product. Hence,

$$\underbrace{\left(\sqrt[q]{x^p}\right)\cdots\left(\sqrt[q]{x^p}\right)}_{r \text{ times}} = \sqrt[q]{x^{rp}}.$$

The above result plus Prop. 1.31 on nested roots yields:

$$\boxed{\begin{aligned}
\left(x^{p/q}\right)^{r/s} &= \sqrt[s]{\sqrt[q]{x^{r\cdot p}}} \\
&= \sqrt[s\cdot q]{x^{r\cdot p}} \\
&= x^{\frac{rp}{sq}},
\end{aligned}} \tag{1.74}$$

which is amazing! Equation 1.74 holds for all rational approximations of the real numbers y and z, as well as for y and z themselves. ■

Proof of Theorem 1.67 Binomial Theorem

The Binomial Theorem states that for any pair of scalar variables x and y, and for any counting number n, the following equation holds:

$$(x+y)^n = \sum_{k=0}^{n} \binom{n}{k} x^{n-k} y^k,$$

where $\binom{n}{k} := \frac{n!}{k!(n-k)!}$ is the binomial coefficient. We will prove this by induction.

Base Case ($n = 1$): When $n = 1$, the left-hand side of the equation is $(x+y)^1 = x + y$. The right-hand side of the equation is $\binom{1}{0}x^1y^0 + \binom{1}{1}x^0y^1 = x + y$, because $\binom{1}{0} = 1$, $\binom{1}{1} = 1$, $x^0 = 1$, and $y^0 = 1$. So, the hypothesis holds for $n = 1$.

Inductive Step: Assume that the theorem holds for some n; that is:

$$(x+y)^n = \sum_{k=0}^{n} \binom{n}{k} x^{n-k} y^k.$$

We need to show that the theorem holds for $(n+1)$; that is:

$$(x+y)^{n+1} = \sum_{k=0}^{n+1} \binom{n+1}{k} x^{n+1-k} y^k.$$

We start with the left-hand side of the equation:

$$(x+y)^{n+1} = (x+y)(x+y)^n.$$

Using the inductive hypothesis, we can expand $(x+y)^n$ as

$$(x+y)^{n+1} = (x+y)\sum_{k=0}^{n} \binom{n}{k} x^{n-k} y^k.$$

We then distribute the $(x+y)$ to obtain

$$(x+y)^{n+1} = \sum_{k=0}^{n} \binom{n}{k} x^{n+1-k} y^k + \sum_{k=0}^{n} \binom{n}{k} x^{n-k} y^{k+1}.$$

We can shift the index of the second sum to align with the first, via

$$(x+y)^{n+1} = \sum_{k=0}^{n} \binom{n}{k} x^{n+1-k} y^k + \sum_{k=1}^{n+1} \binom{n}{k-1} x^{n+1-k} y^k.$$

Now, we can combine the two sums, per

$$(x+y)^{n+1} = \sum_{k=0}^{n+1} \left[\binom{n}{k} x^{n+1-k} y^k + \binom{n}{k-1} x^{n+1-k} y^k \right].$$

Finally, using a property of binomial coefficients, namely $\binom{n}{k} + \binom{n}{k-1} = \binom{n+1}{k}$, yields the result

$$(x+y)^{n+1} = \sum_{k=0}^{n+1} \binom{n+1}{k} x^{n+1-k} y^k.$$

■

Proof of Prop. 1.71 Reducing Powers of Sine and Cosine

For all integers $n > 1$,

$$\cos^n(\theta) = \begin{cases} \frac{1}{2^{n-1}} \cdot \sum_{k=0}^{\frac{n-1}{2}} \binom{n}{k} \cos((n-2k)\theta) & n \text{ odd} \\ \frac{1}{2^n} \binom{n}{\frac{n}{2}} + \frac{1}{2^{n-1}} \cdot \sum_{k=0}^{\frac{n-2}{2}} \binom{n}{k} \cos((n-2k)\theta) & n \text{ even} \end{cases} \tag{1.75}$$

and

$$\sin^n(\theta) = \begin{cases} \frac{(-1)^{\frac{n-1}{2}}}{2^{n-1}} \cdot \sum_{k=0}^{\frac{n-1}{2}} \binom{n}{k} (-1)^k \sin((n-2k)\theta) & n \text{ odd} \\ \frac{1}{2^n} \binom{n}{\frac{n}{2}} + \frac{(-1)^{\frac{n}{2}}}{2^{n-1}} \cdot \sum_{k=0}^{\frac{n-2}{2}} \binom{n}{k} (-1)^k \cos((n-2k)\theta) & n \text{ even} \end{cases} \tag{1.76}$$

The derivation for $\cos^n(\theta)$ was given earlier. The derivation for $\sin^n(\theta)$ follows the same reasoning as the one for $\cos^n(\theta)$, with the key differences being the expansion for $(x - x^{-1})^n$ instead of $(x + x^{-1})^n$, and factoring out $\frac{1}{(2i)^n}$ instead of $\frac{1}{2^n}$ when considering $\sin^n(\theta)$. The derivation follows.

Starting with Euler's formula for $\sin^n(\theta)$, we have:

$$\sin^n(\theta) = \left(\frac{e^{i\theta} - e^{-i\theta}}{2i} \right)^n = \frac{1}{(2i)^n} \cdot \left(e^{i\theta} - e^{-i\theta} \right)^n = \frac{1}{(2i)^n} \cdot \left(x - x^{-1} \right)^n, \tag{1.77}$$

for $x = e^{i\theta}$. The binomial expansion of $\left(x - x^{-1}\right)^n$ is:

$$\left(x - x^{-1}\right)^n = \sum_{k=0}^{n} \binom{n}{k} x^{n-k} (-x^{-1})^k = \sum_{k=0}^{n} \binom{n}{k} (-1)^k x^{2k-n}. \tag{1.78}$$

For n odd, the sum has an even number of terms, allowing us to pair the first term with the last term, and so on, to obtain:

$$\begin{aligned} \left(x - x^{-1}\right)^n &= \sum_{k=0}^{\frac{n-1}{2}} \left(\binom{n}{k} (-1)^k x^{n-2k} + \binom{n}{n-k} (-1)^{n-k} x^{-(n-2k)} \right) \\ &= \sum_{k=0}^{\frac{n-1}{2}} \binom{n}{k} (-1)^k \left(x^{n-2k} - x^{-(n-2k)} \right), \end{aligned} \tag{1.79}$$

where we used $\binom{n}{n-k} = \binom{n}{k}$, and $(-1)^{n-k} = (-1)^n \cdot (-1)^{-k} = -(-1)^k$ (because n is odd and because $(-1)^{-k} = (-1)^k$).

Substituting back into (1.77) for n odd gives

$$\begin{aligned}
\sin^n(\theta) &= \frac{1}{(2i)^n} \cdot \left(x - x^{-1}\right)^n \Big|_{x=e^{\mathrm{i}\theta}} \\
&= \frac{1}{(2i)^n} \cdot \sum_{k=0}^{\frac{n-1}{2}} \binom{n}{k} (-1)^k \left(x^{n-2k} - x^{-(n-2k)}\right) \Big|_{x=e^{\mathrm{i}\theta}} \\
&= \frac{1}{(2i)^{n-1}} \cdot \sum_{k=0}^{\frac{n-1}{2}} \binom{n}{k} (-1)^k \left(\frac{\left(e^{\mathrm{i}\theta}\right)^{n-2k} - \left(e^{\mathrm{i}\theta}\right)^{-(n-2k)}}{2\mathrm{i}} \right) \\
&= \frac{1}{(2i)^{n-1}} \cdot \sum_{k=0}^{\frac{n-1}{2}} \binom{n}{k} (-1)^k \left(\frac{e^{\mathrm{i}(n-2k)\theta} - e^{-(n-2k)\theta}}{2\mathrm{i}} \right) \\
&= \frac{1}{(2i)^{n-1}} \cdot \sum_{k=0}^{\frac{n-1}{2}} \binom{n}{k} (-1)^k \sin\left((n-2k)\theta\right) \\
&= \frac{(-1)^{\frac{n-1}{2}}}{2^{n-1}} \cdot \sum_{k=0}^{\frac{n-1}{2}} \binom{n}{k} (-1)^k \sin\left((n-2k)\theta\right),
\end{aligned}$$

where on the last step, we used the fact that n odd implies $(n-1)$ is even, and hence

$$(i)^{n-1} = (i^2)^{\frac{n-1}{2}} = (-1)^{\frac{n-1}{2}}.$$

For n even, the sum has an odd number of terms and hence the middle term remains unpaired.

$$\begin{aligned}
\left(x - x^{-1}\right)^n &= \sum_{k=0}^{\frac{n-2}{2}} \left(\binom{n}{k} (-1)^k x^{n-2k} + \binom{n}{n-k} (-1)^{n-k} x^{-(n-2k)} \right) + (-1)^{\frac{n}{2}} \binom{n}{n/2} \\
&= \sum_{k=0}^{\frac{n-2}{2}} \binom{n}{k} (-1)^k \left(x^{n-2k} + x^{-(n-2k)}\right) + (-1)^{\frac{n}{2}} \binom{n}{n/2}.
\end{aligned}$$

where we again used $\binom{n}{n-k} = \binom{n}{k}$, and $(-1)^{n-k} = (-1)^n \cdot 1^{-k} = 1^{-k} = (-1)^k$ (because n is even and because $(-1)^{-k} = (-1)^k$).

Substituting back into (1.77) for n even gives

$$\begin{aligned}
\sin^n(\theta) &= \frac{1}{(2i)^n}\cdot\left(x-x^{-1}\right)^n\Big|_{x=e^{\mathrm{i}\,\theta}}\\
&= (-1)^{\frac{n}{2}}\cdot\frac{1}{(2i)^n}\binom{n}{n/2}+\frac{1}{(2i)^n}\cdot\sum_{k=0}^{\frac{n-2}{2}}\binom{n}{k}(-1)^k\left(x^{n-2k}+x^{-(n-2k)}\right)\Big|_{x=e^{\mathrm{i}\,\theta}}\\
&= (-1)^{n/2}\cdot\frac{1}{(2i)^n}\binom{n}{r}+\frac{1}{2^{n-1}}\cdot\frac{1}{i^n}\cdot\sum_{k=0}^{\frac{n-2}{2}}\binom{n}{k}(-1)^k\left(\frac{\left(e^{\mathrm{i}\,\theta}\right)^{n-2k}+\left(e^{\mathrm{i}\,\theta}\right)^{-(n-2k)}}{2}\right)\\
&= (-1)^{r}\cdot\frac{1}{(2i)^n}\binom{n}{r}+\frac{1}{2^{n-1}}\cdot\frac{1}{i^n}\cdot\sum_{k=0}^{\frac{n-2}{2}}\binom{n}{k}(-1)^k\left(\frac{e^{\mathrm{i}(n-2k)\theta}+e^{-(n-2k)\theta}}{2}\right)\\
&= (-1)^{n/2}\cdot\frac{1}{(2i)^n}\binom{n}{n/2}+\frac{1}{2^{n-1}}\cdot\frac{1}{i^n}\cdot\sum_{k=0}^{\frac{n-2}{2}}\binom{n}{k}(-1)^k\cos\left((n-2k)\theta\right)\\
&= \frac{1}{2^n}\binom{n}{n/2}+\frac{(-1)^{n/2}}{2^{n-1}}\cdot\sum_{k=0}^{\frac{n-1}{2}}\binom{n}{k}(-1)^k\cos\left((n-2k)\theta\right),
\end{aligned}$$

where in the last step, we used the fact that for n even, $(i)^n=(i^2)^r=(-1)^r$ and $(-1)^r\cdot(-1)^r=(-1)^n=1$.

■

Chapter 2

Calculus Foundations: Proofs, Finite Sums, Limits at Infinity, and Geometric Sums

Learning Objectives

By the end of this chapter, the student should be able to:

- Understand the significance of mathematical proofs and articulate their importance in calculus and its applications.
- Master the technique of proof by induction and apply it to demonstrate the validity of mathematical statements.
- Comprehend the concept of limits and see some of their initial uses in calculus.

Outcomes

Upon successful completion of this chapter, students will be able to:

- Apply logical rules to construct (simple) mathematical arguments.
- Learn that there are at least two kinds of infinity.
- Analyze sums of powers of integers to build a foundational understanding of integration.
- Calculate limits at infinity for functions that are significant in calculus.
- Acquire additional knowledge about Euler's number and its unique properties.

2.1 What Is a Mathematical Proof? And Why Are Mathematical Proofs Important?

We ask an expert, **ChatGPT**:

A *mathematical proof* is a logical argument that establishes the truth of a mathematical statement. It is a demonstration that if certain statements (called axioms or premises) are assumed to be true, then another statement necessarily follows from these assumptions.

Proofs use the rules of logic combined with the axioms of the mathematical system under consideration to make their arguments. There are different types of proofs, such as direct proof, proof by contradiction, and proof by induction.

A *direct proof* begins with the assumptions and applies the rules of logic and other mathematical truths that have already been established to arrive at the conclusion.

Proof by induction is a method often used to prove statements about all natural numbers. It involves proving a base case (usually for the number 1 or 0) and then proving that if the statement is true for an arbitrary natural number $k > 0$, then it is also true for the number $k + 1$.

Proof by contradiction (also known as *reductio ad absurdum*) assumes that the statement to be proved is actually false, and then shows that this assumption leads to an absurdity. This then means that the statement cannot be false, and hence must actually be true because we are using "binary logic" (any statement is either **T** or **F**; there is nothing in between).

Mathematical proofs are crucial for a number of reasons:

(a) *Certainty:* Unlike in many other fields, in mathematics, knowledge can be established beyond a shadow of a doubt. A proof guarantees that, given the acceptance of the axioms, the conclusion must be true.

(b) *Building Blocks:* Each proven statement becomes a tool that mathematicians can use in further proofs. Mathematics is built up from layers upon layers of proven statements.

(c) *Understanding:* The process of creating a proof often leads to a deeper understanding of the mathematical structure being studied.

(d) *Discovery:* Sometimes, the process of proving a statement can lead to the discovery of new mathematical facts or even whole areas of mathematics.

(e) *Rigor:* Proofs provide a standard of rigor that helps prevent errors in mathematical reasoning. This rigor is crucial in areas such as computer programming and algorithm design, where a small error in reasoning can lead to significant problems.

Mathematical Proofs Promote Understanding

In short, mathematical proofs are the cornerstone of mathematical knowledge, providing certainty, depth of understanding, and a rigorous basis for further discovery. We'll add one more reason as to why they are important: some things that seem "obviously false", even to the larger mathematical community, can turn out to be true, and vice versa! In a later section, we'll use a famous example to accomplish two things: (1) to learn about countable and uncountable sets and (2) to underline that you can be right even when everyone tells you that you are wrong. A sad example of the latter is Georg Cantor, whose surprising discovery about the nature of infinity led to public opposition and personal attacks. Mathematics being as controversial as evolution? You cannot make this stuff up! Please read on.

Related Videos:

- Don't let it fool you by **3Blue1Brown** (moral of the story: patterns often break).
- Proofs Are Awesome! Here's Why! Math Time With Professor Prime (a short low-key pep talk).
- Is math discovered or invented? a TED-Ed talk by Jeff Dekofsky (asks the question, "does math exist independent of humans").

2.2 Vocabulary and Helpful Notation

The following definitions borrow liberally from Math 299 at Michigan State University.

Terminology:

- *Definition:* A precise explanation of the mathematical meaning of a term or concept.
- *Axiom:* A principle or assumption that is accepted as true without proof in a mathematical system.
- *Theorem:* A mathematical statement that has been proven to be true.

- *Lemma:* A mathematical statement that is proven true and used as a stepping stone in the proof of other statements.
- *Claim:* A mathematical statement that is made as a step towards proving a theorem or to highlight an important property of a mathematical object. When used as a stepping stone to prove something else, Claims are essentially equivalent to Lemmas. When a Claim is used to highlight an important property, it is a "baby" Proposition.
- *Proposition:* A mathematical statement that is less central but nonetheless true and interesting.
- *Corollary:* A mathematical statement that is a straightforward deduction from a theorem or proposition.
- *Proof:* The logical reasoning that demonstrates why a statement is true.
- *Conjecture:* A mathematical statement that is believed to be true, but for which no proof has been established. Or, a communication of a potentially true statement that the author hopes to prove or that the mathematical community seeks to prove. Fermat's Last Theorem was a Conjecture until it was proven by Wiles in 1995.

We will mostly use just definitions and propositions so that we do not get tied up in the hierarchy of theorems, lemmas, etc. We will not take this to extremes, by calling the very well-known "Pythagorean Theorem" the "Pythagorean Proposition", or awkwardly referring to the "Binomial Proposition".

Crib Notes for Proofs: where P and Q are logical statements. They can be either true (denoted **T**) or false (denoted **F**). Because we use binary logic, there is no in-between value, such as "maybe". Statements are either **T** or **F**.

In the following, you might have in mind that $P:\ \det(A) \neq 0$ and $Q: A$ is a square invertible matrix, or, $P:$ the rows of A are linearly independent and $Q: Ax = b$ has a minimum norm solution.

- $\neg$ or $\sim$ denotes the logical operation "NOT", where $\neg\mathbf{T} = \mathbf{F}$ and $\neg\mathbf{F} = \mathbf{T}$. Both symbols will be used interchangeably in this book.
- $P \implies Q$ represents "if statement P is true, then statement Q is also true".
- $P \iff Q$ indicates that "P is true if and only if Q is true". Although P iff Q is a valid expression, it will be seldom used in these notes.
- $P \iff Q$ is logically equivalent to:

  (a) $P \implies Q$ and

  (b) $Q \implies P$.
- The *contrapositive* of $P \implies Q$ is denoted as $\neg Q \implies \neg P$.
- The *converse* of $P \implies Q$ is expressed as $Q \implies P$. It is crucial to note that $P \implies Q$ does not generally imply $Q \implies P$ and vice versa. If they were equivalent, there would be no need for $P \iff Q$.
- The relation $(P \implies Q) \iff (\neg Q \implies \neg P)$ indicates that these two statements are logically equivalent. Although they are equivalent, one may be easier to use in a proof than the other.
- *Logical AND*: $P_1 \wedge P_2$ is read as "P_1 AND P_2". It holds true when both P_1 and P_2 are true.
- *Logical OR*: $P_1 \vee P_2$ is interpreted as "P_1 OR P_2". It is true when at least one of P_1 or P_2 is true. In this book, we use the "inclusive OR", which means $\mathbf{T} \vee \mathbf{T} = \mathbf{T} \vee \mathbf{F} = \mathbf{F} \vee \mathbf{T} = \mathbf{T}$, where **T** stands for true and **F** stands for false.
- Q.E.D. is an abbreviation of the Latin phrase "*quod erat demonstrandum*", which translates to "that which was to be demonstrated". It is used to indicate the completion of a proof. Nowadays, you more frequently see □ or ■ instead of Q.E.D.

Warning: In the beginning, it is quite frequent that students confuse the meanings of *contrapositive* and *converse*. Just be careful. With practice, it becomes second nature. The fact that they both start with "con-" is not helpful!

2.3 Proofs by Induction, Some Finite Sums, and Their Applications

2.3.1 Two Finite Sums via Induction

Logic Principle 2.1: First Principle of Induction (Standard Induction)

Let $P(n)$ denote a statement about the counting numbers with the following properties:

(a) **Base case:** $P(1)$ is true.

(b) **Induction hypothesis:** For all $k \geq 1$, if $P(k)$ is true, then $P(k+1)$ is true.

Then $P(n)$ is true for all $n \geq 1$.

Remark 2.2. *Suppose the base case involves an integer $k_0 \neq 1$. Then we can re-index the set and reduce it to the base case having $k_0 = 1$. Alternatively, we assume that $P(k)$ is true for $k \geq k_0$ implies $P(k+1)$ is true, and then we conclude that $P(n)$ is true for all $n \geq k_0$. A common mistake is not to use the correct base case. For example, see "how to prove by induction that all horses are the same color."*

Example 2.3. *Prove the* ***Claim:*** *For all $n \geq 1$,*

$$1 + 2 + 3 + \cdots + n = \frac{n(n+1)}{2}. \tag{2.1}$$

Proof:

- **Step 0:** Write down $P(n)$: $1 + 2 + 3 + \cdots + n = n(n+1)/2$, the statement that is to be proved.
- **Step 1:** Check the base case, $P(1)$: For $n = 1$, we have $1 = 1$; for the left side, the sum of the first integer is one, and for the right side, $n(n+1)/2 = 1 \cdot (1+1)/2 = 2/2 = 1$. Because $1 = 1$, the base case is true.
- **Step 2:** Show that the induction hypothesis is true. That is, assuming that $P(k)$ is true, show that $P(k+1)$ is also true:

  $$P(k+1) : (1 + 2 + \cdots + k) + (k+1) = (k+1) \cdot (k+2)/2.$$

  Often, showing that $P(k) = \mathbf{T} \implies P(k+1) = \mathbf{T}$ involves re-writing the terms in $P(k+1)$ as a sum of terms that show up in $P(k)$, plus another term.

For the induction step, we assume that

$$P(k) : 1 + 2 + 3 + \cdots + k = k \cdot (k+1)/2$$

is true. Next, we note there are terms in $P(k+1)$ that are common to $P(k)$, namely

$$(1 + 2 + \cdots + k) + (k+1) = \underbrace{(1 + 2 + \cdots + k)}_{k \cdot (k+1)/2} + (k+1) = k \cdot (k+1)/2 + (k+1),$$

because we assumed that $P(k)$ holds. Then, by the accepted rules of arithmetic,

$$k(k+1)/2 + (k+1) = (k+1) \cdot (k/2 + 1) = (k+1) \cdot (k+2)/2,$$

and we then deduce that

$$(1+2+\cdots+k)+(k+1)=(k+1)\cdot(k+2)/2,$$

and therefore, $P(k+1)$ is true. Because we have shown that $P(1)$ is true and for all $k \geq 1$, $P(k) \implies P(k+1)$, by the Principle of Induction, we conclude that for all $n \geq 1$,

$$1+2+3+\cdots+n=n(n+1)/2,$$

and therefore (2.1) holds. ■

Remark 2.4. *If we run into a case where the base case $P(1)$ seems to be so totally trivial that we are unsure whether there is anything to show, it's OK to check $P(2)$, just to convince ourselves that it is true too. In our case, we would have checked $P(2): 1+2=2\cdot 3/2$; since this is easily established to be true, we may now be more confident of our proof. If we do one more, $P(3): 1+2+3=3\cdot 4/2$, we are now on a roll and ready to attack the general case by induction. Bottom line: Initially, it's natural to be tentative when we write a proof.* **It takes practice to learn the art of proving things.** ***In this book, writing proofs is not a core skill we are trying to develop.*** **Instead, we want to establish that proofs are important because they explain why something is true**. *They turn a mathematical result from a "black box" (opaque, unclear, pure manipulation of symbols) to a "white box" (clear, transparent, understandable).*[1].

Example 2.5. *Prove the* ***Claim:*** *For all $n \geq 1$,*

$$1^2+2^2+\cdots+n^2=\frac{n(n+1)(2n+1)}{6}. \tag{2.2}$$

Proof: Proofs can be written in many different ways. Let us approach this proof a bit differently by defining:

$$\begin{aligned} L(n) &:= 1^2+2^2+\cdots+n^2, \text{ the left-hand side of the Claim,} \\ R(n) &:= n(n+1)(2n+1)/6, \text{the right-hand side of the Claim.} \end{aligned} \tag{2.3}$$

We want to show that all counting numbers $n \geq 1$,

$$P(n) \text{ is true, that is, } L(n)=R(n).$$

- **Step 0:** Write down $P(k)$: $L(k)=R(k)$.
- **Step 1:** Check the base case, $P(1)$: For $k=1$, we have $L(1)=1$ and $R(1)=1\cdot(1+1)\cdot(2+1)/6=1$, and hence the base case is true.
- **Step 2:** Let us check one more just so we get the hang of it, $P(2)$: we have $L(2)=1^2+2^2=5$ and $R(2)=2\cdot(2+1)(2\cdot 2+1)/6=2\cdot(15)/6=5$, and hence $L(2)=R(2)$, showing that $P(2)$ holds. Our confidence is higher!
- **Step 3:** Show that the induction hypothesis is true. That is, using the fact that $P(k)$ is true for some $k \geq 1$, show that $P(k+1)$ also is true.

[1]The terms "black box" and "white box" are commonly used in various fields, including systems engineering, computing, and business, with their exact meaning depending on the context.

- **Black Box Testing/Modeling:** In black box testing or modeling, the internal workings of the system or model are not known or considered. It is called a "black box" because the system is opaque; we can't see what is happening inside. Instead, we focus on the input and output. We provide certain inputs, and then test to see if the output is as expected. In the context of AI, a black box model is a system where we can observe the input and output, but not the internal decision-making process. For example, deep learning models are often considered black boxes because, although they can make highly accurate predictions, it is difficult to understand how they arrived at a particular decision.
- **White Box Testing/Modeling:** In white box testing or modeling, we do have access to the internal workings of the system or model. This type of testing is sometimes called "clear box" or "glass box" testing. We not only examine input and output but also the processes that lead from one to the other. For example, a decision tree can be considered a white box model because it is possible to follow the path from beginning to end to understand exactly how a decision is made.

In a similar sense, a mathematical proof provides the steps that show us how to go from the hypotheses (i.e., input) to the claimed result (i.e., output). Without proofs, we are taking results on faith.

When we assume that $P(k)$ is true, we are assuming that

$$L(k) = R(k).$$

To see if this implies $L(k+1) = R(k+1)$, we first evaluate

$$L(k+1) = \underbrace{1^2 + 2^2 \cdots + k^2}_{L(k)} + (k+1)^2 = L(k) + (k+1)^2.$$

Using the induction hypothesis, namely, $L(k) = R(k) = k(k+1)(2k+1)/6$, we compute

$$\begin{aligned} L(k+1) &= L(k) + (k+1)^2 \\ &= \underbrace{k(k+1)(2k+1)/6}_{L(k)} + (k+1)^2 \\ &= \underbrace{(2k^3 + 3k^2 + k)/6}_{k(k+1)(2k+1)/6} + \underbrace{(k^2 + 2k + 1)}_{(k+1)^2} \\ &= (k^3/3 + k^2/2 + k/6) + (k^2 + 2k + 1) \\ &= k^3/3 + 3k^2/2 + 13k/6 + 1. \end{aligned} \tag{2.4}$$

Next, we plug $(k+1)$ into the definition of $R(k)$ to evaluate

$$\begin{aligned} R(k+1) &= (k+1) \cdot ((k+1) + 1) \cdot (2(k+1) + 1)/6 \\ &= (k+1) \cdot (k+2) \cdot (2k+3)/6 \\ &= (k+1) \cdot (2k^2 + 7k + 6)/6 \\ &= k^3/3 + 3k^2/2 + 13k/6 + 1. \end{aligned} \tag{2.5}$$

Comparing (2.4) and (2.5), we observe that indeed, $L(k+1) = R(k+1)$, and therefore $P(k+1)$ is true. Because we have shown that $P(1)$ is true, and for all $k \geq 1$, $P(k) \implies P(k+1)$, by the Principle of Induction, we conclude that for all $n \geq 1$,

$$1^2 + 2^2 + \cdots + n^2 = n(n+1)(2n+1)/6.$$

■

This following example, from Prove that $(n^3 + 11n)$ is divisible by 6 by PrimeNewtons, is included to show a different style of proof by induction, always with the goal of enhancing our understanding. A counting number (positive integer) $i \in \mathbb{N}$ is divisible by 6 if there exists another integer $m \in \mathbb{N}$ such that $i = 6 \cdot m$; in other words, $\frac{i}{6}$ is also a counting number.

Example 2.6. *Use induction to show that for all $n \in \mathbb{N}$, $n^3 + 11n$ is divisible by 6.*

Solution: We want to show that for all counting numbers $n \geq 1$, $P(n) : n^3 + 11n = 6 \cdot m$, for some counting number m.

Base Case: $n = 1$: $1^3 + 11 \cdot 1 = 12 = 2 \cdot 6$, and hence the base case is true.

Induction Step: We show that if $P(k)$ is true for some $k \geq 1$, then $P(k+1)$ is true.

We first use basic algebra to arrive at

$$\begin{aligned} (k+1)^3 + 11 \cdot (k+1) &= \left(k^3 + 3k^2 + 3k + 1\right) + (11k + 11) \\ &= \left(k^3 + 11k\right) + 3 \cdot \left(k^2 + k + 4\right). \end{aligned} \tag{2.6}$$

Because $P(k)$ holds, there exists $m_1 \in \mathbb{N}$ such that $\left(k^3 + 11k\right) = 6m_1$. We prove below that $\left(k^2 + k + 4\right)$ is even for all counting numbers k, and hence there exists m_2 such that $\left(k^2 + k + 4\right) = 2m_2$. Substituting these facts into (2.6) gives:

$$\begin{aligned}(k+1)^3 + 11 \cdot (k+1) &= \left(k^3 + 11k\right) + 3 \cdot \left(k^2 + k + 4\right) \\ &= 6m_1 + 3 \cdot 2m_2 \\ &= 6 \cdot (m_1 + m_2)\,.\end{aligned}$$

This completes the proof because $m_1 + m_2$, the sum of two counting numbers, is another counting number. ■

Example 2.7. *$k^2 + k + 4$ is even for all counting numbers k.*

Proof: We shall use basic properties of counting numbers, whose proofs we can either provide or seek online. If k is odd, then so is k^2. Because the sum of two odd numbers is even, we have that $k^2 + k$ is even. Finally, adding the even number 4 to the even number $k^2 + k$ gives an even number. Next, if k is even, so is k^2. Because the sum of two even numbers is even, we have that $k^2 + k$ is even. Finally, adding the even number 4 to the even number $k^2 + k$ gives an even number. ■

Note: The information we learned from $P(k)$, namely that $k^3 + 11k$ is divisible by 6, needed to be augmented with an additional fact about counting numbers, namely that $k^2 + k + 4$ is always even.

- Basic: Proof by induction | Sequences, series and induction | Precalculus by Khan Academy.
- Basic: Mathematical Induction Practice Problems by the Organic Chemistry Tutor.
- More Advanced: Prove by induction, Sum of the first n cubes, $1^3 + 2^3 + 3^3 + \cdots + n^3$ by **BlackPenRedPen**.
- Quite Advanced: Introduction to Proof by Induction [Discrete Math Class] by Mathematical Visual Proofs.

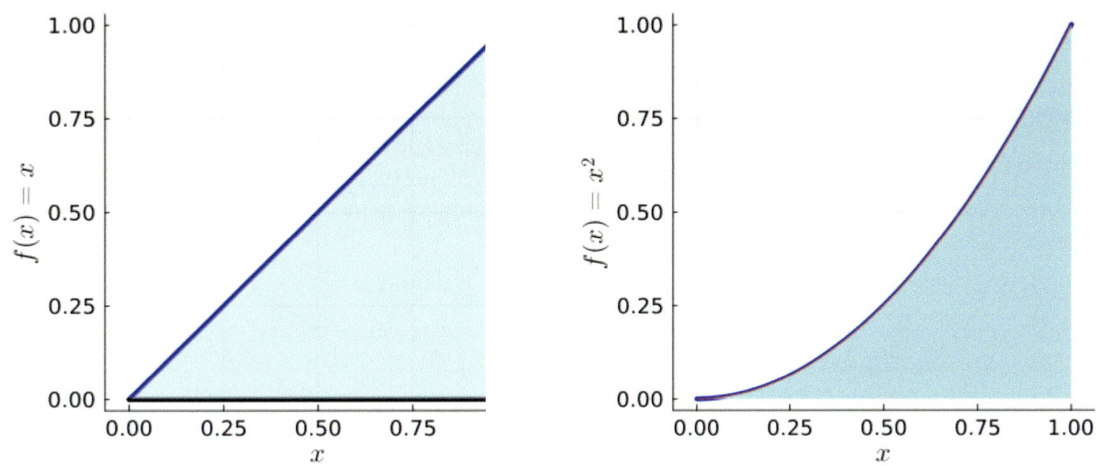

(a) (b)

Figure 2.1: Areas under curves. In blue, the areas under two curves between $x = 0$ and $x = 1$ are shown. (a) For a triangle, we know the formula for its area, $\frac{1}{2}b \cdot h$, where b is the base and h is the height. (b) For the parabola $y = x^2$, we do not have a simple formula for its area. We will develop a method to estimate lower and upper bounds for the computation of areas. We will first show how our estimate works for the triangle, and then apply it to the parabola.

2.3.2 The Amazing Power of the Rectangle: Finding Area under a Curve

In this subsection, we will use what we have learned so far to lay the foundations for computing areas under the simple "curves" shown in **Fig. 2.1**, namely a line and a parabola, **assuming only knowledge of the area of a rectangle!** How can we do this? We will follow Archimedes' basic plan: create both a lower bound and an upper bound for our desired quantity (for Archimedes, it was the circumference of the circle; here, it is the area under the curve), and then show that we can make the lower and upper bounds approach one another. An example of this strategy was presented in Chapter 1 in connection with **Fig. 1.1**.

The curves in **Fig. 2.1** are monotonically increasing, making it particularly easy to formulate upper and lower bounds for their areas by summing the areas of carefully chosen rectangles. We already know a formula for the area of a triangle, namely, one-half base times height, $\frac{1}{2}b \cdot h$, giving us 0.5 for the area in **Fig. 2.1-(a)**. What is the area for the parabola in **Fig. 2.1-(b)**?

Method 2.8: Lower and Upper Bounds for Area under an Increasing Function

Approximating Area: We seek to estimate the area under an increasing function $f : [a,b] \to \mathbb{R}$, where $a < b$. We first divide the interval $[a,b]$ into $n > 1$ evenly spaced subintervals, $[x_i, x_{i+1}]$, where

$$\begin{aligned} \Delta x &:= \frac{b-a}{n} \\ x_i &:= a + (i-1) \cdot \Delta x, \end{aligned} \tag{2.7}$$

so that

$$a = x_1 < x_2 < \cdots < x_n < x_{n+1} = b \tag{2.8}$$

$$x_{i+1} = x_i + \Delta x, \ 1 \le i \le n \tag{2.9}$$

$$f(x_1) \le f(x_2) \le \cdots \le f(x_n) \le f(x_{n+1}). \tag{2.10}$$

- We underestimate the area between $[x_i, x_{i+1}]$ by a rectangle of height $h_i^{\text{Low}} := f(x_i)$ and base $b_i := \Delta x$. This is an **underapproximation** because $h_i^{\text{Low}} := f(x_i) \le f(x)$ for all $x \in [x_i, x_{i+1}]$. We underestimate the total area between $[a,b]$ by summing up the underapproximations,

$$\text{Area}_n^{\text{Low}} := \sum_{i=1}^{n} h_i^{\text{Low}} \cdot b_i = \sum_{i=1}^{n} f(x_i) \cdot \Delta x. \tag{2.11}$$

- We overestimate the area between $[x_i, x_{i+1}]$ by a rectangle of height $h_i^{\text{Up}} := f(x_{i+1})$ and base $b_i := \Delta x$. This is an **overapproximation** because $h_i^{\text{Up}} := f(x_{i+1}) \ge f(x)$ for all $x \in [x_i, x_{i+1}]$. We overestimate the total area between $[a,b]$ by summing up the overapproximations,

$$\text{Area}_n^{\text{Up}} := \sum_{i=1}^{n} h_i^{\text{Up}} \cdot b_i = \sum_{i=1}^{n} f(x_{i+1}) \cdot \Delta x. \tag{2.12}$$

- See **Figs. 2.2** and **2.3**.

Then, just as with Archimedes' Method, we have that for all $n > 1$,

$$\text{Area}_n^{\text{Low}} \le \text{Area} \le \text{Area}_n^{\text{Up}},$$

where Area is the true area under the curve between the values $x = a$ and $x = b$. Moreover, as we make n larger and larger, the lower and upper bounds will approach one another, giving us the ability to approximate the area with arbitrary accuracy.

Area under a Triangle: Let us carry out the method outlined above for $f(x) = x$, $a = 0$ and $b = 1$. Then with

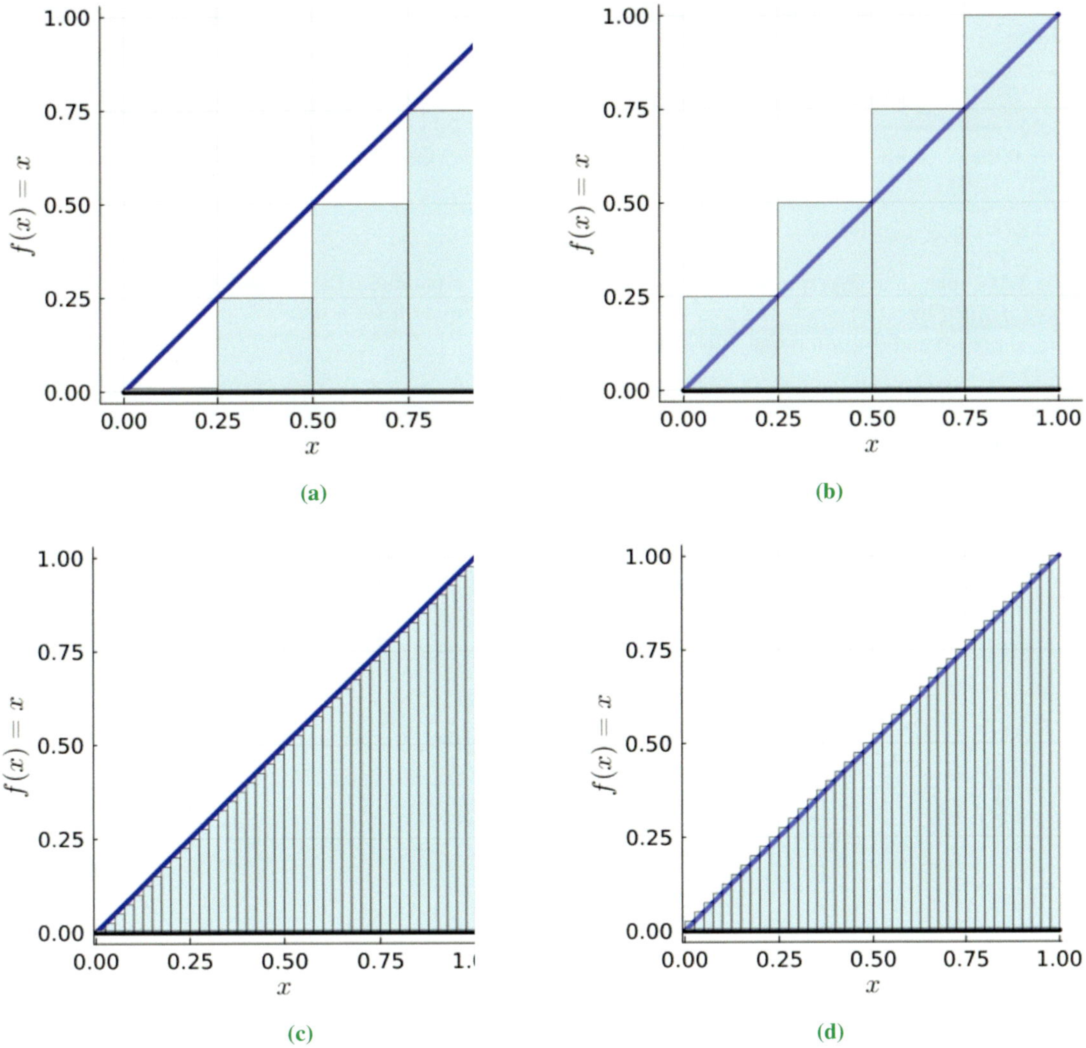

Figure 2.2: Under- and overapproximations for the area of a triangle defined by a line of slope one through the origin. (a) An underapproximation for the triangle defined by $f(x) = x$, $0 \leq x \leq 1$. (b) An overapproximation for the triangle defined by $f(x) = 0$, $0 \leq x \leq 1$. For (a) and (b), $n = 4$ and thus $\Delta x = 0.25$, while for (c) and (d), $n = 40$ and $\Delta x = 0.025$.

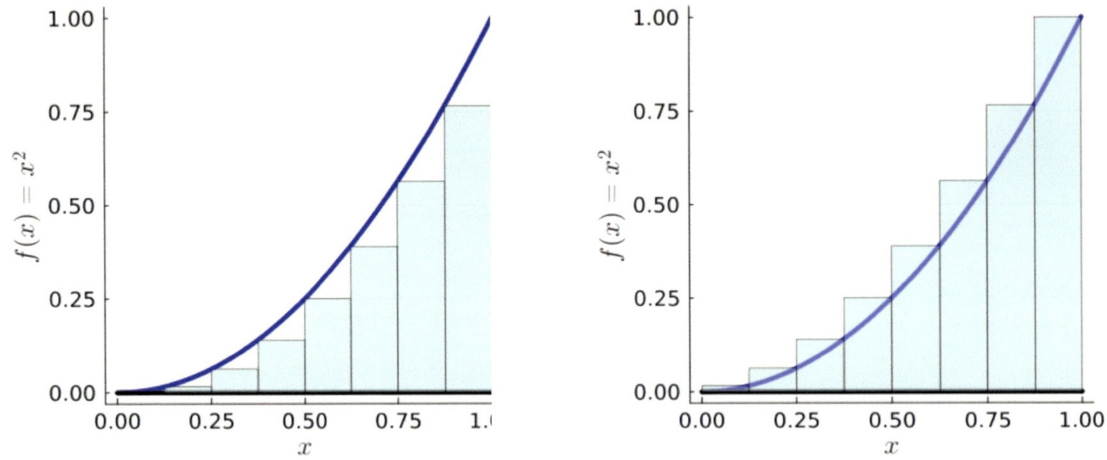

(a) (b)

Figure 2.3: Under- and overapproximations for the area under a parabola. (a) An underapproximation for a parabola defined by $f(x) = x^2$, $0 \le x \le 1$. (b) An overapproximation for a parabola defined by $f(x) = x^2$, $0 \le x \le 1$. For (a) and (b), $n = 8$ and thus $\Delta x = 0.125$.

$n > 1$, we have $\Delta x = \frac{1}{n}$, $x_i = (i-1)\Delta x$, $f(x_i) = (i-1)\Delta x$, and

$$\begin{aligned}
\text{Area}_n^{\text{Low}} &:= \sum_{i=1}^{n} f(x_i) \cdot \Delta x \\
&= \sum_{i=1}^{n} ((i-1)\Delta x) \cdot \Delta x \\
&= \sum_{i=1}^{n} \left((i-1)\frac{1}{n}\right) \cdot \frac{1}{n} \\
&= \frac{1}{n^2} \cdot \sum_{i=1}^{n} (i-1) \quad \text{(because } n \text{ is a constant, we take it out of the sum)} \\
&= \frac{1}{n^2} \cdot \sum_{i=1}^{n-1} i \quad \text{(change of index)} \\
&= \frac{1}{n^2} \cdot \frac{(n-1)(n)}{2} \quad \text{(by Example 2.3)} \\
&= \frac{1}{2} - \frac{1}{2n},
\end{aligned} \tag{2.13}$$

Similarly,

$$\begin{aligned}
\text{Area}_n^{\text{Up}} &:= \sum_{i=1}^{n} f(x_{i+1}) \cdot \Delta x \\
&= \sum_{i=1}^{n} (i\Delta x) \cdot \Delta x \\
&= \sum_{i=1}^{n} \left(i\frac{1}{n}\right) \cdot \frac{1}{n} \\
&= \frac{1}{n^2} \cdot \sum_{i=1}^{n} i \quad \text{(because } n \text{ is a constant, we take it out of the sum)} \\
&= \frac{1}{n^2} \cdot \frac{n(n+1)}{2} \quad \text{(by Example 2.3)} \\
&= \frac{1}{2} + \frac{1}{2n}.
\end{aligned} \tag{2.14}$$

Hence, for all $n > 1$,

$$\text{Area}_n^{\text{Low}} = \frac{1}{2} - \frac{1}{2n} \leq \text{Area} \leq \frac{1}{2} + \frac{1}{2n} = \text{Area}_n^{\text{Up}}. \tag{2.15}$$

This is a useful result, because we know that the true area of the triangle is $\frac{1}{2}$, and we know that we can make $\frac{1}{2n}$ as small as we like by taking n sufficiently large. Because

$$\frac{\text{Area}_n^{\text{Up}} + \text{Area}_n^{\text{Low}}}{2} = \frac{1}{2}$$

and

$$\frac{\text{Area}_n^{\text{Up}} - \text{Area}_n^{\text{Low}}}{2} = \frac{1}{n},$$

we have

$$\boxed{\text{Area}_n^{\text{est}} = \frac{1}{2} \pm \frac{1}{n}.} \tag{2.16}$$

Area under a Parabola: Next, let us carry out our Archimedes-inspired approximation method for $f(x) = x^2$, $a = 0$, and $b = 1$. Again, with $n > 1$, we have $\Delta x = \frac{1}{n}$, $x_i = (i-1)\Delta x$, $f(x_i) = ((i-1)\Delta x)^2$, and

$$\begin{aligned}
\text{Area}_n^{\text{Low}} &:= \sum_{i=1}^{n} f(x_i) \cdot \Delta x \\
&= \sum_{i=1}^{n} ((i-1)\Delta x)^2 \cdot \Delta x \\
&= \sum_{i=1}^{n} \left((i-1)\frac{1}{n}\right)^2 \cdot \frac{1}{n} \\
&= \frac{1}{n^3} \cdot \sum_{i=1}^{n} (i-1)^2 \quad \text{(because } n \text{ is a constant, we take it out of the sum)} \\
&= \frac{1}{n^3} \cdot \sum_{i=1}^{n-1} i^2 \quad \text{(change of index)} \\
&= \frac{1}{n^3} \cdot \frac{(n-1)(n-1+1)(2(n-1)+1)}{6} \quad \text{(by Example 2.5)} \\
&= \frac{1}{n^3} \cdot \frac{2n^3 - 3n^2 + n}{6} \\
&= \frac{1}{3} - \frac{1}{2n} + \frac{1}{6n^2}.
\end{aligned} \tag{2.17}$$

Similarly,

$$\begin{aligned}
\text{Area}_n^{\text{Up}} &:= \sum_{i=1}^{n} f(x_{i+1}) \cdot \Delta x \\
&= \sum_{i=1}^{n} (i\Delta x)^2 \cdot \Delta x \\
&= \sum_{i=1}^{n} \left(i\frac{1}{n}\right)^2 \cdot \frac{1}{n} \\
&= \frac{1}{n^3} \cdot \sum_{i=1}^{n} i^2 \quad \text{(because } n \text{ is a constant, we take it out of the sum)} \\
&= \frac{1}{n^3} \cdot \frac{n(n+1)(2n+1)}{6} \quad \text{(by Example 2.5)} \\
&= \frac{1}{n^3} \cdot \frac{2n^3 + 3n^2 + n}{6} \\
&= \frac{1}{3} + \frac{1}{2n} + \frac{1}{6n^2}.
\end{aligned} \tag{2.18}$$

Hence, for all $n > 1$,

$$\text{Area}_n^{\text{Low}} = \frac{1}{3} - \frac{1}{2n} + \frac{1}{6n^2} \leq \text{Area} \leq \frac{1}{3} + \frac{1}{2n} + \frac{1}{6n^2} = \text{Area}_n^{\text{Up}}. \tag{2.19}$$

Because

$$\frac{\text{Area}_n^{\text{Up}} + \text{Area}_n^{\text{Low}}}{2} = \frac{1}{3} + \frac{1}{6n^2}$$

and

$$\frac{\text{Area}_n^{\text{Up}} - \text{Area}_n^{\text{Low}}}{2} = \frac{1}{n},$$

we have

$$\boxed{\text{Area}_n^{\text{est}} = \left(\frac{1}{3} + \frac{1}{6n^2}\right) \pm \frac{1}{n}.} \tag{2.20}$$

Moreover, as n gets larger and larger, both $\frac{1}{6n^2}$ and $\frac{1}{n}$ become smaller and smaller, leading us to the conclusion that the area under the parabola is $\frac{1}{3}$.

Summary: To estimate the areas, all we did was add up the areas of a carefully selected set of rectangles. Had we been performing the calculations in Julia, we would not have needed any analytical results at all. We would have naturally taken n large and obtained lower and upper bounds on our desired quantity. In order to obtain analytical results, all we needed were Examples 2.3 and 2.5, which address the sum of the first n integers raised to power "1" and power "2", respectively.

Area under a Triangle and a Parabola, Generalized Expressions

Instead of computing the areas under the curves over the interval $[0,1]$, let us consider $[0,x]$, for $x > 0$. We will not repeat the steps for the triangle, because the formula $\frac{1}{2}b \cdot h$ still applies, giving us $x^2/2$ for the area. What about the parabola? For the lower bound, with $n > 1$ rectangles, we have $\Delta x = \frac{x}{n}$, $x_i = (i-1)\Delta x$, $f(x_i) = ((i-1)\Delta x)^2$, and

$$\begin{aligned}
\text{Area}_n^{\text{Low}} &:= \sum_{i=1}^{n} f(x_i)\cdot\Delta x \\
&= \sum_{i=1}^{n} ((i-1)\Delta x)^2 \cdot \Delta x \\
&= \sum_{i=1}^{n} \left((i-1)\frac{x}{n}\right)^2 \cdot \frac{x}{n} \\
&= \frac{x^3}{n^3}\cdot\sum_{i=1}^{n}(i-1)^2 \quad \text{(because both } n, x \text{ are independent of } i, \text{ we take them out of the sum)} \\
&= \frac{x^3}{n^3}\cdot\sum_{i=1}^{n-1} i^2 \quad \text{(change of index)} \\
&= \frac{x^3}{n^3}\cdot\frac{(n-1)(n-1+1)(2(n-1)+1)}{6} \quad \text{(by Example 2.5)} \\
&= \frac{x^3}{n^3}\cdot\frac{2n^3-3n^2+n}{6} \\
&= \frac{x^3}{3} - \frac{x^3}{2n} + \frac{x^3}{6n^2}.
\end{aligned} \tag{2.21}$$

Repeating the calculation for the upper bound leads to

$$\text{Area}_n^{\text{Up}} = \frac{x^3}{3} + \frac{x^3}{2n} + \frac{x^3}{6n^2}.$$

Combining the lower and upper bounds gives:

$$\boxed{\text{Area}_n^{\text{est}} = \left(\frac{x^3}{3} + \frac{x^3}{6n^2}\right) \pm \frac{x^3}{2n}.} \tag{2.22}$$

Moreover, for x fixed (i.e., held constant), as n gets larger and larger, we can verify numerically, or via Fig. 2.4, that both $\frac{x^3}{6n^2}$ and $\frac{x^3}{2n}$ become smaller and smaller, leading us to the conclusion that the area under

Function	Interval	Area
$f(x) = x$	$[0,1]$	$1/2$
$f(x) = x$	$[0,x]$	$x^2/2$
$f(x) = x^2$	$[0,1]$	$1/3$
$f(x) = x^2$	$[0,x]$	$x^3/3$

Table 2.1: Areas under two simple curves, following a method inspired by Archimedes. The only "leap of faith" required so far is that terms of the form $\frac{\text{constant}}{n}$ and $\frac{\text{constant}}{n^2}$ become arbitrarily small (i.e., negligible) as n becomes very large, as illustrated in **Fig. 2.4**.

the parabola is $\frac{x^3}{3}$. For $x > 0$, the area under the parabola $f(x) = x^2$ for the interval $[0, x]$ is equal to $\frac{x^3}{3}$. In other words, we have obtained a closed-form expression for the area for a general interval $[0, x]$, with very little effort.

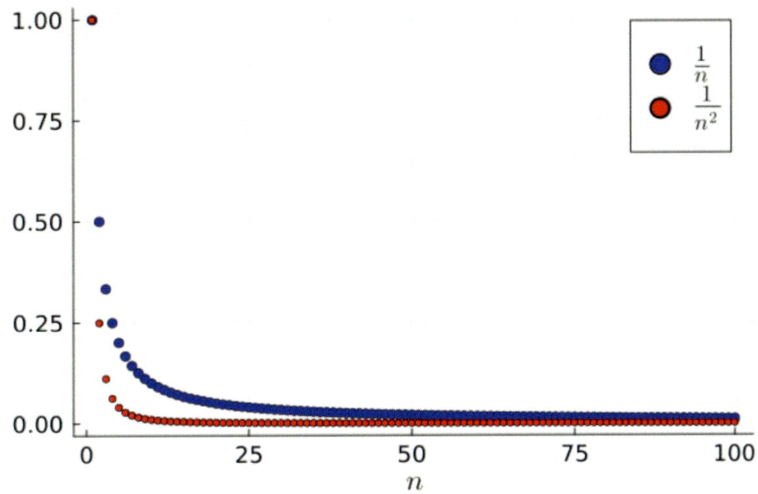

Figure 2.4: Graphs of $1/n$ and $1/n^2$ versus n, for $1 \leq n \leq 100$.

Observation 2.9: Ratios of Polynomials of n as n Tends to Infinity

When approximating our areas with a sum of n rectangles, we were faced with terms of the form

- Triangle ($f(x) = x$): $\frac{1+2+\cdots+n}{n^2} = \frac{an^2+bn}{n^2}$
- Parabola ($f(x) = x^2$): $\frac{1^2+2^2+\cdots+n^2}{n^3} = \frac{an^3+bn^2+cn}{n^3}$

and we can imagine that for the area under x^k, $k > 2$, we would be faced with a term of the form

$$\frac{1^k + 2^k + \cdots + n^k}{n^{k+1}} = \frac{an^{k+1} + bn^k + \cdots + \text{lower powers of } n}{n^{k+1}}.$$

In each case, we want to let n become large so that our system of under- and overapproximations through adding up rectangles becomes more accurate. In the limit, we would like to take an "unbounded" number of "infinitely thin" rectangles, which brings us to the notion of "limits at infinity", the topic of Sections 2.4 and 2.5. In preparation, we will address the subject of sums of powers of integers for arbitrary $k \geq 1$.

Proposition 2.10: Power Sums

For all $n \geq 1$ and $k \geq 1$, the following holds,

$$\begin{aligned}\sum_{i=1}^{n} i^k &:= 1^k + 2^k + 3^k + \cdots + n^k \\ &= \frac{n^{k+1}}{k+1} + \frac{n^k}{2} + \mathrm{O}(n,\ k-1),\end{aligned} \tag{2.23}$$

where $\mathrm{O}(n,\ k-1)$ stands for a polynomial in the variable n with degree less than or equal to $(k-1)$ (i.e., all of the lower-order terms after n^k).

Remark 2.11. *Explicit value of the sum* $\sum_{i=1}^{n} i^k$ *for* $k = 1$ *to* $k = 5$*:*

$$\begin{aligned}\sum_{i=1}^{n} i &= \frac{1}{2}(n^2 + n) = \frac{n^2}{2} + \frac{n}{2}, \\ \sum_{i=1}^{n} i^2 &= \frac{1}{6}(2n^3 + 3n^2 + n) = \frac{n^3}{3} + \frac{n^2}{2} + \mathrm{O}(n,1), \\ \sum_{i=1}^{n} i^3 &= \frac{1}{4}(n^4 + 2n^3 + n^2) = \frac{n^4}{4} + \frac{n^3}{2} + \mathrm{O}(n,2), \\ \sum_{i=1}^{n} i^4 &= \frac{1}{30}(6n^5 + 15n^4 + 10n^3 - n) = \frac{n^5}{5} + \frac{n^4}{2} + \mathrm{O}(n,3), \\ \sum_{i=1}^{n} i^5 &= \frac{1}{12}(2n^6 + 6n^5 + 5n^4 - n^2) = \frac{n^6}{6} + \frac{n^5}{2} + \mathrm{O}(n,4).\end{aligned} \tag{2.24}$$

A general formula is given by Faulhaber.

2.4 Limits at Infinity: The Hard Way

Here, we take our first cut at the concept of a limit. We will first lay out the concept using "everyday language", and then we will formalize it. To make sure we understand the formalization, the examples will be worked out in "excruciating detail". Once we really understand how the process works, we will be able to compute most limits "by inspection".

Finite Limits at ∞

Conceptual Idea: Suppose $f : (0, \infty) \to \mathbb{R}$ is a function and $L \in \mathbb{R}$ is a (finite) real number. Then we say **the limit of $f(x)$ as x approaches infinity equals L**, denoted,

$$\lim_{x \to \infty} f(x) = L,$$

if we can make $f(x)$ stay "arbitrarily close" to L by taking x "sufficiently large in the positive direction."

The formal definition makes precise what we mean by "arbitrarily close" and "sufficiently large."

Definition 2.12. *For the definition of finite limit at infinity, suppose* $f : (0, \infty) \to \mathbb{R}$ *is a function and* $L \in \mathbb{R}$ *is a (finite) real number. Then we say*

$$\lim_{x \to \infty} f(x) = L$$

if, for all $\epsilon > 0$ *(no matter how small), there exists* $N < \infty$*, such that, for all* $x \geq N$*, it holds that* $|f(x) - L| \leq \epsilon$.

Notes:

- For all $x \geq N$ defines "x sufficiently large".
- $|f(x) - L| \leq \epsilon$ defines "arbitrarily close". You can think of $\epsilon > 0$ as specifying a level of precision. How precisely is $f(x)$ approximated by the constant L for x sufficiently large.
- The definition can also be written as: for all $\epsilon > 0$, there exists $N < \infty$, such that, $x \geq N \implies |f(x) - L| \leq \epsilon$.
- It is emphasized that if there is a single point $x \geq N$ such that $|f(x) - L| > \epsilon$, then the "closeness (aka precision)" criterion has failed. We have found a point x where $f(x)$ is not well approximated by the constant L.

Additional Remarks:

- We note that as we take ϵ smaller and smaller (aka our precision requirement becomes finer and finer), we typically need to take N larger and larger. In other words, N typically depends on ϵ. It is sometimes useful to express this dependence as $N(\epsilon)$ or N_ϵ.
- Another way to interpret a limit at infinity is that as we take N larger and larger, the function restricted to $[N, \infty)$ becomes closer and closer to having a constant value equal to L.
- We will treat the definition of a limit as x approaches $-\infty$ later.

We begin with two examples that isolate the respective roles of ϵ and N_ϵ in the definition of a limit.

Example 2.13 (Understanding the Role of ϵ). *Consider the constant function, $f(x) = \pi$ for all $x \geq 0$. Clearly,*

$$\lim_{x \to \infty} f(x) = \pi.$$

Because the function never changes with x, this example isolates the role of the precision parameter $\epsilon > 0$.

Answer the following three questions:

- *Why is the limit not $L = 3$?*
- *Why is the limit not $L = 3.1$?*
- *Why is the limit not $L = 3.14$?*

Solution:

Why is the limit not $L = 3$? Ans. We compute

$$|f(x) - 3| = |\pi - 3| \approx 0.14159.$$

The definition of a limit requires that

$$|f(x) - L| \leq \epsilon \quad \text{for all } \epsilon > 0.$$

ϵ	Does $\|f(x) - 3\| \leq \epsilon$ hold?
1	Yes
0.2	Yes
0.1	No

Because the inequality fails once ϵ is made sufficiently small, the limit cannot be 3.

Why is the limit not $L = 3.1$? Ans. We compute

$$|f(x) - 3.1| = |\pi - 3.1| \approx 0.04159.$$

ϵ	Does $\lvert f(x) - 3.1\rvert \le \epsilon$ hold?
0.1	Yes
0.01	No

Again, the inequality fails once the precision requirement is tightened.

Why is the limit not $L = 3.14$**? Ans.** We compute

$$|f(x) - 3.14| = |\pi - 3.14| \approx 0.00159.$$

ϵ	Does $\lvert f(x) - 3.14\rvert \le \epsilon$ hold?
0.01	Yes
0.001	No

■

Key ideas: ϵ specifies the required precision. A value L can only be the limit of a constant function if the inequality

$$|f(x) - L| \le \epsilon$$

holds for every $\epsilon > 0$, no matter how small. If the inequality fails for even one sufficiently small ϵ, the value L cannot be the limit. We next analyze a function that varies with x so that we can bring in the role of N_ϵ

We next examine how the required precision $\epsilon > 0$ interacts with the settling distance, $N_\epsilon < \infty$.

Example 2.14 (Precision and Settling Distance)**.** *Consider the function* $f : [0, \infty) \to \mathbb{R}$ *defined by* $f(x) = \frac{\pi x}{1+x}$, *for which we take the proposed limit as* $L = \pi$. *For each value of* $\epsilon \in \{1.0, 0.5, 0.1, 0.01\}$, *determine* $N_\epsilon < \infty$ *such that* $|f(x) - L| \le \epsilon$ *for all* $x \ge N_\epsilon$.

Solution: To apply the definition of a limit at infinity, we compute

$$f(x) - \pi = -\frac{\pi}{1+x} \quad \text{and} \quad |f(x) - \pi| = \frac{\pi}{1+x}.$$

To satisfy the definition of a limit, we require

$$\frac{\pi}{1+x} \le \epsilon.$$

Solving for x gives

$$x \ge \frac{\pi}{\epsilon} - 1.$$

Thus, a valid choice is

$$N_\epsilon = \frac{\pi}{\epsilon} - 1,$$

for which substituting in the various values of ϵ gives the table:

ϵ	N_ϵ
1.0	≈ 2.14
0.5	≈ 5.28
0.1	≈ 30.42
0.01	≈ 312.16

Interpretation.

- ϵ sets the required precision around the limiting value.
- N_ϵ sets how far along the x–axis one must go before that precision is guaranteed.
- Tighter precision requirements force larger values of N_ϵ.

The previous example was particularly simple because an appropriate value of N_ϵ could be computed in closed form. Next, we work an example where numerical computations are necessary. Moreover, the chosen function is no longer monotonically increasing as can be seen from its graph in **Fig. 2.5**.

Example 2.15 (Non-closed Form Computation). *Consider $f : [0, \infty) \to \mathbb{R}$ by $f(x) = \pi + xe^{-x}$ for which we again take the proposed limit as $L = \pi$. For each value of $\epsilon \in \{10^{-1}, 10^{-2}, 10^{-3}, 10^{-4}\}$, determine $N_\epsilon < \infty$ such that $|f(x) - L| \le \epsilon$ for all $x \ge N_\epsilon$.*

Solution:

$$f(x) - L = \pi + xe^{-x} - \pi = xe^{-x}$$
$$\Downarrow$$
$$f(x) - L = xe^{-x}$$
$$\Downarrow$$
$$|f(x) - L| = xe^{-x} \quad \text{for all } x \ge 0.$$

We require

$$|f(x) - L| = xe^{-x} \le \epsilon.$$

The inequality $xe^{-x} \le \epsilon$ cannot be solved by elementary algebra, but we can determine N_ϵ numerically. For example, applying the Bisection Algorithm to $xe^{-x} - \epsilon = 0$ will yield the largest possible value for N_ϵ satisfying the inequality, subject to the tolerance we set in the algorithm. Following this process produces the table below

ϵ	N_ϵ (approx.)
10^{-1}	3.58
10^{-2}	6.47
10^{-3}	9.21
10^{-4}	11.61

The results are further illustrated in **Fig. 2.5**.

■

We now present a few examples that are more typical of how limits at infinity are to be computed using the definition.

Example 2.16. *Compute $\lim\limits_{x\to\infty} f(x)$, if it exists, for the following functions, shown as graphs in* **Fig. 2.6**.

(a) $f : (0, \infty) \to \mathbb{R}$ *by* $f(x) = \frac{1}{x}$.

(b) $f : (0, \infty) \to \mathbb{R}$ *by* $f(x) = \frac{1}{x^k}, k > 1$.

(c) $f : \mathbb{R} \to \mathbb{R}$ *by* $f(x) = \frac{x^2}{1+3x^2}$.

(d) $f : \mathbb{R} \to \mathbb{R}$ *by* $f(x) = \frac{x}{1+3x^2}$.

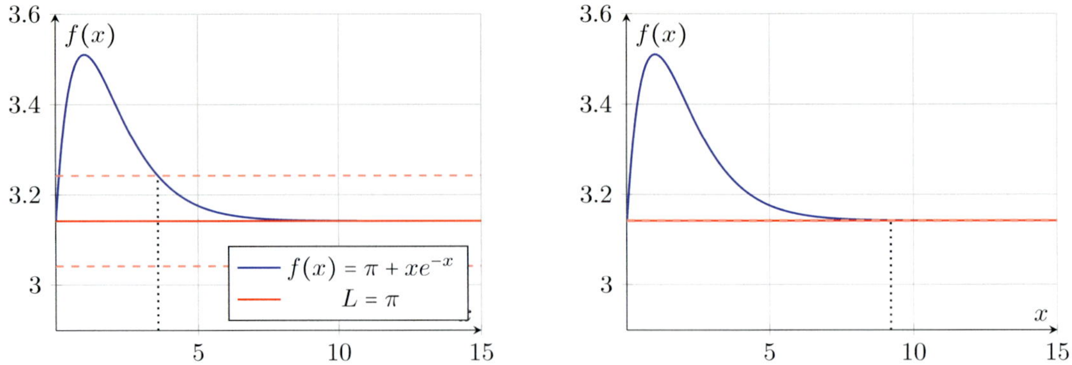

$\epsilon = 0.1 \Rightarrow N_\epsilon \approx 3.58$ $\qquad$ $\epsilon = 0.001 \Rightarrow N_\epsilon \approx 9.21$

Figure 2.5: Comparison of how ϵ and N_ϵ interact for $f(x) = \pi + xe^{-x}$.

Solutions:

(a) $f(x) = \frac{1}{x}$ for $x > 0$. **Ans.** $\lim_{x\to\infty} f(x) = 0.0$.

To show this result, set $L = 0.0$ and choose $\epsilon > 0$ but otherwise arbitrary. Then for all $x \geq \frac{1}{\epsilon}$, we have

$$|f(x) - L| = \left|\frac{1}{x} - 0.0\right| = \frac{1}{x} \leq \frac{1}{\frac{1}{\epsilon}} = \epsilon,$$

where we used $\frac{1}{x} = \frac{1}{\text{a bigger number}} \leq \frac{1}{\text{a smaller number}} = \frac{1}{\frac{1}{\epsilon}}$, when $\boxed{x \geq \frac{1}{\epsilon}}$.

Note: Because $N(\epsilon) := \frac{1}{\epsilon}$, when the bound ϵ on $|f(x) - L|$ is made small, we have at the same time $N(\epsilon) = \frac{1}{\epsilon}$ becoming large, imposing that $x \geq N$ is large.

(b) $f(x) = \frac{1}{x^k}$ for $x > 0$. **Ans.** $\lim_{x\to\infty} f(x) = 0.0$.

To show this result, set $L = 0.0$ and choose $\epsilon > 0$ but otherwise arbitrary. Then, for all $x \geq \frac{1}{\sqrt[k]{\epsilon}}$, we have

$$|f(x) - L| = \left|\frac{1}{x^k} - 0.0\right| = \frac{1}{x^k} \leq \frac{1}{\left(\frac{1}{\sqrt[k]{\epsilon}}\right)^k} = \frac{1}{\frac{1}{\epsilon}} = \epsilon.$$

Notes: The remarks from the previous problem apply here, with the exception that now $N(\epsilon) := \frac{1}{\sqrt[k]{\epsilon}}$. It is still true that making ϵ small forces $N(\epsilon)$ to be large.

(c) $f : \mathbb{R} \to \mathbb{R}$ by $f(x) = \frac{x^2}{1+3x^2}$. **Ans.** $\lim_{x\to\infty} f(x) = \frac{1}{3}$.

To show this result, set $L = 1/3$ and choose $\epsilon > 0$, but otherwise arbitrary. We rewrite the function as, for all $x \neq 0$,

$$f(x) = \frac{x^2}{1+3x^2} \cdot \frac{\frac{1}{x^2}}{\frac{1}{x^2}} = \frac{1}{\frac{1}{x^2} + 3}.$$

Then,

$$|f(x) - L| = \left|\frac{1}{\frac{1}{x^2} + 3} - \frac{1}{3}\right| = \frac{1}{3} - \frac{1}{\frac{1}{x^2} + 3}.$$

Now, let us set $\delta = \frac{1}{x^2}$ and solve for $\delta > 0$ such that $\frac{1}{3} - \frac{1}{\delta+3} < \epsilon$. Hence, after some algebra, we arrive at

$$\frac{1}{3} - \frac{1}{\delta + 3} = \frac{\delta}{3(\delta + 3)} < \epsilon \iff \delta < 3\epsilon(\delta + 3) \iff (1 - 3\epsilon)\delta < 9\epsilon.$$

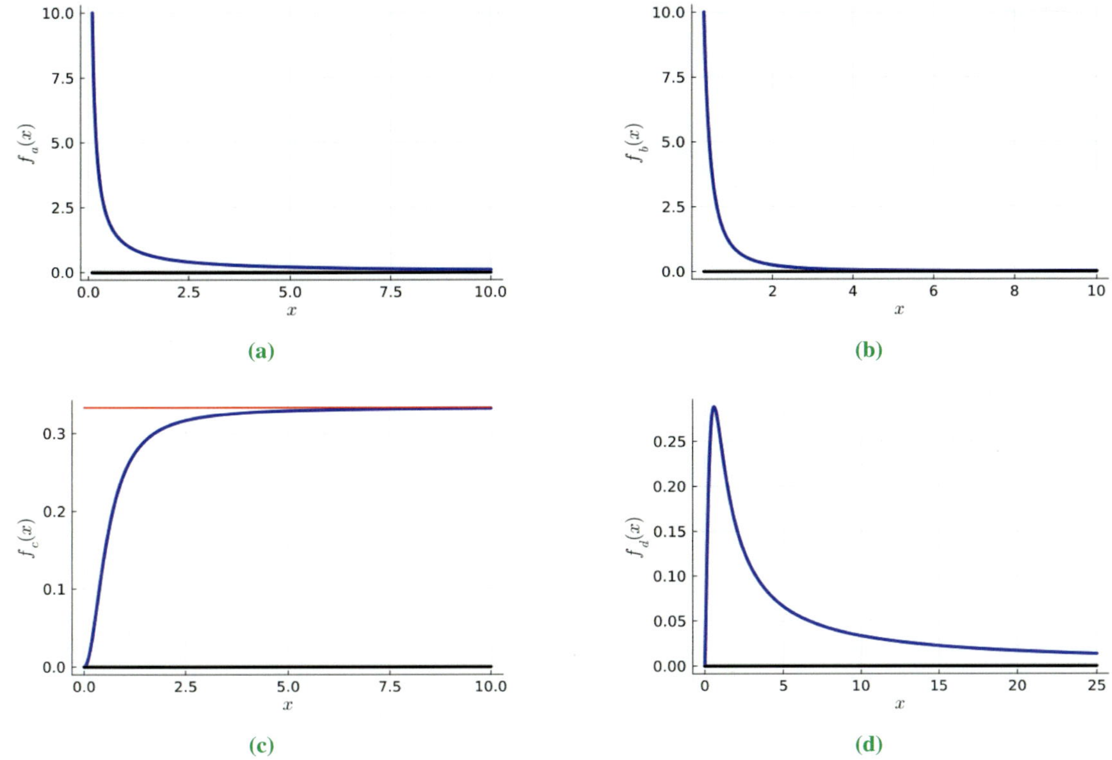

Figure 2.6: Graphs of the four functions in Example 2.16. The red line in (c) shows the nonzero value to which the function is converging.

Because $(1 - 3\epsilon) < 1$ for all $\epsilon > 0$, we conclude that $(1 - 3\epsilon)\delta < \delta$, and

$$\delta < 9\epsilon \implies \left| \frac{1}{\delta + 3} - \frac{1}{3} \right| < \epsilon.$$

Hence, replacing δ by $\frac{1}{x^2}$, we have $\frac{1}{x^2} < 9\epsilon \implies \left| \frac{1}{\frac{1}{x^2}+3} - \frac{1}{3} \right| < \epsilon$, from which we deduce that

$$x > \frac{1}{3\sqrt{\epsilon}} \implies \left| \frac{1}{\frac{1}{x^2} + 3} - \frac{1}{3} \right| < \epsilon,$$

and hence we have shown that $\lim_{x\to\infty} f(x) = \dfrac{1}{3}$.

Notes:

- The above examples are what computing limits by hand typically look like: **a blaze of symbols, multiple lines of computations, each of which is a likely opportunity for making errors**.
- While in this example, the messy algebra and bounding of terms were actually done back in Example 1.8, there is no easy way to remember them.
- **Bottom line:** Finding situations where we can write down the limits by inspection, while avoiding the messy algebra, would be nice. We will do that shortly.

(d) $f(x) = \frac{x}{1+3x^2}$. **Ans.** $\lim_{x\to\infty} f(x) = 0.0$.

To show this result, set $L = 0.0$ and choose $\epsilon > 0$ but otherwise arbitrary. We rewrite the function as, for all $x \neq 0$,

$$f(x) = \frac{x}{1 + 3x^2} \cdot \frac{\frac{1}{x}}{\frac{1}{x}} = \frac{1}{\frac{1}{x} + 3x}.$$

Then, for all $x > 0$,

$$|f(x) - L| = \left| \frac{1}{\frac{1}{x} + 3x} - 0.0 \right| < \frac{1}{0 + 3x}.$$

We observe that

$$\frac{1}{3x} < \epsilon \iff 3x > \frac{1}{\epsilon} \iff x > \frac{1}{3\epsilon}.$$

Hence, here, $N(\epsilon) = \frac{1}{3\epsilon}$.

There is more than one way to arrive at these estimates. For example, if we had instead written, for all $x > 0$,

$$f(x) = \frac{x}{1 + 3x^2} \cdot \frac{\frac{1}{x^2}}{\frac{1}{x^2}} = \frac{\frac{1}{x}}{\frac{1}{x^2} + 3},$$

then, for all $x > 1$,

$$|f(x) - L| = \left| \frac{\frac{1}{x}}{\frac{1}{x^2} + 3} - 0.0 \right| < \frac{\frac{1}{x}}{0 + 3} = \frac{1}{3x},$$

yielding the same result we had before.

■

Strategy for Bounding Ratios of Functions: To find an upper bound for a fraction, $\frac{n(x)}{d(x)}$, you replace the numerator with an upper bound and the denominator with a lower bound for the inequality to hold. In symbols, suppose that $0 \le n(x) \le N(x)$. Then,

$$\frac{n(x)}{d(x)} \le \frac{N(x)}{d(x)},$$

because having a larger number in the numerator gives a larger fraction. Next, suppose that $0 < D(x) \le d(x)$. Then,

$$\frac{n(x)}{d(x)} \le \frac{N(x)}{d(x)} \le \frac{N(x)}{D(x)},$$

because having a smaller number in the denominator yields a larger fraction. **While these are "obvious" facts, applying them successfully when you are first learning limits can be hard to master!**

Conversely, to find a lower bound for a fraction, $\frac{n(x)}{d(x)}$, you replace the numerator with a lower bound and the denominator with an upper bound for the inequality to hold.

In other words, if $n(x) \ge N(x)$ and $0 < d(x) \le D(x)$, then $\frac{n(x)}{d(x)} \ge \frac{N(x)}{D(x)}$. This latter bound is useful for showing a limit is equal to infinity.

Infinite Limit at ∞

Conceptual Idea: A function $f : (0, \infty) \to \mathbb{R}$ converges to infinity as x tends to ∞ if for all finite values of L, "$f(x)$ eventually equals or exceeds L and does not drop below L".

The formal definition makes precise what we mean by "eventually equals or exceeds L and does not drop below it."

Definition 2.17. ***Unbounded Limits at Infinity.*** *The function $f : (0, \infty) \to \mathbb{R}$ has an* ***unbounded positive limit at infinity*** *if, for every finite $L > 0$, there exists $N < \infty$ such that*

$$x \ge N \implies f(x) \ge L.$$

That is, for all $x \in [N, \infty)$, $f(x) \geq L$. We emphasize that there cannot be any $x \in [N, \infty)$ such that $f(x) < L$; the function has to equal or exceed L and not drop below it.

The function $f : (0, \infty) \to \mathbb{R}$ has an ***unbounded negative limit at infinity*** *if, for every finite $L < 0$, there exists $N < \infty$ such that*

$$x \geq N \implies f(x) \leq L.$$

That is, for all $x \in [N, \infty)$, $f(x) \leq L$. We emphasize that there cannot be any $x \in [N, \infty)$ such that $f(x) > L$; the function has to be less than or equal to L and not rise above it.

We denote these limits by $\lim_{x\to\infty} f(x) = \infty$ and $\lim_{x\to\infty} f(x) = -\infty$, respectively.

Note: For the definition of $\lim_{x\to\infty} f(x) = -\infty$, you may be worried that $L < 0$ does not "look big and negative". However, $L = -10^9$ is big and negative. We allow L to be any negative real number.

Example 2.18. *Compute $\lim_{x\to\infty} f(x)$, if it exists, for the following functions, shown as graphs in* Fig. 2.7.

(a) $f : \mathbb{R} \to \mathbb{R}$ *by* $f(x) = \frac{20x^3+11x^2+14}{1+x^2}$.

(b) $f : \mathbb{R} \to \mathbb{R}$ *by* $f(x) = -x^3$.

(c) $f : \mathbb{R} \to \mathbb{R}$ *by* $f(x) = x \cdot \sin(x)$.

(d) $f : \mathbb{R} \to \mathbb{R}$ *by* $f(x) = \frac{-4x^3+11x+14}{1+x^4}$.

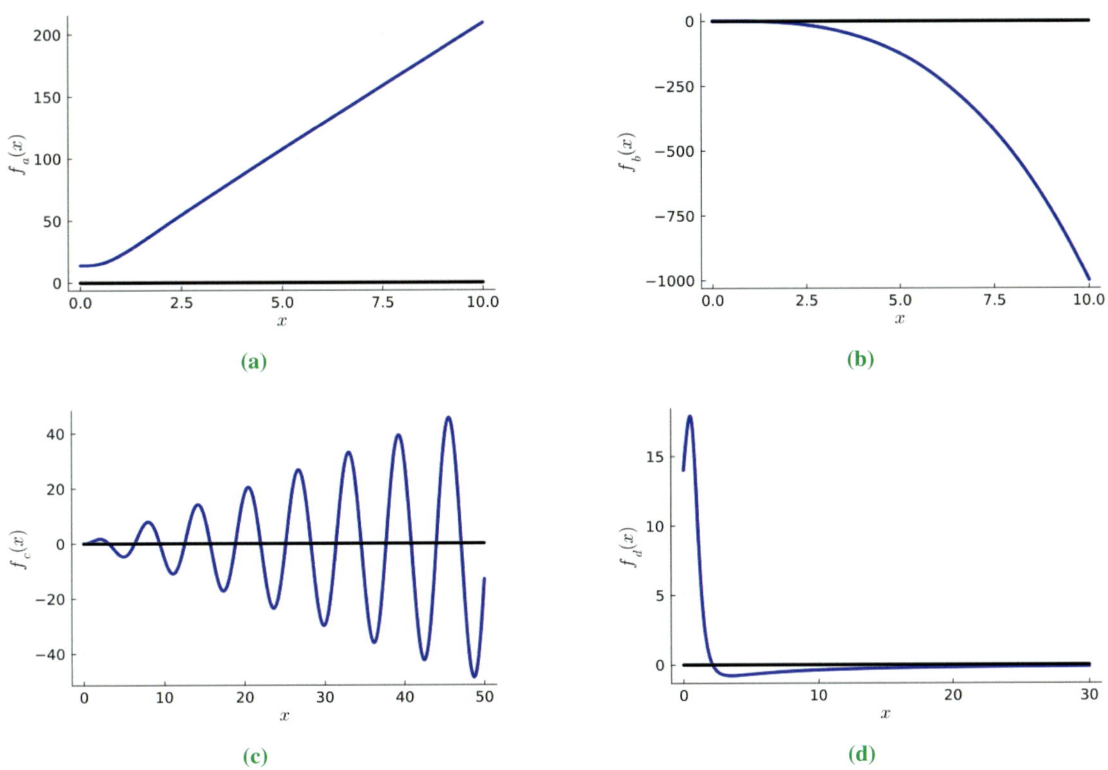

Figure 2.7: Graphs of the four functions in Example 2.18 The plot of $x \cdot \sin(x)$ given in (c) shows that the function becomes unbounded without converging to anything, not even infinity.

Solutions:

(a) $f(x) = \frac{20x^3+11x^2+14}{1+x^2}$. **Ans.** $\lim\limits_{x\to\infty} f(x) = \infty$.

To show this result, we first rewrite the function as we have done several times before, for $x > 1$,

$$f(x) = \frac{20x^3 + 11x^2 + 14}{1 + x^2} = \frac{20x^3 + 11x^2 + 14}{1 + x^2} \cdot \frac{\frac{1}{x^2}}{\frac{1}{x^2}} = \frac{20x + 11 + \frac{14}{x^2}}{\frac{1}{x^2} + 1} \geq \frac{20x + 11}{1 + 1} \geq 10x.$$

Next, let $L > 0$ be otherwise arbitrary and let $N \geq L/10$. Then, for all $x \geq N$, $f(x) \geq L$.

(b) $f(x) = -x^3$. **Ans.** $\lim\limits_{x\to\infty} f(x) = -\infty$.

To show this, let $L < -1$ be otherwise arbitrary and set $N = |L|$. Then $x \geq N \implies f(x) = -x^3 \leq -N^3 \leq -|L|^3 \leq L$. You could also have taken $N = \sqrt[3]{|L|}$ and made similar estimates. All that matters is that you show that you find a region of $[N, \infty) \subset \mathbb{R}$ such that $f(x) \leq L$ for $x \in [N, \infty)$.

(c) $f(x) = x \cdot \sin(x)$. **Ans.** $\lim\limits_{x\to\infty} f(x)$ does not exist.

The issue is that $x \cdot \sin(x)$ becomes arbitrarily large in both the positive direction and the negative direction as $x \to \infty$. To show this, we first note that

$$\sin(x) = \begin{cases} +1 & x = \frac{\pi}{2} + 2k\pi,\ k \in \mathbb{N}, \\ -1 & x = -\frac{\pi}{2} + 2k\pi,\ k \in \mathbb{N}, \end{cases}$$

where we used the $+1$ to emphasize the positive value. Hence, for arbitrary L and $N > 0$, taking $k = \max\{|L|, N\}$ results in $f(\frac{\pi}{2} + 2k\pi) = \frac{\pi}{2} + 2k\pi > L$ and $f(-\frac{\pi}{2} + 2k\pi) = \frac{\pi}{2} - 2k\pi < L$. We cannot make the function stay above L or below L for all $x > N$.

(d) $f(x) = \frac{-4x^3+11x+14}{1+x^4}$. **Ans.** $\lim\limits_{x\to\infty} f(x) = 0$.

To show this, we first rewrite the function as we have done several times before, for $x > 1$,

$$f(x) = \frac{-4x^3 + 11x + 14}{1 + x^4} = \frac{-4x^3 + 11x + 14}{1 + x^4} \cdot \frac{\frac{1}{x^4}}{\frac{1}{x^4}} = \frac{-4\frac{1}{x} + 11\frac{1}{x^3} + 14\frac{1}{x^4}}{\frac{1}{x^4} + 1}.$$

Next, let $L = 0$ and for all $x > 1$,

$$|f(x) - L| = |f(x) - 0| = \frac{|-4\frac{1}{x} + 11\frac{1}{x^3} + 14\frac{1}{x^4}|}{|\frac{1}{x^4} + 1|} \leq \frac{1}{x}\frac{|4 + 11 + 14|}{0 + 1} \leq 29\,\frac{1}{x}.$$

Hence, for all $x \geq \frac{29}{\epsilon} > 1$, $|f(x) - 0| \leq \epsilon$, which shows that the limit is zero.

■

Neither Finite nor Infinite Limit at ∞ Exists

We need to work out what it means to fail a definition. If you are super good at logic, all you have to do is negate the definition! At this stage of your mathematics education, that is easier said than done, so we'll take it step by step. The main mental hurdle is that a definition typically requires several conditions to hold, and hence it fails if any ONE OF THEM does not hold. Logically, this arises from the negation of "and" being "or". For example, if we have **MyDef is True** if **P1** and **P2** (both) hold, then **MyDef is False** if either of **P1** or **P2** does not hold.

Conceptual Ideas: We say that $f : (0, \infty) \to \mathbb{R}$ **does not have a finite limit** as x tends to infinity if, for all candidate finite limits L, it is impossible to ensure that $f(x)$ remains near L by taking x sufficiently large. Similarly, $f(x)$ does not converge to infinity if for some finite $L > 0$, the function keeps "dipping" below L.

The formal anti-definitions clarify what "impossible to keep $f(x)$ near L" or "dipping below L" really mean!

Definition 2.19. ***No Finite Limit at Infinity.*** *Suppose $f : (0, \infty) \to \mathbb{R}$ is a function. Then $f(x)$ does not have a finite limit at infinity if for every choice of $L \in \mathbb{R}$, there exists some $\epsilon > 0$ such that one of the following conditions hold:*

- *There does not exist $N < \infty$ so that $x \geq N \implies |f(x) - L| \leq \epsilon$.*
- *For all $N < \infty$, no matter how large, it is not true that $x \geq N \implies |f(x) - L| \leq \epsilon$.*
- *For all $N < \infty$, no matter how large, there (always) exists some $x > N$ such that $|f(x) - L| > \epsilon$.*

The last three statements are all equivalent. We have simply "unwrapped" the negations one by one. They all capture the idea that "for every choice of $L \in \mathbb{R}$, we are unable to keep the function's values near L by taking x sufficiently large". The notion of "$f(x)$ is not near L" is codified by the existence of some positive value ϵ being used to bound the "distance of $f(x)$ from L," via $|f(x) - L| > \epsilon$.

Definition 2.20. ***No Infinite Limit at Infinity.*** *Suppose $f : (0, \infty) \to \mathbb{R}$ is a function. Then $f(x)$ does not have an unbounded limit at infinity in the positive direction if for some choice of $L > 0$ and every choice of $0 < N_L < \infty$ (no matter how large), there exists $x \geq N_L$ such that $f(x) < L$. Similarly, $f(x)$ does not have an unbounded limit at infinity in the negative direction if for some choice of $L < 0$ and every choice of $0 < N_L < \infty$ (no matter how large), there exists $x \geq N_L$ such that $f(x) > L$.*

***A classic example** is shown in **Fig. 2.7-(c)**. Yes, the function grows without bound, but it fails to stay large in the positive direction (or alternatively, the negative direction), as do the curves in **Fig. 2.8-(a)** and **-(b)**.*

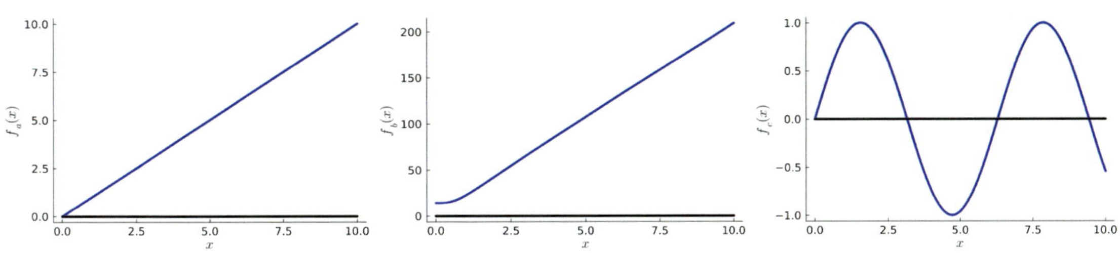

(a) **(b)** **(c)**

Figure 2.8: Graphs of the three functions in Example 2.21.

Example 2.21. *Compute $\lim_{x \to \infty} f(x)$, if a finite limit exists, for the following functions, shown as graphs in* **Fig. 2.8:**

(a) $f : \mathbb{R} \to \mathbb{R}$ *by* $f(x) = x$.

(b) $f : \mathbb{R} \to \mathbb{R}$ *by* $f(x) = \frac{20x^3 + 11x + 14}{1 + x^2}$.

(c) $f : \mathbb{R} \to \mathbb{R}$ *by* $f(x) = \sin(x)$.

Solutions:

(a) $f(x) = x$ does not have a finite limit as x becomes arbitrarily large. While this seems patently obvious, how would one actually show it?

- Solution 1: We use Condition 1 of our "anti-definition", namely, for every choice of $L \in \mathbb{R}$, and for every choice of $N < \infty$, there is always an $x > N$ such that $f(x) > L$. This is easy to show because we can let $L \in \mathbb{R}$ and $0 < N < \infty$ be arbitrary. Then, for $x > \max\{L, N\}$, $f(x) = x > L$.
- Solution 2: We use Condition 4 of our "anti-definition", namely, for every choice of $L \in \mathbb{R}$, there exists some $\epsilon > 0$ such that, for all $N < \infty$, no matter how large, there (always) exists some $x > N$ such that $|f(x) - L| \geq \epsilon$. To show this, let L be arbitrary and choose $\epsilon = 1$. Then, for all $x > L + \epsilon$,

$$|x - L| = x - L > L + \epsilon - L = \epsilon.$$

Hence, no matter what L we choose, and for all N finite, we have $x > \max\{N, L + \epsilon\} \implies |f(x) - L| > \epsilon$.

(b) $f(x) = \frac{20x^3+11x+14}{1+x^2}$ does not have a finite limit as x becomes arbitrarily large. We observe that the numerator is growing at "rate" x^3, while the denominator is growing at rate x^2. Hence, the numerator should "win". How do we formalize this? Let's rewrite, for all $x \neq 0$,

$$\frac{20x^3 + 11x + 14}{1 + x^2} = \frac{20x^3 + 11x + 14}{1 + x^2} \cdot \frac{\frac{1}{x^2}}{\frac{1}{x^2}} = \frac{20x + \frac{11}{x} + \frac{14}{x^2}}{\frac{1}{x^2} + 1}.$$

Then, for all $x > 1$,

$$f(x) = \frac{20x + \frac{11}{x} + \frac{14}{x^2}}{\frac{1}{x^2} + 1} > \frac{20x}{1 + 1} \geq 10x,$$

which grows unbounded as x tends to infinity, meaning there is no finite limit.

(c) $f(x) = \sin(x)$ has no limit as $x \to \infty$ because no matter how large N becomes, $\sin : [N, \infty) \to \mathbb{R}$ still oscillates between -1 and 1. While that really is "enough said", let us examine the details.

(1) Case 1: The candidate limit $L \notin [-1, 1]$ and hence $|L| > 1$. Let $\delta = |L| - 1 > 0$. For all $y \in [-1, 1]$, it follows that $|y| \leq 1$ and hence $|\,|L| - |y|\,| \geq \delta$. By the reverse triangle inequality in Prop. 1.4, we have $|L - y| \geq \delta$. Hence, for all $0 < \epsilon < \delta$ and $x \in \mathbb{R}$, $|L - \sin(x)| = |\sin(x) - L| \geq \delta > \epsilon$, and thus L cannot be a limit.

(2) Case 2: The candidate limit $L = 0$. Then for $\epsilon = 0.5$ and for all $N < \infty$, it follows that $x = N\frac{\pi}{2}$ satisfies $x > N$ and $|\sin(x) - L| = |1 - 0| = 1 > 0.5 = \epsilon$, and hence L is not a limit.

(3) Case 3: The candidate limit $L \in [-1, 1]$ and $L \neq 0$. Then there exists $x \in [-\pi, \pi]$ such that $f(x) = L$ and $f(-x) = -L$. Moreover, because $\sin(-x + 2k\pi) = \sin(-x) = -\sin(x)$, we have that $f(-x + 2k\pi) = -L$. Hence, we can take $\epsilon = |L| > 0$ and note that for all $1 < N < \infty$, $-x + 2N\pi > N$ and $|f(-x + 2N\pi) - L| = |\sin(-x + 2N\pi) - L| = |\sin(-x) - L| = 2|L| > \epsilon$.

■

Secrets of the Arcane 2.22: Why Do We Care about Limits Anyway?

Limits seem hard. Can't calculus exist without them?

1. **Limits Are the Building Blocks of Calculus:** Calculus without limits is like a castle floating in the air. Limits are the bedrock of this mathematical fortress.
2. **Infinity Demystified:** Limits turn the concept of infinity from a perplexing puzzle into a manageable marvel, allowing us to reason about the infinitely small and large.
3. **Solving the Unsolvable:** Limits provide a method to solve problems that might otherwise be

unsolvable. For example, they can be used to determine the value of a function at a point where it's undefined. To see what this might mean, what values should we assign at...

- $x = 1$ for $\frac{x^3-6x^2+11x-6}{x-1}$ (were we not taught NEVER to divide by zero)?
- $t = 0$ for $\frac{\sin(t)}{t}$ (shows up in signal processing when building low-pass filters)?
- $h = 0$ for $\frac{e^h-1}{h}$ (rate of change of the natural exponential at the origin)?
- $n \to \infty$ for $\frac{n^2+n}{2n^2}$ (area under a unit parabola between zero and one)?
- $t \to \infty$ for $t \cdot e^{-t} \cdot \sin(3t)$ (infinity times zero times an oscillation)? Shows up in feedback control!

2.5 Limits at Infinity: The Easy Way

Now that we have computed limits by applying formal methodology, we can make a few observations and profit from them.

2.5.1 Rational Functions

In Examples 2.16, 2.18, and 2.21 we had many **rational functions**; that is, functions consisting of the ratio of two polynomials. Recall that the **degree of a polynomial (aka "order" of the polynomial)** is the highest power of the variable in the polynomial expression. The coefficient multiplying the term of the highest degree is called the **leading coefficient**; by definition, it cannot be zero. In all of our examples, the nature of the limit of a rational function could be determined by the degrees of the numerator and denominator polynomials, and their corresponding leading coefficients.

Intuition for the limit of a rational function of x: With a little algebra, we can rewrite a general rational function of x as:

$$\frac{b_m x^m + b_{m-1}x^{m-1} + \cdots + b_0}{a_n x^n + a_{n-1}x^{n-1} + \cdots + a_0} = \boxed{\frac{x^m}{x^n}} \cdot \frac{b_m + b_{m-1}\frac{1}{x} + \cdots + b_0\frac{1}{x^m}}{a_n + a_{n-1}\frac{1}{x} + \cdots + a_0\frac{1}{x^n}}.$$

Because all terms of the form $\frac{1}{x^k}$ go to zero as $|x|$ tends to infinity, the "messy fraction" on the right has a simple limit, namely,

$$\lim_{x\to\pm\infty} \frac{\boldsymbol{b_m} + b_{m-1}\frac{1}{x} + \cdots + b_0\frac{1}{x^m}}{\boldsymbol{a_n} + a_{n-1}\frac{1}{x} + \cdots + a_0\frac{1}{x^n}} = \boxed{\frac{b_m}{a_n}}.$$

Hopefully, you see it. If not, play with the limit in Julia. Once we have replaced the messy term,

$$\frac{b_m + b_{m-1}\frac{1}{x} + \cdots + b_0\frac{1}{x^m}}{a_n + a_{n-1}\frac{1}{x} + \cdots + a_0\frac{1}{x^n}},$$

with the constant $\frac{b_m}{a_n}$, all we need to do is understand

$$\lim_{x\to\pm\infty} \frac{b_m}{a_n} \cdot \frac{x^m}{x^n} = \frac{b_m}{a_n} \cdot \lim_{x\to\pm\infty} \frac{x^m}{x^n},$$

which seems "infinitely easier"! **This is Prop. 2.23 in a nutshell.**

Proposition 2.23: Limit of a Rational Function as x Tends to Plus or Minus Infinity

Consider a rational function
$$\frac{b_m x^m + b_{m-1}x^{m-1} + \dots + b_0}{a_n x^n + a_{n-1}x^{n-1} + \dots + a_0},$$
where b_m and a_n are nonzero. Then the limit as x tends to plus or minus infinity is equal to
$$\lim_{x\to\pm\infty} \frac{b_m x^m + b_{m-1}x^{m-1} + \dots + b_0}{a_n x^n + a_{n-1}x^{n-1} + \dots + a_0} = \frac{b_m}{a_n} \cdot \lim_{x\to\pm\infty} x^{m-n}. \tag{2.25}$$

In particular,
$$\lim_{x\to\infty} \frac{b_m x^m + b_{m-1}x^{m-1} + \dots + b_0}{a_n x^n + a_{n-1}x^{n-1} + \dots + a_0} = \begin{cases} \operatorname{sign}\left(\frac{b_m}{a_n}\right) \cdot \infty & m > n, \\ \frac{b_m}{a_n} & m = n, \\ 0 & m < n. \end{cases} \tag{2.26}$$

and
$$\lim_{x\to-\infty} \frac{b_m x^m + b_{m-1}x^{m-1} + \dots + b_0}{a_n x^n + a_{n-1}x^{n-1} + \dots + a_0} = \begin{cases} \operatorname{sign}\left(\frac{b_m}{a_n}\right) \cdot (-1)^{m-n} \cdot \infty & m > n, \\ \frac{b_m}{a_n} & m = n, \\ 0 & m < n. \end{cases} \tag{2.27}$$

Note: If either of m and n is odd and the other is even, then $(-1)^{m-n} = -1$; otherwise, it equals $+1$. Equation (2.25) states that the **ratio of the highest degree terms in the numerator and denominator determines the limit**:
$$\begin{aligned} &\lim_{x\to\pm\infty} \frac{b_m x^m + b_{m-1}x^{m-1} + \dots + b_1 x + b_0}{a_n x^n + a_{n-1}x^{n-1} + \dots + a_1 x + a_0} \\ &= \lim_{x\to\pm\infty} \frac{x^m \cdot (b_m + b_{m-1}\frac{1}{x} + \dots + b_1 \frac{1}{x^{m-1}} + b_0 \frac{1}{x^m})}{x^n \cdot (a_n + a_{n-1}\frac{1}{x} + \dots + a_1 \frac{1}{x^{n-1}} + a_0 \frac{1}{x^n})} \\ &= \lim_{x\to\pm\infty} \frac{b_m x^m}{a_n x^n}. \end{aligned}$$

The video "How to Find the Limit at Infinity" by Nancy Pi is relevant to this topic, as is the video by Jennifer McCracken.

Example 2.24. *Where applicable, use Prop. 2.23 to compute $\lim_{x\to\pm\infty} f(x)$ for the following functions:*

(a) $f : \mathbb{R} \to \mathbb{R}$ by $f(x) = \frac{20x^2+11x+14}{1+4x^2}$.

(b) $f : \mathbb{R} \to \mathbb{R}$ by $f(x) = -x^3$.

(c) $f : \mathbb{R} \to \mathbb{R}$ by $f(x) = x \cdot \sin(x)$.

(d) $f : \mathbb{R} \to \mathbb{R}$ by $f(x) = \frac{-4x^3+11x+14}{1+x^4}$.

(e) $f : \mathbb{R} \to \mathbb{R}$ by $f(x) = \frac{4x^4+11x+14}{1+2x^3}$.

Solutions:

(a) $f(x) = \frac{20x^2+11x^2+14}{1+4x^2}$. **Ans.** $\lim_{x\to\infty} f(x) = 5$ and $\lim_{x\to-\infty} f(x) = 5$.

- $\lim_{x\to\infty} f(x) = \frac{20}{4} \cdot \lim_{x\to\infty} x^{2-2} = 5$.
- $\lim_{x\to-\infty} f(x) = \frac{20}{4} \cdot \lim_{x\to-\infty} x^{2-2} = 5$.

(b) $f(x) = -x^3$. **Ans.** $\lim_{x\to\infty} f(x) = -\infty$ and $\lim_{x\to-\infty} f(x) = +\infty$.

- $\lim_{x\to\infty} f(x) = \frac{-1}{1} \cdot \lim_{x\to\infty} x^{3-0} = -\infty$.
- $\lim_{x\to-\infty} f(x) = \frac{-1}{1} \cdot \lim_{x\to-\infty} x^{3-0} = +\infty$.

(c) $f(x) = x \cdot \sin(x)$. **Ans.** Prop. 2.23 is not applicable because $f(x)$ is not a rational function.

(d) $f(x) = \frac{-4x^3+11x+14}{1+x^4}$. **Ans.** $\lim_{x\to\infty} f(x) = 0$ and $\lim_{x\to-\infty} f(x) = 0$.

- $\lim_{x\to\infty} f(x) = \frac{-4}{1} \cdot \lim_{x\to\infty} x^{3-4} = \dfrac{-4}{1} \cdot \lim_{x\to\infty} \dfrac{1}{x} = 0$.
- $\lim_{x\to-\infty} f(x) = \frac{-4}{1} \cdot \lim_{x\to-\infty} x^{3-4} = \dfrac{-4}{1} \cdot \lim_{x\to-\infty} \dfrac{1}{x} = 0$.

(e) $f(x) = \frac{4x^4+11x+14}{1+2x^3}$. **Ans.** $\lim_{x\to\infty} f(x) = +\infty$ and $\lim_{x\to-\infty} f(x) = -\infty$.

- $\lim_{x\to\infty} f(x) = \dfrac{4}{2} \cdot \lim_{x\to\infty} x^{4-3} = \dfrac{4}{2} \cdot \lim_{x\to\infty} x = +\infty$.
- $\lim_{x\to-\infty} f(x) = \dfrac{4}{2} \cdot \lim_{x\to-\infty} x^{4-3} = \dfrac{4}{2} \cdot \lim_{x\to-\infty} x = -\infty$.

■

Sometimes, a change of variable can turn a hard problem into an easy one. We illustrate this next.

Example 2.25. *Evaluate*[2] $\lim_{x\to\infty} \dfrac{2+x}{x^{1/3}-4}$.

Solution: We introduce a change of variable $y := x^{1/3}$. We do this because then

$$\frac{2+x}{x^{1/3}-4} = \frac{2+y^3}{y-4},$$

which is a simple rational function of y. We note that $x \to \infty \iff y \to \infty$, and hence,

$$\lim_{x\to\infty} \frac{2+x}{x^{1/3}-4} = \lim_{y\to\infty} \frac{2+y^3}{y-4} = \infty.$$

Transformations/substitutions/changes of variables are used a lot in calculus to simplify problems. ■

2.5.2 Limit as an Integer Tends to Infinity

Our original motivation for studying limits, namely the area computations in Section 2.3.2, resulted in functions that depended on a counting number, $n \in \mathbb{N}$. All of the definitions and results of Section 2.4 and Section 2.5.1, on limits for a real variable x, apply equally well to limits as an integer $n \in \mathbb{Z}$ tends to plus or minus infinity.

[2]The problem is motivated by Solving Limits with Change of Variable by All Things Mathematics. Here, we take the limit as x tends to infinity. The YouTube video is looking at finite limits, which we treat in Chapter 4.

Taking limits for integer n as $n \to \pm\infty$ is no different than taking limits as $x \to \pm\infty$. While we formalize this next, all we are doing is changing variable names to protect the guilty. If you are good with limits as $x \to \pm\infty$, you can skim a few of the examples and then move on.

Fact 2.26: Limits of Functions of Integers

Definition 2.27. ***Limits at Infinity.*** *Suppose $f : \mathbb{Z} \to \mathbb{R}$ is a function and $L \in \mathbb{R}$ is a (finite) real number. Then:*

(a) $\lim_{n\to\infty} f(n) = L$ *if, for all $\epsilon > 0$ (no matter how small), there exists $N < \infty$, such that, for all $n \geq N$, it holds that $|f(n) - L| \leq \epsilon$.*

(b) $\lim_{n\to-\infty} f(n) = L$ *if, for all $\epsilon > 0$ (no matter how small), there exists $N > -\infty$, such that, for all $n \leq N$, it holds that $|f(n) - L| \leq \epsilon$.*

(c) $\lim_{n\to\infty} f(n) = \infty$ *if, for all $M > 0$ (no matter how large), there exists $N < \infty$, such that, for all $n \geq N$, it holds that $f(n) \geq M$.*

(d) $\lim_{n\to-\infty} f(n) = \infty$ *if, for all $M > 0$ (no matter how large), there exists $N > -\infty$, such that, for all $n \leq N$, it holds that $f(n) \geq M$.*

(e) $\lim_{n\to\infty} f(n) = -\infty$ *if, for all $M \leq 0$ (no matter how large in the negative direction), there exists $N < \infty$, such that, for all $n \geq N$, it holds that $f(n) \leq M$.*

(f) $\lim_{n\to-\infty} f(n) = -\infty$ *if, for all $M < 0$ (no matter how large in the negative direction), there exists $N > -\infty$, such that, for all $n \leq N$, it holds that $f(n) \leq M$.*

Note: *Just as with limits of a real variable x tending to infinity, we examine the behavior of the function on sets of the form $\{n \in \mathbb{N} \mid n \geq N\}$ or $\{n \in \mathbb{Z} \mid n \leq N\}$. We want to know whether the function becomes essentially constant as N becomes arbitrarily large in the positive (or, negative) direction, giving us a finite limit, or can we make the function always stay above (or, below) an arbitrarily large positive (or, negative) number, giving us an infinite limit. Of course, sometimes neither of these conditions hold, in which case a limit as n tends to infinity does not exist.* ***Rational functions will always have either a finite limit or an infinite limit.***

Example 2.28. *Where applicable, use Fact 2.26 and Prop. 2.23 to compute* $\lim_{n\to\infty} f(n)$ *for the following functions:*

(a) $f : \mathbb{Z} \to \mathbb{R}$ *by* $f(n) = \frac{1}{3} + \frac{1}{6n^2}$.

(b) $f : \mathbb{Z} \to \mathbb{R}$ *by* $f(n) = \frac{1}{n^2} \cdot \frac{n(n+1)}{2}$.

(c) $f : \mathbb{Z} \to \mathbb{R}$ *by* $f(n) = \frac{x^3}{n^3} \cdot \frac{2n^3-3n^2+n}{6}$, *where x is a constant not depending upon n.*

(d) $f : \mathbb{Z} \to \mathbb{R}$ *by* $f(n) = \frac{1}{n^{k+1}} \cdot \sum_{i=1}^{n} i^k$, *for $k \geq 1$.*

Solutions:

(a) $f(n) = \frac{1}{3} + \frac{1}{6n^2}$. **Ans.** $\lim_{n\to\infty} f(n) = \dfrac{1}{3}$.

To apply Prop. 2.23, we should rewrite $f(n)$ as $f(n) = \frac{2n^2+1}{6n^2}$ and then compute

$$\lim_{n\to\infty} f(n) = \frac{2}{6} = \frac{1}{3}.$$

(b) $f(n) = \frac{1}{n^2} \cdot \frac{n(n+1)}{2}$. **Ans.** $\lim_{n\to\infty} f(n) = \frac{1}{2}$.

$f(n) = \frac{n^2+n}{2n^2}$ and hence $\lim_{n\to\infty} f(n) = \frac{1}{2}$.

(c) $f(n) = \frac{x^3}{n^3} \cdot \frac{2n^3-3n^2+n}{6}$. **Ans.** $\lim_{n\to\infty} f(n) = \frac{x^3}{3}$.

$f(n) = \frac{x^3}{n^3} \cdot \frac{2n^3-3n^2+n}{6} = \frac{2x^3n^3-3x^3n^2+x^3n}{6n^3}$ and hence $\lim_{n\to\infty} f(x) = \frac{2x^3}{6} = \frac{x^3}{3}$.

(d) $f(n) = \frac{1}{n^{k+1}} \cdot \sum_{i=1}^{n} i^k$, for $k \geq 1$. **Ans.** $\lim_{n\to\infty} f(x) = \frac{1}{k+1}$.

$f(n) = \frac{1}{n^{k+1}} \cdot \sum_{i=1}^{n} i^k = \frac{1}{n^{k+1}} \cdot \left(\frac{n^{k+1}}{k+1} + \frac{n^k}{2} + \mathrm{O}(n,\ k-1)\right)$ by Prop. 2.10. Given what we know so far, we should work the problem as follows: $\mathrm{O}(n,\ k-1)$ is some polynomial of degree $(k-1)$ and hence, $\mathrm{O}(n,\ k-1) = a_{k-1}n^{k-1} + a_{k-2}n^{k-2} + \cdots + a_1 n + a_0$. Hence, we have that

$$f(n) = \frac{\frac{n^{k+1}}{k+1} + \frac{n^k}{2} + a_{k-1}n^{k-1} + a_{k-2}n^{k-2} + \cdots + a_1 n + a_0}{n^{k+1}},$$

and thus $\lim_{n\to\infty} f(x) = \frac{\frac{1}{k+1}}{1} = \frac{1}{k+1}$.

■

Is it possible to take limits term-by-term and not have to rewrite everything as "one giant rational function" before taking the limit? This question is addressed in Section 2.5.5.

Example 2.29. *Compute the following limits:*

(a) $\displaystyle\lim_{N\to\infty} \frac{(x+\frac{1}{N})^2 - x^2}{\frac{1}{N}}$.

(b) $\displaystyle\lim_{N\to\infty} \frac{(x+\frac{1}{N})^3 - x^3}{\frac{1}{N}}$.

(c) $\displaystyle\lim_{N\to\infty} \frac{(x+\frac{1}{N})^n - x^n}{\frac{1}{N}}$, *for* $n \in \mathbb{N}$, $n \geq 3$.

(d) $\lim_{N\to\infty} A_N^{\mathrm{Low}}$ *for the strictly increasing function* $f(y) = y^4$ *over the interval* $[0, x]$, *where* $x > 0$; *see Section 2.3.2.*

(e) $\lim_{N\to\infty} A_N^{\mathrm{Up}}$ *for the strictly increasing function* $f(y) = y^4$ *over the interval* $[0, x]$, *where* $x > 0$; *see Section 2.3.2.*

Solution:

(a) **Ans.** $\lim_{N\to\infty} \frac{(x+\frac{1}{N})^2 - x^2}{\frac{1}{N}} = 2x.$

$$\begin{aligned}\frac{(x+\frac{1}{N})^2 - x^2}{\frac{1}{N}} &= \frac{(x^2 + \frac{2x}{N} + \frac{1}{N^2}) - x^2}{\frac{1}{N}} \\ &= N\cdot\left(\frac{2x}{N} + \frac{1}{N^2}\right) \\ &= \frac{2xN+1}{N}.\end{aligned}$$

Hence,

$$\begin{aligned}\lim_{N\to\infty} \frac{(x+\frac{1}{N})^2 - x^2}{\frac{1}{N}} &= \lim_{N\to\infty} \frac{2xN+1}{N} \\ &= 2x.\end{aligned}$$

Note: $2xN$ is the leading term of the numerator if, and only if, $x \neq 0$. Hence, to be 100% totally correct, we really have two cases to consider here: (i) $x \neq 0$ and (ii) $x = 0$. In Case (i), we have already computed the answer to be $2x$. In Case (ii), the leading coefficient of the numerator is 1, and we have $\lim_{N\to\infty} (1/N) = 0$. Because $2x = 0$ when $x = 0$, being "super careful" versus "not so careful" has not changed the answer.

(b) **Ans.** $\lim_{N\to\infty} \frac{(x+\frac{1}{N})^3 - x^3}{\frac{1}{N}} = 3x^2.$

From Table 1.10, we have:

$$\begin{aligned}\frac{(x+\frac{1}{N})^3 - x^3}{\frac{1}{N}} &= \frac{(x^3 + 3\frac{x^2}{N} + 3\frac{x}{N^2} + \frac{1}{N^3}) - x^3}{\frac{1}{N}} \\ &= N\cdot\left(3\frac{x^2}{N} + 3\frac{x}{N^2} + \frac{1}{N^3}\right) \\ &= \frac{3x^2N^2 + 3xN + 1}{N^2}.\end{aligned}$$

Hence,

$$\begin{aligned}\lim_{N\to\infty} \frac{(x+\frac{1}{N})^3 - x^3}{\frac{1}{N}} &= \lim_{N\to\infty} \frac{3x^2N^2 + 3xN + 1}{N^2} \\ &= 3x^2.\end{aligned}$$

(c) **Ans.** $\lim_{N\to\infty} \frac{(x+\frac{1}{N})^n - x^n}{\frac{1}{N}} = nx^{n-1}.$

For this problem, we need to use the Binomial Theorem 1.67. To avoid losing track of the term $-x^n$, we move it to the front and compute:

$$\begin{aligned}-x^n + \left(x + \frac{1}{N}\right)^n &= -x^n + \binom{n}{0}x^n\cdot\left(\frac{1}{N}\right)^0 + \binom{n}{1}x^{n-1}\cdot\left(\frac{1}{N}\right)^1 + \sum_{k=2}^{n}\binom{n}{k}x^{n-k}\cdot\left(\frac{1}{N}\right)^k \\ &= -x^n + x^n + nx^{n-1}\cdot\left(\frac{1}{N}\right)^1 + \sum_{k=2}^{n}\binom{n}{k}x^{n-k}\cdot\left(\frac{1}{N}\right)^k \\ &= nx^{n-1}\cdot N^{-1} + \sum_{k=2}^{n}\binom{n}{k}x^{n-k}\cdot N^{-k} \\ &= nx^{n-1}\cdot N^{-1} + \sum_{k=2}^{n-1}\binom{n}{k}x^{n-k}\cdot N^{-k} + x^0\cdot N^{-n} \text{ multiply by } \frac{N^n}{N^n} \\ &= \frac{nx^{n-1}N^{n-1} + \sum_{k=2}^{n-1}\binom{n}{k}x^{n-k}\cdot N^{n-k} + 1}{N^n}.\end{aligned}$$

Hence,

$$\lim_{N\to\infty} \frac{(x+\frac{1}{N})^n - x^n}{\frac{1}{N}} = \lim_{N\to\infty} N \cdot \frac{nx^{n-1}N^{n-1} + \sum_{k=2}^{n-1}\binom{n}{k}x^{n-k}\cdot N^{n-k} + 1}{N^n}$$
$$= \lim_{N\to\infty} \frac{nx^{n-1}N^{n-1} + \sum_{k=2}^{n-1}\binom{n}{k}x^{n-k}\cdot N^{n-k} + 1}{N^{n-1}}$$
$$= nx^{n-1}.$$

(d) **Ans.** $\lim_{N\to\infty} A_N^{\text{Low}} = \lim_{N\to\infty} \frac{x^5(6(N-1)^5 + 15(N-1)^4 + 10(N-1)^3 - (N-1))}{30N^5} = \frac{x^5}{5}.$

For $N \geq 2$, define $\Delta x = \frac{x}{N}$, $x_i = (i-1)\Delta x$, $f(x_i) = ((i-1)\Delta x)^4$,

$$\begin{aligned}
\text{Area}_N^{\text{Low}} &:= \sum_{i=1}^{N} f(x_i)\cdot\Delta x \\
&= \sum_{i=1}^{N} ((i-1)\Delta x)^4 \cdot \Delta x \\
&= \sum_{i=1}^{N} \left((i-1)\frac{x}{N}\right)^4 \cdot \frac{x}{N} \\
&= \frac{x^5}{N^5}\cdot\sum_{i=1}^{N}(i-1)^4 \quad \text{(because both } n, x \text{ are independent of } i, \text{ we take them out of the sum)} \\
&= \frac{x^5}{N^5}\cdot\sum_{i=1}^{N-1} i^4 \quad \text{(change of index)} \\
&= \frac{x^5}{N^5}\cdot\frac{6(N-1)^5 + 15(N-1)^4 + 10(N-1)^3 - (N-1)}{30} \quad \text{(by Remark 2.11)} \\
&= \frac{x^5(6(N-1)^5 + 15(N-1)^4 + 10(N-1)^3 - (N-1))}{30N^5}.
\end{aligned}$$

Taking the limit as $N \to \infty$ completes the result, or does it? Do we ‘ need to expand $(N-1)^5$, $(N-1)^4$, and $(N-1)^3$ using the Binomial Theorem so as to identify the leading coefficients? The answer is that NO, we do not need to do that. The leading coefficient comes from $(N-1)^5$, so we can ignore the other two. Morever, from the Binomial Theorem,

$$\begin{aligned}
(N-1)^5 &= \binom{5}{0}N^5 + \text{terms of degree 4 and lower in } N \\
&= N^5 + \mathrm{O}(N, 4).
\end{aligned}$$

Hence, the limit is equal to $\lim_{N\to\infty} \frac{6x^5N^5}{30N^5} = \frac{x^5}{5}.$

(e) **Ans.** $\lim_{N\to\infty} A_N^{\text{Up}} = \lim_{N\to\infty} \frac{x^5(6N^5 + 15N^4 + 10N^3 - N)}{30N^5} = \frac{x^5}{5}.$

For $N \geq 2$, define $\Delta x = \frac{x}{N}$, $x_i = (i-1)\Delta x$, $f(x_{i+1}) = (i\Delta x)^4$,

$$\begin{aligned}
\text{Area}_N^{\text{Up}} := &\sum_{i=1}^{N} f(x_{i+1}) \cdot \Delta x \\
= &\sum_{i=1}^{N} (i\Delta x)^4 \cdot \Delta x \\
= &\sum_{i=1}^{N} \left(i\frac{x}{N}\right)^4 \cdot \frac{x}{N} \\
= &\frac{x^5}{N^5} \cdot \sum_{i=1}^{N} i^4 \quad \text{(because both } N \text{ and } x \text{ are independent of } i, \text{ we take them out of the sum)} \\
= &\frac{x^5}{N^5} \cdot \frac{6N^5 + 15N^4 + 10N^3 - N}{30} \quad \text{(by Remark 2.11)} \\
= &\frac{x^5(6N^5 + 15N^4 + 10N^3 - N)}{30N^5}.
\end{aligned}$$

Taking the limit as $N \to \infty$ completes the result.

■

2.5.3 Products and Ratios of Exponential and Monomial Terms

It is relatively easy to determine the limits of products and ratios of exponentials (a^x) and monomials (x^k), where k can be positive or negative.

Intuition for the limits of $x^k \cdot a^x$ and $\frac{a^x}{x^k}$: The picture to have in mind is that exponentials a^x (for $a > 1$, $x > 0$) **grow infinitely faster** than any monomial x^k, and hence, the exponential term dominates the monomial even while the monomial in the denominator is exploding to infinity. (It's pretty clear that the product of an exploding exponential and an exploding monomial is going to infinity.)

Similarly, exponentials $a^{-x} = \left(\frac{1}{a}\right)^x$ (for $a > 1$, $x > 0$) **decay infinitely faster** than any monomial, and hence, the exponential term dominates the monomial even when the monomial in the numerator is exploding to infinity. (It's pretty clear that the product of a decaying exponential a^{-x} and a decaying monomial $\frac{1}{x^k}$, $k > 0$ goes to zero as $x \to \infty$.)

While the exponential always dominates the tug of war, **there is a subtlety with negative limits**, that is, when you look at limits as $x \to -\infty$: then the monomial can change the sign. That's the only catch. Otherwise, the same story holds: the exponential wins, but the monomial determines the sign.

The above is Prop. 2.30, Corollary 2.31, and Prop. 2.32 in a nutshell.

Proposition 2.30: Exponentials Grow Faster Than Any Monomial Function

The following limits hold for real variables x and integers m:

(a) For all $a > 1$ and integers $0 \le m < \infty$, $\lim\limits_{x\to\infty} \frac{a^x}{x^m} = \infty$.

(b) For all $a > 1$ and integers $0 \le m < \infty$, $\lim\limits_{x\to\infty} \frac{x^m}{a^x} = 0$.

(c) For all $a > 1$ and integers $0 \le m < \infty$, $\lim\limits_{x\to\infty} x^m \cdot a^x = \infty$.

Note: The proof is given in Section 2.9. When using this result, we should keep in mind that

$x^{-m} := \frac{1}{x^m}$ and $a^{-x} := \frac{1}{a^x}$, which allow us to rewrite a function with negative exponents into one with positive exponents.

By noting that $0 < b < 1$ implies that $a := \frac{1}{b} > 1$, and that $a^x = \frac{1}{b^x}$, the limits in Prop. 2.30 can also be applied to exponentials with bases between zero and one. For consistency and ease of use, we restate them in Corollary 2.31 for $0 < a < 1$.

Corollary 2.31: Exponentials Also Decay Faster Than Any Monomial Function

The following limits hold for real variables x and for integers m:

(a) For all $0 < a < 1$ and integers $0 \le m < \infty$, $\lim_{x\to\infty} x^m \cdot a^x = 0$.

(b) For all $0 < a < 1$ and integers $0 \le m < \infty$, $\lim_{x\to\infty} \frac{x^m}{a^x} = \infty$.

(c) For all $0 < a < 1$ and integers $0 \le m < \infty$, $\lim_{x\to\infty} \frac{1}{x^m \cdot a^x} = \infty$.

Proposition 2.32: Relating Limits at Negative Infinity to Limits at (Positive) Infinity

The following relations hold for real variables x and for integers m:

(a) For all $a > 0$, $a \neq 1$ and integers $0 \le m < \infty$, $\lim_{x\to-\infty} x^m \cdot a^x = (-1)^m \cdot \lim_{x\to+\infty} \frac{x^m}{a^x}$.

(b) For all $a > 0$, $a \neq 1$ and integers $0 \le m < \infty$, $\lim_{x\to-\infty} \frac{x^m}{a^x} = (-1)^m \cdot \lim_{x\to+\infty} x^m \cdot a^x$.

(c) For all $a > 0$, $a \neq 1$ and integers $0 \le m < \infty$, $\lim_{x\to-\infty} \frac{a^x}{x^m} = (-1)^m \cdot \lim_{x\to+\infty} \frac{1}{x^m \cdot a^x}$.

Note: Because $1^x = 1$ and $1^n = 1$ for all $x \in \mathbb{R}$ and $n \in \mathbb{Z}$, it is not strictly necessary to make an exception for $a = 1$ in the above limits. It remains true that:

$$\lim_{x\to-\infty} x^m = (-1)^m \cdot \lim_{x\to+\infty} x^m \text{ and } \lim_{x\to-\infty} \frac{1}{x^m} = (-1)^m \cdot \lim_{x\to+\infty} \frac{1}{x^m}.$$

However, because $a = 1$ is excluded from Prop. 2.30 and Corollary 2.31, we wanted to avoid misapplication of those results.

Example 2.33. *Where applicable, use Prop. 2.30, Corollary 2.31, and Prop. 2.32 to compute* $\lim_{x\to\pm\infty} f(x)$ *for the following functions:*

(a) $f : \mathbb{R} \to \mathbb{R}$ *by* $f(x) = x^3 \cdot (0.5)^x$.

(b) $f : \mathbb{R} \to \mathbb{R}$ *by* $f(x) = \frac{2^x}{x^{27}}$.

(c) $f : \mathbb{R} \to \mathbb{R}$ *by* $f(x) = e^x \cdot \sin(x)$.

(d) $f : \mathbb{R} \to \mathbb{R}$ *by* $f(x) = (1.01)^x \cdot \frac{-4x^3+11x+14}{1+x^4}$.

(e) $f : \mathbb{R} \to \mathbb{R}$ *by* $f(x) = \frac{x^4}{5^{\frac{x}{2}}}$.

Solutions:

(a) $f(x) = x^3 \cdot (0.5)^x$. **Ans.** $\lim_{x\to\infty} f(x) = 0$ and $\lim_{x\to-\infty} f(x) = -\infty$.

- $a = 0.5 \implies 0 < a < 1$. Hence, by Corollary 2.31-(a), $\lim_{x\to\infty} f(x) = 0$.
- $a = 0.5 \implies 0 < a < 1$ and m is odd. Hence, by Prop. 2.32-(a),

$$\lim_{x\to-\infty} f(x) = (-1)^3 \cdot \lim_{x\to+\infty} \frac{x^3}{(0.5)^x} = -\infty.$$

(b) $f(x) = \frac{2^x}{x^{27}}$. **Ans.** $\lim_{x\to\infty} f(x) = \infty$ and $\lim_{x\to-\infty} f(x) = 0$.

- $a = 2 \implies a > 1$. Hence, by Prop. 2.30-(a), $\lim_{x\to\infty} f(x) = \infty$.
- $a = 2 \implies a > 1$ and m is odd.
  Hence, by Prop. 2.32-(c),

$$\lim_{x\to-\infty} f(x) = (-1)^{27} \cdot \lim_{x\to+\infty} \frac{1}{x^{27} \cdot 2^x} = 0.$$

(c) $f(x) = e^x \cdot \sin(x)$. **Ans.** Due to the $\sin(x)$ term, Prop. 2.30, Corollary 2.31, and Prop. 2.32 are not applicable. Later, we will learn that $\lim_{x\to-\infty} e^x \cdot \sin(x) = 0$, while $\lim_{x\to+\infty} e^x \cdot \sin(x)$ does not exist.

(d) $f(x) = (1.01)^x \cdot \frac{-4x^3+11x+14}{1+x^4}$. **Ans.** Because the exponential term $(1.01)^x$ is multiplying a rational function instead of a monomial, Prop. 2.30, Corollary 2.31, and Prop. 2.32 are not (directly) applicable. However, Prop. 2.23 encourages us to approximate a rational function by:

$$\frac{b_m x^m + b_{m-1}x^{m-1} + \cdots + b_0}{a_n x^n + a_{n-1}x^{n-1} + \cdots + a_0} \approx \frac{b_m}{a_n} \cdot x^{m-n} \text{ for } |x| \text{ sufficiently large.}$$

Doing so for $\frac{-4x^3+11x+14}{1+x^4}$ results in, for $|x|$ large, $f(x)$ behaving like $g(x) := -4\frac{(1.01)^x}{x}$. For this function, we can apply Prop. 2.30, Corollary 2.31, and Prop. 2.32 to obtain:

- $\lim_{x\to\infty} (1.01)^x \cdot \dfrac{-4x^3 + 11x + 14}{1 + x^4} = \lim_{x\to\infty} g(x) = \lim_{x\to\infty} -4\dfrac{(1.01)^x}{x} = -\infty.$
- $\lim_{x\to-\infty} (1.01)^x \cdot \dfrac{-4x^3 + 11x + 14}{1 + x^4} = \lim_{x\to-\infty} g(x) = \lim_{x\to-\infty} -4\dfrac{(1.01)^x}{x} = 0.$

(e) $f(x) = \frac{x^4}{5^{\frac{x}{2}}}$ **Ans.** $\lim_{x\to\infty} f(x) = 0$ and $\lim_{x\to-\infty} f(x) = \infty$.

We rewrite $f(x) = \frac{x^4}{5^{\frac{x}{2}}} = \frac{x^4}{(\sqrt{5})^x}$

- $a = \sqrt{5} \implies a > 1$. Hence, by Prop. 2.30-(b), $\lim_{x\to\infty} f(x) = 0$.
- $a = \sqrt{5} \implies a > 1$ and m is even. Hence, by Prop. 2.32-(b),

$$\lim_{x\to-\infty} f(x) = \lim_{x\to-\infty} x^4(\sqrt{5})^{-x} = (-1)^4 \cdot \lim_{x\to+\infty} x^4 \cdot \left(\frac{\sqrt{5}}{5}\right)^x = 0,$$

because $\frac{\sqrt{5}}{5} < 1$.

■

Example 2.34. *From Late New Year Problem by Math Curiosity, solve for $a \in \mathbb{N}$, if one exists, such that:*

$$\lim_{n\to\infty} \frac{n^{2022}}{n^a - (n-1)^a} = \frac{1}{2023}.$$

Solution By the Binomial Theorem,

$$n^a - (n-1)^a = n^a - \left[n^a + \binom{a}{1} n^{a-1}(-1)^1 + \mathrm{O}(n, a-2)\right]$$
$$= an^{a-1} - \mathrm{O}(n, a-2).$$

Hence,

$$\frac{n^{2022}}{n^a - (n-1)^a} = \frac{n^{2022}}{an^{a-1} - \mathrm{O}(n, a-2)}.$$

Applying Prop. 2.23 and Fact 2.26, we have a finite limit if, and only if, $a \geq 2023$, and we have a nonzero limit if, and only if, $a \leq 2023$. Hence, our only sensible solution is $a = 2023$. Indeed, we then have

$$\lim_{n\to\infty} \frac{n^{2022}}{n^a - (n-1)^a} = \lim_{n\to\infty} \frac{n^{2022}}{n^{2023} - (n-1)^{2023}} = \lim_{n\to\infty} \frac{n^{2022}}{2023n^{2022} - \mathrm{O}(n, 2021)} = \frac{1}{2023}.$$

■

2.5.4 Algebraic Operations with Infinity: Determinate and Indeterminate Forms

In Definition 1.3, we emphasized that the symbol "∞" is a concept and not a number. In particular:

- ∞ represents that which is larger than any real number, meaning for all $\alpha \in \mathbb{R}$, $\infty > \alpha$ (aka $\alpha < \infty$), while
- $-\infty$ represents that which is smaller than any real number, meaning for all $\alpha \in \mathbb{R}$, $-\infty < \alpha$ (aka $\alpha > -\infty$).

Nevertheless, as with numbers, **some algebraic operations with infinity make sense**.[3] For example,

- for all $\alpha \in \mathbb{R}$, $\alpha + \infty = \infty$ (a finite quantity plus an infinite quantity is still infinite). Its negative is:
- for all $\alpha \in \mathbb{R}$, $\alpha - \infty = -\infty$, which can also be written as $-\infty + \alpha = -\infty$.
- The above can be extended by $\infty + \infty = \infty$ (the sum of two infinitely "positive" quantities is infinitely positive, and the negative of this is:
- $-\infty - \infty = -\infty + (-\infty) = -\infty$ (the sum of two infinitely negative quantities).

What about multiplication and division? Similar rules can be defined:

- For all $\alpha > 0$, $\alpha \cdot \infty = \infty$ (an infinite quantity scaled by a positive number is still infinite).
- For all $\alpha < 0$, $\alpha \cdot \infty = -\infty$ (an infinite quantity scaled by a negative number yields minus infinity). This has to hold if we accept that $(-1) \cdot \infty := -\infty$. Moreover:
- For all $\alpha \in \mathbb{R}$, $\frac{\alpha}{\pm\infty} = 0$.

In the above, we have carefully excluded the following cases, called **indeterminate forms**, because assigning a fixed value to the operations will lead to logical fallacies.

Indeterminate Forms

Definition 2.35. *The following are **indeterminate forms** (aka **operations that are undefined and must remain undefined**):*

(a) $\infty - \infty$

(b) $\frac{\infty}{\infty}$

[3] Determinate is the opposite of indeterminate. Legitimate algebraic operations with infinity are sometimes called **determinate** forms.

(c) $0 \cdot \infty$

(d) $\frac{0}{0}$

For completeness, we mention three additional (less important) indeterminate forms:

(e) 1^∞

(f) 0^0

(g) ∞^0

In the following, we illustrate why the above forms are "indeterminate".

- $\infty - \infty$ can arrive through $\lim_{x\to\infty}(x - x^3) = -\infty$, $\lim_{x\to\infty}(x^3 - x) = \infty$, and $\lim_{x\to\infty}(x - x) = 0$. Without knowing "the rate at which each infinity" arises, it is impossible to know which value to assign!
- $\frac{\infty}{\infty}$ can arrive through $\lim_{x\to\infty}\frac{x^3}{x+1} = \infty$, $\lim_{x\to\infty}\frac{x}{x^3+1} = 0$, and $\lim_{x\to\infty}\frac{x}{x+1} = 1$. Without knowing "the rate at which each infinity" arises, it is impossible to know which value to assign!
- $0 \cdot \infty$ can arrive from $\lim_{x\to\infty} e^{-x} \cdot x = 0$, $\lim_{x\to\infty} 2^{-x} \cdot 3^x = \infty$, and $\lim_{x\to\infty}\frac{1}{1+x^2} \cdot x^2 = 1$. Without knowing "the rate at which zero and infinity" arise, it is impossible to know which value to assign!
- $\frac{0}{0}$ can arrive from $\lim_{x\to\infty}\frac{e^{-x^2}}{e^{-x}} = 0$, $\lim_{x\to\infty}\frac{2^{-x}}{3^{-x}} = \infty$, and $\lim_{x\to\infty}\frac{1-e^{-\frac{1}{x}}}{\frac{1}{x}} = 1$. Without knowing "the rate at which each zero" arises, it is impossible to know which value to assign!

Associated Video: Indeterminate: the hidden power of 0 divided by 0 by Mathologer.

2.5.5 Limit Algebra

It would seem much more efficient to take limits term by term instead of what was done in the solutions to Example 2.28. In fact, such an approach is quite valid **so long as we avoid the indeterminate forms listed above**.

Proposition 2.36: Limits of Linear Combinations

Suppose that for all $1 \le i \le n$, $f_i : (0,\infty) \to \mathbb{R}$ has a finite limit at infinity; that is, $\lim_{x\to\infty} f_i(x)$ exists and is finite. Then, for all coefficients $\alpha_i \in \mathbb{R}$,

$$\lim_{x\to\infty}\left(\sum_{i=1}^{n}\alpha_i f_i(x)\right) = \sum_{i=1}^{n}\left(\alpha_i \lim_{x\to\infty} f_i(x)\right). \tag{2.28}$$

Thus, the limit of the linear combination exists and equals the linear combination of the limits.

Note: The assumption of the individual limits existing and being finite allows us to avoid the dreaded "$\infty - \infty$".

Finite limits of products and ratios can be also computed sequentially. It is even straightforward to allow one of the limits to be unbounded and still draw a conclusion. Below, we avoid the dreaded $\frac{\infty}{\infty}$, $0 \cdot \infty$, and $\frac{0}{0}$ by forcing $g(x)$ to have a finite nonzero limit.

Proposition 2.37: Limits of Products and Ratios

Suppose that $g : (0,\infty) \to \mathbb{R}$ has a limit at infinity of one, i.e., $\lim_{x\to\infty} g(x) = 1.0$. Then, for any function $f : (0,\infty) \to \mathbb{R}$,

$$\begin{aligned} \lim_{x\to\infty} f(x)\cdot g(x) &= \lim_{x\to\infty} f(x), \text{ and} \\ \lim_{x\to\infty} \frac{f(x)}{g(x)} &= \lim_{x\to\infty} f(x). \end{aligned} \tag{2.29}$$

Useful special cases:

(a) If $\lim_{x\to\infty} f(x) = L$, for $L \in \mathbb{R}$, then $\lim_{x\to\infty} f(x)\cdot g(x) = L$ and $\lim_{x\to\infty} \frac{f(x)}{g(x)} = L$.

(b) If $\lim_{x\to\infty} f(x) = \pm\infty$, then $\lim_{x\to\infty} f(x)\cdot g(x) = \pm\infty$ and $\lim_{x\to\infty} \frac{f(x)}{g(x)} = \pm\infty$.

(c) If $\lim_{x\to\infty} f(x)$ does not exist, then $\lim_{x\to\infty} f(x)\cdot g(x)$ and $\lim_{x\to\infty} \frac{f(x)}{g(x)}$ do not exist.

Note: If $\lim_{x\to\infty} g(x) = M \neq 1$, and $M \neq 0$, then $\frac{g(x)}{M}$ has limit one. Hence, using

$$f(x)\cdot g(x) = (Mf(x))\cdot \frac{g(x)}{M}$$

and

$$\frac{f(x)}{g(x)} = \frac{\frac{1}{M}f(x)}{\frac{1}{M}g(x)}$$

reduces the general case of the problem to the given (special) conditions.

Example 2.38. *Compute the indicated limits using Propositions 2.36 and 2.37:*

(a) $\lim_{x\to\infty} \frac{x^2}{3-x^3} + \frac{4+x+3x^2}{x^2-6x+11}$.

(b) $\lim_{x\to\infty} \frac{x^3}{3-x^3} \cdot \frac{4+x+3x^2}{x^2-6x+11}$.

(c) $\lim_{x\to\infty} \frac{\frac{x^2}{3-x^3}}{\frac{4+x+3x^2}{x^2-6x+11}}$.

(d) $\lim_{x\to\infty} \frac{x\cdot\sin(x)}{1+e^{-x}+e^{-2x}}$.

Solution:

(a) **Ans.** $\lim_{x\to\infty} \frac{x^2}{3-x^3} + \frac{4+x+3x^2}{x^2-6x+11} = 3.0$

Why: By Prop. 2.23, $\lim_{x\to\infty} \frac{x^2}{3-x^3} = 0$ and $\lim_{x\to\infty} \frac{4+x+3x^2}{x^2-6x+11} = 3.0$. Hence, by Prop. 2.36, the limits add, and we obtain the given answer.

(b) **Ans.** $\lim_{x\to\infty} \frac{x^2}{3-x^3} \cdot \frac{4+x+3x^2}{x^2-6x+11} = 0.0$

Why: By Prop. 2.23, $\lim_{x\to\infty} \dfrac{x^2}{3-x^3} = 0$ and $\lim_{x\to\infty} \dfrac{4+x+3x^2}{x^2-6x+11} = 3.0$. Hence, by Prop. 2.37, the limits multiply, and we obtain the given answer.

(c) **Ans.** $\lim_{x\to\infty} \dfrac{\frac{x^3}{3-x^3}}{\frac{4+x+3x^2}{x^2-6x+11}} = -\dfrac{1}{3}$.

Why: By Prop. 2.23, $\lim_{x\to\infty} \dfrac{x^3}{3-x^3} = -1.0$ and $\lim_{x\to\infty} \dfrac{4+x+3x^2}{x^2-6x+11} = 3.0$ Hence, by Prop. 2.37, the limits divide, and we obtain the given answer.

(d) **Ans.** $\lim_{x\to\infty} \dfrac{x\cdot\sin(x)}{1+e^{-x}+e^{-2x}}$ does not exist.

Why: By Corollary 2.31 and Prop. 2.36, the denominator limit, $\lim_{x\to\infty} 1+e^{-x}+e^{-2x} = 1.0$, while the numerator limit, $\lim_{x\to\infty} x\cdot\sin(x)$, does not exist due to the oscillating sinusoid. Hence, by Prop. 2.37, the limit of the ratio does not exist.

■

2.6 Geometric Sums

A geometric sequence is a sequence of numbers where each term after the first term is found by multiplying the previous term by a fixed, nonzero number r called the **common ratio**. The general form of a **geometric sequence** is:

$$a_0 = a, \quad a_1 = a\cdot r, \quad a_2 = a\cdot r^2, \quad a_3 = a\cdot r^3, \; \dots, \; a_i = a\cdot r^i,$$

where a is the first term, and r is the **common ratio**. We are more interested in a **geometric sum**, which is the finite sum of the form

$$S_n := \sum_{i=0}^{n} a\cdot r^i = a + a\cdot r + a\cdot r^2 + \cdots + a\cdot r^n. \tag{2.30}$$

Intuition for Geometric Sums (aka geometric series) is uncommonly visual:

- Three Geometric Series in an Equilateral Triangle (visual proof without words) by @MathVisualProofs.
- Playlist for Geometric Sums by @MathVisualProofs.

For later use, we note that

$$S_{n+1} = a + r\cdot S_n. \tag{2.31}$$

Proposition 2.39: Geometric Sums

For $r \neq 1$,

$$S_n = \frac{a(1-r^{n+1})}{1-r}, \tag{2.32}$$

and if $|r| < 1$,

$$\lim_{n\to\infty} S_n = \frac{a}{1-r}. \tag{2.33}$$

Notes:

(i) If $r = 1$, then $S_n = a(n+1)$ and if $a \neq 0$, its limit is unbounded as $n \to \infty$.

(ii) $\sum_{i=0}^{\infty} ar^i := \lim_{n\to\infty} \sum_{i=0}^{n} ar^i$ is called a **geometric series**.

Proof: The proof can be given by induction. We will set $k = 1$ as the base case, even though we could have set it at $k = 0$.

Base Case: $k = 1$:
$S_1 = a + a \cdot r = a \cdot (1 + r) = \frac{a\cdot(1+r)\cdot(1-r)}{1-r} = \frac{a\cdot(1-r^2)}{1-r}$. As is commonly the case, the solution is quite straightforward.

Induction Step: We define $L_k := S_k := \sum_{i=0}^{n} a \cdot r^i$ and $R_k := \frac{a(1-r^{k+1})}{1-r}$ and seek to show that $L_k = R_k$ being true for some $k \geq 1$ implies it is also true for $k + 1$.

$$\begin{aligned}
L_{k+1} &= S_{k+1} \\
&= a + r \cdot S_k \text{ (by (2.30))} \\
&= a + r \cdot \frac{a(1-r^{k+1})}{1-r} \text{ (by the induction hypothesis)} \\
&= a + \frac{a(r-r^{k+2})}{1-r} \text{ (algebra, multiply by } r) \\
&= a \cdot \left(1 + \frac{r-r^{k+2}}{1-r}\right) \text{ (algebra, factor out } a) \\
&= a \cdot \left(\frac{1-r}{1-r} + \frac{r-r^{k+2}}{1-r}\right) \text{ (rewrite the number 1 as a ratio)} \\
&= a \cdot \left(\frac{1-r^{k+2}}{1-r}\right) \text{ (algebra).}
\end{aligned}$$

By its definition,

$$R_{k+1} = \frac{a(1-r^{k+2})}{1-r},$$

which equals L_{k+1}, so the proof by induction is complete.

Now, we look at using Prop. 2.31 to evaluate the limit. We first note that if $r < 0$, then $r = (-1) \cdot |r|$ and therefore, $r^n = (-1)^n \cdot |r|^n$. Hence, by Prop. 2.31, if $|r| < 1$, then $\lim_{n\to\infty} r^{n+1} = r \cdot \lim_{n\to\infty} r^n = 0$, independent of the sign of r. Therefore,

$$\lim_{n\to\infty} S_n = \lim_{n\to\infty} \frac{a(1-r^{n+1})}{1-r} = \frac{a}{1-r}.$$

■

Example 2.40. *Determine which of the following sums have the form of (2.30) and hence are* ***geometric sums****. For those cases that are geometric sums, use Prop. 2.39 to compute* S_n*, and if it exists, compute* $S_\infty := \lim_{n\to\infty} S_n$.

(a) $S_n = \sum_{i=0}^{n} \frac{1}{2^i}$.

(b) $S_n = 1 + \frac{1}{2} + \frac{1}{2^2} + \cdots + \frac{1}{2^n}$.

(c) $S_n = \sum_{i=0}^{n} (-1)^i$.

(d) $S_n = 1 + \frac{1}{2} + \frac{1}{3} + \cdots + \frac{1}{n}$.

(e) $S_n = 3 + 3e^x + 3e^{2x} + \cdots + 3e^{nx}$.

(f) $S_n = 2 + \frac{1}{3^2} + \frac{1}{3^3} + \frac{1}{3^4} + \cdots + \frac{1}{3^n}$.

Solutions:

(a) **Ans.** $\sum_{i=0}^{n} \frac{1}{2^i} = \sum_{i=0}^{n} \left(\frac{1}{2}\right)^i$ is a geometric sum, with common ratio $r = \frac{1}{2}$ and $a = 1$. Hence,

$$S_n = \frac{a \cdot (1 - r^{n+1})}{1 - r} = \frac{1 \cdot (1 - (\frac{1}{2})^{n+1})}{1 - \frac{1}{2}} = 2 \cdot \left(1 - 2^{-(n+1)}\right) = 2 - 2^{-n},$$

and therefore, $S_\infty := \lim_{n\to\infty} S_n = 2.$

(b) **Ans.** $S_n = 1 + \frac{1}{2} + \frac{1}{2^2} + \cdots + \frac{1}{2^n} = \sum_{i=0}^{n} \frac{1}{2^i}$, the sum given in (a). Hence, the answer is the same.

(c) **Ans.** $S_n = \sum_{i=0}^{n}(-1)^n$ is a geometric sum, with common ratio $r = -1$ and $a = 1$. Hence,

$$S_n = \frac{a \cdot (1 - r^{n+1})}{1 - r} = \frac{1 \cdot (1 - (-1)^{n+1})}{1 - (-1)} = \frac{1 - (-1)^{n+1}}{2} = \begin{cases} 0 & n \text{ odd}, \\ 1 & n \text{ even}. \end{cases}$$

Because $|r| = 1$, $\lim_{n\to\infty} S_n$ does not exist; that is hopefully clear because S_n alternates between 0 and 1, which does not converge.

(d) **Ans.** $S_n = 1 + \frac{1}{2} + \frac{1}{3} + \cdots + \frac{1}{n}$ is NOT a geometric sum. If it were to be a geometric sum, from the first term, we would have $a = 1$, and from the second term, we would deduce that $r = \frac{1}{2}$. The problem comes with the third term because

$$\frac{1}{3} \neq r^2 = \left(\frac{1}{2}\right)^2 = \frac{1}{4}.$$

(e) **Ans.** $S_n = 3 + 3e^x + 3e^{2x} + \cdots + 3e^{nx} = \sum_{i=0}^{n} 3 \cdot (e^x)^i$ is a geometric sum, with common ratio $r = e^x$ and $a = 3$. Hence,

$$S_n = \frac{a \cdot (1 - r^{n+1})}{1 - r} = \frac{3 \cdot (1 - (e^x)^{n+1})}{1 - e^x} = 3\frac{1 - e^{(n+1)x}}{1 - e^x}.$$

For $x < 0$, $|r| = |e^x| = e^x < 1$, while for $x > 0$, $|r| > 1$. Therefore:

$$S_\infty := \lim_{n\to\infty} S_n = \begin{cases} \frac{1}{1-e^x} & x < 0, \\ \infty & x > 0. \end{cases}$$

Finally, for $x = 0$, $S_n = 3 \cdot (n + 1)$ and thus $S_\infty = \infty$.

(f) **Ans.** $S_n = 2 + \frac{1}{3^2} + \frac{1}{3^3} + \frac{1}{3^4} + \cdots + \frac{1}{3^n}$ is NOT a geometric sum. If it were to be a geometric sum, from the first term, we would have $a = 2$, and from the second term, we would deduce that $r = \frac{\frac{1}{3^2}}{2} = \frac{1}{18}$. The problem comes with the third term because

$$\frac{1}{3^3} = \frac{1}{27} \neq a \cdot r^2 = 2\left(\frac{1}{18}\right)^2 = \frac{2}{2^2 \cdot 9^2} = \frac{1}{2} \cdot \frac{1}{81}.$$

However, the given sum can be rewritten as a constant plus a geometric sum, and hence we can still compute it with relative ease. Indeed, let us define

$$\overline{S}_n := 1 + \frac{1}{3} + \frac{1}{3^2} + \frac{1}{3^3} + \frac{1}{3^4} + \cdots + \frac{1}{3^n} = -1 + \frac{1}{3} + S_n.$$

$\overline{S}_n$ is a geometric sum with with common ratio $r = \frac{1}{3}$ and $a = 1$. Hence,

$$\overline{S}_n = \frac{a \cdot (1 - r^{n+1})}{1 - r} = \frac{1 \cdot (1 - (\frac{1}{3})^{n+1})}{1 - \frac{1}{3}} = \frac{3}{2} \cdot \left(1 - 3^{-(n+1)}\right),$$

and therefore, $\overline{S}_\infty := \lim_{n\to\infty} \overline{S}_n = \dfrac{3}{2}.$

It follows that:

$$S_n = \overline{S}_n + \frac{2}{3} = \frac{3}{2} \cdot \left(1 - 3^{-(n+1)}\right) + \frac{2}{3},$$

and $S_\infty = \frac{3}{2} + \frac{2}{3} = \frac{9+4}{6} = \frac{13}{6} = 2 + \frac{1}{6}.$

■

Example 2.40-(f) illustrates that sometimes a sum can be modified slightly so that it becomes a geometric sum. That should be considered a standard "trick of the trade". Real problems do not come prepackaged for directly applying schoolwork in a rote manner. Some creativity is required!

2.7 Countable Sets

Our goal here is to illustrate why proofs are important. Moreover, we wish to do that in a setting where the logic of the proof (its roadmap) may be hard to invent on our own, yet; once it is spelled out to us, it is quite understandable and can be understood without a lot of experience with proofs. Even more, after reading through the material, there is a chance, albeit a small one, that the reader might even find it edifying (i.e., enlightening, educational, informative, or illuminating).

To accomplish all of these goals, we will present a nonintuitive, *even shocking*, result that Georg Cantor proved in 1874. The following is a loosely paraphrased excerpt from Wikipedia.

Secrets of the Arcane 2.41: Renegade Mathematician and Corrupter of Youth? You Be the Judge.

In the heat of the mathematical revolution, Georg Cantor stirred up controversy with his mind-bending theory of transfinite numbers. This audacious idea was met with the equivalent of academic side-eye, even triggering the ire of mathematical heavyweights such as Leopold Kronecker and Henri Poincaré, not to mention the philosophical doubts raised by Ludwig Wittgenstein. The vitriol that Cantor faced was far from mild. Kronecker went so far as to lambaste Cantor publicly, **branding him a "scientific charlatan", a "renegade", and a "corrupter of youth."** The ridicule and condemnation of the mathematical community eventually led Cantor to depression and admission to a mental healthcare facility.

While Cantor's treatment by his peers was unjust, he had his issues as well. "Cantor was paradoxically opposed to theories of infinitesimals of his contemporaries Otto Stolz and Paul du Bois-Reymond, describing [Stolz and du Bois-Reymond] as **"an abomination" and "a cholera bacillus of mathematics."**"

In a certain sense, the concept we will consider is quite simple: how many elements are contained in a set? For example, if we can list the elements of a set, such as

$$S = \{s_1, s_2, s_3, s_4, s_5\},$$

where $s_i \neq s_j$ for $i \neq j$, then we would naturally conclude the set has five elements. In mathematics, the number of elements in a set is called its **cardinality**. If we can index the elements of the set using a finite number of counting numbers, then it has finite cardinality. The listing does not have to use consecutive integers for us to conclude that the set's cardinality is finite; for example, we can consider

$$T := \{2, 4, 11, 10^9\}$$

as already being listed by the numbers 2, 4, 11, and 10^9; because the largest element, 10^9, is finite, there is no need to assign $t_1 := 2,\ t_2 := 4,\ t_3 := 11,\ t_4 := 10^9$. However, sets such as the natural numbers $\mathbb{N}$ and the real numbers $\mathbb{R}$ have an infinite number of elements, and hence their cardinality must be infinity. **Cantor asked the question: are the infinities associated with the natural numbers and the real numbers the same, or are they different in some way?** And that led him to be branded a "corrupter of youth?" To dig a little deeper, we must carefully define what we mean by "counting the number of elements in a set."

What does it mean for a set to be countable?

Definition 2.42. *A set S is* **countable** *if there exists an injective (i.e, 1:1) function $f : S \to \mathbb{N}$. This means that each element of the set is assigned a counting number. Moreover, distinct elements of the set are assigned distinct numbers so that we are not undercounting the number of elements.*

*S is **finitely countable** if the range of f is contained in a bounded subset of the natural numbers. Otherwise, S is called **countably infinite**. Stated in a different way: in a finitely countable set, the process of counting comes to an end, whereas in a countably infinite set, it does not, just as with the counting numbers themselves.*

Notes:

*(a) The **empty set** has no elements and hence, by tradition, is considered finitely countable.*

*(b) If there is no injective function from S to the counting numbers, the set is called **uncountable**.*

*(c) If an injective function $f : S \to \mathbb{N}$ exists and its range is contained in a bounded subset of the natural numbers, then S is called **countably finite or finitely countable** (both are used). Otherwise, S is called **countably infinite**. To be extra clear, this means that the range of $f : S \to \mathbb{N}$ contains an unbounded number of natural numbers (equivalently, the range does not have a largest element).*

Remark 2.43. *Some people prefer the terminology **listable** over countable. The term "listable" is less formal and not universally used in mathematics. However, when it is used, it typically implies the same concept as countable. In other words, if you can devise a method to list all the elements of a set—with no element being missed and each being listed only once— then the set is "listable" or countable.*

Example 2.44. *Determine which of the following sets are countably finite and which are countably infinite. In each case, exhibit a function from the set to a **finite** subset of the counting numbers or the full set of natural numbers.*

(a) $S_1 := \{bus, submarine, carrot\}$

(b) $S_2 :=$ the set of even prime numbers.

(c) $S_3 := \{2k \mid k \in \mathbb{N}\}$, the set of even numbers.

(d) $S_4 :=$ the set of all 3×3 matrices with entries from the set $\{0, 1, 2, \ldots, 9\}$.

*(e) $S_5 :=$ the set of all **invertible** 2×2 matrices with entries from the set $\{1, 2, 3, 4, 5\}$.*

(f) $S_6 := \{x \in \mathbb{R} \mid x > 0, \log_{10}(x) \in \mathbb{N}$, and $\log_{10}(x) > \pi \}$ (deliberately made to appear absurdly complicated).

Solution:

(a) **Ans.** S_1 is finitely countable. We define $f : S_1 \to \{1, 2, 3\}$ by $f(\text{bus}) = 1$, $f(\text{submarine}) = 2$, and $f(\text{carrot}) = 3$. f is clearly injective (1:1).

Note: Our mapping does not have to be surjective (onto). Hence, another solution is $f : S_1 \to \{1, 2, 3, 4, 11, 17\}$ by $f(\text{bus}) = 1$, $f(\text{submarine}) = 2$, and $f(\text{carrot}) = 3$. Yet another solution is $f : S_1 \to \{1, 2, 3, 4, 11, 17\}$ by $f(\text{bus}) = 11$, $f(\text{submarine}) = 3$, and $f(\text{carrot}) = 17$. We can even define $f : S_1 \to \mathbb{N}$ by $f(\text{bus}) = 10$, $f(\text{submarine}) = 200$, and $f(\text{carrot}) = 300$, and then note that the range of f is contained in the finite set $\{1, 2, \ldots, 500\}$.

In the same vein, a non-solution is $f : S_1 \to \{1, 2, 3\}$ by $f(\text{bus}) = 3$, $f(\text{submarine}) = 2$, and $f(\text{carrot}) = 2$, because f is not injective (1:1) with both "carrot" and "submarine" being mapped onto the number two. In this case, the undercounting is not critical, but the map $f : \mathbb{N} \to \mathbb{N}$ by $f(k) = 1$ for all k would result in severe undercounting of the natural numbers. That is why injectivity is important.

(b) **Ans.** S_2 is finitely countable because there is only one even prime number, 2.

(c) **Ans.** S_3 is countably infinite. We can define $f : S_3 \to \mathbb{N}$ by $f(2k) = k$. The function is one-to-one because $f(s) = k \iff s = 2k$. Hence, there is only one element of S_3 assigned to any element of $\mathbb{N}$. We could also have defined $f : S_3 \to \mathbb{N}$ by $f(s) = 10^s$. There is no restriction that the numbers be consecutive, i.e., the function can "skip over" numbers in $\mathbb{N}$. We can check that the second function is also injective.

(d) **Ans.** S_4 is finitely countable. A 3×3 matrix has nine entries, and 3×3 matrices in S_4 have entries drawn from the set $\{0, 1, 2, \dots, 9\}$. Hence, there are 10^9 distinct matrices in S_4. Building a map from $f : S_4 \to \{1, 2, 3, \dots, 10^9\}$ can be done using basic arithmetic. Let $a_1, a_2, \dots, a_9$ denote the entries of the 3×3 matrix, enumerated in any order. We define $f : S_4 \to \{1, 2, 3, \dots, 10^9\}$ by $f(A) = 1 + a_1 + a_2 \cdot 10 + \dots + a_8 \cdot 10^8 + a_9 \cdot 10^9$, which associates a unique base-10 counting number to each element of the set. (The "1" is added so that when f is applied to the matrix with all zero entries, it evaluates to a counting number, namely 1, instead of zero.)

Note: Suppose the matrices in S_4 had n-entries and their values were drawn from the set $\{0, 1, \dots, M-1\}$, where n and M are counting numbers, $M \geq 2$. Then there are M^n distinct matrices, and you can define $f(A) = 1 + \sum_{i=1}^{n} a_i M^{i-1}$, which associates a unique counting number to each element of the set, and injectivity (one-to-one-ness) is easy to prove if you think of the number as being in base M.

(e) **Ans.** S_5 is finitely countable, but why? For all of the other solutions, we constructed an explicit injective mapping from the set to the natural numbers or a subset of the natural numbers. To do that here, it appears that we would have to identify those matrices that result in a zero determinant and eliminate them, and then start assigning numbers to the remaining matrices. Is that the only option?

Well, no, we do not have to do that. We see easily that the set of all 2×2 matrices with entries from the set $\{1, 2, 3, 4, 5\}$ has $5^4 = 625$ distinct elements, which is finite. The set S_5 has fewer elements than that because the matrices with a zero determinant are excluded (e.g., the matrix with ones in all four entries is not invertible). A subset of a finite set is finite because it has at most the same number of elements as the "larger, or parent" set. We formalize this observation later in Prop. 2.45. Hence, S_5 **has at most 625 elements** and is therefore finitely countable.

(f) **Ans.** S_6 is countably infinite. The first natural number greater than π is four. Hence, $x \in S_6 \iff x = 10^k$, with $k \in \{4, 5, 6, \dots\}$. We can define $f : S_6 \to \mathbb{N}$ by:

$$f(x) = k,$$

where $x = 10^k$; in other words, $f(x) = \log_{10}(x)$. We can also define $f(10^k) = k - 3$, for all $k \geq 4$, so that f would also be onto (surjective), but as we discussed earlier, that is not a requirement.

■

Proposition 2.45: Subsets of Countable Sets

A subset of a countable set is countable. In particular, a subset of a finitely countable set is finitely countable, while a subset of a countably infinite set can be **either** finitely countable **or** countably infinite.

Proof: Suppose that $f : S \to \mathbb{N}$ is injective and that $A \subset S$. Then, because $a \in A \implies a \in S$, we have that $f(a) \in \mathbb{N}$. Moreover, if $a_1, a_2 \in A$ then $a_1, a_2 \in S$. Hence, if $a_1 \neq a_2$, the injectivity of $f : S \to \mathbb{N}$ yields $f(a_1) \neq f(a_2)$, so $f : A \to \mathbb{N}$ is injective whenever $f : S \to \mathbb{N}$ in injective. The rest of the proof is left as an exercise. ■

Remark 2.46. *When we **restrict** a function to a subset of its original domain, we are technically creating a new function. Why? Because a function is defined by its domain, codomain, and values. All three are important. When the distinction is not all that important, we often do not change the function's name, just as we used the notation f above when its domain was S and when its domain was $A \subset S$. When we want to call attention to the "new function's (new) domain", we use the notation $f|_A$, which is read "f restricted to A" or the "restriction of f to A". We can also write out the full function, $f : A \to \mathbb{N}$.*

All of this has seemed rather tame, so, again, what got Cantor into so much trouble with the math community? The notion of countability led Cantor to the idea that there is more than one kind of infinity, and the mathematicians of his day found that heretical. To consider some easy cases first, let us consider the following questions:

- How many counting numbers are there? (Hint: It is not a number.)
- How many even numbers are there? Let us denote the even numbers by $\mathbb{E}$.
- How many odd numbers are there? Let us denote the odd numbers by $\mathbb{O}$.
- Then, is the infinity of the even numbers half that of the counting numbers? And the same for the odd numbers?

No, they are all the same. We have:

- $f : \mathbb{N} \to \mathbb{N}$ by $f(k) = k$,
- $f : \mathbb{E} \to \mathbb{N}$ by $f(2k) = k$, for all $k \geq 1$,
- $f : \mathbb{O} \to \mathbb{N}$ by $f(2k-1) = k$, for all $k \geq 1$, and
- $\mathbb{E} \cup \mathbb{O} = \mathbb{N}$ and $\mathbb{E} \cap \mathbb{O} = \varnothing$.

Each set is actually in one-to-one correspondence with the natural numbers. The same is true for all multiples of three or five, for example. So can we say that $\infty + \infty = \infty$ and $\frac{\infty}{2} = \infty$? Yes, that's an acceptable interpretation of the above, so long as you recall that you are doing arithmetic on a concept and not a number. Now, what about a matrix with infinitely many rows and columns? Would that have ∞^2 elements? And is ∞^2, assuming we can make sense out of it, different from ∞? Cantor really did stir up a hornet's nest!

2.7.1 The Rational Numbers Are Countable

Because all counting numbers are rational numbers with denominators equal to one, there are at least as many rational numbers as there are counting numbers. But, it seems like there should be many more. Maybe infinitely more? Recall that we defined the rational numbers by:

$$\boxed{\mathbb{Q} := \left\{ \frac{p}{q} \mid p \in \mathbb{Z},\ q \in \mathbb{N},\ \text{no common factors other than one)} \right\}.} \tag{2.34}$$

Having to check that the gcd (greatest common divisor) of two integers equals one would complicate the counting process, so Cantor employed Prop. 2.45 (just as we did in Example 2.44, when dealing with invertible matrices with integer coefficients). Cantor said that if he could prove that the set of all ratios of counting numbers is countable, with $1/2, 2/4, \ldots, \frac{k}{2k}, \ldots$ **all counted as distinct elements,**[4] then the rational numbers must be countable as well. To demonstrate the validity of Cantor's position, we will focus on the positive rational numbers, because if we can show that they are countable, then all rational numbers are countable following the same reasoning applied to the even and odd numbers.

Consider

$$S := \left\{ \frac{i}{j} \mid i \in \mathbb{N},\ j \in \mathbb{N} \right\},$$

which we can envision as an infinite version of Table 2.2.

The first problem faced by Cantor was how to count these numbers. If we try to count row by row, we hit an immediate snag; the first row never terminates, so we cannot come back and start counting the elements in the second row. The same issue arises if we try to count by columns. So what is a young Cantor to do? He

[4]Mathematically, it is better to think of the "nonreduced" ratios p/q as pairs of numbers, (p, q). This avoids having to even think about $120/240$ as being distinct from $1/2$ because the pair of integers $(120, 240)$ is clearly distinct from the pair $(2, 4)$ and the pair $(1, 2)$, while any pair of integers (p, q) generates a rational number through division, p/q.

$i\backslash j$	1	2	3	4	5	6	7	8	9	10	$\cdots$
1	1/1	1/2	1/3	1/4	1/5	1/6	1/7	1/8	1/9	1/10	$\cdots$
2	2/1	2/2	2/3	2/4	2/5	2/6	2/7	2/8	2/9	2/10	$\cdots$
3	3/1	3/2	3/3	3/4	3/5	3/6	3/7	3/8	3/9	3/10	$\cdots$
4	4/1	4/2	4/3	4/4	4/5	4/6	4/7	4/8	4/9	4/10	$\cdots$
5	5/1	5/2	5/3	5/4	5/5	5/6	5/7	5/8	5/9	5/10	$\cdots$
6	6/1	6/2	6/3	6/4	6/5	6/6	6/7	6/8	6/9	6/10	$\cdots$
7	7/1	7/2	7/3	7/4	7/5	7/6	7/7	7/8	7/9	7/10	$\cdots$
8	8/1	8/2	8/3	8/4	8/5	8/6	8/7	8/8	8/9	8/10	$\cdots$
9	9/1	9/2	9/3	9/4	9/5	9/6	9/7	9/8	9/9	9/10	$\cdots$
10	10/1	10/2	10/3	10/4	10/5	10/6	10/7	10/8	10/9	10/10	$\cdots$
$\vdots$	$\vdots$	$\vdots$	$\vdots$	$\vdots$	$\vdots$	$\vdots$	$\vdots$	$\vdots$	$\vdots$	$\vdots$	$\ddots$

Table 2.2: The set of all positive ratios of natural numbers, now with each row having a constant numerator and the denominator increasing. We can regard it as a matrix with infinite rows and columns.

counts the elements using the reverse diagonals, as shown in Rationals are Countable by Cantor Snake by Existsforall Academy. We do something similar in Table 2.3. We can even write down an explicit function, called the *Cantor Pairing Function*:

$$f : S \to \mathbb{N}, \text{ by } f(i,j) := \frac{(i+j)(i+j-1)}{2} + j. \tag{2.35}$$

In Julia, we can easily verify that this formula allows us to associate a unique counting number to each pair (i,j) of counting numbers! Moreover, we can even play with the inverse of the Cantor Pairing Function. Here is a ChatGPT prompt that will produce the Julia code for you.

> Write two functions in Julia that handle Cantor's pairing function and its inverse, using 1-based indexing for counting numbers:
>
> 1. **Cantor Pairing Function:** Implement a function "cantor_pair(i, j)" that takes two positive integers $i \geq 1$ and $j \geq 1$ and returns the unique Cantor pairing number associated with the pair (i,j). Ensure that the function works with 1-based indexing, which means that i and j are both positive integers starting from 1.
> 2. **Inverse Cantor Function:** Implement a function "inverse_cantor(z)" that takes a Cantor pairing number $z \geq 1$ and returns the original pair (i,j) of positive integers. The function should correctly handle the inverse calculation using 1-based indexing.

Hence, the rational numbers are countable because they are a subset of the countable set consisting of all ratios of natural numbers, where every fraction occurs an infinite number of times! Indeed, if p/q is a fraction in reduced form, then $\frac{kp}{kq}$ shows up in the list of all ratios for all $k \geq 1$ (equivalently, the pair of numbers (kp, kq) shows up in Table 2.2).

2.7.2 The Real Numbers Are Uncountable

Cantor employed a proof by contradiction to show that the real numbers are not countable. If you read how Euclid proved that $\sqrt{2}$ is not rational, then, yes, Cantor employed a similar method. But, of course, there is a super cool twist, and it earned the moniker **Cantor's Diagonal Argument**.

$i\backslash j$	1	2	3	4	5	6	7	8	9	10
1	1	3	6	10	15	21	28	36	45	55
2	2	5	9	14	20	27	35	44	54	65
3	4	8	13	19	26	34	43	53	64	76
4	7	12	18	25	33	42	52	63	75	88
5	11	17	24	32	41	51	62	74	87	101
6	16	23	31	40	50	61	73	86	100	115
7	22	30	39	49	60	72	85	99	114	130
8	29	38	48	59	71	84	98	113	129	146
9	37	47	58	70	83	97	112	128	145	163
10	46	57	69	82	96	111	127	144	162	181

Table 2.3: Counting ratios via a "broken" Cantor Snake! We assign one to the first element of the first row. Then we move to the second row and assign two to the first element of the second row, and then, we work our way up the reverse diagonal, assigning three to the second element of the first row. We then go to the third row and repeat the process for the next reverse diagonal. This allows the entries of every row and column to be visited exactly once. **Note: we are not showing numbers outside of the** 10×10 **"footprint".** We can do more with Julia. For example, using the Cantor Pairing Function in (2.35), we can compute that $(i, j) = (11, 1) \to 56$.

Because every real number $y \in \mathbb{R}$ can be expressed as $y = k + x$, for $k \in \mathbb{Z}$ and $x \in [0, 1)$, it is sufficient to show that the interval $[0, 1)$ is uncountable. We recall that:

$$\boxed{x \in [0,1) \iff x = 0.d_1 d_2 d_3 \ldots,} \tag{2.36}$$

and the decimal expansion does not terminate in an infinite string of nines.

Armed with these facts, **we assume that the real numbers in** $[0, 1)$ **ARE COUNTABLE (i.e., listable)**. Hence, we can enumerate them as in Table 2.4; that is:

$$[0, 1) = \{x_1, x_2, x_3, \ldots\}.$$

We emphasize that if the real numbers between zero and one are indeed countable, then we can *completely list* (i.e., without missing a single one of them) all the real numbers in the interval $[0, 1)$. Cantor next shows that this list must be incomplete by producing a new real number in the interval $[0, 1)$ that is not in the list, contradicting the countability of $[0, 1)$, and hence, of $\mathbb{R}$.

Let $d_i[n]$ denote the i-th digit of the decimal expansion of the n-th real number, x_n. The digits $d_n[n]$ are on the diagonal of Table 2.4. Cantor uses these diagonal elements to construct a new real number that is nowhere to be found in the original list! The new number has the decimal expansion:

$$\sigma := \sum_{n=1}^{\infty} \sigma_n 10^{-n}, \text{ where } \sigma_n := \begin{cases} 0 & \text{if } d_n[n] \in \{1, 2, \ldots, 8\}, \\ 1 & \text{otherwise, i.e., } d_n[n] \text{ equals 0 or 9;} \end{cases}$$

in other words, he takes all the digits on the diagonal of his infinite matrix and constructs a new number that differs from the n-th number in its n-th digit. By construction, this new number[5] is nowhere in the original list, because it is different from the first number in the first digit, it is different from the second number in the second digit, etc. The existence of a number that is not in the list contradicts the list containing all real numbers in $[0, 1)$. Because the new number is built from the diagonal elements of the table, the method is called "**Cantor's Diagonal Argument**". It is pretty clever!

[5]Based on Table 2.4, the first ten digits of Cantor's new number are $\sigma = 0.000100000\ldots$

Secrets of the Arcane 2.47: Infinity Comes in Multiple Sizes!

By Cantor showing that real numbers are not countable, the world was informed that the "infinity" of the real numbers is **Strictly Greater Than** the "infinity" of the counting numbers. While we will stop here, Cantor went on to show that there is an infinite chain of "infinities", each one strictly more "numerous" than the previous one!

$d_i[n]\backslash x_n$	x_1	x_2	x_3	x_4	x_5	x_6	x_7	x_8	x_9	x_{10}	
	$\pi/10$	$\phi/10$	$\sqrt{2}/10$	$e/10$	$\sqrt{3}/10$	$\ln(2)/10$	$\sqrt{5}/10$	$\log_{10}(2)/10$	$\delta/10$	$\gamma/10$	
d_1	3	1	1	2	1	0	2	0	4	0	$\cdots$
d_2	1	6	4	7	7	6	2	3	6	5	$\cdots$
d_3	4	1	1	1	3	9	3	0	6	7	$\cdots$
d_4	1	8	4	8	2	3	6	1	9	7	$\cdots$
d_5	5	0	2	2	0	1	0	0	2	2	$\cdots$
d_6	9	3	1	8	5	4	6	2	0	1	$\cdots$
d_7	2	3	3	1	0	7	7	9	1	5	$\cdots$
d_8	6	9	5	8	8	1	9	9	6	6	$\cdots$
d_9	5	8	6	2	0	8	7	9	0	6	$\cdots$
d_{10}	3	8	2	8	7	0	7	5	9	4	$\cdots$
$\vdots$	$\vdots$	$\vdots$	$\vdots$	$\vdots$	$\vdots$	$\vdots$	$\vdots$	$\vdots$	$\vdots$	$\vdots$	$\ddots$

Table 2.4: A hypothesized listing of the real numbers between zero and one. Shown are the first ten digits of some interesting real numbers in the interval $[0,1)$, with the digits on the diagonal, $d_n[n]$, highlighted in red. You may have heard about the golden ratio, $\phi := (1+\sqrt{5})/2 \approx 1.618033988749895...$ You can also learn about δ and γ, the Feigenbaum constant and the Euler–Mascheroni constant, respectively.

While the notion of countable sets is not crucial for the remainder of the material in this book, the idea that mathematics needs to be done carefully is paramount.

2.8 How to Really Define Euler's Number (Optional Read)

We want to define:

$$e := \lim_{n\to\infty} \underbrace{\left(1+\frac{1}{n}\right)^n}_{a_n}, \tag{2.37}$$

but we do not have enough tools to make sense of the limit. Yet, the numbers $a_n := \left(1+\frac{1}{n}\right)^n$ that Bernoulli discovered through his study of compound interest have some very nice properties: for each integer $n \geq 0$:

1. a_n is rational,
2. $a_n \leq a_{n+1}$ (the terms are monotonically increasing), and
3. $a_n < 3$ (the terms are bounded from above).

The first property is true because $1+\frac{1}{n} = \frac{n+1}{n}$ is rational and $\left(1+\frac{1}{n}\right)^n$ is therefore the product of n rational numbers. The second and third properties were given in Prop. 1.74 and are proved at the end of the chapter.

Because the terms are monotonically increasing and bounded from above, they have to "bunch up." Formally, one can show that for all $\epsilon > 0$, there exists $N_\epsilon < \infty$ such that $|a_n - a_m| \le \epsilon$ for all $n, m \ge N_\epsilon$. If they are "bunching up," are they, in turn, "clustering around something" that has all the properties we want numbers to have (we can add and subtract them, multiply them, and even divide by them when they are nonzero)? The answer is YES. One of the constructions can be paraphrased as follows:

Secrets of the Arcane 2.48: What Is a Real Number Anyway?

Every real number $x \in \mathbb{R}$ can be written as $x = \lim\limits_{n\to\infty} a_n$, where for all $n \ge 1$,

1. a_n is rational (we are building new numbers from existing numbers),
2. $a_n \le a_{n+1}$ (the terms are monotonically increasing),
3. there is a finite integer K (independent of n) such that $a_n \le K$ (the terms are bounded from above).

Note: Hence, Euler's number, like all real numbers, is literally built from the ground up from the rational numbers. As long as we start the bisection algorithm with rational lower and upper bounds (bracketing points), all updates will have rational lower and upper bounds; hence, the construction of $\sqrt{2}$ can be achieved through two monotonic sequences of rational numbers, one that is monotonically increasing and the other decreasing. Archimedes' construction of π cannot be executed using rational numbers. However, it can be done using "algebraic numbers", such as $\frac{\sqrt{2}}{2}$, which are in turn built up from the rational numbers.

2.9 Proofs Associated with the Chapter (Optional Read)

Proof of Prop. 2.30 (Exponentials Grow Faster Than Any Monomial Function)

The following limits hold for real variables x and for integers n replacing x:

(a) For all $a > 1$ and integers $0 \le m < \infty$, $\lim\limits_{x\to\infty} \dfrac{a^x}{x^m} = \infty$.

(b) For all $a > 1$ and integers $0 \le m < \infty$, $\lim\limits_{x\to\infty} \dfrac{x^m}{a^x} = 0$.

(c) For all $a > 1$ and integers $0 \le m < \infty$, $\lim\limits_{x\to\infty} x^m \cdot a^x = \infty$.

We only prove the second statement because the first one is the reciprocal of the second, and the third one is clear because, for $x > 1$, $x^m \cdot a^x \ge a^x$, and hence, $\lim\limits_{x\to\infty} x^m \cdot a^x \ge \lim\limits_{x\to\infty} a^x = \infty$.

Because $a > 1$, we can write $a = 1 + b$ for some $b > 0$. Then, for all $k > m$, and all $(n+1) > x > n > 2k$, the

Binomial Theorem 1.67 and Prop. 1.64 give

$$
\begin{aligned}
0 &\le \frac{x^m}{a^x} \\
&= \frac{x^m}{(1+b)^x} \\
&\le \frac{x^m}{(1+b)^n} \quad \text{(because } n < x \text{ and } 1+b \ge 1\text{)} \\
&\le \frac{x^m}{\sum_{k=0}^{n} \binom{n}{k} b^k} \quad \text{(applying the Binomial Theorem 1.67)} \\
&< \frac{(n+1)^m}{\binom{n}{k} b^k} \quad \text{(keeping only one term, and everything dropped is strictly } > 0\text{)} \\
&\le \frac{x^m}{\frac{n^k}{2^k k!} \cdot b^k} \quad \text{(applying (1.39) from Prop. 1.64)} \\
&\le \frac{2^k k!}{b^k} \cdot \frac{x^m}{n^k} \quad \text{(grouping terms independent of } n \text{ and those dependent on } n\text{)} \\
&\le \frac{2^k k!}{b^k} \cdot \frac{(n+1)^m}{n^k} \quad \text{(because } (n+1) > x\text{)}.
\end{aligned}
\tag{2.38}
$$

Because $k > m$, by Prop. 2.23,

$$
\lim_{n \to \infty} \frac{(n+1)^m}{n^k} = 0,
$$

which completes the proof. ■

Sketch of the Proof of Prop. 2.10 (Power Sums)

For all $n \ge 1$ and $k \ge 1$, the following holds:

$$
\begin{aligned}
\sum_{i=1}^{n} i^k :=& 1^k + 2^k + 3^k + \cdots + n^k \\
=& \frac{n^{k+1}}{k+1} + \frac{n^k}{2} + \mathrm{O}(n,\ k-1),
\end{aligned}
\tag{2.39}
$$

where $\mathrm{O}(n,\ k-1)$ stands for a polynomial in the variable n with degree less than or equal to $k-1$ (i.e., all of the lower-order terms).

The proof is by induction on n with $k \ge 1$ a fixed counting number. We define:

$$
\begin{aligned}
L(n) &:= \sum_{i=1}^{n} i^k \\
&\text{and} \\
R(n) &:= \frac{n^{k+1}}{k+1} + \frac{n^k}{2}.
\end{aligned}
\tag{2.40}
$$

Our goal is to show that if for some $n \ge 1$, $L(n) = R(n) + \mathrm{O}(n,\ k-1)$, then it follows that

$$
L(n+1) = R(n+1) + \mathrm{O}(n+1, k-1). \tag{2.41}
$$

We first evaluate $L(n+1)$. By its definition and the induction hypothesis,

$$\begin{aligned} L(n+1) =& L(n) + (n+1)^k \quad \text{(by the definition of } L(n)\text{)} \\ =& R(n) + O(n,\ k-1) + (n+1)^k \quad \text{(because we assume } L(n) = R(n) + O(n,\ k-1)\text{)} \\ =& \frac{n^{k+1}}{k+1} + \frac{n^k}{2} + O(n,\ k-1) + (n+1)^k \quad \text{(using the definition of } R(n)\text{)} \\ =& \frac{n^{k+1}}{k+1} + \frac{n^k}{2} + O(n,\ k-1) + n^k + kn^{k-1} + O(n,\ k-2) \quad \text{(applying the Binomial Theorem)} \\ =& \frac{n^{k+1}}{k+1} + \frac{3n^k}{2} + O(n,\ k-1), \end{aligned}$$

where we noted that $kn^{k-1} + O(n,\ k-1) + O(n,\ k-2) = O(n,\ k-1)$.

Next, we evaluate $R(n+1)$. Using its definition and expanding out terms,

$$\begin{aligned} R(n+1) :=& \frac{(n+1)^{k+1}}{k+1} + \frac{(n+1)^k}{2} \\ =& \frac{n^{k+1} + (k+1)n^k + O(n, k-1)}{k+1} + \frac{n^k + kn^{k-1} + O(n,\ k-2)}{2} \\ & \text{(applying the Binomial Theorem twice)} \\ =& \frac{n^{k+1}}{k+1} + \frac{3n^k}{2} + O(n,\ k-1), \end{aligned}$$

where we noted that

$$\frac{O(n, k-1)}{k+1} + \frac{kn^{k-1} + O(n,\ k-2)}{2} = O(n,\ k-1).$$

Because[6] $O(n+1, k-1) = O(n,\ k-1)$, we have shown that (2.41) holds, and the proof is complete. ■

Proof of Prop. 2.36 (Limits of Linear Combinations)

Suppose that for all $1 \le i \le n$, $f_i : (0, \infty) \to \mathbb{R}$ has a finite limit at infinity, that is, $\lim_{x\to\infty} f_i(x)$ exists and is finite. Then, for all coefficients $\alpha_i \in \mathbb{R}$,

$$\lim_{x\to\infty} \left(\sum_{i=1}^{n} \alpha_i f_i(x) \right) = \sum_{i=1}^{n} \left(\alpha_i \lim_{x\to\infty} f_i(x) \right); \tag{2.42}$$

the limit of the linear combination exists and equals the linear combination of the limits.

Proof: We can assume without loss of generality that for all $1 \le i \le n$, $|\alpha_i| > 0$, because otherwise, the corresponding term(s) can be dropped from the sum. Let $\epsilon > 0$ be given and define

$$\epsilon_i := \frac{\epsilon}{|\alpha_i| n} > 0, \qquad L_i := \lim_{x\to\infty} f_i(x).$$

Then, because the individual limits exist, there exists N_i finite such that, for all $x \ge N_i$, $|f_i(x) - L_i| \le \epsilon$.

[6]The Binomial Theorem implies that $(n+1)^{k-1}$ is a polynomial in the variable n with highest power $k-1$. For the proof to be firm, we need to note that the terms from the Binomial Theorem that are being added to $O(n,\ k-1)$ have coefficients that depend on k, which stays fixed, and not n, which can get arbitrarily large. In fact, this follows from

$$\binom{k-1}{j} := \frac{(k-1)!}{j!(k-1-j)!}, \quad 0 \le j \le k-1.$$

Next, for $N := \max\{N_1, \ldots, N_n\}$ and $x \geq N$, it follows that:

$$\begin{aligned} \left|\sum_{i=1}^{n} \alpha_i f_i(x) - \sum_{i=1}^{n} \alpha_i L_i\right| &= \left|\sum_{i=1}^{n} \alpha_i \left(f_i(x) - L_i\right)\right| \\ &\leq \sum_{i=1}^{n} |\alpha_i| \, |f_i(x) - L_i| \\ &\leq \sum_{i=1}^{n} |\alpha_i| \, \epsilon_i \\ &\leq \sum_{i=1}^{n} |\alpha_i| \frac{\epsilon}{|\alpha_i| n} \\ &\leq \epsilon, \end{aligned}$$

which proves the result. ■

Sketch of Proof of Prop. 2.37 (Limits of Products and Ratios)

Suppose that $g : (0, \infty) \to \mathbb{R}$ has a limit at infinity of one, i.e., $\lim_{x\to\infty} g(x) = 1.0$. Then for any function $f : (0, \infty) \to \mathbb{R}$,

$$\begin{aligned} \lim_{x\to\infty} f(x) \cdot g(x) &= \lim_{x\to\infty} f(x), \text{ and} \\ \lim_{x\to\infty} \frac{f(x)}{g(x)} &= \lim_{x\to\infty} f(x). \end{aligned}$$

Proof: We will prove the case when both limits exist and are finite, leaving the other cases to be proven by the reader. We define $L := \lim_{x\to\infty} f(x)$ and $M := \lim_{x\to\infty} g(x)$. Then, the key inequality is

$$\begin{aligned} |f(x) \cdot g(x) - L \cdot M| &= |(f(x) - L) \cdot (g(x) - L) + L\,(g(x) - M)) + M\,(f(x) - L)| \\ &\leq |(f(x) - L) \cdot (g(x) - L)| + |L| \cdot |g(x) - M| + |M| \cdot |f(x) - L| \\ &\leq |f(x) - L| \cdot |g(x) - L| + |L| \cdot |g(x) - M| + |M| \cdot |f(x) - L|. \end{aligned}$$

Let $\epsilon > 0$ and choose $N_\epsilon < \infty$ such that, for all $x \geq N_\epsilon$

$$\begin{aligned} |f(x) - L| &\leq \sqrt{\frac{\epsilon}{3}}, \\ |g(x) - M| &\leq \sqrt{\frac{\epsilon}{3}}, \\ |M| \cdot |f(x) - L| &\leq \frac{\epsilon}{3}, \\ |L| \cdot |g(x) - M| &\leq \frac{\epsilon}{3}. \end{aligned}$$

This is possible because of the assumptions on the limits of f and g existing and being finite. Then, by the key inequality, for all $x \geq N_\epsilon$

$$\begin{aligned} |f(x) \cdot g(x) - L \cdot M| &\leq |f(x) - L| \cdot |g(x) - L| + |L| \cdot |g(x) - M| + |M| \cdot |f(x) - L| \\ &\leq \frac{\epsilon}{3} + \frac{\epsilon}{3} + \frac{\epsilon}{3} \\ &\leq \epsilon, \end{aligned}$$

proving the first result.

To prove the second result, we must assume $|M| > 0$. The key inequality is then

$$\begin{aligned}\left|\frac{f(x)}{g(x)} - \frac{L}{M}\right| &= \left|\frac{M\,f(x) - L\,g(x)}{M\,g(x)}\right| \\ &= \left|\frac{M\,(f(x) - L) + ML - L\,(g(x) - M) - ML}{M\,g(x)}\right| \\ &= \left|\frac{M\,(f(x) - L) - L\,(g(x) - M)}{M\,g(x)}\right| \\ &\leq \left|\frac{f(x) - L}{g(x)}\right| + \left|\frac{L}{M}\right| \left|\frac{g(x) - M}{g(x)}\right|.\end{aligned}$$

Let $0 < \epsilon < \left|\frac{M}{2}\right|$ and choose $N_\epsilon < \infty$ such that, for all $x \geq N_\epsilon$

$$|f(x) - L| \leq \frac{\epsilon|M|}{4}$$

$$\left|\frac{L}{M}\right| |g(x) - M| \leq \frac{\epsilon|M|}{4}.$$

Then, from the key inequality for ratios,

$$\begin{aligned}\left|\frac{f(x)}{g(x)} - \frac{L}{M}\right| &\leq \left|\frac{f(x) - L}{g(x)}\right| + \left|\frac{L}{M}\right| \left|\frac{g(x) - M}{g(x)}\right| \\ &\leq \left|\frac{f(x) - L}{M/2}\right| + \left|\frac{L}{M}\right| \left|\frac{g(x) - M}{M/2}\right| \\ &\leq \frac{\epsilon}{2} + \frac{\epsilon}{2} \\ &\leq \epsilon,\end{aligned}$$

proving the result. ■

Analytical Background for Secrets of the Arcane 2.48: Why Euler's Number Exists

Suppose that x_k is a monotonically increasing sequence of real numbers that is bounded from above. Then, for every $\epsilon > 0$, there exists $N_\epsilon < \infty$ such that $m, n \geq N_\epsilon \implies |x_m - x_n| \leq \epsilon$.

In plain language, the terms at the tail of the sequence cluster arbitrarily closely to one another. The terms in the "tail" are defined by $m, n \geq N_\epsilon$, and "cluster together" is defined by $|x_m - x_n| \leq \epsilon$.

Proof: We prove the contrapositive.[7] Assume that $\{x_k\}$ is a monotonically increasing sequence of real numbers and that $\{x_k\}$ is not Cauchy. Because the sequence is monotonically increasing, "not Cauchy" is equivalent to the following statement: there exists an $\epsilon > 0$ such that for every $n \in \mathbb{N}$ there exists an $m > n$ with

$$x_m > x_n + \epsilon.$$

We show that $\{x_k\}$ is unbounded.

[7] The logical proposition that $(A \implies B) \iff (\neg B \implies \neg A)$.

Assumption 1: $\{x_k\}$ is monotonically increasing, that is, for all $k \in \mathbb{N}$,

$$x_{k+1} \geq x_k.$$

Assumption 2: There exists an $\epsilon > 0$ such that for all $n \in \mathbb{N}$ there exists an $m > n$ satisfying

$$x_m > x_n + \epsilon.$$

Claim (proved by induction): For every $n \in \mathbb{N}$, there exists an index $k_n \in \mathbb{N}$ such that

$$x_{k_n} > x_1 + n\epsilon.$$

Base Case: For $n = 1$, Assumption 2 with $n = 1$ provides an index $k_1 > 1$ such that

$$x_{k_1} > x_1 + \epsilon.$$

Inductive Step: Assume that for some $n \in \mathbb{N}$ there exists an index k_n such that

$$x_{k_n} > x_1 + n\epsilon.$$

Apply Assumption 2 with $n := k_n$. Then there exists an index $k_{n+1} > k_n$ such that

$$x_{k_{n+1}} > x_{k_n} + \epsilon.$$

Combining the two inequalities gives

$$x_{k_{n+1}} > x_{k_n} + \epsilon > (x_1 + n\epsilon) + \epsilon = x_1 + (n+1)\epsilon,$$

which completes the inductive step.

Therefore, for every $n \in \mathbb{N}$ there exists an index k_n such that $x_{k_n} > x_1 + n\epsilon$. Hence $\{x_k\}$ is unbounded: given any proposed upper bound B, choose $n \in \mathbb{N}$ such that $x_1 + n\epsilon > B$; then $x_{k_n} > B$. ■

Hence, if a monotonically increasing sequence is bounded from above, it must be Cauchy, and therefore its terms "cluster together" in the sense that for every $\epsilon > 0$ there exists N_ϵ such that

$$m, n \geq N_\epsilon \implies |x_m - x_n| \leq \epsilon.$$

Chapter 3

Definite Integration as the Signed Area Under a Curve

Learning Objectives

By the end of this chapter, the student should be able to:

- Define and explain the concept of a definite integral within the context of calculus.
- Execute accurate computations of definite integrals using appropriate mathematical techniques.
- Describe and apply the basic properties of definite integrals to solve problems.
- Recognize and demonstrate the applications of definite integrals in various engineering scenarios.

Outcomes

Upon successful completion of this chapter, students will be able to:

- Construct and calculate Riemann lower and upper sums to approximate definite integrals.
- Acquire insight into what kinds of functions can be integrated.
- Employ numerical algorithms used in engineering practice to compute definite integrals.
- Utilize integration techniques to infer changes in position from a given velocity function, particularly in robotic applications.
- Identify the mathematical origins of parabolic trajectories in ballistic motion.
- Determine the area enclosed between two functions.
- Compute essential parameters for robotic models, such as total mass and center of mass.
- Learn how to "trick" a single-variable integral into computing the volume of a solid of revolution.

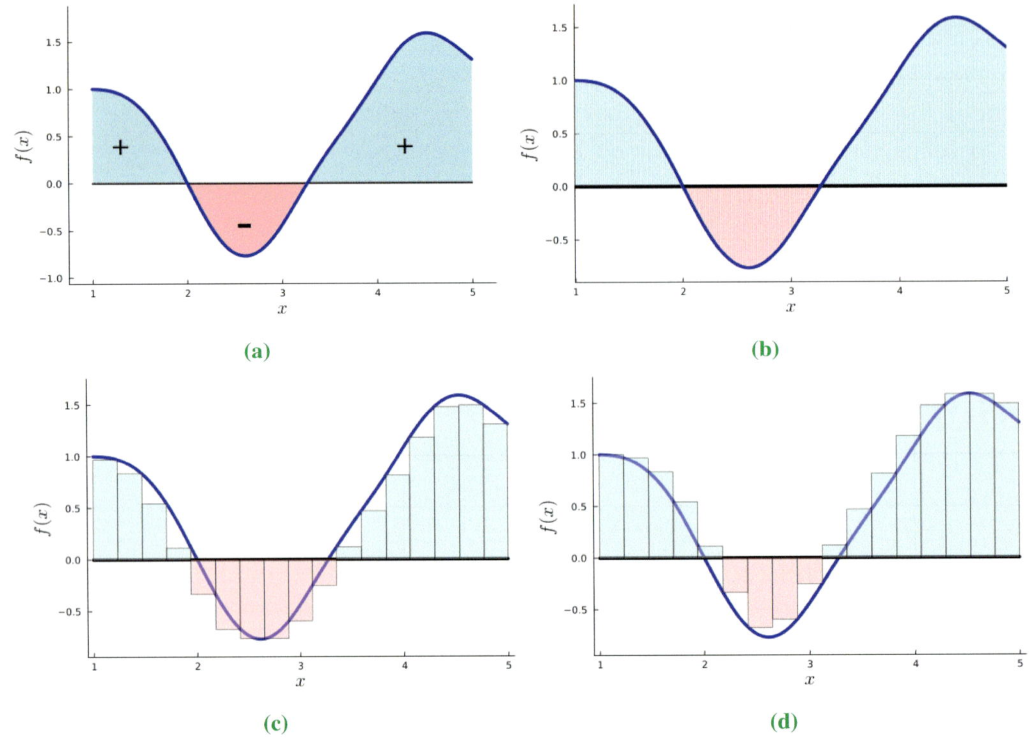

Figure 3.1: The Riemann Integral computes the area under a (decently well-behaved[1]) curve. (a) Shows that where the function is positive, the area is assigned a positive value, and where the function is negative, it is assigned a negative value. The Riemann Integral is the sum of the positive and negative areas. (c) Shows an underapproximation via rectangles, (d) shows an overapproximation, while in (b), we have used so many rectangles that you cannot tell if it is an under- or over-approximation.

The Riemann Integral is the result of applying the Approximation Principle to compute the "signed area under a curve", as illustrated in **Fig. 3.1**. The term "signed area" is used to highlight that we assign a positive number to the area "under" positive values of the function and a negative number to the area "above" negative values of the function. Moving forward, we will use the generic terminology "AREA UNDER the CURVE" to cover both situations. The process of approximating the area should look familiar to us because we carried out a similar process in Section 2.3.2, for monotonically increasing functions, and we only dealt with "positive area."

3.1 The Riemann Integral (aka Riemann-Darboux Integral)

In this chapter, we will address the more general situation of functions that are non-monotonic (aka, "wiggle around") and take on both positive and negative values. For us to even define what we mean by the area under the curve, we will have to impose some conditions on the functions we consider. We will not address the "most general conditions known to humankind", but we will build toward a general enough class of functions for "most engineering applications". The fundamental idea remains the same as we used previously for monotonically increasing functions in Section 2.3.2: we sandwich the area we wish to compute between an underapproximation of the area and an overapproximation of the area. We then seek to analyze when we can make the under- and over-approximations converge to one another in the limit as we take finer and finer rectangular slices. The limiting value is defined to be the area under the curve, that is, the Riemann Integral.

3.1.1 Simple Version of Riemann Lower and Upper Sums

As in Section 2.3.2 and illustrated in **Fig. 3.4** and **Fig. 3.2**, for a function $f : [a,b] \to \mathbb{R}$, where $a < b$, we divide the interval $[a,b]$ into $n > 1$ **evenly spaced subintervals**, $[x_i, x_{i+1}]$, where

$$\begin{aligned} \Delta x &:= \frac{b-a}{n}, \\ x_i &:= a + (i-1)\cdot \Delta x, \end{aligned} \tag{3.1}$$

so that

$$a = x_1 < x_2 < \cdots < x_n < x_{n+1} = b, \tag{3.2}$$

$$x_{i+1} = x_i + \Delta x, \; 1 \le i \le n. \tag{3.3}$$

We still need to define:

$$\boxed{\text{Area}_n^{\text{Low}} := \sum_{i=1}^{n} h_i^{\text{Low}} \cdot b_i \quad \text{and} \quad \text{Area}_n^{\text{Up}} := \sum_{i=1}^{n} h_i^{\text{Up}} \cdot b_i,} \tag{3.4}$$

where $b_i = \Delta x$ is the base of the rectangle and h_i^{Low} and h_i^{Up} are the heights. To form the "best" underapproximation we can, we want to assign to h_i^{Low} the **maximum** (aka, largest) value we can so that the rectangle fits under the function, and similarly, to form the "best" overapproximation we can, we want to assign to h_i^{Up} the **minimum** (aka, smallest) value we can so that the rectangle fits over the function.

For monotonically increasing continuous functions, this is fairly trivial because we can set $h_i^{\text{Low}} = f(x_i)$, the function evaluated at the lower end of the interval, because $f(x_i) \le f(x)$ for all $x \in [x_i, x_{i+1}]$, and we can set $h_i^{\text{Up}} = f(x_{i+1})$, the function evaluated at the upper end of the interval, because $f(x_{i+1}) \ge f(x)$ for all $x \in [x_i, x_{i+1}]$. In symbols, for monotonically increasing continuous functions, we used

$$\boxed{h_i^{\text{Low}} := \min_{x \in [x_i, x_{i+1}]} f(x) = f(x_i) \quad \text{and} \quad h_i^{\text{Up}} := \max_{x \in [x_i, x_{i+1}]} f(x) = f(x_{i+1}).} \tag{3.5}$$

When the function wiggles up and down, we have to search for (or compute) the minimum and maximum values of $f : [x_i, x_{i+1}] \to \mathbb{R}$ instead of just using the function values at the endpoints. But with this one exception, the process remains the same.

Riemann Lower and Upper Sums

Definition 3.1. *(Riemann (Approximating) Sums) For $a < b$ (finite) real numbers, suppose that $f : [a,b] \to \mathbb{R}$ is continuous. For $n \ge 1$, define $\Delta x := \frac{b-a}{n}$ and $x_i := a + (i-1)\cdot \Delta x$, for $1 \le i \le n+1$. Then,*

(a) $\text{Area}_n^{\text{Low}} := \sum_{i=1}^{n} h_i^{\text{Low}} \cdot \Delta x$, *for* $h_i^{\text{Low}} := \min_{x \in [x_i, x_{i+1}]} f(x)$ *is a **Riemann lower sum**, and*

(b) $\text{Area}_n^{\text{Up}} := \sum_{i=1}^{n} h_i^{\text{Up}} \cdot \Delta x$, *for* $h_i^{\text{Up}} := \max_{x \in [x_i, x_{i+1}]} f(x)$ *is a **Riemann upper sum**.*

Notes: Riemann's original formulation was more complicated and more general than the above. We'll get to a more general formulation in due time. The simplified formulation we have followed mirrors the formulation of Darboux, though, as a Calculus novice, you "could (and should) care less!"

The division of $[a,b]$ as $a = x_1 < x_2 < \cdots < x_n < x_{n+1} = b$ is called a **partition**. We have been using a **uniform partition** where $x_{i+1} - x_i = \Delta x$, for all $1 \le i \le n$. Darboux allowed for non-uniform partitions by requiring that

$$\max_{1 \le i \le n} (x_{i+1} - x_i) \underset{n \to \infty}{\longrightarrow} 0.$$

This is immediate for a uniform partition because $\max_{1 \le i \le n}(x_{i+1} - x_i) = \Delta x = \frac{b-a}{n} \underset{n\to\infty}{\longrightarrow} 0$. Some theorems about Riemann Integrals only work when non-uniform partitions are allowed.

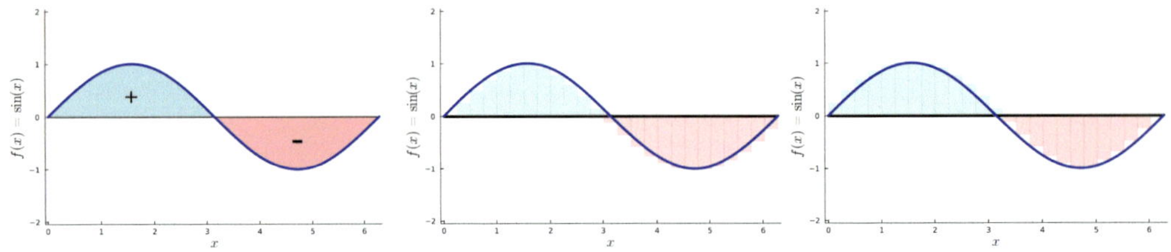

(a) (b) (c)

Figure 3.2: An illustration of Riemann Lower and Upper Sums for $\sin : [0, 2\pi] \to \mathbb{R}$. (a) Shows that the area we are approximating should equal to zero, because the area in pink appears to be the negative of the area in blue. A lower sum is shown in (b) and an upper sum is shown in (c). The number of intervals is $n = 25$.

Below is a Julia implementation to estimate the Riemann Lower and Upper Sums. They are estimations because we are NOT computing the exact minima and maxima of the function $\sin[x_i, x_{i+1}] \to \mathbb{R}$. Instead, we introduce `nRefine` additional points in between $[x_i, x_{i+1}]$, evaluate the function at the array of points

```
xRefine = range(x[i], x[i+1], length=ceil(Int64,nRefine)),
```

and then apply Julia's `minimum` and `maximum` functions to the vector `f(xRefine)`, which contains the function's values at the two endpoints, x_i and x_{i+1}, plus any additional points specified by `nRefine` ≥ 2.

We should review (1.2) for how the estimate and the error bound are defined.

```
function riemannLowerUpperSums(f, a, b, N=100, nRefine=4, aTol=1e-6)

    if abs(a-b) < aTol
        println("abs(a-b) < $aTol. Estimated integral is zero.")
        return 0.0
    end

    if nRefine < 2
        nRefine = 2
        println("nRefine was set to its minimum value of $nRefine")
    end

    # make sure N is an integer greater or equal to 2
    N = max(ceil(Int64, N), 2)

    # Define Deltax
    Deltax = (b-a)/N

    # Both sums start at zero
    LowerSum = 0.0
    UpperSum = 0.0

    # Initialize x_i
    xi = a

    for i = 1:N

        # x_{i+1}
        xip1 = xi + Deltax

```

```
        # Add extra points so that we can estimate
        # the max and min on [x_i, x_{i+1}]
        xRefine = range(xi, xip1,
            length = ceil(Int64, nRefine))
        fRefined = f.(xRefine)

        fMax = maximum(fRefined)
        fMin = minimum(fRefined)

        # Now compute upper and lower sums
        UpperSum = UpperSum + fMax*Deltax
        LowerSum = LowerSum + fMin*Deltax

        # Update x_i
        xi = xip1
    end

    # @show b-xi # for debugging

    estIntegral = (UpperSum + LowerSum)/2.0
    pmError = (UpperSum - LowerSum)/2

    return (
        LowerSum = LowerSum,
        UpperSum = UpperSum,
        estIntegral = estIntegral,
        pmError = pmError
    )

end
```

Output

```
riemannLowerUpperSums (generic function with 4 methods)
```

Two illustrations follow.

```
f(x) = sin(x)
a = 0.0
b = 2*pi
N=25
F=riemannLowerUpperSums(f, a, b, N)
@show F.LowerSum
@show F.UpperSum;
```

Output

```
F.LowerSum = -0.5021037670817627
F.UpperSum = 0.502103767081769
```

Note the lower and upper Riemann sums are not close to one another.

```
f(x) = sin(x)
a = 0.0
```

```
b = 2*pi
N=250
F=riemannLowerUpperSums(f, a, b, N)
@show F.LowerSum
@show F.UpperSum;
```

Output

```
F.LowerSum = -0.05026327759307434
F.UpperSum = 0.05026327759309964
```

The lower and upper Riemann sums are now closer to one another. With even more rectangles, $N = 2{,}500$, we obtain `F.LowerSum ≈ -0.00503` and `F.UpperSum ≈ 0.00503`.

3.1.2 Riemann Integral of a Continuous Function over a Bounded Interval

Definition 3.2. *Riemann Integral for a Continuous Function over a (Bounded) Interval* $[a,b]$, $a < b$: *If* $\lim\limits_{n\to\infty} \text{Area}_n^{\text{Low}}$ *and* $\lim\limits_{n\to\infty} \text{Area}_n^{\text{Up}}$ *both exist, are finite, and equal one another, then the* ***Riemann Integral*** *of* $f:[a,b]\to\mathbb{R}$ *exists and is equal to the common limit. In other symbols,*

$$\int_a^b f(x)\,dx := \begin{cases} \lim_{n\to\infty} \text{Area}_n^{\text{Low}} \\ \lim_{n\to\infty} \text{Area}_n^{\text{Up}} \end{cases} \quad \textit{provided both limits exist, are finite, and are equal.} \tag{3.6}$$

Notes:

- The symbol $\int$ is an old-fashioned "lazy S" and comes from the fact that an integral is a sum.
- The symbol $\int_a^b$ stands for "the sum from a to b" and is read "**the integral from a to b**".
- a is called the **lower limit** and b the **upper limit**. Together, they are called **limits of integration**.
- The function being integrated is called the **integrand**.
- The Riemann integral exists for bounded continuous functions on bounded intervals.
- The bounded intervals can be open (a,b) "with endpoints excluded", semi-open $(a,b]$ or $[a,b)$ "with one endpoint excluded", or closed $[a,b]$ "with both endpoints included". In all cases, $\int_a^b f(x)\,dx$ gives the same value.
- More general functions admitting a Riemann Integral are discussed in Section 4.7 "When does a Function have a Definite Riemann Integral?".

```
f(x) = sin(x)
a = 0.0
b = 2*pi
N=250
F=riemannLowerUpperSums(f, a, b, N)
```

Output

```
(LowerSum = -0.05026327759307434, UpperSum = 0.05026327759309964,
estIntegral = 1.2649603586822877e-14, pmError = 0.05026327759308699)
```

```
f(x) = sin(x)
a = 0.0
b = 2*pi
N=2500
F=riemannLowerUpperSums(f, a, b, N)
```

Output

```
(LowerSum = -0.005026548246017541, UpperSum = 0.005026548245468393 5,
estIntegral = -2.74573766501085e-13, pmError = 0.005026548245742967)
```

In case you did not review (1.2) for how the estimate and error bound are defined, they are defined by

- estIntegral = (upperSum + lowerSum)/2,
- pmError = (upperSum − lowerSum)/2.

Even though the estimated integral `estIntegral` may appear to be amazingly good, we must examine the error bound `pmError` to understand how much confidence to attach to the answer. The error bound tells us that with 250 rectangles, `pmError` $\approx \pm 0.05$, which is not so great, while with 2,500 rectangles, the error bound is approximately ± 0.005.

It is straightforward to numerically integrate complicated functions that we are hesitant to integrate by hand.

```
f(x) = cos(exp(-x)*sin(x)/(1 + x + 5x^2)) + log2( (1 + x)/(cos(x) + 5x^2) )
a = 0.0
b = 10
N = 5000
F = riemannLowerUpperSums(f, a, b, N)
```

Output

```
(LowerSum = -28.837938067489223, UpperSum = -28.826646469106663,
estIntegral = -28.83229226829794, pmError = 0.005645799191279721)
```

$$\int_0^{10} \cos\left(\frac{\exp(-x)\sin(x)}{1+x+5x^2}\right) + \log_2\left(\frac{1+x}{\cos(x)+5x^2}\right) dx \approx -28.832292 \pm 0.005646. \quad (3.7)$$

3.1.3 Illustration of the Riemann Indefinite Integral of a Monomial

If we replace one of the limits of integration with a variable, say x, which means we are integrating over $[a, x]$ or $[x, b]$, the resulting integral produces a function of x. Because x is not a "definite real number", the integral is called an **indefinite integral**. In other words, if $f : [a, b] \to \mathbb{R}$ is Riemann integrable, then $g : [a, b] \to \mathbb{R}$, defined by

$$g(x) := \int_a^x f(y)\, dy,$$

is called an **indefinite integral**.

Proposition 3.3: Integrating x^k

For all counting numbers k and for all $x > 0$ finite,

$$\int_0^x y^k\,dy = \frac{x^{k+1}}{k+1}. \tag{3.8}$$

Note: So far, we only know how to integrate functions that can be built from monomials. In Prop. 4.24, we will have enough tools to show that $\int_0^x e^y\,dy = e^x - 1$.

Proof: We will work this one out in full detail so as to further ingrain the **Approximation Principle** that underlies most of Calculus. We let $n > 1$ be an integer and define $\Delta y := \frac{x-0}{n}$ along with $y_i := a + (i-1)\Delta y$, yielding

$$a = y_1 < y_2 < \cdots < y_n < y_{n+1} = b.$$

Because $f(y) = y^k$ is increasing, $f(y_i) \le f(y) \le f(y_{i+1})$ for $y \in [y_i, y_{i+1}]$. This makes computing the min and max of the function over subintervals very easy.

Riemann Lower Sum: We underestimate the total area between $[0, x]$ by summing up the underapproximations, $f(y_i) \cdot \Delta y$, for each interval $[y_i, y_{i+1}]$,

$$\begin{aligned}
\text{Area}_n^{\text{Low}} &:= \sum_{i=1}^{n} f(y_i)\cdot\Delta y \\
&= \sum_{i=1}^{n} [(i-1)\Delta y]^k \cdot \Delta y \quad \text{(substituting in terms)} \\
&= \sum_{i=1}^{n} (i-1)^k \cdot (\Delta y)^k \cdot \Delta y \quad \text{(algebra)} \\
&= (\Delta y)^{k+1} \cdot \sum_{i=1}^{n} (i-1)^k \quad \text{(take terms that do not depend on } i \text{ outside the sum)} \\
&= \left(\frac{x}{n}\right)^{k+1} \cdot \sum_{i=1}^{n-1} i^k \quad \text{(change of index and algebra)} \\
&= \frac{x^{k+1}}{n^{k+1}} \cdot \left(\frac{(n-1)^{k+1}}{k+1} + \frac{(n-1)^k}{2} + \mathrm{O}(n-1, k-1)\right) \quad \text{(Prop. 2.10).}
\end{aligned} \tag{3.9}$$

From our results in Section 2.5.1, we have that

$$\begin{aligned}
\lim_{n\to\infty} \frac{x^{k+1}}{n^{k+1}} \cdot \left(\frac{(n-1)^{k+1}}{k+1} + \frac{(n-1)^k}{2} + \mathrm{O}(n-1, k-1)\right) &= x^{k+1} \cdot \lim_{n\to\infty} \left(\frac{(n-1)^{k+1}}{(k+1)\cdot n^{k+1}} + \frac{(n-1)^k}{2\cdot n^{k+1}} + \right. \\
&\qquad \left. + \frac{\mathrm{O}(n-1, k-1)}{n^{k+1}}\right) \\
&= \frac{x^{k+1}}{k+1}.
\end{aligned} \tag{3.10}$$

We next repeat this process for the upper sum.

Riemann Upper Sum: We overestimate the total area between $[0, x]$ by summing up the overapproximations, $f(y_{i+1}) \cdot \Delta y$, for each interval $[y_i, y_{i+1}]$,

$$
\begin{aligned}
\text{Area}_n^{\text{Up}} :&= \sum_{i=1}^{n} f(y_{i+1}) \cdot \Delta y \\
&= \sum_{i=1}^{n} [i\Delta y]^k \cdot \Delta y \;\; \text{(substituting in terms)} \\
&= \sum_{i=1}^{n} i^k \cdot (\Delta y)^k \cdot \Delta y \;\; \text{(algebra)} \\
&= (\Delta y)^{k+1} \cdot \sum_{i=1}^{n} i^k \;\; \text{(take terms that do not depend on } i \text{ outside the sum)} \\
&= \left(\frac{x}{n}\right)^{k+1} \cdot \sum_{i=1}^{n} i^k \;\; \text{(algebra)} \\
&= \frac{x^{k+1}}{n^{k+1}} \cdot \left(\frac{n^{k+1}}{k+1} + \frac{n^k}{2} + \mathrm{O}(n, k-1) \right) \;\; \text{(Prop. 2.10)}.
\end{aligned}
\tag{3.11}
$$

From our results in Section 2.5.1, we have that

$$
\lim_{n\to\infty} \frac{x^{k+1}}{n^{k+1}} \cdot \left(\frac{n^{k+1}}{k+1} + \frac{n^k}{2} + \mathrm{O}(n, k-1) \right) = \frac{x^{k+1}}{k+1}. \tag{3.12}
$$

Because the limits of the Upper and Lower Riemann Sums both exist and are equal, by Def. 3.2, the Riemann Integral of y^k over the interval $[0, x]$ is defined and equals $\frac{x^{k+1}}{k+1}$; that is:

$$
\int_0^x y^k \, dy = \frac{x^{k+1}}{k+1}. \tag{3.13}
$$

■

3.1.4 Are All Functions Riemann Integrable?

The answer is NO! Consider this discontinuous function, $f : [0, 1] \to \mathbb{R}$ by

$$
f(x) := \begin{cases} 1 & x \text{ is rational} \\ 0 & x \text{ is irrational.} \end{cases} \tag{3.14}
$$

Then, for any interval $0 \le x_i < x_{i+1} \le 1$,

$$
\min_{x\in[x_i, x_{i+1}]} f(x) = 0.0 \;\text{ and }\; \max_{x\in[x_i, x_{i+1}]} f(x) = 1.0 \tag{3.15}
$$

It follows that for all n, the Riemann Lower Sum, $\text{Area}_n^{\text{Low}} = 0.0$ (is identically equal to zero), while for all n, the Riemann Upper Sum, $\text{Area}_n^{\text{Up}} = 1.0$. (is identically one). Hence, even though both limits exist, they are not equal. The function is, therefore, not Riemann Integrable. In the next chapter, we will establish a better understanding of what types of functions have Riemann Integrals.

3.2 A Few Properties of the Riemann Integral

Here are a few properties that will prove useful when evaluating Riemann Integrals.

3.2.1 A Basic Additivity Property of Area Under a Curve

Figure 3.3 illustrates a general additivity property of area under a curve: if the interval is broken up into disjoint pieces, then the individual signed areas can be added up to obtain the total signed area. This translates into a useful property of the Riemann Integral.

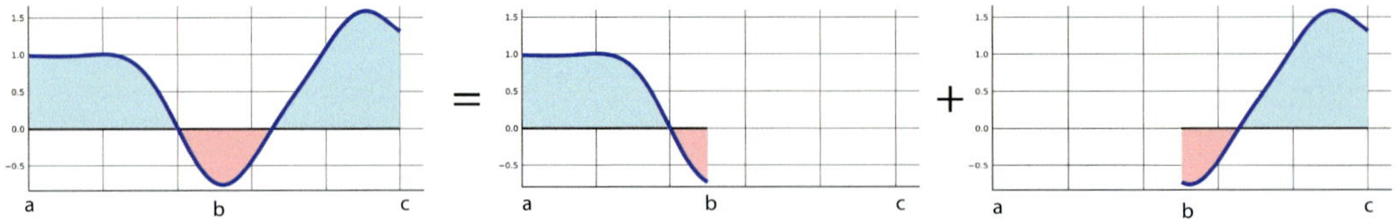

Figure 3.3: The signed area under the curve from $x = a$ to $x = c$ is equal to the signed area from $x = a$ to $x = b$ plus the signed area from $x = b$ to $x = c$. Hence, when the Riemann Integral exists, it can be computed as $\int_a^c f(x)\,dx = \int_a^b f(x)\,dx + \int_b^c f(x)\,dx$.

Proposition 3.4: First Additivity Property of the Riemann Integral

Suppose that $a < b < c$ and $f : [a,c] \to \mathbb{R}$ is continuous. Then the three Riemann Integrals $\int_a^c f(x)\,dx$, $\int_a^b f(x)\,dx$, and $\int_b^c f(x)\,dx$ all exist and satisfy

$$\int_a^c f(x)\,dx = \int_a^b f(x)\,dx + \int_b^c f(x)\,dx. \tag{3.16}$$

Remark 3.5. *There is nothing to stop us from using a different number of rectangles when approximating $\int_a^b f(x)\,dx$ versus $\int_b^c f(x)\,dx$. In fact, "professional" packages for numerically approximating integrals use "adaptive refinement" where they selectively refine some subintervals $[x_i, x_{i+1}]$ more than others when approximating an area under a curve. We discuss later in (3.31) a criterion that can be used for deciding where to use a finer partition.*

3.2.2 Dummy Variables of Integration

A variable that appears in a calculation only as a placeholder and disappears completely in the final result is called a **dummy variable**. As an example, in (2.20) we computed an estimate of the area under a parabola $y = x^2$ for the interval $[0,1]$ using n rectangles and obtained $\text{Area}_n^{\text{est}} = \left(\frac{1}{3} + \frac{1}{6n^2}\right) \pm \frac{1}{n}$. Nowhere does the variable x appear in the final answer. Hence, it is a dummy variable. We could have similarly computed the area under the parabola $\beta = \alpha^2$, and we would have obtained the same answer, $\text{Area}_n^{\text{est}} = \left(\frac{1}{3} + \frac{1}{6n^2}\right) \pm \frac{1}{n}$. Similarly, when we computed the Riemann Integral of y^k over the interval $[0,x]$, we obtained an answer that depended on x, the upper limit, but not on the variable y.

Proposition 3.6: Dummy Variables

Suppose that $a < b$ and $f : [a,b] \to \mathbb{R}$ is continuous. Then,

$$\int_a^b f(x)\,dx = \int_a^b f(y)\,dy = \int_a^b f(\lambda)d\lambda, \tag{3.17}$$

for any choice of real variables x, y, and λ. The Riemann Integral depends on the function $f : [a,b] \to \mathbb{R}$ and the lower and upper limits. The integral does not depend on what we name the variable we are plugging into f. In other words, the variable under the integral sign is a **dummy variable**.

3.2.3 Integrating Linear Combinations

Proposition 3.7: Linearity of the Riemann Integral

Suppose that $a < b$ and each of the functions $1 \le i \le N$, $f_i : [a,b] \to \mathbb{R}$, is Riemann integrable (i.e., their Riemann Integral exists and is finite). Then for constants $\alpha_i \in \mathbb{R}$, the function

$$f(x) := \alpha_1 \cdot f_1(x) + \alpha_2 \cdot f_2(x) + \cdots + \alpha_N \cdot f_N(x),$$

defined as a "linear combination of the functions $\{f_1, f_2, \ldots, f_N\}$", is Riemann integrable and, moreover,

$$\int_a^b f(x)\,dx = \alpha_1 \int_a^b f_1(x)\,dx + \alpha_2 \int_a^b f_2(x)\,dx + \cdots + \alpha_N \int_a^b f_N(x)\,dx. \tag{3.18}$$

Note: Because we know how to integrate monomials and linear combinations, we now know how to integrate any polynomial.

3.2.4 Making Sense of an Integral When Its Lower Limit Is Greater Than Its Upper Limit

How to make sense of $\int_a^b f(x)\,dx$ when $a > b$? The interval $[a,b]$ does not make sense because $\{x \in \mathbb{R} \mid a \le x \le b\} = \emptyset$; so what gives? By convention, when $a > b$, the Riemann Integral from a to b is defined as the negative of the integral from b to a. This is compatible with our approach to integration via Riemann sums because when $a > b$, $\Delta x := \frac{b-a}{n} = -\frac{a-b}{n} < 0$, and $x_i := a + (i-1)\Delta x$ satisfy $b = x_{n+1} < x_n < \cdots < x_2 < x_1 = a$. In symbols, we have, if $b < a$ and $f : [b,a] \to \mathbb{R}$ is continuous, then

$$\boxed{\int_a^b f(x)\,dx := -\int_b^a f(x)\,dx.} \tag{3.19}$$

3.2.5 Generalized First Additivity Property of Integrals

We build on Prop. 3.4, which we restate as: for $a < c < b$ and $f : [a,b] \to \mathbb{R}$ continuous,

$$\int_a^b f(x)\,dx = \int_a^c f(x)\,dx + \int_b^c f(x)\,dx.$$

It is useful to know that the relative order of a, b, and c is irrelevant (aka, does not matter).

Corollary 3.8: Generalized First Additivity Property

For a, b, and c finite real numbers in no particular order, define $\alpha := \min\{a,b,c\}$ to be the smallest of the three numbers, and $\beta := \max\{a,b,c\}$ to be the largest. If $f : [\alpha, \beta] \to \mathbb{R}$ is continuous, then

$$\int_a^b f(x)\,dx = \int_a^c f(x)\,dx + \int_c^b f(x)\,dx. \tag{3.20}$$

Example 3.9. *For a and b both finite real numbers, determine $\int_a^b y^k\,dy$.*

Solutions: Because $(\bullet)^k : \mathbb{R} \to \mathbb{R}$ is continuous, Corollary 3.8 yields:

$$\begin{aligned}\int_a^b y^k\,dy &= \int_a^0 y^k\,dy + \int_0^b y^k\,dy \\ &= -\int_0^a y^k\,dy + \int_0^b y^k\,dy \\ &= -\frac{a^{k+1}}{k+1} + \frac{b^{k+1}}{k+1} = \frac{b^{k+1} - a^{k+1}}{k+1}.\end{aligned}$$

■

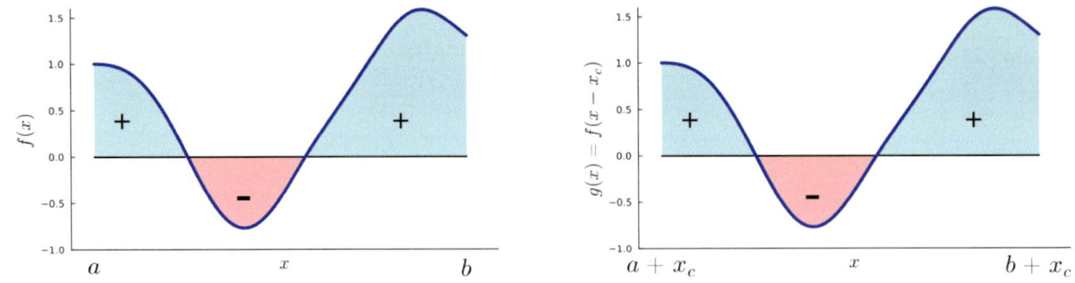

(a) (b)

Figure 3.4: Shifting does not change the area: A plot of $f : [a, b] \to \mathbb{R}$ is shown in (a) while (b) shows a plot of a shifted version of f, namely, $g : [a + x_c, b + x_c] \to \mathbb{R}$, where $g(x) := f(x - x_c)$. Clearly, the area under each function is the same, **when the limits of integration shift with the function**, as shown in the plots. Note that $g(a + x_c) = f(a + x_c - x_c) = f(a)$; similarly, $g(b + x_c) = f(b)$.

3.2.6 Shifting and Integration

Section 1.16 introduced the idea of shifting a function along the x-axis. It also noted that the areas enclosed by rectangles and triangles do not change under shifts (their areas are **invariant** under the operation of shifting). Because the Riemann-Darboux integral is based on the areas of rectangles, it inherits this invariance property, as illustrated in Fig. 3.4.

Proposition 3.10: Shift Property

Suppose that $a < b$, $f : [a, b] \to \mathbb{R}$ is continuous, and $x_c \in \mathbb{R}$. Then $g : [a + x_c, b + x_c] \to \mathbb{R}$, given by $g(x) := f(x - x_c)$, is also continuous and $\int_a^b f(x)\,dx = \int_{a+x_c}^{b+x_c} g(x)\,dx$. That is,

$$\int_a^b f(x)\,dx = \int_{a+x_c}^{b+x_c} f(x - x_c)\,dx. \tag{3.21}$$

Note: By defining $\alpha := a + x_c$ and $\beta := b + x_c$, the above result can be rewritten as:

$$\boxed{\int_\alpha^\beta f(x - x_c)\,dx = \int_{\alpha - x_c}^{\beta - x_c} f(x)\,dx.} \tag{3.22}$$

The latter form is usually easier to apply.

Example 3.11. *Evaluate the following definite integrals.*

(a) $\int_0^1 (x - 2)^2\,dx$

(b) $\int_0^1 (x - x_c)^2\,dx$

(c) $\int_0^1 \mathrm{rect}(x - 3)\,dx$

(d) $\int_3^5 \mathrm{tri}(x - 4)\,dx$

Solution:

(a) **Ans.** $\int_0^1 (x-2)^2\,dx = 2\frac{1}{3}$.

Applying Prop. 3.10,

$$\begin{aligned}\int_0^1 (x-2)^2\,dx &= \int_{0-2}^{1-2} x^2\,dx \\ &= \int_{-2}^{-1} x^2\,dx \\ &= \frac{x^3}{3}\Big|_{-2}^{-1} \\ &= \frac{(-1)^3}{3} - \frac{(-2)^3}{3} \\ &= -\frac{1}{3} + \frac{8}{3} \\ &= 2\frac{1}{3}.\end{aligned}$$

The solution looks long because we gave every step. Normally, two or three lines is all one would need!

(b) **Ans.** $\int_0^1 (x-x_c)^2\,dx = \frac{1}{3} - x_c + x_c^2$.

$$\begin{aligned}\int_0^1 (x-x_c)^2\,dx &= \int_{-x_c}^{1-x_c} x^2\,dx \\ &= \frac{x^3}{3}\Big|_{-x_c}^{1-x_c} \\ &= \frac{(1-x_c)^3}{3} - \frac{(-x_c)^3}{3} \\ &= \frac{1}{3} - x_c + x_c^2.\end{aligned}$$

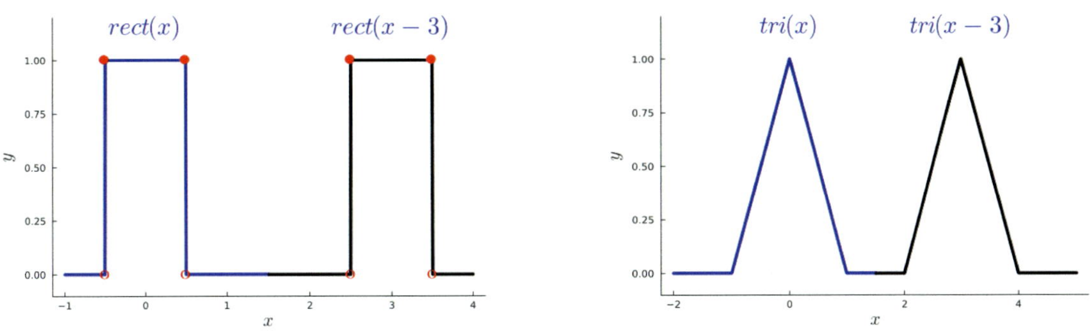

(a) (b)

Figure 3.5: Shifting two common functions: (a) Rectangle functions and (b) triangle functions.

(c) **Ans.** $\int_0^1 \text{rect}(x-3)\,dx = 0$.

A plot of the shifted rectangle function is shown in **Fig. 3.5-(a)**. It is zero over the range of integration, $[0,1]$, and therefore, its integral is zero.

Applying Prop. 3.10,

$$\int_0^1 \operatorname{rect}(x-3)\,dx = \int_{-3}^{-2} \operatorname{rect}(x)\,dx$$
$$= 0$$

because the function $\operatorname{rect}(x)$ is zero over the range of integration, $[-3,-2]$

(d) **Ans.** $\int_3^5 \operatorname{tri}(x-3)\,dx = \frac{1}{2}$

A plot of the shifted (unit) triangle function is shown in **Fig. 3.5-(b)**. The range of integration, $[3,5]$, captures the right half of the triangle, and hence its integral is one half.

Applying Prop. 3.10,

$$\int_3^5 \operatorname{tri}(x-3)\,dx = \int_0^2 \operatorname{tri}(x)\,dx$$
$$= 0.5.$$

Because the function $\operatorname{tri}(x)$ is centered at the origin, the (shifted) range of integration, $[0,2]$, excludes the left half of the triangle and includes its right half.

■

3.2.7 Scaling and Integration

Scaling, introduced in Section 1.16, causes most learners (and some instructors) more "issues" than shifting. This is because the area scales inversely proportional to the scale factor, as long as the limits of integration are scaled as well. **Figures 3.6** and **3.7** illustrate how $f(2x)$ encloses half the area of $f(x)$, and similarly, $f(\frac{x}{2})$ encloses double the area. The easiest way to keep this straight is to sketch the first half-periods of $\sin(x)$ and $\sin(2x)$. We are unlikely to forget that $\sin(2x)$ has twice the frequency and hence, half the period, of $\sin(x)$, and hence encloses half the area.

Proposition 3.12: Scaling Property

Suppose that $a < b$, $f : [a,b] \to \mathbb{R}$ is continuous, and $\omega_0 > 0$ is a real number. Then $g : [\frac{a}{\omega_0}, \frac{b}{\omega_0}] \to \mathbb{R}$, given by $g(x) := f(\omega_0 \cdot x)$ is also continuous and

$$\int_{\frac{a}{\omega_0}}^{\frac{b}{\omega_0}} g(x)\,dx = \frac{1}{\omega_0} \cdot \int_a^b f(x)\,dx. \tag{3.23}$$

Similarly, if $\omega_0 < 0$, then $g : [\frac{b}{\omega_0}, \frac{a}{\omega_0}] \to \mathbb{R}$ by $g(x) := f(\omega_0 \cdot x)$ is also continuous and

$$\int_{\frac{b}{\omega_0}}^{\frac{a}{\omega_0}} g(x)\,dx = \frac{-1}{\omega_0} \cdot \int_a^b f(x)\,dx = \frac{1}{|\omega_0|} \cdot \int_a^b f(x)\,dx. \tag{3.24}$$

Note: Assume $\omega_0 \neq 0$. By defining $\alpha := \frac{a}{\omega_0}$ and $\beta := \frac{b}{\omega_0}$, the above results can be rewritten as

$$\boxed{\int_\alpha^\beta f(\omega_0 \cdot x)\,dx = \frac{1}{\omega_0} \cdot \int_{\alpha \cdot \omega_0}^{\beta \cdot \omega_0} f(x)\,dx.} \tag{3.25}$$

This latter form is often easier to apply and is valid for $\omega_0 < 0$ and $\omega_0 > 0$.

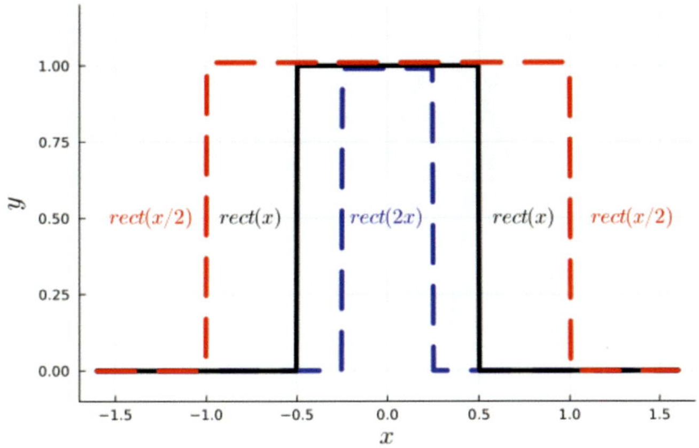

Figure 3.6: Scaling does change the area: Plots for three rectangles are shown each with a different scaling factor. The area scales inversely with the scale factor used in the function when the integration limits are appropriately adjusted. A unit rectangle has area 1.0. A unit rectangle scaled by $2x$ has area 0.5, while a unit rectangle scaled by $x/2$ has area 2.0.

Example 3.13. *Evaluate the following definite integrals.*

(a) $\int_0^1 (2x)^2\, dx$

(b) $\int_0^2 \text{rect}(\frac{x}{2})\, dx$

(c) $\int_0^1 \text{rect}(\omega_0\, x)\, dx,$ *for* $\omega_0 > 0$

(d) $\int_3^5 \text{tri}(\frac{x-2}{2})\, dx$ *(has both a shift and a scale).*

Solution:

(a) **Ans.** $\int_0^1 (2x)^2\, dx = \frac{4}{3}$.

The most obvious way to solve this problem is to note that $(2x)^2 = 4x^2$, and hence

$$\int_0^1 (2x)^2\, dx = 4\int_0^1 x^2\, dx = 4\left.\frac{x^3}{3}\right|_0^1 = 4\left(\frac{1}{3} - 0\right) = \frac{4}{3}.$$

Applying Prop. 3.12 gives:

$$\begin{aligned}\int_0^1 (2x)^2\, dx &= \frac{1}{2}\int_0^2 x^2\, dx \\ &= \left.\frac{1}{2}\frac{x^3}{3}\right|_0^2 \\ &= \frac{1}{2}\left(\frac{8}{3} - 0\right) \\ &= \frac{4}{3}.\end{aligned}$$

(b) **Ans.** $\int_0^2 \text{rect}(\frac{x}{2})\, dx = 1.0$.

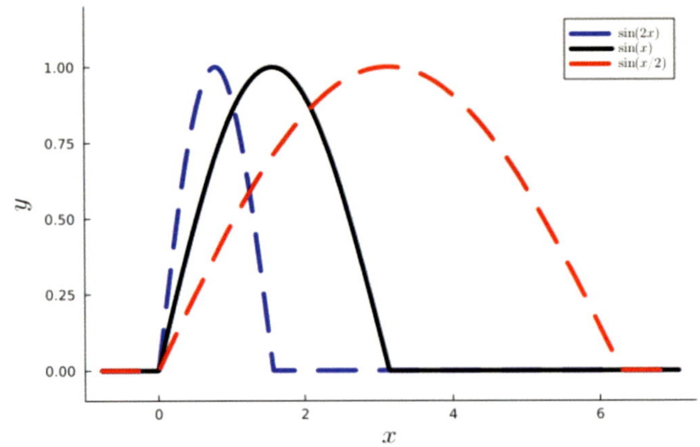

Figure 3.7: Scaling does change the area, take 2: Similarly, a half-period of $\sin(x)$ (integrated over $[0, \pi]$) has area 2.0, whereas a half-period of $\sin(2x)$ (integrated over $[0, \frac{\pi}{2}]$) has area 1.0 and a half-period of $\sin(\frac{x}{2})$ (integrated over $[0, 2\pi]$) has area 4.0. **If we had integrated the sinusoids over a full period, all of the answers would have been zero.**

Recall that

$$\operatorname{rect}(x) = \begin{cases} 1 & |x| \le 0.5, \\ 0 & \text{otherwise.} \end{cases}$$

Applying Prop. 3.12 gives:

$$\begin{aligned} \int_0^2 \operatorname{rect}(\frac{x}{2})\, dx &= 2 \int_0^1 \operatorname{rect}(x)\, dx \\ &= 2 \int_0^{0.5} 1\, dx + 2 \int_{0.5}^1 0\, dx \\ &= 2 \cdot \frac{1}{2} = 1. \end{aligned}$$

(c) **Ans.** For $\omega_0 > 0$, $\int_0^1 \operatorname{rect}(\omega_0\, x)\, dx = \frac{\min\{0.5, \omega_0\}}{\omega_0} = \begin{cases} 1 & 0 < \omega_0 \le 0.5, \\ \frac{1}{2\omega_0} & \omega_0 > 0.5. \end{cases}$

Applying Prop. 3.12 gives:

$$\begin{aligned} \int_0^1 \operatorname{rect}(\omega_0\, x)\, dx &= \frac{1}{\omega_0} \int_0^{\omega_0} \operatorname{rect}(x)\, dx \\ &= \frac{1}{\omega_0} \int_0^{\min\{0.5,\omega_0\}} 1\, dx + \frac{1}{\omega_0} \int_{\min\{0.5,\omega_0\}}^{\omega_0} 0\, dx \\ &= \frac{1}{\omega_0} \cdot \min\{0.5, \omega_0\}. \end{aligned}$$

(d) **Ans.** $\int_3^5 \operatorname{tri}(\frac{x-2}{2})\, dx = \frac{3}{4}$.

Recall that

$$\operatorname{tri}(x) = \begin{cases} |x| & |x| \le 1, \\ 0 & \text{otherwise.} \end{cases}$$

Applying Propositions 3.10 and 3.12 gives:

$$\begin{aligned}
\int_3^5 \operatorname{tri}\left(\frac{x-2}{2}\right) dx &= \int_1^3 \operatorname{tri}\left(\frac{x}{2}\right) dx \quad \text{(remove the shift first)} \\
&= 2 \int_{1/2}^{3/2} \operatorname{tri}(x)\, dx \quad \text{(then take care of the scale)} \\
&= 2 \int_{1/2}^{1} x\, dx + 2 \int_{1}^{3/2} 0\, dx \quad \text{(apply def. of triangle function)} \\
&= 2 \left.\frac{x^2}{2}\right|_{1/2}^{1} + 0 \\
&= 2 \left(\frac{1}{2} - \frac{1}{8}\right) \\
&= 2 \cdot \frac{3}{8} = \frac{3}{4}.
\end{aligned}$$

■

3.3 Numerical Methods for Approximating Riemann Integrals

Rectangles are not always the most efficient shape for estimating area, and we can imagine that other geometric shapes might better "fill in" the area. One popular shape is a trapezoid, as shown in **Fig. 3.8**, which caps a rectangle with a right triangle; this gives rise to the **Trapezoidal Rule** for approximating the area under a curve. A trapezoid results from approximating a function over a "small" interval $[x_i, x_{i+1}]$ by a (straight) line segment connecting $f(x_i)$ and $f(x_{i+1})$. Another popular idea is to approximate a function over $[x_i, x_{i+1}]$ by a quadratic, which gives rise to **Simpson's Rule**.

3.3.1 Trapezoidal Rule

The area under the trapezoids shown in **Fig. 3.8-(a)** and **-(b)** is given by

$$\text{trapArea}_i = \frac{f(x_i) + f(x_{i+1})}{2} \cdot (x_{i+1} - x_i), \tag{3.26}$$

which is the average (or, mean) of the two rectangles used in the Riemann lower and upper sums, when the function is increasing. For a partition, $a = x_1 < x_2 < \cdots < x_n < x_{n+1} = b$, the area under the curve over the interval $[a, b]$ is approximated by:

$$\text{Area}_n^{\text{Trap}} := \sum_{i=1}^{n} \text{trapArea}_i = \sum_{i=1}^{n} \frac{f(x_i) + f(x_{i+1})}{2} \cdot (x_{i+1} - x_i). \tag{3.27}$$

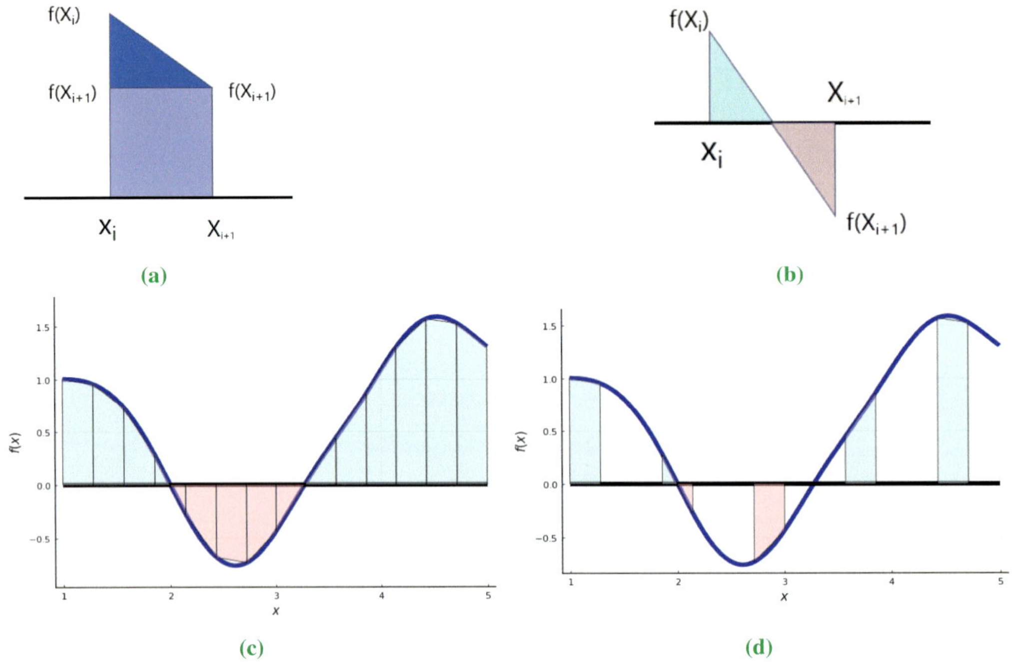

Figure 3.8: Using trapezoids to approximate the area of a function $f : [1, 5] \to \mathbb{R}$. Trapezoids are polygons with four corners, parallel sides, and a flat base, like a rectangle, but where the two sides can have different heights. (a) Illustration that a trapezoid can also be viewed as a rectangle plus a right triangle, where the rectangle is determined by the smallest height, the base of the right triangle is the top of the rectangle, and the side of the right triangle is determined by the difference in the two heights. (b) When $f(x_i)$ and $f(x_{i+1})$ have opposite signs, the trapezoid becomes two right triangles, one "pointing up" and one "pointing down". Part (c) shows 14 trapezoids approximating the area under the curve defined by f. Part (d) shows every third trapezoid so that the individual shapes can be more clearly seen.

Proposition 3.14: Trapezoidal Rule for Computing a Riemann Integral

Suppose that $a < b$ and $f : [a, b] \to \mathbb{R}$ is continuous. Then the Riemann Integral $\int_a^b f(x)\,dx$ exists and it can be evaluated by the **Trapezoidal Rule**. That is,

$$\int_a^b f(x)\,dx = \lim_{n\to\infty} \mathrm{Area}_n^{\mathrm{Trap}} = \lim_{n\to\infty} \sum_{i=1}^{n} \frac{f(x_i) + f(x_{i+1})}{2} \cdot (x_{i+1} - x_i) = \lim_{n\to\infty} \sum_{i=1}^{n} \frac{f(x_i) + f(x_{i+1})}{2} \cdot \Delta x, \tag{3.28}$$

where, $\Delta x := \frac{b-a}{n}$ and $x_i := a + (i-1)\Delta x$. In the above, the limits mean that for all $\epsilon > 0$, there exists $N < \infty$ such that $n \geq N \implies \left| \int_a^b f(x)\,dx - \sum_{i=1}^{n} \frac{f(x_i)+f(x_{i+1})}{2} \cdot (x_{i+1} - x_i) \right| \leq \epsilon$.

Remark 3.15. *There is no obvious "Lower Trapezoidal Sum" or "Upper Trapezoidal Sum" that can be used to bound the error in the approximation for a given, finite value of n, as we had with the method of rectangles. A common approach is to form the sum for an initial value of n, and then recompute it for $2n$, $3n$, etc., until either the absolute change or the relative change in the calculation is less than a given* `aTol`*; that is, for $k = 1, 2, \ldots$, stop when*

$$\left| \mathrm{Area}_{(k+1)\cdot n}^{\mathrm{Trap}} - \mathrm{Area}_{k\cdot n}^{\mathrm{Trap}} \right| \leq \mathtt{aTol} \quad (\textit{absolute change, or})$$

$$\left| \frac{\mathrm{Area}_{(k+1)\cdot n}^{\mathrm{Trap}} - \mathrm{Area}_{k\cdot n}^{\mathrm{Trap}}}{\mathrm{Area}_{k\cdot n}^{\mathrm{Trap}}} \right| \leq \mathtt{aTol} \quad (\textit{relative change}).$$

Neither of these measures of the error provides the firm bound we had previously,

$$\left|\text{Area}^{\text{True}} - \text{Area}_n^{\text{Approx}}\right| \leq \texttt{aTol}.$$

Is there something more that can be done? Yes.

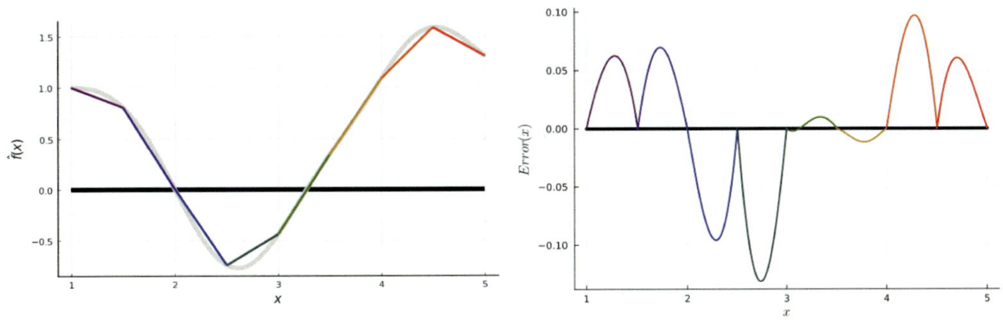

(a) (b)

Figure 3.9: The Trapezoidal Rule can be viewed as a piecewise linear approximation of the function $f : [1,5] \to \mathbb{R}$ over intervals $[x_i, x_{i+1}]$. (a) Shows an approximation of f using eight linear segments, one for each of the intervals $[1, 1.5]$, $[1.5, 2]$, etc., to $[4.5, 5]$. (b) Shows the error in the fit. The peak values in the error over each interval $[x_i, x_{i+1}]$ can be used to bound the total error in the approximated integral.

We formally define a **piecewise linear approximation** to the function $f : [1,5] \to \mathbb{R}$ by

$$\widehat{f}(x) := m_i(x - x_i) + f(x_i) \text{ for } x \in [x_i, x_{i+1}), \tag{3.29}$$

where

$$m_i := \frac{f(x_{i+1}) - f(x_i)}{x_{i+1} - x_i} \quad \text{(rise over run)}. \tag{3.30}$$

Figure 3.9-(a) shows a plot of a typical $\widehat{f}$. The **approximation uses the term "piecewise"** because a different approximation (in this case, linear) is used for each subinterval $[x_i, x_{i+1}]$. The Trapezoidal Rule integrates exactly the linear approximation. The error in the Riemann Integral of the original function f arises from the difference, $\text{Error}(x) := f(x) - \widehat{f}(x)$. A typical error plot is shown in **Fig. 3.9-(b)**. We can use this to form an estimate of the error in the Riemann Integral by

$$\text{Error}_i^{\max} := \max_{x_i \leq x \leq x_{i+1}} |f(x) - \widehat{f}(x)|. \tag{3.31}$$

The error in the computation of the Riemann Integral can then be upper-bounded by:

$$\text{Error}^{\text{RI}} \leq \sum_{i=1}^{n} \text{Error}_i^{\max} \cdot \Delta x = \Delta x \cdot \sum_{i=1}^{n} \max_{x_i \leq x \leq x_{i+1}} |f(x) - \widehat{f}(x)|. \tag{3.32}$$

For any interval $[x_i, x_{i+1}]$ where $\text{Error}_i^{\max}$ in (3.31) is larger than some preset tolerance, we can subdivide the interval and attempt to reduce the error.

```
function trapezoidalRule(f, a, b, N, aTol=1e-6)
    if abs(a-b) < aTol
        println("abs(a-b) < $aTol. Estimated integral is zero.")
            return 0.0
    end
    # make sure N is an integer greater or equal to 2
```

```
    N=max(ceil(Int64,N),2)
    # Define \Delta x
    Deltax = (b-a)/N
    # Initialize Sum
    estIntegralTrapezoidal = 0.0
    # Initialize x_i
    xi=a
    for i = 1:N
        # define x_{i+1}
        xip1 = xi + Deltax
        estIntegralTrapezoidal = estIntegralTrapezoidal + Deltax*(f(xi) + f(xip1))/2.0
        # update xi
        xi = xip1
    end
    return estIntegralTrapezoidal
end
```

Output

```
trapezoidalRule (generic function with 2 methods)
```

```
f(x) = sin(x)
a = 0.0
b = 2*pi
N = 25
trapezoidalRule(f, a, b, N)
```

Output

```
2.373101715136272e-15
```

Recall that we know that the area under the function $\sin : [0, 2\pi] \to \mathbb{R}$ is zero. The Trapezoidal Rule has returned a value very close to zero, 2.37×10^{-15}, but how accurate is it? To find out, we implement the error estimate in (3.32).

```
function trapezoidalRuleWithErrorBounds(f, a, b, N=100, aTol=1e-6)
    if abs(a-b) < aTol
        println("abs(a-b) < $aTol. Estimated integral is zero.")
            return 0.0
    end
    # make sure N is an integer greater or equal to 2
    N=max(ceil(Int64,N),2)
    # Define \Delta x
    Deltax = (b-a)/N
    # Initialize Sum
    estIntegralTrapezoidal = 0.0
    # Initialize Error Estimate
    pmError  = 0.0
    nRefine = 10
    # Initialize x_i
    xi=a
    for i = 1:N
        # define x_{i+1}
        xip1 = xi + Deltax
        y1 = f(xi)
```

```
        y2 = f(xip1)
        estIntegralTrapezoidal = estIntegralTrapezoidal + Deltax*(y1 + y2)/2.0
        xRefined = range(xi, xip1, length=nRefine)
        fRefined = f.(xRefined)
        g(x) = y1 + (x-xi)*(y2-y1)/Deltax
        gRefined = g.(xRefined)
        errorLocal = fRefined-gRefined
        errorChoice = "max"
        if errorChoice == "max"
            maxError = maximum(abs.(fRefined-gRefined))
            pmError = pmError + Deltax * maxError
        elseif errorChoice == "avg"
            avgError = sum(abs.(fRefined-gRefined))/length(errorLocal)
            pmError = pmError + Deltax * avgError
        else # mean error
            meanError=sum(errorLocal)/length(errorLocal)
            pmError = pmError + Deltax * meanError
        end
        # update xi
        xi = xip1
    end
    return (estIntegralTrapezoidal=estIntegralTrapezoidal, pmError=pmError)
end
```

Output

```
trapezoidalRuleWithErrorBounds (generic function with 3 methods)
```

```
f(x) = sin(x)
a = 0.0
b = 2*pi
N = 25
G=trapezoidalRuleWithErrorBounds(f, a, b, N)
display(G)
```

Output

```
(estIntegralTrapezoidal = 2.373101715136272e-15,
pmError = 0.03127822283578252)
```

If we increase the number of line segments, the result improves, as we would expect.

```
f(x) = sin(x)
a = 0.0
b = 2*pi
N = 250
G=trapezoidalRuleWithErrorBounds(f, a, b, N)
display(G)
```

Output

```
(estIntegralTrapezoidal = 1.2034589904480475e-14,
pmError = 0.0003120774612516434)
```

We can also test it on the complicated function in (3.7).

```
f(x) = cos(exp(-x)*sin(x)/(1 + x + 5x^2)) + log2( (1 + x)/(cos(x) + 5x^2) )
a = 0.0
b = 10
N = 5000
G=trapezoidalRuleWithErrorBounds(f, a, b, N)
display(G)
```

Output

```
(estIntegralTrapezoidal = -28.832292268297902,
pmError = 2.701204661411364e-6)
```

The error is now more than a thousand times smaller than with the Riemann Upper and Lower Sums, using the same number of subintervals,

$$\int_0^{10} \cos\left(\frac{\exp(-x)\sin(x)}{1+x+5x^2}\right) + \log_2\left(\frac{1+x}{\cos(x)+5x^2}\right) dx \approx -28.832292 \pm 2.701205e\text{-}6, \tag{3.33}$$

versus the previous error bound of $\approx \pm 5e\text{-}3$.

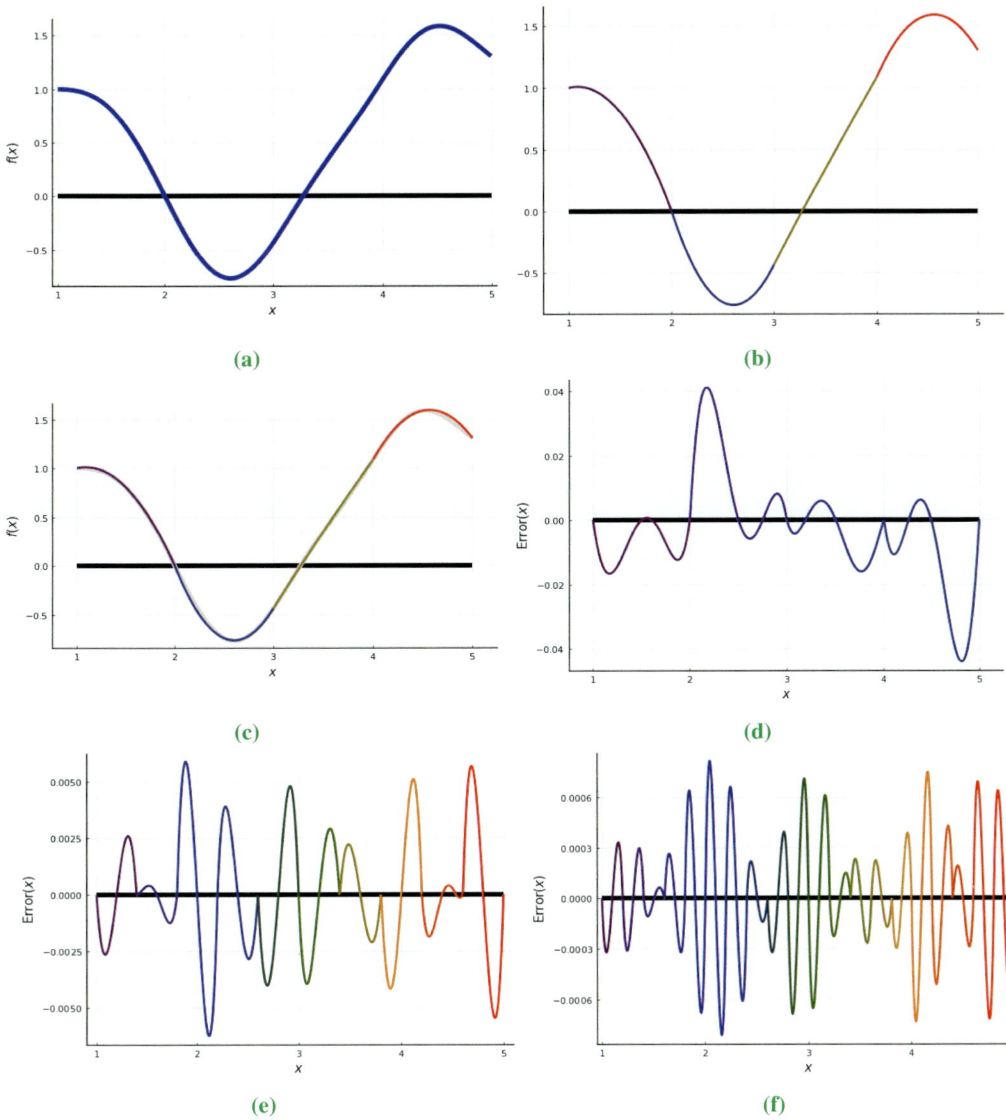

Figure 3.10: Simpson's Rule uses piecewise quadratic functions to approximate the function $f : [1, 5] \to \mathbb{R}$ over intervals $[x_i, x_{i+1}]$. (a) The function f that is to be integrated. Part (b) shows an approximation of f using only four quadratics, one for each of the intervals $[1, 2]$, $[2, 3]$, $[3, 4]$, and $[4, 5]$. See Prop. 3.17 for how to design the quadratic functions. Part (c) shows an overlay of the approximation via four quadratics onto the original function; it is difficult to discern the difference. Part (d) shows the error in the fit, once again, using only four quadratics. Part (e) shows the error in the fit using ten quadratics. Part (f) shows the error in the fit using twenty quadratics. You need to look at the scales to appreciate that the magnitude of the error is being reduced by a factor of ten each time.

3.3.2 Simpson's Rule

Simpson's Rule is inspired by the thought, "If we can make headway by piecewise approximating a function using lines, why not make even more headway by piecewise approximating a function using quadratics?" From Prop. 3.3 and Prop. 3.4, we obtain that:

$$\boxed{\begin{aligned} \int_a^b 1\,dx &= b-a \\ \int_a^b x\,dx &= \frac{b^2-a^2}{2} \\ \int_a^b x^2\,dx &= \frac{b^3-a^3}{3}. \end{aligned}} \tag{3.34}$$

When combined with Prop. 3.7, we now know how to integrate exactly a general quadratic function as the sum of the integrals of the functions 1, x, and x^2. **Figure 3.10** illustrates how the use of piecewise quadratic approximations may yield higher-fidelity numerical integration than the piecewise linear approximations used in the Trapezoidal Rule.

Interpolating a Quadratic Function through a Set of Three Points

Definition 3.16. *Consider an interval $[x_i, x_{i+1}] \subset \mathbb{R}$ and a function $f : [x_i, x_{i+1}] \to \mathbb{R}$. Let $x_c := \frac{x_i + x_{i+1}}{2}$ be the midpoint of the interval and $\Delta x := x_{i+1} - x_i > 0$ be its length. The quadratic function*

$$q(x) := \alpha(x - x_c)^2 + \beta(x - x_c) + \gamma \tag{3.35}$$

*is said to be **centered** about x_c. The function q is said to **interpolate the function** f if the two functions agree on the three points $\{x_i, x_c, x_{i+1}\}$; that is:*

$$\begin{aligned} q(x_i) &= f(x_i) \\ q(x_c) &= f(x_c) \\ q(x_{i+1}) &= f(x_{i+1}). \end{aligned}$$

Intuition for Simpson's Rule: The quadratic function $q(x) := \alpha(x - x_c)^2 + \beta(x - x_c) + \gamma$ has three unknowns, α, β, and γ. We need, therefore, three equations to determine the unknowns. The equations come from matching the quadratic to $f : [x_i, x_{i+1}] \to \mathbb{R}$ at the two endpoints x_i and x_{i+1}, and at the center point $x_c = \frac{x_{i+1} + x_i}{2}$. These are the interpolation conditions given in the above definition.

Matching $q(x)$ and $f(x)$ at the midpoint gives $\gamma = f(x_c)$. The algebra to do the match at the endpoints is given in the proof of Prop. 3.17. Integrating the quadratic, linear, and constant terms in $q(x)$ over the interval $[x_i, x_{i+1}]$ then gives the exact area under the quadratic function, $q : [x_i, x_{i+1}] \to \mathbb{R}$, and a "good approximation" of the area under the given function, $f : [x_i, x_{i+1}] \to \mathbb{R}$. There is nothing magical about using a quadratic function. We can use any function where solving for its unknown coefficients is straightforward, and where we know how to compute its Riemann integral in closed form.

Proposition 3.17: Basis of Simpson's Rule: Integral of an Interpolated Quadratic Function

The quadratic function given by (3.35) and centered about the midpoint $x_c = x_i + \frac{\Delta x}{2}$ interpolates

$f : [x_i, x_{i+1}] \to \mathbb{R}$ at the points $x_i < x_c < x_{i+1}$ if, and only if,

$$\begin{aligned}
\alpha &= \frac{2}{(\Delta x)^2}\left(f(x_i) + f(x_{i+1}) - 2f(x_c)\right) \\
\beta &= \frac{1}{\Delta x}\left(f(x_{i+1}) - f(x_i)\right) \\
\gamma &= f(x_c).
\end{aligned} \tag{3.36}$$

Moreover, the Riemann integral of the interpolating quadratic function is equal to:

$$\begin{aligned}
\int_{x_i}^{x_{i+1}} q(x)\,dx &= \alpha \cdot \frac{(\Delta x)^3}{12} + \gamma \cdot \Delta x \\
&= \frac{\Delta x}{6} \cdot \left(f(x_i) + 4f(x_c) + f(x_{i+1})\right),
\end{aligned} \tag{3.37}$$

where $x_c := \frac{x_i + x_{i+1}}{2}$.

Proof: Because $q(x_c) = \gamma$, we immediately obtain $\gamma = f(x_c)$. The other two coefficients are obtained from the following (2×2) system of linear equations:

$$\begin{bmatrix} \left(\frac{\Delta x}{2}\right)^2 & -\frac{\Delta x}{2} \\ \left(\frac{\Delta x}{2}\right)^2 & \frac{\Delta x}{2} \end{bmatrix} \begin{bmatrix} \alpha \\ \beta \end{bmatrix} = \begin{bmatrix} f(x_i) - f(x_c) \\ f(x_{i+1}) - f(x_c) \end{bmatrix},$$

where we have used $\Delta x = x_{i+1} - x_i$.

To obtain Simpson's formula, we apply Prop. 3.10 to $q(x)$ to obtain:

$$\int_{x_i}^{x_{i+1}} q(x)\,dx = \int_{x_i - x_c}^{x_{i+1} - x_c} q(x + x_c)\,dx = \int_{-\frac{\Delta x}{2}}^{\frac{\Delta x}{2}} \left(\alpha x^2 + \beta x + \gamma\right)\,dx = \alpha \cdot \frac{(\Delta x)^3}{12} + \gamma \cdot \Delta x.$$

■

Remark 3.18. *The "integration" formula is amazingly simple: two-thirds of the function's value at the midpoint of the interval* $[x_i, x_{i+1}]$*, plus one-sixth for each of the two endpoints; we note that* $\frac{2}{3} + 2 \cdot \frac{1}{6} = 1.0$. *Let us keep in mind the road we have traveled: Riemann Sums use one value of the function for each interval to approximate area via a rectangle, the Trapezoidal Rule uses two values,* $f(x_i)$ *and* $f(x_{i+1})$ *with weights of one half, and Simpson's Rule uses three values,* $f(x_i)$, $f(x_{i+1})$, *with weight one-sixth and* $f(x_c)$ *with weight two thirds. If we approximate with a cubic, we will use four function values and different weights. In general, we can fit with an* n*-th degree polynomial, but eventually, we are faced with* ***Diminishing Returns***.

```
function simpsonRule(f, a, b, N=100, aTol=1e-6)
    # Uses a centered quadratic
    if abs(a-b) < aTol
        println("abs(a-b) < $aTol. Estimated integral is zero.")
            return 0.0
    end
    # make sure N is an integer greater or equal to 2
    N=max(ceil(Int64,N),2)
    # Define \Delta x
    Deltax = (b-a)/N
    # Initialize Sum
    estIntegralSimpson = 0.0
    # Initialize x_i
```

```
    xi=a
    for i = 1:N
        xip1 = xi + Deltax
        a1 = f(xi)
        a2 = f(xip1)
        ac = f( (xi + xip1)/2 )
        estIntegralSimpson = estIntegralSimpson + Deltax*(a1 + a2)/6.0 + 2.0*Deltax*
    ac/3.0
        # update xi
        xi = xip1
    end
    return estIntegralSimpson
end
```

Output

```
simpsonRule (generic function with 3 methods)
```

Here is a version with error estimation.

```
function simpsonRuleWithErrorBounds(f, a, b, N=100, aTol=1e-6)
    # Uses a centered quadratic
    if abs(a-b) < aTol
        println("abs(a-b) < $aTol. Estimated integral is zero.")
            return 0.0
    end
    # make sure N is an integer greater or equal to 2
    N=max(ceil(Int64,N),2)
    # Define \Delta x
    Deltax = (b-a)/N
    # Initialize Sum
    estIntegralSimpson = 0.0
    # Initialize total error
    pmError = 0.0
    # Initialize x_i
    xi=a
    for i = 1:N
        xip1 = xi + Deltax
        xc = (xi + xip1)/2
        a1 = f(xi)
        a2 = f(xip1)
        ac = f(xc)
        estIntegralSimpson = estIntegralSimpson + Deltax*(a1 + a2)/6.0
                                        + 2.0*Deltax*ac/3.0
        nRefine = 50
        xRefined = range(xi, xip1, length=nRefine)
        fRefined = f.(xRefined)
        # Coefficients of the centered quadratic
        alpha = (2/Deltax^2)*(a1 + a2 - 2*ac)
        beta = (1/Deltax)*(a2-a1)
        gamma = ac
        # centered quadratic
        g(x) = alpha*(x-xc)^2 + beta*(x-xc) + gamma
        gRefined = g.(xRefined)
        errorLocal = fRefined-gRefined
        errorChoice = "max"
```

```
            if errorChoice == "max"
                maxError = maximum(abs.(fRefined-gRefined))
                pmError = pmError + Deltax * maxError
            elseif errorChoice == "avg"
                avgError = sum(abs.(fRefined-gRefined))/length(errorLocal)
                pmError = pmError + Deltax * avgError
            else # mean error
                meanError=sum(errorLocal)/length(errorLocal)
                pmError = pmError + Deltax * meanError
            end
            # update xi
            xi = xip1
        end
        return (estIntegralSimpson=estIntegralSimpson, pmError=pmError)
end
```

Output

```
simpsonRuleWithErrorBounds (generic function with 3 methods)
```

```
f(x) = sin(x)
a = 0.0
b = 2*pi
N = 25
G = simpsonRuleWithErrorBounds(f, a, b, N)
display(G)
```

Output

```
(estIntegralSimpson = 3.021888295151598e-15,
pmError = 0.0005179561770644536)
```

```
f(x) = sin(x)
a = 0.0
b = 2*pi
N = 250
G = simpsonRuleWithErrorBounds(f, a, b, N)
display(G)
```

Output

```
(estIntegralSimpson = 1.3069054302303207e-14,
pmError = 5.096276651101236e-7)
```

```
f(x) = cos(exp(-x)*sin(x)/(1 + x + 5x^2)) + log2( (1 + x)/(cos(x) + 5x^2) )
a = 0.0
b = 10
N = 5000
G = simpsonRuleWithErrorBounds(f, a, b, N)
display(G)
```

Output

```
(estIntegralSimpson = -28.832291734252568,
pmError = 1.1358825205876213e-9)
```

The error is now more than a thousand times smaller than with the Trapezoidal Rule, using the same number of subintervals,

$$\int_0^{10} \cos\left(\frac{\exp(-x)\sin(x)}{1+x+5x^2}\right) + \log_2\left(\frac{1+x}{\cos(x)+5x^2}\right) dx \approx -28.832292 \pm 1.135883e\text{-}9, \tag{3.38}$$

versus the previous error bound of $\approx\ \pm 2.701205e\text{-}6$. And the error is two and a half million times (2.5×10^6) smaller than when we estimated the area with rectangles.

Even though, for many real problems, such a degree of accuracy is overkill, it is still fun to see how it all works out. In practice, most applications are satisfied with an error of 10^{-3} and would use a far smaller n to achieve it with Simpson's Rule than with Riemann Lower and Upper Sums.

3.3.3 Julia Packages

QuadGK is a numerical integration function found in languages like MATLAB and Julia, which uses the Gauss-Kronrod quadrature rule to calculate definite integrals. Recall that Simpson's Rule,

$$\int_{x_i}^{x_i+\Delta x} f(x)\,dx \approx \frac{\Delta x}{6}\cdot\left(f(x_i)+f(x_{i+1})\right) + \frac{2\cdot\Delta x}{3}\cdot f(x_c),$$

works by forming a weighted sum of $f : [x_i, x_{i+1}] \to \mathbb{R}$ evaluated at x_i, $x_c := \frac{x_i+x_{i+1}}{2}$, and x_{i+1}, and then adding up all of these terms to form an approximation of area. QuadGK is similar, except it uses a more sophisticated choice of points and weights discovered by Gauss and improved on by Kronrod. The estimates over each interval $[x_i, x_{i+1}]$ are then summed for the final result. Moreover, the package QuadGK refines estimates iteratively until they are within a threshold, making it adaptive and able to handle various functions, including difficult cases with discontinuities. See the documentation for more information.

Example 3.19. *Use the function* `quadgk` *to evaluate the following definite integrals taken from YouTube. Each integral is supposed to be uncommonly difficult for students to integrate.*

(a) $\int_1^e \frac{1+x^2\cdot\ln(x)}{x+x^2\cdot\ln(x)}\,dx$ *from Cipher.*

(b) $\int_0^\pi \frac{2\cdot x\cdot\sin(x)}{3+\cos(2\cdot x)}\,dx$ *from Lets Solve Math Problems, 2018 Stanford Math Tournament.*

(c) $\int_{0.1}^{0.5} \left(\sqrt{1-x^2} + \sqrt{x^2+1} + x^2\cdot\ln(x) + \frac{\operatorname{atan}(x)}{x^2} + x^{\frac{\sqrt{x}}{\ln(x)}} + \frac{x^2+1}{x^3-2x} + \sin(x)^4\cdot\cos(x)^2 + \frac{1}{2+\cos(x)} + \tan(x)^{10} + x^x\right)^2 dx$ *from "when the Calculus teacher promises there will be only one integral on the test" by* **BlackPenRedPen**.

(d) $\int_{-\frac{1}{2}}^{\frac{1}{2}} \sqrt{x^2+1+\sqrt{x^4+x^2+1}}\,dx$ *from MathTuts, solving integrals from the 2023 MIT Integration Bee Finals: Q3.*

Solutions:

(a)

```julia
using QuadGK
f(x) = ( 1 + x^2 * log(x) )/( x + x^2*log(x) )
area, error = quadgk(f, 1, exp(1))
```

Output

```
(1.4050201409408223, 2.707664759071804e-9)
```

(b)

```
f(x) = ( 2.0*x*sin(x) )/( 3 + cos(2*x) )
area, error = quadgk(f, 0, pi)
```

Output

```
(2.4674011002723395, 4.898666333685853e-10)
```

(c)

```
f(x) = ( sqrt(1 - x^2) + sqrt(x^2+1)+ x^2 * log(x) + atan(x)/x^2 +
     x^( sqrt(x)/log(x) ) + (x^2+1)/(x^3-2x) + sin(x)^4 * cos(x)^2 +
          1/( 2 + cos(x) ) + tan(x)^10 + x^x )^2
F = quadgk(f, 0.1, 0.5)
G = simpsonRuleWithErrorBounds(f, 0.1, 0.5, 3000)
display(F)
display(G);
```

Output

```
(16.53669833564563, 6.188844992038867e-10)

(estIntegralSimpson = 16.5366983356456,
pmError = 4.499301236175278e-10)
```

(d)

```
f(x) = sqrt( x^2 + 1 + sqrt( x^4 + x^2 + 1 ) )
F = quadgk(f, -0.5, 0.5)
G = simpsonRuleWithErrorBounds(f, -0.5, 0.5, 3000)
display(F)
display(G);
```

Output

```
(1.4586629142874117, 7.910228028151778e-11)

(estIntegralSimpson = 1.458662914287415,
pmError = 3.0790407260876096e-14)
```

Optional Read: **ApproxFun** is a second interesting package for numerical integration. The syntax is hard to grasp at first, but it even allows the solution of "integral equations[2]", a generalization of the Ordinary Differential Equations we will introduce in Chapter 9. This package first takes a function $f : [a,b] \to \mathbb{R}$ defined on a bounded interval $[a,b]$, $-\infty < a < b < \infty$, and then transforms it to an equivalent function on the interval $[-1,1]$ through shifting and scaling. It then approximates the function by a linear combination of the form,

$$\widehat{f}(x) = \sum_{k=0}^{N} f_k \cdot \cos(k \cdot \arccos(x)),$$

where the functions $\cos(k \cdot \arccos(x))$ are **orthogonal** on $-1 \le x \le 1$ with respect to the weight $\frac{1}{\sqrt{1-x^2}}$. The orthogonality condition is

$$\int_{-1}^{1} \cos(j \cdot \arccos(x)) \cdot \cos(k \cdot \arccos(x)) \frac{1}{\sqrt{1-x^2}}\, dx = 0,$$

for all $j \neq k$. The number of terms N is adjusted so that the approximation error $f(x) - \widehat{f}(x)$ is close to "machine epsilon". Hence, the results are very accurate.

[2]Think differential equations, but replace derivatives with integrals.

The downside of the ApproxFun package is that one needs more than basic Calculus to really understand what it is doing, whereas the concepts behind QuadGK are not dissimilar to Simpson's Rule.

We will repeat two of the previous examples so that we can see how the syntax goes when using ApproxFun.

```
using SpecialFunctions, ApproxFun
f(x)= ( 1 + x^2 * log(x) )/( x + x^2*log(x) )

# Define the approximating function
g =  Fun(f,1..exp(1))
# Define the operation of computing a Definite Integral
Sigma = DefiniteIntegral()
# Apply the Definite Integral operator to g
Sigma*g
```

Output

```
1.4050201409408223 anywhere
```

```
f(x) = ( sqrt(1 - x^2) + sqrt(x^2+1)+ x^2 * log(x) + atan(x)/x^2 +
     x^( sqrt(x)/log(x) ) + (x^2+1)/(x^3-2x) + sin(x)^4 * cos(x)^2 +
         1/( 2 + cos(x) ) + tan(x)^10 + x^x )^2
F = quadgk(f, 0.1, 0.5)
G = simpsonRuleWithErrorBounds(f, 0.1, 0.5, 3000)
display(F)
display(G)

# Define the approximating function
g = Fun(f,0.1..0.5)
# Define the operation of computing a Definite Integral
Sigma = DefiniteIntegral()
# Apply the Definite Integral operator to g
Sigma*g
```

Output

```
(16.53669833564563, 6.188844992038867e-10)

(estIntegralSimpson = 16.5366983356456,
pmError = 4.499301236175278e-10)

16.536698335645635 anywhere
```

3.3.4 Optional Read: Even and Odd Functions as Great Test Cases

When using a numerical integration package on a challenging function, it is always useful to be able to generate "test cases," wherein we use functions that are related to an important problem we are trying to solve, and yet, we know ahead of time what the integral of the function is, exactly.

Even and Odd Functions, and Parts of Functions

Definition 3.20.

(a) $f : \mathbb{R} \to \mathbb{R}$ *is* ***even*** *if* $f(-x) = f(x)$.

(b) $f : \mathbb{R} \to \mathbb{R}$ *is* ***odd*** *if* $f(-x) = -f(x)$.

Notes:

- Even functions are modeled on $f(x) = x^2$, while odd functions are modeled on $f(x) = x^3$.
- Common even functions include x^{2k} for $k \geq 0$, $\cos(x)$, and $\cosh(x)$.
- Common odd functions include x^{2k+1} for $k \geq 0$, $\sin(x)$, and $\sinh(x)$.
- More examples can be found at Math is Fun and Khan Academy.

Most functions are neither even nor odd. That is what makes this class of functions very useful to us.

Definition 3.21. *Consider* $f : \mathbb{R} \to \mathbb{R}$.

(a) $f_{\mathrm{e}} : \mathbb{R} \to \mathbb{R}$ *by* $f_{\mathrm{e}}(x) := \frac{f(x)+f(-x)}{2}$ *is called the* ***even part of f***.

(b) $f_{\mathrm{o}} : \mathbb{R} \to \mathbb{R}$ *by* $f_{\mathrm{o}}(x) := \frac{f(x)-f(-x)}{2}$ *is called the* ***odd part of f***.

Notes:

- $f_{\mathrm{e}}(-x) := \frac{f(-x)+f(-(-x))}{2} = \frac{f(-x)+f(x)}{2} = \frac{f(x)+f(-x)}{2} = f_{\mathrm{e}}(x)$, and hence f_{e} is an even function.
- $f_{\mathrm{o}}(-x) := \frac{f(-x)-f(-(-x))}{2} = \frac{f(-x)-f(x)}{2} = -\frac{f(x)-f(-x)}{2} = f_{\mathrm{o}}(x)$, and hence f_{o} is an odd function.
- $f_{\mathrm{e}}(x) + f_{\mathrm{o}}(x) = \frac{f(x)+f(-x)}{2} + \frac{f(x)-f(-x)}{2} = \frac{f(x)+f(x)}{2} + \frac{f(-x)-f(-x)}{2} = f(x)$
- **Hence, every function is the sum of its even and odd parts.**

Example 3.22. *Compute the even and odd parts of* e^x.

Solution:

We define $f(x) := e^x$. Then,

$$f_{\mathrm{e}}(x) = \frac{f(x) + f(-x)}{2}$$

$$= \frac{e^x + e^{-x}}{2}$$

$$= \cosh(x).$$

Similarly,

$$\begin{aligned} f_o(x) &= \frac{f(x) - f(-x)}{2} \\ &= \frac{e^x - e^{-x}}{2} \\ &= \sinh(x). \end{aligned}$$

■

Proposition 3.23: Integrals of Even and Odd Functions over Symmetric Intervals

Suppose that $f : \mathbb{R} \to \mathbb{R}$ is a function and that for $a > 0$, the definite integrals $\int_{-a}^{0} f(x)\,dx$ and $\int_0^a f(x)\,dx$ both exist and are finite. Then the following hold:

(a) If f is an even function,

- $\int_{-a}^{0} f(x)\,dx = \int_0^a f(x)\,dx$, and
- $\int_{-a}^{a} f(x)\,dx = 2\int_0^a f(x)\,dx$.

(b) If f is an odd function,

- $\int_{-a}^{0} f(x)\,dx = -\int_0^a f(x)\,dx$, and
- $\int_{-a}^{a} f(x)\,dx = 0$.

(c) For all $[b, c] \subset [-a, a]$,

- $\int_b^c f(x)\,dx = \int_b^c f_e(x)\,dx + \int_b^c f_o(x)\,dx$.

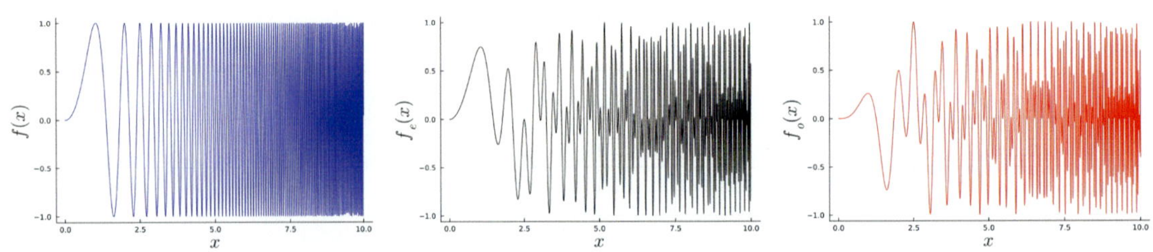

Figure 3.11: Even and Odd Parts to the Rescue: Parth (a) shows a complicated function that may be hard to integrate. Parts (b) and (c) show the even and odd parts, respectively. While they still look hard to integrate, Prop. 3.23 shows how they can be used to check the accuracy of an integration routine.

Example 3.24. Figure 3.11 *shows a complicated function to integrate. We are concerned about the ability of* `QuadGK` *to integrate the function. One means of checking its accuracy is*

- *Integrate* $f(x)$.
- *Integrate the odd part of* $f(x)$, $f_o(x)$.
- *Integrate the even part of* $f(x)$, $f_e(x)$.
- *Check that* $\int_a^b f(x)\,dx = \int_a^b f_o(x)\,dx + \int_a^b f_e(x)\,dx$.

They match to 4.385×10^{-15}, *which is either outstanding or* `QuadGK` *actually integrated the even and odd parts separately and added them together!*

Solution:

```
using QuadGK

f(x) = sin(x^2 + .5*x^3)
fo(x) = ( f(x) - f(-x) )/2.0
fe(x) = ( f(x) + f(-x) )/2.0

xmax = 10.0
xmin = 0

area, error = quadgk(f, xmin, xmax);            @show area, error
area_o, error_o =  quadgk(fo, xmin, xmax);   @show area_o, error_o
area_e, error_e =  quadgk(fe, xmin, xmax);   @show area_e, error_e

area - (area_o + area_e)
```

Output

```
(area, error) = (0.4817407024672528, 6.724935180040881e-9)
(area_o, error_o) = (0.08683542238431138, 1.235326109499035e-9)
(area_e, error_e) = (0.39490528008293707, 5.824010571660494e-9)

4.385380947269368e-15
```

■

Figure 3.12: A small mobile robot used in various Michigan Robotics courses. The user commands wheel speed. If both wheels move at the same speed, the robot moves in a straight line. If one wheel turns faster than the other, the robot turns toward the side of the slower wheel.

3.4 Applications of the Definite Integral

Now that we have delved into the theoretical and computational underpinnings of the definite integral, it is time to witness its practical prowess. This cornerstone of calculus is not just an abstract mathematical concept; it is a versatile tool that brings clarity to complex problems across diverse fields. In robotics and mechanical design, its value becomes immediately apparent. Consider, for example, a mobile robot traversing a landscape: by integrating its velocity, we can pinpoint its exact change in position, offering a detailed account of its journey. When we are faced with determining the area sandwiched between two functions, the definite integral steps in, quantifying the shared space with precision. This is especially vital in robot link design, where discerning properties like the center of mass is paramount for ensuring both stability and accurate movement. Venturing further, the definite integral plays a starring role in the realm of solids of revolution, allowing us to both visualize and calculate the volume of shapes birthed from rotating curves around an axis. As we work through these various applications, the definite integral's depth and adaptability will shine through, underscoring why it was invented at the precise time in human history when we were trying to decode the intricacies of the world and the solar system.

3.4.1 From Speed to Change in Position

When we drive a car, fly an airplane, or pilot a boat, with the goal of going from point A to point B, we typically do not set new position destinations (i.e., waypoints) every second along a planned path. Instead, we command speed and direction. Here, we will explore the relationship between speed and change in position in the context of a small mobile ground robot, such as the one in **Fig. 3.12**.

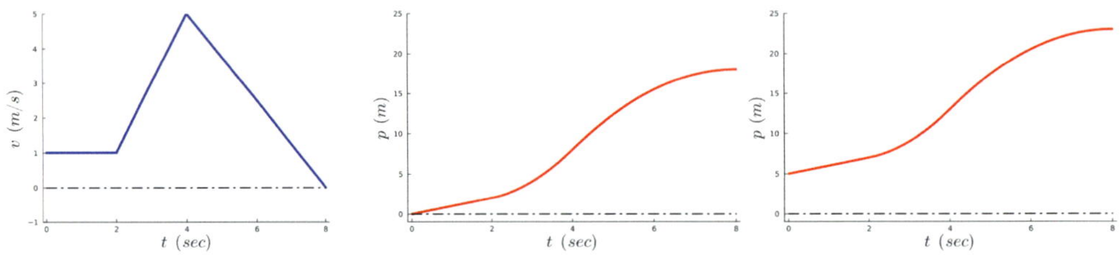

(a) **(b)** **(c)**

Figure 3.13: How is position related to speed of a moving body? Part (a) shows a simple speed profile given by (3.39). Part (b) shows the resulting position profile when the body starts at the point $p_0 = 0.0$, while part (c) shows the position profile when the body starts at a point $p_0 = 5.0$ meters, instead of at the origin. Varying p_0 simply shifts the curve up or down.

Consider the speed profile,

$$v(t) = \begin{cases} 1 & 0 \le t < 2, \\ 1 + 2 \cdot (t-2) & 2 \le t < 4, \\ 5 - 1.25 \cdot (t-4) & 4 \le t \le 8, \end{cases} \tag{3.39}$$

as plotted in **Fig. 3.13-(a)**. For the first two seconds, the robot is moving at 1 m/s (i.e., meters per second). Hence, after 1 s, it has traveled 1 m, and after 2 s, it has traveled 2 m. That is simple to comprehend because the speed is constant over the first two seconds. What about the next two seconds, where the speed is ramping up with a slope of 2 m/s^2? Well, we could compute the change in position by approximating the speed over a small interval of time as being constant, then multiply the constant by the time interval and, finally, add the displacement to the previous speed position. That is the same procedure we followed when defining the Riemann Integral! It follows that **position is related to speed** by[3]

$$\boxed{p(t) = p_0 + \int_{t_0}^{t} v(\tau) d\tau,} \tag{3.40}$$

where t_0 is the initial time, $p_0 := p(t_0)$ is the initial position of the moving body at time t_0, and τ is a dummy variable of integration. **Fig. 3.13-(b)** and **(c)** show the position profile for two different initial conditions, namely, $p(0) = 0.0$ m and $p(0) = 5.0$ m. The integral itself computes the **change in position** or **relative displacement** of the robot. We have to add to the relative displacement the initial position of the robot to obtain the true position of the robot.

In the following, we show an inefficient but effective means of computing $p_0 + \int_{t_0}^{t} v(t)\, dt$, as a function of time, for $t \ge t_0$.

[3]So as not to confuse the variable being integrated with the time variable t, we used τ as the "dummy variable".

```
# Speed profile
function v(t)
    if t<2
        v = 1
    elseif t < 4
        v = 1 + 2(t-2)
    else
        v = 5 - 1.25(t-4)
    end
    return v
end

# How to integrate a generic function f from t0 to t
function integral_t(f, t0, t)
    result, err = quadgk(f, t0, t)
    return result
end

# Set an initial position
p0=5.0

# Define the integral as a function so that we can plot it
int(t) = p0 + integral_t(v, 0, t)

p1 = plot(int, 0, 8, linewidth=4, color=:red, label=false,
    xlabel=L"$t ~~ ({\rm sec})$", ylabel=L"$p ~~(m)$",
    yguidefontsize=18, xguidefontsize=18)
plot!(g,0, 8, linewidth=4, color=:black, label=:false)
xlims!(p1, (-0.1, 8.1))
ylims!(p1, (-1, 25))
display(p1)
png(p1, "SimpleDisplacementProfilOffset")
```

Output: The plots in Fig. 3.13-(b) and (c).

Let us now expand our geometry by considerig a robot moving in the plane $\mathbb{R}^2$, instead of along a line. We will model a velocity profile as a vector, $[v_x(t), v_y(t)]$. Because each component of the velocity depends on a single variable, t, we can perform the Riemann Integral separately for each component, as in:

$$p(t) = \begin{bmatrix} p_x(t) \\ p_y(t) \end{bmatrix} = \begin{bmatrix} p_x(t_0) + \int_{t_0}^{t} v_x(\tau)d\tau \\ \\ p_y(t_0) + \int_{t_0}^{t} v_y(\tau)d\tau \end{bmatrix}. \tag{3.41}$$

Example 3.25. *Figure 3.14 shows a patch of $\mathbb{R}^2$ with some obstacles in it. We want to steer our robot so that it starts at Point A, goes to Point B, and avoids the obstacles shown by the colored sections. Which of the following velocity profiles will work?*

$$(i) \quad \begin{bmatrix} v_y(t) \\ v_y(t) \end{bmatrix} = \begin{bmatrix} 0.00807179 \\ -1.97588 \end{bmatrix} + \begin{bmatrix} 0.677259 \\ 2.01576 \end{bmatrix} \cdot t + \begin{bmatrix} 0.437326 \\ 0.645312 \end{bmatrix} \cdot t^2 + \begin{bmatrix} -0.197531 \\ -0.592593 \end{bmatrix} \cdot t^3; \text{ see Fig. 3.14-(b).}$$

$$(ii) \quad \begin{bmatrix} v_y(t) \\ v_y(t) \end{bmatrix} = \begin{bmatrix} 0.00807179 \\ -1.97596 \end{bmatrix} + \begin{bmatrix} 0.677259 \\ 2.00199 \end{bmatrix} \cdot t + \begin{bmatrix} 0.437326 \\ 0.0767879 \end{bmatrix} \cdot t^2 + \begin{bmatrix} -0.197531 \\ -0.335802 \end{bmatrix} \cdot t^3; \text{ see Fig. 3.14-(c)}$$

Solution

We first use the velocity profile from (i).

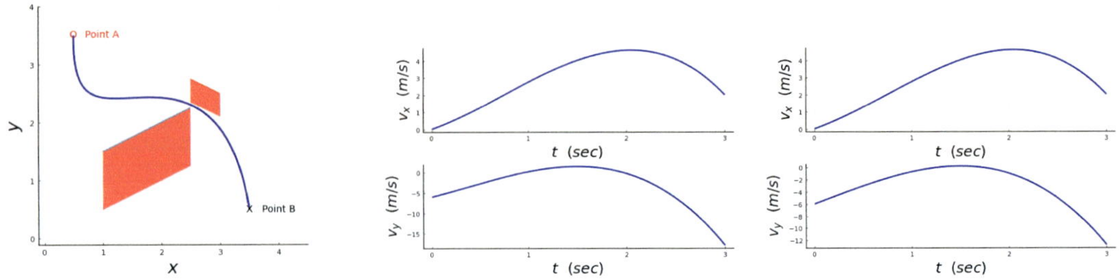

(a) (b) (c)

Figure 3.14: Our objective is to steer from Point A to Point B, without hitting the obstacles in (a). Among the speed profiles shown in (b) and (c), which one will get the job done? They look pretty similar.

```
# Show the parameters being used
@show P0
@show P4
@show beta

# Define the proposed velocity profiles and so that we can integrate
# them
Velx(t) = beta[1,1] + t*beta[2,1] + t^2*beta[3,1] + t^3*beta[4,1]
Vely(t) = beta[1,2] + t*beta[2,2] + t^2*beta[3,2] + t^3*beta[4,2]

# How to integrate a generic function f from t0 to t
function integral_t(f, t0, t)
    # Could just as easily use our version of Simpson's Rule
    result, err = quadgk(f, t0, t)
    return result
end

# Set an initial position
px0 = P0[1]
py0 = P0[2]

# Define the integral as a function so that we can plot it
intx(t) = px0 + integral_t(Velx, 0, t)
inty(t) = py0 + integral_t(Vely, 0, t)

# Use integration to compute the Robot's Path in the Plane
N=1000
t = range(0, 3, length=N)
posRobotx = intx.(t)
posRoboty = inty.(t)

# Plot the Obstacles
ob1(x)=1 + x/2
ob2(x)=0.0 + x/2
x=range(1, 2.5, length=N)
p1=plot(x, ob1.(x), fillrange=ob2.(x), fillalpha=0.6, fillcolor=:red, label=false)
ob1(x)=4 - x/2
ob2(x)=3.6 - x/2
x=range(2.5, 3.0, length=N)
p1=plot!(x, ob1.(x), fillrange=ob2.(x), fillalpha=0.6, fillcolor=:red, label=false)

```

```
# Plot the Robot's path in the Plane with Obstacles
plot!(posRobotx, posRoboty, legend=false, linewidth=2, color=:blue, label=false,
    aspect_ratio=1,xlabel=L"$x$", ylabel=L"$y$",
    yguidefontsize=18, xguidefontsize=18)
annotate!(P0[1], P0[2], text("O", :red, :center, 10))
annotate!(P0[1] + 0.5, P0[2], text("Point A", :red, :center, 10))
annotate!(P4[1], P4[2], text("X", :black, :center, 10))
annotate!(P4[1] + 0.5, P4[2], text("Point B", :black, :center, 10))
xlims!(p1, (-0.1, 4.5))
ylims!(p1, (-0.1, 4))
```

Output

```
P0 = [0.5, 3.5]
P4 = [3.5, 0.5]

beta = [0.008071793840493424 -1.9758808107720176;
0.6772585064197526 2.015759503265411;
0.4373262151028052 0.6453119786307883;
-0.19753086419795485 -0.592592592589968]
```

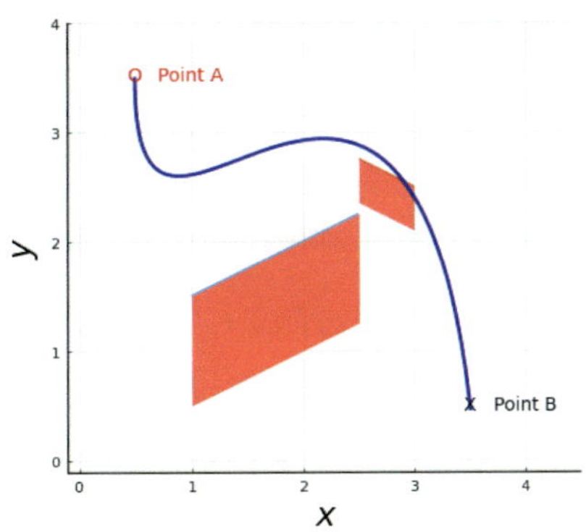

A plot of the path corresponding to the velocity profile (i) in Example 3-7 is shown above. While it ALMOST misses the upper obstacle, it does, in fact, hit it, as shown in the plot. The correct answer is velocity profile (ii), which produces the path shown in Fig. 3.14-(a). ■

3.4.2 Ballistic Motion

Ballistic motion, also known as **projectile motion**, refers to the motion of objects that are launched, thrown, or projected into the air and subject to only the force of gravity; see Fig. 3.15. This type of motion combines both vertical and horizontal movements. When an object is launched, it follows a curved path known as a parabola, assuming there is no air resistance. The analysis of ballistic motion allows us to predict the trajectory, range, maximum height, and time of flight of the projectile, making it an important concept in various applications, from sports to engineering.

Similarly to how (3.40) allows us to go from velocity $v(t)$ to position $p(t)$, we can also go from acceleration $a(t)$ to velocity $v(t)$ via

$$\boxed{v(t) = v_0 + \int_{t_0}^{t} a(\tau)d\tau,} \tag{3.42}$$

and then apply (3.40) to obtain position. For ballistic motion in the absence of air resistance, the acceleration is constant,

$$a(t) = \begin{bmatrix} 0.0 \\ -g \end{bmatrix}, \tag{3.43}$$

Figure 3.15: **Ballistic Motion:** All Hail San Antonio's Coyote, Inventor of the T-shirt Cannon? An example of ballistic motion in everyday life.

making it quite easy to obtain velocity:

$$v(t) = \begin{bmatrix} v_x(t) \\ v_y(t) \end{bmatrix} = \begin{bmatrix} v_x(t_0) + \int_{t_0}^{t} a_x(\tau)\, d\tau \\ v_y(t_0) + \int_{t_0}^{t} a_y(\tau)\, d\tau \end{bmatrix} = \begin{bmatrix} v_x(t_0) + \int_{t_0}^{t} 0.0\, d\tau \\ v_y(t_0) - \int_{t_0}^{t} g\, d\tau \end{bmatrix} = \begin{bmatrix} v_x(t_0) \\ v_y(t_0) - g\cdot(t - t_0) \end{bmatrix}. \tag{3.44}$$

Integrating once more to obtain position yields the well-known equations:

$$p(t) = \begin{bmatrix} p_x(t_0) + \int_{t_0}^{t} v_x(\tau)\, d\tau \\ p_y(t_0) + \int_{t_0}^{t} v_y(\tau)\, d\tau \end{bmatrix} = \begin{bmatrix} p_x(t_0) + \int_{t_0}^{t} v_x(t_0)\, d\tau \\ p_y(t_0) + \int_{t_0}^{t} \left(v_y(t_0) - g\cdot(\tau - t_0)\right)\, d\tau \end{bmatrix} \tag{3.45}$$

$$= \begin{bmatrix} p_x(t_0) + v_x(t_0)\cdot(t - t_0) \\ p_y(t_0) + v_y(t_0)\cdot(t - t_0) - \frac{1}{2} g\cdot(t - t_0)^2 \end{bmatrix}. \tag{3.46}$$

In the above equation, t_0, $p(t_0)$, and $v(t_0)$ are initial conditions for the motion, whereas g is a physical parameter.

(a) (b)

Figure 3.16: Ballistic motion with emotion! (a) University of Michigan's Crisler Center. (b) Jalen Rose is manning the t-shirt cannon.

Example 3.26. *Similar to the depiction in Fig. 3.16, suppose that Jalen Rose of Fab Five fame wants to pop a t-shirt to Jimmy King, whom he has spotted hiding up in the stands. If Jimmy is at a position 20 meters longitudinally and 10 meters vertically from the t-shirt cannon, at what angle should Jalen tilt the cannon if* *the exit velocity of the t-shirt is 25 m/s?*

Additional Data: *The mass of a t-shirt is 140 g or 0.14 Kg; zero degrees corresponds to the cannon's barrel being parallel to the ground, and* $g \approx 9.81$ *m/s*^{2}.

Solution: Ans. ≈ 0.6395 rad $\approx 37°$.

We can assume $t_0 = 0$. And because we are given the desired end position of the t-shirt relative to its exit point from the cannon, we have $p_x(t_0) = p_y(t_0) = 0.0$. Next, we resolve the t-shirt's exit velocity from the cannon barrel into its x and y components:

$$\begin{bmatrix} v_x(0) \\ v_y(0) \end{bmatrix} = \begin{bmatrix} 25\cos(\theta) \\ 25\sin(\theta) \end{bmatrix}.$$

Let T be the time of flight (the time the t-shirt lands in Jimmy King's lap, a quantity we need to solve for). The ballistic equation given by (3.45) simplifies to:

$$p(T) = \begin{bmatrix} 25\cos(\theta)\cdot T \\ \\ 25\sin(\theta)\cdot T - \frac{1}{2}(9.81)T^2 \end{bmatrix} = \begin{bmatrix} 20 \\ \\ 10 \end{bmatrix}. \tag{3.47}$$

Solving the first equation for T yields $T = \frac{20}{25\cos(\theta)}$, which when substituted into the second equation gives

$$10 = 20\tan(\theta) - 4.905\cdot\left(\frac{4}{5\cos(\theta)}\right)^2 \implies \theta \approx 0.6395 \text{ rad} \approx 36.6°,$$

which we solve with the bisection algorithm. From the Julia code, we also obtain that it takes 0.997 s for the t-shirt to arrive in Jimmy's lap.

```

# make the algorithm handy
function bisection(f, a, b, tol)
    if f(a) * f(b) > 0
        error("f(a) and f(b) must have opposite signs")
        return Na
    end
    kmax = 1e5 # max number of loops so that the algorithm terminates
    k=1
    c=NaN #define c out of the loop so we can access it
    while ( (b - a) / 2 > tol || abs(f(c))> tol ) && k < kmax
        c = (a + b) / 2
        #
        if f(a) * f(c) < 0
            b = c
        else
            a = c
        end
        k = k+1
    end
    pmErrorBound = (b - a)/2
   println("Number of iterations = ",k)
    return (est=(a + b) / 2, low=a, up=b, error=pmErrorBound)
end

# Set up the problem
px0 = 0 # meters
py0 = 0 # meters
```

```
v0 = 25
0  # m/s

pxT = 20 # meters
pyT = 10 # meters

gG = 9.81

# pxT =  v0*cos(theta)*T
T(theta) = pxT/(v0*cos(theta))
# py =  v0*sin(theta)*T
px(theta) = v0*cos(theta)*pxT/(v0*cos(theta))

#f(theta) = -pyT + v0*sin(theta)*T - 0.5*gG*T^2
f(theta) = -pyT + v0*sin(theta)*pxT/(v0*cos(theta)) -
                0.5*gG*(pxT/(v0*cos(theta)))^2

thetaStar = bisection(f, pi/16, pi/4, 1e-6)

@show theta_rad = thetaStar.est

@show theta_deg = theta_rad*180/pi

Tstar = T(thetaStar.est)
println("Duration of the tShirts toss = $Tstar  seconds")
```

Output

```
Number of iterations = 22
theta_rad = thetaStar.est = 0.6395180791349615
theta_deg = (theta_rad * 180) / pi = 36.64168685674669
Duration of the tShirts toss = 0.9970295179594787  seconds
```

3.4.3 Area between Two Functions

When an area of interest is bounded by two functions instead of one, the calculation for area changes slightly, as illustrated in Fig. 3.17. The area formula becomes:

$$\text{Signed Area}_g^f := \int_a^b (f(x) - g(x))\, dx. \tag{3.48}$$

As described in the figure, this is the signed area, just as with the normal Riemann Integral. If we want the pink-shaded regions to be counted as positive areas, then we would need to evaluate:

$$\text{Unsigned Area}_g^f := \int_a^b |f(x) - g(x)|\, dx. \tag{3.49}$$

It is a fact that if $f : [a,b] \to \mathbb{R}$ and $g : [a,b] \to \mathbb{R}$ are both continuous, then so are $f - g : [a,b] \to \mathbb{R}$ and $|f - g| : [a,b] \to \mathbb{R}$, and thus their Riemann Integrals exist.

Example 3.27. *Compute the area bounded by the two functions in* Fig. 3.18, *defined by*

$$f(x) = \begin{cases} 2 - |x| & |x| \le 2 \\ 0 & \textit{otherwise} \end{cases} \quad \textit{and} \quad g(x) = \begin{cases} |x| - 1 & |x| \le 1 \\ 0 & \textit{otherwise.} \end{cases} \tag{3.50}$$

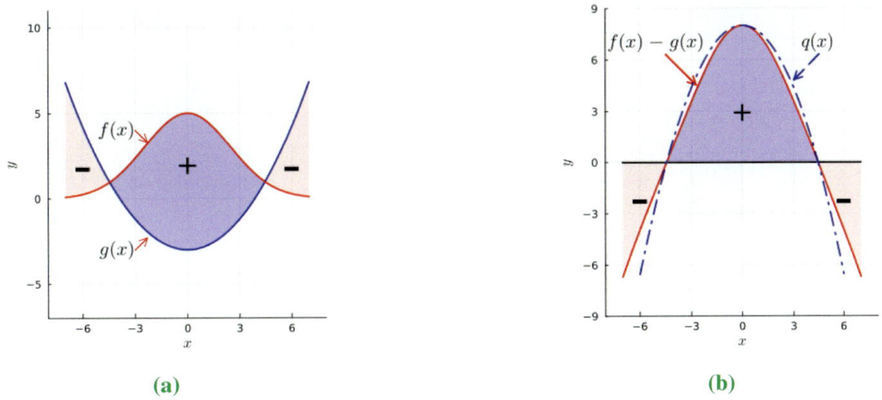

Figure 3.17: The two images have the same scale: the y-axis in image (a) extends from −7.0 to +11.0, and in (b), it extends from −9.0 to +9.0. We defined the Riemann Integral as the (signed) area under a curve. In reality, we were looking at the area between a function $f(x)$ and the x-axis. When the function was positive, i.e., above the x-axis, the area was positive, and when the function was below the x-axis, the area was negative. More generally, it is common for a shape to be defined by more than one function. In (a), we treat the case of two functions, $f(x)$ and $g(x)$. The (signed) area is the Riemann Integral of $f(x) - g(x)$ over the region of interest. Here, we emphasize that when $f(x) - g(x) < 0$, the area "between the curves" is negative. If you want to count that as a positive area, then you would integrate $|f(x) - g(x)|$, the absolute value of $f(x) - g(x)$. In (b), we plot the function $f(x) - g(x)$. The corresponding shaded areas in each figure have the same area. Moreover, the plot in (b) looks like an inverted parabola, but it is not. The dashed line is the function $q(x) = -0.404022x^2 + 8.0$, which matches $f(x) - g(x)$ at its peak and its two zero crossings. Because a parabola is determined by its values at three distinct points, this shows that $f(x) - g(x)$ is not a parabola.

Solution: The area in Fig. 3.18-(a) is the sum of two triangles,

$$\text{Area}_a = \frac{1}{2} \cdot 4 \cdot 2 + \frac{1}{2} \cdot 2 \cdot 1 = 5.0.$$

The area in Fig. 3.18-(b) is the sum of three triangles and a rectangle,

$$\text{Area}_b = \underbrace{\frac{1}{2} \cdot 2 \cdot 2}_{\text{top triangle}} + \underbrace{2 \cdot \frac{1}{2} \cdot 1 \cdot 1}_{\text{two side triangles}} + \underbrace{2 \cdot 1}_{\text{rectangle}} = 5.0.$$

Hence, the two areas are equal. ■

Example 3.28. *For the functions $f(x) := 5 \cdot \exp(-x^2/12)$ and $g(x) := -3 + 0.2 \cdot x^2$ plotted in* Fig. 3.17, *compute the following areas:*

(a) Signed area between $f(x)$ and $g(x)$ for the interval $[a, b] := [-6, 6]$.

(b) Unsigned area between $f(x)$ and $g(x)$ for the interval $[a, b] := [-6, 6]$.

(c) The area indicated by the blue patch in Fig. 3.17.

Solutions:

(a) **Ans.** 37.461.

(b) **Ans.** 49.587

(c) **Ans.** 43.524

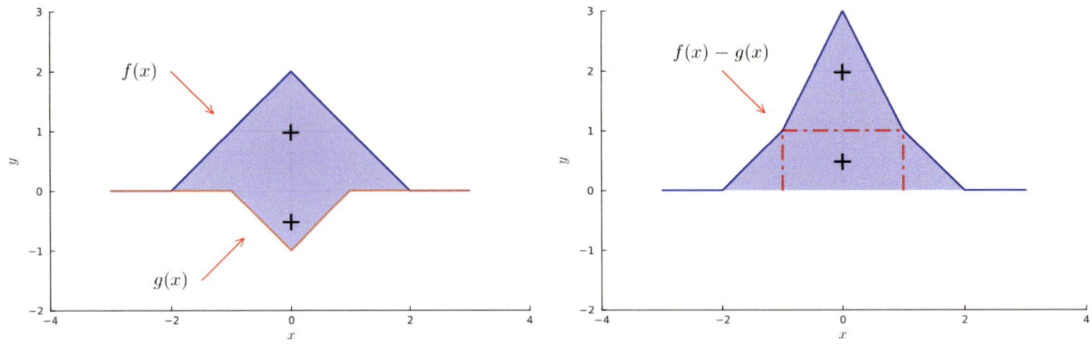

(a) (b)

Figure 3.18: Triangles provide a simpler situation in which to understand that the area bounded by two functions is the same as the area defined by the difference of the two functions. The functions are defined in (3.50). (a) From the image, we compute the area of the top triangle as 4, and the bottom triangle as 1, for a total area of 5 units squared. (b) Decomposing the shape into a top triangle, a rectangle, and two side triangles, we arrive at its area as $1 + 2 + 1 + 1 = 5$. They are the same, whether they look it or not!

```
# Problem Solutions

f(x) = 5*exp(-x^2/12)
g(x) = -3 + 0.2*x^2

h(x) = f(x) - g(x)
absh(x) = abs(h(x))
a = -6
b = 6

# For bounding the blue-shaded region. we use the
# Bisection Algorithm to find where h(x)=0
xRoot1 = bisection(h, -5, -4, 1e-6)
xRoot1 = xRoot1.est
xRoot2 = bisection(h, 2.5, 5, 1e-6)
xRoot2 = xRoot2.est

# We compute the various areas for parts (a), (b), and (c)

@show Ans_a, error_a = quadgk(h, a, b)
println(" ")
@show Ans_b, error_b = quadgk(absh, a, b)
println(" ")
@show [xRoot1, xRoot2]
@show Ans_c, error_c = quadgk(h, xRoot1, xRoot2);
```

Output

```
(Ans_a, error_a) = quadgk(h, a, b) = (37.46061361388632,
1.2664415294239006e-8)

(Ans_b, error_b) = quadgk(absh, a, b) = (49.58734405494167,
1.984501950419177e-7)

[xRoot1, xRoot2] = [-4.4498186111450195, 4.449818730354309]
```

```
(Ans_c, error_c) = quadgk(h, xRoot1, xRoot2) = (43.52397876865702,
1.0885337076160795e-9)
```

■

Example 3.29. *Use numerical methods to solve this problem from a Putnam Exam: Find c so that the two areas defined by the curves $y = 2x - 3x^3$ and $y = c$ are equal.*

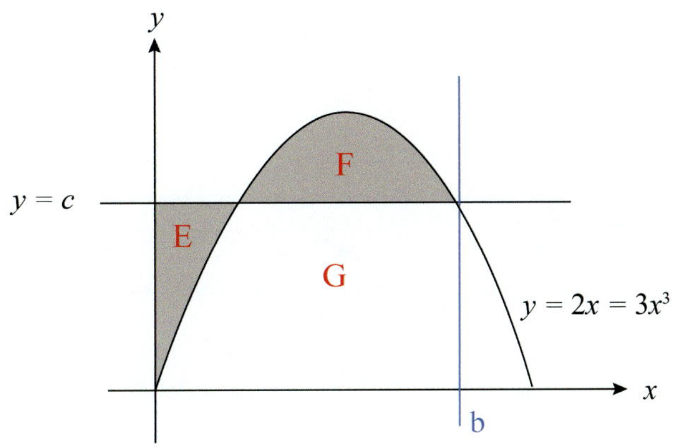

Solution:

The peak of the curve $y = 2x - 3x^3$ occurs at $x = \frac{\sqrt{2}}{3}$. For $c \in \mathbb{R}$ such that the horizontal line $y = c$ intersects the curve $y = 2x - 3x^3$ in two points, let b be the rightmost intersection. Then, areas E and F are equal if, and only if, $\int_0^b (\underbrace{2x - 3x^3}_{f(x)} - \underbrace{c}_{g(x)})\, dx = 0$. We use the Bisection Algorithm to solve for b and c.

```
f(x) = 2x - 3x^3

function Int(c)
    h(x) = f(x)-c
    # For c given, solve for b such that f(b)-c = 0.
    # Bisection is great for finding roots
    Ans = bisection(h, sqrt(2)/3, 2, 1e-6)
    b = Ans.est
    # Compute the integral from zero to b. We seek the value of c
    # such that the integral from 0 to b equals zero
    # Hence, we can use Bisection for that too.
    value, error = quadgk(h, 0, b)
    return Area=value
end
```

Output

```
Int (generic function with 1 method)
```

```
Ans = bisection(Int, .3, .5, 1e-9)
@show c = Ans.est

# let's find b, even though it is not part of the answer
```

```
h(x) = f(x)-c
Ans = bisection(h, sqrt(2)/3, 2, 1e-6)
@show b = Ans.est;
```

Output

```
c = Ans.est = 0.4444444440305233
b = Ans.est = 0.6666664352015328
```

Note: $\frac{4}{9} = 0.444...$ and $\frac{2}{3} = 0.666...$ Hence, we guess that the analytical answers are $c = \frac{4}{9}$ and $b = \frac{2}{3}$. ■

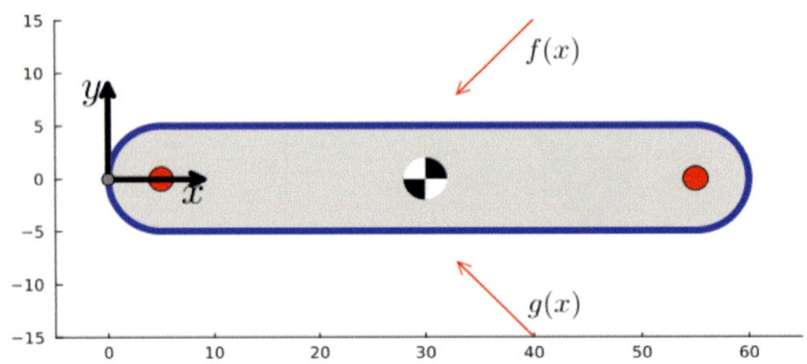

Figure 3.19: A planar version of a very basic robot link in units of centimeters (cm). A Cartesian coordinate system is attached to the link, with its origin at the left end of the link. In these coordinates, the leftmost point on the link is at $(0, 0)$, and the rightmost point is at $(60, 0)$. The red dots indicate the location of two revolute joints, located at $(0, 5)$ and $(0, 55)$. The symbol in the middle indicates the location of the link's **center of mass**, a term that will be defined in the text.

Helpful Videos:

- Area Between Curves by Integration (Easy Way) by Quoc Dat Phung has high-quality drawings to support the discussion.
- Area Between Curves - Integrating with Respect to Y by Quoc Dat Phung has high-quality drawings to support the discussion.

3.4.4 Using the Power of the Rectangle to Find the Center of Mass of a (Planar) Robot Link

In physics and robotics, we are constantly leaning upon foundational principles to explore new frontiers. One of those is Newton's second law, $F = ma$, a seemingly simple equation that elegantly describes the motion of a (single) point mass. Yet, we quickly realize that a real robot link is not just a solitary point but an entire body comprised of infinitely many point masses, all moving in tandem. This realization is not just a curiosity of nature but a profound insight that echoes the essence of Calculus. By viewing a body as an infinite collection of points, we unveil deeper layers of understanding. The **Total Mass** becomes more than just a summation; it is an integral, a continuous accumulation of infinitesimal masses. The **Center of Mass** transforms from a mere geometric point to a gravitational anchor, summarizing the collective pull of gravity on every particle, as seen in this video. And the **Moment of Inertia**, discussed later, encapsulates a body's resistance to change in its rotational state, just as total mass encapsulates a body's resistance to change in its horizontal and vertical positions. The more mass a body has, the harder it is to push, and similarly, the more inertia a body has, the more effort it takes to change its angle of rotation.

Because this is not a mechanics course, we will make two simplifying assumptions:

- Each of our robot's links has a constant thickness in the z-direction, which will not be shown in our link drawings as they only give the outline in the x-y plane.

- The density of a link's material is constant throughout the link.

Here is some useful notation to get us going.

Notation:

- A is the (total) area of a link.

- h is the thickness (or height in the z-direction) of a link.

- V is the (total) volume of a link, and hence equals $A \cdot h$.

- ρ is the material density of a link. Because the volume of a typical robot link in m^3 is tiny, we use units of grams per centimeter cubed (g/cm^3) for the density. For reference, the density of aluminum is $\rho_0 := 2.7$ g/cm^3.

- M_{Total}, or M_{T} for short, is the total mass of a link. Hence,

$$\boldsymbol{M_{\text{T}} = \rho \cdot V = \rho \cdot h \cdot A},$$

because of our assumptions of uniform link thickness and density. Hence, for us, computing total mass is tantamount to computing the area, which we know how to do.

Total Mass:

From Section 3.4.3, if $f : [x_{\text{min}}, x_{\text{max}}] \to \mathbb{R}$ and $g : [x_{\text{min}}, x_{\text{max}}] \to \mathbb{R}$ provide the upper and lower boundaries of an object, such as the robot link shown in **Fig. 3.19**, then,

$$M_{\text{T}} = \rho \cdot h \cdot \int_{x_{\text{min}}}^{x_{\text{max}}} (f(x) - g(x))\, dx. \tag{3.51}$$

This is very straightforward due to our assumptions of constant density and thickness. We next move on to an object's center of mass.

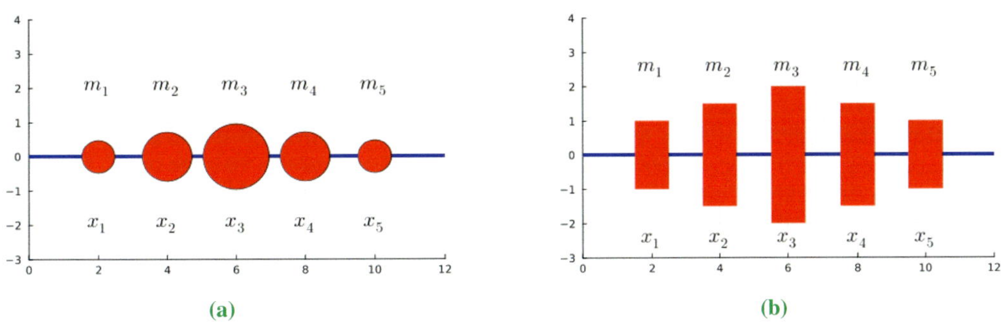

Figure 3.20: The **x-center of mass** is the point where the collection of masses would balance, taking into account both the position and magnitude of each mass. (a) Representation of a system of discrete point masses m_i located at positions x_i along the x-axis. The size of each red ball is proportional to its corresponding mass. (b) Visualization of the same system of discrete masses as rectangles. Each rectangle's height corresponds to the mass m_i and is centered at position x_i. This representation bridges the concept from discrete point masses to continuous mass distributions, emphasizing the idea of the center of mass as the system's balance point in a continuous setting. The depiction with rectangles should inspire us to recall that the Riemann integral is, after all, a sum of infinitely many, infinitely thin rectangles.

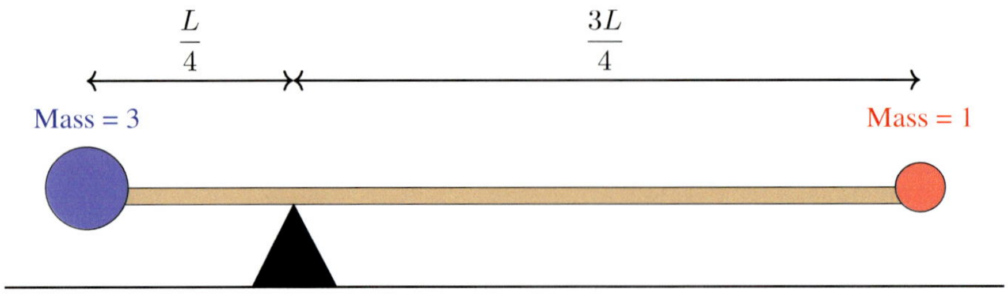

Figure 3.21: To be in balance, the sum of the moments about the fulcrum must be equal. If x_c is the position of the fulcrum and x_1, x_2 denote the positions of the masses m_1 and m_2, then the condition for balance is $m_1 \cdot (x_1 - x_c) + m_2 \cdot (x_2 - x_c) = 0$.

Center of Mass, First for Point Masses:

You may have encountered this concept in a physics course for a discrete set of masses as shown in Figs. 3.20. Given the symmetric distribution of the masses in the figure, the center of mass is computed by eye to be the position of the middle mass, x_3. But how does one determine it in general, such as in Fig. 3.21?

Consider a collection of N point masses $m_1, m_2, \ldots, m_N$ of various sizes located at various positions $x_1, x_2, \ldots, x_N$, respectively, along the x-axis. The **total mass** is

$$M_{\text{T}} := \sum_{i=1}^{N} m_i.$$

For the case of point masses lying along the x-axis (i.e., their y-coordinate is identically zero) the moment about the origin and the moment about the y-axis are the same. Because we will soon consider masses with nonzero y-coordinates, we use the term **Moment about the y-axis**, denoted M_y.

For the i-th mass, its contribution to M_y is the product of its mass and its distance from the y-axis, namely

$$M_{y,i} := m_i \cdot x_i.$$

The **Total Moment about the y-axis** is the sum of the individual moments, leading to

$$M_y := \sum_{i=1}^{N} M_{y,i} = \sum_{i=1}^{N} m_i \cdot x_i.$$

The **x-coordinate of the center of mass**, denoted $\boldsymbol{x_c}$, is the total y-moment divided by the total mass of the system,

$$x_c := \frac{M_y}{M_{\text{T}}} = \frac{\sum_{i=1}^{N} m_i \cdot x_i}{\sum_{i=1}^{N} m_i}. \tag{3.52}$$

Interpretation: The formula for x_c is the weighted average of the x-positions of the point masses, where each position is weighted by its corresponding mass. The center of mass is the "balance point" of the system, and the formula reflects this by taking into account both the position and mass of each point in the system. More information is available at Equation for Center of Mass by Khan Academy.

Source of the Equation for Center of Mass. From physics, for a "plank" (or teeter-totter) to be in equilibrium, the sum of the "moment arms" must be zero. That is,

$$\sum_{i=1}^{N} m_i \cdot (x_i - x_c) = 0.$$

This yields,

$$\sum_{i=1}^{N} m_i \cdot x_i = \sum_{i=1}^{N} m_i \cdot x_c = x_c \cdot \sum_{i=1}^{N} m_i.$$

Solving this equation for x_c yields the formulas given by (3.52).

Example 3.30. *For Fig. 3.20, the masses and x-locations are:*

$$(m_1, m_2, m_3, m_4, m_5) = (2, 3, 4, 3, 2)$$

and

$$(x_1, x_2, x_3, x_4, x_5) = (2, 4, 6, 8, 10).$$

Compute the center of mass, x_c.

Solution: The total mass is $M_{\mathrm{T}} = 14$ and the total y-moment equals 84. Hence,

$$x_c := \frac{84}{14} = 6 = x_3.$$

To be in equilibrium, a fulcrum would be placed exactly in the middle at x_3.

```
x_positions = [2, 4, 6, 8, 10]
masses = [2, 3, 4, 3, 2]
@show M_T = sum(masses)
@show M_y = sum(masses.*x_positions)
@show x_c = M_y / M_T;
```

Output

```
M_T = sum(masses) = 14
M_y = sum(masses .* x_positions) = 84
x_c = M_y / M_T = 6.0
```

Example 3.31. *Suppose we make the mass at the far end truly massive,*

$$(m_1, m_2, m_3, m_4, m_5) = (2, 3, 4, 3, 200)$$

and leave the x values the same,

$$(x_1, x_2, x_3, x_4, x_5) = (2, 4, 6, 8, 10).$$

Compute the new center of mass, x_c.

Solution: The total mass is $M_{\mathrm{T}} = 212$ and the total y-moment equals 2064. Hence,

$$x_c := \frac{2064}{212} = 9.736 \approx x_5.$$

To be in equilibrium, a fulcrum would have to be placed almost directly under the huge mass at x_5.

```
x_positions = [2, 4, 6, 8, 10]
masses = [2, 3, 4, 3, 200]
@show M_T = sum(masses)
@show totalMomentx = sum(masses.*x_positions)
@show x_c = totalMomentx / M_T;
```

Output

```
M_T = sum(masses) = 212
totalMomentx = sum(masses .* x_positions) = 2064
x_c = totalMomentx / M_T = 9.735849056603774
```

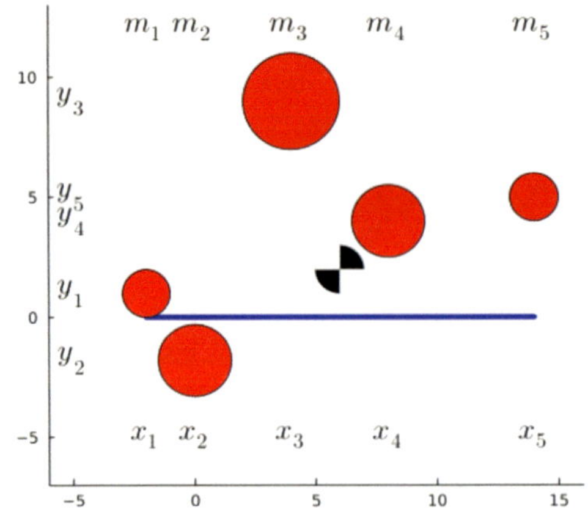

Figure 3.22: Five masses of various sizes and locations.

Example 3.32. *As in Fig. 3.22, suppose the masses also have nonzero y-locations in addition to their x-locations. Compute both the x and y components of the center of mass. Note that it is harder to eyeball its location.*

Data:

$$(m_1, m_2, m_3, m_4, m_5) = (2, 3, 4, 3, 2)$$

$$(x_1, x_2, x_3, x_4, x_5) = (-2, 0, 4, 8, 14),$$

and

$$(y_1, y_2, y_3, y_4, y_5) = (1, -1.8, 9, 4, 5).$$

Compute the center of mass, (x_c, y_c).

Solution: The total mass is $M_{\mathrm{T}} = 14$ and the total y-moment equals 64. Hence,

$$x_c := \frac{64}{14} = 4.571.$$

The x-moment equals 54.6. Hence,

$$y_c := \frac{54.6}{14} = 3.9.$$

It follows that if the masses were attached to a massless support and we were to drill a finger hole precisely at

$$(x_c, y_c) = (4.571, 3.9),$$

the overall object would balance perfectly on our fingertip without any tendency to rotate. Hopefully, this makes physical sense to you, but if not, Fig. 3.23 or the video Center of Mass along x y plane by Ian Page, may help.

Our closing remark is that the center of mass of a "curved" object does not have to lie within the object itself; consider, for example, an annulus or a doughnut.

```
# Balls with nonzero y_positions
x_positions = [-2, 0, 4, 8, 14]
y_positions = [1, -1.8, 9, 4, 5]
masses = [2, 3, 4, 3, 2]

@show M_T = sum(masses)
@show M_y = sum(masses.*x_positions)
@show M_x = sum(masses.*y_positions)
@show x_c = M_y / M_T
@show y_c = M_x  / M_T;
```

Output

```
M_T = sum(masses) = 14
M_y = sum(masses .* x_positions) = 64
M_x = sum(masses .* y_positions) = 54.6
x_c = M_y / M_T = 4.571428571428571
y_c = M_x / M_T = 3.9
```

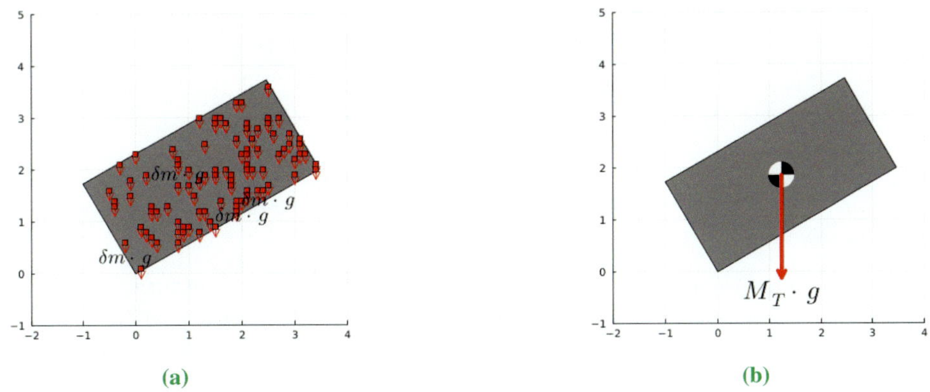

Figure 3.23: Physical Significance of the Center of Mass: Calculus allows us to correctly apply Newton's Laws to macroscopic objects. Image (a) illustrates that a physical object is comprised of a myriad of tiny particles of mass δm, each one undergoing the effects of gravity. Image (b) shows that once the **Center of Mass** is computed, we recover the Second Law of Newton, $F = M_{\rm T} \cdot g$. **There is one more important feature about the center of mass**: if the object has zero initial rotation (spin), placing the force of gravity at the center of mass will not produce any "torque" on the object, and hence it will not rotate under the action of gravity, which is what Newton observed (his famous apple was not spinning as it fell from the tree; that would have complicated his explanation immensely!). If we were to place $M_{\rm T} \cdot g$ at a corner of the rectangle, it would cause the rectangle to rotate, which would go counter to experimental evidence. M_y and M_x are capturing that effect.

Center of Mass for a (Continuous) Planar Robot Link

Center of Mass in the Plane

In the following, we assume that $f : [x_{\rm min}, x_{\rm max}] \to \mathbb{R}$ and $g : [x_{\rm min}, x_{\rm max}] \to \mathbb{R}$ provide the upper and lower boundaries of a planar object, such as the robot link shown in Fig. 3.19.

Definition 3.33. *The following mechanical parameters are defined for a planar object:*

(a) ***Total Moment about the y-axis:***

$$M_y = \rho \cdot h \cdot \int_{x_{\rm min}}^{x_{\rm max}} x \cdot (f(x) - g(x))\, dx.$$

(b) ***Total Moment about the x-axis:***

$$M_x = \rho \cdot h \cdot \int_{x_{\rm min}}^{x_{\rm max}} \frac{1}{2} \cdot \left(f^2(x) - g^2(x)\right) dx.$$

The formula for M_y is directly analogous to what we had for point masses, while the formula for M_x seems totally different. The justification is given later in an optional reading section. Once we get over the hurdle for M_x, the definition of the Cartesian center of mass is straightforward, namely

(c) x-Coordinate of Center of Mass:

$$x_c := \frac{M_y}{M_{\rm T}} = \frac{\int_{x_{\min}}^{x_{\max}} x \cdot (f(x) - g(x))\, dx}{\int_{x_{\min}}^{x_{\max}} (f(x) - g(x))\, dx}, \tag{3.53}$$

and

(d) y-Coordinate of Center of Mass:

$$y_c := \frac{M_x}{M_{\rm T}} = \frac{\int_{x_{\min}}^{x_{\max}} \frac{1}{2} \cdot \left(f^2(x) - g^2(x)\right) dx}{\int_{x_{\min}}^{x_{\max}} (f(x) - g(x))\, dx}, \tag{3.54}$$

where the term $\rho \cdot h$ cancels in the numerators and denominators. Hence, (x_c, y_c) is related to the ***centroid*** *of an object. The two notions are ONLY the same when the mass is uniformly distributed throughout the object (e.g., ρ is a constant, independent of (x, y)).*

Notes: $x \cdot (f(x) - g(x))$ and $\left(f^2(x) - g^2(x)\right)$ have the same units, namely distance squared. This is because x, $f(x)$, and $g(x)$ are each measured in meters or centimeters. Consequently, the units work out fine for M_x and M_y, and most importantly for us, the units on x_c and y_c are distance.

A good source of SIMPLE examples is How to find the centroid of simple composite shapes by **Engineer4Free**.

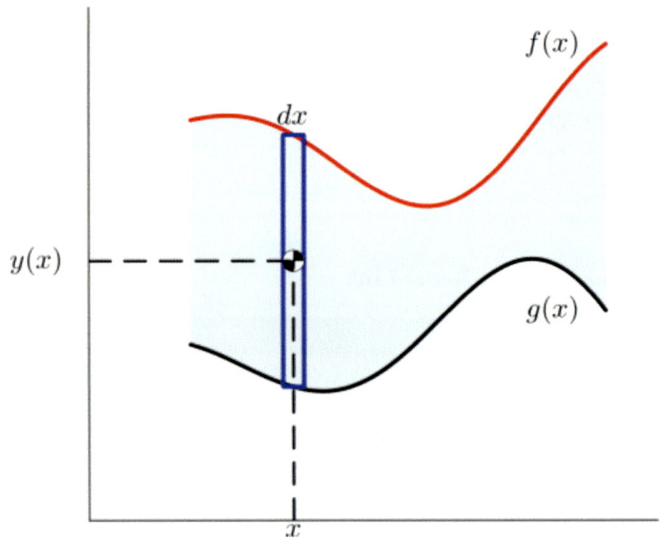

Figure 3.24: The surprising power of the rectangle is on full display. The (differential) area of the rectangle is height × width or $dA(x) = (f(x) - g(x))\, dx$. Its center (of mass) is located at $(x, \frac{f(x)+g(x)}{2})$, where the y-value is the mean of the top and bottom of the rectangle. Calculus tells us how to use an integral to add up the values associated with each rectangle to compute the object's overall mass and center of mass. The details are given in the text. The same process can be followed for moments of inertia.

Source of the Equation for Center of Mass. Figure 3.24 shows a body defined by two functions, $f : [a, b] \to \mathbb{R}$ and $g : [a, b] \to \mathbb{R}$. Moreover, for all $x \in [a, b]$ we have $f(x) > g(x)$. Let (x_c, y_c) denote the Cartesian position of the center of mass, and define:

- **Differential Element and Mass:** Choose a vertical differential element of width dx at a distance x from the y-axis. The mass of this strip is

$$dm = \rho \cdot h \cdot (f(x) - g(x))\, dx,$$

(density times z-thickness times height times differential width).

- **Condition of Balance from Physics for x_c:**

$$\int_a^b (x - x_c) \cdot \rho \cdot h \cdot (f(x) - g(x))\, dx = 0,$$

which is equivalent to

$$\int_a^b x \cdot \rho \cdot h \cdot (f(x) - g(x))\, dx = \int_a^b x_c \cdot \rho \cdot h \cdot (f(x) - g(x))\, dx = x_c \cdot \int_a^b \rho \cdot h \cdot (f(x) - g(x))\, dx$$

because x_c is a constant and can be taken out of the integral. Solving for x_c leads to the expression given by (3.57).

Similarly, the balance condition for y_c is

- **Condition of Balance from Physics for y_c:**

$$\int_a^b (y(x) - y_c) \cdot \rho \cdot h \cdot (f(x) - g(x))\, dx = 0,$$

which is equivalent to

$$\int_a^b y(x) \cdot \rho \cdot h \cdot (f(x) - g(x))\, dx = \int_a^b y_c \cdot \rho \cdot h \cdot (f(x) - g(x))\, dx = y_c \cdot \int_a^b \rho \cdot h \cdot (f(x) - g(x))\, dx$$

because y_c is a constant and can be taken out of the integral. Things become a bit more interesting when we note that the $y(x)$-center of mass position of the infinitesimally small rectangle is its midpoint, namely, $y(x) = \frac{f(x)+g(x)}{2}$, as shown below.

Derivation of x_c:

- **Moment of the Differential Element along the x-axis:** This is given by:

$$dM_y = \rho \cdot h \cdot x \cdot (f(x) - g(x))\, dx.$$

- **Total Moment about the y-axis:**

$$M_y = \int_a^b x \cdot \rho \cdot h \cdot (f(x) - g(x))\, dx. \tag{3.55}$$

- **Total Mass of the Body:**

$$M_T = \int_a^b \rho \cdot h \cdot (f(x) - g(x))\, dx. \tag{3.56}$$

- **x-coordinate of the Center of Mass:**

$$x_c := \frac{M_y}{M_T} = \frac{\int_a^b x \cdot (f(x) - g(x))\, dx}{\int_a^b (f(x) - g(x))\, dx}. \tag{3.57}$$

Interpretation: The formula for x_c is the weighted average of the x-positions of the differential elements, where each position is weighted by its corresponding mass. The center of mass is the "balance point" of the planar body.

Derivation of y_c:

- **Differential Element and Mass:** We use the same vertical differential element as before; it has width dx at a distance x from the y-axis. The height of this strip is $f(x) - g(x)$ and its differential mass is

$$\text{dm} = \rho \cdot h \cdot (f(x) - g(x))\, dx. \tag{3.58}$$

- **y-Center of Mass for the Differential Element:** is the middle of the rectangle,

$$\boldsymbol{y(x)} = \frac{f(x) + g(x)}{2}. \tag{3.59}$$

- **Moment of the Differential Element about the x-axis:**

$$dM_x = \rho \cdot h \cdot \boldsymbol{y(x)} \cdot (f(x) - g(x))\, dx \tag{3.60}$$

$$= \rho \cdot h \cdot \frac{f(x) + g(x)}{2} \cdot (f(x) - g(x))\, dx \tag{3.61}$$

$$= \rho \cdot h \cdot \frac{(f(x) - g(x))^2}{2} dx. \tag{3.62}$$

- **Total Moment about the x-axis:**

$$M_x = \int_a^b \rho \cdot h \cdot \frac{(f(x) - g(x))^2}{2} dx. \tag{3.63}$$

- **Total Mass of the Body:**

$$M_{\rm T} = \int_a^b \rho \cdot h \cdot (f(x) - g(x))\, dx. \tag{3.64}$$

- **y-coordinate of the Center of Mass:**

$$y_c = \frac{M_x}{M_{\rm T}} = \frac{1}{2} \frac{\int_a^b (f(x) - g(x))^2\, dx}{\int_a^b (f(x) - g(x))\, dx}, \tag{3.65}$$

  where, because the density ρ and thickness h are constant, they cancel out.

There are many derivations available on YouTube. Here is one by **BlackPenRedPen**: Centroid formulas of a region bounded by two curves.

Example 3.34. *Confirm the position of the center of mass for the standard link in* **Fig. 3.19** *and determine the center of mass for the objects in* **Fig. 3.25**. *For all of the calculations, use Julia and quadgk.*

Note: *For the basic link in* **Fig. 3.19**,

$$f(x) = \begin{cases} \sqrt{|5^2 - (x-5)^2|} & 0 \le x \le 5 \\ 5 & 5 < x \le 55 \\ \sqrt{|5^2 - (55-x)^2|} & 55 < x \le 60 \end{cases}$$

and $g(x) = -f(x)$.

Solutions:

Basic Robot Link:

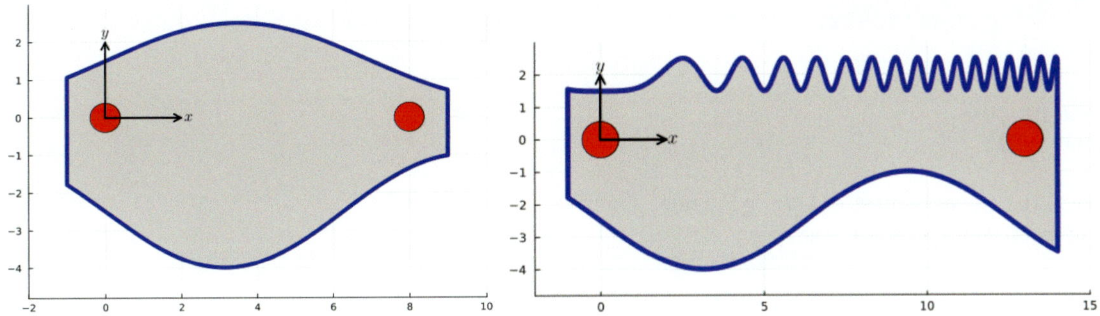

(a)

(b)

Figure 3.25: We will compute the center of mass of these two planar objects, where (a) could be a "hip" and (b) could be an "old-fashioned saw"? Our point is that doing the calculations by hand would be extremely tedious, while in Julia they are totally manageable. For today's robots, the mechanical parameters are determined numerically.

```
# Basic Link Top
function f(x)
     #
     # Basic Link
     #
     # Recall origin is placed at left end of the link
     if x < 5
          # Left end semicircle
          # (x-r)^2 + y^2 = r^2
          y = sqrt(abs(5^2 - (x-5)^2))
     elseif x < 55
          y =  5
     else
          # Right end semicircle
          # (r + distancebetweenJoints - x)^2 + y^2 = r^2
          y = sqrt(abs(5^2 - (5 + 50 - x)^2))
     end
     return y
end
g(x) = -f(x)

xmin  = 0.0
xmax = 60.0

(xc, yc) = computeCoM(f, g, xmin, xmax)
```

Output

```
(30.000000027228815, 0.0)
```

Hip:

```
# Hip
#
# Bounding functions having sinusoidal waves
```

```
f(x) = sin(0.45x) + 1.75
g(x) = -1.5sin(0.50x) - 2.5
L = 10.0

xmin = 0.0
xmax = L

(xc, yc) = computeCoM(f, g, xmin, xmax)
center_of_mass_symbol((xc,yc), .5, 0.2)
```

Output

```
(4.1845164408139786, -0.4255811707111333)
```

Saw:

```
# Saw
#
# Functions having crazier sinusoidal waves
f(x) = sin(0.25x^2)^2 + 1.5
g(x) = -1.5sin(0.50x) - 2.5
L = 15.0

xmin = 0.0
xmax = L

(xc, yc) = computeCoM(f, g, xmin, xmax)
```

Output

```
(7.192793764557787, -0.43051124387170386)
```

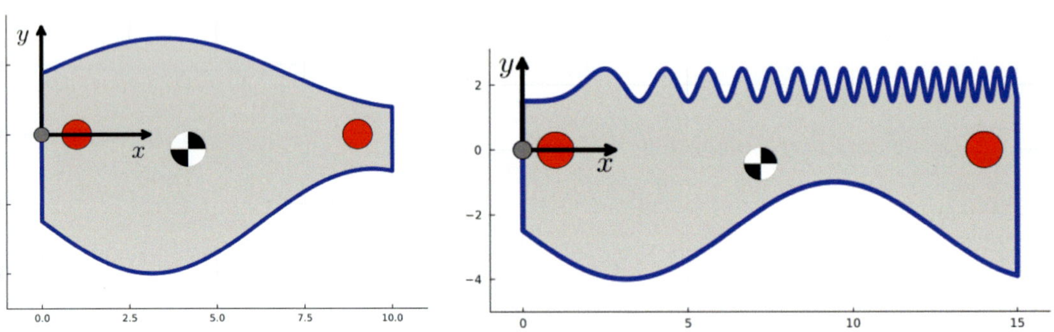

(a) (b)

```
# Compute center of mass
using QuadGK

function computeCoM(f, g, xmin, xmax)
    #
    # f = upper bounding function
    # g = lower bounding function
```

```
    # [xmin, xmax] = domain of f and g

    # Set up functions for calling quadgk
    areaIntegrand(x) = f(x) - g(x)
    MyIntegrand(x)   = x * areaIntegrand(x)          # moment about y-axis
    MxIntegrand(x)   = 0.5 * (f(x)^2 - g(x)^2)         # moment about x-
axis

    # Apply quadgk
    A, = quadgk(areaIntegrand, xmin, xmax)   # Area
    My, = quadgk(MyIntegrand, xmin, xmax)      # Total y-axis moment
    Mx, = quadgk(MxIntegrand, xmin, xmax)      # Total x-axis moment

    # (xc, yc) of object
    xc = My / A
    yc = Mx / A

    # return values
    return (xc, yc)
end
```

Output

```
computeCoM (generic function with 1 method)
```

■

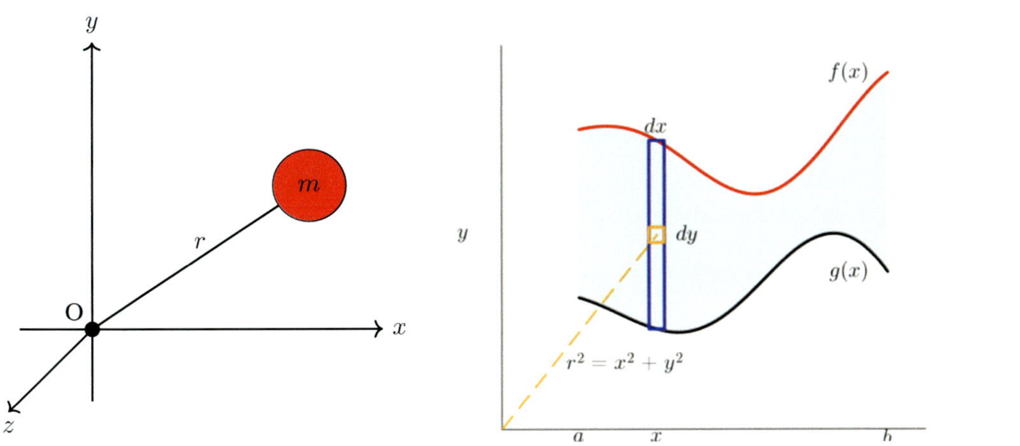

(a) Physics tells us the moment of inertia for a point mass.

(b) Computing the moment of inertia for a planar area uses Calculus.

Figure 3.27: Illustrations of the moment of inertia: (a) for a point mass about the z-axis and (b) for a planar area using differential elements. The amazing power of the rectangle is on even fuller display for the computation of the moment of inertia. We want to compute the moment of inertia about the z-axis for a continuous body, based on what we know about the moment of inertia of a point mass, m, namely, $I_z = mr^2$, where r is the distance of the point mass from the origin. We apply this knowledge to compute the moment of inertia of the pseudo point mass defined by the differential orange rectangle centered at (x, y), where $\rho \cdot h \cdot dx \cdot dy$ plays the role of a point mass and $\sqrt{x^2 + y^2}$ is the distance from the origin. We use Calculus to sum up the differential inertias along the y-direction, yielding a formula for the differntial inertia of a very thin rectangle (its width is dx). We use Calculus a second time to add up the inertia associated with each infinitesimally thin rectangle to compute the object's overall moment of inertia.

3.4.5 Optional Read: Using Again the Power of the Rectangle to Compute the Inertia of a (Planar) Robot Link

This is the first and only time in the textbook that we compute a two-dimensional integral. It is perhaps the ultimate illustration of a Superpower of Calculus: adding up Infinitesimal Objects in 2D instead of 1D! We define the differential mass contained in the orange rectangle of **Fig. 3.27-b** by

$$dM(x,y) := \rho \cdot h \cdot dy \cdot dx, \tag{3.66}$$

which is density, times thickness, times height, times width. The distance squared of the differential mass from the origin is

$$r^2 = x^2 + y^2. \tag{3.67}$$

To compute the differential inertia dI_z of the entire rectangle of height $(f(x) - g(x))$, we need to compute an integral in the y-direction, namely

$$\begin{aligned} dI_x &= \int_{g(x)}^{f(x)} r^2 \cdot dM(x,y) \\ &= \int_{g(x)}^{f(x)} (x^2 + y^2) \cdot \rho \cdot h \cdot dy \cdot dx \text{ (factor out constants)} \\ &= \rho \cdot h \cdot dx \cdot \int_{g(x)}^{f(x)} (x^2 + y^2) \cdot dy \text{ (dx is treated as a constant when integrating with respect to y)} \\ &= \rho \cdot h \cdot dx \cdot (x^2 \cdot y + \frac{1}{3} y^3)\Big|_{g(x)}^{f(x)} \text{ (x is also treated as a constant when integrating with respect to y)} \\ &= \rho \cdot h \cdot \left[x^2 \cdot (f(x) - g(x)) + \frac{1}{3} \cdot \left(f^3(x) - g^3(x)\right)\right] \cdot dx \text{ (moved dx to the end).} \end{aligned} \tag{3.68}$$

This result pertains to an infinitesimally thin rectangle. Now, to compute the entire moment of inertia about the z-axis, we use Calculus to sum up all of the differential moments of inertia along the x-direction, yielding,

$$I_z = \int_a^b dI_z = \rho \cdot h \cdot \int_a^b \left[x^2 \cdot (f(x) - g(x)) + \frac{1}{3} \cdot \left(f^3(x) - g^3(x)\right)\right] dx. \tag{3.69}$$

The integral can be evaluated in closed form in simple examples. However, for most robot links of interest, numerical methods (such as QuadGK) are used to evaluate the integral.

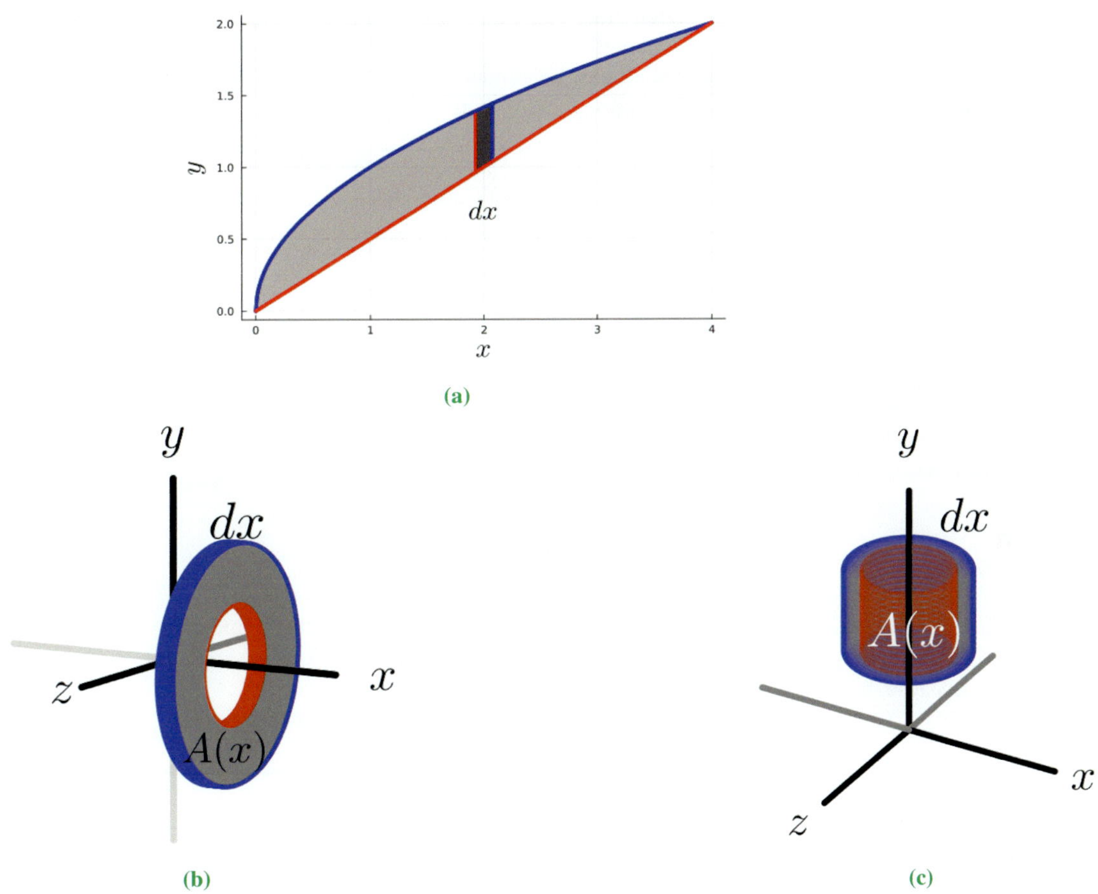

Figure 3.28: Differential volumes dV in the form of disk/washers and shells. The area upper-bounded by $f : [a, b] \to \mathbb{R}$ in blue and lowerbounded by $g : [a, b] \to \mathbb{R}$ in red is shown in (a), for $[a, b] = [0, 4]$. The area enclosed is shaded gray. A solid of revolution results by rotating the indicated area about either the x-axis or the y-axis. Using Calculus, we can compute the volume of the solid of rotation, whether or not we can visualize the resulting solid in our mind. The key point is the following: The Riemann integral comes down to understanding narrow rectangles. Within the area enclosed in (a) is a dark gray "rectangle" of width dx, the same rectangle we would use to compute the area enclosed by f and g. The result of rotating the narrow rectangle about the x-axis is the **washer** shown in (b); a washer is a thin disk with a hole in it, having thickness dx. Because its thickness is nonzero, the washer has a small, but nonzero volume dV. The same dark gray rectangle is rotated about the y-axis in (c). The blue cylinder is traced out by the right side of the rectangle, and the red cylinder is traced out by the left side of the rectangle. The result is a **shell**, namely a cylinder with thickness dx. Due to its nonzero thickness, the shell has a small but nonzero volume dV. Calculus allows us to add up the infinitesimal volumes dV to obtain the overall volume.

3.4.6 Solids of Revolution or More on the Power of the Rectangle

Consider a shape in the plane enclosing a nonzero area, as illustrated in Fig. 3.28-(a). A **solid of revolution** is the result of rotating the shape under consideration about either the x-axis or the y-axis. When rotated about the x-axis, Fig. 3.28-(a) becomes the shape in a blue parabolic-shaped vase with a red cone-shaped interior lying on its side. When rotated about the y-axis, the shape in Fig. 3.28-(a) generates an upright red cone-shaped object with an anti-parabolic-(square-root)-shaped blue interior. To compute the volume of these two shapes, we do not need to be able to sketch them. As shown in Fig. 3.28, rotating a rectangle of width dx about the x-axis generates a washer (aka a disk with a hole in the middle) of thickness dx, while rotating that same rectangle about the y-axis produces a shell, i.e., a cylinder with thickness dx. Each of these objects has a differential volume equal to $A(x)\,dx$. The total volume is then:

$$\boxed{V_{\text{solid}} := \int_a^b dV(x) := \int_a^b A(x)\,dx.} \tag{3.70}$$

Computing the Volume

- **Area and Differential Volume of the Disk:** The outer radius is $f(x)$ and the inner radius is $g(x)$, and its thickness is dx. Hence,
$$A(x) := \pi\left(f^2(x) - g^2(x)\right)$$
and
$$dV_{\text{x-axis}} := \pi\left(f^2(x) - g^2(x)\right)dx.$$

- **Area and Differential Volume of the Cylinder/Shell:** The top edge of the cylinder/shell is $f(x)$ and its bottom edge is $g(x)$, giving a height of $h := (f(x) - g(x))$. The radius is $r := x$, and its thickness is dx. Because the surface area of a cylinder is $2\pi \cdot r \cdot h$, we have
$$A(x) := 2\pi x\,(f(x) - g(x)) \text{ and } dV_{\text{y-axis}} := 2\pi x\,(f(x) - g(x))\,dx.$$

Proposition 3.35: Volume of a Solid of Revolution

Key Assumptions: The area lies totally in the first-quadrant of the (x,y)-plane, that is, the quadrant where $x \geq 0$ and $y \geq 0$. Moreover, the area is delimited by two functions $f : [a,b] \to \mathbb{R}$ and $g : [a,b] \to \mathbb{R}$, and for all $0 \leq a \leq x \leq b$, $f(x) \geq g(x)$. The last condition means that f is the upper bound on the area, g is the lower bound, and the two functions do not enclose any "negative area".

Under the above assumptions, the volumes of the two standard solids of revolution are computed as follows:

$$V_{\text{x-axis}} := \int_a^b dV_{\text{x-axis}} = \int_a^b \pi\left(f^2(x) - g^2(x)\right)dx, \tag{3.71}$$

and it is called the **disc/washer method** for rotation about the x-axis.

$$V_{\text{y-axis}} := \int_a^b dV_{\text{y-axis}} = \int_a^b 2\pi x\,(f(x) - g(x))\,dx, \tag{3.72}$$

and it is called the **shell method** for rotation about the y-axis. The disc/washer and shell are shown in Fig. 3.28.

Remark 3.36. *On YouTube, there are many videos that illustrate these two methods. You need to be cognizant of two points:*

- *We are applying the disc-washer method to a solid formed by rotating an area about the x-axis. If you apply the disc-washer method when rotating the area about the y-axis, you will have to use the inverse functions, $x = f^{-1}(y)$ and $x = g^{-1}(y)$. Watch out for this in the videos. Also, what do you do when the inverse functions do not exist? Ouch!*

- *We are applying the shell method to a solid formed by rotating an area about the y-axis. If you apply the shell method when rotating the area about the x-axis, you will have to use the inverse functions, $x = f^{-1}(y)$ and $x = g^{-1}(y)$. Watch out for this in the videos. Also, what do you do when the inverse functions do not exist? Ouch! You apply our method!*

- *Our suggestion: Avoid situations where you need inverse functions; they are usually difficult to compute. The computational methods introduced in this book do not use inverse functions.*

- ***Optional Read: A minor point:** On rare occasions, an area is delimited by $x = F(y)$ and $x = G(y)$, and thus the functions are parameterized by y instead of x. This means the roles of the x and y axes have been flipped. In this case, we would draw the dark rectangle as wide in the x-direction and short in the y-direction, flipping the axes that produce a disc/washer versus a shell. Hence, we have to flip the formulas in (3.71) and (3.72) for the disc/washer and shell methods.*

Justification of (3.71): Consider Fig. 3.28-(a). When the darkly shaded rectangle of width dx is rotated about the x-axis, it traces out the **washer** shown in Fig. 3.28-(b). The outside radius of the washer is $r_2 = f(x)$, the inside radius is $r_1 = g(x)$, and its width is dx. It then follows that the area of the washer is $A(x) = \pi(r_2^2 - r_1^2) = \pi\left(f^2(x) - g^2(x)\right)$. The washer is the solid of revolution traced out by the small rectangle when it is rotated about the x-axis. The **differential volume** of the washer is its area times width (or thickness), that is, $dV = A(x)\,dx$. Hence, the total volume of the solid of revolution about the x-axis is the sum of the differential volumes of revolution. That is, $V = \int_a^b dV = \int_a^b A(x)dx$; we simply add up the differential volumes $dV := A(x)\,dx$ of each washer from the lower limit of x to its upper limit.

Justification of (3.72): Consider again Fig. 3.28-(a). When the darkly shaded rectangle of width dx is rotated about the y-axis, it traces out the **shell** shown in Fig. 3.28-(c). A "shell" is a cylinder with a thin but nonzero thickness; in our case, the thickness is dx. The area of the cylinder is $2\pi \cdot r \cdot h$, where its radius $r = x$ and height $h = f(x) - g(x)$. Therefore, the surface area of the shell is $A(x) = 2\pi \cdot x \cdot (f(x) - g(x))$, and the differential volume of the resulting solid of revolution is $dV = A(x)\,dx = 2\pi \cdot x \cdot (f(x) - g(x))\,dx$. Hence, the total volume of the solid of revolution about the y-axis is the sum of the differential volumes of revolution. That is, $V = \int_a^b dV = \int_a^b A(x)\,dx$, we simply add up the differential volumes $dV := A(x)\,dx$ of each shell from the lower limit of x to its upper limit.

Example 3.37. *For the two functions $f : [0,4] \to \mathbb{R}$ by $f(x) = \sqrt{x}$ and $g : [0,4] \to \mathbb{R}$ by $g(x) = 0.5x$, compute the volumes for the resulting solids of revolution about the x- and y-axes. After this example, we will learn how to sketch the volumes.*

Rotation about x-axis:

```
# Volume of revolution about the x-axis
using QuadGK

f(x) = sqrt(x)
g(x) = 0.5x
xmin = 0.0
xmax = 4.0

# Area of disc
areaDiscWasher(x) = pi * ( (f(x))^2 - (g(x)^2) )

Volume_x, Error = quadgk(areaDiscWasher, xmin, xmax)
```

Output

```
(8.377580409572783, 0.0)
```

Rotation about y-axis:

```
# Volume of revolution about the y-axis
using QuadGK

f(x) = sqrt(x)
g(x) = 0.5x
xmin = 0.0
xmax = 4.0

# Area of disc
areaShell(x) = 2*pi * x* ( f(x) - g(x) )

Volume_y, Error = quadgk(areaShell, xmin, xmax)
```

Output

```
(13.40412865472173, 1.244413129087435e-7)
```

Note: The volumes are not the same.

How to Sketch the Volumes

Figures 3.29 and 3.30 walk us through the steps for sketching a solid of revolution about the x-axis. The same procedure works for revolving about the y-axis. To get the hang of the process, seeing "live" examples worked out step-by-step is extremely helpful. **Recommended Video:** Disk Washer Method to Find the Volume of Solids of Revolution by Quoc Dat Phung, an engineering student in Canada. The video has excellent graphics to support the discussion. The first two examples, which take 10 minutes total at 1X, are adequate to get you going. Quoc Dat uses the term "antiderivative". Don't let it bother you; we will cover the concept in an upcoming chapter. Roughly speaking, it's another term for the "indefinite integral" that we covered in Prop. 3.3: said another way, the antiderivative of x^k is $\frac{x^{k+1}}{k+1}$. The first two examples introduced by Quoc Dat are for rotations about the x-axis, and he applies the disk/washer method.

ChatGPT Prompt for Students: "Using the disk/washer method, determine the volume of the solid of revolution formed by rotating about the x-axis the region bounded by $y = \sqrt{x}$ and $y = \frac{1}{2}x$ for x between zero and four. In addition, using the shell method, determine the volume of the solid of revolution formed by rotating about the y-axis the region bounded by $y = \sqrt{x}$ and $y = \frac{1}{2}x$ for x from 0 to 4. Provide well-documented Julia code for each method, explaining each step in its solution. Additionally, generate supporting plots to illustrate the methods explained in the code."

Example 3.38. ***Volume of a Basketball:*** *Use the disk/washer method or the shell method to compute the volume of an NBA basketball.*

Solution: According to Wikipedia, an NBA basketball has a circumference of 75 centimeters (cm), meaning its radius is $r = \frac{75}{2\pi} = 11.9366$ cm. We define $f : [0, r] \to \mathbb{R}$ by $f(x) = \sqrt{r^2 - 1}$ and $g : [0, r] \to \mathbb{R}$ by $g(x) = 0$. The graph of f is a quarter of a circle, as shown in Fig. 3.32.

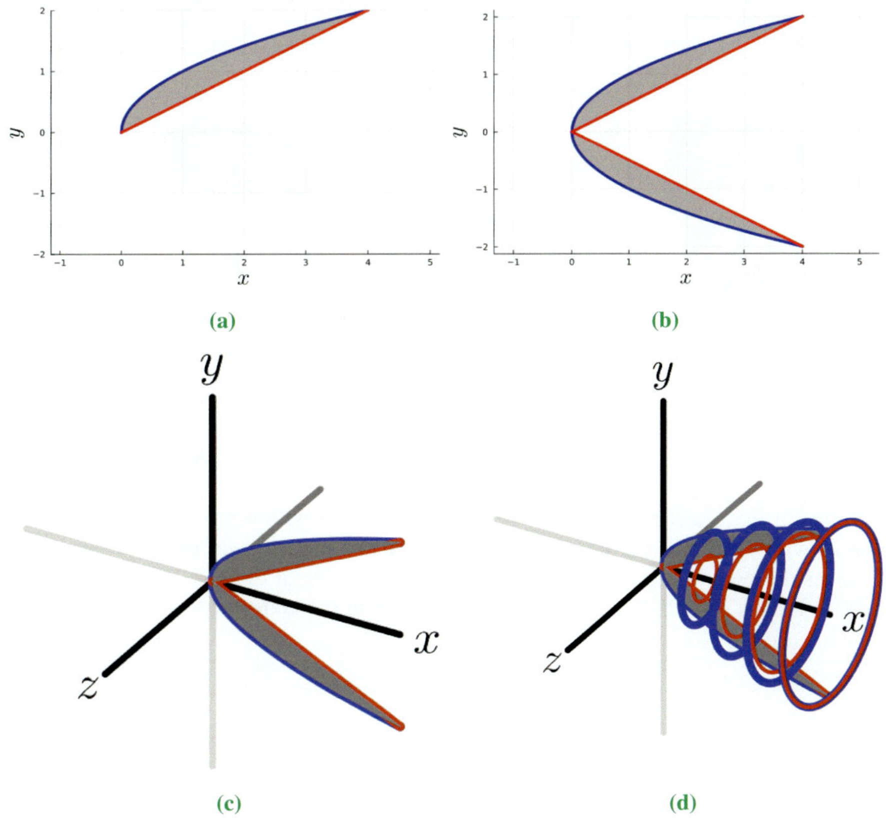

Figure 3.29: How to Sketch a Solid of Revolution. Part (a) shows the area of interest in the (x, y)-plane. To form a rotation about the x-axis, make a mirror image of the area below the x-axis, as in (b); this is really the area rotated by 180 degrees about the x-axis. Next, as in (c), add in a z-axis to form a 3D object. The final step is to realize that each point (x, y) in the original area forms a circle in the (y, z)-plane when rotated about the x-axis. The blue circles in (d) delineate the outer surface of the solid of revolution, and the red circles delineate the inner surface of the solid.

```
# Volume of revolution about the x-axis
using QuadGK

r = 75/(2*pi)

f(x) = sqrt(r^2 - x^2)
g(x) = 0.0x
xmin = 0
xmax = r

# Area of disc
areaDiscWasher(x) = pi * ( (f(x))^2 - (g(x)^2) )

VolumeSemisphere_x, Error = quadgk(areaDiscWasher, xmin, xmax)

VolumeBasketball = 2 * VolumeSemisphere_x
```

Output

```
7124.145724851874
```

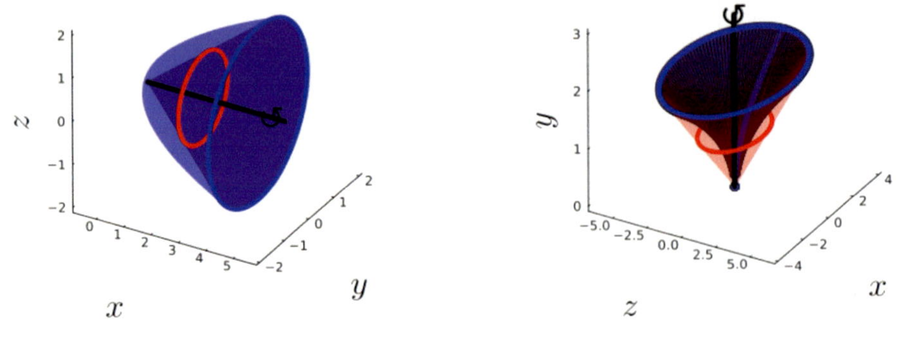

(a) (b)

Figure 3.30: Solids of Revolution for Example 3.37 and **Fig. 3.29** with bounding functions $f : [0,4] \to \mathbb{R}$ and $g : [0,4] \to \mathbb{R}$, with $f(x) \geq g(x)$. (a) Rotation of the area about the x-axis. (b) Rotation of the area about the y-axis. f forms the outer boundary when the area is rotated about the x-axis, while it forms the inner boundary when the area is rotated about the y-axis. The opposite is true for g.

```
# Volume of revolution about the y-axis
using QuadGK

r = 75/(2*pi)

f(x) = sqrt(r^2 - x^2)
g(x) = 0.0x
xmin = 0
xmax = r

# Area of disc
areaShell(x) = 2*pi * x* ( f(x) - g(x) )

VolumeSemisphere_y, Error = quadgk(areaShell, xmin, xmax)

VolumeBasketball = 2 * VolumeSemisphere_y
```

Output

```
7124.145729241054
```

Our estimate is, therefore, Volume = 7124.145727 cm^3 ± 4.39e-06 ■.

Other Recommended Videos:

- Cylindrical Shells to Find the Volume of Solids of Revolution by Quoc Dat Phung has excellent graphics to support what he is teaching.
- Disk/Washer vs. Cylindrical Shell...when to use which? by Quoc Dat Phung has excellent graphics to support what he is teaching. **Note:** For us, the answer is clear for functions $y = f(x)$ and $y = g(x)$: when rotating about the x-axis, use the Disk/Washer Method, and when rotating about the y-axis, use the Shell Method. This avoids having to compute inverse functions, as illustrated in **Fig. 3.31**.
- Disc/Washer Method vs. Shell Method (rotated about different lines) by **BlackPenRedPen** has clear, hand-drawn graphics so that you too can make nice sketches.

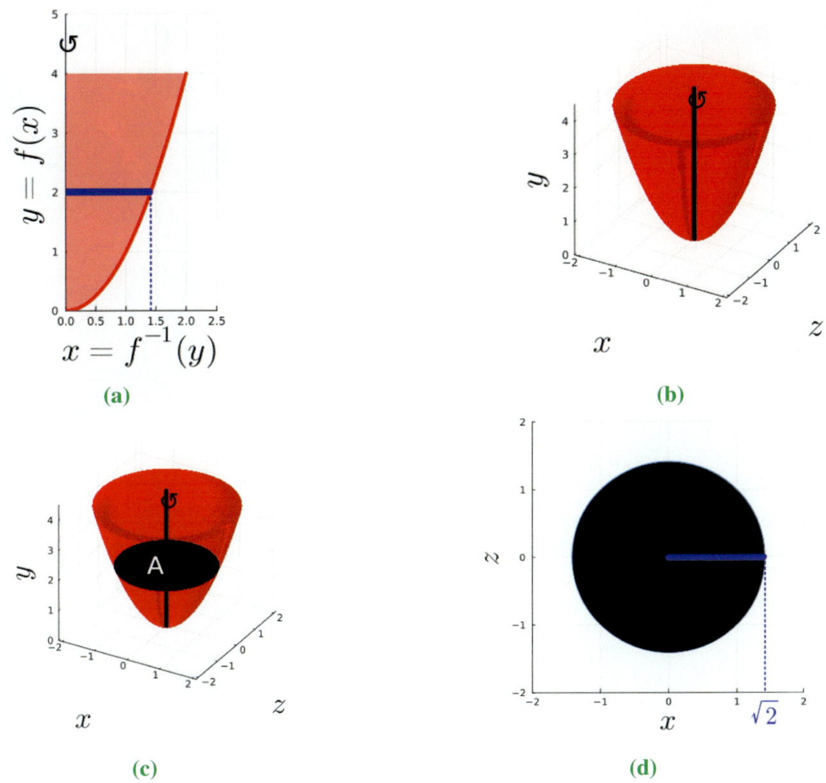

Figure 3.31: Applying the disk/washer method to a solid of revolution about the y-axis requires inverse functions. The area between the line $y = 4$ and $y = x^2$ has been shaded in (a). In (b), the shaded area has been rotated around the y-axis. This means that each point $(x = f^{-1}(y), y)$ in the shaded area will trace out a circle of radius $f^{-1}(y)$ in the (x, z)-plane at a height of $y = f(x)$. In particular, the blue bar in (a), which has thickness dy and radius $r(y) = f^{-1}(y)$, traces out the disk in (c). In (d), the disk has been projected to the (x, z)-plane, confirming it is a "perfect" circle. Hence, the area in (c) is $A = \pi r^2$, where $r = f^{-1}(y)$. It is wise to apply the shell method in cases like this one to avoid having to compute inverses of functions.

- Shell method for volume of revolution (rotated about different axis and lines) by **BlackPenRedPen** has clear, hand-drawn graphics so that you too can make nice sketches.

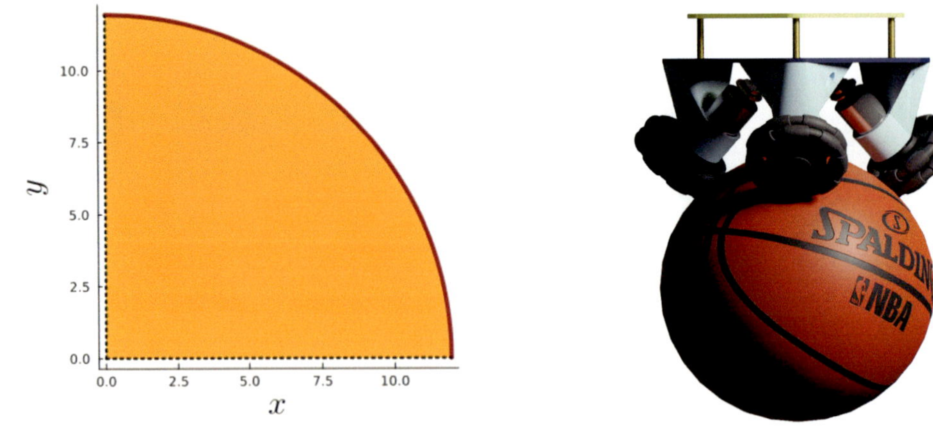

(a) (b)

Figure 3.32: We can revolve the quarter circle shown in (a) around either axis to form a hemisphere. The volume of a sphere, such as the basketball shown in (b), will be twice that of a hemisphere.

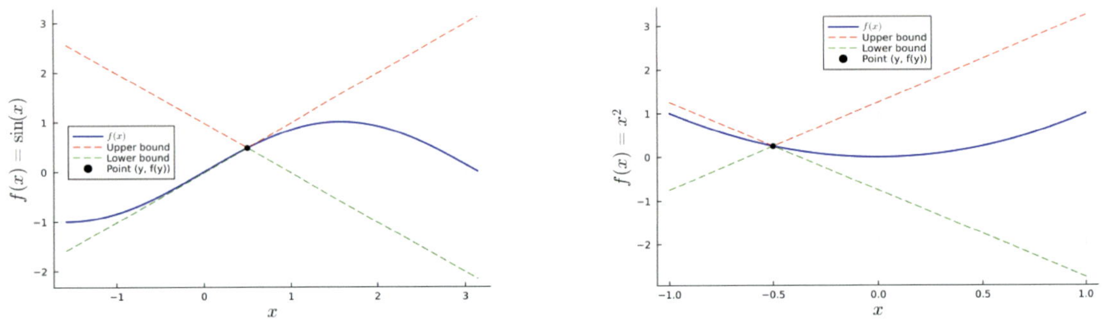

(a) (b)

Figure 3.33: **Lipschitz Continuity** is more restrictive than "normal continuity". We can think of Lipschitz continuity as a rule that imposes a bound on how quickly a function can change as we move from one point to another along its graph, forcing the function to behave as if it were "locally linear". This means the function does not spike too steeply or drop too suddenly, ensuring a certain smoothness in its graph.

3.5 Optional Read: Proofs Associated with the Chapter

We open this section by providing proof of the existence of the Riemann-Darboux Integral for functions that satisfy a stronger condition than the "usual notion of continuity" we used so far in the book.

Lipschitz Continuity

Definition 3.39. *A function $f : [a, b] \to \mathbb{R}$ is **Lipschitz continuous** if there exists a finite constant $L < \infty$ such that for all $x, y \in [a, b]$,*

$$|f(x) - f(y)| \leq L \cdot |x - y|. \tag{3.73}$$

***Note:** Essentially, Lipschitz continuity means the function cannot grow faster than a linear function with slope L. More precisely, for any point $y \in [a, b]$ and all points $x \in [a, b]$,*

$$f(y) - L|x - y| \leq f(x) \leq f(y) + L|x - y|, \tag{3.74}$$

as illustrated in Fig. 3.33. Near each point $y \in [a,b]$, the local change in f as a function of $(x-y)$ is linear with slope L. Indeed, for $x > y$, $|x-y| = x-y$, and we have $f(x) \le f(y) + L(x-y)$ (a constant plus a slope times the change of x with respect to y), while for $x < y$, $|x-y| = -(x-y)$, and we have $f(x) \ge f(y) + L(x-y)$ (one again, a constant plus a slope times the change of x with respect to y).

Note: From Prop. 1.4-(c), the "no-name property of inequalities", (3.73) is equivalent to

$$-L\cdot|x-y| \le f(x) - f(y) \le L\cdot|x-y|.$$

Adding $f(y)$ to each term gives,

$$f(y) - L\cdot|x-y| \le f(x) \le f(y) + L\cdot|x-y|.$$

Note: Due to the inequality (3.74), a Lipschitz continuous function cannot have a "jump", and hence every Lipschitz continuous function is also "continuous" in the sense that we have been using the term in this textbook. Indeed, considering Fig. 1.2-(b) or (c), drawing a discontinuous function requires a line with an infinite slope (i.e., a vertical line) or lifting your pen from the paper. In the next chapter, we will be more precise about the notion of continuity. ■

Proposition 3.40: (Existence of the Riemann-Darboux Integral for Lipschitz Continuous Functions

Suppose that $f : [a,b] \to \mathbb{R}$ is Lipschitz continuous. Then its Riemann-Darboux integral $\int_a^b f(x)\,dx$ exists and is finite.

Proof: We must check that the conditions in Definition 3.2 are satisfied. To do that, we recall some of the details. For $a < b$ (finite) real numbers, and $n \ge 1$, define $\Delta x := \frac{b-a}{n}$ and $x_i := a + (i-1)\cdot\Delta x$, for $1 \le i \le n+1$. Then,

(a) $\text{Area}_n^{\text{Low}} := \sum_{i=1}^n h_i^{\text{Low}}\cdot\Delta x$, for $h_i^{\text{Low}} := \min\limits_{x\in[x_i,x_{i+1}]} f(x)$ is the **Riemann lower sum**, and

(b) $\text{Area}_n^{\text{Up}} := \sum_{i=1}^n h_i^{\text{Up}}\cdot\Delta x$, for $h_i^{\text{Up}} := \max\limits_{x\in[x_i,x_{i+1}]} f(x)$ is the **Riemann upper sum**.

We need to show that

- $\lim_{n\to\infty} \text{Area}_n^{\text{Low}}$ exists and is finite,
- $\lim_{n\to\infty} \text{Area}_n^{\text{Up}}$ exists and is finite, and
- the two limits are equal.

We first show the last property: if the two limits exist and are finite, then they converge to the same value.

The key step is to estimate the max and min of $f : [x_i, x_{i+1}] \to \mathbb{R}$. From (3.74), for all $x \in [x_i, x_{i+1}]$

$$f(x_i) - L|x - x_i| \le f(x),$$

and hence

$$h_i^{\text{Low}} := \min_{x\in[x_i,x_{i+1}]} f(x) \ge \min_{x\in[x_i,x_{i+1}]} \left(f(x_i) - L|x-x_i|\right) = f(x_i) - L|x_{i+1} - x_i| = f(x_i) - L\Delta x.$$

Similarly, from (3.74), for all $x \in [x_i, x_{i+1}]$,

$$f(x) \le f(x_i) + L|x - x_i|,$$

and hence

$$h_i^{\text{Up}} := \max_{x\in[x_i,x_{i+1}]} f(x) \le \max_{x\in[x_i,x_{i+1}]} (f(x_i) + L|x - x_i|) = f(x_i) + L|x_{i+1} - x_i| = f(x_i) + L\Delta x.$$

In case the math in these steps is unclear, the bounds also can be deduced from **Fig. 3.33**.

From these inequalities, it follows that

$$\begin{aligned}
\text{Area}_n^{\text{Up}} - \text{Area}_n^{\text{Low}} &= \sum_{i=1}^{n} \left(h_i^{\text{Up}} - h_i^{\text{Low}}\right)\Delta x \\
&\le \sum_{i=1}^{n} \left(h_i^{\text{Up}} - h_i^{\text{Low}}\right)\Delta x \\
&\le \sum_{i=1}^{n} (f(x_i) + L\Delta x - f(x_i) + L\Delta x)\,\Delta x \\
&\le \sum_{i=1}^{n} 2L\Delta x \cdot \Delta x \\
&\le n2L\left(\frac{b-a}{n}\right)^2 \\
&\le 2L\frac{(b-a)^2}{n}.
\end{aligned}$$

Hence,

$$\lim_{n\to\infty} \left(\text{Area}_n^{\text{Up}} - \text{Area}_n^{\text{Low}}\right) \le \lim_{n\to\infty} 2L\frac{(b-a)^2}{n} = 0.$$

Because $\text{Area}_n^{\text{Up}} - \text{Area}_n^{\text{Low}} \ge 0$, we deduce that

$$\lim_{n\to\infty} \left(\text{Area}_n^{\text{Up}} - \text{Area}_n^{\text{Low}}\right) = 0,$$

and therefore, if the two limits exist and are finite, then they converge to the same value.

To show the convergence of the Riemann Upper or Lower Sum, we need to usee a result from Real Analysis that is beyond the scope of this book: the Monotone Convergence Theorem. The theorem states that if a sequence is monotonically increasing and bounded from above, then it has a limit, and the limit is finite; similarly, if a sequence is monotonically decreasing and bounded from below, then it has a limit, and the limit is finite.

For us, the sequences of interest are $\text{Area}_n^{\text{Up}}$ and $\text{Area}_n^{\text{Low}}$, for n a power of two, that is, $n = 2^k$. **Why Powers of 2?** When n is chosen as a power of 2, each step from n to $2n$ involves splitting each subinterval into exactly two equal parts. This ensures that every new subinterval's maximum cannot exceed the maximum of the "parent interval" in the previous partition. To illustrate this, consider a subinterval $[x_i, x_i + \Delta x]$ and divide it into two equal halves,

$$[x_i, x_i + \Delta x] = [x_i, x_i + \frac{\Delta x}{2}] \bigcup [x_i + \frac{\Delta x}{2}, x_i + \Delta x].$$

Then, upon defining,

$$h_i^{\rm Up} := \max_{x \in [x_i, x_i + \Delta x]} f(x)$$

$$\bar{h}_{2i}^{\rm Up} := \max_{x \in [x_i, x_i + \frac{\Delta x}{2}]} f(x)$$

$$\bar{h}_{2i+1}^{\rm Up} := \max_{x \in [x_i + \frac{\Delta x}{2}, x_i + \Delta x]} f(x),$$

we have

$$h_i^{\rm Up} = \max\{\bar{h}_{2i}^{\rm Up}, \bar{h}_{2i+1}^{\rm Up}\},$$

and therefore

$$h_i^{\rm Up} \cdot \Delta x \geq \bar{h}_{2i}^{\rm Up} \cdot \frac{\Delta x}{2} + \bar{h}_{2i+1}^{\rm Up} \cdot \frac{\Delta x}{2},$$

which shows that the sequence, $\text{Area}_n^{\rm Up}$, is monotonically decreasing. Using Lipschitz continuity, the sequence is lower bounded by

$$(f(a) - L|b-a|) \cdot (b-a),$$

showing that all of the conditions of the Monotone Convergence Theorem are met.

We can repeat the above argument (with small changes) to show that $\text{Area}_n^{\rm Low}$ is monotonically increasing and bounded from above, and hence also satisfies the conditions of the Monotone Convergence Theorem. ■

Proof of Prop. 3.4 Additivity Property of the Riemann Integral

Suppose that $a < b < c$ and $f : [a, c] \to \mathbb{R}$ is continuous. Then the three Riemann Integrals $\int_a^c f(x)\,dx$, $\int_a^b f(x)\,dx$, and $\int_b^c f(x)\,dx$ all exist and satisfy

$$\int_a^c f(x)\,dx = \int_a^b f(x)\,dx + \int_b^c f(x)\,dx.$$

Proof: For many people, Fig. 3.3 is proof enough! If we were using the more general Darboux partitions, instead of the uniform partitions $x_i := a + (i-1) \cdot \Delta x$, then the proof would be immediate as well. Due to our use of simplified partitions, the best way to organize the proof is to wait for the integrability results in Section 4.7 on piecewise continuous functions. Define

$$f_{ab}(x) := f(x),\ a \leq x < b,\ \text{zero elsewhere},$$
$$f_{bc}(x) := f(x),\ b \leq x \leq c,\ \text{zero elsewhere},$$

and note that $f(x) = f_{ab}(x) + f_{bc}(x)$. Finally, we can then apply Prop. 3.7. ■

Proof of Prop. 3.7 Linearity of the Riemann Integral

Suppose that $a < b$ and each of the functions $1 \leq k \leq N$, $f_k : [a, b] \to \mathbb{R}$, is Riemann-integrable (i.e., their Riemann Integral exists and is finite). Then for constants $\alpha_k \in \mathbb{R}$, the function

$$f(x) := \alpha_1 \cdot f_1(x) + \alpha_2 \cdot f_2(x) + \cdots + \alpha_N \cdot f_N(x),$$

defined as a "linear combination of the functions $\{f_1, f_2, \ldots, f_N\}$", is Riemann-integrable, and moreover

$$\int_a^b f(x)\,dx = \alpha_1 \int_a^b f_1(x)\,dx + \alpha_2 \int_a^b f_2(x)\,dx + \cdots + \alpha_N \int_a^b f_N(x)\,dx.$$

Proof: The proof uses the Riemann Lower and Upper Sums and needs a few key facts, namely

$$\max_{x\in[x_i,x_{i+1}]}(\alpha_1\cdot f_1(x)+\cdots+\alpha_N\cdot f_N(x))\le\max_{x\in[x_i,x_{i+1}]}(\alpha_1\cdot f_1(x))+\cdots+\max_{x\in[x_i,x_{i+1}]}(\alpha_N\cdot f_N(x)),\quad(3.75a)$$

$$\min_{x\in[x_i,x_{i+1}]}(\alpha_1\cdot f_1(x)+\cdots+\alpha_N\cdot f_N(x))\ge\min_{x\in[x_i,x_{i+1}]}(\alpha_1\cdot f_1(x))+\cdots+\min_{x\in[x_i,x_{i+1}]}(\alpha_N\cdot f_N(x)).\quad(3.75b)$$

In words, the max of a sum is less than or equal to the sum of the maxes, while the min of a sum is greater than or equal to the sum of the mins.

Without loss of generality, we can assume that the coefficients[4] $\alpha_k\ge0$. Indeed, if for some k, we have $\alpha_k<0$, then the substitutions $f_k(x)\to-f_k(x)$ and $\alpha_k\to-\alpha_k$ result in the sum being unchanged and all of the coefficients being nonnegative. With this condition, it is now true that for all $1\le k\le n$,

$$\alpha_k\cdot\min_{x\in[x_i,x_{i+1}]}f_k(x)=\min_{x\in[x_i,x_{i+1}]}(\alpha_k\cdot f_k(x))$$

and

$$\alpha_k\cdot\max_{x\in[x_i,x_{i+1}]}f_k(x)=\max_{x\in[x_i,x_{i+1}]}(\alpha_k\cdot f_k(x)).$$

When combined with (3.75), we have

$$\max_{x\in[x_i,x_{i+1}]}(\alpha_1\cdot f_1(x)+\cdots+\alpha_N\cdot f_N(x))\le\alpha_1\cdot\max_{x\in[x_i,x_{i+1}]}f_1(x)+\cdots+\alpha_N\cdot\max_{x\in[x_i,x_{i+1}]}f_N(x),\quad(3.76a)$$

$$\min_{x\in[x_i,x_{i+1}]}(\alpha_1\cdot f_1(x)+\cdots+\alpha_N\cdot f_N(x))\ge\alpha_1\cdot\min_{x\in[x_i,x_{i+1}]}f_1(x)+\cdots+\alpha_N\cdot\min_{x\in[x_i,x_{i+1}]}f_N(x).\quad(3.76b)$$

We now follow our general (simplified) procedure for Riemann Lower and Upper Sums from Section 3.1.1. Let $n\ge2$, and define:

$$\begin{aligned}\Delta x&:=\frac{b-a}{n},\\ x_i&:=a+(i-1)\cdot\Delta x,\\ h_i^{\rm Low}&:=\min_{x\in[x_i,x_{i+1}]}f(x),\\ h_i^{\rm Up}&:=\max_{x\in[x_i,x_{i+1}]}f(x),\\ h_{i,j}^{\rm Low}&:=\min_{x\in[x_i,x_{i+1}]}f_j(x),\\ h_{i,j}^{\rm Up}&:=\max_{x\in[x_i,x_{i+1}]}f_j(x),\\ {\rm Area}_n^{\rm Low}&:=\sum_{i=1}^n h_i^{\rm Low}\cdot\Delta x,\\ {\rm Area}_n^{\rm Up}&:=\sum_{i=1}^n h_i^{\rm Up}\cdot\Delta x,\\ {\rm Area}_{n,j}^{\rm Low}&:=\sum_{i=1}^n h_{i,j}^{\rm Low}\cdot\Delta x,\\ {\rm Area}_{n,j}^{\rm Up}&:=\sum_{i=1}^n h_{i,j}^{\rm Up}\cdot\Delta x.\end{aligned}\quad(3.77)$$

Then, from (3.76), we have

$$\sum_{j=1}^N\alpha_j{\rm Area}_{n,j}^{\rm Low}\le{\rm Area}_n^{\rm Low}\le{\rm Area}_n^{\rm Up}\le\sum_{j=1}^N\alpha_j{\rm Area}_{n,j}^{\rm Up}.\quad(3.78)$$

[4]If minus signs were involved, then we would have to deal with $\max\, f(x)=-\min(-f(x))$, etc., making the argument clumsy. In math, whenever possible, we like to reduce things to a simpler case that still shows the result we seek.

Then, because N is finite,

$$\lim_{n\to\infty}\sum_{j=1}^{N}\alpha_j \text{Area}_{n,j}^{\text{Low}} = \sum_{j=1}^{N}\alpha_j \lim_{n\to\infty}\text{Area}_{n,j}^{\text{Low}}, \tag{3.79a}$$

$$\lim_{n\to\infty}\sum_{j=1}^{N}\alpha_j \text{Area}_{n,j}^{\text{Up}} = \sum_{j=1}^{N}\alpha_j \lim_{n\to\infty}\text{Area}_{n,j}^{\text{Up}}. \tag{3.79b}$$

Because $\{f_1, \cdots, f_N\}$ are Riemann Integrable, the above limits exist, are finite, and are equal. Hence, by the Squeeze Theorem for Limits,

$$\lim_{n\to\infty}\text{Area}_n^{\text{Low}} = \lim_{n\to\infty}\text{Area}_n^{\text{Up}} = \sum_{j=1}^{N}\alpha_j \lim_{n\to\infty}\text{Area}_{n,j}^{\text{Up}} = \sum_{j=1}^{N}\alpha_j \int_a^b f_j(x)\,dx.$$

This proves the result. ■

Proof of Cor. 3.8 Generalized First Additivity Property

For a, b, and c finite real numbers in no particular order, define $\alpha := \min\{a,b,c\}$ to be the smallest of the three numbers, and $\beta := \max\{a,b,c\}$ the largest. If $f : [\alpha,\beta] \to \mathbb{R}$ is continuous, then

$$\int_a^b f(x)\,dx = \int_a^c f(x)\,dx + \int_c^b f(x)\,dx. \tag{3.80}$$

Proof:

From Prop. 3.4, if $a < b < c$ and $f : [a,c] \to \mathbb{R}$ is continuous, then the three Riemann Integrals $\int_a^c f(x)\,dx$, $\int_a^b f(x)\,dx$, and $\int_b^c f(x)\,dx$ all exist and satisfy

$$\int_a^c f(x)\,dx = \int_a^b f(x)\,dx + \int_b^c f(x)\,dx.$$

Hence, $\int_a^b f(x)\,dx = \int_a^c f(x)\,dx - \int_b^c f(x)\,dx$. If we apply (3.19) on swapping the order of the limits of the integral, we obtain

$$\int_a^b f(x)\,dx = \int_a^c f(x)\,dx + \int_c^b f(x)\,dx,$$

which shows that the additivity property holds when c is larger than both limits and not just for c between the two limits.

Similarly,

$$\int_b^c f(x)\,dx = -\int_a^b f(x)\,dx + \int_a^c f(x)\,dx,$$

by the First (Basic) Additivity Property, and, if we apply (3.19), we obtain

$$\int_b^c f(x)\,dx = \int_b^a f(x)\,dx + \int_a^c f(x)\,dx,$$

which shows that the additivity property holds when c is smaller than both limits. This is half[5] of what we need to show.

If we negate the First (Basic) Additivity Property, we obtain:

$$-\int_a^c f(x)\,dx = -\int_a^b f(x)\,dx - \int_b^c f(x)\,dx,$$

which, upon flipping all three sets of limits of integration, yields:

$$\int_c^a f(x)\,dx = \int_b^a f(x)\,dx + \int_c^b f(x)\,dx.$$

From this result, the remaining cases follow easily, and hence the order of a, b, and c is irrelevant. ■

[5]The integrals on the left of our three cases all have a lower limit that is smaller than the upper limit. What happens when the reverse is true?

Proof of Prop. 3.10 Shift Property

Suppose that $a < b$ and $f : [a,b] \to \mathbb{R}$ is continuous and let $x_c \in \mathbb{R}$. Then $g : [a+x_c, b+x_c] \to \mathbb{R}$ by $g(x) := f(x - x_c)$ is also continuous and $\int_a^b f(x)\,dx = \int_{a+x_c}^{b+x_c} g(x)\,dx$.

A graphical proof is provided by the Khan Academy and in **Fig. 3.4**.

Proof: If we want a formal proof with Riemann Sums, we can develop that too! For simplicity and clarity, we assume the function is monotonically increasing and we use the lower Riemann sum:

- $\Delta x := \frac{b-a}{n}$
- $x_i = a + (i-1)\Delta x$
- $\text{Area}_n^{\text{Low}} := \sum_{i=1}^n f(x_i) \cdot \Delta x$ is the lower Riemman sum.

For $g : [a+x_c, b+x_c] \to \mathbb{R}$ by $g(x) := f(x - x_c)$, we construct its lower Riemann sum, via

- $\Delta \bar{x} := \frac{(b+x_c)-(a+x_c)}{n} = \frac{b-a}{n} = \Delta x$
- $\bar{x}_i = a + x_c + (i-1)\Delta x = x_i + x_c$
- $\overline{\text{Area}}_n^{\text{Low}} := \sum_{i=1}^n g(\bar{x}_i) \cdot \Delta \bar{x} = \sum_{i=1}^n g(x_i + x_c) \cdot \Delta x$ is the lower Riemann sum for g.

Because $g(x_i + x_c) := f(x_i + x_c - x_c) = f(x_i)$, the two lower Riemann sums are identical, and hence, if the integrals exist, they are also identical.

■

Proof of Prop. 3.12 Scaling Property

Suppose that $a < b$, $f : [a,b] \to \mathbb{R}$ is continuous, and $\omega_0 > 0$ is a real number. Then $g : [\frac{a}{\omega_0}, \frac{b}{\omega_0}] \to \mathbb{R}$ by $g(x) := f(\omega_0 \cdot x)$ is also continuous and $\int_a^b f(x)\,dx = \omega_0 \cdot \int_{\frac{a}{\omega_0}}^{\frac{b}{\omega_0}} g(x)\,dx$.

Proof:
We reuse the proof strategy for the Shifting Property. For $g : [\frac{a}{\omega_0}, \frac{b}{\omega_0}] \to \mathbb{R}$ by $g(x) := f(\omega_0 \cdot x)$, we construct its lower Riemann sum, via

- $\Delta \bar{x} := \frac{\frac{b}{\omega_0} - \frac{a}{\omega_0}}{n} = \frac{1}{\omega_0} \cdot \frac{b-a}{n} = \frac{1}{\omega_0} \cdot \Delta x$
- $\bar{x}_i = \frac{a}{\omega_0} + (i-1)\Delta \bar{x} = \frac{a}{\omega_0} + (i-1)\frac{1}{\omega_0} \cdot \Delta x = \frac{x_i}{\omega_0}$
- $\overline{\text{Area}}_n^{\text{Low}} := \sum_{i=1}^n g(\bar{x}_i) \cdot \Delta \bar{x} = \sum_{i=1}^n g(\frac{x_i}{\omega_0}) \cdot \frac{1}{\omega_0} \cdot \Delta x$ is the lower Riemman sum for g.

Because $g(\frac{x_i}{\omega_0}) := f(\omega_0 \cdot \frac{x_i}{\omega_0}) = f(x_i)$, the two lower Riemann sums are identical, up to the scale factor of $\frac{1}{\omega_0}$, and hence, if the integrals exist, they are also identical up to the scale factor of $\frac{1}{\omega_0}$.

For the case $\omega_0 < 0$, the same proof is valid and is not repeated. ■

Proof of Prop. 3.14 Trapezoidal Rule for Computing a Riemann Integral

Suppose that $a < b$ and $f : [a,b] \to \mathbb{R}$ is continuous. Then the Riemann Integral $\int_a^b f(x)\,dx$ exists and it can be evaluated by the **Trapezoidal Rule**. That is,

$$\int_a^b f(x)\,dx = \lim_{n\to\infty} \text{Area}_n^{\text{Trap}} = \lim_{n\to\infty} \sum_{i=1}^{n} \frac{f(x_i) + f(x_{i+1})}{2} \cdot (x_{i+1} - x_i),$$

where, $\Delta x := \frac{b-a}{n}$ and $x_i := a + (i-1)\Delta x$. In the above, the limits mean that for all $\epsilon > 0$, there exists $N < \infty$ such that $n \geq N \implies \left|\int_a^b f(x)\,dx - \sum_{i=1}^{n} \frac{f(x_i)+f(x_{i+1})}{2} \cdot (x_{i+1} - x_i)\right| \leq \epsilon.$

Proof: We offer a proof for functions that are monotonically increasing. For the general case, we must appeal to the uniform continuity property of continuous functions over closed bounded intervals.

If the integral $\int_a^b f(x)\,dx$ exists, then the lower and upper Riemann sums for f both exist and converge to the same number, which defines $\int_a^b f(x)\,dx$. By Prop. 2.36, the average of the lower and upper Riemann sums must also converge to $\int_a^b f(x)\,dx$. For monotonic functions, the trapezoidal rule is precisely the average of the upper and lower Riemann sums. Hence, the result is proved. ∎

Proof of Prop. 3.23 Integrals of Even and Odd Functions over Symmetric Intervals

Suppose that $f : \mathbb{R} \to \mathbb{R}$ is a function and that for $a > 0$, the definite integrals $\int_{-a}^0 f(x)\,dx$ and $\int_0^a f(x)\,dx$ both exist and are finite. Then the following hold:

(a) If f is an even function,

- $\int_{-a}^0 f(x)\,dx = \int_0^a f(x)\,dx$, and
- $\int_{-a}^a f(x)\,dx = 2\int_0^a f(x)\,dx$.

(b) If f is an odd function,

- $\int_{-a}^0 f(x)\,dx = -\int_0^a f(x)\,dx$, and
- $\int_{-a}^a f(x)\,dx = 0$.

(c) For all $[b,c] \subset [-a,a]$,

- $\int_b^c f(x)\,dx = \int_b^c f_e(x)\,dx + \int_b^c f_o(x)\,dx$.

Proof:

(a) If f is an even function, $\int_{-a}^0 f(x)\,dx = \int_0^a f(x)\,dx$ follows from Prop. 3.12 by taking $\omega_0 = -1$. The second statement is immediate from the first statement.

(b) If f is an odd function, $\int_{-a}^0 f(x)\,dx = -\int_0^a f(x)\,dx$ follows from Prop. 3.12 by taking $\omega_0 = -1$. The second statement is immediate from the first statement.

(c) For all $[b,c] \subset [-a,a]$, $\int_b^c f(x)\,dx = \int_b^c f_e(x)\,dx + \int_b^c f_o(x)\,dx$ follows from Prop. 3.7 because $f(x) = f_e(x) + f_o(x)$ by the construction of the odd and even parts.

∎

Chapter 4

Properties of Functions: Left and Right Limits, Types of Continuity, Boundedness, and Generalizations of Max and Min

Remark: This chapter reveals some of the mathematical backbone of calculus. Traditionally, much of this material is placed earlier in a calculus course. Because many of the topics are abstract and very technical, we delayed them until this chapter to make sure that the user is ready to combine programming and numerical calculations when learning mathematical concepts.

Learning Objectives

By the end of this chapter, the student should be able to:

- Appreciate the foundational role of calculus in mathematical modeling and problem-solving, integrating programming and numerical methods to solidify these concepts.
- Analyze the behavior of functions at specific points using the concept of one-sided limits.
- Thoroughly understand the nature of function continuity and discover that continuity comes in more than one flavor.
- Derive closed-form expressions for the integrals of exponential functions and apply these techniques to integrate trigonometric functions.
- Determine the boundedness of functions and understand the implications for mathematical analysis and applications.
- Recognize that while maximum and minimum values are important, they are not the full story.

Outcomes

Upon successful completion of this chapter, students will be able to:

- Acquire both intuitive and formal understandings of one-sided limits and their calculation.
- Apply numerical methods to estimate one-sided limits and assess the continuity of functions using the epsilon-delta definition.

- Recognize when it is permissible to take limits within functions, and apply this knowledge to the integration of exponentials.
- Identify and analyze piecewise continuous functions and extend this understanding to a broader class of Riemann integrable functions.
- Review and apply the concepts of maximum and minimum function values, and explore alternative behaviors when these extrema do not exist.

4.1 Intuition for Limits from the Left and the Right

To help with developing an intuitive feel for one-sided limits, consider the function

$$f(x) := \begin{cases} 0.0 & x < 0.0 \\ 0.5 & x = 0.0 \\ 1.0 & 0.0 < x < 2.0 \\ 0.75 & x = 2.0 \\ 1.0 - (x - 2.0) & 2.0 < x \leq 3.0 \\ 0.0 & x > 3.0 \end{cases} \tag{4.1}$$

as plotted in **Fig. 4.1** (blue lines and open red circles). We have defined the value of the function at the origin to be neither zero nor one; in fact, $f(0) = 0.5$. However, if we approach the origin from the left side, namely through negative values of x, the value of the function is constant and equal to zero. Hence, we say that the limit from the left at $x = 0.0$ equals zero, and we denote this by

$$\lim_{x \to 0^-} f(x) = 0. \tag{4.2}$$

Here, the minus sign as a superscript on the right side of a value of x, currently $x_0 = 0$, indicates the limit is taken with values of x that are strictly less than $x_0 = 0$, are never equal to x_0, but become "arbitrarily close to x_0". Similarly, if we approach the origin from the right side, namely through small positive values of x, the value of the function is constant and equal to one. Hence, we say that the limit from the right at $x_0 = 0.0$ equals one, and we denote this by

$$\lim_{x \to 0^+} f(x) = 1. \tag{4.3}$$

Here, the plus sign as a superscript on the right side of a value of x, say x_0, indicates the limit is taken with values of x that are strictly greater than $x_0 = 0$, are never equal to x_0, but become "arbitrarily close to x_0". Before we move on, let us note that the limit of the function from the left at the origin equals zero, its limit from the right equals one, and the value of the function at the origin is 0.5. The function is clearly discontinuous at the origin because there is a jump.

Let us next consider the behavior of the function near $x_0 = 2$. If we approach $x_0 = 2$ from its left side, namely through values of x strictly less than two, the value of the function is constant and equal to one. Hence, we say that the limit from the left at $x = 2.0$ equals one, and we denote this by:

$$\lim_{x \to 2^-} f(x) = 1. \tag{4.4}$$

Again, the minus sign as a superscript on the right side of a value of $x_0 = 2$ indicates the limit is taken with values of x that are strictly less than $x_0 = 2$, are never equal to x_0, but become arbitrarily close to x_0. Similarly, if we approach $x_0 = 2$ from the right side, namely through values of x strictly greater than two, the value of the function is less than one but becomes arbitrarily close to one as x approaches two. Hence, we say that the limit from the right at $x = 2$ equals one, and we denote this by

$$\lim_{x \to 2^+} f(x) = 1. \tag{4.5}$$

Again, the plus sign as a superscript on the right side of $x_0 = 2$ indicates the limit is taken with values of x that are strictly greater than x_0, are never equal to x_0, but become arbitrarily close to x_0. Before we move on, let us note that the limit of the function from the left at $x_0 = 2$ equals one, its limit from the right equals one,

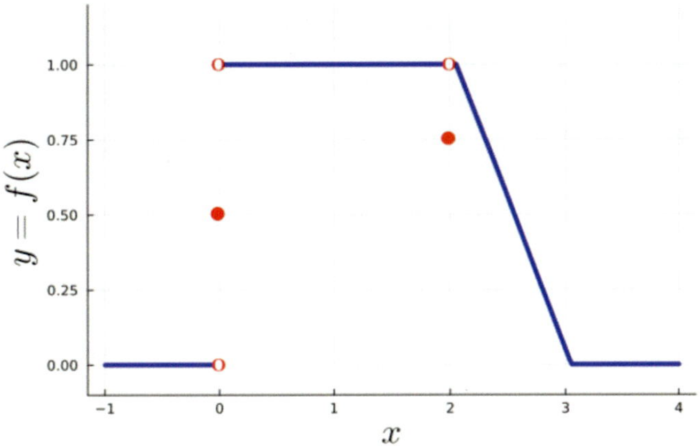

Figure 4.1: This oddly defined function will help us understand the notions of limit from the left and limit from the right at a point. The value of the function at $x = 0$ is 0.5, and its value at $x = 2$ is 0.75.

and the value of the function at $x_0 = 2$ is equal to 0.75. The function is discontinuous at $x_0 = 2$ because its value jumps from 1.0 to 0.75 at the point $x_0 = 2$.

For our final case, let us consider the behavior of the function near $x_0 = 3$. If we approach $x_0 = 3$ from its left side, namely through values of x strictly less than three, the value of the function decreases to zero. Hence, we say that the limit from the left at $x = 3$ equals zero, and we denote this by:

$$\lim_{x \to 3^-} f(x) = 0. \tag{4.6}$$

As before, the minus sign as a superscript on the right side of a value of $x_0 = 3$ indicates the limit is taken with values of x that are strictly less than $x_0 = 3$, are never equal to x_0, but become arbitrarily close to x_0. Similarly, if we approach $x_0 = 3$ from the right side, namely through values of x strictly greater than three, the value of the function is constant and equal to zero. Hence, we say that the limit from the right at $x = 3$ equals zero, and we denote this by

$$\lim_{x \to 3^+} f(x) = 0. \tag{4.7}$$

Again, the plus sign as a superscript on the right side of $x_0 = 3$ indicates the limit is taken with values of x that are strictly greater than $x_0 = 3$, are never equal to x_0, but become arbitrarily close to x_0. Before we move on, let us note that the limit of the function from the left at $x_0 = 3$ equals zero, its limit from the right equals zero, and the value of the function at $x_0 = 3$ is equal to zero. The function is continuous at $x_0 = 3$ because its value does not undergo a jump as x is varied near $x_0 = 3$. This is "mathematically certified" by the limit from the left, the limit from the right, and the function all having the same value at the point $x_0 = 3$. In fact, this property, namely, the two one-sided (left and right) limits existing and agreeing with the value of the function, holds for all $-1 < x \leq 0$, $0 < x < 2$, and $2 < x < 4$; in other symbols, for all $x_0 \in \{(-1, 4) \mid x \neq 0 \text{ and } x \neq 2\}$,

$$\lim_{x \to x_0^-} f(x) = f(x_0) \text{ and } \lim_{x \to x_0^+} f(x) = f(x_0),$$

and this analysis agrees with our intuitive notion of the continuity of the function at those points.

Fact 4.1: Relation to Limits at $\pm\infty$

We connect the one-sided limits to the material in Sections 2.4 and 2.5. The limit from the right is the same as the following (positive) infinite limit:

$$\lim_{x \to x_0^+} f(x) = \lim_{\eta \to +\infty} f(x_0 + \frac{1}{\eta})$$

and the limit from the left is the same as the corresponding negative infinite limit

$$\lim_{x \to x_0^-} f(x) = \lim_{\eta \to -\infty} f\left(x_0 + \frac{1}{\eta}\right).$$

The key points are:

- $\frac{1}{\eta}$ is never exactly equal to zero, so we never evaluate the function at x_0;
- as $\eta \to \infty$, $x_0 + \frac{1}{\eta}$ approaches x_0 from the **right** because $x_0 + \frac{1}{\eta} > x_0$ for all $\eta > 0$; and
- as $\eta \to -\infty$, $x_0 + \frac{1}{\eta}$ approaches x_0 from the **left** because $x_0 + \frac{1}{\eta} < x_0$ for all $\eta < 0$.

Note: If we make the substitution $h := \frac{1}{\eta}$, then $\eta \to +\infty \iff h \to 0^+$ (because h decreases to zero through positive values, that is, h approaches zero from the right) and $\eta \to -\infty \iff h \to 0^-$ (because h increases to zero through negative values, that is, h approaches zero from the left).

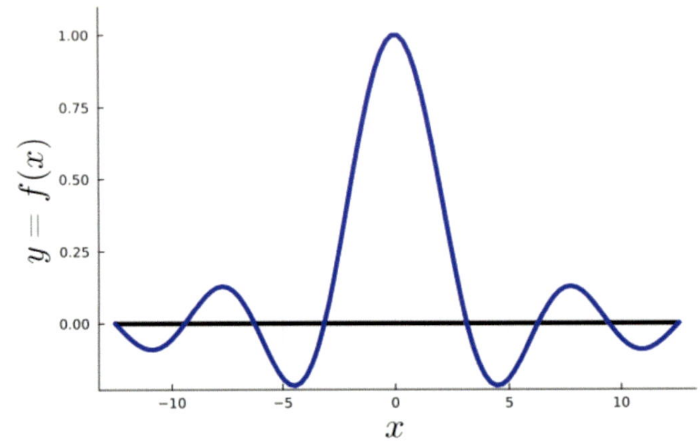

Figure 4.2: This is a plot of $\sin(x)/x$, a function that plays a starring role in signal processing. How is it possible to make sense of its value at the origin, where we have $0/0$? Inquiring minds want to know!

Secrets of the Arcane 4.2: How Might Limits Be Useful

The function

$$\operatorname{sinc}(x) := \begin{cases} \frac{\sin(x)}{x} & x \neq 0 \\ 1 & x = 0 \end{cases}$$

(pronounced "sink" or "sync") plays a big role in signal processing because it can be used to create low-pass filters (i.e., filters that remove high-frequency "noise" from a data signal). Why is it defined to be one at the origin? **Limits give us a way to understand and analyze such conundrums as $\mathbf{0/0}$.** The function's graph is displayed in Fig. 4.2.

This chapter puts us on the road to understanding how to deal with "0/0," while L'Hôpital's Rule in the next chapter should clear up any remaining doubts.

4.2 Formal Definition of Limits from the Left and the Right

We build off the observations presented in Fact 4.1.

One-Sided Limits at Zero

Definition 4.3. *Suppose that $f : A \to \mathbb{R}$ is a real-valued function, $A \subset \mathbb{R}$, and both L and R are (finite) real numbers.*

*(a) The **limit of f from the left at** 0 is equal to L if, for all $\epsilon > 0$ (no matter how small), there exists $\delta > 0$ such that the open interval $(-\delta, 0) \subset A$ (i.e., it is contained in the domain of definition of the function), and $-\delta < h < 0 \implies |f(h) - L| \le \epsilon$. The left limit is denoted as*

$$\lim_{h\to 0^-} f(h) = L.$$

*(b) The **limit of f from the right at** 0 is equal to R if, for all $\epsilon > 0$ (no matter how small), there exists $\delta > 0$ such that the open interval $(0, \delta) \subset A$ (i.e., it is contained in the domain of definition of the function), and $0 < h < \delta \implies |f(h) - R| \le \epsilon$. The right limit is denoted as*

$$\lim_{h\to 0^+} f(h) = R.$$

Notes:

- In the above, h is a dummy variable. We can just as easily write $\lim\limits_{x\to 0^-} f(x)$, $\lim\limits_{y\to 0^+} f(y)$, etc.
- As we make ϵ smaller and smaller, we typically need to make δ smaller and smaller. In other words, δ usually depends on ϵ. It is sometimes useful to express this dependence as $\delta(\epsilon)$ or δ_ϵ, just as we did with $N(\epsilon)$ and N_ϵ when doing limits at infinity.
- **Similar to limits as x tends to infinity, most one-sided limits in Engineering can be computed by inspection, as in Section 2.5, or with minor computations. This will become more clear when we get to L'Hôpital's Rule. For now, we focus on the concept of how the function behaves as we approach the origin from the right or the left.**

Example 4.4. *Compute the left and right limits at zero for the following functions:*

(a) $f(y) = y^2, y \in \mathbb{R}$

(b) $f(z) = \begin{cases} -1 + z^4 & z < 0 \\ 0 & z = 0 \\ 1 + z^2 & z > 0 \end{cases}$

(c) $f(h) = \begin{cases} \frac{h}{|h|} & h \neq 0 \\ \textit{undefined} & h = 0 \end{cases}$

(d) $f(x) = \begin{cases} \sin(\frac{1}{x}) & x > 0 \\ 0 & x \le 0 \end{cases}$

Solutions: We first give solutions that use the epsilon-delta-definitions, and then provide a pointer for doing the limits by inspection.

(a) **Ans.** $\lim\limits_{y\to 0^-} f(y) = 0$ and $\lim\limits_{y\to 0^+} f(y) = 0$.

Let $\epsilon > 0$ be given and set $\delta := \sqrt{\epsilon}$. Then, for all $-\delta < y < 0$, we have $|f(y) - L| = |y^2 - 0| = |y^2| = y^2 < (-\delta)^2 = \epsilon$, and hence the definition of the limit from the left is verified.

Similarly, for all $0 < y < \delta$, we have $|f(y) - R| = |f(y) - 0| = |y^2| = y^2 < (\delta)^2 = \epsilon$, and hence the definition of the limit from the right is verified. In this case, we could use the same δ for both sides of the limit.

(b) **Ans.** $\lim_{z \to 0^-} f(z) = -1$ and $\lim_{z \to 0^+} f(z) = 1$.

Let $\epsilon > 0$ be given and set $\delta := \sqrt[4]{\epsilon}$. Then for all $-\delta < z < 0$, we have $|f(z) - L| = |-1 + z^4 - (-1)| = |z^4| = z^4 < (-\delta)^4 = \epsilon$, and hence the definition of the limit from the left is verified.

Let $\epsilon > 0$ be given and set $\delta := \sqrt{\epsilon}$ for all $0 < z < \delta$, we have $|f(z) - R| = |1 + z^2 - 1| = |z^2| = z^2 < (\delta)^2 = \epsilon$, and hence the definition of the limit from the right is verified.

(c) **Ans.** $\lim_{h \to 0^-} f(h) = -1$ and $\lim_{h \to 0^+} f(h) = 1$.

Let $\epsilon > 0$ be given and set $\delta := 27$. Then, for all $-\delta < h < 0$, we have $|f(h) - L| = |\frac{h}{|h|} - (-1)| = |-1 - (-1)| = 0 < \epsilon$, and hence the definition of the limit from the left is verified. In this case, we can take δ to be any positive real number. It did not have to be small. The graph of $f(h)$ is shown in Fig. 4.3.

Similarly, for all $0 < h < \delta$, we have $|f(h) - R| = |\frac{h}{|h|} - 1| = |1 - 1| = 0 < \epsilon$, and hence the definition of the limit from the right is verified.

(d) **Ans.** $\lim_{x \to 0^-} f(x) = 0$ and $\lim_{x \to 0^+} f(x)$ does not exist.

Here is the insight: $\lim_{x \to 0^+} \sin\left(\frac{1}{x}\right) = \lim_{y \to +\infty} \sin(y)$, and we know that $\sin(y)$ oscillates forever between -1 and $+1$ as $y \to \infty$; hence, the limit from the right at zero does not exist. Can we show that this violates the existence of the limit, using epsilons and deltas? Of course!

Let $R \in \mathbb{R}$, the candidate right limit, be arbitrary. Set $\epsilon = 0.25$ and let $\delta > 0$ be arbitrary. We seek to show that there always exists $0 < x < \delta$ such that $|f(x) - R| > \epsilon$.

Let k be an even integer such that $0 < \frac{1}{k} < \delta$. Then $x_1 := \frac{1}{k\pi}$ and $x_2 := \frac{1}{(k+0.5)\pi}$ satisfy $0 < x_2 < x_1 < \delta$. These numbers were chosen because

$$f(x_1) = \sin(k\pi) = 0 \text{ and } f(x_2) = \sin((k + 0.5)\pi) = \sin\left(\frac{\pi}{2}\right) = 1.$$

If $|f(x_1) - R| = |0 - R| = |R| > \epsilon$, then R is not the right limit. Hence, suppose that $|R| \le \epsilon = 0.25$. Then, by the Reverse Triangle Inequality, $|f(x_2) - R| \ge \big|\, |f(x_2)| - |R| \,\big| \ge 1 - 0.25 = 0.75 > \epsilon$, and thus R is not the right limit. Because R was arbitrary, the right limit does not exist at zero. A zoom of the function's graph near the origin is shown below.

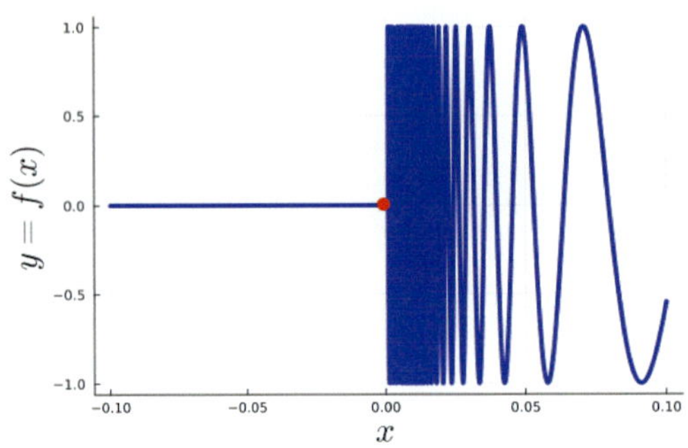

Because the limit from the left is "obvious," this completes the analytical solutions. **The solutions by inspection** proceed as in Section 4.1, using Fig. 4.3. We can trace the graph of each function to the left and right of zero to determine the limits.

■

We can also estimate the limits numerically as follows: define $h[i] = \frac{1}{i}$, $0 < i_{\min} \le i \le i_{\max}$, for example, and then compute $f(h)$ and $f(-h)$ on the vector of values specified by h. This is illustrated in the next code block.

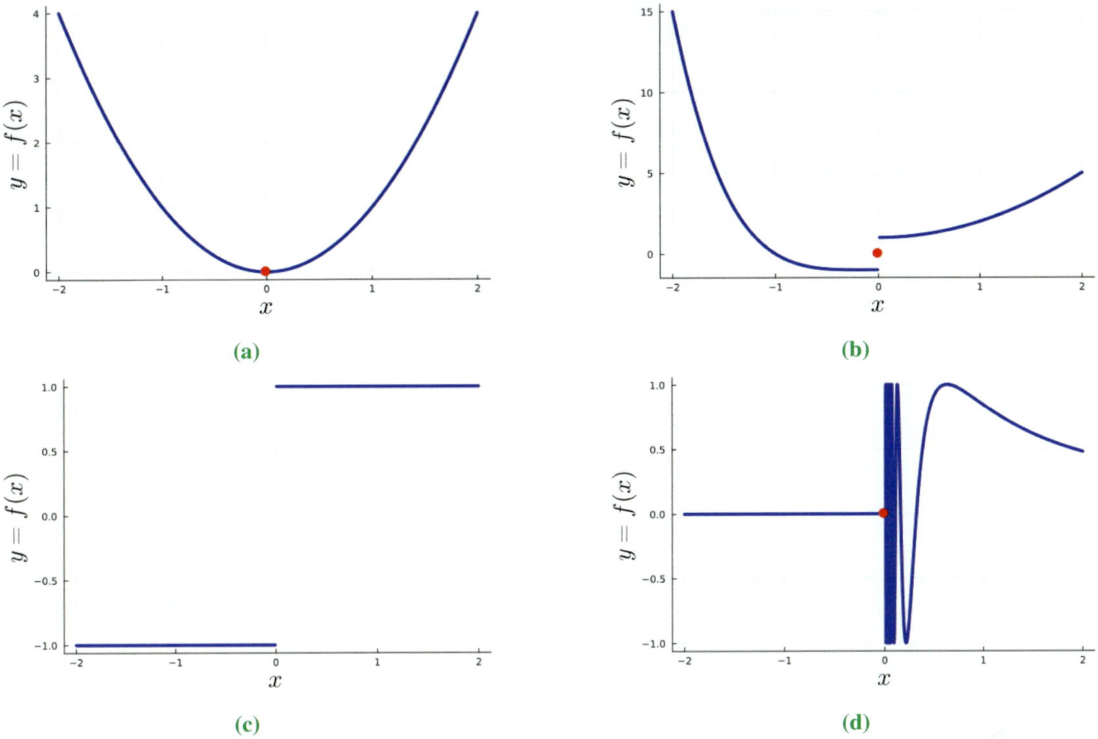

Figure 4.3: Graphs of the functions defined in Example 4.4, which treats left and right limits at zero. (a) The left and right limits are both equal to zero. (b) The left limit equals -1 and the right limit equals $+1$. (c) The left limit equals -1 and the right limit equals $+1$. (d) The left limit equals 0 and the right limit is undefined because, from the right side of zero, the function oscillates between -1 and $+1$ an infinite number of times for $0 < x < \delta$, for any $\delta > 0$, no matter how small.

```
eta = 100:1:1e4
h_Right = 1.0 ./eta
k = 2
y_Right = f.(h_Right,k)
p1 = plot(h_Right,y_Right, linewidth=4, color=:blue, label=false, guidefont=20)
plot!(xlabel=L"$x$", ylabel=L"$y=f(x)$")
h_Left = -h_Right
y_Left = f.(h_Left,k)
p1 = plot!(h_Left,y_Left, linewidth=4, color=:green, label=false, guidefont=20)
if k !=3
    annotate!(0, f(0,k), text(L"$\bullet$", :red, :center, 20))
end
ylims!(p1, (minimum(y_minus), maximum(y_Right)))

display(p1)

@show LeftLimit = y_Left[end]
@show RightLimit = y_Right[end]

png(p1, "LeftRightLimitAtZeroBSecondMethod")
```

Output

```
LeftLimit = y_Left[end] = -0.9999999999999999
```

```
RightLimit = y_Right[end] = 1.00000001
```

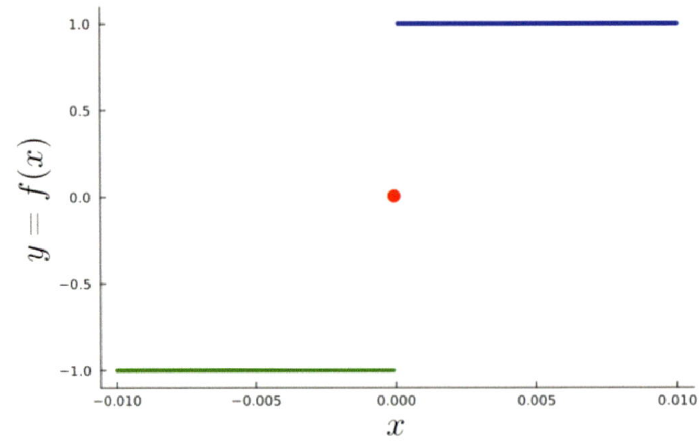

As illustrated in Fig. 4.4, if the function diverges to infinity as we approach zero from the left or right, then finite limits will certainly not exist. We take care of this, just as we did for limits at infinity in Definition 2.17.

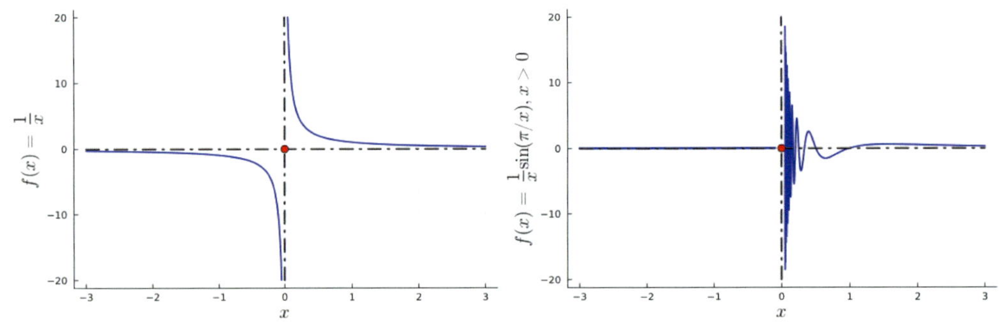

(a) (b)

Figure 4.4: Functions can also blow up as the origin is approached from the left or the right. (a) Both one-sided limits of $1/x$ exist but are $\pm\infty$. (b) The limit from the left exists and is bounded, while the limit from the right does not exist for $(1/x) \cdot \sin(\pi/x), x > 0$.

Unbounded One-Sided Limits at 0

Definition 4.5. *Suppose that $A \subset \mathbb{R}$ and $f : A \to \mathbb{R}$ is a real-valued function.*

(a) *The **limit of f from the left at** 0 is equal to ∞ if, for all $0 < N < \infty$ (no matter how large), there exists $\delta > 0$ such that the open interval $(-\delta, 0) \subset A$ (i.e., it is contained in the domain of definition of the function) **and** $-\delta < h < 0 \implies f(h) > N$. The left limit is then denoted as:*

$$\lim_{h \to 0^-} f(h) = \infty.$$

*Similarly, the **limit of f from the left at** 0 is equal to $-\infty$ if, for all $0 < N < \infty$ (no matter how large), there exists $\delta > 0$ such that the open interval $(0, \delta) \subset A$ (i.e., it is contained in the domain of definition of the function), **and** $-\delta < h < 0 \implies f(h) < -N$. The left limit is then denoted as:*

$$\lim_{h \to 0^-} f(h) = -\infty.$$

*(b) The **limit of f from the right at** 0 is equal to ∞ if, for all $0 < N < \infty$ (no matter how large), there exists $\delta > 0$ such that the open interval $(0, \delta) \subset A$ (i.e., it is contained in the domain of definition of the function) **and** $0 < h < \delta \implies f(h) > N$. The right limit is then denoted as:*

$$\lim_{h \to 0^+} f(h) = \infty.$$

*Similarly, the **limit of f from the right at** 0 is equal to −∞ if, for all $0 < N < \infty$ (no matter how large), there exists $\delta > 0$ such that the open interval $(0, \delta) \subset A$ (i.e., it is contained in the domain of definition of the function), **and** $0 < h < \delta \implies f(h) < -N$. The right limit is then denoted as*

$$\lim_{h \to 0^+} f(h) = -\infty.$$

Example 4.6. *For $f : \mathbb{R} \to \mathbb{R}$ given by*

$$f(x) = \begin{cases} \frac{1}{x} & x \neq 0 \\ 0 & x = 0, \end{cases} \tag{4.8}$$

compute the left and right limits at the origin.

Solution: We provide first an analytical solution and then a numerical solution.

Ans. $\lim_{h \to 0^-} f(h) = -\infty$. To prove this answer using the definition, let $0 < N < \infty$ be arbitrary. Then for $0 < \delta < \frac{1}{N}$,

$$-\delta < x < 0 \implies f(x) = \frac{1}{x} \leq -\frac{1}{\delta} \leq -\frac{1}{\frac{1}{N}} = -N.$$

Ans. $\lim_{h \to 0^+} f(h) = \infty$. To prove this answer using the definition, let $0 < N < \infty$ be arbitrary. Then for $0 < \delta < \frac{1}{N}$,

$$0 < x < \delta \implies f(x) = \frac{1}{x} \geq \frac{1}{\delta} \geq \frac{1}{\frac{1}{N}} = N.$$

```
using Plots, LaTeXStrings

# Create a logarithmic scale of h values
k = 7 #
h = 2.0 .^ range(-1, -k, step=-1)

f(x) = 1/x

# Limit from the right
p1 = scatter(h, f.(h), legend=false, markersize = 4, color=:red, guidefont=20)
plot!(xlabel = "h", ylabel = "f(h) = 1/h")
# Limit from the left
scatter!(-h, f.(-h), markersize = 4, color=:blue)
```

Output The plot for $f(x)$ as defined by (4.8) is shown below.

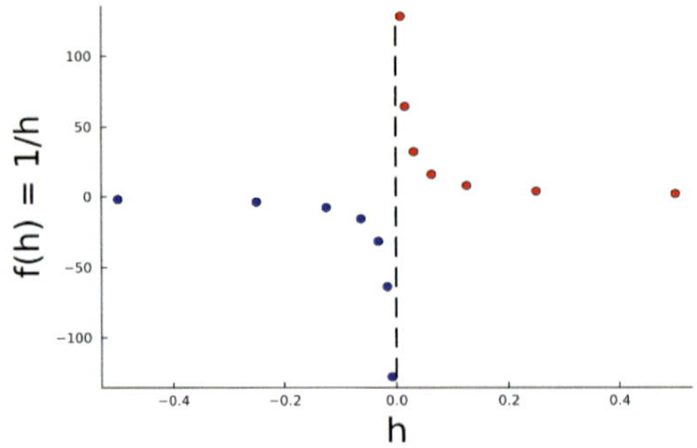

One-Sided Limits at a Point

Definition 4.7. *Suppose that $f : A \to \mathbb{R}$ is a real-valued function, $A \subset \mathbb{R}$, $x_0 \in \mathbb{R}$, and there exists $\delta > 0$ such that the two open intervals $(x_0 - \delta, x_0)$ and $(x_0, x_0 + \delta)$ are contained in A, the domain of definition of the function.*

(a) *The* ***limit from the left at*** x_0 *is*

$$\lim_{x \to x_0^-} f(x) := \lim_{h \to 0^-} f(x_0 + h).$$

(b) *The* ***limit from the right at*** x_0 *is*

$$\lim_{x \to x_0^+} f(x) := \lim_{h \to 0^+} f(x_0 + h).$$

Notes:

- The above is valid for finite limits and unbounded limits.
- $(x_0 - \delta, x_0) \subset A$ and $(x_0, x_0 + \delta) \subset A$ ensure that considering the limits from the left and right makes sense.

Example 4.8. *Let us reuse the function that we studied in (4.1) and* **Fig. 4.1**, *namely,*

$$f(x) := \begin{cases} 0.0 & x < 0.0 \\ 0.5 & x = 0.0 \\ 1.0 & 0.0 < x < 2.0 \\ 0.75 & x = 2.0 \\ 1.0 - (x - 2.0) & 2.0 < x \le 3.0 \\ 0.0 & x > 3.0. \end{cases}$$

Determine the limits from the left and right for

(a) $x_0 = 0$

(b) $x_0 = 2$, *and*

(c) $x_0 = 3$.

Solution: We can take a numerical approach by plotting the function near the specified values of x_0, but we will do something a bit more quantitative. We will define a (finite) vector of h values that terminates in very small values, e.g., on the order of 10^{-6}. We will also compute the mean and (sample) standard deviation of the function evaluated at $x_0 \pm h$, namely:

- $h[i] = \frac{1}{i+i_{\min}}$, $1 \le i \le N$, $i_{\min} > 0$ a constant.
- $y_R[i] := f(x_0 + h[i])$
- $\mu_R := \frac{1}{N} \cdot \sum_{i=1}^{N} y_R[i]$ (mean value)
- $\sigma_R := \sqrt{\frac{1}{N-1} \cdot \sum_{i=1}^{N} (y_R[i] - \mu_R)^2}$ (sample standard deviation)

and similarly for the left-sided limit. If the function has a right limit at x_0, then y_R should be nearly a constant for N large. Hence, when

- $\mu_R \approx y_R[N]$ and
- $\sigma_R \approx 0$,

the numerical evidence **suggests** that $y_R[N]$ is a good estimate of the limit.

```
function f(x)
    if x < 0
        y = -1.0
    elseif abs(x)<1e-12
        y = 0.0
    elseif (0 < x < 2)
        y = 1.0
    elseif abs(x-2.0)<1e-12
        y = 0.75
    elseif (2 < x <= 3.0)
        y = 1.0 - (x - 2.0)
    else
        y = 0.0
    end
    return y
end
#
using Statistics
eta = 1e3:1e3:1e6
h_Right = (1.0)./ eta
h_Left = -h_Right
x0Vec = [0.0 2.0 3.0]

delta = h_Right[end]

@show delta
println(" ")

for k = 1:length(x0Vec)
    x0 = x0Vec[k]
    y_Left = f.(x0 .+ h_Left)
    y_Right = f.(x0 .+ h_Right)
    #
    @show x0
    @show limitLeft = y_Left[end]
    @show meanLeft = mean(y_Left)
    @show stdDevLeft = std(y_Left)
    println("--------")
    @show limitRight = y_Right[end]
    @show meanRight = mean(y_Right)
```

```
    @show stdDevRight = std(y_Right)
    println(" ")
    println(" ")
    #
end
```

Output

```
delta = 1.0e-6

x0 = 0.0
limitLeft = y_Left[end] = -1.0
meanLeft = mean(y_Left) = -1.0
stdDevLeft = std(y_Left) = 0.0
--------
limitRight = y_Right[end] = 1.0
meanRight = mean(y_Right) = 1.0
stdDevRight = std(y_Right) = 0.0


x0 = 2.0
limitLeft = y_Left[end] = 1.0
meanLeft = mean(y_Left) = 1.0
stdDevLeft = std(y_Left) = 0.0
--------
limitRight = y_Right[end] = 0.9999989999999999
meanRight = mean(y_Right) = 0.9999925145291394
stdDevRight = std(y_Right) = 3.9868430925503734e-5


x0 = 3.0
limitLeft = y_Left[end] = 1.000000000139778e-6
meanLeft = mean(y_Left) = 7.4854708605545105e-6
stdDevLeft = std(y_Left) = 3.986843092550373e-5
--------
limitRight = y_Right[end] = 0.0
meanRight = mean(y_Right) = 0.0
stdDevRight = std(y_Right) = 0.0
```

We see that the estimated numerical limits correspond well to what we computed previously in Section 4.1, when all we knew to do was to read values from a graph. In each case, $y[N] \approx \mu$ and $\sigma \approx 0$. ■

Let us now try the more challenging example of $\sin(\frac{1}{x}), x > 0$ and zero otherwise.

Example 4.9. *Determine the limits from the left and right for* $g(x) := \sin(\frac{1}{x}), x > 0$ *and* $g(x) = 0, x \leq 0,$ *for* $x_0 \in \{0.0, \frac{1}{\pi}, \frac{2}{\pi}\}$.

Solution:

```
eta = 1e3:1e3:1e6
h_Right = (1.0)./ eta
h_Left = -h_Right
x0Vec = [0.0 1.0/pi 2.0/pi]

g(x) = x > 0 ? sin(1/x) : 0.0

delta = h_Right[end]

```

```
@show delta
println(" ")

for k = 1:length(x0Vec)
    x0 = x0Vec[k]
    y_Left = g.(x0 .+ h_Left)
    y_Right = g.(x0 .+ h_Right)
    #
    @show x0
    @show limitLeft = y_Left[end]
    @show meanLeft = mean(y_Left)
    @show stdDevLeft = std(y_Left)
    println("--------")
    @show limitRight = y_Right[end]
    @show meanRight = mean(y_Right)
    @show stdDevRight = std(y_Right)
    println(" ")
    println(" ")

    println(" ")
    #
end
```

Output

```
delta = 1.0e-6

x0 = 0.0
limitLeft = y_Left[end] = 0.0
meanLeft = mean(y_Left) = 0.0
stdDevLeft = std(y_Left) = 0.0
--------
limitRight = y_Right[end] = -0.34999350217129294
meanRight = mean(y_Right) = -0.00011524369884409064
stdDevRight = std(y_Right) = 0.7075827861802105



x0 = 0.3183098861837907
limitLeft = y_Left[end] = -9.869635406884313e-6
meanLeft = mean(y_Left) = -7.392953161190493e-5
stdDevLeft = std(y_Left) = 0.00039441035871027 89
--------
limitRight = y_Right[end] = 9.86957339451198e-6
meanRight = mean(y_Right) = 7.382758963882898e-5
stdDevRight = std(y_Right) = 0.0003925579978142143



x0 = 0.6366197723675814
limitLeft = y_Left[end] = 0.999999999996956
meanLeft = mean(y_Left) = 0.9999999949842888
stdDevLeft = std(y_Left) = 1.0037167987310087e-7
--------
limitRight = y_Right[end] = 0.999999999996956
meanRight = mean(y_Right) = 0.9999999950072798
stdDevRight = std(y_Right) = 9.976893600410653e-8
```

As we suspected from Example 4.4-(d), the left limit at the $x_0 = 0$ exists and is equal to zero, but the right limit does not exist. This shows up in the numerical calculations by:

- $|\mu_R - y_R[N]| \approx 0.35$ is "large"; and
- $\sigma_R \approx 0.708$ is also large.

These observations are compatible with the extreme oscillations on the right side of the origin, as seen in Fig. 4.3. All other one-sided limits exist. ■

We next illustrate unbounded one-sided limits at a point.

Example 4.10. *Determine the limits from the left and right at $x_0 \in \{0.0, -\frac{\pi}{2}, -\pi\}$ for $g(x) := \tan(x)$, $(x \mod \pi) \neq \frac{\pi}{2}$ (i.e., x is not an odd multiple of $\pi/2$) and $g(x) = 0$ otherwise. The plot for this function is shown below.*

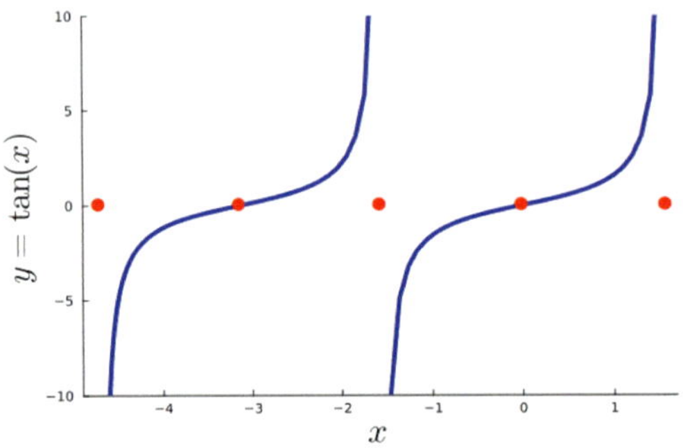

Solution: Because $(x \mod \pi) = \frac{\pi}{2} \implies \sin(x) = \pm 1$ and $\cos(x) = 0$, we expect unbounded limits at these points (numerator is non-zero and the denominator is equal to zero). Similarly, because $(x \mod \pi) = 0 \implies \sin(x) = 0$ and $\cos(x) = \pm 1$, we expect the limits to be zero. Indeed, based on the plot, we can see the one-sided limits at $x_0 = 0$ and $x_0 = -\pi$ all exist and are equal to zero, which is confirmed by the numerical calculations below. Moreover, the limit from the left at $-\frac{\pi}{2}$ appears to be $+\infty$, while the limit from the right appears to be $-\infty$, consistent with the "analytical" study of the problem.

```
using Statistics
eta = 1e3:1e3:1e6
h_Right = (1.0)./ eta
h_Left = -h_Right
x0Vec = [0.0 -pi/2 -pi]

g(x) = abs(x % pi) >1e-10  ? tan(x) : 0.0

delta = h_Right[end]

@show delta
println(" ")

for k = 1:length(x0Vec)
    x0 = x0Vec[k]
    y_Left = g.(x0 .+ h_Left)
    y_Right = g.(x0 .+ h_Right)
    #
    @show x0
    @show limitLeft = y_Left[end]
    @show meanLeft = mean(y_Left)
    @show stdDevLeft = std(y_Left)
```

```
    println("--------")
    @show limitRight = y_Right[end]
    @show meanRight = mean(y_Right)
    @show stdDevRight = std(y_Right)
    println(" ")
    println(" ")

    println(" ")
    #
end
```

Output

```
delta = 1.0e-6

x0 = 0.0
limitLeft = y_Left[end] = -1.0000000000003333e-6
meanLeft = mean(y_Left) = -7.485471261235949e-6
stdDevLeft = std(y_Left) = 3.986843990838751e-5
--------
limitRight = y_Right[end] = 1.0000000000003333e-6
meanRight = mean(y_Right) = 7.485471261235949e-6
stdDevRight = std(y_Right) = 3.986843990838751e-5



x0 = -1.5707963267948966
limitLeft = y_Left[end] = 1.0000000001431657e6
meanLeft = mean(y_Left) = 500500.000018453
stdDevLeft = std(y_Left) = 288819.43611785874
--------
limitRight = y_Right[end] = -1.000000000020701e6
meanRight = mean(y_Right) = -500499.9999775702
stdDevRight = std(y_Right) = 288819.4360824532



x0 = -3.141592653589793
limitLeft = y_Left[end] = -1.0000000000176466e-6
meanLeft = mean(y_Left) = -7.485471261117652e-6
stdDevLeft = std(y_Left) = 3.986843990838485e-5
--------
limitRight = y_Right[end] = 1.000000000262576e-6
meanRight = mean(y_Right) = 7.485471261362585e-6
stdDevRight = std(y_Right) = 3.986843990838485e-5
```

■

4.3 The Power of Continuity Done Right

In Section 1.5.1, we provided an intuitive notion of continuity, namely, "a function is continuous if we can draw its graph without lifting our pencil from the paper." Using the concept of a limit, we next provide an analytical definition. **This more careful definition of continuity at a point will provide a significant benefit to us when it comes to evaluating limits and computing integrals.**

4.3.1 Continuity at a Point and Continuous Everywhere on the Domain of Definition

Continuity at a Point and Continuous Everywhere

Definition 4.11. *Suppose that* $I \subset \mathbb{R}$ *is an open interval, that is,* I *is either* $(a,b), (a,\infty), (-\infty,b)$, *or* $(-\infty,\infty)$ *as in Definition 1.3.*

(a) A function $f : I \to \mathbb{R}$ *is* ***continuous at*** $x_0 \in I$ *if the two one-sided limits at* x_0 *exist and agree with the value of the function at* x_0. *In other symbols,*

$$\begin{aligned} \lim_{x\to x_0^-} f(x) &= f(x_0), \text{ and} \\ \lim_{x\to x_0^+} f(x) &= f(x_0). \end{aligned} \tag{4.9}$$

For intervals containing an endpoint, such as $[a,b]$, $[a,b)$, $[a,\infty)$, $(a,b]$, *and* $[b,\infty)$, *we have to treat the endpoints differently, but all other points in the interval are treated as above.*

(b) For I *a closed or half-open interval as highlighted above and* $x_0 \in I$ ***not an endpoint***, f *is continuous at* x_0 *if the conditions in (4.9) hold. At the endpoints of the interval, only one of the one-sided limits makes sense. Hence,* f *is* ***continuous at*** $x_0 = a$***, a lower endpoint,*** *if*

$$\lim_{x\to a^+} f(x) = f(a),$$

and f *is* ***continuous at*** $x_0 = b$***, an upper endpoint,*** *if*

$$\lim_{x\to b^-} f(x) = f(b).$$

(c) A function f *that is not continuous at* x_0 *is said to be* ***discontinuous at*** x_0 *or* x_0 *is said to be* ***a point of discontinuity*** *of the function.*

(d) If f *is continuous at each point* $x_0 \in I$, *its domain of definition, then* f ***is simply said to be continuous***.

(e) If f *is not continuous at all points of its domain of definition, then it is* ***discontinuous***. *It only takes one point of discontinuity to make a function discontinuous.*

Remark 4.12. *For* $x_0 \in \mathbb{R}$, *it can* never *be the case that* $f(x_0) = \pm\infty$ *because* ***the codomains of our functions contain only real numbers***. *We recall that* ∞ *is a concept and not a number. Limits can equal* ∞ *because we allow functions to grow without bound, such as the function* $f : \mathbb{R} \to \mathbb{R}$ *by* $f(x) = x$ *or* $f : (0,\infty) \to (0,\infty)$ *by* $f(x) = \frac{1}{x}$.

Equation 4.9 can be rearranged as

$$\begin{gathered} \lim_{x\to x_0^-} (f(x) - f(x_0)) = 0 \\ \text{and} \\ \lim_{x\to x_0^+} (f(x) - f(x_0)) = 0. \end{gathered} \tag{4.10}$$

We will use this equivalence in Example 4.13.

Example 4.13. *For the following functions, determine, if any exist, their points of discontinuity. If there are no points of discontinuity, then state that the function is continuous everywhere in its domain of definition.*

(a) $f : \mathbb{R} \to \mathbb{R}$ *by* $f(x) = x^3$.

(b) $f : [2, \infty) \to \mathbb{R}$ *by* $f(x) = x^2$.

(c) $f : \mathbb{R} \to \mathbb{R}$ *by* $f(x) = \begin{cases} x^3 & x \geq 0 \\ x^2 & x < 0. \end{cases}$

(d) $f : \mathbb{R} \to \mathbb{R}$ *by* $f(x) = \begin{cases} 1 + x & x \geq 0 \\ x^2 & x < 0. \end{cases}$

(e) $f : [0, \infty) \to \mathbb{R}$ *by* $f(x) = \begin{cases} \sin(\frac{1}{x}) & x > 0 \\ 0 & x = 0 \end{cases}$

(f) $f : (0, \infty) \to \mathbb{R}$ *by* $f(x) = \log_a(x)$, *where* $a > 1$.

Solutions:

(a) **Ans.** x^3 is continuous at all points of its domain of definition. We show this via the definition. Let $x_0 \in \mathbb{R}$ be arbitrary. Then,

$$\lim_{x \to x_0^-} x^3 := \lim_{h \to 0^-} (x_0 + h)^3 = \lim_{h \to 0^-} \left(x_0^3 + 3hx_0^2 + 3h^2 x_0 + h_0^3 \right) = x_0^3 + 0 + 0 + 0 = x_0^3$$

and

$$\lim_{x \to x_0^+} x^3 := \lim_{h \to 0^+} (x_0 + h)^3 = \lim_{h \to 0^+} \left(x_0^3 + 3hx_0^2 + 3h^2 x_0 + h_0^3 \right) = x_0^3 + 0 + 0 + 0 = x_0^3.$$

Because x_0 was arbitrary, x^3 is continuous at all points of its domain of definition. **Note that we determined the limits by inspection after some simple algebra (namely, the Binomial Theorem).**

(b) **Ans.** $(\bullet)^2 : [2, \infty) \to \mathbb{R}$ is continuous at all points of its domain of definition. We show this via the definition. We first consider $2 < x_0 < \infty$, but otherwise arbitrary. Then,

$$\lim_{x \to x_0^-} x^2 := \lim_{h \to 0^-} (x_0 + h)^2 = \lim_{h \to 0^-} \left(x_0^2 + 2hx_0 + h_0^2 \right) = x_0^2 + 0 + 0 = x_0^2$$

and

$$\lim_{x \to x_0^+} x^2 := \lim_{h \to 0^+} (x_0 + h)^2 = \lim_{h \to 0^+} \left(x_0^2 + 2hx_0 + h_0^2 \right) = x_0^2 + 0 + 0 = x_0^2.$$

Let us now set $x_0 = 2$. We only check the limit from the right, namely,

$$\lim_{x \to 2^+} x^2 := \lim_{h \to 0^+} (2 + h)^2 = \lim_{h \to 0^+} \left(4 + 4h + h_0^2 \right) = 4 + 0 + 0 = 4 = x_0^2.$$

Because we verified continuity at all points of its domain, $(\bullet)^2 : [2, \infty) \to \mathbb{R}$ is continuous at all points of its domain of definition. **Once again, the limits themselves were computed by inspection.**

(c) **Ans.** $f(x) = \begin{cases} x^3 & x \geq 0 \\ x^2 & x < 0. \end{cases}$ is continuous at all points of its domain of definition. We solve this one a bit differently, because continuity for $x > 0$ and $x < 0$ can be checked the same as we did for parts (a) and (b). The only point that seems a bit different is the origin, but because x^3 and x^2 both vanish at the origin, there should not be a problem; to be sure, we evaluate

$$\lim_{x \to 0^-} x^2 := \lim_{h \to 0^-} (0 + h)^2 = \lim_{h \to 0^-} h^2 = 0 = 0^2$$

and

$$\lim_{x \to 0^+} x^3 := \lim_{h \to 0^+} (0 + h)^2 = \lim_{h \to 0^+} h^3 = 0 = 0^3.$$

Hence, the definition of continuity at a point is met for all points of the domain of definition of the function.

(d) **Ans.** $f(x) = \begin{cases} 1+x & x \geq 0 \\ x^2 & x < 0. \end{cases}$ is discontinuous at $x_0 = 0$ and continuous everywhere else. Once again, checking the limits for all $x_0 \neq 0$ is straightforward. For $x_0 = 0$, we see that

$$\lim_{x \to 0^-} x^2 := \lim_{h \to 0^-} (0+h)^2 = \lim_{h \to 0^-} h^2 = 0 \neq f(0) = 1$$

and

$$\lim_{x \to 0^+} 1 + x := \lim_{h \to 0^+} (1 + 0 + h) = \lim_{h \to 0^+} 1 + h = 1 = f(0).$$

Because the left limit at zero is not equal to the value of the function at zero, the function is discontinuous at the origin. **It was not necessary to compute the limit from the right; we simply did it for extra practice.**

(e) **Ans.** $f(x) = \begin{cases} \sin(\frac{1}{x}) & x > 0 \\ 0 & x = 0 \end{cases}$ is discontinuous at the origin, and continuous everywhere else in its domain of definition, meaning, it is continuous for all $x > 0$. The function is discontinuous at $x_0 = 0$ because we showed earlier, in Example 4.4, that $\lim_{x \to 0^+} \sin(1/x)$ does not exist. Hence, the limit from the right cannot equal the value of the function, which is zero.

For $x > 0$, set $y := \frac{1}{x}$. Then $y > 0 \iff x > 0$ reduces us to whether $\sin(y)$ is continuous or not for $y > 0$. But we "know" from experience that it is continuous. A formal proof can be given using trigonometric identities and the bounds given in **Fig. 4.7**.

(f) **Ans.** $f(x) = \log_a(x)$, where $a > 1$ is continuous everywhere in its domain of definition, which we know is $(0, \infty)$. We first show continuity at $x_0 = 1$ and then use the pattern established for this special case to handle the general case. Recall from Prop. 1.49 that for all $a > 1$, $\log_a(1) = 0$. Hence, we need to show that both

$$\lim_{h \to 0^-} \log_a(1+h) = 0$$

and

$$\lim_{h \to 0^+} \log_a(1+h) = 0.$$

This limit is less obvious than the previous limits; hence, we give an epsilon-delta proof: Let $\epsilon > 0$ be arbitrary. We seek $\delta > 0$ such that

$$-\delta \leq h < 0 \implies |\log_a(1+h)| \leq \epsilon$$

and

$$0 < h \leq \delta \implies |\log_a(1+h)| \leq \epsilon,$$

where we note that we can use different values for δ in the two cases. From Prop. 1.4-(c), we have that

$$|\log_a(1+h)| < \epsilon \iff -\epsilon < \log_a(1+h) < \epsilon.$$

Because, for $a > 1$, a^x is strictly monotonically increasing on $(-\infty, \infty)$, taking the exponential preserves the inequalities; hence,

$$-\epsilon < \log_a(1+h) < \epsilon \iff a^{-\epsilon} < 1 + h < a^{\epsilon}.$$

Therefore,[1]

$$|\log_a(1+h)| < \epsilon \iff \underbrace{a^{-\epsilon} - 1}_{<0} < h < \underbrace{a^{\epsilon} - 1}_{>0},$$

which implies that

$$-\delta < h < 0 \implies |\log_a(1+h)| \leq \epsilon, \text{ when } \delta = 1 - a^{-\epsilon}$$

and

$$0 < h < \delta \implies |\log_a(1+h)| \leq \epsilon, \text{ when } \delta = a^{\epsilon} - 1,$$

[1]For all $a > 1$, $a^x < 1 \iff x < 0$.

establishing continuity at the origin. To show continuity at a general $x > 0$, we want to show that the left and right limits as h tends to zero of $\log_a(x+h) - \log_a(x) = 0$. Using log properties,

$$\log_a(x+h) - \log_a(x) = \log_a\left(\frac{x+h}{x}\right) = \log_a\left(1 + \frac{h}{x}\right).$$

Because $x > 0$, if we define $\bar{h} := \frac{h}{x}$, then $\bar{h} \to 0^- \iff h \to 0^-$ and $\bar{h} \to 0^+ \iff h \to 0^+$. Hence,

$$\lim_{h\to 0^-} \log_a(1 + \frac{h}{x}) = \lim_{\bar{h}\to 0^-} \log_a(1 + \bar{h})$$

and

$$\lim_{h\to 0^+} \log_a(1 + \frac{h}{x}) = \lim_{\bar{h}\to 0^+} \log_a(1 + \bar{h}),$$

reducing the problem to the case we have already considered, namely, continuity at $x_0 = 1$.

■

4.3.2 Two-Sided Limits and Continuous Functions

If the limit from the left and the right both exist at a point, then we can combine them into a single limit.

Two-sided Limit at a Point

To set the stage, consider a function $f : A \to \mathbb{R}$ where A is a subset of $\mathbb{R}$, $x_0 \in \mathbb{R}$, and there exists $\delta > 0$ such that both $(x_0 - \delta, x_0) \subset A$ and $(x_0, x_0 + \delta) \subset A$. Then $x_0 + h \in A$ for all $0 < |h| < \delta$, meaning that we can evaluate the left and right limits:

$$\lim_{h\to 0^-} f(x_0 + h)$$

and

$$\lim_{h\to 0^+} f(x_0 + h).$$

Definition 4.14. *Suppose that $f : A \to \mathbb{R}$ as above and $x_0 \in A$. Then the limit as x approaches x_0 exists and is equal to L, a finite real number, if for all $\epsilon > 0$, there exists $\delta > 0$, such that $0 < |h| \leq \delta \implies |f(x_0 + h) - L| \leq \epsilon$. The limit is denoted as:*

$$\lim_{x\to x_0} f(x) = L.$$

"Two-sided limits" are typically called "limits," with the adjective "two-sided" dropped. It is only in the case of a one-sided limit that we need to state its chirality ("sidedness" or handedness of the limit).

Fact 4.15: Equivalence between a Two-sided Limit and Two One-sided Limits

Two of one, versus one of the other: the following are equivalent statements for $|L| < \infty$:

(a) $\lim\limits_{x\to x_0^-} f(x) = L$ and $\lim\limits_{x\to x_0^+} f(x) = L$.

(b) $\lim\limits_{x\to x_0} f(x) = L$.

Notes: Statements being equivalent means $(a) \iff (b)$, where we note that **(a) is true** if, and only if, **both one-sided limits exist and are equal to L**. Just as with a one-sided limit, we never evaluate the function at x_0 when evaluating a two-sided limit. **In theory, we can replace every "limit" with two one-sided limits. However, this is not common practice.**

Using the above Fact, continuity of a function at a point can be checked with a single (two-sided) limit instead of two one-sided limits. What we will do instead is to give the classical epsilon-delta definition of a function being continuous at a point and then **relate all of the (commonly seen) definitions of continuity at a point**.

Classical Definition of Continuity at a Point

Definition 4.16. *Suppose that $f : I \to \mathbb{R}$, where I is an interval and $x_0 \in A$. Then f is continuous at x_0 if for all $\epsilon > 0$, there exists $\delta > 0$, such that*

$$\{x \mid |x - x_0| \le \delta\} \subset I \textbf{ and } |x - x_0| \le \delta \implies |f(x) - f(x_0)| \le \epsilon.$$

Note: *If $\{x \in \mathbb{R} \mid |x - x_0| \le \delta\} \not\subset I$, then it does not make sense to evaluate the function at all of these points because some of them (at least) are not in the domain of the function. This observation explains why the definition is written the way it is.*

Fact 4.17: Equivalent Ways to Define Continuity at a Point

The following statements are equivalent for a function $f : I \to \mathbb{R}$, where I is an interval:

(a) f is continuous at x_0 by Def. 4.16.

(b) $\lim_{x \to x_0^-} f(x) = f(x_0)$ and $\lim_{x \to x_0^+} f(x) = f(x_0)$, meaning that Def. 4.11 of continuity at a point holds.

(c) $\lim_{x \to x_0} f(x) = f(x_0)$ (equivalent to the above because two one-sided limits are always equivalent to a single one-sided limit.)

More examples of using limits and checking continuity with the modified formulations would be repetitive. Instead, let us list some common continuous functions.

Proposition 4.18: Common Continuous Functions

The following functions are continuous for all $x \in \mathbb{R}$ unless indicated otherwise:

Polynomial Functions

- Constant function: $f(x) = c$
- Linear function: $f(x) = ax + b$
- Quadratic function: $f(x) = ax^2 + bx + c$
- Higher-degree polynomials: $f(x) = a_n x^n + a_{n-1} x^{n-1} + \cdots + a_1 x + a_0$

Trigonometric Functions

- Sine: $f(x) = \sin(x)$
- Cosine: $f(x) = \cos(x)$
- Tangent (continuous except at odd multiples of $\pm\frac{\pi}{2}$): $f(x) = \tan(x)$

Power Function

- Power (continuous for $x > 0$ and $y \in \mathbb{R}$): $f(x) = x^y$

Exponential and Logarithmic Functions

- Exponential: $f(x) = a^x$ (where $a > 1$)

- Natural Exponential: $f(x) = e^x$
- Logarithm (continuous for $x > 0$): $f(x) = \log_a(x)$ (where $a > 1$)
- Natural Logarithm (continuous for $x > 0$): $f(x) = \ln(x)$

Rational Functions (subject to domain restrictions)

- $f(x) = \frac{p(x)}{q(x)}$ where $p(x)$ and $q(x)$ are polynomials and $q(x) \neq 0$ (means, rational functions are continuous for all $x \in \mathbb{R}$ such that $q(x) \neq 0$)

Root Functions

- Square Root (continuous for $x \geq 0$): $f(x) = \sqrt{x}$
- Cube Root: $f(x) = \sqrt[3]{x}$
- n-th Root (continuous for $x \geq 0$ if $n \in \mathbb{N}$ is even, all $x \in \mathbb{R}$ if $n \in \mathbb{N}$ is odd): $f(x) = \sqrt[n]{x}$

Other Special Functions

- Absolute Value: $f(x) = |x|$
- Triangle: $f(x) = \mathrm{tri}(x)$
- Radial Basis Function: $f(x) = e^{-\frac{(x-x_c)^2}{s^2}}$
- Gaussian (Normal Distribution) with mean μ and standard deviation σ: $f(x) = \frac{1}{\sqrt{2\pi\sigma^2}}e^{-\frac{(x-\mu)^2}{2\sigma^2}}$

4.3.3 Key Properties of Continuous Functions

For the purpose of developing and understanding the fundamentals of calculus, the following Proposition ranks among the most important results in this book.

Proposition 4.19: Taking Limits Inside a Function

Consider two functions $g : X \to Y$ and $f : Y \to Z$, where X, Y, and Z are subsets of $\mathbb{R}$. Suppose that f is continuous at a point $y_0 \in Y$. Then, the following hold:

(a) If $x_0 \in \mathbb{R}$ and $\lim_{x \to x_0} g(x) = y_0$ (i.e., the limit exists and equals y_0), then $\lim_{x \to x_0} f(g(x)) = f(\lim_{x \to x_0} g(x)) = f(y_0)$. **Thus, the limit can be taken "inside" the function f.**

(b) Similarly, if $\lim_{x \to \infty} g(x) = y_0$, then $\lim_{x \to \infty} f(g(x)) = f(\lim_{x \to \infty} g(x)) = f(y_0)$.

(c) If $\lim_{x \to -\infty} g(x) = y_0$, then $\lim_{x \to -\infty} f(g(x)) = f(\lim_{x \to -\infty} g(x)) = f(y_0)$.

In short, if $g(x)$ has a limit (as x approaches a finite point or as x approaches $\pm\infty$), and f is continuous at the limit of g (i.e., y_0), then, when evaluating a limit of $f(g(x)) = f \circ g(x)$, the limit can be taken inside the function f and applied directly to g.

Recommended video: Mu Prime Math "When can we switch the limit and function?"

Example 4.20. *Use Prop. 4.19 to evaluate the following limits. In each case, as part of the solution, identify the functions $g : X \to Y$ and $f : Y \to Z$, as well as the limit point of g, which we have been calling y_0.*

(a) $\lim_{x \to 0} \sin\left(\frac{x}{1 + x^2}\right)$

(b) $\lim_{x \to \frac{\pi}{2}} \frac{3 + \sin^3(x)}{\sin^2(x) + 3\sin(x) + 4}$

(c) $\lim_{x \to \infty} \ln(e^{-2x} + e^{-x} + 1)$

Solutions:

(a) **Ans.** $\lim_{x \to 0} \sin\left(\frac{1+x}{1+x^2}\right) = \sin(1) \approx 0.8415.$

We define $f : \mathbb{R} \to \mathbb{R}$ by $f(y) = \sin(y)$, $g : \mathbb{R} \to \mathbb{R}$ by $g(x) = \frac{1+x}{1+x^2}$, and note that

$$\lim_{x \to 0} g(x) = \lim_{x \to 0} \frac{1+x}{1+x^2} = 1 =: y_0.$$

The limit exists because g is continuous everywhere on its domain of definition per Prop. 4.18, Common Continuous Functions. Finally, also by Prop. 4.18, $f(y)$ is continuous everywhere and hence at the point $y_0 := 1$.

(b) **Ans.** $\lim_{x \to \frac{\pi}{2}} \frac{3 + \sin^3(x)}{\sin^2(x) + 3\sin(x) + 4} = \frac{1}{2}$ (because $\sin(\pi/2) = 1$).

We define $f : \mathbb{R} \to \mathbb{R}$ by $f(y) = \frac{3+y^3}{y^2+3y+4}$, $g : \mathbb{R} \to \mathbb{R}$ by $g(x) = \sin(x)$, and note that

$$\lim_{x \to \frac{\pi}{2}} g(x) = \lim_{x \to \frac{\pi}{2}} \sin(x) = 1.0 =: y_0.$$

The limit exists because g is continuous everywhere on its domain of definition per Prop. 4.18, Common Continuous Functions. Finally, also by Prop. 4.18, $f(y)$ is continuous everywhere and hence at the point $y_0 := 1$.

(c) **Ans.** $\lim_{x \to \infty} \ln(e^{-2x} + e^{-x} + 1) = 0.$

Given what we know at this point in the chapter, this problem requires two applications of Prop. 4.19. We will write $\ln(e^{-2x} + e^{-x} + 1)$ as the composition of three functions, $\ln(e^{-2x} + e^{-x} + 1) =: f(g(h(x)))$, where

$$\begin{aligned} f : (0, \infty) \to \mathbb{R} \ \text{ by } \ f(z) = \ln(z), \\ g : (0, \infty) \to (1, \infty) \ \text{ by } \ g(y) = y^2 + y + 1, \\ h : \mathbb{R} \to (0, \infty) \ \text{ by } \ h(x) = e^{-x}, \end{aligned}$$

where we note that $g(h(x)) = (e^{-x})^2 + e^{-x} + 1 = e^{-2x} + e^{-x} + 1$ because $(a^\alpha)^\beta = a^{\alpha \cdot \beta}$ for all $a > 0$, and real numbers α and β.

All three functions are continuous everywhere on their domains of definition. Moreover,

$$\begin{aligned} \lim_{x \to \infty} h(x) &= \lim_{x \to \infty} e^{-x} = 0 =: y_0, \\ g(y_0) &= g(0) = 1 =: z_0, \\ f(z_0) &= f(1) = \ln(1) = 0. \end{aligned}$$

Therefore,

$$\begin{aligned}\lim_{x\to\infty}\ln(e^{-2x}+e^{-x}+1) &= \lim_{x\to\infty} f(g(h(x)))\\ &= f(\lim_{x\to\infty} g(h(x)))\\ &= f(g(\underbrace{\lim_{x\to\infty} h(x)}_{y_0=0}))\\ &= f(\underbrace{g(0)}_{z_0=1})\\ &= f(1)\\ &= \ln(1)\\ &= 0.\end{aligned}$$

Note: All of these limits can be approximated numerically using the same techniques we have already demonstrated earlier.

■

The above problem was mainly an exercise in writing a "complicated" function as a composition of simpler functions. The next two problems are more representative of how Prop. 4.19 is typically used. Their solutions will illustrate how a change of variable can make an apparently intractable problem become tractable.

Example 4.21. *Using the fact that* $\lim_{n\to\infty}\left(1+\frac{1}{n}\right)^n =: e$*, compute the following limits:*

(a) $\lim_{n\to\infty}\left(1-\frac{1}{n}\right)^{-n}$;

(b) $\lim_{n\to\infty}\left(1+\frac{x}{n}\right)^n$*, where* $x\in\mathbb{R}$ *is arbitrary; and*

(c) $\lim_{h\to 0}\frac{\ln(x+h)-\ln(x)}{h}$*, where* $x>0$.

Hints: For (a), note that $\left(1-\frac{1}{n}\right)^{-1} = \left(\frac{n-1}{n}\right)^{-1} = \left(\frac{n}{n-1}\right) = \left(\frac{n-1+1}{n-1}\right) = 1+\frac{1}{n-1}$. Also, Prop. 2.37 is helpful. Finally, remember the power rules!

For (b), when $x=0$, the limit is clearly one, leaving two cases to investigate, $x>0$ and $x<0$. First, assume $x>0$, and then use the substitution $m := \frac{n}{x}$ and look for the power function $f:(0,\infty)\to(0,\infty)$ by $f(y)=y^x$. Note that here, the variable is y, and x is a fixed positive constant. Follow a similar pattern for $x<0$.

For (c), recall the logarithm rules for $\ln(x)\pm\ln(y)$ and $\ln(x^y)$. Try the substitution $n := \frac{1}{h}$.

Solutions:

(a) **Ans.** $\lim_{n\to\infty}\left(1-\frac{1}{n}\right)^{-n} = e$.

Using the Hint, we have:

$$\begin{aligned}\lim_{n\to\infty}\left(1-\frac{1}{n}\right)^{-n} &= \lim_{n\to\infty}\left(\left(1-\frac{1}{n}\right)^{-1}\right)^n\\ &= \lim_{n\to\infty}\left(1+\frac{1}{n-1}\right)^n.\end{aligned}$$

This is very close to the limit for e, but we are not quite there yet. We make a change of variable $m := n - 1$ so that

$$n \to \infty \iff m \to \infty.$$

Hence,

$$\begin{aligned}\lim_{n\to\infty}\left(1-\frac{1}{n}\right)^{-n} &= \lim_{n\to\infty}\left(1+\frac{1}{n-1}\right)^{n}\\ &= \lim_{m\to\infty}\left(1+\frac{1}{m}\right)^{m+1}\\ &= \lim_{m\to\infty}\underbrace{\left(1+\frac{1}{m}\right)^{m}}_{f(m)}\cdot\underbrace{\left(1+\frac{1}{m}\right)}_{g(m)},\end{aligned}$$

setting us up to apply Prop. 2.37, because

$$\lim_{m\to\infty} g(m) = 1 \text{ and } \lim_{m\to\infty} f(m) = e.$$

We conclude therefore that $\lim_{n\to\infty}\left(1-\frac{1}{n}\right)^{-n} = e.$

Note: Can we show that for all integers $k \geq 1$,

$$\lim_{n\to\infty}\left(1+\frac{1}{n\pm k}\right)^{n} = \lim_{n\to\infty}\left(1+\frac{1}{n}\right)^{n\pm k} = e\,?$$

The key point is that k is a fixed (finite) value when applying Prop. 2.37.

(b) **Ans.** $\lim_{n\to\infty}\left(1+\frac{x}{n}\right)^{n} = e^{x}.$

Using the hint, we assume $x > 0$ and set $m := \frac{n}{x}$ so that

$$n = m\cdot x \quad \text{and } m \to \infty \iff n \to \infty.$$

Making the substitution, we have

$$\begin{aligned}\lim_{n\to\infty}\left(1+\frac{x}{n}\right)^{n} &= \lim_{m\to\infty}\left(1+\frac{1}{m}\right)^{m\cdot x}\\ &= \lim_{m\to\infty}\left(\left(1+\frac{1}{m}\right)^{m}\right)^{x}.\end{aligned}$$

We define $f(y) := y^x$ and note that it is continuous for all $y > 0$. Hence,

$$\begin{aligned}\lim_{n\to\infty}\left(1+\frac{x}{n}\right)^{n} &= \lim_{m\to\infty}\left(\left(1+\frac{1}{m}\right)^{m}\right)^{x}\\ &= \left(\lim_{m\to\infty}\left(1+\frac{1}{m}\right)^{m}\right)^{x} \quad \text{(take limit inside } f\text{)}\\ &= (e)^{x} \quad \text{(recognizing the limit)}\\ &= e^{x}.\end{aligned}$$

We still have $x < 0$ to consider, but the pattern is now very clear. Note that because $x < 0$, $x = -|x|$ and $|x| \neq 0$. Hence, we set $m := \frac{n}{|x|}$ so that:

$$n = m\cdot|x| \quad \text{and } m \to \infty \iff n \to \infty.$$

Making the indicated substitution, we have

$$\begin{aligned}\lim_{n\to\infty}\left(1+\frac{x}{n}\right)^{n} &= \lim_{n\to\infty}\left(1-\frac{|x|}{n}\right)^{n}\\ &= \lim_{m\to\infty}\left(1-\frac{1}{m}\right)^{m\cdot|x|}\\ &= \lim_{m\to\infty}\left(\left(1-\frac{1}{m}\right)^{-m}\right)^{-|x|},\end{aligned}$$

where in the last step we used our power rules. We define $f(y) := y^x = y^{-|x|}$ and note that it is a continuous function of y for all $y > 0$. Hence,

$$\begin{aligned}
\lim_{n\to\infty}\left(1+\frac{x}{n}\right)^n &= \lim_{m\to\infty}\left(\left(1-\frac{1}{m}\right)^{-m}\right)^{-|x|} \\
&= \left(\lim_{m\to\infty}\left(1-\frac{1}{m}\right)^{-m}\right)^{-|x|} \quad \text{(take limit inside } f\text{)} \\
&= (e)^{-|x|} \quad \text{(recognizing the limit)} \\
&= e^{-|x|} \\
&= e^x \quad (\text{because } x = -|x| \text{ for } x < 0).
\end{aligned}$$

(c) **Ans.** For all $x > 0$, $\lim_{h\to 0}\dfrac{\ln(x+h)-\ln(x)}{h} = \dfrac{1}{x}$.

Using our logarithm rules,

$$\begin{aligned}
\frac{\ln(x+h)-\ln(x)}{h} &= \frac{1}{h}\ln\left(\frac{x+h}{x}\right) \\
&= \ln\left(\left(\frac{x+h}{x}\right)^{\frac{1}{h}}\right) \\
&= \ln\left(\left(1+\frac{h}{x}\right)^{\frac{1}{h}}\right).
\end{aligned}$$

We make the substitution $n := \frac{1}{h}$ so that

$$n \to \infty \iff h \to 0^+.$$

Using this substitution, we have

$$\begin{aligned}
\lim_{h\to 0^+}\frac{\ln(x+h)-\ln(x)}{h} &= \lim_{h\to 0^+}\ln\left(\left(1+\frac{h}{x}\right)^{\frac{1}{h}}\right) \\
&= \lim_{n\to\infty}\ln\left(\left(1+\frac{1}{nx}\right)^{n}\right) \\
&= \ln\left(\lim_{n\to\infty}\left(1+\frac{1}{nx}\right)^{n}\right) \quad \text{(taking the limit inside because } \ln(x) \text{ is continuous)} \\
&= \ln\left(e^{\frac{1}{x}}\right) \quad \text{(recognizing the limit by part (b) of this problem)} \\
&= \frac{1}{x} \quad \text{(using logarithm rules).}
\end{aligned}$$

In theory, we should now consider $\lim_{h\to 0^-}\dfrac{\ln(x+h)-\ln(x)}{h}$, but we leave it to the reader to do.

■

This following (challenging) limit is the key to understanding the definite integral $\int_a^b e^x dx$.

Example 4.22. *Evaluate* $\lim_{h\to 0}\dfrac{e^h-1}{h}$.

Solution: **Ans.** $\lim_{h\to 0}\dfrac{e^h-1}{h} = 1$.

At first blush, none of the methods we have covered so far seems applicable. This is more typical of how real engineering problems go. Initially, we are totally convinced that none of the tools we know are useful for the problem at hand, but then, all of a sudden, we see how the problem can be tackled if we look at it the right way. Here, we need to make a change of variable and apply logarithm rules.

The power move is to define $u := e^h - 1$ so that $e^h = u + 1 \iff h = \ln(1+u)$. We note that $\ln(x)$ is continuous everywhere on $(0, \infty)$, and that $h \to 0 \iff u \to 0$. With this substitution, we have:

$$\lim_{h\to 0} \frac{e^h - 1}{h} = \lim_{u\to 0} \frac{u}{\ln(1+u)} = \lim_{u\to 0} \frac{1}{\frac{1}{u} \cdot \ln(1+u)}.$$

From our logarithm properties, we recall that

$$y \cdot \ln(x) = \ln(x^y),$$

and hence, $\frac{1}{u} \cdot \ln(1+u) = \ln\left((1+u)^{\frac{1}{u}}\right)$, yielding

$$\lim_{h\to 0} \frac{e^h - 1}{h} = \lim_{u\to 0} \frac{1}{\ln\left((1+u)^{\frac{1}{u}}\right)}.$$

If we define $n := \frac{1}{u}$, we would have

$$(1+u)^{\frac{1}{u}} = \left(1 + \frac{1}{n}\right)^n \underset{n\to\infty}{\longrightarrow} e,$$

which is Euler's number, and we know that $\ln(e) = 1$. How can these observations be useful to us?

We write

$$\frac{1}{\ln\left((1+\frac{1}{n})^n\right)} = f(g(k(n))),$$

where

$$\begin{aligned} f(\beta) &:= \frac{1}{\beta}, \quad \text{continuous for all } \beta > 0, \\ g(\alpha) &:= \ln(\alpha), \quad \text{continuous for all } \alpha > 0, \\ k(n) &:= \left(1 + \frac{1}{n}\right)^n \underset{n\to\infty}{\longrightarrow} e. \end{aligned}$$

Then, as in Example 4.20, we have:

$$\begin{aligned} \lim_{h\to 0^+} \frac{e^h - 1}{h} &= \lim_{n\to\infty} \frac{1}{\ln\left((1+\frac{1}{n})^n\right)} \quad \text{(after two changes of variable)} \\ &= \lim_{n\to\infty} f(g(k(n))) \quad \text{(using our defined functions, two being continuous and one having a known limit)} \\ &= f\left(\lim_{n\to\infty} g(k(n))\right) \quad \text{(take limit inside } f\text{)} \\ &= f\left(g\left(\lim_{n\to\infty} k(n)\right)\right) \quad \text{(take limit inside } g\text{)} \\ &= f\left(g\left(\lim_{n\to\infty} \left(1+\frac{1}{n}\right)^n\right)\right) \quad \text{(identify } k(n)\text{)} \\ &= f\left(g\left(e\right)\right) \quad \text{(use known limit)} \\ &= f\left(\ln\left(e\right)\right) \quad \text{(now we are on autopilot)} \\ &= f\left(1\right) \\ &= \frac{1}{1} \\ &= 1. \end{aligned}$$

Note: Because we showed every step, the solution looks impossibly arduous. With practice, we can skip most of the steps because "they become obvious with sufficient practice." The key ideas were to make a useful substitution and subsequently recognize when limits can be taken inside functions. ■

Example 4.23. *Determine the limit* $\lim_{n\to\infty}\left(\frac{n+3}{n}\right)^n$.

Solution: We first note that $\frac{n+3}{n} = 1 + \frac{3}{n}$, and hence, $\left(\frac{n+3}{n}\right)^n = \left(1+\frac{3}{n}\right)^n$. We recognize immediately that this has something to do with Euler's number, e. Indeed, from Example 4.21-(b), we have

$$\lim_{n\to\infty}\left(\frac{n+3}{n}\right)^n = \lim_{n\to\infty}\left(1+\frac{3}{n}\right)^n = e^3.$$

■

How is your limit intuition coming along? **BlackPenRedPen** has a video to find out! No epsilons and deltas. He uses continuity and qualitative reasoning. The answer pivots on the difference between 0^- and 0^+. Intrigued?

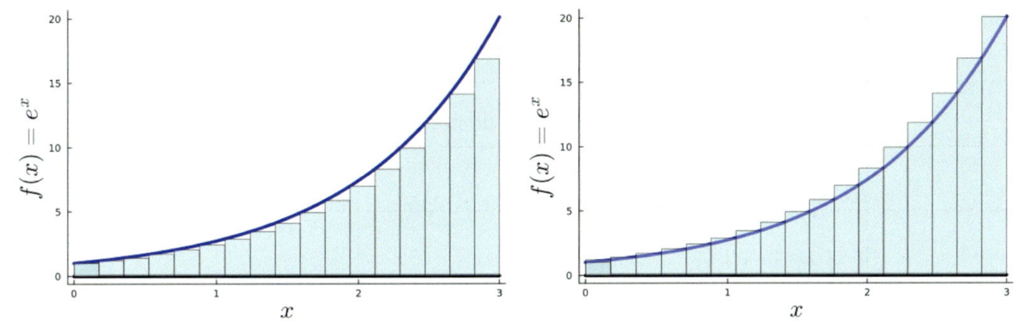

(a) (b)

Figure 4.5: Under- and overapproximations for the area under e^x. (a) An underapproximation for the area. (b) An overapproximation for the area.

4.3.4 Application to Integrating the Natural Exponential, e^x

Proposition 4.24: Integrating e^x

For $x > 0$ finite,

$$\int_0^x e^y\,dy = e^x - 1. \tag{4.11}$$

Proof: In Section 3.1.3, we developed closed-form expressions for integrating monomials; namely, for $k \in \mathbb{N}$,

$$\int_0^x y^k dy = \frac{y^{k+1}}{k+1}.$$

Here, we follow the same pattern to discover the closed-form expression for exponentials, namely

- Form the Riemann Lower and Upper Sums as indicated in Fig. 4.5.
- By taking $\Delta y := \frac{x}{n}$, we will compute the lower and upper bounds of (4.11), $\text{Area}_n^{\text{Low}}$ and $\text{Area}_n^{\text{Up}}$, that depend on n.

- We will seek an exact value for (4.11) by evaluating $\lim_{n\to\infty} \text{Area}_n^{\text{Low}}$ and $\lim_{n\to\infty} \text{Area}_n^{\text{Up}}$.
- If the two limits exist and are equal, then we will have determined the definite integral.

Solution: We work this out in full detail so as to review the **Approximation Principle** that undergirds most of calculus. We let $n > 1$ be an integer, and because we are working on the interval $[0, x]$ for $x > 0$, we define $\Delta y := \frac{x-0}{n}$ along with $y_i := a + (i-1)\Delta y$, yielding

$$0 =: a = y_1 < y_2 < \cdots < y_n < y_{n+1} = b := x.$$

Because $f(y) = e^y$ is increasing, $f(y_i) \le f(y) \le f(y_{i+1})$ for $y \in [y_i, y_{i+1}]$, it is easy to determine the min and max values of the function over subintervals.

Important Observation: $e^{k\Delta y} = \left(e^{\Delta y}\right)^k$ by Prop. 1.45-(g).

Riemann Lower Sum: We underestimate the total area between $[0, x]$ by summing up the underapproximations, $f(y_i) \cdot \Delta y$, for each interval $[y_i, y_{i+1}]$,

$$\begin{aligned}
\text{Area}_n^{\text{Low}} &:= \sum_{i=1}^{n} f(y_i) \cdot \Delta y \\
&= \sum_{i=1}^{n} e^{(i-1)\Delta y} \cdot \Delta y \ \ \text{(substituting in terms)} \\
&= \sum_{i=1}^{n} \left(e^{\Delta y}\right)^{(i-1)} \cdot \Delta y \ \ \text{(by the observation)} \\
&= \Delta y \cdot \sum_{i=1}^{n} \left(e^{\Delta y}\right)^{(i-1)} \ \ \text{(take terms that do not depend on } i \text{ outside the sum)} \\
&= \frac{x}{n} \cdot \sum_{i=0}^{n-1} \left(e^{\Delta y}\right)^{i} \ \ \text{(change of index)} \\
&= \frac{x}{n} \cdot \left(\frac{1-r^n}{1-r}\right) \ \ \text{(Prop. 2.39)},
\end{aligned} \tag{4.12}$$

where

$$r := e^{\Delta y}. \tag{4.13}$$

Hence, after noting that $r^n = e^{n\Delta y} = e^x$, we have

$$\text{Area}_n^{\text{Low}} = \frac{x}{n} \cdot \frac{1-e^x}{1-e^{\frac{x}{n}}} = \frac{x}{n} \cdot \frac{e^x - 1}{e^{\frac{x}{n}} - 1}. \tag{4.14}$$

Therefore,

$$\begin{aligned}
\text{Area}^{\text{Low}} &:= \lim_{n\to\infty} \text{Area}_n^{\text{Low}} \\
&= \lim_{n\to\infty} \frac{x}{n} \cdot \frac{e^x - 1}{e^{\frac{x}{n}} - 1} \\
&= (e^x - 1) \cdot \lim_{n\to\infty} \frac{x}{n} \cdot \frac{1}{e^{\frac{x}{n}} - 1} \ \ (e^x - 1 \ \ \text{does not depend on n, so factor out of the limit}) \\
&= (e^x - 1) \cdot \lim_{h\to 0^+} h \cdot \frac{1}{e^h - 1} \ \ \left(\text{change of variable } h := \frac{x}{n} \iff n = \frac{x}{h}\right) \\
&= (e^x - 1) \cdot \lim_{h\to 0^+} \frac{1}{\frac{e^h - 1}{h}} \\
&= (e^x - 1) \cdot \frac{1}{\lim_{h\to 0^+} \frac{e^h - 1}{h}} \ \ \text{(by Prop. 4.19)} \\
&= (e^x - 1) \cdot \frac{1}{1} \\
&= e^x - 1,
\end{aligned} \tag{4.15}$$

because, by Example 4.22, $\lim_{h\to 0^+} \frac{e^h - 1}{h} = 1$. Next, we repeat this process for the Riemann Upper Sum[2].

Riemann Upper Sum: We overestimate the total area between $[0, x]$ by summing up the overapproximations, $f(y_{i+1}) \cdot \Delta y$, for each interval $[y_i, y_{i+1}]$,

$$
\begin{aligned}
\text{Area}_n^{\text{Up}} := & \sum_{i=1}^{n} f(y_{i+1}) \cdot \Delta y \\
= & \sum_{i=1}^{n} e^{i\Delta y} \cdot \Delta y \quad \text{(substituting in terms)} \\
= & \sum_{i=1}^{n} \left(e^{\Delta y}\right)^i \cdot \Delta y \quad \text{(by the observation)} \\
= & \Delta y \cdot \sum_{i=1}^{n} \left(e^{\Delta y}\right)^i \quad \text{(take terms that do not depend on } i \text{ outside the sum)} \qquad (4.16) \\
= & \frac{x}{n} \cdot \left(\sum_{i=0}^{n} \left(e^{\Delta y}\right)^i - 1 \right) \quad \text{(adding and subtracting one for the geometric sum)} \\
= & \frac{x}{n} \cdot \left(\frac{1 - r^{n+1}}{1 - r} - 1 \right) \quad \text{(Prop. 2.39)} \\
= & \frac{x}{n} \cdot r \cdot \left(\frac{r^n - 1}{r - 1} \right) \quad \text{(algebra)}
\end{aligned}
$$

where, as before, $r = e^{\Delta y} = e^{\frac{x}{n}}$.

Hence, after noting that $r^n = \left(e^{\Delta y}\right)^n = e^{n\Delta y} = e^x$, we have

$$
\text{Area}_n^{\text{Up}} = \frac{x}{n} \cdot e^{\frac{x}{n}} \cdot \left(\frac{e^x - 1}{e^{\frac{x}{n}} - 1} \right). \qquad (4.17)
$$

Therefore,

$$
\begin{aligned}
\text{Area}^{\text{Up}} := & \lim_{n\to\infty} \text{Area}_n^{\text{Up}} \\
= & \lim_{n\to\infty} \frac{x}{n} \cdot e^{\frac{x}{n}} \cdot \left(\frac{e^x - 1}{e^{\frac{x}{n}} - 1} \right) \\
= & \left(e^x - 1\right) \cdot \lim_{n\to\infty} \frac{x}{n} \cdot e^{\frac{x}{n}} \cdot \left(\frac{1}{e^{\frac{x}{n}} - 1} \right) \\
= & \left(e^x - 1\right) \cdot \lim_{h\to 0^+} h \cdot e^h \cdot \frac{1}{e^h - 1} \quad \left(\text{change of variable } h := \frac{x}{n}\right) \qquad (4.18) \\
= & \left(e^x - 1\right) \cdot \lim_{h\to 0^+} \frac{1}{\frac{e^h - 1}{h}} \cdot \lim_{h\to 0^+} e^h \quad \text{(by Prop. 4.19)} \\
= & \left(e^x - 1\right) \cdot 1 \cdot 1 \\
= & e^x - 1,
\end{aligned}
$$

because,

- by Example 4.22, $\lim_{h\to 0^+} \frac{e^h - 1}{h} = 1$, and
- because e^x is continuous, $\lim_{h\to 0^+} e^h = e^{\left(\lim_{h\to 0^+} h\right)} = e^0 = 1$.

The limits of the Riemann Lower and Upper Sums both exist and are equal. Hence, our proof is complete. ■

[2] Because e^x is a continuous function, we know the Riemann integral exists. Hence, the two Riemann sums converge to a common finite value, meaning that we really only have to do one of them to compute the value of the integral. Here, we do both because it is our last opportunity to practice the definition of the Riemann integral!

Proposition 4.25: Integrating $e^{\alpha x}$

For $x > 0$ finite and $\alpha \in \mathbb{R}$ non-zero,

$$\int_0^x e^{\alpha y}\, dy = \frac{1}{\alpha} \cdot (e^{\alpha x} - 1). \tag{4.19}$$

Proof: By Prop. 3.12, Integration and Scaling,

$$\begin{aligned}\int_0^x e^{\alpha y}\, dy &= \frac{1}{\alpha} \int_{\alpha \cdot 0}^{\alpha \cdot x} e^y\, dy \\ &= \frac{1}{\alpha} e^y \Big|_{\alpha \cdot 0}^{\alpha \cdot x} \\ &= \frac{1}{\alpha} (e^{\alpha x} - 1)\end{aligned}$$

■

Example 4.26. *Evaluate the limit as $n \to \infty$ of the finite sum,*

$$\frac{1}{n} \sum_{k=1}^{n} e^{\frac{k}{n}} = \frac{\sqrt[n]{e^1} + \sqrt[n]{e^2} + \cdots + \sqrt[n]{e^n}}{n}.$$

Solution: See the video "The trick you need to know for limits", by Michael Penn.

■

Fact 4.27: Euler's Formula for Integrating Sine and Cosine

Recall that by Euler's formula,

$$\cos(\omega_0 x) = \frac{e^{\mathbf{i}\omega_0 x} + e^{-\mathbf{i}\omega_0 x}}{2} \tag{4.20a}$$

and

$$\sin(\omega_0 x) = \frac{e^{\mathbf{i}\omega_0 x} - e^{-\mathbf{i}\omega_0 x}}{2\mathbf{i}}. \tag{4.20b}$$

While we have not proved it, Prop. 4.25 also holds for $\alpha \in \mathbb{C}$ non-zero. Hence,

$$\int_0^x e^{\mathbf{i}\omega_0 y}\, dy = \frac{1}{\mathbf{i}\omega_0} \left(e^{\mathbf{i}\omega_0 x} - 1\right) \tag{4.21a}$$

and

$$\int_0^x e^{-\mathbf{i}\omega_0 y}\, dy = \frac{-1}{\mathbf{i}\omega_0} \left(e^{-\mathbf{i}\omega_0 x} - 1\right). \tag{4.21b}$$

After some algebra, we obtain

$$\int_0^x \cos(\omega_0 y)\, dy = \frac{1}{\omega_0} \sin(\omega_0 x) \tag{4.22a}$$

and

$$\int_0^x \sin(\omega_0 y)\, dy = \frac{1}{\omega_0} (1 - \cos(\omega_0 x)). \tag{4.22b}$$

We will not belabor this point here because in Chapter 7, we will find another means to arrive at the same result. Nevertheless, it is very satisfying to see how things tie together.

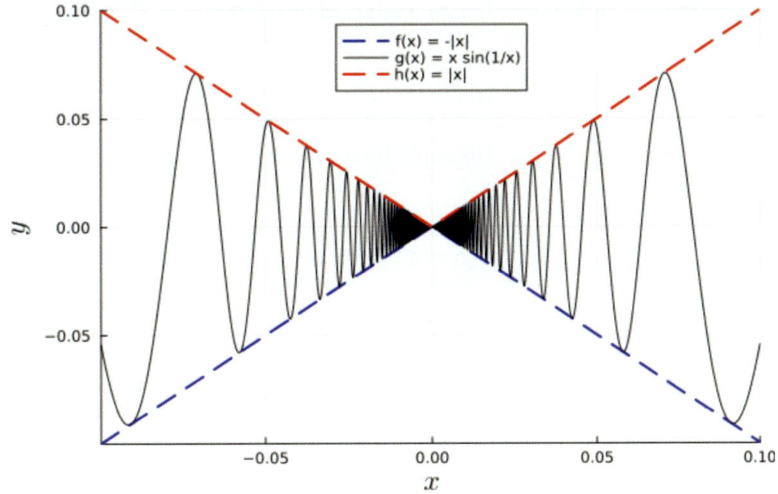

Figure 4.6: **(Squeeze or Sandwich Theorem)** If a function is sandwiched between two functions that have a common limit, then the function itself also has the same limit.

4.4 The Squeeze Theorem

Proposition 4.28: Squeeze or Sandwich Theorem

Suppose that $a < x_0 < b$ and that f, g, and h are functions from the interval (a, b) to $\mathbb{R}$, except possibly being undefined at x_0 itself (the usual situation we face when computing limits). If for every $x \in (a, b)$, $x \neq x_0$, it holds that

$$f(x) \leq g(x) \leq h(x),$$

and if

$$\lim_{x \to x_0} f(x) = \lim_{x \to x_0} h(x) = L,$$

then

$$\lim_{x \to x_0} g(x) = L.$$

Note: The result also holds for limits from the left or right at x_0. To see this, simply redefine all three functions to the limit from the left for $x > x_0$ (or the limit from the right, for $x < x_0$). A proof is given in Wikipedia, which also has some good examples.

Example 4.29. *Use the Squeeze Theorem to evaluate the following limits.*

(a) $\lim_{x \to 0} x \sin\left(\frac{1}{x}\right)$.

(b) $\lim_{x \to 0} \frac{\sin(x)}{x}$.

(c) $\lim_{x \to 0} \frac{1 - \cos(x)}{x}$.

Solution:

(a) **Ans.** $\lim_{x \to 0} x \sin\left(\frac{1}{x}\right) = 0$.

The intuition is illustrated in Fig. 4.6. For an analytical proof, because $-1 \le \sin(\frac{1}{x}) \le 1$ for all $x \neq 0$, we have that

$$-x \le x \sin(\frac{1}{x}) \le x \quad \text{(multiply both sides by } x > 0\text{)},$$
$$-x \ge x \sin(\frac{1}{x}) \ge x \quad \text{(multiply both sides by } x < 0 \text{ and note the flipped inequalities)},$$

and, therefore,

$$x \le x \sin(\frac{1}{x}) \le -x \quad \text{(rewrite the previous line from right to left, still for } x < 0\text{)}.$$

From the first and last lines, we have

$$-|x| \le x \sin\left(\frac{1}{x}\right) \le |x|,$$

as illustrated in Fig. 4.6. Hence, we can take $f(x) := -|x|$, $g(x) := x \sin(\frac{1}{x})$, and $h(x) := |x|$. Because

$$\lim_{x\to 0} f(x) = \lim_{x\to 0} h(x) = 0,$$

the Squeeze Theorem gives the stated conclusion.

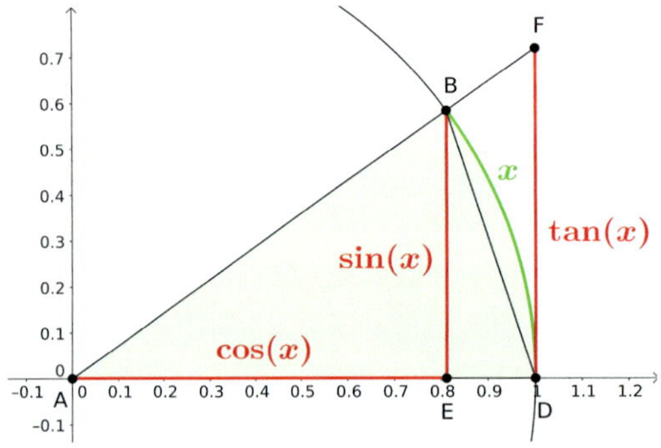

(a)

	Area($\triangle ADB$)	$\le$	Area(sector ADB)	$\le$	Area($\triangle ADF$)
$\Longrightarrow$	$\frac{1}{2} \cdot \sin(x) \cdot 1$	$\le$	$\frac{x}{2\pi} \cdot \pi$	$\le$	$\frac{1}{2} \cdot \tan(x) \cdot 1$
$\Longrightarrow$	$\sin(x)$	$\le$	x	$\le$	$\frac{\sin(x)}{\cos(x)}$
$\Longrightarrow$	$\frac{\cos(x)}{\sin(x)}$	$\le$	$\frac{1}{x}$	$\le$	$\frac{1}{\sin(x)}$
$\Longrightarrow$	$\cos(x)$	$\le$	$\frac{\sin(x)}{x}$	$\le$	1

(b)

Figure 4.7: Image and table from Wikipedia. The image shows a right triangle $\triangle ADB$, a sector of the unit circle, sector ADB having perimeter x radians, and a right triangle $\triangle ADF$. The table notes their respective areas and relative sizes. From the formulas for area, useful implications ($\Longrightarrow$) are obtained for bounding $\sin(x)$ and $\cos(x)$.

(b) **Ans.** $\lim_{x\to 0} \frac{\sin(x)}{x} = 1.$

The hard work is done in Fig. 4.7, which shows that we can take $f(x) := \cos(x)$, $g(x) := \frac{\sin(x)}{x}$, and $h(x) := 1$. Because

$$\lim_{x\to 0} f(x) = \lim_{x\to 0} h(x) = 1,$$

the Squeeze Theorem gives the stated conclusion.

(c) **Ans.** $\lim_{x\to 0} \dfrac{1-\cos(x)}{x} = 0.$

This is the most challenging problem. For $x > 0$,

$$
\begin{aligned}
\cos(x) \le & \frac{\sin(x)}{x} \le 1 \\
& \Downarrow \\
\cos(x) \le & \frac{\sqrt{1-\cos^2(x)}}{x} \le 1 \quad \text{(Basic trig identity)} \\
& \Downarrow \\
\cos^2(x) \le & \frac{1-\cos^2(x)}{x^2} \le 1 \quad \text{(Square "all" sides)} \\
& \Downarrow \\
x\cos^2(x) \le & \frac{1-\cos^2(x)}{x} \le x \quad \text{(Multiply "all" sides by } x>0\text{)} \\
& \Downarrow \\
x\cos^2(x) \le & \frac{(1-\cos(x))\,(1+\cos(x))}{x} \le x \quad \text{(Factor the middle term).}
\end{aligned}
\tag{4.23}
$$

Applying the Squeeze Theorem to the last line of (4.23), we obtain

$$\lim_{x\to 0+} \frac{(1-\cos(x))\,(1+\cos(x))}{x} = 0.$$

And because $\lim_{x\to 0+} (1+\cos(x)) = 1$, it follows that $\lim_{x\to 0+} \frac{(1-\cos(x))}{x} = 0$.

To complete the problem, we need to compute the limit from the left, which we leave for the reader. The key issue is that we used $x > 0$ in the fourth line of (4.23). When multiplying by $x < 0$, the inequalities in that line need to be flipped. ■

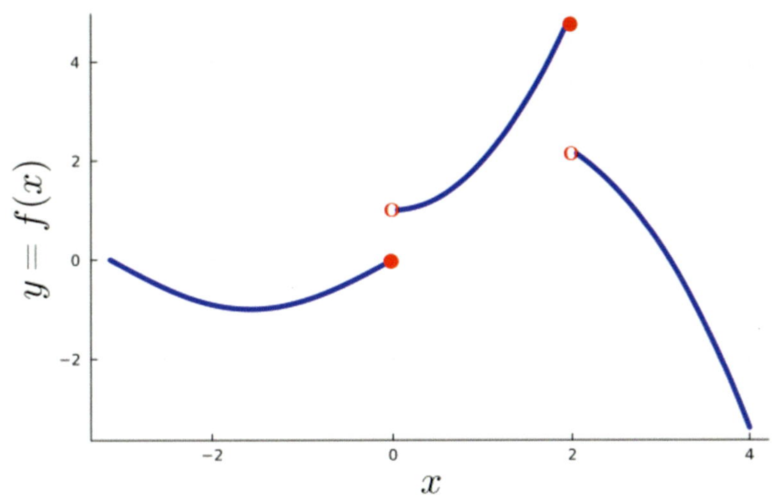

Figure 4.8: The function $f : [-\pi, 4] \to \mathbb{R}$ is an example of what we will call a **piecewise continuous function**. If we look at the function on the open interval $(-\pi, 0)$, the leftmost section, the function is continuous and has finite one-sided limits at $-\pi$ and 0 (i.e., the two ends of the interval); the function is also continuous on the middle interval $(0, 2)$ and has finite one-sided limits at 0 and 2; finally, the function is continuous on the open interval $(2, 4)$ and has finite one-sided limits at 2 and 4.

4.5 Piecewise Continuity

When a function is discontinuous, it may be continuous on subintervals of its domain of definition, as illustrated in Fig. 4.8. Such functions occur in Engineering when we are controlling the speed of a robot, for example. When we update a planned speed profile, it is not always convenient to make the command continuous. Allowing for jumps gives us more freedom in directing the robot.

When we studied definite integrals, we noted that a continuous function on a closed interval $[a, b]$ could always be integrated. We will define piecewise continuous functions in such a way that they can be integrated as well.

Piecewise Continuity

Definition 4.30. *A real-valued function $f : [a, b] \to \mathbb{R}$, $a < b$ both finite, is* ***piecewise continuous*** *if there exists a partition of $[a, b]$,*

$$a =: a_1 < a_2 < \cdots < a_{n+1} := b, \tag{4.24}$$

such that for all $1 \le i \le n$,

(a) $f(a_i, a_{i+1}) \to \mathbb{R}$ is continuous (i.e., f is continuous on the open interval (a_i, a_{i+1}),

(b) $\lim_{x \to a_i^+} f(x)$ and $\lim_{x \to a_{i+1}^-} f(x)$ both exist and are finite, and

(c) $\lim_{x \to a+} f(x) = f(a)$ and $\lim_{x \to b^-} f(x) = f(b)$ (matches the function definition at the two endpoints).

A function $f : I \to \mathbb{R}$, for I an ***unbounded interval,*** *is* ***piecewise continuous*** *if for all (bounded) intervals $[a, b] \subset I$, $a < b$ finite,*

$$\bar{f} : [a, b] \to \mathbb{R} \text{ by } \bar{f}(x) := f(x)$$

satisfies the above conditions for piecewise continuity.

Notes:

- We have broken the interval $[a, b]$ into pieces, hence the adjective "**piecewise**," and we impose properties on the function for each of the pieces. Since we are defining "piecewise continuity," the condition in (a) is evidently important. What is the role of the two limits in (b)? They prevent the function from "exploding at the points a_i, $2 \le i \le n$" as we will illustrate shortly. You can take a look back at Example 4.10 to see what this means!
- What about functions defined on an open interval, $f : (a, b) \to \mathbb{R}$? If the two limits $\lim_{x \to a+} f(x)$ and $\lim_{x \to b-} f(x)$ exist and are finite, then we simply define $f_e : [a, b] \to \mathbb{R}$ by

$$f_e(x) := \begin{cases} \lim_{x \to a+} f(x) & x = a \\ f(x) & a < x < b \\ \lim_{x \to b-} f(x) & x = b. \end{cases}$$

  We can then apply the definition to the extended function, f_e. If one of the two one-sided limits does not exist or is infinite, then the function is automatically not piecewise continuous.
- Some books do not include the requirement that the function has finite limits at the "boundary points," a_i. We will always require that the two limits exist and are finite, but in other contexts, you have to be careful.
- If $f : [a, b] \to \mathbb{R}$ is continuous, then it is also piecewise continuous. The converse is evidently false, as exemplified in Fig. 4.8.
- There is no such thing as "piecewise continuous at a point."

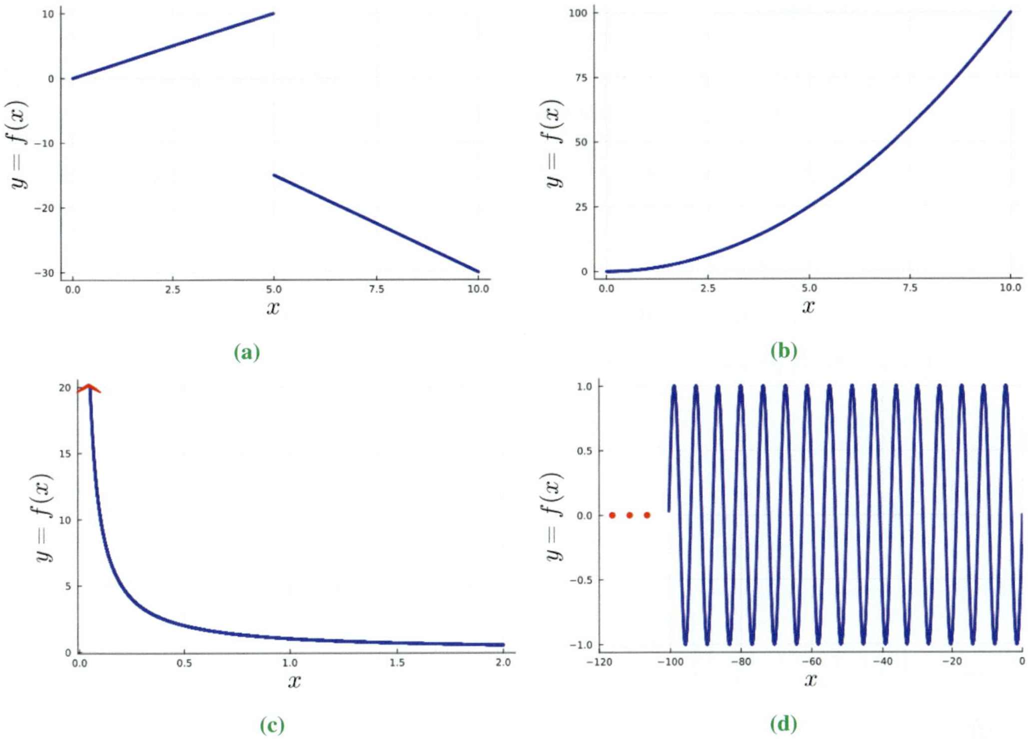

Figure 4.9: Graphs for Example 4.31. In (c), the graph heads to ∞ as $x \to 0^+$. In (d), the graph extends infinitely to the left, oscillating between -1 and $+1$ as sine waves like to do! The graphs in (a) and (b) show all the values of the functions.

Example 4.31. *Determine which, if any, of the following functions are piecewise continuous.*

(a) $f : [0, 10] \to \mathbb{R}$ *by* $f(x) = \begin{cases} 2x & 0 \le x < 5 \\ -3x & 5 \le x \le 10 \end{cases}$.

(b) $f : [0, 10) \to \mathbb{R}$ *by* $f(x) = x^2$.

(c) $f : (0, 2] \to \mathbb{R}$ *by* $f(x) = \frac{1}{x}$.

(d) $f : (-\infty, 0) \to \mathbb{R}$ *by* $f(x) = \sin(x)$.

Solutions: Plots are provided in **Fig. 4.9**; analytical reasoning is provided below.

(a) **Ans.** Piecewise continuous.

We break the interval $[0, 10]$ into two pieces, $A_1 := (0, 5)$ and $A_2 := (5, 10)$, with

$$f_1 : (0, 5) \to \mathbb{R}, \text{ by } f_1(x) = 2x \text{ and } f_2 : (5, 10) \to \mathbb{R}, \text{ by } f_2(x) = -3x.$$

Each of these functions is continuous and

$$\lim_{x \to 0^+} f_1(x) = 0 = f(0), \lim_{x \to 5-} f_1(x) = 10, \lim_{x \to 5+} f_2(x) = -15, \text{ and } \lim_{x \to 10^-} f_2(x) = -30 = f(10).$$

Hence, all of the conditions are met.

(b) **Ans.** Piecewise continuous.

$f(x) = x^2$ has a finite left limit at $x = 10$ and it is equal to $(10)^2$. Therefore, we have $f_e : [0, 10] = x^2$, which we know is continuous everywhere, and hence it is piecewise continuous.

(c) **Ans.** Not piecewise continuous because $\lim_{x \to 0^+} \frac{1}{x} = \infty$. Our definition of piecewise continuity requires finite limits.

(d) **Ans.** Piecewise continuous, because

- the function has a finite left limit at zero, namely, $\lim_{x \to 0^-} \sin(x) = 0$, and
- the function is continuous everywhere for all bounded intervals $[a, 0)$, for $a < 0$ finite.

■

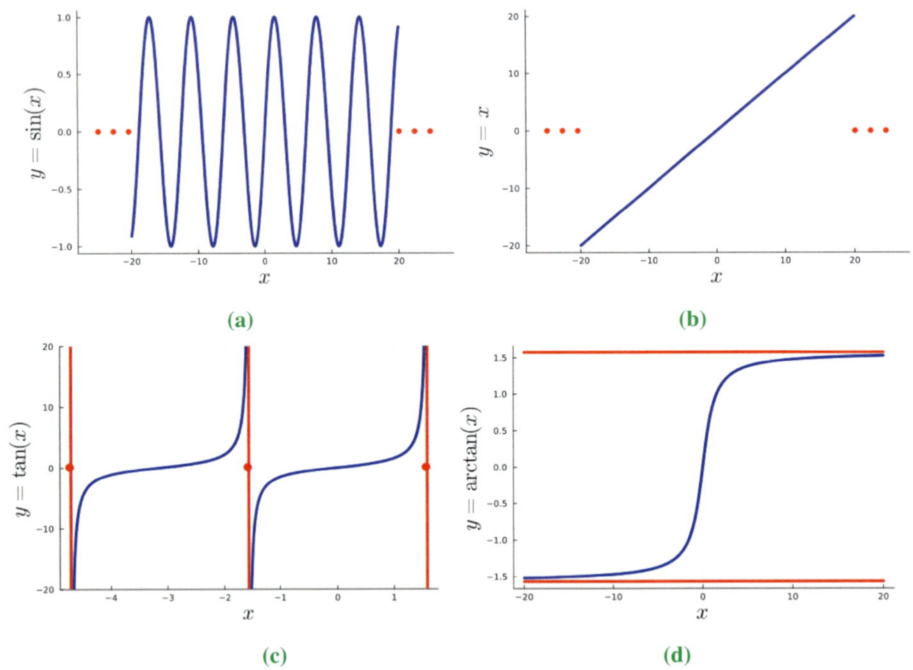

Figure 4.10: Functions can be bounded or unbounded, have horizontal or vertical asymptotes, or not! (a) Bounded, with no asymptotes of any kind. (b) Represents $f(x) = x$ and is unbounded with no asymptotes of any kind (c) Unbounded with vertical asymptotes at odd multiples of $\frac{\pi}{2}$ indicated in red. (d) Bounded with horizontal asymptotes at $\pm\frac{\pi}{2}$ indicated in red

4.6 Bounded vs. Unbounded Functions and Asymptotes

Figure 4.10 illustrates further characteristics of functions that, when present, often aid in their analysis or, in the case of functions of time (aka trajectories), aid in the analysis of the system that produces them. Bounded functions, such as the sine function, are confined within a fixed range, offering predictability and stability, qualities essential in fields like engineering and physics where maintaining variables within limits is crucial. On the other hand, asymptotes describe the behavior of functions as they approach specific values or infinity, as seen in the tangent and arc-tangent functions. These concepts are not just abstract mathematical ideas but are pivotal in understanding growth rates, stability, and constraints in real-world applications ranging from economic models to biological processes.

Bounded Functions

Definition 4.32. *Let $A \subset \mathbb{R}$ be a subset of $\mathbb{R}$ or $\mathbb{Z}$, and $f : A \to \mathbb{R}$ a function.*

*(a) f is **bounded from above** if there exists $M < \infty$ such that $f(x) \leq M$ for all $x \in A$.*

*(b) f is **bounded from below** if there exists $M > -\infty$ such that $f(x) \geq M$ for all $x \in A$.*

*(c) f is **bounded** if it is both bounded from above and from below. This is equivalent to there exists $M < \infty$ such that $|f(x)| \leq M$ for all $x \in A$.*

*(d) f is **unbounded** if it is **not bounded** (that is, either not bounded from above or not bounded from below). This is equivalent to for all $M < \infty$, there $x \in A$, such that $|f(x)| \geq M$. In particular, a function that is bounded from above but not from below is unbounded, and similarly, a function that is bounded from below but not from above is unbounded.*

(e) Recall that when $A \subset \mathbb{N}$, then $a_n := f(n)$ defines a sequence $\{a_n\}_{n \in A}$.

Illustration: See Fig. 4.10-(a) and (b).

Notes: The notion of "boundedness" appears in:

- **Numerical Methods**: Bounded functions are easier to approximate numerically. For example, numerical integration techniques often rely on the function being bounded to guarantee a certain level of accuracy.
- **Stability**: In control theory and engineering, bounded functions are often desirable because they indicate a system's stability. A bounded output in response to a bounded input usually signifies a stable system.
- **Optimization**: Bounded functions are often easier to optimize because you can be certain that the optimum lies within a specific range. This is particularly useful in machine learning algorithms.
- **Convergence**: In mathematics and computer science, bounded functions are important for analyzing the convergence of sequences and series.

Before we dive into examples, it is worthwhile to understand the related notion of **asymptotes**.

Asymptotes Come in Two Flavors: Horizontal and Vertical

Definition 4.33. *Let $A \subset \mathbb{R}$ be a subset of $\mathbb{R}$ and $f : A \to \mathbb{R}$ a function. A **horizontal asymptote** is a horizontal line $y = b$ where the function f approaches b as x approaches positive infinity (∞) or negative infinity ($-\infty$). In other symbols, a horizontal line $y = b$ is a **horizontal asymptote** if at least one of the following two conditions holds:*

1. $\lim_{x \to \infty} f(x) = b$;
2. $\lim_{x \to -\infty} f(x) = b$.

Illustration: See Fig. 4.10-(c) and (d).

Definition 4.34. *Let $A \subset \mathbb{R}$ be a subset of $\mathbb{R}$ and $f : A \to \mathbb{R}$ a function. A **vertical asymptote** is a vertical line at $x = x_0$ where the function f approaches $\pm\infty$ as x approaches x_0 from the left or the right. In other symbols, a vertical line $\{(x_0, y) \mid y \in \mathbb{R}\}$ is a **vertical asymptote** if at least one of the following two conditions holds:*

1. $\lim_{x \to x_0^+} f(x) = \pm\infty$ *(the limit exists and is unbounded);*

2. $\lim_{x \to x_0^-} f(x) = \pm\infty$ *(the limit exists and is unbounded).*

Note: *At least one of the limits needs to exist for x_0 to be a vertical asymptote. As an example, the function*

$$f(x) := \frac{\sin(\frac{1}{x})}{x^2}$$

"blows up" as x approaches the origin from the left or from the right, but neither limit exists. Hence, $x_0 = 0$ is not a vertical asymptote.

Notes: The notion of "asymptotes" is useful because:

- **Behavior at Infinity**: Asymptotes help us understand the behavior of functions as points in their domain approach infinity or a finite point. This is crucial in calculus for evaluating limits and integrals.
- **Graphical Interpretation**: Asymptotes provide a way to graphically represent the behavior of functions, making it easier to visualize and understand their properties.
- **Modeling**: In physics, engineering, and economics, asymptotic behavior can model saturation effects. For example, the law of diminishing returns in economics and saturation current in electrical engineering can be modeled using functions with horizontal asymptotes.

As illustrated in Fig. 4.10, a function can have at most two horizontal asymptotes, while it can have multiple (even infinite in number) vertical asymptotes or none at all.

Example 4.35. *Determine which of the following functions are bounded versus unbounded. Determine also if they have any asymptotes or not.*

(a) $f : \mathbb{R} \to \mathbb{R}$ *by* $f(x) = e^x$.

(b) $f : \mathbb{R} \to \mathbb{R}$ *by* $f(x) = x^2$.

(c) $f : \mathbb{R} \to \mathbb{R}$ *by* $f(x) = \frac{.001x^2 + x}{1 + x^2}$.

(d) $f : \mathbb{R} \to \mathbb{R}$ *by* $f(x) = \begin{cases} \frac{1}{(x-1)(x+1)} & |x| \neq 1 \\ 0 & \textit{otherwise} \end{cases}$.

(e) $f : \mathbb{R} \to \mathbb{R}$ *by* $f(x) = \begin{cases} \frac{x-1}{x^2-1} & |x| \neq 1 \\ 0 & \textit{otherwise} \end{cases}$. ***Hint:*** *Compare to the previous problem.*

(f) $f : \mathbb{N} \to \mathbb{R}$ *by* $f(n) = (1 - 1/n)^{-n}$. ***Hint:*** *See Example 4.21-(a).*

(g) $f : \mathbb{N} \to \mathbb{R}$ *by* $f(n) = \sum_{k=0}^{n} \frac{1}{k!}$. ***Hint:*** *See Section 1.18.1.*

Solutions:

(a) **Ans.** $f(x) = e^x$ is unbounded. It has a horizontal asymptote at $y = 0$, corresponding to $\lim_{x \to -\infty} e^x = 0$.

(b) **Ans.** $f(x) = x^2$ is unbounded and has no asymptotes.

(c) **Ans.** $f(x) = \frac{-0.001x^2+x}{1+x^2}$ is bounded and has a horizontal asymptote at $y = -10^{-3}$, corresponding to $\lim_{x\to\pm\infty} f(x) = -10^{-3}$.

(d) **Ans.** $f(x) = \begin{cases} \frac{1}{(x-1)(x+1)} & |x| \neq 1 \\ 0 & \text{otherwise} \end{cases}$ is unbounded with two vertical asymptotes. One is at $x_0 = -1$, corresponding to $\lim_{x\to -1^-} f(x) = +\infty$ and $\lim_{x\to -1^+} f(x) = -\infty$; the other is at $x_0 = +1$, corresponding to $\lim_{x\to 1^-} f(x) = -\infty$ and $\lim_{x\to 1^+} f(x) = +\infty$. If we were to plot the function, the answer would jump out at us. Here, we numerically approximate the limits.

```
f(x) = 1/(x^2-1)
del= 1e-5
X0 = [-1, 1]
Data = Array{Float64}(undef, 0, 2)
for k = 1:2
    x=X0[k]-del
    Data = [Data; x f(x)]
    x=X0[k]+del
    Data = [Data; x f(x)]
end
Data
```

Output The first two rows are for $x_0 = -1$ and the last two are for $x_0 = 1$.

```
4×2 Matrix{Float64}:
 -1.00001    49999.8
 -0.99999   -50000.3
  0.99999   -50000.3
  1.00001    49999.8
```

(e) **Ans.** $f(x) = \begin{cases} \frac{x-1}{x^2-1} & |x| \neq 1 \\ 0 & \text{otherwise} \end{cases}$ has a vertical asymptote at $x_0 = -1$, but does not have an additional vertical asymptote at $x_0 = 1$ because the "apparent singularity" at +1 is canceled by the term in the numerator. We can look at the problem in two ways:

First: Because $x^2 - 1 = (x-1)(x+1)$, $\frac{x-1}{x^2-1} = \frac{x-1}{(x-1)(x+1)} = \frac{1}{x+1}$.

Second: Numerically approximate the limits as in the previous problem.

```
f(x) = (x-1)/(x^2-1)
del= 1e-5
X0 = [-1, 1]
Data = Array{Float64}(undef, 0, 2)
for k = 1:2
    x=X0[k]-del
    Data = [Data; x f(x)]
    x=X0[k]+del
    Data = [Data; x f(x)]
end
Data
```

Output The first two rows are for $x_0 = -1$ and show that the right and left limits diverge to $\pm\infty$; the last two are for $x_0 = 1$, and indicate that the left and right limits converge to a half.

```
4×2 Matrix{Float64}:
```

```
-1.00001   -100000.0
-0.99999    100000.0
 0.99999          0.500003
 1.00001          0.499998
```

Hence, $\lim_{x \to 1^-} f(x) = 0.5$ and $\lim_{x \to 1^+} f(x) = 0.5$.

Both methods are valid.

(f) **Ans.** $f : \mathbb{N} \to \mathbb{R}$ by $f(n) = (1 - 1/n)^{-n}$ has a horizontal asymptote at e because $\lim_{n \to \infty} (1 - 1/n)^{-n} = e$, Euler's number. We computed the limit in Example 4.21-(a). We illustrate the limit in the following plot.

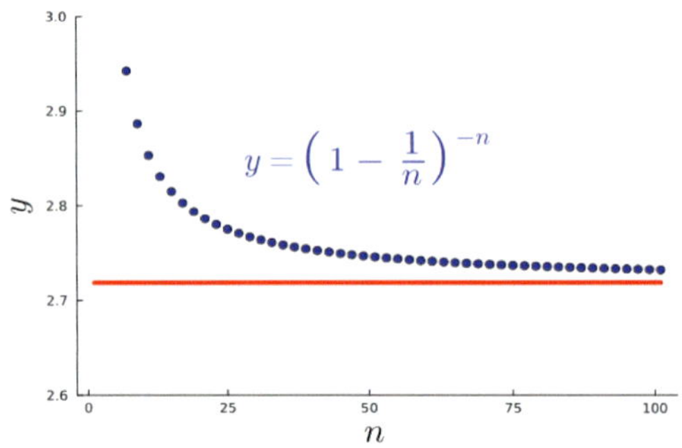

(g) **Ans.** $f : \mathbb{N} \to \mathbb{R}$ by $f(n) = \sum_{k=0}^{n} \frac{1}{k!}$ has a horizontal asymptote at e. We illustrate this via a plot. Note how fast the summation approaches e. Euler really was smart!

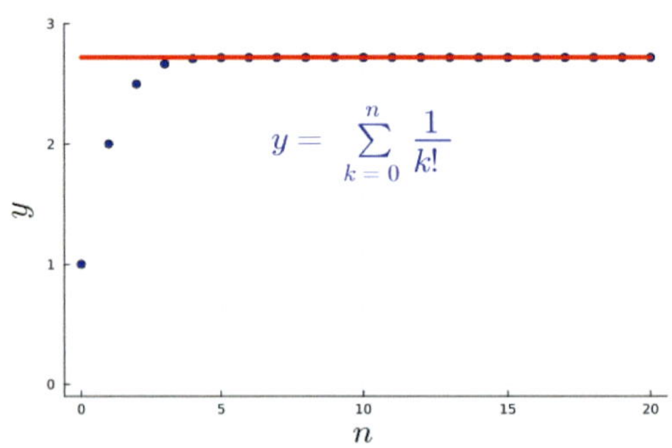

■

4.7 When Does a Function Have a Definite Riemann Integral?

We developed the Riemann Integral in Chapter 3 as the signed area under a curve. Our original development was for continuous functions $f : [a, b] \to \mathbb{R}$, where $[a, b]$ is a closed bounded interval. We now understand what a continuous function is. We also now understand other properties of functions, such as piecewise continuous functions, bounded functions, and functions that have vertical asymptotes. How do these properties relate to the Riemann Integral?

Proposition 4.36: Handy List of Functions Having a Riemann Integral

Functions Commonly Encountered by Engineers:

- **Continuous Functions:** All continuous functions on a closed and bounded interval $[a, b]$

are Riemann integrable. This is the most straightforward class of functions that are Riemann integrable. Riemann integrability may fail for continuous functions on an open and bounded interval (a, b) or half-open intervals. It is very important that $a < b$ are finite real numbers and the function is continuous on $[a, b]$.

- **Piecewise Continuous Functions:** A function that is continuous on $[a, b]$ except for a finite number of discontinuities, while having finite left and right limits at the discontinuities, is also Riemann integrable. If the one-sided limits at points of discontinuity either do not exist or are unbounded, then the result can be a function that is not Riemann integrable. Hence, the presence of vertical asymptotes, while not a deal breaker, signals that caution is needed.

- **Bounded Functions with a Finite Number of Discontinuities (not so common, but interesting):** More generally, any function that is bounded on $[a, b]$ and has a finite number of discontinuities is Riemann integrable. The function does not have to be continuous or even piecewise continuous. It does not have to possess one-sided limits at the discontinuities. This result requires graduate-level mathematics to prove. *Mostly, when an engineer encounters such functions, they are piecewise continuous. An example that is not piecewise continuous would be* $f(0) := 0$, and for $0 < x < 1$, $f(x) := \sin(1/x)$. The function $f : [0, 1] \to [-1, 1]$ is bounded between ± 1 and has a single discontinuity, namely at $x_0 = 0$. Amazingly, it is Riemann integrable.

- **Composite Functions:** If $f : [a, b] \to \mathbb{R}$ is continuous and $g : [a, b] \to \mathbb{R}$ is Riemann integrable, then $f(g(x))$ is also Riemann integrable. The function g does not have to be continuous or piecewise continuous. It could be one of the more exotic functions mentioned elsewhere in this list.

Functions $f : (a, b) \to \mathbb{R}$ with Vertical Asymptotes or Functions Defined on Unbounded Intervals:

- Such integrals are called **Improper Integrals**, and we treat them in Chapter 8.
- Sometimes, we can make sense of them, and sometimes not, as indicated above.
- Interesting examples include $\int_0^\infty e^{-x}\, dx$ (finite), $\int_0^\infty x^k \cdot e^{-x}\, dx$ (finite), and $\int_0^1 \frac{1}{x}\, dx$ (infinite), all of which are encountered frequently.

More Exotic Functions, Beyond the Scope of this Course:

- **Monotonic Functions:** Functions that are monotonic (either entirely non-increasing or entirely non-decreasing) on a closed and bounded interval $[a, b]$ are Riemann integrable. **Such functions could have a countably infinite number of discontinuities and still be Riemann integrable.** For open bounded intervals, (a, b), half-open bounded intervals $[a, b)$ and $(a, b]$, or unbounded intervals, it is NOT true that all monotonic functions are guaranteed to be Riemann integrable.

- **Functions Equivalent to Continuous Functions:** If a function $f : [a, b] \to \mathbb{R}$ is (i) bounded and (ii) its points of discontinuity have Lebesgue measure zero, then f is Riemann integrable. Moreover, if (iii), f differs from a continuous function $g : [a, b] \to \mathbb{R}$ on a set of Lebesgue measure zero, then $f(x)$ is Riemann integrable, and its integral is the same as that of $g(x)$. The Lebesgue measure is treated in graduate-level courses on Functional Analysis or Measure Theory.

- **Functions with Bounded Variation:** Functions of bounded variation on a closed and bounded interval are Riemann-Stieltjes integrable with respect to any other function of bounded variation. It is noted that the Lebesgue integral generalizes the Riemann integral and can integrate a broader class of functions. However, every Riemann integrable function is also Lebesgue integrable, and the two integrals will agree on the value for such functions.

4.8 Maximum and Minimum Values of Sets and Their Generalizations

This is the first time we come close to the "nitty gritty" of the real numbers!

Max and Min Values of a Set

Definition 4.37. *Let $A \subset \mathbb{R}$ be a nonempty subset of the real numbers.*

*(a) a^* is a **maximum element** of the set A if (i) $a^* \in A$ and (ii) $a^* \geq a$ for all $a \in A$. The notation is*

$$a^* = \max\{A\}.$$

*(b) a_* is a **minimum element** of the set A if (i) $a_* \in A$ and (ii) $a_* \leq a$ for all $a \in A$. The notation is*

$$a_* = \min\{A\}.$$

Notes:

- A maximum or minimum element must be in the set.
- A finite set always has both maximum and minimum elements. Here, "finite set" means a finite number of elements.
- A set with an infinite number of elements may not have a maximum or minimum element.

Example 4.38. *Indicate the maximum and minimum elements of the following sets, if they exist.*

(a) $A := [3,4]$.

(b) $B := (3,4]$.

(c) $C := [3,4)$.

(d) $D := (3,4)$.

(e) $E := (0,1) \cap \mathbb{Q}$, *the set of rational numbers in the open interval* $(0,1)$.

(f) $F := \{x \in [0,1] \mid x \notin \mathbb{Q}\}$, *the set of irrational numbers in the closed interval* $[0,1]$.

Solutions:

(a) **Ans.** $\min\{A\} = 3$ and $\max\{A\} = 4$.

While we take this as obvious, it would be good practice to try writing out a proof by appealing to the definition.

(b) **Ans.** $\min\{B\}$ does not exist while $\max\{B\} = 4$.

To prove that the set B does not have a minimum element, let $x \in B$ be arbitrary and define $\delta := x - 3$. Then $\delta > 0$ because $3 < x \leq 4$. Hence, we have

$$3 < 3 + \delta/2 < x,$$

and thus $3 + \delta/2$ is both smaller than x and an element of B; in symbols, $3 + \delta/2 < x$ and $(3 + \delta/2) \in B$. Hence, x cannot be a minimum, and because x was arbitrary, no element of B can be a minimum.

(c) **Ans.** $\max\{C\}$ does not exist while $\min\{C\} = 3$.

To prove that the set C does not have a maximum element, let $x \in C$ be arbitrary and define $\delta := 4 - x$. Then $\delta > 0$ because $4 \notin C$. Hence, we have

$$x < x + \delta/2 < 4,$$

and thus $x + \delta/2 > x$ and $(x + \delta/2) \in C$. Hence, x cannot be a maximum, and because x was arbitrary, no element of C can be a maximum.

(d) **Ans.** Neither $\min\{D\}$ nor $\max\{D\}$ exist. The reasoning is the same as in the previous two parts of this Example.

(e) **Ans.** Neither $\min\{E\}$ nor $\max\{E\}$ exist.

Let $r \in E$ be an arbitrary rational number in the open interval $(0,1)$ and define $\delta := 1 - r$. Then $\delta > 0$ (because $1 \notin E$) and δ is rational (because it is the difference of two rational numbers)

Then we have

$$r < r + \delta/2 < 1$$

and thus $r + \delta/2 > r$ and $(r + \delta/2) \in E$ because $r + \delta/2$ is also rational. Hence, r cannot be a maximum. Similar reasoning shows that r cannot be a minimum.

(f) **Ans.** Neither $\min\{F\}$ nor $\max\{F\}$ exist. The reasoning is the same as in the previous part of this Example, because, if $x \in F$ is irrational, then $1-x$, $1-x/2$, and $x/2$ are all irrational numbers belonging to F.

■

Sup and Inf Values of a Set

Definition 4.39. *Let $A \subset \mathbb{R}$ be a nonempty subset of the real numbers.*

*(a) a^* is the **least upper bound (aka, supremum)** of the set A if (i) $a^* \in \mathbb{R}$, (ii) $a^* \geq a$ for all $a \in A$, and (iii) if $b \in \mathbb{R}$ satisfies $b \geq a$ for all $a \in A$, then $a^* \leq b$. The notation is*

$$a^* = \sup\{A\},$$

*where "sup" is short for "supremum". Note that (ii) says that a^* is an **upper bound** and (iii) says that, among all upper bounds, it is the **least upper bound**.*

*(b) a_* is the **greatest lower bound (aka, infimum)** of the set A if (i) $a_* \in \mathbb{R}$, (ii) $a_* \leq a$ for all $a \in A$, and (iii) if $b \in \mathbb{R}$ satisfies $b \leq a$ for all $a \in A$, then $a_* \geq b$. The notation is*

$$a^* = \inf\{A\},$$

*where "inf" is short for "infimum". Note that (ii) says that a_* is a **lower bound** and (iii) says that, among all lower bounds, it is the **greatest lower bound**.*

Remark 4.40. *Properties of the infimum and supremum.*

- *The **key difference between a maximum or minimum element and the corresponding supremum or infimum element** is where they live! $\max\{A\}$ and $\min\{A\}$ must belong to A, whereas $\sup\{A\}$ and $\inf\{A\}$ belong to $\mathbb{R}$. In fact,*

  $$\sup\{\mathbf{A}\} = \max\{\mathbf{A}\} \iff \sup\{\mathbf{A}\} \in \mathbf{A} \text{ and } \inf\{\mathbf{A}\} = \min\{\mathbf{A}\} \iff \inf\{\mathbf{A}\} \in \mathbf{A}.$$

- **Every bounded set of real numbers has a supremum and an infimum, whereas we noticed in the previous example many bounded sets of real numbers do not have a maximum or a minimum.**

- **The existence of the supremum and the infimum for any bounded set of real numberrs is actually part of the definition of the real numbers.**
- **If a set is unbounded from above, such as $[0, \infty)$, it is common to define the supremum as ∞. Similarly, if a set is unbounded from below, it is common to define the infimum as $-\infty$. In this way, all subsets of the real numbers have an infimum and a supremum.**

Example 4.41. *Find the supremum and infimum for the following sets, if they exist. If a supremum or infimum is also a maximum or minimum, indicate that as well.*

(a) $A := [3, 4]$.

(b) $B := (3, 4]$.

(c) $C := [3, 4)$.

(d) $D := (3, 4)$.

(e) $E := (0, 1) \cap \mathbb{Q}$, *the set of rational numbers in the open interval* $(0, 1)$.

(f) $F := \{x \in \mathbb{R} \mid x \notin \mathbb{Q}\}$, *the set of irrational numbers.*

Solutions:

(a) **Ans.** $\inf\{A\} = 3$ and $\sup\{A\} = 4$. Because $3 = \inf\{A\} \in A$, it is also a minimum. Similarly, because $4 = \sup\{A\} \in A$, it is also a maximum.

(b) **Ans.** $\inf\{B\} = 3$ and $\sup\{B\} = 4$. Because $3 = \inf\{B\} \notin B$, it is NOT a minimum. However, because $4 = \sup\{B\} \in B$, it is also a maximum.

(c) **Ans.** $\inf\{C\} = 3$ and $\sup\{C\} = 4$. Because $3 = \inf\{C\} \in C$, it is also a minimum. However, because $4 = \sup\{C\} \notin C$, it is NOT a maximum.

(d) **Ans.** $\inf\{D\} = 3$ and $\sup\{D\} = 4$. Because $3 = \inf\{D\} \notin D$, it is NOT a minimum. Similarly, because $4 = \sup\{D\} \notin D$, it is NOT a maximum.

(e) **Ans.** $\inf\{E\} = 0$ and $\sup\{E\} = 1$. Because $0 = \inf\{E\} \notin E$, it is NOT a minimum. Similarly, because $1 = \sup\{E\} \notin E$, it is NOT a maximum.

(f) **Ans.** $\inf\{F\} = -\infty$ and $\sup\{F\} = \infty$. Because $-\infty = \inf\{F\} \notin F$, it is NOT a minimum. Similarly, because $\infty = \sup\{F\} \notin F$, it is NOT a maximum. Recall that $\pm\infty$ are concepts, not real numbers.

■

4.9 Maximum and Minimum Values of Functions and Their Generalizations

In some sense, this material should be super easy: it is a repeat of the previous definitions of max, min, sup, and inf applied to the range of a function! In practice, it seems more difficult to grasp for most learners. Perhaps this is because the range of a function can be very complicated? In any case, we will rearrange things by first presenting the sup and inf, before moving on to max and min.

Extremal Values of a Function (or a Function's Range)

Definition 4.42. *Let $f : A \to \mathbb{R}$ be a function, with its domain A being nonempty. Define $Y := f(A) := \{y \in \mathbb{R} \mid y = f(x) \text{ for some } x \in A\}$ to be the* ***range*** *of f.*

(a) $f^* := \sup\{Y\}$ *is the* ***supremum*** *of the range of f, though it is most commonly called the supremum of f. The notation is*

$$f^* = \sup_{x \in A} f(x), \text{ or } f^* = \sup\big\{\{y = f(x) \mid x \in A\}\big\},$$

which provides good insight: f^* is the least upper bound of $f(x)$ as x varies over the entire domain, A.

(b) $f_* := \inf\{Y\}$ *is the* ***infimum*** *of the range of f, though it is most commonly called the infimum of f. The notation is*

$$f_* = \inf_{x \in A} f(x), \text{ or } f^* = \inf\big\{\{y = f(x) \mid x \in A\}\big\},$$

which provides good insight: f_* is the greatest lower bound of $f(x)$ as x varies over the entire domain, A.

Recalling that the key difference between sup and max is "where they live", we define the max and min of a function as follows.

(c) $f^* := \sup\{Y\}$ *is the* ***maximum of*** *f if $f^* \in Y$, that is, there exists $\bar{a} \in A$ such that $f(\bar{a}) = f^*$. The notation is*

$$f^* = \max_{x \in A} f(x),$$

which provides good insight: f^* is the maximum element of $f(x)$ as x varies over the entire domain, A.

(d) $f_* := \inf\{Y\}$ *is the* ***minimum of*** *f if $f_* \in Y$, that is, there exists $\underline{a} \in A$ such that $f(\underline{a}) = f^*$. The notation is*

$$f_* = \min_{x \in A} f(x),$$

which provides good insight: f_* is the minimum element of $f(x)$ as x varies over the entire domain, A.

Notes:

- For the maximum (resp., minimum) of a function to exist, there must be an element in the domain that achieves the maximum (resp., minimum) value of the function. This is not the case of the supremum (resp., infimum).
- $\sup_{x \in A} f(x)$ and $\inf_{x \in A} f(x)$ always exist, while $\max_{x \in A} f(x)$ and $\min_{x \in A} f(x)$ sometimes exist, and sometimes do not.

Proposition 4.43: Extreme Value Theorem Due to Bolzano-1830

Let $f : [a, b] \to \mathbb{R}$ be a **continuous function on a closed and bounded interval** $[a, b]$. Then f always achieves its extreme values (aka, maximum and minimum values). In other symbols, there always exist $x^* \in [a, b]$ and $x_* \in [a, b]$ such that

$$f(x^*) = \max_{x \in [a,b]} f(x) \text{ and } f(x_*) = \min_{x \in [a,b]} f(x).$$

Example 4.44. *For the following functions, determine their sup (resp., inf) values, and if they exist, also give the max (resp. min) values.*

(a) $f : (-\infty, \infty) \to \mathbb{R}$ *by* $f(x) = e^{-x}$.

(b) $f : [0, 10] \to \mathbb{R}$ *by* $f(x) = \frac{e^x}{x^2+1}$.

(c) $f : (-\infty, \infty) \to \mathbb{R}$ *by* $f(x) = e^{-x^2} \sin(x)$

(d) $f : [0, \infty) \to \mathbb{R}$ *by* $f(x) = \frac{2x+1}{x^2+1}$.

(e) $f : [0, 1] \to \mathbb{R}$ *by* $f(x) = \frac{\tan(x)-x^3}{x^2+2x\sin(x)+10}$.

Solutions:

(a) **Ans.** $\inf\limits_{-\infty<x<\infty} e^{-x} = 0$ and $\sup\limits_{-\infty<x<\infty} e^{-x} = \infty$. The min and max of f do not exist.

e^{-x} is strictly monotonically decreasing. Hence, $\inf\limits_{-\infty<x<\infty} e^{-x} = \lim\limits_{x\to\infty} e^{-x} = 0$ and $\sup\limits_{-\infty<x<\infty} e^{-x} = \lim\limits_{x\to-\infty} e^{-x} = \infty$.

(b) **Ans.** $\min\limits_{x\in[0,10]} \dfrac{e^x}{x^2+1} = 1$ and $\max\limits_{x\in[0,10]} \dfrac{e^x}{x^2+1} \approx 218.1$.

Because the denominator never vanishes, $f : [0, 10] \to \mathbb{R}$ is a continuous function on the closed and bounded interval, $[0, 10]$. Hence, the max and min values do exist. If the function is strictly monotonically increasing, then the extreme values will be at the endpoints of the interval. We do not know enough calculus at the moment to check whether the function is monotonic, so we'll determine the extreme values numerically; see the plot below. The code below also provides values of x that achieve the max and min of f.

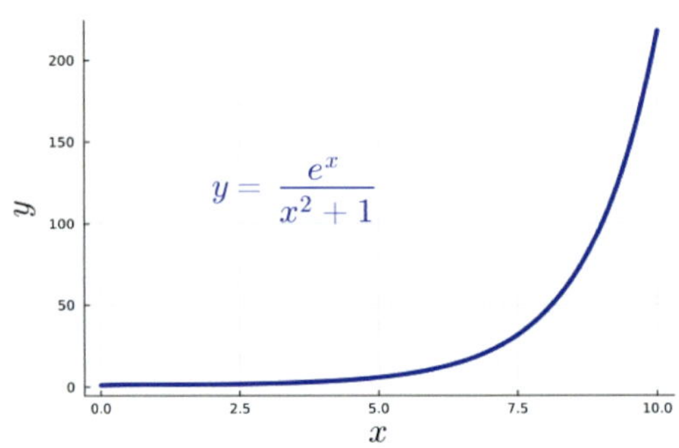


```
xmin = 0
xmax = 10
xVals=xmin:.01:xmax

f(x) = exp(x)/(1 + x^2)

y = f.(xVals)

p1 = plot(xVals,y, linewidth=4, color=:blue, label=false)
plot!(xlabel=L"$x$", ylabel=L"$y=\frac{e^{x}}{x^2 + 1}$")


@show MaxF = maximum(y)
```

```
@show MinF = minimum(y)

max_value, index = findmax(y)
println("Maximum value: ", max_value)
println("x^*: ", xVals[index] )

min_value, index = findmin(y)
println("Minimum value: ", min_value,)
println("x_*: ", xVals[index] )

display(p1)

png(p1, "MaxMinB")
```

Output

```
MaxF = maximum(y) = 218.08381975056156
MinF = minimum(y) = 1.0
Maximum value: 218.08381975056156
x^*: 10.0
Minimum value: 1.0
x_*: 0.0
```

(c) **Ans.** $\min_{x\in\mathbb{R}} e^{-x^2}\sin(x) \approx -0.397$ and $\max_{x\in\mathbb{R}} e^{-x^2}\sin(x) \approx 0.397$.

Because the domain of the function is unbounded, Prop. 4.43 does not apply. **However, the function is continuous everywhere on its domain, and it is easy to see that the extreme values do not occur at $\pm\infty$; in fact, they occur within the closed bounded set $[-2, 2]$, where the Proposition does apply**. How did we see all of that? Well, $\lim_{x\to\pm\infty} e^{-x^2}\sin(x) = 0$. Yet, we know that e^{-x^2} is always positive, while $\sin(x)$ takes on positive and negative values. Hence, $f(x)$ takes on positive and negative values, and, therefore, the infimum will be negative, and the supremum will be positive. We then need to graph the function to find a range where the extreme values of the function will occur; see the following plot.

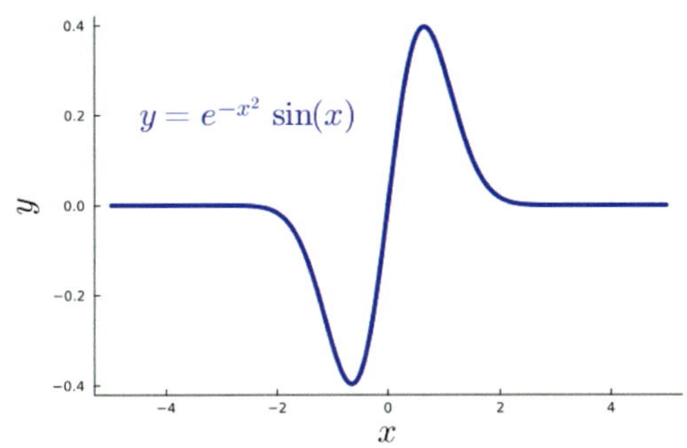


```
xmin = -5
xmax = 5
xVals=xmin:.01:xmax

f(x) = exp(-x^2)*sin(x)

y = f.(xVals)

p1 = plot(xVals,y, linewidth=4, color=:blue, label=false)
plot!(xlabel=L"$x$", ylabel=L"$y=e^{-x^2} ~ \sin(x)$")
```

```
@show MaxF = maximum(y)
@show MinF = minimum(y)

max_value, index = findmax(y)
println("Maximum value: ", max_value)
println("x^*: ", xVals[index] )

min_value, index = findmin(y)
println("Minimum value: ", min_value,)
println("x_*: ", xVals[index] )

display(p1)

png(p1, "MaxMinC")
```

Output

```
MaxF = maximum(y) = 0.3966429553529812
MinF = minimum(y) = -0.3966429553529812
Maximum value: 0.3966429553529812
x^*: 0.65
Minimum value: -0.3966429553529812
x_*: -0.65
```

(d) **Ans.** $\min_{x\in\mathbb{R}} \dfrac{2x+1}{x^2+1} \approx 0.423$ and $\max_{x\in\mathbb{R}} \dfrac{2x+1}{x^2+1} \approx 1.618$.

Because the domain of the function is unbounded, Prop. 4.43 does not apply. **However, the function is continuous everywhere on its domain, and it is easy to see that the extreme values do not occur at ∞; in fact, they occur within the closed bounded set $[0, 5]$, where the Proposition does apply.**

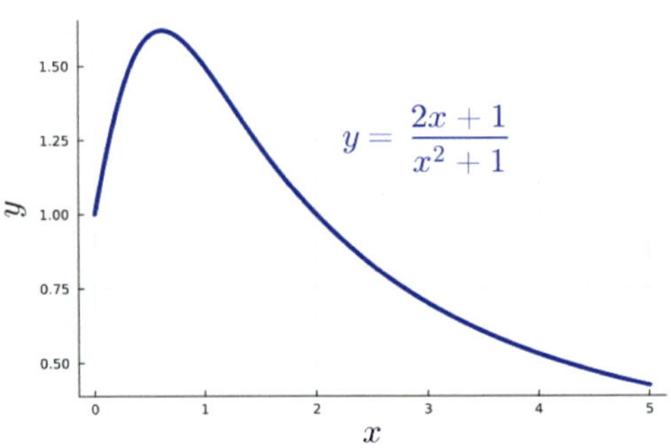


```
xmin = 0
xmax = 5
xVals=xmin:.01:xmax

f(x) = (2x + 1)/(x^2 + 1)

y = f.(xVals)

p1 = plot(xVals,y, linewidth=4, color=:blue, label=false)
plot!(xlabel=L"$x$", ylabel=L"$y= \frac{2x + 1}{x^2 + 1}$")

@show MaxF = maximum(y)
```

```
@show MinF = minimum(y)

max_value, index = findmax(y)

println("Maximum value: ", max_value)
println("x^*: ", xVals[index] )

display(p1)

png(p1, "MaxMinD")
```

Output

```
MaxF = maximum(y) = 1.6180294712510837
MinF = minimum(y) = 0.4230769230769231
Maximum value: 1.6180294712510837
x^*: 0.62
```

(e) **Ans.** $\min_{x\in[0,1]} \dfrac{\tan(x) - x^3}{x^2 + 2x\sin(x) + 10} = 0$ and $\max_{x\in[0,1]} \dfrac{\tan(x) - x^3}{x^2 + 2x\sin(x) + 10} \approx 0.044.$

Because the denominator never vanishes and $1 < \pi/2$, $f : [0,1] \to \mathbb{R}$ is a continuous function on the closed and bounded interval, $[0,1]$. Hence, the max and min values do exist. How do we know the denominator does not vanish? We complete the square,

$$x^2 + 2x\sin(x) + 10 = x^2 + 2x\sin(x) + 10 + \sin^2(x) - \sin^2(x) = (x + \sin(x))^2 + 10 - \sin^2(x),$$

and we note that, for all x, $10 - \sin^2(x) \geq 9$ and $(x + \sin(x))^2 \geq 0$. We also generate the plot below.

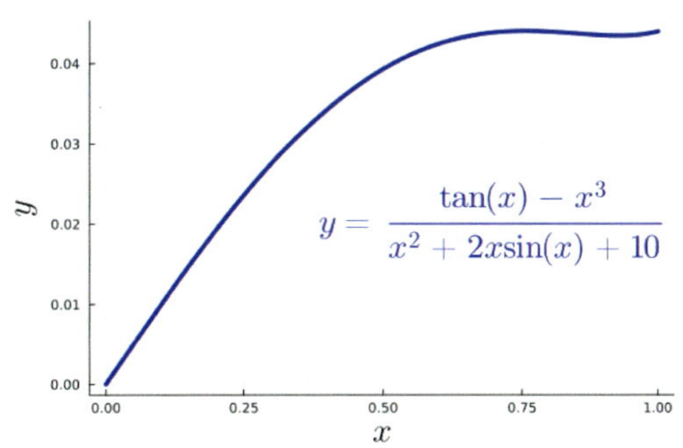


```
xmin = 0
xmax = 1
xVals=xmin:.001:xmax

f(x) = (tan(x) - x^3)/(x^2 + 2x * sin(x) + 10)

y = f.(xVals)

p1 = plot(xVals,y, linewidth=4, color=:blue, label=false)
plot!(xlabel=L"$x$", ylabel=L"$y=\frac{\tan(x)-x^3}{x^2+2x\sin(x)+10}$")

@show MaxF = maximum(y)
@show MinF = minimum(y)

max_value, index = findmax(y)

```

```
println("Maximum value: ", max_value)
println("x^*: ", xVals[index] )

display(p1)

png(p1, "MaxMinE")
```

Output

```
MaxF = maximum(y) = 0.043999918166140344
MinF = minimum(y) = 0.0
Maximum value: 0.043999918166140344
x^*: 0.755
```

■

4.10 (Optional Read) Intermediate Value Theorem and the Mean Value Theorem

The Intermediate Value Theorem is a fundamental result in calculus and real analysis. It asserts that for any continuous function defined on a closed interval, the function takes on every value between its minimum and maximum on that interval. This theorem has important implications in various areas of mathematics and its applications. We used it to establish the soundness of the Bisection Algorithm.

Proposition 4.45: Intermediate Value Theorem

Let $f : [a, b] \to \mathbb{R}$ be a continuous function on the closed interval $[a, b]$, where $a < b$. If d is any number between $f(a)$ and $f(b)$, then there exists at least one $c \in [a, b]$ such that $f(c) = d$.

Proof: Without loss of generality, assume $f(a) < d < f(b)$. Define the set $S = \{x \in [a, b] \mid f(x) \leq d\}$.

Because $f(a) \leq d$, we have that $a \in S$, and hence the set S is non-empty. Moreover, the set S is bounded above by b because $S \subset [a, b]$. Finally, by the completeness property of the real numbers, S has a supremum. Define

$$c := \sup S.$$

We now want to show that $f(c) = d$. Because $c = \sup S$ and $c \in [a, b]$, c is the least upper bound of S. Therefore, for any $\epsilon > 0$, there exists an $x \in S$ such that $c - \epsilon < x \leq c$.

- Since $x \in S$, $f(x) \leq d$. Given f is continuous, we have $\lim_{x \to c^-} f(x) = f(c) \leq d$.
- By definition of c being the supremum, for any $\epsilon > 0$, there exists a point $y \in (c, c + \epsilon]$ such that $y \notin S$. Thus, $f(y) > d$.
- Since f is continuous at c, we have $\lim_{y \to c^+} f(y) = f(c) \geq d$.

Combining the inequalities $f(c) \leq d$ and $f(c) \geq d$, we obtain $f(c) = d$, and hence, there exists at least one $c \in [a, b]$ such that $f(c) = d$.

■

The Mean Value Theorem is a central result in calculus and real analysis. It states that for any function continuous on a closed interval and differentiable on the open interval, there exists at least one point in the open interval where the derivative of the function equals the average rate of change over the closed interval.

Proposition 4.46: Mean Value Theorem

Let $f : [a,b] \to \mathbb{R}$ be a continuous function on the closed interval $[a,b]$ and differentiable on the open interval (a,b), where $a < b$. Then, there exists at least one $c \in (a,b)$ such that

$$f'(c) = \frac{f(b) - f(a)}{b-a}.$$

Proof: Define the function $g(x) = f(x) - \frac{f(b)-f(a)}{b-a}(x-a)$. Note that g is continuous on $[a,b]$ and differentiable on (a,b).

We first observe that:

$$g(a) = f(a) - \frac{f(b) - f(a)}{b-a}(a-a) = f(a),$$

and

$$g(b) = f(b) - \frac{f(b) - f(a)}{b-a}(b-a) = f(b) - (f(b) - f(a)) = f(a).$$

Thus, $g(a) = g(b)$. By Rolle's Theorem, since g is continuous on $[a,b]$, differentiable on (a,b), and $g(a) = g(b)$, there exists at least one $c \in (a,b)$ such that $g'(c) = 0$. Next, we compute $g'(x)$:

$$g'(x) = f'(x) - \frac{f(b) - f(a)}{b-a}.$$

At $x = c$, we have $g'(c) = 0$, which gives us:

$$f'(c) - \frac{f(b) - f(a)}{b-a} = 0,$$

or equivalently,

$$f'(c) = \frac{f(b) - f(a)}{b-a}.$$

Thus, there exists at least one $c \in (a,b)$ such that

$$f'(c) = \frac{f(b) - f(a)}{b-a}.$$

■

Related Videos:

- Intermediate Value Theorem by The Organic Chemistry Tutor
- Mean Value Theorem by @MindSphereYT.

4.11 (Optional Read) Proofs Associated with the Chapter

We first state and prove a few lemmas, and then we get into the proofs.

Lemma (Product of Two Continuous Functions Is Continuous): Let $f : [a,b] \to \mathbb{R}$ and $g : [a,b] \to \mathbb{R}$ be continuous functions at a point $x_0 \in [a,b]$. Then, the function $h(x) := f(x)g(x)$ is also continuous at x_0.

Proof: To prove that $h(x)$ is continuous at x_0, we need to show that for every $\epsilon > 0$, there exists a $\delta > 0$ such that if $|x - x_0| < \delta$, then $|h(x) - h(x_0)| < \epsilon$.

Because f and g are continuous at x_0, for any $\epsilon_1 > 0$, there exists $\delta_1 > 0$ such that if $|x - x_0| < \delta_1$, then $|f(x) - f(x_0)| < \epsilon_1$. Similarly, for any $\epsilon_2 > 0$, there exists $\delta_2 > 0$ such that if $|x - x_0| < \delta_2$, then $|g(x) - g(x_0)| < \epsilon_2$.

Let $0 < \epsilon < 1$. Choose $\epsilon_1 = \dfrac{\epsilon}{2(|g(x_0)| + 1)}$ and $\epsilon_2 = \dfrac{\epsilon}{2(|f(x_0)| + 1)}$. By the continuity of f and g, there exist δ_1 and δ_2 corresponding to ϵ_1 and ϵ_2, respectively. Let $\delta = \min\{\delta_1, \delta_2\}$. Then, for $|x - x_0| < \delta$, we have,

$$\begin{aligned}
|f(x)g(x) - f(x_0)g(x_0)| &= |f(x)g(x) - f(x)g(x_0) + f(x)g(x_0) - f(x_0)g(x_0)| \\
&\leq |f(x)||g(x) - g(x_0)| + |g(x_0)||f(x) - f(x_0)| \\
&< |f(x)|\epsilon_2 + |g(x_0)|\epsilon_1 \\
&\leq (|f(x_0)| + 1)\epsilon_2 + (|g(x_0)| + 1)\epsilon_1 \\
&= \frac{\epsilon}{2} + \frac{\epsilon}{2} \\
&= \epsilon.
\end{aligned}$$

Therefore, $h(x) = f(x)g(x)$ is continuous at x_0. ■

Lemma (Sum of Two Continuous Functions Is Continuous): Let $f : [a, b] \to \mathbb{R}$ and $g : [a, b] \to \mathbb{R}$ be continuous functions on the interval $[a, b]$. Then, the function $h(x) := f(x) + g(x)$ is also continuous on $[a, b]$.

Proof: To prove that $h(x)$ is continuous on $[a, b]$, we need to show that for every point $x_0 \in [a, b]$ and for every $\epsilon > 0$, there exists a $\delta > 0$ such that if $|x - x_0| < \delta$, then $|h(x) - h(x_0)| < \epsilon$.

Since f and g are continuous at x_0, for any $\epsilon > 0$, there exist $\delta_f > 0$ and $\delta_g > 0$ such that:

- If $|x - x_0| < \delta_f$, then $|f(x) - f(x_0)| < \frac{\epsilon}{2}$.
- If $|x - x_0| < \delta_g$, then $|g(x) - g(x_0)| < \frac{\epsilon}{2}$.

Let $\delta = \min\{\delta_f, \delta_g\}$. Now, for $|x - x_0| < \delta$, we have by the triangle inequality,

$$\begin{aligned}
|h(x) - h(x_0)| &= |(f(x) + g(x)) - (f(x_0) + g(x_0))| \\
&= |(f(x) - f(x_0)) + (g(x) - g(x_0))| \\
&\leq |f(x) - f(x_0)| + |g(x) - g(x_0)| \\
&< \frac{\epsilon}{2} + \frac{\epsilon}{2} \\
&= \epsilon.
\end{aligned}$$

Thus, $h(x) = f(x) + g(x)$ is continuous at x_0. Since x_0 was arbitrary, $h(x)$ is continuous on the entire interval $[a, b]$. ■.

Lemma (Composition of Two Continuous Functions Is Continuous): Let $f : X \to Y$ and $g : Y \to Z$ be continuous functions, where X, Y, and Z are subsets of $\mathbb{R}$. Then, the composition $g \circ f : X \to Z$, defined by $(g \circ f)(x) := g(f(x))$, is continuous.

Proof:

To show that $g \circ f$ is continuous at any point $x_0 \in X$, we need to verify that for every $\varepsilon > 0$, there exists a $\delta > 0$ such that if $|x - x_0| < \delta$, then $|(g \circ f)(x) - (g \circ f)(x_0)| < \varepsilon$.

Because g is continuous at $f(x_0) \in Y$, for every $\varepsilon > 0$, there exists an $\eta > 0$ such that if $|y - f(x_0)| < \eta$, then $|g(y) - g(f(x_0))| < \varepsilon$, where $y = f(x)$. Furthermore, since f is continuous at x_0, for the $\eta > 0$ found above, there exists a $\delta > 0$ such that if $|x - x_0| < \delta$, then $|f(x) - f(x_0)| < \eta$.

Combining these two observations, if $|x - x_0| < \delta$, then $|f(x) - f(x_0)| < \eta$, which implies $|g(f(x)) - g(f(x_0))| < \varepsilon$. This means that $|(g \circ f)(x) - (g \circ f)(x_0)| < \varepsilon$. Therefore, by the ε-δ definition of continuity, $g \circ f$ is continuous at x_0. Since x_0 was arbitrary, $g \circ f$ is continuous on X. ■

We leave unproven that all three of these results can be extended to $n > 2$ functions. You may wish to prove at least one of them for yourself. Each of the extensions can be proven by induction.

Proof of Prop. 4.18 (Common Continuous Functions

We'll prove continuity for a few of the dozen or more functions listed in Prop. 4.18: the following functions are continuous for all $x \in \mathbb{R}$ unless indicated otherwise:

(a) Monomials: $f(x) = x^k$

(b) Polynomials: $f(x) = a_n x^n + a_{n-1} x^{n-1} + \cdots + a_1 x + a_0$

(c) Sine: $f(x) = \sin(x)$

(d) Natural Exponential: $f(x) = e^x$

(e) Natural Logarithm (continuous for $x > 0$): $f(x) = \ln(x)$

(f) Power (continuous for $x > 0$ and $y \in \mathbb{R}$): $f(x) = x^y$

(g) Square Root (continuous for $x \geq 0$): $f(x) = \sqrt{x}$

(a) Monomials: $f(x) = x^k$.

We prove by induction that the function $f(x) = x^k$ is continuous for all positive integers k.

Base Cases:

- For $k = 0$, $f(x) = x^0 = 1$, a constant function, which is continuous everywhere.
- For $k = 1$, $f(x) = x$, the identity function, which is also continuous everywhere.

Inductive Step: Assume $f(x) = x^k$ is continuous. We need to show that x^{k+1} is continuous as well. Given $f(x) = x^k$ is continuous by our inductive hypothesis, and knowing that $g(x) = x$ is continuous, consider $h(x) := x^{k+1} = f(x) \cdot g(x) = x^k \cdot x$. Since the product of two continuous functions is continuous, it follows that x^{k+1} is continuous.

Therefore, by mathematical induction, $f(x) = x^k$ is continuous for all positive integers k. ■

(b) Polynomials: $f(x) = a_n x^n + a_{n-1} x^{n-1} + \cdots + a_1 x + a_0$

This follows immediately from the above result for monomials and the lemma on the sum of continuous functions.

(c) Sine: $f(x) = \sin(x)$

To prove continuity at $x_0 \in \mathbb{R}$, we need to show that for every $\epsilon > 0$, there exists a $\delta > 0$ such that if $|x - x_0| < \delta$, then $|\sin(x) - \sin(x_0)| < \epsilon$.

Using the trigonometric identity for the difference of two sines,

$$|\sin(x)-\sin(x_0)| = |2\cos\left(\frac{x+x_0}{2}\right)\sin\left(\frac{x-x_0}{2}\right)|.$$

Because $|\cos(\theta)| \le 1$ for all θ, and $|\sin(\theta)| \le |\theta|$ for all θ, we can further simplify the inequality to:

$$|\sin(x)-\sin(x_0)| \le |2\sin\left(\frac{x-x_0}{2}\right)| \le |x-x_0|.$$

To ensure $|\sin(x)-\sin(x_0)| < \epsilon$, it suffices to choose $\delta = \epsilon$. Thus, for $|x-x_0| < \delta$, we indeed have $|\sin(x)-\sin(x_0)| < \epsilon$.

Therefore, by the ϵ-δ criterion of continuity, $f(x) = \sin(x)$ is continuous at x_0. ■

(d) Natural Exponential: $f(x) = e^x$

Let x_0 be an arbitrary point in $\mathbb{R}$. To prove that $f(x) = e^x$ is continuous at x_0, consider any $\varepsilon > 0$. We aim to show that there exists a $\delta > 0$ such that if $|x-x_0| < \delta$, then $|e^x - e^{x_0}| < \varepsilon$.

Given $\varepsilon > 0$, define $\theta > 1$ such that

$$e^{x_0} - \varepsilon < \frac{e^{x_0}}{\theta} < e^{x_0} < \theta e^{x_0} < e^{x_0} + \varepsilon.$$

It follows from the definition of θ that

$$\frac{e^{x_0}}{\theta} < e^{x_0} < \theta e^{x_0}.$$

Let $\delta = \ln(\theta)$. Then, for $x \in (x_0-\delta, x_0+\delta)$, we have

$$e^{x_0-\delta} < e^x < e^{x_0+\delta}.$$

This implies

$$\frac{e^{x_0}}{\theta} < e^x < \theta e^{x_0},$$

or equivalently,

$$e^{x_0} - \varepsilon < e^x < e^{x_0} + \varepsilon.$$

Therefore, $|e^x - e^{x_0}| < \varepsilon$ whenever $|x-x_0| < \delta$, proving that $f(x) = e^x$ is continuous at x_0. Since x_0 was arbitrary, $f(x) = e^x$ is continuous for all $x_0 \in \mathbb{R}$. ■

(e) Natural Logarithm (continuous for $x > 0$): $f(x) = \ln(x)$

Let $x_0 > 0$ be an arbitrary point. To prove that $f(x) = \ln(x)$ is continuous at x_0, consider any $\varepsilon > 0$. We must show that there exists a $\delta > 0$ such that if $|x-x_0| < \delta$, then $|\ln(x) - \ln(x_0)| < \varepsilon$.

Given $\varepsilon > 0$, consider the interval $(e^{-\varepsilon}x_0, e^{\varepsilon}x_0)$. For x in this interval, we have $e^{-\varepsilon} < \frac{x}{x_0} < e^{\varepsilon}$, implying that $-\varepsilon < \ln\left(\frac{x}{x_0}\right) < \varepsilon$, which simplifies to $-\varepsilon < \ln(x)-\ln(x_0) < \varepsilon$, or equivalently, $|\ln(x)-\ln(x_0)| < \varepsilon$.

Let $\delta = \min\{x_0 - e^{-\varepsilon}x_0, e^{\varepsilon}x_0 - x_0\}$. Then, for $|x-x_0| < \delta$, x lies within the interval $(e^{-\varepsilon}x_0, e^{\varepsilon}x_0)$, ensuring that $|\ln(x) - \ln(x_0)| < \varepsilon$. Therefore, by the ε-δ definition of continuity, $f(x) = \ln(x)$ is continuous at x_0. Since x_0 was arbitrary and $x_0 > 0$, $f(x) = \ln(x)$ is continuous for all $x_0 > 0$. ■

(f) Power (continuous for $x > 0$ and $y \in \mathbb{R}$): $f(x) = x^y$

From $x > 0$ and logarithm properties, we have that $x^y = e^{y\ln x}$, which is the composition of the continuous functions, e^x and $y\ln x$. ■

(g) Square Root (continuous for $x \geq 0$): $f(x) = \sqrt{x}$

This is a special case of the power function, with $y = \frac{1}{2}$. ■

Proof of Prop. 4.19 (Taking Limits Inside a Function)

Consider two functions $g : X \to Y$ and $f : Y \to Z$, where X, Y, and Z are subsets of $\mathbb{R}$. Suppose that f is continuous at a point $y_0 \in Y$. Then, the following hold:

(a) If $x_0 \in \mathbb{R}$ and $\lim_{x \to x_0} g(x) = y_0$ (i.e., the limit exists and equals y_0), then $\lim_{x \to x_0} f(g(x)) = f(\lim_{x \to x_0} g(x)) = f(y_0)$. **We say the limit can be taken "inside" the function f.**

(b) Similarly, if $\lim_{x \to \infty} g(x) = y_0$, then $\lim_{x \to \infty} f(g(x)) = f(\lim_{x \to \infty} g(x)) = f(y_0)$.

(c) If $\lim_{x \to -\infty} g(x) = y_0$, then $\lim_{x \to -\infty} f(g(x)) = f(\lim_{x \to -\infty} g(x)) = f(y_0)$.

In short, if $g(x)$ has a limit (as x approaches a finite point or as x approaches $\pm\infty$), and f is continuous at the limit of g (i.e., y_0), then, when evaluating a limit of $f(g(x)) = f \circ g(x)$, the limit can be taken inside the function f and applied directly to g.

Proof:

(a) Given that $\lim_{x \to x_0} g(x) = y_0$, we know that for every $\varepsilon > 0$ there exists a $\delta > 0$ such that if $0 < |x - x_0| < \delta$, then $|g(x) - y_0| < \varepsilon$. Since f is continuous at y_0, for this ε, there exists a $\delta_f > 0$ such that $|y - y_0| < \delta_f$ implies $|f(y) - f(y_0)| < \varepsilon$ for all y in the domain of f.

Since $g(x)$ approaches y_0 as x approaches x_0, and f maps values near y_0 to values near $f(y_0)$, it follows that as x approaches x_0, $f(g(x))$ approaches $f(y_0)$. Hence, $\lim_{x \to x_0} f(g(x)) = f(y_0)$.

(b) The proofs for parts (b) and (c) follow a similar reasoning to part (a), but considering the limits as x approaches ∞ and $-\infty$, respectively. The key is the continuous behavior of f at y_0 and the fact that $g(x)$ approaches y_0 in each scenario. By the continuity of f, the image under f of the values of $g(x)$ that approach y_0 will approach $f(y_0)$, thus allowing the limit to be taken inside the function f.

(c) See above.

In conclusion, if $g(x)$ converges to a point y_0 as x approaches a finite value or $\pm\infty$, and f is continuous at y_0, the composition $f \circ g(x)$ behaves such that the limit operation can be directly applied after the composition, yielding $f(\lim_{x \to x_0} g(x)) = f(y_0)$, where x_0 represents a finite value, ∞, or $-\infty$.

■

Proof of Prop. 4.36 (Handy List of Functions Having a Riemann Integral)

Functions Commonly Encountered by Engineers:

(a) **Continuous Functions:** All continuous functions on a closed and bounded interval $[a, b]$ are Riemann integrable. This is the most straightforward class of functions that are Riemann integrable. Riemann integrability may fail for continuous functions on an open and bounded interval (a, b) or half-open intervals. It is very important that $a < b$ are finite real numbers and the function is continuous on $[a, b]$.

(b) **Piecewise Continuous Functions:** A function that is continuous on $[a, b]$ except for a finite number of discontinuities, while having finite left and right limits at the discontinuities, is also

Riemann integrable. If the one-sided limits at points of discontinuity either do not exist or are unbounded, then the result can be a function that is not Riemann integrable.

Proofs: We realize this is far from the entire list given in Prop. 4.36, but even these two are on the edge of what we can handle in the course.

(a) **Continuous Functions:** Because f is continuous on $[a,b]$, it is also uniformly continuous on this interval due to the interval being closed and bounded. This means that for any $\epsilon > 0$, there exists a $\delta > 0$ such that for all $x, y \in [a,b]$, if $|x-y| < \delta$, then $|f(x) - f(y)| < \frac{\epsilon}{b-a}$. We have not covered this "flavor" of continuity in the textbook.

Uniformly partition the interval $[a,b]$ into n subintervals $\{[x_i, x_{i+1}]\}_{i=1}^n$, where $x_1 = a$ and $x_{n+1} = b$, and each subinterval has a width equal to $\Delta x := \frac{b-a}{n} \leq \delta$. This ensures that the supremum and infimum of f in each subinterval,

$$h_i^{\text{Up}} := \sup_{x \in [x_i, x_{i+1}]} f(x)$$

$$h_i^{\text{Low}} := \inf_{x \in [x_i, x_{i+1}]} f(x)$$

satisfy

$$\omega_i := h_i^{\text{up}} - h_i^{\text{low}} \leq \frac{\epsilon}{b-a}.$$

The difference between the Riemann upper sum and lower sum for the partition can be expressed as the sum of the "oscillations", ω_i, over all subintervals, weighted by the width of each subinterval:

$$\sum_{i=1}^{n} \omega_i \Delta x \leq \sum_{i=1}^{n} \frac{\epsilon}{b-a} \cdot \frac{b-a}{n} = \sum_{i=1}^{n} \frac{\epsilon}{n} = \epsilon.$$

This shows that the difference between the upper and lower Riemann sums for f over $[a,b]$ can be made less than any given $\epsilon > 0$, which proves that f is Riemann integrable on $[a,b]$.

(b) **Piecewise Continuous Functions:** By the way we have defined piecewise continuous functions, we can write them as a finite sum of functions that are continuous on disjoint sub-intervals of the form (a_i, b_i), have bounded limits at the endpoints, and are zero everywhere else. Next, the proof of Prop. 3.4 (First Additivity Property) can be adapted to cover any jumps at the endpoints of the intervals.

■

Chapter 5

Differentiation

Learning Objectives

By the end of this chapter, the student should be able to:

- Understand two conceptual views of a single-variable derivative.
- Appreciate the effectiveness of various software tools for computing derivatives.
- Examine real problems where single-variable derivatives are important in engineering.
- Make the leap to partial derivatives, which are single-variable derivatives applied to multivariable functions.

Outcomes

Upon successful completion of this chapter, students will be able to:

- Apply the definition of the derivative as rise over run in the limit.
- Understand the centrality of the derivative for linear approximation of a function.
- Compute a few derivatives via the rise over run definition.
- Develop a sense of when common functions are differentiable and when they are not.
- Learn how to compute derivatives with various software packages.
- Master the rules of differentiation and understand their origin.
- Apply single-variable derivatives to determine speed from position.
- Apply the fact that a strictly positive derivative implies the function is strictly monotonically increasing.
- Use l'Hôpital's Rule for limits of indeterminate form.
- Compute Jacobians, gradients, and Hessians, with examples using software.
- Understand and use the total derivative.

5.1 The Derivative as the Local Slope of a Function

Figure 5.1 illustrates the notion of the **derivative of a function at a point** as the limiting **slope**, or **rise over run**, as the run approaches zero (from both sides).

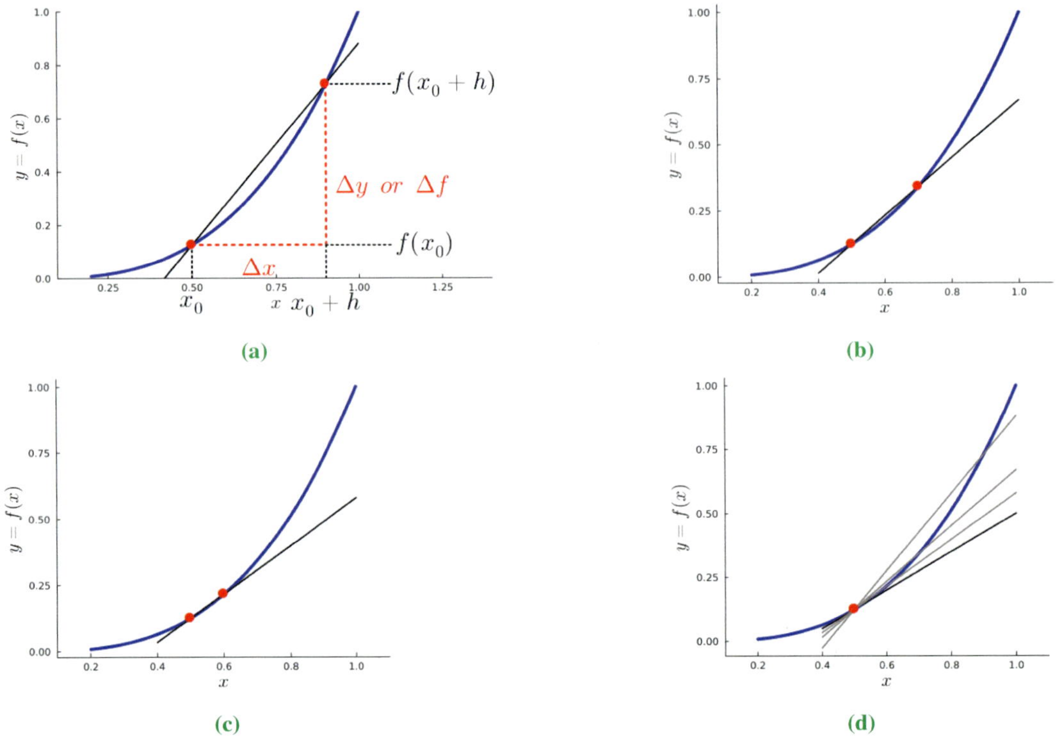

Figure 5.1: The derivative of $f : (a, b) \to \mathbb{R}$ at $x_0 \in (a, b)$ is the limiting value of the "rise over run," as $h \to 0$. (a) Graphical definition of all the terms in (5.1). The "rise" is $\Delta f := f(x_0 + h) - f(x_0)$, and the "run" is $\Delta x := (x_0 + h) - x_0 = h$; the slope, of course, is their ratio, $\frac{\Delta f}{\Delta x}$. A line passing through the pair of points $(x_0, f(x_0))$ and $(x_0 + h, f(x_0 + h))$ is called a **secant line**, where the term "secant" is Latin for "cutting" (it cuts the function at the two indicated points, as opposed to a "tangent line", which touches the function at only one point). (b) and (c) show that as the run, $\Delta x := (x_0 + h) - x_0 = h$, gets smaller and smaller, so does the rise, $\Delta f := f(x_0 + h) - f(x_0)$. (d) Shows the "limiting case" where the secant line becomes a tangent line (indicated by the dark line), the slope of which is the value of the derivative at x_0. The lighter lines are the previous secant lines so that we can see the convergence process playing out as $h \to 0$.

Derivative at a Point

Definition 5.1. *The derivative of $f : (a, b) \to \mathbb{R}$ at a point $x_0 \in (a, b)$ is defined as the limiting value of the rise over run,*

$$\frac{df(x_0)}{dx} := \lim_{h \to 0} \frac{f(x_0 + h) - f(x_0)}{(x_0 + h) - x_0} = \lim_{h \to 0} \frac{f(x_0 + h) - f(x_0)}{h}. \tag{5.1}$$

Notes: *Recall that the double-sided limit $\lim_{h \to 0} \frac{f(x_0 + h) - f(x_0)}{h}$ is equivalent to two single-sided limits,*

$$\lim_{h \to 0^-} \frac{f(x_0 + h) - f(x_0)}{h} \text{ and } \lim_{h \to 0^+} \frac{f(x_0 + h) - f(x_0)}{h}.$$

*Both have to exist, be finite, and be equal for the single-sided limit to exist. To show that a function is **not differentiable at** x_0, it is sufficient to show that any one of the following conditions is true:*

- *The two one-sided limits both exist and are finite, but they have different values.*
- *One of the limits is infinite.*
- *One of the limits does not exist (i.e., is undefined).*

If $f : [a,b] \to \mathbb{R}$ is defined on a closed interval, the derivatives at the two endpoints can be defined via single-sided limits, namely

$$\frac{df(a)}{dx} := \lim_{h\to 0^+} \frac{f(x_0+h)-f(x_0)}{h}$$

and (5.2)

$$\frac{df(b)}{dx} := \lim_{h\to 0^-} \frac{f(x_0+h)-f(x_0)}{h}.$$

The derivative at points $a < x_0 < b$ is given by (5.1).

Example 5.2. *Explain why the function $f(x) := |x|$ is not differentiable at $x_0 = 0$.*

Solution: The key is to compute both single-sided limits at the origin. To make this particularly easy, we note that

$$f(x) := |x| = \begin{cases} x & x > 0 \\ -x & x < 0 \\ 0 & x = 0. \end{cases}$$

Hence,

$$\lim_{h\to 0^+} \frac{f(0+h)-f(0)}{h} = \lim_{h\to 0^+} \frac{h}{h} = +1$$

and

$$\lim_{h\to 0^-} \frac{f(0+h)-f(0)}{h} = \lim_{h\to 0^-} \frac{-h}{h} = -1.$$

Both limits exist and are finite, but they are not equal. Hence, $|x|$ is not differentiable at the origin, even though it is differentiable at all other points.

■

Example 5.3. *Is the function*

$$f(x) := \begin{cases} 1+x & x > 0 \\ x & x \le 0. \end{cases}$$

differentiable at $x_0 = 0$?

Solution: **Ans.** No.

The key, once again, is to compute a single-sided limit at the origin. If any one of them does not exist or is unbounded, then the derivative is undefined at that point.

$$\lim_{h\to 0^+} \frac{f(0+h)-f(0)}{h} = \lim_{h\to 0^+} \frac{1+h}{h} = \infty.$$

This shows that f is not differentiable at the origin. For completeness, we compute the other derivative, namely

$$\lim_{h\to 0^-} \frac{f(0+h)-f(0)}{h} = \lim_{h\to 0^+} \frac{h}{h} = 1.$$

Note that the "jump" discontinuity at the origin made the function not differentiable at that point. ■

The following two Propositions are actually equivalent (i.e., they state the same fact).

Proposition 5.4: Never Differentiable at a Point of Discontinuity

Suppose $f : [a,b] \to \mathbb{R}$ is discontinuous at $x_0 \in [a,b]$. Then, f is nondifferentiable[a] at x_0.

Note: x_0 can be a point of discontinuity that arises from a "jump" in the function's value at x_0, as in Example 5.3, or it can be due to something wilder, such as

$$f(x) := \begin{cases} \sin(1/x) & x \neq 0 \\ 0 & x = 0 \end{cases}$$

not being continuous at $x_0 = 0$ due to "infinitely rapid variation" in the function's values near the origin.

[a] I.e., not differentiable.

Proposition 5.5: Differentiable Implies Continuous

Suppose $f : [a,b] \to \mathbb{R}$ is differentiable at $x_0 \in [a,b]$. Then, f is continuous at x_0.

Why are they equivalent? Well, it is the classic rule in Logic: $(A \implies B) \iff (\neg B \implies \neg A)$. In our case, let us select A to be "f is differentiable at x_0" and B to be "f is continuous at x_0". Then $(A \implies B)$ is Prop. 5.5. Continuing, $\neg B$ is "f is discontinuous at x_0" and $\neg A$ is "f is nondifferentiable at x_0". The content of Prop. 5.4 is $(\neg B \implies \neg A)$.

Importantly, the two propositions are NOT telling us that a function is differentiable if, and only if, it is continuous. We have already seen that $|x|$ is not differentiable at the origin even though it is continuous everywhere.

Example 5.6. *Compute the derivatives of the functions below at an arbitrary point $x \in \mathbb{R}$.*

(a) $f : \mathbb{R} \to \mathbb{R}$ *by* $f(x) = x^2$.

(b) $g : \mathbb{R} \to \mathbb{R}$ *by* $g(x) = e^x$.

(c) $k : \mathbb{R} \to \mathbb{R}$ *by* $k(x) = \sin(x)$.

(d) $\ell : (0, \infty) \to \mathbb{R}$ *by* $\ell(x) = \ln(x)$.

Solutions:

(a) $f(x) = x^2$. **Ans.** $\frac{df(x)}{dx} = 2x$.

We can also write the answer so that $\frac{d}{dx}$ looks like an operation being performed on the function: $\frac{d}{dx}\left(x^2\right) = 2x$. The operation is one of taking the derivative (when it exists).

In this problem, as requested, we let the arbitrary point be x instead of x_0. This will help us to understand that the derivative of a function is often another function.

$$\begin{aligned}\frac{df(x)}{dx} &= \lim_{h\to 0}\frac{f(x+h)-f(x)}{h} \text{ (apply definition)}\\ &= \lim_{h\to 0}\frac{(x+h)^2-x^2}{h} \text{ (plug in the function)}\\ &= \lim_{h\to 0}\frac{(x^2+2hx+h^2)-x^2}{h} \text{ (expand the square)}\\ &= \lim_{h\to 0}\frac{2hx+h^2}{h} \text{ (algebra)}\\ &= \lim_{h\to 0} 2x+h \text{ (more algebra)}\\ &= 2x.\end{aligned}$$

(b) $g(x) = e^x$. **Ans.** $\frac{dg(x)}{dx} = e^x$.

The solution to this problem uses Example 4.22, which showed that $\lim_{h\to 0}\frac{e^h-1}{h} = 1$.

$$\begin{aligned}\frac{dg(x)}{dx} &= \lim_{h\to 0}\frac{g(x+h)-g(x)}{h} \text{ (apply definition)}\\ &= \lim_{h\to 0}\frac{e^{(x+h)}-e^x}{h} \text{ (plug in the function)}\\ &= \lim_{h\to 0} e^x\cdot\frac{e^h-1}{h} \text{ (factor out } e^x\text{)}\\ &= e^x\cdot\lim_{h\to 0}\frac{e^h-1}{h} \text{ (move } e^x \text{ outside the limit because it is independent of } h\text{)}\\ &= e^x\cdot 1 \text{ (Example 4.22)}\\ &= e^x.\end{aligned}$$

(c) $k(x) = \sin(x)$. **Ans.** $\frac{dk(x)}{dx} = \cos(x)$.

The solution requires two trigonometric identities, namely,

$$\sin(a+b) = \cos(a)\cdot\sin(b) + \sin(a)\cdot\cos(b)$$

and

$$2\sin^2\left(\frac{c}{2}\right) = 1-\cos(c), \quad (5.3)$$

in addition to the limit $\lim_{h\to 0}\frac{\sin(h)}{h} = 1$ from Example 4.29.

$$\begin{aligned}\frac{dk(x)}{dx} &= \lim_{h\to 0}\frac{k(x+h)-k(x)}{h} \text{ (apply definition)}\\ &= \lim_{h\to 0}\frac{\sin(x+h)-\sin(x)}{h} \text{ (plug in the function)}\\ &= \lim_{h\to 0}\frac{(\cos(x)\cdot\sin(h)+\sin(x)\cdot\cos(h))-\sin(x)}{h} \text{ (use the first identity)}\\ &= \cos(x)\cdot\lim_{h\to 0}\frac{\sin(h)}{h} + \sin(x)\cdot\lim_{h\to 0}\frac{(\cos(h)-1)}{h} \text{ (algebra)}\\ &= \cos(x)\cdot\lim_{h\to 0}\frac{\sin(h)}{h} - \sin(x)\cdot\lim_{h\to 0} 2\cdot\frac{\sin^2\left(\frac{h}{2}\right)}{h} \text{ (use the second identity)}\\ &= \cos(x)\cdot 1 - \sin(x)\cdot\lim_{h\to 0}\sin\left(\frac{h}{2}\right)\cdot\frac{\sin\left(\frac{h}{2}\right)}{\frac{h}{2}} \text{ (limit from Example 4.29)}\\ &= \cos(x),\end{aligned}$$

because from Prop. 2.37, when we have a product of functions where at least one has a finite limit, then the limit of the product is the product of the limits. Moreover,

$$\lim_{h\to 0} \sin\left(\frac{h}{2}\right) = 0 \text{ and } \lim_{h\to 0} \frac{\sin\left(\frac{h}{2}\right)}{\frac{h}{2}} = 1.$$

Therefore,

$$\begin{aligned}\lim_{h\to 0}\left(\sin\left(\frac{h}{2}\right)\right)\cdot\left(\frac{\sin\left(\frac{h}{2}\right)}{\frac{h}{2}}\right) &= \underbrace{\left(\lim_{h\to 0}\sin\left(\frac{h}{2}\right)\right)}_{0}\cdot\underbrace{\left(\lim_{h\to 0}\frac{\sin\left(\frac{h}{2}\right)}{\frac{h}{2}}\right)}_{1} \quad \text{(applying Prop. 2.37 and the above limits)}\\ &= 0\cdot 1\\ &= 0.\end{aligned}$$

Related video The Limit (do not use l'Hôpital's Rule) by **BlackPenRedPen**.

(d) $\ell(x) = \ln(x)$. **Ans.** $\frac{d\ell(x)}{dx} = \frac{1}{x}$.

The solution is available in Example 4.21-(c). ■

That's how derivatives are computed via the definition. As practicing engineers, we are unlikely to be asked to compute a derivative by the definition. Hence, we will not belabor it.

Notation for Differentiation

Definition 5.7. *In Lagrange's notation for the derivative, the primary objects are functions, such as $f(x) = x^2$, and derivatives are written with a prime, as in $f'(x) = 2x$. In Leibniz's notation, the primary objects are relationships, such as $y = x^2$, and derivatives are written as a ratio, as in $\frac{dy}{dx} = 2x$. Both notations are quite common, and we will even see a third notation (based on dots) when we focus on derivatives with respect to time of a trajectory, such as $x(t)$.*

$$\begin{aligned} f'(x) &:= \lim_{h\to 0}\frac{f(x+h)-f(x)}{h} \quad \textit{Lagrange Style}\\ \frac{df(x)}{dx} &:= \lim_{h\to 0}\frac{f(x+h)-f(x)}{h} \quad \textit{Leibniz Style}. \end{aligned} \tag{5.4}$$

Note: *Once again, we could create a perfect bubble in the course where one and only one notation is used, but then you would be at a disadvantage in other courses and life.*

Proposition 5.8: Common Functions, Their Derivatives, and Domains

The following functions are differentiable on the indicated domains. Here, we use Lagrange's notation for the derivative, namely

$$f'(x) := \lim_{h\to 0}\frac{f(x+h)-f(x)}{h}.$$

Function	**Derivative**	**Domain**
$f(x) = c$	$f'(x) = 0$	$x \in \mathbb{R}$
$f(x) = x^n$	$f'(x) = nx^{n-1}$	$x \in \mathbb{R}$
$f(x) = e^x$	$f'(x) = e^x$	$x \in \mathbb{R}$
$f(x) = \ln(x)$	$f'(x) = \frac{1}{x}$	$x > 0$
$f(x) = \sin(x)$	$f'(x) = \cos(x)$	$x \in \mathbb{R}$
$f(x) = \cos(x)$	$f'(x) = -\sin(x)$	$x \in \mathbb{R}$
$f(x) = \tan(x)$	$f'(x) = \sec^2(x)$	$x \neq \frac{\pi}{2} + k\pi,\ k \in \mathbb{Z}$
$f(x) = \tan(x)$	$f'(x) = 1 + \tan^2(x)$	$x \neq \frac{\pi}{2} + k\pi,\ k \in \mathbb{Z}$
$f(x) = \cot(x)$	$f'(x) = -\csc^2(x)$	$x \neq k\pi,\ k \in \mathbb{Z}$
$f(x) = \sec(x)$	$f'(x) = \sec(x)\tan(x)$	$x \neq \frac{\pi}{2} + k\pi,\ k \in \mathbb{Z}$
$f(x) = \csc(x)$	$f'(x) = -\csc(x)\cot(x)$	$x \neq k\pi,\ k \in \mathbb{Z}$
$f(x) = \sinh(x)$	$f'(x) = \cosh(x)$	$x \in \mathbb{R}$
$f(x) = \cosh(x)$	$f'(x) = \sinh(x)$	$x \in \mathbb{R}$
$f(x) = \tanh(x)$	$f'(x) = \operatorname{sech}^2(x)$	$x \in \mathbb{R}$

Note: $1 + \tan^2(x) = \left(\frac{\cos(x)}{\cos(x)}\right)^2 + \left(\frac{\sin(x)}{\cos(x)}\right)^2 = \frac{\sin^2(x)+\cos^2(x)}{\cos^2(x)} = \left(\frac{1}{\cos(x)}\right)^2 = \sec^2(x)$. Both forms are common for the derivative of $\tan(x)$. In a similar manner, $-\csc^2(x) = -\frac{1}{\sin^2(x)} = -\frac{\sin^2(x)+\cos^2(x)}{\sin^2(x)} = -\left(1 + \cot^2(x)\right)$.

5.2 The Derivative as a Local Linear Approximation of a Function

Let $f : (a, b) \to \mathbb{R}$ and let $x_0 \in (a, b)$. Suppose we want to approximate the function by a line, $y = y_0 + m \cdot (x - x_0)$, near the point x_0. **Figure 5.1** suggests that the derivative is doing something like that. Let us explore it a bit more deeply.

We pose

$$f(x) \approx y_0 + m \cdot (x - x_0) \tag{5.5}$$

for $|x - x_0| < \delta$ and some $\delta > 0$. What values should we assign to y_0 and m? Well, if we plug $x = x_0$ into (5.5), we obtain

$$f(x_0) = y_0.$$

Hence,

$$f(x) \approx f(x_0) + m(\cdot x - x_0). \tag{5.6}$$

Next, because we are used to limits, we plug in $x = x_0 + h$ into (5.6), yielding

$$f(x_0 + h) \approx f(x_0) + \underbrace{m \cdot h}_{h = (x_0 + h) - x_0}.$$

A little algebra leads to:

$$\big(f(x_0 + h) - f(x_0) \approx m \cdot h\big) \iff \left(m \approx \frac{f(x_0 + h) - f(x_0)}{h}\right). \tag{5.7}$$

The approximation becomes better and better with smaller and smaller h. In the limit, we have

$$m = \lim_{h \to 0} \frac{f(x_0 + h) - f(x_0)}{h}, \tag{5.8}$$

as long as the limit exists. In other symbols, if f is differentiable at x_0, then

$$f(x) \approx f(x_0) + \frac{df(x_0)}{dx} \cdot (x - x_0) \tag{5.9}$$

is a **local linear approximation** to the function near the point x_0.

(Optional Read) More Formal Definition of a Linear Approximation of a Function Near a Point

Definition 5.9. *Let $f(x)$ be a differentiable function at x_0. The function $y(x) = y_0 + m(x - x_0)$ is said to be a **linear approximation** of $f(x)$ near x_0 if:*

$$\lim_{x \to x_0} \frac{f(x) - (y_0 + m(x - x_0))}{x - x_0} = 0.$$

Roughly speaking, this definition implies that the error between $f(x)$ and its linear approximation $y(x)$ is "quadratic" in $(x - x_0)$.

Claim: The only possible values for y_0 and m such that $y(x) = y_0 + m(x - x_0)$ is a linear approximation of $f(x)$ near x_0 are $y_0 = f(x_0)$ and $m = f'(x_0)$.

Proof: Assume that $y(x) = y_0 + m(x - x_0)$ is a linear approximation of $f(x)$ near x_0. By the definition of linear approximation, we have the condition,

$$\lim_{x \to x_0} \frac{f(x) - (y_0 + m(x - x_0))}{x - x_0} = 0.$$

Expanding the numerator gives:

$$f(x) - (y_0 + m(x - x_0)) = f(x) - y_0 - m(x - x_0).$$

Thus, the condition becomes:

$$\lim_{x \to x_0} \frac{f(x) - y_0 - m(x - x_0)}{x - x_0} = 0.$$

We can rewrite this as:

$$\lim_{x \to x_0} \left(\frac{f(x) - f(x_0)}{x - x_0} + \frac{f(x_0) - y_0}{x - x_0} - m \right) = 0.$$

Using the fact that $\lim_{x \to x_0} \frac{f(x) - f(x_0)}{x - x_0} = f'(x_0)$, we obtain:

$$f'(x_0) + \lim_{x \to x_0} \frac{f(x_0) - y_0}{x - x_0} - m = 0.$$

Now, observe that $\lim_{x \to x_0} \frac{f(x_0) - y_0}{x - x_0}$ is only well defined if $f(x_0) = y_0$; otherwise, the limit would be unbounded. Therefore, we must have:

$$y_0 = f(x_0).$$

Substituting this back into the equation, we obtain:

$$f'(x_0) - m = 0,$$

which implies that:

$$m = f'(x_0).$$

Thus, the only possible values for y_0 and m are $y_0 = f(x_0)$ and $m = f'(x_0)$. ■

Later in this chapter, we will see how derivatives help us to build local quadratic, cubic, ..., nth degree polynomial approximations to functions. These have two names: Maclaurin series, when the approximation is performed at the origin (i.e., $x_0 = 0$), and Taylor Series when $x_0 \neq 0$.

Fact 5.10: (Optional Read) Quality of the Approximation of a Linear Approximation

To approximate a function $f(x)$ at a point x using its behavior near x, we can expand it in terms of a small increment h. The first-order approximation, often referred to as the linear approximation, assumes that $f(x+h)$ can be expressed as:

$$f(x+h) \approx f(x) + f'(x) \cdot h,$$

which implies that the change in f is primarily governed by its derivative at x. This approximation is valid for small values of h, where higher-order terms become negligible.
To rigorously understand the accuracy of this approximation, we consider the error term,

$$\mathrm{Error(h)} := f(x+h) - (f(x) + f'(x) \cdot h).$$

The first-order approximation assumes that this error is of order $o(h)$, meaning that it diminishes faster than h as $h \to 0$. Mathematically, we can write:

$$\lim_{h \to 0} \frac{f(x+h) - f(x) - f'(x) \cdot h}{h} = 0.$$

This statement ensures that the remainder of the approximation is proportional to $o(h)$ and becomes insignificant for small h. Thus, the first-order approximation provides a simple, yet powerful, tool to estimate $f(x+h)$, while the error term analysis assures its accuracy.

5.3 Software Tools for Computing Derivatives

This section is based on the ROB 101 Programming Guide.

5.3.1 Symbolic Derivatives

"Symbolic" means computing derivatives and manipulating functions the way a human would on a sheet of paper. We literally define variables as "symbols" (i.e., they do not take on numerical values, such as x, y, or z), and manipulate them as variables. It is pretty incredible that a machine designed to crunch numbers can do that.

```
using Symbolics # Only need to do this once per session

# Create a custom function to make differentiation convenient
function deriv(f, x)
    return expand_derivatives(Differential(x)(f))
end

# Sample Usage
@variables x # Makes x a symbolic variable
f = x^3*cos(x) + x^2*sin(x)^2 + log(3x) # Note that we did not say
   # f(x) =, just f =
dfdx = deriv(f, x)
```

Output

$$\frac{1}{x} - x^3 \sin(x) + 2\sin^2(x)\,x + 3x^2\cos(x) + 2x^2\cos(x)\sin(x)$$

Recall that in Julia, $\log(x) = \ln(x)$ and all trigonometric functions are in radians.

The above gives us the derivative as a function of x. If we want to evaluate it at a point, say $x_0 = 0.5$, we can do so as follows, with a convenient, custom-made function.

```julia
function evaluate(f, x0)
    return Symbolics.value(substitute(f, (Dict(x => x0))))
end

# Example use
@show dfdx
evaluate(dfdx, .5)
```

Output

```
dfdx = 1 / x + 2x*(sin(x)^2) + 3(x^2)*cos(x) + 2(x^2)*cos(x)*sin(x)
        - (x^3)*sin(x)

3.0384753223601586
```

Note: The output of `substitute(f, (Dict(x => x0)))` has Type `Num`, as you can check with the command `typeof`.

```julia
val_numeric = substitute(dfdx, (Dict(x => 0.5))) # Will be Type Num

@show typeof(val_numeric)

val = Symbolics.value(val_numeric) # converts to Float64
```

Output

```
typeof(val_numeric) = Num

3.0384753223601586
```

It is recommended to use the "evaluate" command created above. It is not, for some reason, a part of the Symbolic.jl package.

Continuing with our demonstration of symbolic differentiation, we can even compute the derivative of the derivative! It goes like so:

```julia
d2fdx2 = deriv(dfdx, x)
```

Output

$$\frac{-1}{x^2} + 2\sin^2(x) + 2\cos^2(x)\,x^2 - x^3\cos(x) + 6x\cos(x) - 2\sin^2(x)\,x^2 - 6x^2\sin(x) + 8x\cos(x)\sin(x)$$

And, we can even put the derivative operation in a loop, like so:

```julia
@variables x
f = x^3*cos(x) + x^2*sin(x)^2 + log(3x)
N=5
temp = f
for k = 1:N
```

```
    temp= deriv(temp, x)
end
dNfdxN = temp
```

Output

$$\frac{-32}{x^5} - 60\cos(x) - 80\cos^2(x)\,x - 4x^3\frac{2}{x^8} - x^3\sin(x) + 60x\sin(x)$$
$$+80\sin^2(x)\,x - 64x^{11}\frac{-1}{x^{16}} + 15x^2\cos(x)$$
$$-160\cos(x)\sin(x) + 32x^2\cos(x)\sin(x)\,.$$

Note that the awkward expression,

$$-4x^3\frac{2}{x^8} - 64x^{11}\frac{-1}{x^{16}},$$

was not simplified. We can fix that as follows:

```
# Added a simplify operation
function deriv(f, x)
    df = Differential(x)(f)
    return simplify(expand_derivatives(df)) # Note the simplify command
end

# Example use
@variables x
f = x^3*cos(x) + x^2*sin(x)^2 + log(3x)
N=5
temp = f
for k = 1:N
    temp= deriv(temp, x)
end
dNfdxN = temp
```

Output

$$\frac{24}{x^5} + 6(-\cos(x) - 4\sin(2x)) - 54\cos(x) - 56\sin(2x) - x^3\sin(x) + 15x^2\cos(x) + 60x\sin(x)$$
$$+ 16x^2\sin(2x) - 80x\cos(2x)$$

The "cost" of the additional simplification is that the function runs slower. Hence, it is our decision whether or not to `simplify`.

Advantages of Symbolic Differentiation

- Symbolic Differentiation allows you to compute exact derivatives up to machine precision.
- Symbolic Differentiation is also a great tool for analysis. Being able to generate closed-form derivatives of a model facilitates the understanding of the dynamics of the model.
- Symbolic Differentiation is also significantly useful for generating closed-form derivatives of functions as code. Closed-loop controllers running on real-time systems often have to run at frequencies in the 1 kHz range. This constrains the control loop to a time budget of 1 millisecond. Given this time budget, operations in the controller (such as computing derivatives and Jacobians) have to be as fast as possible. Being able to generate closed-form parameterized functions for derivatives and Jacobians offline using Symbolic Differentiation that can then be evaluated online in microseconds can be very useful in speeding up the controller and meeting the controller time budget.

Disadvantages of Symbolic Differentiation

- The main disadvantage of symbolic derivatives is that they tend to get complicated and exponentially long very quickly. The phenomenon is called **expression swell**. Consider taking the derivative of a function
$$h(x) = f(x)g(x)$$
where $f(x)$ is itself the product $f(x) = u(x)\, v(x)$, the derivative of $h(x)$ ends up as this challenging function:
$$\frac{d}{dx}h(x) = \left(\frac{d}{dx}u(x)\, v(x) + u(x)\, \frac{d}{dx}v(x)\right) g(x) + u(x)\, v(x)\, \frac{d}{dx}g(x),$$
leading to inefficient code. We will learn the **Product Rule** in Section 5.4.

5.3.2 Numerical Differentiation

Numerical Differentiation is also called **finite differences**.

Numerical Approximations of a Derivative

Common approximations of a derivative are based on "not letting the limit go all the way to zero" but stopping at h small.
$$\frac{df(x_0)}{dx} := \lim_{h\to 0} \frac{f(x_0+h) - f(x_0)}{h}. \tag{5.10}$$
In practice, we will use "small positive values" for h and never compute the exact limit. Hence, we have an **approximation for the derivative** at a point, namely
$$\frac{df(x_0)}{dx} \approx \frac{f(x_0+h) - f(x_0)}{h}, \tag{5.11}$$
which is called a **forward difference approximation to the derivative**.

We can also do a **backward difference approximation to the derivative**,
$$\frac{df(x_0)}{dx} \approx \frac{f(x_0) - f(x_0-h)}{h}, \tag{5.12}$$
and a **symmetric difference approximation**, where we go both forward and backward from the point x_0,
$$\frac{df(x_0)}{dx} \approx \frac{f(x_0+h) - f(x_0-h)}{2h}. \tag{5.13}$$
The forward and backward difference approximations to the derivative are, in fact, exact for linear functions, while the symmetric difference approximation is *exact* for quadratic polynomials. The symmetric difference is also sometimes called a **central difference**.

If the derivative of $f(x)$ at a point x_0 exists, then for h sufficiently small, the forward difference, backward difference, and symmetric difference approximations to the derivative will always agree. If they provide different answers, then the limit in (5.10) does not exist, and the function is said to be *nondifferentiable*.

In Julia, we perform numerical derivatives as follows.

```
using FiniteDiff # Julia package for finite difference approximations of
    derivatives.

# Example function
g(x) = x^3*cos(x) + x^2*sin(x)^2 + log(3x)

```

```
val = FiniteDiff.finite_difference_derivative(g, 0.5)
```

Output

```
3.038475322499568
```

The "true" (symbolic) value is 3.0384753223601586**; their difference is only** 1.394×10^{-10}**.** That seems pretty good! Note that the `FiniteDiff` package chooses the value of "$h > 0$" for you, which is pretty sweet.

Advantages of Numerical Differentiation

- Numerical differentiation is the simplest differentiation method to implement. It only depends on the function values and not on any set of differential calculus rules.
- Hence, numerical differentiation can be applied to any differentiable function to obtain reasonably accurate results.

Disadvantages of Numerical Differentiation

- Finite Difference suffers from *Truncation error*, resulting in inaccurate derivatives when the step size is large.
- Finite Difference can also be expensive to compute for functions with large input dimensions. This becomes important when we get to Jacobians and gradients later in the chapter.
- It is not easy to "iterate" numerical differentiation methods.

5.3.3 Automatic Differentiation

Similar to Symbolic Differentiation, **Automatic Differentiation (aka AD)** uses differential calculus rules to compute derivatives. **However, its goal is to return a numerical solution instead of a single closed-form analytical expression**. As such, Automatic Differentiation keeps track of intermediate variables and their derivatives to speed up computation. It decomposes the function into primitive operations using the "chain rule", evaluates the operations and their derivatives, and stores them in intermediate variables. We cover the **chain rule** in Section 5.4. Automatic Differentiation is sometimes characterized as "the chain rule meets databases"! It is very efficient.

There are two primary forms of automatic differentiation, **Forward Mode AD** and **Reverse Mode AD**.

Forward Mode AD:

- In forward mode AD, the derivative information is propagated from the input variables through to the output of the function.
- This is accomplished by augmenting each intermediate variable in the computation with its derivative (or tangent) with respect to the input variables.
- Forward mode is particularly efficient when the number of input variables is small compared to the number of output variables.

Reverse Mode AD:

- In reverse mode AD, the derivative information is propagated from the output back to the input variables.
- This is accomplished by computing the adjoint or cotangent of each intermediate variable, which represents the rate of change of the output with respect to that intermediate variable.
- The adjoints are then used to compute the gradient of the output with respect to the input variables.

- Reverse mode is particularly efficient when the number of output variables is small compared to the number of input variables. This property makes it especially popular in deep learning, where the objective is typically scalar (e.g., loss) but the input (e.g., parameters of a neural network) is high-dimensional.

Both forward and reverse modes of AD are exact methods (up to machine precision) and are more accurate than numerical differentiation. In addition to these primary forms, there are also hybrid methods that combine aspects of both forward and reverse modes to optimize the computation of derivatives for specific types of functions or computational graphs (e.g., Neural Networks).

To appreciate how automatic differentiation works, let us consider an example, namely the function

$$f(x) = \sin(x^2 + 2x).$$

Using forward mode automatic differentiation, we can compute its derivative by breaking down the function into elementary operations.

Intermediate Variables

$$\begin{aligned} v_1 &= x^2 \\ v_2 &= 2x \\ v_3 &= v_1 + v_2 \\ f(x) &= \sin(v_3) \end{aligned}$$

Derivatives

Here, we are using the "prime" notation for the first derivative of a quantity:

$$\begin{aligned} v_1' &= 2x \quad \text{(derivative of } x^2\text{)} \\ v_2' &= 2 \quad \text{(derivative of } 2x\text{)} \\ v_3' &= v_1' + v_2' \quad \text{(derivative of a sum equals sum of the derivatives)} \\ f'(x) &= \cos(v_3) \cdot v_3' \quad \text{(chain rule, which we have not yet covered)} \end{aligned}$$

Evaluation

For $x = 1$:

$$\begin{aligned} v_1 &= 1 \\ v_2 &= 2 \\ v_3 &= 3 \\ f(1) &= \sin(3) \\ v_1' &= 2 \\ v_2' &= 2 \\ v_3' &= 4 \\ f'(1) &= \cos(3) \cdot 4 \end{aligned}$$

Thus, the value of the derivative of the function at $x = 1$ is $4\cos(3) = -3.9599699864017817$.

By way of an example, we return to our previous function and compute its derivative using Forward-Mode Automatic Differentiation.

```
using ForwardDiff # load the ForwardDiff.jl package as usual
                  # only need to do once per session
```

```
f2(x) = x^3*cos(x) + x^2*sin(x)^2 + log(3x) # define the function

x0 = 0.5 # specific point where we want to evaluate the derivative

df2dx = ForwardDiff.derivative(f2, x0)
```

Output

```
3.0384753223601586
```

5.3.4 Guidelines on Choosing Differentiation Methods

Symbolic Differentiation

- **When to use:** For functions that have a known algebraic form and when we need an explicit formula for the derivative.
- **Advantages:**
  - Provides exact, closed-form expressions for derivatives.
  - Useful for understanding the underlying behavior and properties of a function.
- **Limitations:**
  - Can lead to complex and lengthy expressions for even moderately complicated functions.
  - Not suitable for functions without an explicit mathematical expression.

Numerical Differentiation

- **When to use:** When we have discrete data points or when an analytical solution is difficult to obtain.
- **Advantages:**
  - Simple to implement and understand.
  - Does not require knowledge of the function's underlying algebraic form.
- **Limitations:**
  - Can introduce errors due to truncation and rounding.
  - Accuracy depends on the choice of step size.

Automatic Differentiation (AD)

- **When to use:** For complex functions where symbolic differentiation is impractical, and when high accuracy is needed. Common in computational applications and machine learning.
- **Advantages:**
  - Provides exact derivatives up to machine precision.
  - Can handle a wide range of functions, including those defined by algorithms or programs.
- **Limitations:**
  - Might be overkill for simple functions.
  - Requires specialized software or libraries.

General Advice

For learners just starting with calculus, it's beneficial to begin with symbolic differentiation to grasp the fundamental concepts. As you progress and encounter more complex functions or real-world data, exploring numerical and automatic differentiation methods will be valuable.

5.4 Differentiation Rules that All Engineers Are Expected to Understand

Proposition 5.8 provided the derivatives of many common or elementary functions, but failed to consider more complicated functions, such as

- $e^{\sin(x)}$,
- $\cos(\omega_0 x + \theta_0)$,
- $\sin(x^3) + \tan(x)$,
- $x \cdot \ln(x)$,

which are compositions of elementary functions. Determining their derivatives via the limit definition would be very slow! Instead, as hinted at in our discussion of Automatic Differentiation, there are **Differentiation Rules** that are used to compute the derivatives.

As a practicing engineer, in discussions with colleagues, you will be expected to know the following means of computing derivatives. **More immediately, in most of your engineering courses, you will need competency in the following derivative rules in order to understand lectures, complete homework, and pass exams.** **All of the differentiation rules can be derived from our result in Section 5.2 relating differentiation to linear approximations of functions.** An animated, quite beautiful demonstration of these rules using this idea is provided by **Mathologer: Why is calculus so ... EASY ?**; here, we have jumped ahead in the video to the relevant location. While you are welcome to start it from the beginning, we are not ready for that just yet; you may wish to come back to this video when we do **antiderivatives** in Chapter 7.

5.4.1 Differentiating by the Rules

Proposition 5.11: Differentiation Rules

Here are the rules of differentiation:

(a) **Sum/Difference Rule**: If $f(x) = g(x) + h(x)$ or $f(x) = g(x) - h(x)$, then

$$f'(x) = g'(x) + h'(x) \text{ or } f'(x) = g'(x) - h'(x).$$

(b) **Product Rule**: If $f(x) = g(x) \cdot h(x)$, then

$$f'(x) = g'(x) \cdot h(x) + g(x) \cdot h'(x).$$

In particular, if $g(x) = c$, a constant, then

$$f'(x) = c \cdot h'(x).$$

(c) **Quotient Rule or Ratio Rule**: If $f(x) = \frac{g(x)}{h(x)}$ and $h(x) \neq 0$, then

$$f'(x) = \frac{g'(x) \cdot h(x) - g(x) \cdot h'(x)}{(h(x))^2}.$$

In particular, if $g(x) = c$, a constant, then

$$f'(x) = -c \cdot \frac{h'(x)}{(h(x))^2}.$$

(d) **Chain Rule**: If $f(x) = g(h(x))$, then

$$f'(x) = g'(h(x)) \cdot h'(x).$$

Another way to state it is: decompose $f(x) = g(h(x))$ as $g(y)$ evaluated at $y = h(x)$. Then,

$$\frac{d}{dx}\left(g(h(x))\right) = \left.\frac{dg(y)}{dy}\right|_{y=h(x)} \cdot \frac{dh(x)}{dx},$$

where $\Big|_{y=h(x)}$ means to evaluate y at $h(x)$, or equivalently, substitute in $h(x)$ for y.

Select one of these forms of the Chain Rule and stick with it! The Chain Rule shows up everywhere in calculus.

(e) **Exponential Rule**: If $f(x) = a^x$, where a is a positive constant, then

$$f'(x) = \ln(a) \cdot a^x.$$

In particular, if $a = e$, Euler's constant, then

$$f'(x) = e^x.$$

(f) **Logarithm Rule**: If $f(x) = \log_a(x)$, where a is a positive constant, then

$$f'(x) = \frac{1}{x \ln(a)}.$$

In particular, if $a = e$, Euler's constant, then

$$f'(x) = \frac{1}{x}.$$

(g) **Power Rule**: If $f(x) = x^n$, where $n \in \mathbb{N}$, the counting numbers, then

$$f'(x) = nx^{n-1}$$

and there is no restriction on x.

(h) **Generalized Power Rule**: If $f(x) = x^\alpha$, where $\alpha \in \mathbb{R}$ and $x > 0$, then

$$f'(x) = \alpha x^{\alpha-1}.$$

Here, when the exponent (power) is an arbitrary real number, we need $x > 0$. When the exponent is a counting number, $f'(x) = nx^{n-1}$, and there is no restriction on x.

There are many online sources for the proofs of the differentiation rules. Here are a few.

- Paul's Online Notes
- openstax.org Differentiation Rules
- **Mathologer: Why is calculus so … EASY ?** Recall that Leibniz's notation for the derivative is $\frac{df}{dx}$, an increment in f over an increment in x. The prime notation is due to Lagrange. **Mathologer shows how the $\frac{df}{dx}$ notation eases the derivation of the rules of differentiation.**

  As we show below, it is really all about linear approximations of functions about a point.

5.4.2 The How and Why of the Rules

In this section, some of the examples explain why the rules are true, while other examples illustrate the application of the rules.

Example 5.12. *Derive the Chain Rule for* $(f \circ g(x))'$ *using the result in Section 5.2 relating differentiation to linear approximations of functions.*

Solution: Using the prime notation for derivatives, we write the linear approximations for f and g about points x and y as

$$g(x+h) \approx g(x) + g'(x) \cdot h \text{ for } h \text{ small} \iff \lim_{h \to 0} \frac{g(x+h) - g(x)}{h} = g'(x)$$

and

$$f(y+\delta) \approx f(y) + f'(y) \cdot \delta \text{ for } \delta \text{ small} \iff \lim_{\delta \to 0} \frac{f(y+\delta) - f(y)}{\delta} = f'(y).$$

Shortly, we'll be motivated to set $y = g(x)$ and $\delta = g'(x) \cdot h$.

By the definition of the derivative of $f \circ g$ at x,

$$(f \circ g(x))' = (f(g(x)))' = \lim_{h \to 0} \frac{f(g(x+h)) - f(g(x))}{h}.$$

As Step 1, in the above formula, we replace $g(x+h)$ by its linear approximation about x,

$$(f(g(x)))' = \lim_{h \to 0} \frac{f(g(x) + g'(x) \cdot h)) - f(g(x))}{h}.$$

As Step 2, in the above formula, we identify $f(g(x) + g'(x) \cdot h))$ as $f(y+\delta)$. We can then replace f with its linear approximation about $y = g(x)$ and $\delta = \delta(h) := g'(x) \cdot h$, where we note that $h \to 0$ implies $\delta(h) \to 0$. This gives us,

$$\begin{aligned}(f(g(x)))' &= \lim_{h \to 0} \frac{f(g(x)) + f'(g(x)) \cdot g'(x) \cdot h - f(g(x))}{h} \\ &= \lim_{h \to 0} \frac{f'(g(x)) \cdot g'(x) \cdot h}{h} \\ &= f'(g(x)) \cdot g'(x).\end{aligned}$$

Again, the Mathologer video animates the above proof. The animation even includes the important point that all of this works because of the equivalence between differentiation and linear approximation of a function near a point. ■

Example 5.13. *Derive the Product Rule for* $(f(x) \cdot g(x))'$ *using the result in Section 5.2 relating differentiation to linear approximations of functions.*

Solution: Using the prime notation for derivatives, we write the linear approximations for f and g about a point x as:

$$g(x+h) \approx g(x) + g'(x) \cdot h \text{ for } h \text{ small} \iff \lim_{h \to 0} \frac{g(x+h) - g(x)}{h} = g'(x)$$

and

$$f(x+h) \approx f(x) + f'(x) \cdot h \text{ for } h \text{ small} \iff \lim_{h \to 0} \frac{f(x+h) - f(x)}{h} = f'(x).$$

By the definition of the derivative of $f(x) \cdot g(x)$ at x,

$$(f(x) \cdot g(x))' = \lim_{h \to 0} \frac{f(x+h) \cdot g(x+h) - f(x) \cdot g(x)}{h}.$$

We replace both $f(x+h)$ and $g(x+h)$ by their linear approximations about x, and expand out the multiplication:

$$
\begin{aligned}
(f(x)\cdot g(x))' &= \lim_{h\to 0} \frac{(f(x)+f'(x)\cdot h)\cdot(g(x)+g'(x)\cdot h)-f(x)\cdot g(x)}{h} \\
&= \lim_{h\to 0} \frac{\left(f(x)\cdot g(x)+f(x)\cdot g'(x)\cdot h+f'(x)\cdot g(x)\cdot h+f'(x)\cdot g'(x)\cdot h^2\right)-f(x)\cdot g(x)}{h} \\
&= \lim_{h\to 0} \frac{f(x)\cdot g'(x)\cdot h+f'(x)\cdot g(x)\cdot h+f'(x)\cdot g'(x)\cdot h^2}{h} \\
&= f(x)\cdot g'(x)+f'(x)\cdot g(x)+\lim_{h\to 0}\frac{f'(x)\cdot g'(x)\cdot h^2}{h} \\
&= f(x)\cdot g'(x)+f'(x)\cdot g(x)
\end{aligned}
$$

because

$$
\lim_{h\to 0}\frac{f'(x)\cdot g'(x)\cdot h^2}{h}=\lim_{h\to 0} f'(x)\cdot g'(x)\cdot h = f'(x)\cdot g'(x)\cdot \lim_{h\to 0} h = 0.
$$

Again, the Mathologer video animates the above proof. ■

Remark 5.14. *The Quotient Rule can be derived following the same pattern as in Examples 5.12 and 5.13.*

Fact 5.15: Practice Videos

A great source of practice examples: **A Nancy Pi Playlist on Differentiation**. Below are some of the key videos.

- Derivatives ... How? (basic intro, product rule, quotient rule)
- The Chain Rule ... How? When? **Excellent Video! Be sure to watch to the end.** We will borrow some of Nancy Pi's ideas in the examples that follow.
- More Chain Rule means more practice.

Other sources include Calculus I Tutorial: how to use the product rule, quotient rule, and chain rule by **BlackPenRedPen**, and Differentiation Formulas by The Organic Chemistry Tutor, which basically is a list that is written out on the screen.

Finally, a highly recommended video: HOW TO READ CALCULUS OUT LOUD! LIMITS, DERIVATIVES & INTEGRAL SYMBOLS by **I Need to Pass Calculus ...** .

Function	c	x^n	e^x	$\ln(x)$	$\sin(x)$	$\cos(x)$	$\tan(x)$	$\text{atan}(x)$
Derivative	0	nx^{n-1}	e^x	$\frac{1}{x}$	$\cos(x)$	$-\sin(x)$	$1+\tan^2(x)$	$\frac{1}{1+x^2}$

Table 5.1: Common Elementary Functions and Their Derivatives.

Example 5.16. *Using Table 5.1 and the Rules of Differentiation, compute by hand the derivatives of the following functions.*

(a) $f(x) = 3x^2+2x+4$

(b) $f(x) = \frac{1}{x}$

(c) $f(x) = \frac{1}{x^2}$

(d) $f(x) = \frac{\sin(x)}{x}$

(e) $f(x) = \sin(x^3 + 4x)$

(f) $f(x) = e^{\sin(x^3+4x)}$

(g) $f(x) = e^{-(x-x_c)^2/s^2} = \exp\left(-\frac{(x-x_c)^2}{s^2}\right)$ *(Radial Basis Function)*

It is suggested that the reader check solutions using the `Symbolic` *differentiation tools from Section 5.3.1.*

Solutions:

(a) $f(x) = 3x^2 + 2x + 4$ **Ans.** $f'(x) = 6x + 2$ uses the Sum Rule, Constant Rule $(c)' = 0$, and the Power Rule $\left(x^k\right)' = kx^{k-1}$.

(b) $f(x) = \frac{1}{x}$ **Ans.** $f'(x) = -\frac{1}{x^2}$ uses the Quotient Rule. Here, $g(x) = 1$ and $h(x) = x$. Hence,

$$f'(x) = \frac{g'(x)\cdot h(x) - g(x)\cdot h'(x)}{(h(x))^2} = \frac{0\cdot x - 1\cdot 1}{x^2} = -\frac{1}{x^2}.$$

(c) $f(x) = \frac{1}{x^2}$ **Ans.** $f'(x) = -\frac{2x}{x^4} = -2\frac{1}{x^3}$ uses the Quotient Rule. Here, $g(x) = 1$ and $h(x) = x^2$. Hence,

$$f'(x) = \frac{g'(x)\cdot h(x) - g(x)\cdot h'(x)}{(h(x))^2} = \frac{0\cdot x^2 - 1\cdot 2x}{(x^2)^2} = -2\frac{1}{x^3}.$$

(d) $f(x) = \frac{\sin(x)}{x}$ **Ans.** $f'(x) = \frac{x\cdot\cos(x) - \sin(x)}{x^2}$ uses the Quotient Rule. Here, $g(x) = \sin(x)$ and $h(x) = x$. Hence,

$$f'(x) = \frac{g'(x)\cdot h(x) - g(x)\cdot h'(x)}{(h(x))^2} = \frac{\cos(x)\cdot x - 1\cdot\sin(x)}{(x)^2} = \frac{x\cdot\cos(x) - \sin(x)}{x^2}.$$

(e) $f(x) = \sin(x^3 + 4x)$ **Ans.** $f'(x) = \left(3x^2 + 4\right)\cdot\cos(x^3 + 4x)$ uses the Chain Rule and the Power Rule. Here, $g(x) = \sin(x)$ and $h(x) = x^3 + 4x$. Hence,

$$f'(x) = g'(h(x))\cdot h'(x) = \underbrace{\cos(x^3 + 4x)}_{g'(h(x))}\cdot\underbrace{\left(3x^2 + 4\right)}_{h'(x)} = \left(3x^2 + 4\right)\cdot\cos(x^3 + 4x),$$

where it is traditional to move the polynomial terms in front of the trigonometric terms, but it is not mandatory to do so.

(f) $f(x) = e^{\sin(x^3+4x)}$ **Ans.** $f'(x) = \left(3x^2 + 4\right)\cdot\cos(x^3 + 4x)\cdot e^{\sin(x^3+4x)}$ uses the Chain Rule **TWICE** and the Power Rule. Here, $g(x) = e^x$ and $h(x) = \sin(x^3 + 4x)$. Fortunately for us, we already computed $h'(x) = \left(3x^2 + 4\right)\cdot\cos(x^3 + 4x)$ in the previous part via the Chain Rule, so we only have one application of the Chain Rule to do here! Hence,

$$f'(x) = g'(h(x))\cdot h'(x) = \underbrace{e^{\sin(x^3+4x)}}_{g'(h(x))}\cdot\underbrace{\left(3x^2 + 4\right)\cdot\cos(x^3 + 4x)}_{h'(x)} = \left(3x^2 + 4\right)\cdot\cos(x^3 + 4x)\cdot e^{\sin(x^3+4x)}.$$

(g) $f(x) = e^{-(x-x_c)^2/s^2}$ **Ans.** $f'(x) = -2\frac{(x-x_c)}{s^2}\cdot e^{-(x-x_c)^2/s^2}$ uses the Chain Rule and the Power Rule. Here, $g(x) = e^x$ and $h(x) = -\frac{(x-x_c)^2}{s^2}$. Hence,

$$f'(x) = g'(h(x))\cdot h'(x) = \underbrace{\left(e^{-\frac{(x-x_c)^2}{s^2}}\right)}_{g'(h(x))}\cdot\underbrace{\left(-2\frac{(x-x_c)}{s^2}\right)}_{h'(x)} = -2\frac{(x-x_c)}{s^2}\cdot e^{-(x-x_c)^2/s^2}.$$

You can write the terms in any order.

```
@variables x

fa = 3x^2 + 2x + 4

fb = 1/x

fc = 1/(x^2)

fd = sin(x)/x

fe = sin(x^3 + 4x)

ff = exp( ( sin(x^3 + 4x) ) )

@variables xc s # additional variables in the radial basis function
                # which are treated as constants because we are
                # differentiating with respect to x

fg = exp(-(x-xc)^2 / s^2)

# With the simplify operation
function deriv(f, x)
    df = Differential(x)(f)
    return simplify(expand_derivatives(df)) # Note the simplify command
end
```

Output

```
deriv (generic function with 1 method)
```

```
@show deriv(fa, x)

@show deriv(fb, x)

@show deriv(fc, x)

@show deriv(fd, x)

@show deriv(fe, x)

@show deriv(ff, x)

@show deriv(fg, x);
```

Output

```
deriv(fa, x) = 2 + 6x
deriv(fb, x) = -1 / (x^2)
deriv(fc, x) = -2 / (x^3)
deriv(fd, x) = (x*cos(x) - sin(x)) / (x^2)
deriv(fe, x) = (4 + 3(x^2))*cos(x^3 + 4x)
deriv(ff, x) = (4 + 3(x^2))*cos(x^3 + 4x)*exp(sin(x^3 + 4x))
deriv(fg, x) = (-2(x - xc)*exp(((xc - x)*(x - xc)) / (s^2))) / (s^2)
```

■

We next consider examples that are "similar" to the power rule.

Example 5.17. *The Generalized Power Rule, Exponential Rule, and Logarithm Rules of Differentiation in Prop. 5.11 can all be derived from the Chain Rule and the Logarithm Rules in Prop. 1.49. Use this approach to derive the derivatives of the following functions.*

(a) $f(x) = x^y$, *where* $x > 0$ *and* $y \in \mathbb{R}$ *is a constant.*

(b) $f(x) = \sqrt[a]{x}$, *where* $x > 0$ *and* $a \neq 0$ *is a constant.*

(c) $f(x) = a^x$, *where* $x \in \mathbb{R}$ *and* $a > 0$ *is a constant.*

Hint: *For* $u > 0$ *and* $v \in \mathbb{R}$, *the Logarithm Rules from Prop. 1.49 yield,*

$$u^v = e^{\ln(u^v)} = e^{v\cdot\ln(u)}.$$

This "observation" is key to computing the derivatives of the above functions.

Solutions:

(a) $f(x) = x^y$, where $x > 0$ and $y \in \mathbb{R}$ is a constant.

Ans. $f'(x) = y \cdot x^{y-1}$ uses the Logarithm Rules and the Chain Rule. As in the Hint, we write

$$f(x) = e^{\ln(x^y)} = e^{y\cdot\ln(x)}.$$

We choose $g(x) = e^x$ and $h(x) = y \cdot \ln(x)$. We note that

$$h'(x) = y \cdot (\ln(x))' = \frac{y}{x}$$

because $(\ln(x))' = \frac{d}{dx}\ln(x) = \frac{1}{x}$ and y is a constant. Hence,

$$f'(x) = g'(h(x)) \cdot h'(x) = \underbrace{\left(e^{y\cdot\ln(x)}\right)}_{g'(h(x))} \cdot \underbrace{\left(\frac{y}{x}\right)}_{h'(x)} = \underbrace{(x^y)}_{g'(h(x))} \cdot \underbrace{\left(\frac{y}{x}\right)}_{h'(x)} = y \cdot x^{y-1},$$

where we used $e^{y\cdot\ln(x)} = x^y$.

Note: As long as we are careful about the domain of the function, namely $x > 0$, this works like the Power Rule.

(b) $f(x) = \sqrt[a]{x}$ where $x > 0$ and $a \neq 0$ is a constant.

Ans. $f'(x) = \frac{1}{a} \cdot x^{\frac{1-a}{a}}$ follows directly from part (a) because

$$\sqrt[a]{x} = x^{\frac{1}{a}},$$

and we identify $y = \frac{1}{a}$. The rest is algebra; indeed,

$$y - 1 = \frac{1}{a} - 1 = \frac{1-a}{a}.$$

Note: If we take $a = 2$, we have

$$\boxed{(\sqrt{x})' = \frac{d}{dx}\sqrt{x} = \frac{d\sqrt{x}}{dx} = \frac{1}{2}x^{-\frac{1}{2}} = \frac{1}{2\sqrt{x}},}$$

where we have emphasized the multiple ways in which the derivative operation can be written.

(c) $f(x) = a^x$, where $x \in \mathbb{R}$ and $a > 0$ is a constant.

Ans. $f'(x) = \ln(a) \cdot a^x$ uses the Logarithm Rules and the Chain Rule. As in the Hint, we write

$$f(x) = e^{\ln(a^x)} = e^{x \cdot \ln(a)}.$$

We choose $g(x) = e^x$ and $h(x) = x \cdot \ln(a)$. We note that $h(x)$ is linear in x because a is a constant and, therefore,

$$h'(x) = \ln(a).$$

Hence,

$$f'(x) = g'(h(x)) \cdot h'(x) = \underbrace{\left(e^{x \cdot \ln(a)}\right)}_{g'(h(x))} \cdot \underbrace{\ln(a)}_{h'(x)} = \underbrace{(a^x)}_{g'(h(x))} \cdot \underbrace{\ln(a)}_{h'(x)} = \ln(a) \cdot a^x,$$

where we used $e^{x \cdot \ln(a)} = a^x$.

Note: If $a = e$, Euler's constant, then $\ln(a) = \ln(e) = 1$ and $f'(x) = (e^x)' = e^x$.

```
@variables x y n a # y, n, and a are all treated as constants
                   # because we are taking derivatives with respect
                   # to the variable x

fa = x^y

fb = x^(1/n)

fc = a^x

@show deriv(fa, x)

@show deriv(fb, x)

@show deriv(fc, x);
```

Output

```
deriv(fa, x) = y*(x^(y - 1))

deriv(fb, x) = (x^((1 - n) / n)) / n

deriv(fc, x) = (a^x)*log(a)
```

■

5.4.3 Bonus Examples

Example 5.18. *(Looks Can be Deceiving) From the Generalized Power Rule,*

$$\frac{d}{dx}\sqrt{x} = \frac{d}{dx}x^{\frac{1}{2}} = \frac{1}{2} \cdot x^{\frac{1}{2}-1} = \frac{1}{2} \cdot x^{-\frac{1}{2}} = \frac{1}{2} \cdot \frac{1}{\sqrt{x}}.$$

But what is the solution to:

$$\frac{d}{dx}\left(\sqrt{x}^{\sqrt{x}}\right)?$$

Solution: We give two solutions based on a common technique developed in Example 5.17.

Option 1: The problem looks terrifying until we recall our power rules: $(x^a)^b = x^{a \cdot b}$. For us, $a = \frac{1}{2}$ and $b = \sqrt{x} = x^{\frac{1}{2}}$. Hence, we seek to differentiate

$$\sqrt{x}^{\sqrt{x}} = x^{\frac{1}{2}\sqrt{x}} = e^{\ln\left(x^{\frac{1}{2}\sqrt{x}}\right)} = e^{\frac{1}{2}\sqrt{x} \cdot \ln(x)}.$$

Key Differentiation Fact: For $h(x)$ differentiable, $\frac{d}{dx}e^{h(x)} = e^{h(x)} \cdot h'(x) = e^{h(x)} \cdot \frac{d}{dx}h(x)$.

Hence,

$$\begin{aligned}
\frac{d}{dx}\left(\sqrt{x}^{\sqrt{x}}\right) &= \frac{d}{dx}\left(e^{\frac{1}{2}\sqrt{x} \cdot \ln(x)}\right) \\
&= e^{\frac{1}{2}\sqrt{x} \cdot \ln(x)} \cdot \frac{d}{dx}\left(\frac{1}{2}\sqrt{x} \cdot \ln(x)\right) \\
&= e^{\frac{1}{2}\sqrt{x} \cdot \ln(x)} \cdot \frac{1}{2} \cdot \frac{d}{dx}\left(\sqrt{x} \cdot \ln(x)\right) \\
&\sqrt{x}^{\sqrt{x}} \cdot \frac{1}{2}\left(\frac{1}{2} \cdot \frac{1}{\sqrt{x}} \cdot \ln(x) + \sqrt{x} \cdot \frac{1}{x}\right) \\
&= \sqrt{x}^{\sqrt{x}} \cdot \frac{1}{4\sqrt{x}}\left(\ln(x) + 2\right) \\
&= \frac{1}{4}\left(\ln(x) + 2\right) \cdot \sqrt{x}^{\sqrt{x} - \frac{1}{2}}.
\end{aligned}$$

Option 2: We solve a more general case first! Sometimes, generality can oblige us to focus. What is $\frac{d}{dx}\left(f(x)^{g(x)}\right)$, assuming $f : \mathbb{R} \to (0, \infty)$ and $g : \mathbb{R} \to \mathbb{R}$ are both differentiable? We start with:

$$f(x)^{g(x)} = e^{\ln\left(f(x)^{g(x)}\right)} = e^{g(x) \cdot \ln(f(x))}.$$

Key Differentiation Facts: For $h(x)$ differentiable,

$$\frac{d}{dx}e^{h(x)} = e^{h(x)} \cdot h'(x) = e^{h(x)} \cdot \frac{d}{dx}h(x)$$

and

$$\frac{d}{dx}\ln(h(x)) = \frac{1}{h(x)} \cdot h'(x) \quad (\text{for } h(x) \text{ positive}).$$

Hence, the combination of the Chain Rule and the Product Rule leads to:

$$\begin{aligned}
\frac{d}{dx}e^{g(x) \cdot \ln(f(x))} &= e^{g(x) \cdot \ln(f(x))} \cdot \frac{d}{dx}\left(g(x) \cdot \ln(f(x))\right) \\
&= e^{g(x) \cdot \ln(f(x))} \cdot \left(g'(x) \cdot \ln(f(x)) + g(x) \cdot \frac{1}{f(x)} \cdot f'(x)\right) \\
&= f(x)^{g(x)} \cdot \left(g'(x) \cdot \ln(f(x)) + \frac{g(x)}{f(x)} \cdot f'(x)\right).
\end{aligned}$$

In our particular case, $f(x) = g(x) = \sqrt{x}$, and the problem statement gave us $f'(x) = g'(x) = \frac{1}{2} \cdot \frac{1}{\sqrt{x}}$. Hence,

$$\begin{aligned}
\frac{d}{dx}\left(\sqrt{x}^{\sqrt{x}}\right) &= \sqrt{x}^{\sqrt{x}} \cdot \left(\frac{1}{2} \cdot \frac{1}{\sqrt{x}} \cdot \ln\left(\sqrt{x}\right) + \frac{\sqrt{x}}{\sqrt{x}} \cdot \frac{1}{2} \cdot \frac{1}{\sqrt{x}}\right) \\
&= \frac{1}{2}\sqrt{x}^{\sqrt{x}} \cdot \frac{1}{\sqrt{x}} \cdot \left(\frac{1}{2}\ln(x) + 1\right) \\
&= \frac{1}{4}\left(\ln(x) + 2\right) \cdot \sqrt{x}^{\sqrt{x} - \frac{1}{2}}.
\end{aligned}$$

The problem comes from in-class derivative vs. derivative on the test! by **BlackPenRedPen**. The take-home message is: when confronted with a complicated-looking differentiation problem that involves powers (aka exponents), $x^y = e^{\ln(x^y)}$ can be a powerful opening gambit, as long as the quantity x is positive, thereby ensuring that $\ln(x)$ makes sense. ■.

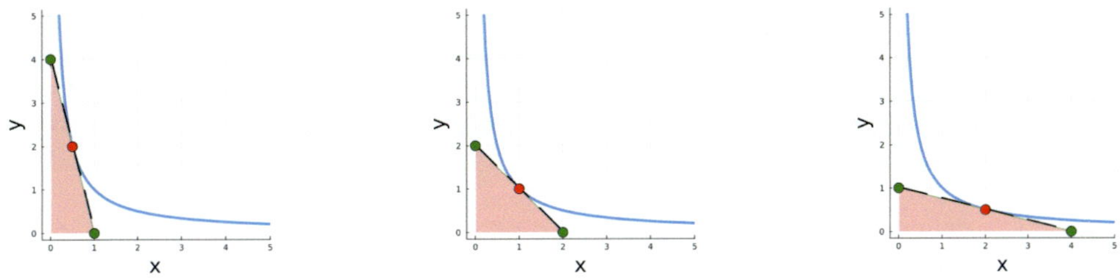

(a) (b) (c)

Figure 5.2: The blue line is the graph of $\frac{1}{x}$. The black dashed line is tangent to the red dot. The two green dots are the x- and y-intercepts for the tangent line. The pink triangle is the area between the tangent line and the (x, y)-axes. Its area is the same for all choices of the tangent point. Why the area under the tangent line of $1/x$ is special by **BlackPenRedPen**.

Example 5.19. *Compute the area highlighted in* Fig. 5.2.

Solution: The formula for the tangent line is precisely the linear approximation of $f(x) = \frac{1}{x}$ at a point $x_0 = a$. Hence, from (5.9),

$$y = f(a) + f'(a) \cdot (x - a).$$

Because $f'(x) = -\frac{1}{x^2}$, we arrive at

$$y = \frac{1}{a} - \frac{1}{a^2}(x - a) = \frac{2}{a} - \frac{1}{a^2}x.$$

Hence, the y-intercept of the line is the point $(0, \frac{2}{a})$ and the x-intercept is the point $(2a, 0)$. The area of the triangle is

$$\frac{1}{2}b \cdot h = \frac{1}{2} \cdot 2a \cdot \frac{2}{a} = 2,$$

independent of the choice of $a > 0$. ■

5.5 Use Cases of the Single-Variable Derivative

Thus far, we have examined the theory for derivatives of a function that depends on a single variable. Before moving on to multivariable functions, we shall illustrate why derivatives are such a useful tool.

Figure 5.3: This is the same small mobile robot that we showed earlier in Section 3.4.1. The user commands wheel speed. If both wheels move at the same speed, the robot moves in a straight line. If one wheel turns faster than the other, the robot turns toward the side of the slower wheel. We are given position trajectories (aka position curves in $\mathbb{R}^2$), instead of velocity. We need to convert the position trajectories to velocity commands. A position trajectory tracking controller for the robot is not yet available, but will be introduced in Chapter 10.

5.5.1 From Position to Velocity

The company **Algorithms in Motion**® has given us the paths in $\mathbb{R}^2$ shown in **Fig. 5.4**. Each path gets us from:

$$\text{Point A} = \begin{bmatrix} 0 \\ 0 \end{bmatrix} \text{ to Point B} = \begin{bmatrix} 10 \\ 10 \end{bmatrix},$$

while avoiding the obstacle shown in red. Shown by a dotted black line is a straight line connecting the two points, which goes right through the obstacle and would destroy the robot. Our job is the following:

> **Task: Select the path that has the lowest peak speed so as to minimize the chance of the wheels slipping while maneuvering**, where **speed is defined to be the Euclidean norm of the velocity vector.** At each point in time, velocity has x- and y-components, making it a 2-vector. The speed is:
>
> $$||(v_x, v_y)|| = \sqrt{(v_x)^2 + (v_y)^2}.$$

Our plan is as follows:

(a) We start with the equations for the position curves, parameterized by time.

(b) We differentiate the position curves (functions) to obtain velocity.

(c) We compute the instantaneous speed of each curve from the instantaneous velocity.

(d) We search over the speed curves to find the peak speed.

(e) We choose the best curve for our robot, the one that demands the lowest peak speed (thereby inducing the least chance of wheel slippage).

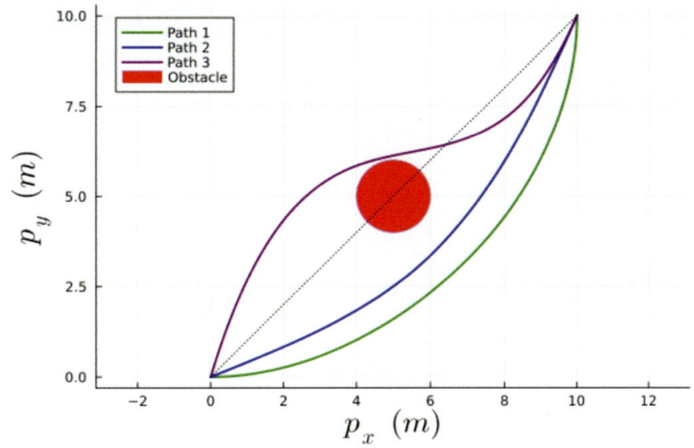

Figure 5.4: Proposed paths for the robot to follow, each of which avoids the obstacle indicated in red. As explained in **Fig. 5.3**, our robot accepts the velocity of the paths. We can use calculus to convert the paths to velocity commands because velocity is the (time) rate of change of position. Hence, $v(t) = \frac{d}{dt}p(t)$.

Below are the path formulas that are given to us.

$$\begin{aligned}
\text{path}_1(t)\begin{bmatrix} x_1(t) \\ y_1(t) \end{bmatrix} &= \begin{bmatrix} 0.37500000t - 0.00007813t^3 \\ 0.00937500t^2 - 0.00007813t^3 \end{bmatrix}, \\
\text{path}_2(t)\begin{bmatrix} x_2(t) \\ y_2(t) \end{bmatrix} &= \begin{bmatrix} -0.00562500t^2 + 0.37500000t + 0.00006250t^3 \\ -0.00187500t^2 + 0.15000000t + 0.00010938t^3 \end{bmatrix}, \\
\text{path}_3(t)\begin{bmatrix} x_1(t) \\ y_1(t) \end{bmatrix} &= \begin{bmatrix} 0.00187500t^2 + 0.22500000t - 0.00003125t^3 \\ -0.03187500t^2 + 0.75000000t + 0.00048438t^3 \end{bmatrix}.
\end{aligned} \tag{5.14}$$

Next, we place the functions in Julia to perform the differentiation line by line using the Power Rule.

```
# Paths from the Planning Team
using LinearAlgebra
@variables t

path1 = [0.375t - 7.812500000000002e-5(t^3),
    0.009375000000000001(t^2) - 7.812500000000002e-5(t^3)]

path2 = [0.375t + 6.250000000000001e-5(t^3) - 0.005625000000000015(t^2),
    0.15000000000000002t + 0.00010937500000000003(t^3) -
   0.0018750000000000004(t^2)]

path3 = [0.0018750000000000004(t^2) + 0.225t - 3.1250000000000001e-5(t^3),
    0.75t + 0.0004843750000000001(t^3) - 0.03187500000000001(t^2)]

# Derivatives of the paths
v1 = deriv(path1, t)
v2 = deriv(path2, t)
v3 = deriv(path3, t)

# Range of t values
t_values = collect(0:0.01:40)
```

```
v1_traj = evaluate(v1, t_values)
v2_traj = evaluate(v2, t_values)
v3_traj = evaluate(v3, t_values)

# The rows are indexed by time. We want to apply
# norm to the columns. That is easy to do in Julia
speed_1 = norm.(eachcol(v1_traj))
speed_2 = norm.(eachcol(v2_traj))
speed_3 = norm.(eachcol(v3_traj))

# Compute max speeds
@show maximum(speed_1)
@show maximum(speed_2)
@show maximum(speed_3)

# Make Speed Plots vs.~time

# Define an array of colors
colors = [:green, :blue, :purple, :orange]

plot1=plot(t_values, speed_1, linewidth=2, linecolor=colors[1], label="Speed 1",
    xlabel=L"$t ~~(s)$", ylabel=L"$Speed ~~(m/s)$",
    yguidefontsize=18, xguidefontsize=18)
plot1=plot!(t_values, speed_2, linewidth=2, linecolor=colors[2], label="Speed 2")
plot1=plot!(t_values, speed_3, linewidth=2, linecolor=colors[3], label="Speed 3")

display(plot1)
png(plot1, "MobileRobotSpeeds")
```

Output

```
maximum(speed_1) = 0.39774756441743303
maximum(speed_2) = 0.5711829829397932
maximum(speed_3) = 0.7830229881682913
```

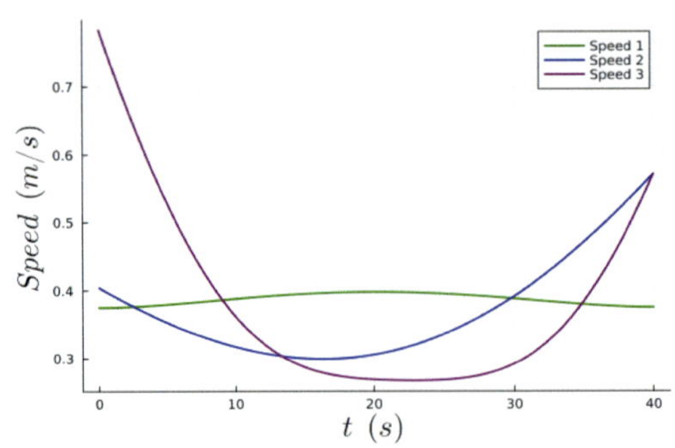

> **Conclusion:** Each of the paths gets us from Point A to Point B in 40 seconds. **Path 1 is the best by our criterion of lowest peak speed.** All of them will likely work.

Here is the "upgraded" `evaluate` function, using Julia's multiple dispatch feature.

```
# evaluate has here been extended to use multiple dispatch
```

```

# Method when f is a symbolic expression and x0 is a Float64
function evaluate(f::Any, x0::Float64)
    vars = Symbolics.get_variables(f)
    return Symbolics.value(substitute(f, (Dict(first(vars) => x0))))
end

# Method when f is a function and x0 is a Float64
function evaluate(f::Function, x0::Float64)
    return f(x0)
end

# Method when f is a vector of symbolic expressions and x0 is a vector
# of Float64
function evaluate(f::Vector{Any}, x0::Vector{Float64})
    vars = Symbolics.get_variables(f[1])
    var = first(vars)
    return hcat([float.(Symbolics.value.(substitute.(f, Ref(var => x_val)))) for x_val
     in x0]...)
end


# Method when f is a vector of symbolic expressions and x0 is a vector
# of Float64
function evaluate(f::Vector{Num}, x0::Vector{Float64})
    vars = Symbolics.get_variables(f[1])
    var = first(vars)
    return hcat([float.(Symbolics.value.(substitute.(f, Ref(var => x_val))))
            for x_val in x0]...)
end

# Method when f is a vector of functions and x0 is a vector of Float64
function evaluate(f::Vector{Function}, x0::Vector{Float64})
    return [fi(xi) for (fi, xi) in zip(f, x0)]
end
```

Output

```
evaluate (generic function with 5 methods)
```

Secrets of the Arcane 5.20: Relations between Differentiation and Integration

In the above example, we differentiated the (x, y)-coordinates of a position trajectory with respect to time and obtained a velocity profile as a function of time. Earlier in Section 3.4.1, we performed the opposite operation: we started with (x, y)-coordinates of a velocity profile and integrated them to obtain position trajectories:

$$v(t) = \begin{bmatrix} v_x(t) \\ v_y(t) \end{bmatrix} = \begin{bmatrix} \frac{dp_x(t)}{dt} \\ \frac{dp_y(t)}{dt} \end{bmatrix}$$

and (5.15)

$$p(t) = \begin{bmatrix} p_x(t) \\ p_y(t) \end{bmatrix} = \begin{bmatrix} p_x(t_0) + \int_{t_0}^{t} v_x(\tau)d\tau \\ p_y(t_0) + \int_{t_0}^{t} v_y(\tau)d\tau \end{bmatrix}.$$

In other words, it seems that we can go back and forth in the sense that:

- If we apply the differentiation operator to a function (or vector of functions), we can "undo it" via integration; and
- if we apply the integration operator to a function (or vector of functions), we can "undo it" via differentiation.

In plain words, one operation is the inverse of the other! This is remarkable. Differentiation takes a very local view of a function to extract its slope at a point, while definite integration focuses on an arbitrarily large (but bounded) region of the function's domain to compute the area under the function.

One possibility is that this is simply an accident of physics ($F = ma$), and for more general functions, the inverse relationships between differentiation and integration fall apart. Another possibility is that they hold very broadly. Which is it?

Chapter 7 shows that the inverse relationships between differentiation and integration hold very broadly. Moreover, these relationships will give us a new way of computing integrals.

Function ($f(x)$)	Derivative ($f'(x)$)	Monotonicity
x^3	$3x^2$	Strictly increasing
$-x^3$	$-3x^2$	Strictly decreasing
$\ln(x)$	$\frac{1}{x}$	Strictly increasing for $x > 0$
e^x	e^x	Strictly increasing
e^{-x}	$-e^{-x}$	Strictly decreasing
$\tan(x)$	$1 + \tan^2(x)$	Strictly increasing (in its domain)
$\frac{1}{1+x^2}$	$-\frac{2x}{(1+x^2)^2}$	Strictly decreasing for $x \geq 0$
$\frac{1}{1+x^2}$	$-\frac{2x}{(1+x^2)^2}$	Strictly increasing for $x \leq 0$
$\frac{x^2}{x^2+1}$	$\frac{2x}{(x^2+1)^2}$	Strictly increasing for $x \geq 0$

Table 5.2: Monotonicity of functions through the lens of their derivatives.

5.5.2 If the Derivative Does Not Change Sign, the Function Is Monotonic

In Section 1.5.5, we learned that a strictly monotonic function (increasing or decreasing) has an inverse as long as the codomain of the function has been chosen properly (it must equal the function's range). We address here how to use a function's derivative to check monotonicity, both strict and non-strict.

Let $I \subset \mathbb{R}$ be an open interval and let $f : I \to \mathbb{R}$ be a function on I. For convenience, we recall the various forms of monotonicity:

- f is **nondecreasing** if $y > x \implies f(y) \geq f(x)$;
- f is **strictly increasing** if $y > x \implies f(y) > f(x)$. Similarly,
- f is **nonincreasing** if $y > x \implies f(y) \leq f(x)$;
- f is **strictly decreasing** if $y > x \implies f(y) < f(x)$.
- **Note:** A constant function is both nondecreasing and nonincreasing but is neither strictly increasing nor strictly decreasing.

Proposition 5.21: Monotonic Functions

Let $I \subset \mathbb{R}$ be an open interval and let $f : I \to \mathbb{R}$ be a differentiable function on I. The following statements are true.

(a) If $\frac{df(x)}{dx} \geq 0$ for all $x \in I$, then f is nondecreasing.

(b) If $\frac{df(x)}{dx} \leq 0$ for all $x \in I$, then f is nonincreasing.

(c) If $\frac{df(x)}{dx} > 0$ for all $x \in I$, then f is strictly increasing.

(d) If $\frac{df(x)}{dx} < 0$ for all $x \in I$, then f is strictly decreasing.

The last two statements can be relaxed just a bit.

(c') If $\frac{df(x)}{dx} = 0$ for a finite number of $x \in I$ and $\frac{df(x)}{dx} > 0$ elsewhere, then f is strictly increasing.

(d') If $\frac{df(x)}{dx} = 0$ for a finite number of $x \in I$ and $\frac{df(x)}{dx} < 0$ elsewhere, then f is strictly decreasing.

These "friendly amendments" are important for a function such as x^3, where the derivative $3x^2$ vanishes at the origin. Otherwise, we would be saying the function is strictly increasing on $(-\infty, 0)$ and $(0, \infty)$, and be perplexed about what is happening for $x = 0$. That would not be good!

Several examples of Prop. 5.21 are given in **Table 5.2**. Traditionally, a proof of Prop. 5.21 is based on the Fundamental Theorem of Calculus, developed later in the book. Fortunately for us, the limit definition of a derivative provides good insight into why it is true! Indeed, consider

$$\frac{df(x)}{dx} = \lim_{h \to 0^+} \frac{f(x+h) - f(x)}{h},$$

where $x + h > x$ plays the role of $y > x$. Because h is greater than zero when we take the limit from the right, for the derivative to be strictly greater than zero at x, it must be true that $f(x+h) > f(x)$. And then $f(x+2h) > f(x+h)$ follows from $\frac{df(x+h)}{dx} > 0$, and so forth. While this reasoning is only qualitative, you get the idea.

Secrets of the Arcane 5.22: Once Upon A Time ...

... you could purchase mathematical results and claim them as your own! L'Hôpital's Rule is named for the French mathematician Guillaume-François-Antoine, Marquis de l'Hôpital, who purchased the formula from his teacher, the Swiss mathematician Johann Bernoulli. Recall that the number e was discovered by the Swiss mathematician Jacob Bernoulli. They were brothers. Can you imagine buying some math results to improve your standing in society?

Note: L'Hôpital is also spelled l'Hospital. In French, the circumflex is used to indicate that an "s" has been dropped. The circumflex is also present in the French word "hôtel", which we use in English as "hotel", but we also have "hostel", as in a "Youth Hostel". The ways we have incorporated French into English can be very confusing. One example: the French word "entrée" means "starter" or "appetizer" on a "carte" (menu)" in France. We, of course, use it as "plat principal", aka the main dish. Yikes!

5.5.3 L'Hôpital's Rule for Evaluating Limits

L'Hôpital's Rule is a powerful tool used to evaluate limits that take the form $\frac{0}{0}$ or $\frac{\pm\infty}{\pm\infty}$. Such forms are called **indeterminate**. Named after the French mathematician Guillaume de l'Hôpital, this rule simplifies the process of finding such limits by differentiating the numerator and the denominator of the function.

Proposition 5.23: L'Hôpital's Rule

Let $I := (a, b)$ be an open interval, $x_0 \in I$ a point in I, and $f : I \to \mathbb{R}$ and $g : I \to \mathbb{R}$ two functions that are differentiable on I except possibly at x_0. If f and g satisfy the additional conditions,

(a) For all $x \in I$, $x \neq x_0$, $g'(x) \neq 0$.

(b) $\lim_{x \to x_0} f(x) = \lim_{x \to x_0} g(x) = 0$ or $|\lim_{x \to x_0} f(x)| = |\lim_{x \to x_0} g(x)| = \infty$ (means the limits of f and g can be $\pm\infty$).

(c) $\lim_{x \to x_0} \dfrac{f'(x)}{g'(x)}$ exists (can be finite or infinite, but it must exist).

Then,

$$\lim_{x \to x_0} \frac{f(x)}{g(x)} = \lim_{x \to x_0} \frac{f'(x)}{g'(x)}. \tag{5.16}$$

Notes:

- In the above we have expressed the result using two-sided limits. It also works for one-sided limits, $x \to x_0^-$ or $x \to x_0^+$, and for $x \to \pm\infty$.
- What if $\lim_{x \to x_0} \dfrac{f'(x)}{g'(x)}$ is also indeterminate? Then we can apply l'Hôpital's Rule to $\frac{f'(x)}{g'(x)}$ and work with the second derivatives, $\frac{f''(x)}{g''(x)}$, etc. This can go on until we run out of steam or get an answer. Usually, one application suffices.

Example 5.24. *Consider the function $\frac{\sin(x)}{x}$ as x approaches 0. Both the numerator and the denominator approach 0, placing us in the indeterminate form $\frac{0}{0}$. Can we apply l'Hôpital's Rule?*

Solution: Differentiating the numerator (aka $f(x) = \sin(x)$), we obtain $\cos(x)$, and differentiating the denominator (aka $g(x) = x$), yields 1, which is nonzero. Therefore, the limit becomes:

$$\lim_{x \to 0} \frac{\cos(x)}{1} = \cos(0) = 1.$$

Because all the conditions for l'Hôpital's Rule are met, we conclude that:

$$\lim_{x \to 0} \frac{\sin(x)}{x} = \lim_{x \to 0} \frac{\cos(x)}{1} = \cos(0) = 1.$$

■

Bottom line: L'Hôpital's Rule is a convenient method for evaluating limits of indeterminate forms. It saves time and simplifies the process by allowing us to differentiate the numerator and the denominator separately to find the limit.

Another **indeterminate form** is $0 \cdot \infty$. Can l'Hôpital's Rule help there as well? Yes, because we can rewrite it in the standard form treated in Prop. 5.23, as we illustrate next.

Example 5.25. *Determine the following limits.*

(a) $\lim_{x \to \infty} x e^{-x}$

(b) $\lim_{x \to \infty} \dfrac{x^2 + 3x + 2}{2x^2 + 4x + 1}$

(c) $\lim_{x \to 0} \dfrac{\ln(1 + x)}{x}$

(d) $\lim_{x\to\infty} x \ln\left(\frac{1}{x}\right)$

(e) $\lim_{x\to 0} \frac{e^x - \cos(x)}{x}$

Note: *The first two problems can be solved using the results of Subsections 2.5.2 and 2.5.3. It is good to see them through the lens of l'Hôpital. Part (d) of the problem can* **easily trip you up**. *Definitely take a look at it for yourself before reading the solutions.*

Solutions:

(a) **Ans.** $\lim_{x\to\infty} xe^{-x} = 0.$

This problem has the indeterminate form $\infty \cdot 0$. We rewrite it in the form $\frac{\infty}{\infty}$, via

$$\begin{aligned}\lim_{x\to\infty} xe^{-x} &= \lim_{x\to\infty} \frac{x}{e^x} \quad \text{(reciprocal property of exponentials)}\\ &= \lim_{x\to\infty} \frac{1}{e^x} \quad \text{(differentiate numerator and denominator)}\\ &= 0.\end{aligned}$$

We could have also rewritten it in the form $\frac{0}{0}$, but as we will see, it leaves us with a harder problem to solve in that case. In other problems, the opposite is true. Hence, there is "art" in the application of mathematical tools! And thank goodness for that! If there were no art involved (i.e., one could just turn a crank), a bot could easily take our jobs as engineers. Consider:

$$\begin{aligned}\lim_{x\to\infty} xe^{-x} &= \lim_{x\to\infty} \frac{e^{-x}}{\frac{1}{x}} \quad (x = \frac{1}{1/x})\\ &= \lim_{x\to\infty} \frac{-e^{-x}}{\frac{-1}{x^2}} \quad \text{(differentiate numerator and denominator)}\\ &= \text{a mess!}\end{aligned}$$

(b) **Ans.** $\lim_{x\to\infty} \frac{x^2+3x+2}{2x^2+4x+1} = \frac{1}{2}$

This problem uses l'Hôpital's Rule twice. Its indeterminate form is $\frac{\infty}{\infty}$, so it falls directly into the l'Hôpital framework.

$$\begin{aligned}\lim_{x\to\infty} \frac{x^2+3x+2}{2x^2+4x+1} &= \lim_{x\to\infty} \frac{2x+3}{4x+4} \quad \text{(differentiate numerator and denominator)}\\ &= \lim_{x\to\infty} \frac{2}{4} \quad \text{(differentiate numerator and denominator once more)}\\ &= \frac{1}{2}.\end{aligned}$$

(c) **Ans.** $\lim_{x\to 0} \frac{\ln(1+x)}{x} = 1.$

The indeterminate form of this problem is $\frac{0}{0}$, so it falls directly into the l'Hôpital framework.

$$\begin{aligned}\lim_{x\to 0} \frac{\ln(1+x)}{x} &= \lim_{x\to 0} \frac{\frac{1}{x+1}}{1} \quad \text{(differentiate numerator and denominator)}\\ &= 1.\end{aligned}$$

(d) **Ans.** $\lim_{x\to\infty} x\ln\left(\frac{1}{x}\right) = -\infty.$

This is not an indeterminate form,[1] and thus l'Hôpital's Rule does not apply. It is still valid to assign $f(x) := x$ and $g(x) := \ln\left(\frac{1}{x}\right)$. Then, we note,

$$\begin{aligned}\lim_{x\to\infty} f(x) &= \lim_{x\to\infty} x \\ &= \infty\end{aligned}$$

$$\begin{aligned}\lim_{x\to\infty} g(x) &= \lim_{x\to\infty} \ln\left(\frac{1}{x}\right) \\ &= \lim_{x\to\infty} -\ln(x) \quad \text{(logarithm rules)} \\ &= -\infty.\end{aligned}$$

Hence,

$$\lim_{x\to\infty} f(x)\cdot g(x) = -\infty,$$

because something large and positive times something large and negative is something (super) large and negative.

> Now, suppose we skipped the above analysis and applied l'Hôpital's Rule **blindly!** We might do something like this
>
> $$\begin{aligned}\lim_{x\to\infty} x\ln\left(\frac{1}{x}\right) &= \lim_{x\to\infty} \frac{\ln\left(\frac{1}{x}\right)}{\frac{1}{x}} \quad \left(x = \frac{1}{1/x}\right) \\ &= \lim_{x\to\infty} \frac{x\cdot\frac{-1}{x^2}}{\frac{-1}{x^2}} \quad \text{(differentiate numerator and denominator with the Chain Rule)} \\ &= \lim_{x\to\infty} x \quad \text{(cancel like terms in the numerator and denominator)} \\ &= \infty.\end{aligned}$$
>
> That means we blew the sign on the limit! Or, we might even say, our error is $-2\cdot\infty$ ($-\infty-\infty$, actual limit minus computed limit). Crazy! But, that is what happens when you misapply l'Hopital's Rule, as ChatGPT4 did for your author! You gotta be careful.

Here is another VALID solution using logarithm rules.

$$\begin{aligned}\lim_{x\to\infty} x\ln\left(\frac{1}{x}\right) &= \lim_{x\to\infty} x\cdot(\ln(1)-\ln(x)) \quad (\ln(a/b) = \ln(a)-\ln(b)) \\ &= \lim_{x\to\infty} -x\cdot\ln(x) \quad (\ln(1)=0) \\ &= -\infty,\end{aligned}$$

because, $\lim_{x\to\infty} x = \infty$ and $\lim_{x\to\infty} \ln(x) = \infty$.

(e) **Ans.** $\lim_{x\to 0} \dfrac{e^x - \cos(x)}{x} = 1$

This problem's indeterminate form is $\frac{0}{0}$, so it falls directly into the l'Hôpital framework.

$$\begin{aligned}\lim_{x\to 0} \frac{e^x-\cos(x)}{x} &= \lim_{x\to 0} \frac{e^x+\sin(x)}{1} \quad \text{(differentiate numerator and denominator)} \\ &= 1.\end{aligned}$$

■

[1] Plus infinity times minus infinity is not indeterminate. To be indeterminate, we need $\frac{0}{0}$ or $\frac{\pm\infty}{\pm\infty}$.

Fact 5.26: Need to Apply l'Hôpital's Rule with Care

All four conditions for l'Hôpital's rule are **necessary** (from Wikipedia):

- **Indeterminacy of form:** $\lim_{x \to x_0} f(x) = \lim_{x \to x_0} g(x) = 0$ or $\pm\infty$;
- **Differentiability of functions:** $f(x)$ and $g(x)$ are differentiable on an open interval I except possibly at a point x_0 contained in I (the same point used in the limit);
- **Non-zero derivative of denominator:** $g'(x) \neq 0$ for all x in I with $x \neq x_0$; and
- **Existence of limit of the quotient of the derivatives:** $\lim_{x \to x_0} \dfrac{f'(x)}{g'(x)}$ exists.

If even one of the above conditions is not satisfied, l'Hôpital's rule is not valid (i.e., it CANNOT be applied).

Example 5.27. *Compute* $\lim_{x \to \infty} \sqrt[x]{x}$.

Solution: 1 in 1 Minute by **BlackPenRedPen**. ■

Want more examples?

- **L'Hôpital's Rule: Ultimate Study guide by BlackPenRedPen**
- **L'Hopital's rule by The Organic Chemistry Tutor.**
- **What is l'Hospital's rule? by Krista King. (She also talks about Johann Bernoulli.)**

5.5.4 Higher-Order Derivatives Lead to Taylor's and Maclaurin's Expansions

In Section 5.2, we showed how the notion of a linear approximation to a function near a point and the derivative of the function at that same point go hand in hand; they are really two sides of the same coin. As hinted at in Fig. 5.5, why stop at a linear approximation of a function? Why not use higher-order derivatives to propose a polynomial approximation that may be able to provide better accuracy? Progress on this question began 700 years ago by the Indian mathematician, Madhava of Sangamagrama, and then accelerated under "the relatively more recent" investigation by Issac Newton in the mid to late 1600's. The general theory came together under Brook Taylor in 1715. The importance of Taylor's work became more clear when Colin Maclaurin used it to characterize maxima, minima, and "other extreme points" of functions in 1725. Today, the uses of Taylor approximations range from the creation of faster and more accurate numerical integration routines to highly efficient ray tracing algorithms that render photo-realistic images in video games.

Fundamental Idea

We start with a very simple proposition that lays bare the essence of Taylor series.

Proposition 5.28: Local Polynomial Approximation

Consider a function$f : (a,b) \to \mathbb{R}$ that is n-times differentiable at $0 \in (a,b)$, the origin. Then the polynomialMaclaurin!polynomial

$$p(x) := a_n \frac{x^n}{n!} + a_{n-1} \frac{x^{(n-1)}}{(n-1)!} + \cdots + a_2 \frac{x^2}{2!} + a_1 x + a_0 \tag{5.17}$$

matches the values of f and its first n derivatives at the origin if, and only if,

$$a_k = f^{(k)}(0) = \frac{d^k f(x)}{dx^k}\Big|_{x=0}. \tag{5.18}$$

Note: We have added factorial terms to the monomials. This is because

$$\frac{d}{dx}\left(\frac{x^k}{k!}\right) = \frac{x^{k-1}}{(k-1)!}, \quad \frac{d^2}{dx^2}\left(\frac{x^k}{k!}\right) = \frac{x^{k-2}}{(k-2)!}, \quad \frac{d^3}{dx^3}\left(\frac{x^k}{k!}\right) = \frac{x^{k-3}}{(k-3)!}, \quad \text{etc.}$$

Note 2: A similar result holds for

$$q(x) := \sum_{k=0}^{n} a_k \frac{(x-x_0)^k}{k!} = a_n \frac{(x-x_0)^n}{n!} + \cdots + a_2 \frac{(x-x_0)^2}{2!} + a_1(x-x_0) + a_0,$$

that is, $p(x)$ shifted by x_0. Then we need $a_k = f^{(k)}(x_0)$.

Proof: It is super informative to look at the second-order case first. We are interested in constructing a polynomial

$$p(x) = a_2 \frac{x^2}{2!} + a_1 x + a_0$$

that matches the function f and its first two derivatives at the origin.

Function Value at the Origin: For the polynomial $p(x)$ to match the value of $f(x)$ at $x = 0$, we must have

$$p(0) = a_0 = f(0).$$

This establishes that the constant term a_0 is equal to $f(0)$, the value of the function at the origin.

First Derivative: The first derivative of $p(x)$ is

$$p'(x) = 2a_2 \frac{x}{2!} + a_1 = a_2 x + a_1.$$

Matching the first derivative of $p(x)$ with that of $f(x)$ at $x = 0$ yields

$$p'(0) = a_1 = f'(0).$$

This shows that the coefficient a_1 is determined by the first derivative of f at the origin.

Second Derivative: The second derivative of $p(x)$ is

$$p''(x) = a_2.$$

Matching the second derivative of $p(x)$ with that of $f(x)$ at $x = 0$ gives

$$p''(0) = a_2 = f''(0).$$

This confirms that the coefficient a_2 is equal to the second derivative of f at the origin.

(Optional Read) Proof by Induction:

Base Case: For $k = 0$, we have $p(0) = a_0$. For the polynomial $p(x)$ to match the value of $f(x)$ at the origin, we must therefore have

$$a_0 = f(0).$$

All other terms in $p(x)$ contain x and hence vanish when $x = 0$.

Induction Hypothesis: Assume that for some $k \geq 0$ the kth derivative of $p(x)$ has the form

$$p^{(k)}(x) = a_k + a_{k+1}\frac{x}{1!} + a_{k+2}\frac{x^2}{2!} + \cdots + a_n \frac{x^{n-k}}{(n-k)!}.$$

Evaluating this expression at $x = 0$ gives

$$p^{(k)}(0) = a_k.$$

Inductive Step: Differentiate the expression for $p^{(k)}(x)$. This yields

$$p^{(k+1)}(x) = a_{k+1} + a_{k+2}\frac{x}{1!} + \cdots + a_n \frac{x^{n-(k+1)}}{(n-(k+1))!}.$$

Evaluating at $x = 0$ gives

$$p^{(k+1)}(0) = a_{k+1}.$$

Thus the formula also holds for $k+1$. By mathematical induction, for every $k = 0, 1, \ldots, n$,

$$p^{(k)}(0) = a_k.$$

Therefore, for the polynomial $p(x)$ to match f and its first n derivatives at 0, the coefficients must satisfy

$$a_k = f^{(k)}(0) = \frac{d^k f(x)}{dx^k}\bigg|_{x=0}.$$

This completes the proof. ■

The Actual Taylor and Maclaurin Polynomials

Maclaurin!polynomial

Taylor and Maclaurin Polynomials

Let $f : (a,b) \to \mathbb{R}$ be a function defined on an open interval (a,b) and let $x_0 \in (a,b)$. Recall that if f is once differentiable at x_0, then

$$f(x) \approx f(x_0) + \frac{df(x_0)}{dx} \cdot (x - x_0) \quad \text{or with the prime notation} \quad f(x) \approx f(x_0) + f'(x_0) \cdot (x - x_0)$$

is the **linear approximation** of the function about x_0.

Definition 5.29. *Suppose that $f : (a,b) \to \mathbb{R}$ is n-times differentiable at x_0. Then*

$$f(x) \approx f(x_0) + f'(x_0)(x - x_0) + f''(x_0) \cdot \frac{(x-x_0)^2}{2!} + \cdots + f^{(n)}(x_0)\frac{(x-x_0)^n}{n!}, \tag{5.19}$$

*is the **n-th degree Taylor expansion of f about the point x_0**. For the special case $x_0 = 0$ (the origin),*

$$f(x) \approx f(0) + f'(0) \cdot x + f''(0) \cdot \frac{x^2}{2!} + \cdots + f^{(n)}(0) \cdot \frac{x^n}{n!} \tag{5.20}$$

*is called an **n-th degree Maclaurin polynomial or Maclaurin expansion**.*

Note: The Taylor expansion is the unique polynomial of degree n that exactly matches the values of f and its first n derivatives at the point x_0.

Note 2: There is a nice theory for quantifying the approximation error when using a Taylor expansion. Here is a very clear written document by the University of South Carolina and here is a video by the Organic Chemistry Tutor.

Taking It to the Limit

Many functions can be differentiated an infinite number of times. For example, e^x is **infinitely differentiable** because $\frac{d}{dx}e^x = e^x$ implies that $\frac{d^k}{dx^k}e^x = e^x$ for all $k \geq 1$. Sine and cosine are also infinitely differentiable because of the simple chain of derivatives,

$$\frac{d}{dx}\sin(x) = \cos(x) \quad \text{and} \quad \frac{d}{dx}\cos(x) = -\sin(x).$$

In fact, the k-th derivative of $\sin(x)$ is

$$\frac{d^k}{dx^k}\sin(x) = \begin{cases} \sin(x) & \text{if } k \equiv 0 \pmod 4 \\ \cos(x) & \text{if } k \equiv 1 \pmod 4 \\ -\sin(x) & \text{if } k \equiv 2 \pmod 4 \\ -\cos(x) & \text{if } k \equiv 3 \pmod 4 \end{cases}$$

and the k-th derivative of $\cos(x)$ is

$$\frac{d^k}{dx^k}\cos(x) = \begin{cases} \cos(x) & \text{if } k \equiv 0 \pmod 4 \\ -\sin(x) & \text{if } k \equiv 1 \pmod 4 \\ -\cos(x) & \text{if } k \equiv 2 \pmod 4 \\ \sin(x) & \text{if } k \equiv 3 \pmod 4. \end{cases}$$

For functions that are infinitely differentiable, one can propose an **infinite Taylor or Maclaurin series** representation of the function.

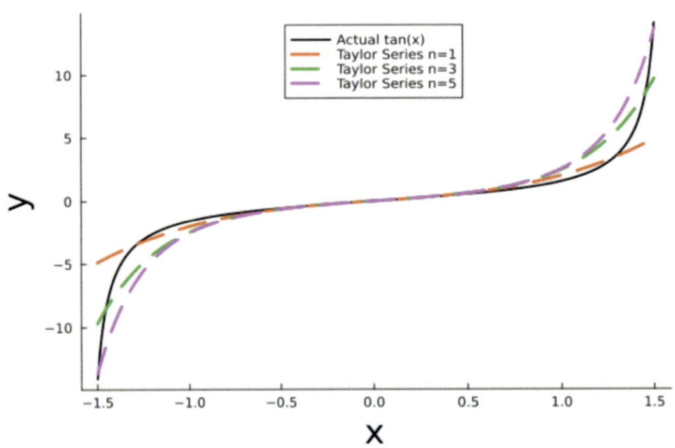

Figure 5.5: (Taylor-Maclaurin Expansion for Tangent) Taylor and Maclaurin studied polynomial approximations of functions based on higher-order derivatives. As such, their work extends the notion of a function's linearization, which only uses the first derivative. Because of Taylor series, people often say "**first-order Taylor approximation**" when they really mean the "**linearization of a function**".

Fact 5.30: (Infinite) Taylor Series Expansion

Maclaurin!series
The Taylor series expansion of an infinitely differentiable function $f(x)$ around a point x_0 is given by,

$$f(x) = f(x_0)+f'(x_0)(x-x_0)+\frac{f''(x_0)}{2!}(x-x_0)^2+\frac{f'''(x_0)}{3!}(x-x_0)^3+\cdots = \lim_{n\to\infty}\sum_{k=0}^{n}\frac{f^{(k)}(x_0)}{k!}(x-x_0)^k, \tag{5.21}$$

where $f^{(k)}(x_0)$ denotes the k-th derivative of f evaluated at the point x_0.

Whether or not the limit in an infinite Taylor series **converges** (i.e., exists at a point $x \in \mathbb{R}$) is a highly technical question that we will not explore. Nevertheless, there are three CONVERGENT infinite Maclaurin series that show up frequently in Engineering and STEM. We highlight them below:

Expansion for e^x

$$e^x = 1 + x + \frac{x^2}{2!} + \frac{x^3}{3!} + \cdots = \lim_{n\to\infty} \sum_{k=0}^{n} \frac{x^k}{k!}, \tag{5.22}$$

Expansion for $\sin(x)$

$$\sin(x) = x - \frac{x^3}{3!} + \frac{x^5}{5!} - \frac{x^7}{7!} + \cdots = \lim_{n\to\infty} \sum_{k=0}^{n} \frac{(-1)^k x^{2k+1}}{(2k+1)!}, \tag{5.23}$$

Expansion for $\cos(x)$

$$\cos(x) = 1 - \frac{x^2}{2!} + \frac{x^4}{4!} - \frac{x^6}{6!} + \cdots = \lim_{n\to\infty} \sum_{k=0}^{n} \frac{(-1)^k x^{2k}}{(2k)!}. \tag{5.24}$$

If we were to add a fourth one, it would be the expansion for $\ln(1+x)$.

Deriving Euler's Formula from Fact 5.30

Euler's formula, $e^{\mathbf{i}\,x} = \cos(x) + \mathbf{i}\sin(x)$, follows from the three Taylor series in Fact 5.30. Here is how we proceed. We first DEFINE $e^{\mathbf{i}\,x}$ by substituting $\mathbf{i}\,x$ for x into (5.22). Yes, otherwise, we do not know how to evaluate the function with a complex argument. However, we know how to perform algebra on terms like $(\mathbf{i}\,x)^k$.

This approach to defining $e^{\mathbf{i}\,x}$ through a power series is in and of itself a powerful use of Taylor series; continuing, we have

$$e^{\mathbf{i}\,x} := 1 + \mathbf{i}\,x + \frac{(\mathbf{i}\,x)^2}{2!} + \frac{(\mathbf{i}\,x)^3}{3!} + \frac{(\mathbf{i}\,x)^4}{4!} + \cdots$$

Then, because $\mathbf{i}^k = \begin{cases} 1 & k = 0 \mod 4 \\ \mathbf{i} & k = 1 \mod 4 \\ -1 & k = 2 \mod 4 \\ -\mathbf{i} & k = 3 \mod 4, \end{cases}$ the above simplifies to,

$$e^{\mathbf{i}\,x} = 1 + \mathbf{i}\,x - \frac{x^2}{2!} - \mathbf{i}\,\frac{x^3}{3!} + \frac{x^4}{4!} + \mathbf{i}\,\frac{x^5}{5!} \cdots$$

Grouping the real and imaginary parts, yields

$$e^{\mathbf{i}\,x} = \left(1 - \frac{x^2}{2!} + \frac{x^4}{4!} - \cdots\right) + \mathbf{i}\left(x - \frac{x^3}{3!} + \frac{x^5}{5!} - \cdots\right)$$

The series within the parentheses are recognized as the Taylor series expansions of $\cos(x)$ and $\sin(x)$, respectively,

$$\cos(x) = 1 - \frac{x^2}{2!} + \frac{x^4}{4!} - \cdots$$
$$\sin(x) = x - \frac{x^3}{3!} + \frac{x^5}{5!} - \cdots$$

Therefore, by direct comparison, we obtain

$$e^{\mathbf{i}\,x} = \cos(x) + \mathbf{i}\sin(x).$$

■

For Further Learning

- Higher order derivatives by 3Blue1Brown.
- Taylor series by 3Blue1Brown.
- Taylor & Maclaurin Polynomials: Intro Khan Academy.
- Taylor Series and Power Series Made Easy (with Pictures): Review of Calculus by Steve Brunton.
- Taylor Series of the Exponential Function and Euler's Formula! by Steve Brunton.

m	Expression for $\mathrm{Im}\left\{(1+i)^m\right\}$	Answer
0	$\mathrm{Im}\left\{(1+i)^0\right\}$	0
1	$\mathrm{Im}\left\{(1+i)^1\right\}$	1
2	$\mathrm{Im}\left\{(1+i)^2\right\}$	2
3	$\mathrm{Im}\left\{(1+i)^3\right\}$	2
4	$\mathrm{Im}\left\{(1+i)^4\right\}$	0
5	$\mathrm{Im}\left\{(1+i)^5\right\}$	-2^2
6	$\mathrm{Im}\left\{(1+i)^6\right\}$	-2^3
7	$\mathrm{Im}\left\{(1+i)^7\right\}$	-2^3

Table 5.3: A Y2K generalization can be found in this table of values for m, expressions for $\mathrm{Im}\left\{(1+i)^m\right\}$, and computed numbers.

5.5.5 (Optional Read) Y2K or BlackPenRedPen Strikes Again

While the original Y2K Problem was of potential major significance for our engineered world, the following problem has zero engineering relevance. Nevertheless, many of us got into engineering because we liked math. Hence, scratching that itch is a worthwhile endeavor.

Problem: Compute $\frac{d^{2000}}{dx^{2000}}\left(e^x \cdot \sin(x)\right)\Big|_{x=0}$. Yes, the Y2K derivative of $e^x \cdot \sin(x)$ evaluated at the origin.

Ans. 1 BCE.[2]

Why: Recall Euler's formula: $e^{\mathrm{i}\,x} = \cos(x) + \mathrm{i}\sin(x)$. Hence, we can write

$$\sin(x) = \mathrm{Im}\{e^{\mathrm{i}\,x}\},$$

where Im stands for the imaginary part. Our problem can, therefore, be written as

$$\frac{d^{2000}}{dx^{2000}}\left(\mathrm{Im}\{e^x \cdot e^{\mathrm{i}\,x}\}\right)\Big|_{x=0} = \mathrm{Im}\left\{\frac{d^{2000}}{dx^{2000}}\left(e^x \cdot e^{\mathrm{i}\,x}\right)\Big|_{x=0}\right\} = \mathrm{Im}\left\{\frac{d^{2000}}{dx^{2000}}\left(e^{(1+\mathrm{i})x}\right)\Big|_{x=0}\right\}.$$

This greatly simplifies things because

$$\frac{d^k}{dx^k}\left(e^{\alpha x}\right)\Big|_{x=0} = \alpha^k \cdot e^{\alpha x}\big|_{x=0} = \alpha^k,$$

[2] The year immediately before 1 CE (Common Era) is 1 BCE (Before Common Era). There is no year 0 in the Gregorian calendar system, so it jumps directly from 1 BCE to 1 CE. Does this mean that 1 BCE, being the year before 1 CE, functions as zero?

even for $\alpha \in \mathbb{C}$. It follows that,

$$\frac{d^{2000}}{dx^{2000}} \left(e^x \cdot \sin(x)\right)\Big|_{x=0} = \mathrm{Im}\{(1+\mathbf{i})^{2000}\},$$

and while this may look painful, it is not! We first note that

$$(1+\mathbf{i})^2 = (1+2\,\mathbf{i}+(\mathbf{i})^2)) = (1+2\,\mathbf{i}-1) = 2\,\mathbf{i}.$$

Then, we note that

$$(1+\mathbf{i})^8 = (2\,\mathbf{i})^4 = 2^4,$$

because $\mathbf{i}^4 = (-1)^2 = 1$. Do you see where this is going? $2000 = 8 \cdot 250$. Hence, using our exponent rules,

$$(1+\mathbf{i})^{2000} = \left((1+\mathbf{i})^8\right)^{250} = \left(2^4\right)^{250} = 2^{1000},$$

a real number. We conclude that

$$\frac{d^{\mathrm{Y2K}}}{dx^{\mathrm{Y2K}}} \left(e^x \cdot \sin(x)\right)\Big|_{x=0} = \frac{d^{2000}}{dx^{2000}} \left(e^x \cdot \sin(x)\right)\Big|_{x=0} = \mathrm{Im}\left\{(1+\mathbf{i})^{2000}\right\} = \mathrm{Im}\left\{2^{1000}\right\} = 0.$$

Here's the link to **BlackPenRedPen**'s video solving the tiebreaker problems from the 2022 Berkeley Math Tournament. The Berkeley version of the problem is ever so slightly different. After watching the video, can you generalize the problem to

$$\textbf{Find:}\quad \frac{d^{\mathrm{Year}}}{dx^{\mathrm{Year}}} \left(e^x \cdot \sin(x)\right)\Big|_{x=0} \;?$$

Hint: Given in the footnote[3] so that you can play with the problem first. Scratch that itch!

5.6 Partial Derivatives

Because you have taken Linear Algebra already, where you solved problems involving thousands of variables, working with functions of a single variable must seem somewhat "tame". The training of modern AI neural networks involves literally billions of variables where their derivatives are computed via Automatic Differentiation (AD); see Section 5.3.3. While we will NOT attempt any examples of this magnitude (look for a course on deep learning), we will treat **multivariable (aka multivariate) functions** in $n > 1$ variables.

5.6.1 Intuition

Imagine you are standing somewhere on the side of a mountain, and you want to know how fast you are ascending or descending. You can use partial derivatives to figure this out. Partial derivatives are a way of measuring how a function changes when one of its inputs changes while keeping all of the other inputs the same. Since you are standing on a mountain, let us take the function as your altitude (aka height) and the inputs as your (x, y) position on the mountain.

If you compute the local slope of your function as x varies while holding y constant, it is called a **partial derivative with respect to x**. **Why partial?** Because you are only considering a "small part of the variables" (one at a time) and not "all of them" (together). The partial derivative with respect to x would tell you the rate of change of your altitude as you move a small amount in an eastward direction (i.e., move to the right), **while keeping y constant**. A partial derivative with respect to y would tell you the rate of change of your altitude as you move a small amount northward (i.e., straight ahead), **while keeping x constant**. It is the "slope" or rate of change taken one variable at a time, a kind of divide-and-conquer approach.

A partial derivative is a little like how a basis for vectors allows you to understand how much each individual basis vector contributes to a given vector, and by adding up all of the contributions, you can reconstruct the entire vector. Recall, if

$$\{u_1, u_2, \ldots, u_n\}$$

[3] Write Year $= 8 \cdot N + m$, where $m \in \{0, 1, 2, \ldots, 7\}$. Then the answer is $2^{4 \cdot N} \cdot \mathrm{Im}\left\{(1+\mathbf{i})^m\right\}$. You are almost there. See also Table 5.3.

is a basis for $\mathbb{R}^n$, then we can write any vector $v \in \mathbb{R}^n$ as a linear combination of the basis vectors, that is,

$$v = \alpha_1 u_1 + \alpha_2 u_2 + \cdots + \alpha_n u_n,$$

for constants $\alpha_i \in \mathbb{R}$. The quantity $\boldsymbol{\alpha_i u_i}$ is the **component of $\boldsymbol{v}$** lying in the **direction $\boldsymbol{u_i}$**.

We will see that partial derivatives are EXACTLY LIKE THIS, and in the best possible way. We use the natural basis vectors, $\{e_1, e_2, \ldots, e_n\}$, to define partial derivatives. For definiteness, we recall that in $\mathbb{R}^n$, the **natural basis vectors (aka canonical basis vectors)** are the columns of the $n \times n$ identity matrix, that is,

$$e_1 := \begin{bmatrix} 1 \\ 0 \\ 0 \\ \vdots \\ 0 \end{bmatrix}, \quad e_2 := \begin{bmatrix} 0 \\ 1 \\ 0 \\ \vdots \\ 0 \end{bmatrix}, \quad e_3 := \begin{bmatrix} 0 \\ 0 \\ 1 \\ \vdots \\ 0 \end{bmatrix}, \ldots, \quad e_n := \begin{bmatrix} 0 \\ 0 \\ 0 \\ \vdots \\ 1 \end{bmatrix}.$$

5.6.2 Partial Derivatives: Single-Variable Derivatives Meet Multivariable Functions

To set the stage, let us note that if $f : \mathbb{R}^n \to \mathbb{R}^m$ is a vector-valued function depending on $x = (x_1, x_2, \ldots, x_n)$, $x_0 \in \mathbb{R}^n$, and $h \in \mathbb{R}$, then $g(h) := f(x_0 + h \cdot e_i)$ is only a function of h; it is a single-variable function. As an example, consider

$$f(x_1, x_2) := \begin{bmatrix} x_1 \cdot x_2 \\ e^{x_1} + \sin(x_2) \end{bmatrix}$$

and let $x_0 = (3, \frac{\pi}{4})$. Then

$$g(h) := f(x_0 + h \cdot e_1) = f\left(\begin{bmatrix} 3 \\ \frac{\pi}{4} \end{bmatrix} + h \cdot \begin{bmatrix} 1 \\ 0 \end{bmatrix}\right) = f(3 + h, \frac{\pi}{4}) = \begin{bmatrix} (3+h) \cdot \frac{\pi}{4} \\ e^{3+h} + \sin\left(\frac{\pi}{4}\right) \end{bmatrix};$$

similarly,

$$\gamma(h) := f(x_0 + h \cdot e_2) = f\left(\begin{bmatrix} 3 \\ \frac{\pi}{4} \end{bmatrix} + h \cdot \begin{bmatrix} 0 \\ 1 \end{bmatrix}\right) = f(3, \frac{\pi}{4} + h) = \begin{bmatrix} 3 \cdot \left(\frac{\pi}{4} + h\right) \\ e^{3} + \sin\left(\frac{\pi}{4} + h\right) \end{bmatrix}.$$

Each of these functions depends on h alone. And where h shows up in the function is dictated by which basis vector e_i is used in forming $f(x_0 + h \cdot e_i)$. The first function depends on h everywhere that f depends on x_1, and the second function depends on h everywhere that f depends on x_2. Once again, this is a divide-and-conquer strategy that works very well. Moreover, because these functions depend on a single variable, we can use our limit concept from Def. 5.1 to define a partial derivative! **Yes, partial derivatives are "rise over run" computed one variable at a time.**

Partial Derivative at a Point

Definition 5.31. *Suppose $f : \mathbb{R}^n \to \mathbb{R}^m$ is a vector-valued function depending on $x = (x_1, x_2, \ldots, x_n)$ and that $x_0 \in \mathbb{R}^n$ is a point (aka a fixed vector). The* ***partial derivative of $\boldsymbol{f}$ with respect to $\boldsymbol{x_i}$ at the point $\boldsymbol{x_0}$*** *is*

$$\begin{aligned} \frac{\partial f(x_0)}{\partial x_i} &:= \lim_{h \to 0} \frac{f(x_0 + h \cdot e_i) - f(x_0)}{h}, \\ \frac{\partial f(x_0)}{\partial \boldsymbol{x_i}} &:= \lim_{h \to 0} \frac{f(x_0 + h \cdot \boldsymbol{e_i}) - f(x_0)}{h} \quad \textit{(same as above, highlighting } \boldsymbol{x_i} \textit{ and } \boldsymbol{e_i}\textit{)}, \end{aligned} \tag{5.25}$$

assuming the limit exists and is finite.

Notes:

- *To distinguish a "partial" derivative from an "ordinary" derivative, we replace* d *with the symbol* ∂*, which is pronounced "partial".*
- *The symbol* $\frac{\partial f(x_0)}{\partial x_i}$ *is called the "partial derivative of* f *with respect to* x_i*, evaluated at* x_0*.*
- *Because the function* $f : \mathbb{R}^n \to \mathbb{R}^m$ *has* m*-components, that is,*

$$f(x_1, x_2, \ldots, x_n) = \begin{bmatrix} f_1(x_1, x_2, \ldots, x_n) \\ f_2(x_1, x_2, \ldots, x_n) \\ \vdots \\ f_m(x_1, x_2, \ldots, x_n) \end{bmatrix},$$

the limit in (5.25) must exist and be finite for each component of the function, that is, for each $f_i(x_1, x_2, \ldots, x_n)$**, or the partial derivative of** f **does not exist.**

Didn't we just stop using limits to compute the derivative with respect to a single variable? Yes! We compute partial derivatives the same way as ordinary derivatives because we are only differentiating with respect to one variable at a time, as in

$$\frac{\partial f(x_0)}{\partial \boldsymbol{x_i}} = \frac{d}{dh} f(x_0 + h \cdot \boldsymbol{e_i})\Big|_{h=0},$$

and hence all of the rules in Prop. 5.11 for ordinary derivatives apply to partial derivatives as well. Moreover, we have Software tools for turning complicated problems into manageable problems. We illustrate all of this next.

Example 5.32. *Compute–by hand–the partial derivatives of*

$$f(x_1, x_2) := \begin{bmatrix} x_1 \cdot x_2 \\ e^{x_1} + \sin(x_2) \end{bmatrix}$$

at a general point, $x \in \mathbb{R}^2$*. In other words, you do not have to evaluate them at a specific point.*

Solution:

$$\frac{\partial}{\partial x_1} f(x_1, x_2) = \frac{\partial}{\partial x_1} \begin{bmatrix} x_1 \cdot x_2 \\ e^{x_1} + \sin(x_2) \end{bmatrix} = \begin{bmatrix} \frac{\partial}{\partial x_1}(x_1 \cdot x_2) \\ \frac{\partial}{\partial x_1}(e^{x_1} + \sin(x_2)) \end{bmatrix} = \begin{bmatrix} x_2 \\ e^{x_1} \end{bmatrix}.$$

Indeed, when taking the derivative with respect to x_1, we treat everything else as a constant,

- $\frac{\partial}{\partial x_1}(x_1 \cdot x_2) = x_2$ (we think of $x_2 = c$, a constant),
- $\frac{\partial}{\partial x_1}(\sin(x_2)) = 0$ (we think of $x_2 = c$, a constant), and
- $\frac{\partial}{\partial x_1}(e^{x_1}) = e^{x_1}$ (because the derivative of e^{x_1} is itself).

Next,

$$\frac{\partial}{\partial x_2} f(x_1, x_2) = \frac{\partial}{\partial x_2} \begin{bmatrix} x_1 \cdot x_2 \\ e^{x_1} + \sin(x_2) \end{bmatrix} = \begin{bmatrix} \frac{\partial}{\partial x_2}(x_1 \cdot x_2) \\ \frac{\partial}{\partial x_2}(e^{x_1} + \sin(x_2)) \end{bmatrix} = \begin{bmatrix} x_1 \\ \cos(x_2) \end{bmatrix}.$$

Indeed, when taking the derivative with respect to x_2, we treat everything else as a constant,

- $\frac{\partial}{\partial x_2}(x_1 \cdot x_2) = x_1$ (we think of $x_1 = c$, a constant),

- $\frac{\partial}{\partial x_2}(e^{x_1}) = 0$ (we think of $x_1 = c$, a constant), and
- $\frac{\partial}{\partial x_2}(\sin(x_2)) = \cos(x_2)$ (because the derivative of $\sin(x_2)$ is $\cos(x_2)$).

■

Remark 5.33. *Why do we sometimes write* $\frac{\partial f(x_1,x_2)}{\partial x_1}$ *and sometimes* $\frac{\partial}{\partial x_1} f(x_1,x_2)$*? Because*

$$\frac{\partial}{\partial x_1}\begin{bmatrix} x_1 \cdot x_2 \\ e^{x_1} + \sin(x_2) \end{bmatrix} \quad \textbf{looks much better than} \quad \frac{\partial \begin{bmatrix} x_1 \cdot x_2 \\ e^{x_1} + \sin(x_2) \end{bmatrix}}{\partial x_1}.$$

Don't you agree? Also, it's good to recognize differentiation as an operator that is applied to functions.

Example 5.34. *Use Julia to compute partial derivatives of the following functions:*

(a) $f_a(x_1,x_2) = \begin{bmatrix} x_1 x_2^2 \\ e^{3x_1} + \cos(4x_2) \end{bmatrix}$

(b) $f_b(x_1,x_2) = e^{3x_1} + \cos(4x_2)$

(c) $f_c(x_1,x_2) = \begin{bmatrix} x_1 x_2^2 \\ x_2 \sin^3(x_1) + 11 \\ 4x_1^2 + (x_2+3)^2 \end{bmatrix}$

(d) $f_d(x_1,x_2,x_3) = \begin{bmatrix} x_1 x_2 x_3^3 \\ x_2 e^{x_1 x_3} \\ x_1 \end{bmatrix}$

Solution:

```julia
# Reminder about our function

using Symbolics

# For scalar functions
function deriv(f::Symbolics.Num, x::Symbolics.Num)
    df = Differential(x)(f)
    return simplify(expand_derivatives(df))
        # Note the simplify command
end

# For vector-valued functions
function deriv(f::Vector{Symbolics.Num}, x::Symbolics.Num)
    df = []
    for component in f
        df_comp = Differential(x)(component)
        push!(df, simplify(expand_derivatives(df_comp)))  # Note the simplify command
    end
    return df
end
```

Output

```
deriv (generic function with 2 methods)
```

```
@variables x1 x2 x3

fa = [x1 * (x2)^2; exp(3*x1) + cos(4*x2)]
fb = exp(3*x1) + cos(4*x2) # not a vector, hence no brackets!
fc = [x1 * (x2)^2; x2*( sin(x1) )^3 + 11; 4*(x1)^2 + ( x2 +3 )^2]
fd = [x1*x2*(x3)^3; x2*exp(x1*x3); x1]



@show dfadx1 = deriv.(fa, x1)
@show dfadx2 = deriv.(fa, x2)
println(" ")
@show dfbdx1 = deriv.(fb, x1)
@show dfbdx2 = deriv.(fb, x2)
println(" ")
@show dfcdx1 = deriv.(fc, x1)
@show dfccx2 = deriv.(fc, x2)
println(" ")
@show dfddx1 = deriv.(fd, x1) # the dfddxi is really dfd/dxi
@show dfddx2 = deriv.(fd, x2)
@show dfddx3 = deriv.(fd, x3);
```

Output

```
dfadx1 = deriv.(fa, x1) = Num[x2^2, 3exp(3x1)]
dfadx2 = deriv.(fa, x2) = Num[2x1*x2, -4sin(4x2)]

dfbdx1 = deriv.(fb, x1) = 3exp(3x1)
dfbdx2 = deriv.(fb, x2) = -4sin(4x2)

dfcdx1 = deriv.(fc, x1) = Num[x2^2, 3x2*(sin(x1)^2)*cos(x1), 8x1]
dfccx2 = deriv.(fc, x2) = Num[2x1*x2, sin(x1)^3, 2(3 + x2)]

dfddx1 = deriv.(fd, x1) = Num[x2*(x3^3), x2*x3*exp(x1*x3), 1]
dfddx2 = deriv.(fd, x2) = Num[x1*(x3^3), exp(x1*x3), 0]
dfddx3 = deriv.(fd, x3) = Num[3x1*x2*(x3^2), x1*x2*exp(x1*x3), 0]
```

■

Explaining the Num Type in Julia

We asked ChatGpt4 to explain the Num Type in the expression Num$[x_2^2, 3 \cdot x_2 \cdot (\sin(x_1)^2) \cdot \cos(x_1), 8x1]$. Here's what it said. "**Num: In the context of Julia and symbolic computation:** Num is a type that holds symbolic numbers or expressions. These are not evaluated until you explicitly ask Julia to do so. They can be variables, standard mathematical functions, or more complex expressions." ChatGPT4 did not say this, but it should have: **The key thing is, you can continue taking derivatives like so,**

```
@show d2fadx1dx1 = deriv.(dfadx1 , x1)
@show d2fadx1dx2 = deriv.(dfadx1 , x2)
println(" ")
@show d2fadx2dx1 = deriv.(dfadx2 , x1)
```

```
@show d2fadx2dx2 = deriv.(dfadx2 , x2)
println(" ")
```

Output

```
d2fadx1dx1 = deriv.(dfadx1, x1) = Num[0, 9exp(3x1)]
d2fadx1dx2 = deriv.(dfadx1, x2) = Num[2x2, 0]

d2fadx2dx1 = deriv.(dfadx2, x1) = Num[2x2, 0]
d2fadx2dx2 = deriv.(dfadx2, x2) = Num[2x1, -16cos(4x2)]
```

where the correct mathematical notation would be

$$\frac{\partial^2 f_a(x_1,x_2)}{\partial x_1^2} = \frac{\partial^2 f_a(x_1,x_2)}{\partial x_1 \partial x_1} = \begin{bmatrix} 0 \\ 9e^{3x_1} \end{bmatrix} \text{ and } \frac{\partial^2 f_a(x_1,x_2)}{\partial x_1 \partial x_2} = \begin{bmatrix} 2x_2 \\ 0 \end{bmatrix},$$

$$\frac{\partial^2 f_a(x_1,x_2)}{\partial x_2 \partial x_1} = \begin{bmatrix} 2x_2 \\ 0 \end{bmatrix} \text{ and } \frac{\partial^2 f_a(x_1,x_2)}{\partial x_2^2} = \frac{\partial^2 f_a(x_1,x_2)}{\partial x_2 \partial x_2} = \begin{bmatrix} 2x_1 \\ -16\cos(4x_2) \end{bmatrix}.$$

And then we take derivatives of those expressions as well; see Introduction to Higher Order Partial Derivatives Notation and Example by the Math Sorcerer.

5.7 Packaging Partial Derivatives to Form Jacobians, Gradients, and Hessians

Jacobians and gradients are matrix and vector forms, respectively, of partial derivatives. They collect all of the "local slope information" into a single object, just as we learned to do for linear equations, $Ax = b$. The Hessian is the Jacobian of the gradient.

5.7.1 The Jacobian

Consider a multivariable vector-valued function $f : \mathbb{R}^n \to \mathbb{R}^m$. The notation introduced in Section 5.6.2 for partial derivatives allows us to make the following very compact definition.

Jacobian of a Function

Definition 5.35. *The **Jacobian** of $f : \mathbb{R}^n \to \mathbb{R}^m$ is*

$$\frac{\partial f(x)}{\partial x} := \begin{bmatrix} \frac{\partial f(x)}{\partial x_1} & \frac{\partial f(x)}{\partial x_2} & \cdots & \frac{\partial f(x)}{\partial x_n} \end{bmatrix}, \tag{5.26}$$

assuming the indicated partial derivatives exist. The Jacobian of a function is formed by packaging the column vectors $\frac{\partial f(x)}{\partial x_j}$ into a matrix. We need to keep in mind that, for each value of $x \in \mathbb{R}^n$, the compact and innocent-looking object

$$\frac{\partial f(x)}{\partial x}$$

is really an $m \times n$ matrix: because there are n columns of the form $\frac{\partial f(x)}{\partial x_j}$, and each column is an m-vector. There are software tools that compute the Jacobian in one fell swoop.

***Note:** The Jacobian of a function has two common notations (i) $\frac{\partial f(x)}{\partial x}$ as we used above, and (ii) $J_f(x)$.*

Just for the record, we write out $\frac{\partial f(x)}{\partial x}$ as an $m \times n$ matrix. Suppose

$$f(x) = \begin{bmatrix} f_1(x) \\ f_2(x) \\ \vdots \\ f_m(x) \end{bmatrix} = \begin{bmatrix} f_1(x_1, x_2, \dots, x_n) \\ f_2(x_1, x_2, \dots, x_n) \\ \vdots \\ f_m(x_1, x_2, \dots, x_n) \end{bmatrix}.$$

Writing out all of the entries in the $m \times n$ Jacobian matrix gives

$$\frac{\partial f(x)}{\partial x} := \begin{bmatrix} \frac{\partial f(x)}{\partial x_1} & \frac{\partial f(x)}{\partial x_2} & \cdots & \frac{\partial f(x)}{\partial x_n} \end{bmatrix} = \begin{bmatrix} \frac{\partial f_1(x)}{\partial x_1} & \frac{\partial f_1(x)}{\partial x_2} & \cdots & \frac{\partial f_1(x)}{\partial x_n} \\ \frac{\partial f_2(x)}{\partial x_1} & \frac{\partial f_2(x)}{\partial x_2} & \cdots & \frac{\partial f_2(x)}{\partial x_n} \\ \vdots & \vdots & \ddots & \vdots \\ \frac{\partial f_m(x)}{\partial x_1} & \frac{\partial f_m(x)}{\partial x_2} & \cdots & \frac{\partial f_m(x)}{\partial x_n} \end{bmatrix}, \tag{5.27}$$

which is much more intimidating than (5.26). In other symbols, the ij component of $J_f(x) = \frac{\partial f(x)}{\partial x}$ is

$$\begin{aligned} \left[\frac{\partial f(x)}{\partial x}\right]_{ij} &= \frac{\partial f_i(x)}{\partial x_j}, \\ J_{ij}(x) &= \frac{\partial f_i(x)}{\partial x_j}. \end{aligned} \tag{5.28}$$

If you want to ignore all of the benefits of vector-matrix notation, you can compute each entry of the Jacobian matrix one by one, as in (5.28). When doing hand computations for simple examples, such as $m = n = 2$, the above very explicit 1×1 scalar (i.e., non-vectorized) calculations can be informative.

Proposition 5.36: Linear Approximation at a Point for Multivariable Functions

The linear approximation of a (nonlinear) function $f : \mathbb{R}^n \to \mathbb{R}^m$ at a point x_0 is defined to be

$$f(x) \approx f(x_0) + A(x - x_0) := f(x_0) + \frac{\partial f(x_0)}{\partial x}(x - x_0), \tag{5.29}$$

where the $m \times n$ matrix $A := J_f(x_0) = \frac{\partial f(x_0)}{\partial x}$ is the Jacobian of f at the point x_0. We note that the dimensions make sense because $f(x_0)$ is $m \times 1$, $(x - x_0)$ is $n \times 1$, and therefore $J_f(x_0)(x - x_0) = \frac{\partial f(x_0)}{\partial x}(x - x_0)$ is $(m \times n) \cdot (n \times 1) = m \times 1$.

Note: The linear approximation is also called the **first-order Taylor expansion of f about the point x_0.**

Proof: We pose

$$f(x) \approx y_0 + A \cdot (x - x_0) \tag{5.30}$$

for $||x - x_0|| < \delta$ and some $\delta > 0$. What values should we assign to y_0 and A? Just as in the case of a scalar-valued function of a single variable, if we plug $x = x_0$ into (5.30), we obtain

$$f(x_0) = y_0.$$

Hence,

$$f(x) \approx f(x_0) + A \cdot (x - x_0). \tag{5.31}$$

Next, we plug in $x = x_0 + h \cdot e_i$ into (5.31), yielding

$$\begin{aligned} f(x_0 + h \cdot e_i) &\approx f(x_0) + \underbrace{A \cdot h \cdot e_i}_{h \cdot e_i = (x_0 + h \cdot e_i) - x_0} \\ &\approx f(x_0) + h \cdot A \cdot e_i \\ &\approx f(x_0) + h \cdot A_i, \end{aligned}$$

where we note that $A \cdot e_i = A_i$, the i-th column of A. A little algebra gives us

$$\big(f(x_0 + h \cdot e_i) - f(x_0) \approx h \cdot A_i\big) \iff \left(A_i \approx \frac{f(x_0 + h \cdot e_i) - f(x_0)}{h}\right). \tag{5.32}$$

The approximation should be better and better with smaller and smaller h. In the limit, we have

$$A_i = \lim_{h \to 0} \frac{f(x_0 + h \cdot e_i) - f(x_0)}{h} =: \frac{\partial f(x_0)}{\partial x_i}, \tag{5.33}$$

as long as the limit exists. In other symbols, if f is differentiable at x_0, once we assemble the columns of

$$A = \begin{bmatrix} \frac{\partial f(x_0)}{\partial x_1} & \frac{\partial f(x_0)}{\partial x_2} & \cdots & \frac{\partial f(x_0)}{\partial x_n} \end{bmatrix},$$

we have

$$f(x) \approx f(x_0) + \frac{\partial f(x_0)}{\partial x} \cdot (x - x_0) \tag{5.34}$$

is a **local linear approximation** to the function near the point x_0. ■

Secrets of the Arcane 5.37: Use of First-Order Taylor Expansions in Engineering

The use of **first-order Taylor expansions**, often referred to as **linear approximations**, is pervasive in various fields of engineering. While it's difficult to quantify how much of engineering is based on this concept, it's safe to say that it's a fundamental tool used frequently for analysis, design, and optimization. Here are some areas where it's commonly used:

Feedback Control Systems: In control theory, nonlinear systems are often linearized around an operating point to make them easier to analyze and control. The linearized models are then used to design controllers. **We will make use of linearization when we design feedback controllers for the BallBot.** A cool thing about the use of linearizations in feedback control is that **the controller is trying to keep an error term small**, and this error is often $(x - x_0)$. In other words, the controller is helping to validate the assumptions required for using a linear approximation in the first place.

Structural Engineering: In the analysis of structures like beams, bridges, and buildings, linear approximations are often used to simplify complex equations describing deformation and stress.

Electrical Engineering: In circuit analysis, devices like transistors are often modeled using linear approximations around an operating point, especially for small-signal analysis.

Fluid Mechanics: In fluid dynamics, linear approximations are used to simplify the Navier-Stokes equations under certain conditions, leading to more tractable models like potential flow.

Thermodynamics: In thermodynamics, properties like enthalpy and entropy are often linearized to analyze processes like isothermal and adiabatic expansions.

Signal Processing: In signal processing, systems are often linearized to make them easier to analyze using Fourier and Laplace transforms.

Optimization: In optimization problems, the objective function and constraints are often linearized to find a local minimum or maximum.

Economics and Operations Research: These fields also make extensive use of linear approximations for problems like supply chain optimization, market equilibrium analysis, etc.

Limitations: It's important to note that linear approximations are only valid "near the point of expansion"[a] and may not capture the behavior of a nonlinear function (or system model) accurately over a large domain. Engineers must exercise caution when using these approximations, especially for systems that exhibit "strong"[b] nonlinearities.

[a] In practice, it is an art to understand what "near" means in any given application. Once again, this is great news as it keeps domain experts, like you, employed!

[b] Nonlinear terms, such as x^2 or e^x that "blow up quickly". Functions that are bounded, such as $\sin(x)$, are considered "mild nonlinearities".

Example 5.38. *For the following functions $f : \mathbb{R}^n \to \mathbb{R}^m$, compute their Jacobians at a general point $x \in \mathbb{R}^n$* ***and*** *compute their linearization about the given point.*

(a) $f(x_1, x_2) := \begin{bmatrix} x_1 \cdot (x_2)^2 \\ e^{3x_1} + \cos(4x_2) \end{bmatrix}$ *and* $x_0 = \begin{bmatrix} \sqrt{2} \\ \pi \end{bmatrix}$.

(b) $f(x_1, x_2) := e^{3x_1} + \cos(4x_2)$ *and* $x_0 = \begin{bmatrix} 0.77 \\ 11.6 \end{bmatrix}$.

(c) $f(x_1, x_2) := \begin{bmatrix} x_1 \cdot (x_2)^2 \\ x_2 \cdot \sin^3(x_1) + 11 \\ 4(x_1)^2 + (x_2 + 3)^2 \end{bmatrix}$ *and* $x_0 = \begin{bmatrix} \sqrt{7} \\ 1.0 \end{bmatrix}$.

(d) $f(x_1, x_2, x_3) := \begin{bmatrix} x_1 \cdot x_2 \cdot (x_3)^3 \\ x_2 \cdot e^{x_1 \cdot x_3} \\ x_1 \end{bmatrix}$ *and* $x_0 = \begin{bmatrix} 0 \\ 0 \\ 0 \end{bmatrix}$.

Solutions: These are the same functions from Example 5.38. We could assemble the various partial derivatives we have already computed by hand and verified with the `Symbolics` package. Instead, we will do everything in software and use the functions `Symbolics.jacobian` and `ForwardDiff.jacobian`. Would you accept anything else?

```
using Symbolics
# Define variables
@variables x1 x2 x3

fa = [x1 * (x2)^2; exp(3*x1) + cos(4*x2)]
fb = [exp(3*x1) + cos(4*x2)]
fc = [x1 * (x2)^2; x2*( sin(x1) )^3 + 11; 4*(x1)^2 + ( x2 +3 )^2]
fd = [x1*x2*(x3)^3; x2*exp(x1*x3); x1]

# Symbolic Jacobians

#a)
Ja = Symbolics.jacobian(fa, [x1, x2])
```

```
println(" (a)  ")
display(Ja)
println(Ja)
println("  ")

#b)
Jb = Symbolics.jacobian(fb, [x1, x2]) # The Jacobian function x to be
                                     # a vector; hence, the brackets
println(" (b)  ")
display(Jb)
println(Jb)
println("  ")

#c)
Jc = Symbolics.jacobian(fc, [x1, x2])
println(" (c)  ")
display(Jc)
println(Jc)
println("  ")

#d)
Jd = Symbolics.jacobian(fd, [x1, x2, x3])
println(" (d)  ")
display(Jd)
println(Jd)
println("  ")
```

Output

```
(a)
```

$$\begin{bmatrix} x2^{2.0000} & 2.0000x1x2 \\ 3.0000e^{3.0000x1} & -4.0000\sin(4.0000x2) \end{bmatrix}$$

```
Num[x2^2 2x1*x2; 3exp(3x1) -4sin(4x2)]
```

```
(b)
```

$$\begin{bmatrix} 3.0000e^{3.0000x1} & -4.0000\sin(4.0000x2) \end{bmatrix}$$

```
Num[3exp(3x1) -4sin(4x2)]
```

```
(c)
```

$$\begin{bmatrix} x2^{2.0000} & 2.0000x1x2 \\ 3.0000\sin^{2.0000}(x1)x2\cos(x1) & \sin^{3.0000}(x1) \\ 8.0000x1 & 2.0000\,(3.0000 + x2) \end{bmatrix}$$

```
Num[x2^2 2x1*x2; 3x2*(sin(x1)^2)*cos(x1) sin(x1)^3; 8x1 2(3 + x2)]
```

```
(d)
```

$$\begin{bmatrix} x3^{3.0000}x2 & x3^{3.0000}x1 & 3.0000x3^{2.0000}x1x2 \\ x2x3e^{x1x3} & e^{x1x3} & x1x2e^{x1x3} \\ 1.0000 & 0.0000 & 0.0000 \end{bmatrix}$$

```
Num[x2*(x3^3) x1*(x3^3) 3x1*x2*(x3^2); x2*x3*exp(x1*x3) exp(x1*x3) x1*x2*exp(x1*x3); 1 0 0]
```

Next, we evaluate the Jacobians at a point.

```
# Example of converting from Symbolic to Numerical Functions
using Symbolics, ForwardDiff
```

```
#a)
x0 = [sqrt(2); pi]
# Build a numerical function from the symbolic expression
fa_num = build_function(fa, [x1, x2])
fa_num = eval(fa_num[1])
# Compute the Jacobian of the new function at the test point
Jaca = ForwardDiff.jacobian(fa_num, x0)
println("(a) ")
display(Jaca)
println(" ")

#b)
x0 = [0.77; 11.6]
fb_num = build_function(fb, [x1, x2])
fb_num = eval(fb_num[1])
# Compute the Jacobian of the new function at the test point
Jacb = ForwardDiff.jacobian(fb_num, x0)
println("(b) ")
display(Jacb)
println(" ")

#c)
x0 = [sqrt(7); 1.0]
fc_num = build_function(fc, [x1, x2])
fc_num = eval(fc_num[1])
# Compute the Jacobian of the new function at the test point
Jacc = ForwardDiff.jacobian(fc_num, x0)
println("(c) ")
display(Jacc)
println(" ")

#d)
x0 = [0.0; 0; 0]
# Build a numerical function from the symbolic expression
fd_num = build_function(fd, [x1, x2, x3])
fd_num = eval(fd_num[1])
# Compute the Jacobian of the new function at the test point
Jacd = ForwardDiff.jacobian(fd_num, x0)
println("(d) ")
display(Jacd)
println(" ")
```

Output

```
(a)
2×2 Matrix{Float64}:
   9.8696  8.88577
 208.774   1.95943e-15

(b)
1×2 Matrix{Float64}:
 30.2233  -2.64922

(c)
3×2 Matrix{Float64}:
  1.0       5.2915
```

```
 -0.597294  0.107695
 21.166     8.0
```

```
(d)
3×3 Matrix{Float64}:
 0.0  0.0  0.0
 0.0  1.0  0.0
 1.0  0.0  0.0
```

In each case, the first-order linear approximation is $f(x_0) + \frac{\partial f(x_0)}{\partial x} \cdot (x - x_0)$. All of the Jacobians have been evaluated.

Converting Symbolic Functions to Numerical Functions in Julia

The various Julia packages for computing derivatives do not yet play so nicely with one another. They will someday. The following method was found by applying numerous prompts to ChatGPT4. After many false starts, it is actually pretty clean and is what we used above. In the following, the code is explained.

```julia
# Example of converting from Symbolic to Numerical Functions
using Symbolics, ForwardDiff

@variables x1 x2 x3
fd = [x1*x2*(x3)^3; x2*exp(x1*x3); x1]

# Build a numerical function from the symbolic expression
fd_num = build_function(fd, [x1, x2, x3])
fd_num = eval(fd_num[1])

# Test the new function
x0 = [1.0, 2.0, 3.0]
result = fd_num(x0)
println("Result of fd_num(x0): ", result)

# Compute the Jacobian of the new function at the test point
Jac = ForwardDiff.jacobian(fd_num, x0)
println("Jacobian: ", Jac)
```

The above code snippet demonstrates how to convert a symbolic function to a numerical function in Julia using the Symbolics and ForwardDiff packages. The symbolic function is defined with variables x_1, x_2, x_3 and consists of three components. The numerical function is then tested at a point $x_0 = [1.0, 2.0, 3.0]$, and its Jacobian is computed at this point.

Importing Packages

```julia
using Symbolics, ForwardDiff
```

The Symbolics package is used for symbolic computations, and ForwardDiff is used for automatic differentiation.

Defining Symbolic Variables and Function

```julia
@variables x1 x2 x3
fd = [x1*x2*(x3)^3; x2*exp(x1*x3); x1]
```

Symbolic variables x_1, x_2, x_3 are defined, and a symbolic function fd is created as a vector of three components.

Building the Numerical Function

```
fd_num = build_function(fd, [x1, x2, x3])
fd_num = eval(fd_num[1])
```

The `build_function` method from the Symbolics package is used to convert the symbolic function into a numerical function. The `eval` function is then used to make it callable.

Testing the Numerical Function

```
x0 = [1.0, 2.0, 3.0]
result = fd_num(x0)
println("Result of fd_num(x0): ", result)
```

The numerical function is tested at a point $x_0 = [1.0, 2.0, 3.0]$.

Computing the Jacobian

```
Jac = ForwardDiff.jacobian(fd_num, x0)
println("Jacobian: ", Jac)
```

The Jacobian of the numerical function at the point x_0 is computed using ForwardDiff's `jacobian` function.

Output

```
Result of fd_num(x0): [54.0, 40.171073846375336, 1.0]
Jacobian: [54.0 27.0 54.0; 120.51322153912601 20.085536923187668
           40.171073846375336; 1.0 0.0 0.0]
```

You may prefer the following alternative method. **It is simpler but has some manual intervention.**

```
# Exercise in textbook
using Symbolics
# define variables
@variables x[1:3]

fa = [x[1] * (x[2])^2; exp(3*x[1]) + cos(4*x[2])]
fb = [exp(3*x[1]) + cos(4*x[2])]
fc = [x[1] * (x[2])^2; x[2]*( sin(x[1]) )^3 + 11; 4*(x[1])^2 +
        ( x[2] +3 )^2]
fd = [x[1]*x[2]*(x[3])^3; x[2]*exp(x[1]*x[3]); x[1]]

# Compute Symbolic Jacobians
#a)
Ja = Symbolics.jacobian(fa, x[1:2])
println(" (a) ")
display(Ja)
println(Ja)
println(" ")

#b)

Jb = Symbolics.jacobian(fb, x[1:2])
println(" (b) ")
display(Jb)
println(Jb)
println(" ")

#c)
```

```
Jc = Symbolics.jacobian(fc, x[1:2])
println(" (c) ")
display(Jc)
println(Jc)
println(" ")

#d)
Jd = Symbolics.jacobian(fd, x)
println(" (d) ")
display(Jd)
println(Jd)
println(" ")
```

Output

(a)

$$\begin{bmatrix} x_2^{2.0000} & 2.0000x_1x_2 \\ 3.0000e^{3.0000x_1} & -4.0000\sin(4.0000x_2) \end{bmatrix}$$

```
Num[x[2]^2 2x[1]*x[2]; 3exp(3x[1]) -4sin(4x[2])]
```

(b)

$$\begin{bmatrix} 3.0000e^{3.0000x_1} & -4.0000\sin(4.0000x_2) \end{bmatrix}$$

```
Num[3exp(3x[1]) -4sin(4x[2])]
```

(c)

$$\begin{bmatrix} x_2^{2.0000} & 2.0000x_1x_2 \\ 3.0000\sin^{2.0000}(x_1)\cos(x_1)x_2 & \sin^{3.0000}(x_1) \\ 8.0000x_1 & 2.0000(3.0000 + x_2) \end{bmatrix}$$

```
Num[x[2]^2 2x[1]*x[2]; 3(sin(x[1])^2)*cos(x[1])*x[2] sin(x[1])^3; 8x[1] 2(3 + x[2])]
```

(d)

$$\begin{bmatrix} x_3^{3.0000}x_2 & x_3^{3.0000}x_1 & 3.0000x_3^{2.0000}x_1x_2 \\ e^{x_1x_3}\,x_2x_3 & e^{x_1x_3} & e^{x_1x_3}\,x_1x_2 \\ 1.0000 & 0.0000 & 0.0000 \end{bmatrix}$$

```
Num[(x[3]^3)*x[2] (x[3]^3)*x[1] 3(x[3]^2)*x[1]*x[2]; exp(x[1]*x[3])*x[2]*x[3] exp(x[1]*x[3]) exp(x[1]*x[3])*x[1]*x[2]; 1 0
0]
```

```
using ForwardDiff

# Copy and paste to create regular Julia functions. The names are
# slightly altered so as not to have a conflict with the Symbolic
# functions fa, fb, fc, fd

fA(x) = [x[1] * (x[2])^2; exp(3*x[1]) + cos(4*x[2])]
fB(x) = [exp(3*x[1]) + cos(4*x[2])]
fC(x) = [x[1] * (x[2])^2; x[2]*( sin(x[1]) )^3 + 11; 4*(x[1])^2 +
            ( x[2] +3 )^2]
fD(x) = [x[1]*x[2]*(x[3])^3; x[2]*exp(x[1]*x[3]); x[1]];

#a)
x0 = [sqrt(2); pi]
Ja = ForwardDiff.jacobian(fA, x0)
println(" (a) ")
display(Ja)
println(" ")

#b)
x0 = [0.77; 11.6]
```

```
Jb = ForwardDiff.jacobian(fB, x0)
println("(b) ")
display(Jb)
println(" ")

#c)
x0 = [sqrt(7); 1.0]
Jc = ForwardDiff.jacobian(fC, x0)
println("(c) ")
display(Jc)
println(" ")

#d)
x0 = [0.0; 0; 0]
Jd = ForwardDiff.jacobian(fD, x0)
println("(d) ")
display(Jd)
println(" ")
```

Output

```
(a)
2×2 Matrix{Float64}:
   9.8696  8.88577
 208.774   1.95943e-15

(b)
1×2 Matrix{Float64}:
 30.2233  -2.64922

(c)
3×2 Matrix{Float64}:
  1.0       5.2915
 -0.597294  0.107695
 21.166     8.0

(d)
3×3 Matrix{Float64}:
 0.0  0.0  0.0
 0.0  1.0  0.0
 1.0  0.0  0.0
```

One more example, just for fun.

```
@variables t x y z a b

# Define a gnarly function with variables t, x, y, z, a, b and
# 3 components
f_gnarly = [
    sin(a*x^3) + cos(b*exp(y)) + t*z^2 + exp(a - b),
    a*x^2*y - b*t*z + log(x + y + z + 1),
    a*b*t*x*y*z + sin(t) + cos(a) + (atan(exp(b))+11)^z]

# Compute the Jacobian symbolically
Jacobian_gnarly = Symbolics.jacobian(f_gnarly, [t, x, y, z, a, b])

@show f_gnarly
```

Output To save space, we replace $f_{\text{gnarly}}(t,x,y,z,a,b)$ with $f(...)$, and note that

$$\text{Jac}f(...) = \left[\begin{array}{cccccc} \frac{\partial f(...)}{\partial t} & \frac{\partial f(...)}{\partial x} & \frac{\partial f(...)}{\partial y} & \frac{\partial f(...)}{\partial z} & \frac{\partial f(...)}{\partial a} & \frac{\partial f(...)}{\partial b} \end{array}\right],$$

where

$$\frac{\partial f(...)}{\partial t} = \begin{bmatrix} z^2 \\ -bz \\ abxyz + \cos(t) \end{bmatrix}, \frac{\partial f(...)}{\partial x} = \begin{bmatrix} 3x^2 a\cos(x^3 a) \\ \frac{1}{1+x+y+z} + 2axy \\ abtyz \end{bmatrix}, \frac{\partial f(...)}{\partial y} = \begin{bmatrix} -be^y \sin(be^y) \\ x^2 a + \frac{1}{1+x+y+z} \\ abtxz \end{bmatrix}$$

$$\frac{\partial f(...)}{\partial z} = \begin{bmatrix} 2tz \\ \frac{1}{1+x+y+z} - bt \\ (11+\arctan(e^b))^z \log(11+\arctan(e^b)) + abtxy \end{bmatrix}$$

$$\frac{\partial f(...)}{\partial a} = \begin{bmatrix} x^3\cos(x^3 a) + e^{a-b} \\ x^2 y \\ -\sin(a) + btxyz \end{bmatrix}, \frac{\partial f(...)}{\partial b} = \begin{bmatrix} -e^{a-b} - e^y \sin(be^y) \\ -tz \\ \frac{(11+\arctan(e^b))^{-1+z} z e^b}{1+(e^b)^2} + atxyz \end{bmatrix}.$$

That would be hard to do by hand. ■

5.7.2 The Gradient

The **gradient** is the special name given to the **transpose of the Jacobian** of a function $f:\mathbb{R}^n \to \mathbb{R}$, (i.e., for each $x \in \mathbb{R}^n$, $f(x) \in \mathbb{R}$ is a scalar). In Section 6.2.2, where we deal with optimization, we will show that the gradient of a function $f:\mathbb{R}^n \to \mathbb{R}$ **points in the direction of maximum increase of the function**, and hence the negative of the gradient points in the **direction of maximum decrease**. All of the marvelous AI tools at our disposal today rely on the gradient for *training* the neural networks on which the tools are based. **That makes the gradient one of the most important objects in all of calculus** ... and yet, it is just a collection of partial derivatives stacked into a column vector. Who could have imagined that?

The Gradient

Definition 5.39. *The* ***gradient*** *of $f:\mathbb{R}^n \to \mathbb{R}$ is the partial derivatives of f with respect to x_i arranged to form a* ***column vector****, instead of a row vector,*

$$\nabla f(x_0) := \begin{bmatrix} \frac{\partial f(x_0)}{\partial x_1} \\ \frac{\partial f(x_0)}{\partial x_2} \\ \vdots \\ \frac{\partial f(x_0)}{\partial x_n} \end{bmatrix}. \tag{5.35}$$

*The cool symbol ∇ is usually pronounced as "**grad**" and one says "**grad f**" when speaking of ∇f.*

Remark 5.40. *The Jacobian of $f:\mathbb{R}^n \to \mathbb{R}$ is the row vector, $\frac{\partial f(x)}{\partial x} := \left[\begin{array}{cccc} \frac{\partial f(x)}{\partial x_1} & \frac{\partial f(x)}{\partial x_2} & \cdots & \frac{\partial f(x)}{\partial x_n} \end{array}\right]$. It is a $1 \times n$ row vector because each entry, $\frac{\partial f(x)}{\partial x_i}$, is 1×1 due to f having codomain $\mathbb{R}$, and it has n columns because $x \in \mathbb{R}^n$.*

The gradient is the transpose of the Jacobian. Hence the linearization of $f:\mathbb{R}^n \to \mathbb{R}$ can be written *as*

$$f(x) \approx f(x_0) + (\nabla f(x_0))^\top \cdot (x - x_0)$$
$$\approx f(x_0) + \nabla f(x_0) \bullet (x - x_0), \tag{5.36}$$

where $\nabla f(x_0) \bullet (x - x_0)$ is the ***dot product*** *of the $n \times 1$ gradient vector with the $n \times 1$ vector $(x - x_0)$.*

Example 5.41. *Compute the gradient of $f : \mathbb{R}^2 \to \mathbb{R}$, for $f(x_1, x_2) = x_1 \cos(x_2)$ and $x_0 = [2 \;\; \pi/4]^\top$.*

Solution: We compute both $\nabla f(x)$ and $\nabla f(x_0)$ in software.

```
using Symbolics

@variables x[1:2]
f = x[1] * cos(x[2])
grad_f = Symbolics.gradient(f, [x[1], x[2]])

@show f
@show grad_f

x0 = [2; pi/4]
f_num = build_function(f , x)
f_num = eval(f_num)
# Compute the Jacobian of the new function at the test point
Grad_f_num= ForwardDiff.gradient(f_num, x0)
println(" ")
display(Grad_f_num)
println(" ")
```

Output

$$\nabla\left(x_1 \cos(x_2)\right) = \begin{bmatrix} \frac{\partial}{\partial x_1}\left(x_1 \cos(x_2)\right) \\ \frac{\partial}{\partial x_2}\left(x_1 \cos(x_2)\right) \end{bmatrix} = \begin{bmatrix} \cos(x_2) \\ -x_1 \sin(x_2) \end{bmatrix},$$

$$\nabla\left(x_1 \cos(x_2)\right)\big|_{x_0} = \begin{bmatrix} 0.70710678 \\ -1.41421356 \end{bmatrix}.$$

■

5.7.3 A Handy Product Rule for the Gradient and the Jacobian

On the one hand, because partial derivatives are ordinary derivatives taken one variable at a time while holding the others constant, all of the rules for differentiation presented in Section 5.4 apply to individual partial derivatives, $\frac{\partial}{\partial x_i}$. On the other hand, applying these rules in a matrix-vector form typically requires tensor notation, which we do not want to touch in this textbook. There is, however, one useful case where things work out nicely, and we treat that here.

Proposition 5.42: Gradient and Jacobian Meet the Dot Product

Suppose $f : \mathbb{R}^n \to \mathbb{R}^m$ and $g : \mathbb{R}^n \to \mathbb{R}^m$ are vector-valued functions and define $h : \mathbb{R}^n \to \mathbb{R}$ by

$$h(x) := (f(x))^\top \cdot g(x) = f(x) \bullet g(x), \tag{5.37}$$

the dot product of f and g. If f and g are both differentiable, then so is h and

$$J_h(x) := \begin{bmatrix} \frac{\partial h(x)}{\partial x_1} & \frac{\partial h(x)}{\partial x_2} & \cdots & \frac{\partial h(x)}{\partial x_n} \end{bmatrix} = [g(x)]^\top \cdot \frac{\partial f(x)}{\partial x} + [f(x)]^\top \cdot \frac{\partial g(x)}{\partial x} \tag{5.38}$$

and because the gradient is the transpose of the Jacobian, and the transpose of a product equals the product of the transposes in reverse order, we have

$$\nabla h(x) := \begin{bmatrix} \frac{\partial h(x)}{\partial x_1} \\ \frac{\partial h(x)}{\partial x_2} \\ \vdots \\ \frac{\partial h(x)}{\partial x_n} \end{bmatrix} = \left[\frac{\partial g(x)}{\partial x}\right]^\top \cdot f(x) + \left[\frac{\partial f(x)}{\partial x}\right]^\top \cdot g(x). \tag{5.39}$$

Proof: The function $h(x)$ is equal to

$$h(x) = \sum_{j=1}^{m} f_j(x) g_j(x).$$

Taking the partial derivative of h with respect to a specific component x_i, we can apply the product rule for single-variable derivatives to obtain:

$$\frac{\partial h(x)}{\partial x_i} = \sum_{j=1}^{m} \left(\frac{\partial f_j(x)}{\partial x_i} g_j(x) + f_j(x) \frac{\partial g_j(x)}{\partial x_i} \right)$$

(next, rewrite each sum as a row vector times a column vector)

$$= \begin{bmatrix} g_1(x) & g_2(x) & \cdots & g_m(x) \end{bmatrix} \cdot \begin{bmatrix} \frac{\partial f_1(x)}{\partial x_i} \\ \frac{\partial f_2(x)}{\partial x_i} \\ \vdots \\ \frac{\partial f_m(x)}{\partial x_i} \end{bmatrix} + \begin{bmatrix} f_1(x) & f_2(x) & \cdots & f_m(x) \end{bmatrix} \cdot \begin{bmatrix} \frac{\partial g_1(x)}{\partial x_i} \\ \frac{\partial g_2(x)}{\partial x_i} \\ \vdots \\ \frac{\partial g_m(x)}{\partial x_i} \end{bmatrix}$$

$$= [g(x)]^\top \cdot \left[\frac{\partial f(x)}{\partial x}\right]_i + [f(x)]^\top \cdot \left[\frac{\partial g(x)}{\partial x}\right]_i,$$

where $\left[\frac{\partial f(x)}{\partial x}\right]_i$ and $\left[\frac{\partial g(x)}{\partial x}\right]_i$ denote the ith columns of the Jacobian matrices. Hence, putting it all together,

$$J_h(x) := \begin{bmatrix} \frac{\partial h(x)}{\partial x_1} & \frac{\partial h(x)}{\partial x_2} & \cdots & \frac{\partial h(x)}{\partial x_n} \end{bmatrix} = [g(x)]^\top \cdot \frac{\partial f(x)}{\partial x} + [f(x)]^\top \cdot \frac{\partial g(x)}{\partial x}.$$

■

5.7.4 (Optional Read) The Hessian

Assume that the gradient of the scalar-valued function $f : \mathbb{R}^n \to \mathbb{R}$ exists. Then $\nabla f : \mathbb{R}^n \to \mathbb{R}^n$ is an $(n \times 1)$ vector-valued function. If its partial derivatives with respect to $x = (x_1, x_2, \ldots, x_n)$ exist, then we can compute its Jacobian, which will be an $n \times n$ matrix. The name Hessian for this object is in honor of its inventor, Ludwig Otto Hesse.

Hessian of a Function

Definition 5.43. *The **Hessian** of $f : \mathbb{R}^n \to \mathbb{R}$ is*

$$H_f(x) := \frac{\partial \nabla f(x)}{\partial x} := \frac{\partial}{\partial x}(\nabla f(x)) = \begin{bmatrix} \frac{\partial \nabla f(x)}{\partial x_1} & \frac{\partial \nabla f(x)}{\partial x_2} & \cdots & \frac{\partial \nabla f(x)}{\partial x_n} \end{bmatrix}, \tag{5.40}$$

assuming the indicated partial derivatives exist. The Hessian of a function is formed by packaging the column vectors $\frac{\partial \nabla f(x)}{\partial x_j}$ into a matrix. We need to keep in mind that, for each value of $x \in \mathbb{R}^n$, the *compact and awesome-looking object*

$$H_f(x) := \frac{\partial \nabla f(x)}{\partial x}$$

is really an $n \times n$ matrix: because there are n columns of the form $\frac{\partial \nabla f(x)}{\partial x_j}$, and each column is an n-vector. ***There are software tools that compute the Hessian in one fell swoop. You do not need to first compute the gradient and then the Jacobian, as in the definition.***

Just for the record, we write out $H_f(x) := \frac{\partial \nabla f(x)}{\partial x}$ as an $n \times n$ matrix. Suppose $f(x) = f(x_1, x_2, \ldots, x_n)$. Then,

$$\nabla f(x) = \begin{bmatrix} \frac{\partial f(x)}{\partial x_1} \\ \frac{\partial f(x)}{\partial x_2} \\ \vdots \\ \frac{\partial f(x)}{\partial x_n} \end{bmatrix}.$$

Computing the Jacobian of the gradient and writing out all of the entries in their $n \times n$ matrix glory gives

$$H_f(x) = \frac{\partial \nabla f(x)}{\partial x} := \begin{bmatrix} \frac{\partial \nabla f(x)}{\partial x_1} & \frac{\partial \nabla f(x)}{\partial x_2} & \cdots & \frac{\partial \nabla f(x)}{\partial x_n} \end{bmatrix} = \begin{bmatrix} \frac{\partial^2 f(x)}{\partial x_1 \partial x_1} & \frac{\partial^2 f(x)}{\partial x_2 \partial x_1} & \cdots & \frac{\partial^2 f(x)}{\partial x_n \partial x_1} \\ \frac{\partial^2 f(x)}{\partial x_1 \partial x_2} & \frac{\partial^2 f(x)}{\partial x_2 \partial x_2} & \cdots & \frac{\partial^2 f(x)}{\partial x_n \partial x_2} \\ \vdots & \vdots & \ddots & \vdots \\ \frac{\partial^2 f(x)}{\partial x_1 \partial x_n} & \frac{\partial^2 f(x)}{\partial x_2 \partial x_n} & \cdots & \frac{\partial^2 f(x)}{\partial x_n \partial x_n} \end{bmatrix}. \tag{5.41}$$

In other symbols, the ij component of $H_f(x) := \frac{\partial \nabla f(x)}{\partial x}$ is

$$\left[\frac{\partial \nabla f(x)}{\partial x}\right]_{ij} = \frac{\partial^2 f(x)}{\partial x_j \partial x_i} \left(= \frac{\partial^2 f(x)}{\partial x_i \partial x_j} \text{ when the partial derivatives are continuous} \right). \tag{5.42}$$

5.7.5 (Optional Read) More Matrix-Vector Calculus

We begin by highlighting the very useful site: Matrix Calculus, an online tool that computes vector and matrix derivatives (matrix calculus), so that we do not have to memorize all the formulas. The use of this site brings up an issue we mentioned previously: there are different conventions for whether $\frac{\partial}{\partial x}$ applied to a scalar-valued function $f : \mathbb{R}^m \to \mathbb{R}$ is interpreted as a row vector, which would be a Jacobian, or a column vector, which would be a gradient. We discuss the "why" of this below, in case you are interested. Let us note, though, that in spoken and written language, you deal with contradictory meanings and conventions all the time and accept that as part of being fluent. It's no different in math. **Context usually settles any possible confusion**.

With these comments out of the way, we will list a few key formulas and derive them so that you have confidence when using matrix-vector calculus.

Proposition 5.44: Key Matrix-Vector Calculus Formulas

Most useful formulas for engineers:

(a) For C an $m \times n$ matrix and $x \in \mathbb{R}^n$, $\frac{\partial}{\partial x}(Cx) = C$.

(b) For $x \in \mathbb{R}^n$, $\frac{\partial}{\partial x}(x^\top x) = 2x^\top$.

(c) For $x \in \mathbb{R}^n$, $\nabla(x^\top x) = 2x$.

(d) For Q an $n \times n$ symmetric matrix and $x \in \mathbb{R}^n$, $\frac{\partial}{\partial x}\left(x^\top Q x\right) = 2x^\top \cdot Q$.

(e) For Q an $n \times n$ symmetric matrix and $x \in \mathbb{R}^n$, $\nabla\left(x^\top Q x\right) = 2Qx$.

(f) For Q an $n \times n$ symmetric matrix and $x \in \mathbb{R}^m$, $\frac{\partial \nabla(x^\top Q x)}{\partial x} = 2Q$ (yes, this is the Hessian of $x^\top Qx$). You will also see it written as $\frac{\partial}{\partial x}\left(\frac{\partial(x^\top Q x)}{\partial x}\right) = 2Q$, especially when using the `Symbolics` package or `SymPy`.

Note: You will often see a $\frac{1}{2}$ inserted into $x^\top x$ and $x^\top Qx$ so that $\nabla\left(\frac{1}{2}x^\top x\right) = x$ and $\nabla\left(\frac{1}{2}x^\top Qx\right) = Qx$. There has to be a two or a one-half somewhere! It's your call.

Row vs. Column? How to Know: The Jacobian matrix provides a linear approximation of a multivariable function. In this context, if $f : \mathbb{R}^n \to \mathbb{R}^m$ and $m \geq 2$, then the Jacobian of f has to be an $m \times n$ matrix. However, when $m = 1$, all bets are off! **Because the gradient is just the transpose of the Jacobian, in many contexts, it seems redundant to use both operations**. The choice to represent either the gradient or Jacobian of $f : \mathbb{R}^n \to \mathbb{R}$ as a row vector or a column vector, therefore, depends on conventions, contexts, and the operations you're planning to perform with it. Here's how it usually plays out:

(i) When $f : \mathbb{R}^n \to \mathbb{R}^m$, $m > 1$, then the Jacobian $\frac{\partial f(x)}{\partial x}$ is an $m \times n$ matrix.

(ii) When $f : \mathbb{R}^n \to \mathbb{R}$, (i.e., $m = 1$), then it's 25–75% whether the Jacobian $\frac{\partial f(x)}{\partial x}$ is a $1 \times n$ row vector (25%) or $n \times 1$ column vector (75%), even in this textbook. Whenever possible, we will write the Jacobian as a row vector and use the gradient when we want a column vector. However, we cannot fly in the face of tradition and rewrite famous equations in a form that would be unrecognizable to most users. When treating Lagrange Multipliers in Section 6.3, it is very natural to use the gradient for a column of partial derivatives, and we will. When treating Lagrangian Dynamics in Section 6.4, indicating a column of partial derivatives with a gradient instead of a Jacobian would make the equations unrecognizable.

(iii) Hence, for us, the Jacobian of $x^\top x$ is a row vector. The extremely useful site, Matrix Calculus, the online tool that computes vector and matrix derivatives so that you do not have to, makes the Jacobian of $x^\top x$ a column vector. Outside of this book, when writing out the Jacobian or gradient, always be clear about which convention you're using, especially if your colleagues might be accustomed to a different convention.

(iv) When the gradient of $f : \mathbb{R}^n \to \mathbb{R}$ is represented as a column vector, it is the transpose of the Jacobian matrix of f,

$$\nabla f(x) = \begin{bmatrix} \frac{\partial f(x)}{\partial x_1} \\ \\ \frac{\partial f(x)}{\partial x_2} \\ \vdots \\ \frac{\partial f(x)}{\partial x_n} \end{bmatrix} = \left[J_f(x)\right]^\top.$$

This is the more common representation, especially in optimization contexts.

(v) The gradient of $f : \mathbb{R}^n \to \mathbb{R}$ can also be represented as a row vector, in which case it's exactly the Jacobian matrix for f,

$$\nabla f(x) = \begin{bmatrix} \frac{\partial f(x)}{\partial x_1} & \frac{\partial f(x)}{\partial x_2} & \cdots & \frac{\partial f(x)}{\partial x_n} \end{bmatrix} = J_f(x).$$

In conclusion, while mathematical conventions can guide the choice of representation, the ultimate decision often depends on the specific context and operations in which the gradient or Jacobian is used. Always prioritize clarity and correctness in your work. **Always ensure that the dimensionality matches the operations you intend to perform. That's what we mean by context!**

Proofs of Prop. 5.44

In the following, $x = \begin{bmatrix} x_1 \\ x_2 \\ \vdots \\ x_n \end{bmatrix}$.

(a) For C an $m \times n$ matrix and $x \in \mathbb{R}^n$, $\frac{\partial}{\partial x}(Cx) = C$. Express $C = \begin{bmatrix} C_1 & C_2 & \cdots & C_n \end{bmatrix}$ in terms of its columns. Then $Cx = x_1C_1 + x_2C_2 + \cdots + x_nC_n$. Hence, $\frac{\partial(Cx)}{\partial x_i} = C_i$, the i-th column of C. Therefore,

$$\frac{\partial}{\partial x}(Cx) = \begin{bmatrix} \frac{\partial Cx}{\partial x_1} & \frac{\partial Cx}{\partial x_2} & \cdots & \frac{\partial Cx}{\partial x_n} \end{bmatrix} = C.$$

Notice what we did: we broke it down to first computing $\frac{\partial}{\partial x_i}$ and then stacked things into rows or columns as required. Keep this in mind when making vector computations with derivatives.

(b) For $x \in \mathbb{R}^n$, $\frac{\partial}{\partial x}(x^\top x) = 2x^\top$.

$x^\top x = x_1^2 + x_2^2 + \cdots + x_n^2$. Hence, $\frac{\partial}{\partial x_i}(x^\top x) = 2x_i$. Therefore,

$$\frac{\partial(x^\top x)}{\partial x} = \begin{bmatrix} \frac{\partial(x^\top x)}{\partial x_1} & \frac{\partial(x^\top x)}{\partial x_2} & \cdots & \frac{\partial(x^\top x)}{\partial x_n} \end{bmatrix} = \begin{bmatrix} 2x_1 & 2x_2 & \cdots & 2x_n \end{bmatrix} = 2x^\top.$$

Alternative proof based on the Product Rule in Prop. 5.42. We identify $f(x) = x$ and $g(x) = x$. Hence, $\frac{\partial f(x)}{\partial x} = \frac{\partial g(x)}{\partial x} = \frac{\partial x}{\partial x} = I_n$, where I_n is the $n \times n$ identity matrix. This gives the result immediately.

(c) For $x \in \mathbb{R}^n$, $\nabla(x^\top x) = 2x$.

In this textbook, the gradient of a scalar-valued function is always the transpose of the Jacobian. The result is immediate from the previous part or from Prop. 5.42.

(d) For Q an $n \times n$ symmetric matrix and $x \in \mathbb{R}^n$, $\frac{\partial}{\partial x}(x^\top Qx) = 2x^\top \cdot Q$.

This part follows from the next part by taking transposes and noting that Q symmetric implies that $Q^\top = Q$.

(e) For Q an $n \times n$ symmetric matrix and $x \in \mathbb{R}^n$, $\nabla(x^\top Qx) = 2Qx$.

We apply Prop. 5.42 by identifying $f(x) = x$ and $g(x) = Qx$. Hence, $\frac{\partial f(x)}{\partial x} = I_n$, and by part (a) of this problem, $\frac{\partial Qx}{\partial x} = Q$. Hence,

$$\nabla\left(x^\top Qx\right) = I_n^\top \cdot Qx + Q^\top \cdot x = I_n \cdot Qx + Q^{\cdot} x = 2Qx,$$

because both I_n and Q are symmetric.

(f) For Q an $n \times n$ symmetric matrix and $x \in \mathbb{R}^n$, $\frac{\partial \nabla(x^\top Qx)}{\partial x} = 2Q$ (yes, this is the Hessian of $x^\top Qx^\top$).

Apply part (a) to part (e).

5.8 The Total Derivative and the Chain Rule

Our journey into the realm of differentiation began with the study of functions of a single variable. It was a straightforward world where changes took place in a one-dimensional landscape. But as with most explorations, our innate curiosity led us to broaden our horizons. We ventured into the more intricate domain of functions of several variables, acquainting ourselves with the powerful tool of Jacobians.

Now, imagine combining the simplicity of our initial one-dimensional world with the complexity of the multi-dimensional one. Enter the **total derivative**. This handy tool allows us to efficiently compute the derivative of a function of a single scalar variable, t, that emerges from the composition of two functions: $f : \mathbb{R}^n \to \mathbb{R}^m$ and $g : \mathbb{R} \to \mathbb{R}^n$. In essence, the total derivative marries our initial knowledge with our further explorations, showcasing the true power of differentiation and meriting its place as the last section in this Chapter.

Here's the intuition: Suppose we are given a scalar-valued function of three variables, $f(x, y, z)$, where each of the variables in turn depends on time, denoted as $x(t)$, $y(t)$, and $z(t)$. Then the **total derivative of f with respect to t is**

$$\frac{df}{dt} = \frac{\partial f}{\partial x}\frac{dx}{dt} + \frac{\partial f}{\partial y}\frac{dy}{dt} + \frac{\partial f}{\partial z}\frac{dz}{dt}, \tag{5.43}$$

which we recognize as being (the rate of change of f with respect to x) **times** (the rate of change of x with respect to t), plus dot dot dot, **plus** (the rate of change of f with respect to z) **times** (the rate of change of z with respect to t). Hence, (5.43) is a vector version of the Chain Rule given in Prop. 5.11 Differentiation Rules, parts (d) and (e). Because (5.43) **adds up all of the individual rates of change**, it is given the name **total derivative**.

You may be thinking, "Couldn't we have simply plugged in $x(t)$, $y(t)$, and $z(t)$ into f and then differentiated with respect to t?" and you would be absolutely correct. It is equivalent, and in some instances, that would be the easier thing to do. But just as it is easier to compute

$$\frac{d}{dx}\left(\sin(x^2)\right) = \left(\cos(x^2)\right) \cdot (2x) = 2x \cdot \cos(x^2)$$

with the Chain Rule than it is to memorize a ginormous table of derivatives that includes the above function, it can be easier to compute the total derivative using a formula such as (5.43). That is all that is going on here.

This discussion is formalized in the following definition,which also treats the case of vector-valued multivariable functions.

Total Derivative of a Multivariable Function

Suppose that $f : \mathbb{R}^n \to \mathbb{R}^m$ is a differentiable function and $x : \mathbb{R} \to \mathbb{R}^n$ by $x(t) = g(t)$ is also differentiable.

Definition 5.45. *The **total derivative** of $f(g(t))$ is $\frac{d}{dt}f(g(t))$, which can be computed as*

$$\frac{d}{dt}f(g(t)) = \left.\frac{\partial f(x)}{\partial x}\right|_{x=g(t)} \cdot \frac{dg(t)}{dt}. \tag{5.44}$$

Notes:

- $\left.\frac{\partial f(x)}{\partial x}\right|_{x=g(t)}$ *is the Jacobian of $f(x)$ evaluated at $x = g(t)$.*
- $\frac{dg(t)}{dt}$ *is the ordinary derivative of $g(t)$ with respect to t. When g is a vector of functions of t, you simply differentiate each of its components.*
- *When $f(x)$ is scalar-valued, that is $f : \mathbb{R}^n \to \mathbb{R}$, then expanding out the terms in $\frac{\partial f(x)}{\partial x}$ yields a formula like (5.43).*
- *Indeed, to obtain a formula similar to (5.43) for vector-valued functions, we rewrite (5.44) as follows:*
  - *let $J_f(x) := \frac{\partial f(x)}{\partial x}$ denote the Jacobian of $f(x)$ and*
  - *let $g'(t) := \frac{dg(t)}{dt}$ be the ordinary derivative of g with respect to t;*
  - *then, the total derivative becomes*

$$\frac{d}{dt}f(g(t)) = J_f(g(t)) \cdot g'(t).$$

We will illustrate the total derivative on (i) two basic examples to show the mechanics of its computation and (ii) on an example that is much closer to its real use for us: an application to an equation with derivatives, aka a differential equation. In Section 6.4 we begin laying the groundwork for differential equations through the study of Lagrange's powerful reformulation of Newtonian mechanics (i.e., $F = ma$), allowing us to derive dynamical models of multi-link robots and other things that move and go bump in the night.

Example 5.46. *Mechanics of the Total Derivative: Suppose that* $f : \mathbb{R}^2 \to \mathbb{R}$ *by*

$$f(x_1, x_2) = x_1 \cdot \sin(x_2)$$

and

$$g(t) = \begin{bmatrix} g_1(t) \\ g_2(t) \end{bmatrix} = \begin{bmatrix} t^2 \\ t^3 \end{bmatrix}.$$

Compute $\frac{df(t)}{dt} := \frac{d}{dt} f(g(t))$.

Solution: $\frac{df(t)}{dt} = 2t \cdot \sin\left(t^3\right) + 3t^4 \cdot \cos\left(t^3\right)$.

Using the Chain Rule,

$$\frac{df(g(t))}{dt} = \left.\frac{\partial f(x)}{\partial x}\right|_{x=g(t)} \cdot \begin{bmatrix} \frac{dg_1(t)}{dt} \\ \frac{dg_2(t)}{dt} \end{bmatrix}.$$

We compute that

$$\frac{\partial f(x)}{\partial x} = \begin{bmatrix} \sin(x_2) & x_1 \cos(x_2) \end{bmatrix},$$

$$\left.\frac{\partial f(x)}{\partial x}\right|_{x=g(t)} = \begin{bmatrix} \sin(t^2) & t^2 \cdot \cos(t^3) \end{bmatrix},$$

and

$$\begin{bmatrix} \frac{dg_1(t)}{dt} \\ \frac{dg_2(t)}{dt} \end{bmatrix} = \begin{bmatrix} 2t \\ 3t^2 \end{bmatrix}.$$

Putting it all together, we obtain

$$\frac{df(g(t))}{dt} = \left.\frac{\partial f(x)}{\partial x}\right|_{x=g(t)} \cdot \begin{bmatrix} \frac{dg_1(t)}{dt} \\ \frac{dg_2(t)}{dt} \end{bmatrix} = \begin{bmatrix} \sin(t^2) & t^2 \cdot \cos(t^3) \end{bmatrix} \cdot \begin{bmatrix} 2t \\ 3t^2 \end{bmatrix} = 2t \sin\left(t^3\right) + 3t^4 \cos\left(t^3\right).$$

■

You will need to do examples like this by hand in various courses. In code, to check your work, you could do the following:

```
using Symbolics

# Create a custom function to make differentiation convenient
function deriv(f, x)
    return expand_derivatives.(Symbolics.Differential(x)(f))
end

@variables x1, x2, t

f = x1*sin(x2)

```

```
# Compute the Jacobian of f with respect to x1, x2
Jac = Symbolics.jacobian([f], [x1, x2])

# Define the functions x1(t), x2(t)
x1_def = t^2
x2_def = t^3


# Compute the derivatives of x1(t) and x2(t) with respect to t
derivatives = [deriv(x1_def, t), deriv(x2_def, t)]

# Compute the total derivative of f using the chain rule
# (matrix multiplication)
df_dt_chain = Jac * derivatives

# Substitute in for x1(t) and x2(t)
df_dt = substitute(df_dt_chain, Dict(x1 => x1_def, x2 => x2_def))

# The result of df_dt_chain = Jac * derivatives is a 1 x 1 column
# vector.
# In Julia, we need to extract the first (and only) element of this
# vector using df_dt_chain[1] to produce a scalar-valued function. Why?
# The simplify command in Symbolics expects a scalar and not a vector.
# You can also use broadcasting to make it work
#
df_dt = Symbolics.simplify(df_dt[1],expand=true) # extract scalar


println("\nThe total derivative of f with respect to t is:\n\n", df_dt)
```

Output

```
The total derivative of f with respect to t is:

2t*sin(t^3) + 3(t^4)*cos(t^3)
```

Example 5.47. *Mechanics of the Total Derivative: Suppose that $f : \mathbb{R}^3 \to \mathbb{R}$ by*

$$f(x,y,z) = x^3 + y^3 + z^3$$

and

$$\begin{bmatrix} x(t) \\ y(t) \\ z(t) \end{bmatrix} = \begin{bmatrix} e^t \\ e^t \sin(t) \\ e^t \cos(t) \end{bmatrix}.$$

Compute $\frac{df(t)}{dt} := \frac{d}{dt} f(x(t), y(t), z(t))$.

Solution: $\frac{df(t)}{dt} = 3e^{3t} - 3\cos^2(t) \cdot \sin(t) + 3\cos(t) \cdot \sin^2(t)$.

We solve this one using non-vector notation so that you can find a preference: Using the Chain Rule and leaving out the arguments on the functions, we have

$$\frac{df}{dt} = \frac{\partial f}{\partial x}\frac{dx}{dt} + \frac{\partial f}{\partial y}\frac{dy}{dt} + \frac{\partial f}{\partial z}\frac{dz}{dt}.$$

We compute that

$$\frac{\partial f}{\partial x} = 3x^2, \quad \frac{\partial f}{\partial y} = 3y^2, \quad \text{and} \quad \frac{\partial f}{\partial z} = 3z^2,$$

and

$$\frac{dx}{dt} = e^t, \quad \frac{dy}{dt} = e^t \sin(t) + e^t \cos(t), \quad \text{and} \quad \frac{dz}{dt} = e^t \cos(t) - e^t \sin(t).$$

Combining these values and doing the required substitutions, we obtain

$$\begin{aligned}
\frac{df}{dt} &= \underbrace{3\left(e^t\right)^2}_{\frac{\partial f}{\partial x}} \cdot \underbrace{\left(e^t\right)}_{\frac{dx}{dt}} + \underbrace{3\left(e^t \sin(t)\right)^2}_{\frac{\partial f}{\partial y}} \cdot \underbrace{\left(e^t \sin(t) + e^t \cos(t)\right)}_{\frac{dy}{dt}} + \underbrace{3\left(e^t \cos(t)\right)^2}_{\frac{\partial f}{\partial z}} \cdot \underbrace{\left(e^t \cos(t) - e^t \sin(t)\right)}_{\frac{dz}{dt}} \\
&= \textbf{(lol, where's Julia?)} \\
&= 3e^{3t} - 3\cos^2(t) \cdot \sin(t) + 3\cos(t) \cdot \sin^2(t).
\end{aligned}$$

Some Perspective: The above illustrates the "grind of calculus". Everyone dislikes it. You will have to do some messy calculations in your science and engineering courses. Once again, this is just part of the learning process, akin to running wind sprints in sports—they're no fun, but they make you better in the game! As a practicing STEM professional, you will use "productivity tools", such as Julia or MATLAB. As a student, you can always check your HW with Julia!

```julia
using Symbolics

# Create a custom function to make differentiation convenient
function deriv(f, x)
    return expand_derivatives.(Symbolics.Differential(x)(f))
end

@variables x, y, z, t

f = [x^3 + y^3 + z^3]

# Compute the Jacobian of f with respect to x, y, z
Jac = Symbolics.jacobian(f, [x, y, z])

# Define the functions x(t), y(t), and z(t)
x_def = exp(t)
y_def = sin(t)
z_def = cos(t)

# Compute the derivatives of x(t), y(t), and z(t) with respect to t
derivatives = [deriv(x_def, t), deriv(y_def, t), deriv(z_def, t)]

# Compute the total derivative of f using the chain rule
# (matrix multiplication)
df_dt_chain = Jac * derivatives

# Substitute in for x(t), y(t), and z(t)
df_dt = substitute(df_dt_chain, Dict(x => x_def, y => y_def, z => z_def))

# The result of df_dt_chain = Jac * derivatives is a 1 x 1 column
# vector.
# In Julia, we need to extract the first (and only) element of this
# vector using df_dt_chain[1] to produce a scalar-valued function. Why?
# The simplify command in Symbolics expects a scalar and not a vector.
# You can also use broadcasting to make it work
```

```
#
df_dt = Symbolics.simplify.(df_dt,expand=true) # broadcasting, note the dot

println("\nThe total derivative of f with respect to t using the chain rule is:\n\n",
    df_dt)
```

Output

```
The total derivative of f with respect to t using the chain rule is:

3exp(3t) - 3(cos(t)^2)*sin(t) + 3cos(t)*(sin(t)^2)
```

Setting the stage for the next example: The following is called a linear system of differential equations,

$$\begin{bmatrix} \frac{dx_1(t)}{dt} \\ \frac{dx_2(t)}{dt} \end{bmatrix} = \begin{bmatrix} 0 & 1 \\ -1 & -1 \end{bmatrix} \begin{bmatrix} x_1(t) \\ x_2(t) \end{bmatrix} = \begin{bmatrix} x_2(t) \\ -x_1(t) - x_2(t) \end{bmatrix}. \tag{5.45}$$

Physically, this system of equations, with derivatives on the left side and unknown (aka to be found) functions of time on the right side, represents a linear approximation of a pendulum with friction at the pivot: $x_1(t)$ is the angle of the pendulum and $x_2(t)$ is the angular velocity of the pendulum.

We do not yet know how to find the unknown functions, $x_1(t)$ and $x_2(t)$. When we come to this later in the book, we'll understand that a solution to this system of equations is not so different than what is required for solving any system of equations: we need to find functions that, when substituted into the left side (which involves taking their derivatives), equals the right side (which involves only substituting in the functions). That's why they are called equations: an equality holds.

The next example asks (and answers) an intriguing question: without knowing the solutions to the above differential equations, can we prove that the solutions do not explode? That is,

$$\lim_{t\to\infty} (x_1(t))^2 + (x_2(t))^2 \tag{5.46}$$

is finite? And just maybe, that it even equals zero? Huh? We do not know the solution, but we seek to determine some of its properties anyway? That seems like a power mindset! **To answer these questions, we will use the total derivative.**

Example 5.48. *For the differential equations in (5.45):*

(a) Show that the limit in (5.46) does not blow up.

(b) Bonus: Give a plausible argument that the limit is, in fact, equal to zero.

Solution: Define a function $V(x_1, x_2) := (x_1)^2 + (x_2)^2$, which is the norm squared of the vector (x_1, x_2). We compute its total derivative with respect to time, $\frac{dV(x_1(t), x_2(t))}{dt}$, using the Chain Rule,

$$\begin{aligned}
\frac{d}{dt}V(x_1(t), x_2(t)) &= \frac{\partial V(x_1(t), x_2(t))}{\partial x_1} \cdot \frac{dx_1(t)}{dt} + \frac{\partial V(x_1(t), x_2(t))}{\partial x_2} \cdot \frac{dx_2(t)}{dt} \\
&= \underbrace{2x_1(t)}_{\frac{\partial V}{\partial x_1}} \cdot \frac{dx_1(t)}{dt} + \underbrace{2x_2(t)}_{\frac{\partial V}{\partial x_2}} \cdot \frac{dx_2(t)}{dt} \\
&= 2x_1(t) \cdot \underbrace{x_2(t)}_{\frac{dx_1(t)}{dt}} + 2x_2(t) \cdot \underbrace{(-x_1(t) - x_2(t))}_{\frac{dx_2(t)}{dt}} \quad \text{(plugging in from (5.45))} \\
&= -2(x_2(t))^2.
\end{aligned}$$

Because

$$\frac{d}{dt}V(x_1(t),x_2(t)) = -2(x_2(t))^2 \leq 0 \tag{5.47}$$

for all t, Prop. 5.21 implies that the function $V(x_1(t),x_2(t))$ is nonincreasing! We do not know the actual time-function $V(t) = V(x_1(t),x_2(t))$, but we know that its initial value at time $t = 0$ is its largest possible value (otherwise, it could not be nonincreasing). Therefore, the limit in (5.46), if it exists, cannot go to infinity. **That settles the first part of the question, which is already an awesome demonstration of the total derivative in action**. What about the second part of the question?

Bonus: The limit likely goes to zero, and the reasoning is as follows. If a pendulum is not at rest at the downward position, it oscillates. That's what pendulums are designed to do. Now, as long as the angular velocity $x_2(t)$ of the pendulum is nonzero, the time rate of change of $V(x_1(t),x_2(t))$ is negative, so $V(x_1(t),x_2(t))$ continues to decrease. The angular velocity of the pendulum goes to zero when the pendulum reaches one of its two "extreme swing angles" and reverses course. So, as long as the pendulum is not at rest in its downward position, $V(x_1(t),x_2(t))$ is decreasing. It cannot decrease below zero because $(x_1(t))^2 + (x_2(t))^2$ represents the norm squared of the vector $(x_1(t),x_2(t))$. Hence $V(x_1(t),x_2(t))$ likely converges to zero as $t \to \infty$.

```
using Symbolics

# Create a custom function to make differentiation convenient
function deriv(f, x)
    return expand_derivatives.(Differential(x)(f))
end

@variables x1 x2
x=[x1 x2]

# Define the differential equations as a vector
dx_dt = [x2; -x1 - x2]

# Define a function V(x)
V = x1^2 + x2^2

# Compute the Jacobian of V with respect to x
Jac_V = Symbolics.jacobian([V], x)

# Compute the total derivative of V using the chain rule
dV_dt = Jac_V * dx_dt

# The result of the matrix multiplication Jac_V * dx_dt
# is a 1 x 1column vector.
# Since we are interested in the scalar value of the derivative,
# we extract the first (and only) element of this vector using dV_dt[1].
#
# Simplify the result and expand polynomial expressions
#
dV_dt = Symbolics.simplify(dV_dt[1], expand=true)

println("The total derivative of V with respect to t using the chain rule is:\n \n",
    dV_dt)
```

Output

```
The total derivative of V with respect to t using the chain rule is:

-2(x2^2)
```

Bottom Line:

- **Multidimensional Chain Rule:** The total derivative can be thought of as the chain rule's multidimensional counterpart. While the partial derivative captures the rate of change of a multivariable function with respect to a single variable, the total derivative encapsulates the compounded effect of each variable's rate of change on the function. It's the chain rule, but on a grander scale.

- **Versatility:** While plugging in $x(t)$, $y(t)$, and $z(t)$ directly into f and differentiating is a straightforward approach, the total derivative offers a more structured method. This becomes especially valuable when dealing with functions where the relationship between variables isn't immediately clear (such as our differential equation) or when the function's form makes direct differentiation less intuitive.

- **Efficiency:** Just as the chain rule can simplify differentiation for certain composite functions, the total derivative can be more efficient for functions of multiple variables. Instead of working out derivatives for every possible combination of functions, the total derivative provides a systematic approach that breaks the computation down into simpler pieces. Recall that Automatic Differentiation, heavily used in machine learning, also uses the Chain Rule to decompose a complicated derivative into simpler parts.

Secrets of the Arcane 5.49: Let's Ask ChatGPT4

Question to ChatGPT4: What do you think is the most important mathematical tool used for training your neural networks?

Answer: The training of neural networks, like mine, relies on a combination of mathematical concepts, algorithms, and techniques. If we had to pinpoint the most fundamental mathematical tool, it would be:

1. **Calculus and the Chain Rule:** The backpropagation algorithm, which is central to training deep neural networks, is essentially an application of the chain rule from calculus. It computes the **gradient of the loss function with respect to each weight** by propagating the gradient backward through the network.

Other important mathematical tools and concepts include:

2. **Linear Algebra:** Neural networks involve a lot of matrix multiplications and transformations, especially in architectures like convolutional neural networks (CNNs) and recurrent neural networks (RNNs).

3. **Probability and Statistics:** These are crucial for understanding concepts like regularization, dropout, and batch normalization. They also play a role in the design of loss functions and the interpretation of neural network outputs, especially in classification tasks.

4. **Optimization:** Gradient descent and its variants (like stochastic gradient descent, Adam, RMSprop, etc.) are optimization algorithms used to adjust the weights of the network to minimize the loss function.

5. **Differential Equations:** Especially relevant in the context of understanding and designing recurrent neural networks and more recent architectures like Neural ODEs.

6. **Information Theory:** Concepts like entropy and cross-entropy, which are foundational to certain loss functions (e.g., the cross-entropy loss for classification tasks).

While the chain rule and calculus form the backbone of the training process, the successful training and deployment of neural networks require a holistic understanding and application of various mathematical tools and concepts.

13 October 2023, Saint-Rémy-de-Provence, France

5.9 (Optional Read) Proofs Associated with the Chapter

Proof of Prop. 5.5 (Differentiable at x_0 Implies Continuous at x_0)

Suppose $f : [a, b] \to \mathbb{R}$ is differentiable at $x_0 \in [a, b]$. Then, f is continuous at x_0.

Proof: Because f is differentiable at x_0,

$$\lim_{h \to 0} \frac{f(x_0 + h) - f(x_0)}{h} = L$$

for some finite number L.

For $h \neq 0$, we rewrite the difference

$$f(x_0 + h) - f(x_0) = h \cdot \frac{f(x_0 + h) - f(x_0)}{h}$$

as the product of two terms, each of which has a finite limit as $h \to 0$, so that

$$\lim_{h \to 0} (f(x_0 + h) - f(x_0)) = \lim_{h \to 0} \left(h \cdot \frac{f(x_0 + h) - f(x_0)}{h} \right).$$

From Prop. 2.37, we conclude that

$$\lim_{h \to 0} \left(h \cdot \frac{f(x_0 + h) - f(x_0)}{h} \right) = \left(\lim_{h \to 0} h \right) \left(\lim_{h \to 0} \frac{f(x_0 + h) - f(x_0)}{h} \right) = 0 \cdot L = 0,$$

and hence

$$\lim_{h \to 0} f(x_0 + h) = f(x_0).$$

It follows that by Definition 4.11, f is continuous at x_0. ∎

Chapter 6

Engineering Applications of the Derivative

Learning Objectives

By the end of this chapter, the student should be able to:

- Analyze and model engineering problems using the principles of calculus, specifically through the application of derivatives to understand system behaviors.
- Develop strategies for solving optimization problems, both with and without constraints.
- Critically assess the role of derivatives in determining system dynamics and understand the significance of dynamic equations in engineering contexts.
- Apply the concept of energy to the analysis of mechanical systems using the framework of Lagrangian mechanics.

Outcomes

Upon successful completion of this chapter, students will be able to:

- Calculate path length and arc length for given paths.
- Solve engineering problems involving root finding and minimization.
- Use gradient descent to find local minima of functions and understand its limitations.
- Apply second-derivative tests to determine the nature of critical points in functions.
- Solve optimization problems involving equality and inequality constraints using Lagrange multipliers.
- Derive and apply Lagrange's equations to solve problems in dynamics.
- Compute kinetic and potential energy for mechanical systems.
- Model and analyze the motion of complex systems like multi-link manipulators using Lagrange's formalism.
- Determine moments of inertia for various bodies and understand their effects on rotational motion.

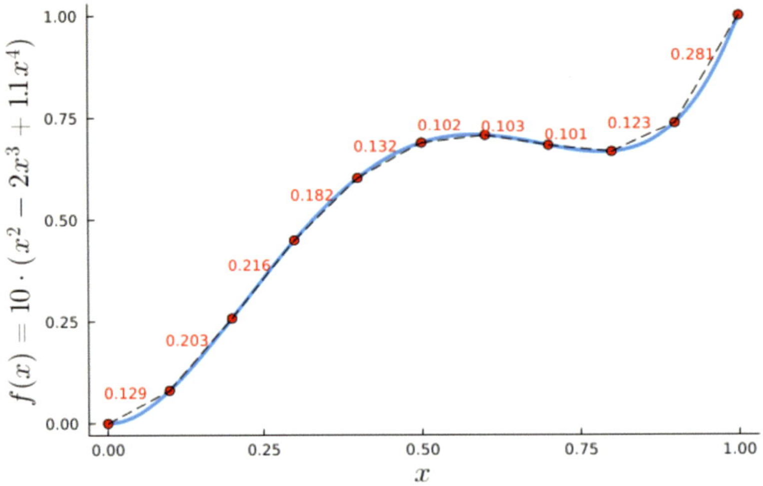

Figure 6.1: Path length is defined by chopping up the curve into short segments, connecting the endpoints of the segments by straight lines, and then adding up the lengths of the lines. As the segment size tends to zero, the true path length is obtained. Simple? Yes. Useful? Very! It can be used to formulate an important problem such as: determine a path of shortest length for a robot to travel without encountering obstacles. Table 6.1 shows that the ten segment lengths in the plot add up to 1.572; the true answer, obtained in the limit as the segment lengths shrink to zero and become infinite in number, is $1.5852977678599067 \pm 6.654127165939272e\text{-}9$. The approximation with ten segments is already quite good!

6.1 Path Length or Arc Length

It is easy to understand what we mean by the length of a line segment in a plane and the length of the circumference of a circle. But for more complicated curves, what is the length from a point a to a point b? **Figure 6.1** and Table 6.1 illustrate the concept of **arc length (aka path length)**, which is fundamental in engineering because it allows us to quantify the "length of a curve" in the plane $\mathbb{R}^2$, which can be defined by a function $f : (a, b) \to \mathbb{R}$ or parametric equations $x(t), y(t)$, $t_1 \le t \le t_2$; the arc length is the distance traveled along the curve between two points.

The integral formulas for arc length arise from the Pythagorean theorem. By dividing the curve into infinitesimally small segments, each segment can be approximated by a straight line whose length can be calculated using the Pythagorean theorem as the hypotenuse of a right triangle with sides dx and dy. As we sum up the lengths of all these segments (a process achieved by integrating), we obtain the total arc length of the curve.

Arc Length = Path Length

Definition 6.1. *Given a differentiable function $f : [a, b] \to \mathbb{R}^2$, the arc length S of the curve described by $y = f(x)$ from $x = a$ to $x = b$ is given by the integral, $S := \int_a^b \sqrt{dx^2 + dy^2}$. After factoring out dx, we obtain the standard formula,*

$$S = \int_a^b \sqrt{1 + \left(\frac{dy}{dx}\right)^2}\, dx. \tag{6.1}$$

For a curve defined parametrically by $x(t)$ and $y(t)$ where t ranges from t_1 to t_2, the arc length is given by:

$$S = \int_{t_1}^{t_2} \sqrt{\left(\frac{dx}{dt}\right)^2 + \left(\frac{dy}{dt}\right)^2}\, dt. \tag{6.2}$$

Table 6.1: Path length is easy to understand but tedious to compute by hand,
$S := \int_a^b \sqrt{dx^2 + dy^2} = \int_a^b \sqrt{1 + \left(\frac{dy}{dx}\right)^2}\, dx$ (after factoring out dx)

x_i	y_i	$\sqrt{(x_i - x_{i-1})^2 + (y_i - y_{i-1})^2}$
0.0	0.0	0.0
0.1	0.0811	0.129
0.2	0.2576	0.203
0.3	0.4491	0.216
0.4	0.6016	0.182
0.5	0.6875	0.132
0.6	0.7056	0.102
0.7	0.6811	0.103
0.8	0.6656	0.101
0.9	0.7371	0.123
1.0	1.0	0.281
$\int_a^b \sqrt{dx^2 + dy^2} \approx \sum_{i=2}^{11} \sqrt{(x_i - x_{i-1})^2 + (y_i - y_{i-1})^2} =$		1.572

Notes: Equation (6.2) is the more general formula for computing path length. When the path is given by a function, as in $(x, y(x))$, (6.1) follows from (6.2) as shown below:

$$S = \int_{a=x(t_1)}^{b=x(t_2)} \sqrt{\left(\frac{dx}{dt}\right)^2 + \left(\frac{dy}{dt}\right)^2}\, dt \text{ (factor out } \left(\frac{dx}{dt}\right)^2\text{)}$$

$$= \int_{a=x(t_1)}^{b=x(t_2)} \sqrt{1 + \left(\frac{\frac{dy}{dt}}{\frac{dx}{dt}}\right)^2}\, \left(\frac{dx}{dt}\right) dt \text{ (cancel the } dt's\text{)}$$

$$= \int_a^b \sqrt{1 + \left(\frac{dy}{dx}\right)^2}\, dx.$$

Source of the formula for arc length: Figure 6.1 and Table 6.1 provide the basic idea. Let Δx and Δy be incremental changes in x and y along a curve. Then the distance covered is given by the Pythagorean Theorem, $\Delta S := \sqrt{(\Delta x)^2 + (\Delta y)^2}$. Taking the limit as Δx goes to zero, we obtain $dS = \sqrt{(dx)^2 + (dy)^2}$, and hence

$$S = \int_a^b dS = \int_a^b \sqrt{(dx)^2 + (dy)^2}.$$

We have not previously encountered an integral of this form. To address it, suppose that the (x, y) coordinates of the curve are parameterized by time. Then, using the linearization (aka first-order Taylor expansion), we have:

$$\Delta x = x(t + \Delta t) - x(t) \underset{\Delta t \to 0}{\longrightarrow} \left(\frac{dx(t)}{dt}\right) \cdot dt,$$

$$\Delta y = y(t + \Delta t) - y(t) \underset{\Delta t \to 0}{\longrightarrow} \left(\frac{dy(t)}{dt}\right) \cdot dt$$

Substituting into the previous equation and factoring out dt gives the result.

Example 6.2. *(Classical example because the integral is easy to evaluate). A circle of radius* $r > 0$ *about the origin can be parameterized by* $(x(t) = r\sin(t), y(t) = r\cos(t))$*, for* $0 \le t \le 2\pi$*. Compute the arc length from* $t_1 = \pi/3$ *to* $t_2 = \pi$*.*

Solution: Ans. $S = 2r\frac{\pi}{3}$.

We apply (6.2):

- $x(t) = r\sin(t) \implies \frac{dx}{dt} = r\cos(t)$.
- $y(t) = r\cos(t) \implies \frac{dy}{dt} = -r\sin(t)$.
- $\sqrt{\left(\frac{dx}{dt}\right)^2 + \left(\frac{dy}{dt}\right)^2} = r$.

Hence,

$$S = \int_{t_1}^{t_2} \sqrt{\left(\frac{dx}{dt}\right)^2 + \left(\frac{dy}{dt}\right)^2}\, dt = \int_{\pi/3}^{\pi} r\, dt = r \cdot t\Big|_{\pi/3}^{\pi} = r\left(\pi - \pi/3\right) = 2r\pi/3.$$

■

Example 6.3. *Compute the path length of the curve defined by* $y = f(x) = x^2$ *from* $x = 0$ *to* $x = 1$*.*

Solution: Ans. $S = 1.479$.

The derivative $\frac{dy}{dx}$ is $2x$. Hence, the arc length is calculated as:

$$S = \int_0^1 \sqrt{1 + (2x)^2}\, dx. \tag{6.3}$$

Setting up the integral is the main skill to be mastered here. In Chapter 7, we will learn how to compute closed-form solutions[1] for such integrals when the function defining the path is simple enough, such as $f(x) = x^2$. For now, we seek a numerical answer, which is what real engineers must do in 99.9% of cases.

```
using QuadGK

# Define the function for the curve
f(x) = x^2

# Define the derivative of the function
df(x) = 2x

# Define the integrand for arc length
function arc_length_integrand(x)
    return sqrt(1 + df(x)^2)
end

# Compute the arc length using QuadGK
integral, error = quadgk(arc_length_integrand, 0, 1)

println("The arc length of the curve from x = 0 to x = 1
is approximately: ", integral)
println("The estimated error of the computation is: ", error)
```

[1] $S = \int_0^1 \sqrt{1 + (2x)^2}\, dx = \int_0^2 \sqrt{1 + u^2}\, \frac{du}{2} = \frac{1}{2}\int_0^2 \sqrt{1 + u^2}\, du = \text{asinh}(u)\Big|_0^2 = \frac{1}{2}\text{asinh}(2)$,

Output

```
The arc length of the curve from x = 0 to x = 1 is approximately:
1.4789428575446002

The estimated error of the computation is: 1.0424397123287577e-8
```

■

Example 6.4. *Compute the arc length of the curve defined by* $y = f(x) = 10(x^2 - 2x^3 + 1.1x^4)$ *from* $x = 0$ *to* $x = 1$.

Solution: Ans. $S = 1.585$.

The derivative $\frac{dy}{dx}$ is $10(2x - 6x^2 + 4.4x^3)$. Hence, the arc length is calculated as:

$$S = \int_0^1 \sqrt{1 + (10(2x - 6x^2 + 4.4x^3))^2}\, dx. \tag{6.4}$$

```
using QuadGK

# Define the function for the curve
f(x) = 10*(x^2 - 2*x^3 + 1.1*x^4)

# Define the derivative of the function
df(x) = 10*(2x - 6x^2 + 4.4x^3)

# Define the integrand for arc length
function arc_length_integrand(x)
    return sqrt(1 + df(x)^2)
end

# Compute the arc length using QuadGK
integral, error = quadgk(arc_length_integrand, 0, 1)

println("The arc length of the curve from x = 0 to x = 1 is
approximately: ", integral)

println("The estimated error of the computation is: ", error)
```

Output

```
The arc length of the curve from x = 0 to x = 1 is approximately:
1.5852977678599067

The estimated error of the computation is: 6.654127165939272e-9
```

■

Example 6.5. Figure 6.2 *depicts a figure eight parameterized by:*

$$x(t) = \pi \cdot \frac{\cos\left(\frac{t \cdot 2\pi}{4}\right)}{1 + \sin^2\left(\frac{t \cdot 2\pi}{4}\right)}$$

$$y(t) = \pi \cdot \frac{\sin\left(\frac{t \cdot 2\pi}{4}\right) \cdot \cos\left(\frac{t \cdot 2\pi}{4}\right)}{1 + \sin^2\left(\frac{t \cdot 2\pi}{4}\right)}$$

for $0 \le t \le 4$. *Compute its arc length.*

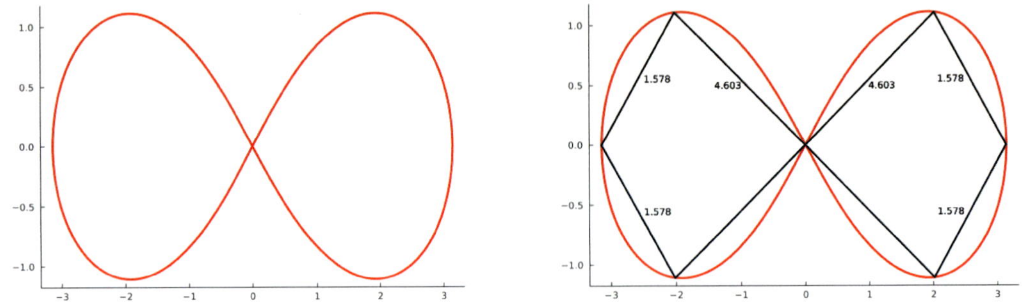

(a) (b)

Figure 6.2: A figure eight is clearly not a function from x to y. (a) Rather, it is a parameterized curve in the plane. Example 6.5 explains what is going on. Part (b) shows a very crude straight line underapproximation of the arc length yielding $S \approx 4 \cdot (1.578) + 2 \cdot (4.603) = 15.518$. How far off is the approximation? The two lines of length 4.603 cross through the origin to hit the other side of the figure.

Solution: Ans. $S = 16.47$.

This time, the path in $\mathbb{R}^2$ is not specified by a function, but instead by a parameterized curve $(x(t), y(t))$, $0 \leq t \leq 4$. This is more general than a function because the curve can cross itself, fail the vertical line test, and do all sorts of crazy and wonderful things, just like a drone flying in the air. To compute arc length, we need to apply the formula:

$$S = \int_{t_1}^{t_2} \sqrt{\left(\frac{dx}{dt}\right)^2 + \left(\frac{dy}{dt}\right)^2}\, dt = \int_{t_1}^{t_2} \sqrt{(x'(t))^2 + (y'(t))^2}\, dt.$$

We compute the derivatives using software and obtain:

$$x'(t) = -\frac{\pi \cdot 2\pi \cdot \sin\left(\frac{2\pi t}{4}\right)}{4 \cdot \left(\sin^2\left(\frac{2\pi t}{4}\right) + 1\right)} - \frac{\pi \cdot 4\pi \cdot \sin\left(\frac{2\pi t}{4}\right) \cdot \cos^2\left(\frac{2\pi t}{4}\right)}{4 \cdot \left(\sin^2\left(\frac{2\pi t}{4}\right) + 1\right)^2},$$

$$y'(t) = -\frac{\pi \cdot 2\pi \cdot \sin^2\left(\frac{2\pi t}{4}\right)}{4 \cdot \left(\sin^2\left(\frac{2\pi t}{4}\right) + 1\right)} + \frac{\pi \cdot 2\pi \cdot \cos^2\left(\frac{2\pi t}{4}\right)}{4 \cdot \left(\sin^2\left(\frac{2\pi t}{4}\right) + 1\right)} - \frac{\pi \cdot 4\pi \cdot \sin^2\left(\frac{2\pi t}{4}\right) \cdot \cos^2\left(\frac{2\pi t}{4}\right)}{4 \cdot \left(\sin^2\left(\frac{2\pi t}{4}\right) + 1\right)^2}.$$

We are now all set to do the integral numerically.

```
using QuadGK, Plots, LaTeXStrings

function x_position(t)
    a = pi; T = 4
    a * cos(t*2*π/T) / (1 + sin(t*2*π/T)^2)
end
function y_position(t)
    a = pi; T = 4
    a * sin(t*2*π/T) * cos(t*2*π/T) / (1 + sin(t*2*π/T)^2)
end

# Compute the arc length

# Derivatives
function dx_dt(t)
    a = pi; T = 4
    -2*π*a*sin(2*π*t/T)/(T*(sin(2*π*t/T)^2 + 1)) -
    4*π*a*sin(2*π*t/T)*cos(2*π*t/T)^2/(T*(sin(2*π*t/T)^2 + 1)^2)
```

```
end

function dy_dt(t)
    a = pi; T = 4
    -2*π*a*sin(2*π*t/T)^2/(T*(sin(2*π*t/T)^2 + 1)) +
    2*π*a*cos(2*π*t/T)^2/(T*(sin(2*π*t/T)^2 + 1)) -
    4*π*a*sin(2*π*t/T)^2*cos(2*π*t/T)^2/(T*(sin(2*π*t/T)^2 + 1)^2)
end

# Define the integrand for arc length
function arc_length_integrand(t)
    return sqrt(dx_dt(t)^2 + dy_dt(t)^2)
end

# Compute the arc length using QuadGK
integral, error = quadgk(arc_length_integrand, 0, 4)

println("The arc length of the curve from x = 0 to x = 4
is approximately: ", integral)

println("The estimated error of the computation is: ", error)

# make a nice plot

dt = 0.01
t = collect(0.0:dt:4.0)
xvalues = x_position.(t)
yvalues = y_position.(t)

p1 = plot(xvalues, yvalues, lw=3, color=:red, apect_ratio = 1,
guidefont = 15, label=false)
```

Output

```
The arc length of the curve from x = 0 to x = 4 is approximately:
16.474873499707456

The estimated error of the computation is: 1.5850782530968388e-7
```

Hence, the very crude approximation in **Figure 6.2**-(b) is off by about 10%. ■

Helpful videos: Keep in mind that setting up the integral is the main task to master. Once you have the problem properly formulated, you can then employ software tools to find the answer.

- Arc Length (formula explained) by **BlackPenRedPen**.
- Arc length intro | Applications of definite integrals by the Khan Academy.
- Find the Arc Length of a Function with Calculus by Quoc Dat Phung. Positive: He has strong graphics. Negative: He uses manual integration instead of numerical integration. We cover how to manually compute similar integrals in Chapter 7. For now, just watch how he sets up the problems.
- Arc Length Calculus Problems by the Organic Chemistry Tutor.

6.2 Root Finding and Unconstrained Minimization

The methods of this section are grounded in the linear approximation of a nonlinear function developed in Prop. 5.36. Root finding uses

$$f(x) \approx f(x_0) + \frac{\partial f(x_0)}{\partial x} \cdot (x - x_0), \quad f : \mathbb{R}^n \to \mathbb{R}^n,$$

while minimization uses the equivalent form developed in (5.36),

$$f(x) \approx f(x_0) + \nabla f(x_0) \bullet (x - x_0), \quad f : \mathbb{R}^n \to \mathbb{R},$$

where the heavy dot, $\bullet$, denotes the dot product of two column vectors.

6.2.1 Root Finding

We consider a function $f : \mathbb{R}^n \to \mathbb{R}^n$ and seek a **root, that is, a point $\mathbf{x_0} \in \mathbb{R}^n$ such that $\mathbf{f(x_0) = 0}$**. Note that the domain and range are both $\mathbb{R}^n$, and thus, this is the nonlinear equivalent of solving a square linear system of equations, $Ax - b = 0$. We recall that $\det(A) \neq 0$ is the necessary and sufficient condition for the existence and uniqueness of solutions to $Ax - b = 0$.

Suppose that x_k is the current guess for a root. The **Newton-Raphson Algorithm** uses the linear approximation of f about the point x_k to choose x_{k+1} so that $f(x_{k+1}) \approx 0$. In other symbols, x_{k+1} should satisfy

$$0 = f(x_{k+1}) \approx f(x_k) + \frac{\partial f(x_k)}{\partial x} \cdot (x_{k+1}) - x_k), \tag{6.5}$$

which is equivalent to

$$0 \approx f(x_k) + \frac{\partial f(x_k)}{\partial x} \cdot (x_{k+1} - x_k). \tag{6.6}$$

If the Jacobian $\frac{\partial f(x_k)}{\partial x}$ is invertible, and the approximation sign is replaced with an equals sign, a unique value for x_{k+1} is obtained. Solving this equation iteratively yields the Newton-Raphson Algorithm.

Method 6.6: Newton-Raphson Algorithm

Based on (6.6), we define[a] $x_{k+1} := x_k + \Delta x_k$, where Δx_k is our update to x_k. We can then break the algorithm into two steps:

$$\left(\frac{\partial f(x_k)}{\partial x}\right) \Delta x_k = -f(x_k) \quad \text{(solve for } \Delta x_k) \tag{6.7}$$

and

$$x_{k+1} = x_k + \Delta x_k \ \text{ (use } \Delta x_k \text{ to update our estimate of the root)}. \tag{6.8}$$

While for small-scale problems we can use the matrix inverse to solve (6.7) for Δx_k, for larger problems, we should use LU Factorization or QR Factorization. Once (6.7) has been solved, x_{k+1} is updated in (6.8) and the process repeats.

A **damped Newton-Raphson Algorithm** is obtained by replacing (6.8) with

$$x_{k+1} = x_k + \epsilon \Delta x_k, \tag{6.9}$$

for some $\epsilon > 0$. The validity of the Newton-Raphson Algorithm rests upon:

- the function f being differentiable;
- the Jacobian $\frac{\partial f(x_k)}{\partial x}$ having a nonzero determinant at points generated by (6.7) and (6.8); and
- the linear equation $f_{\text{lin}}(x) = f(x_k) + \frac{\partial f(x_k)}{\partial x}(x - x_k)$ being a good approximation to the nonlinear function, $f(x)$.

[a]Note that $\Delta x_k = x_{k+1} - x_k$.

Here, the focus is on Julia Packages; we suggest NLsolve.jl, which is based on Newton-Raphson and important variations of the same.

using NLsolve

```
# Import the NLsolve package
using NLsolve

# This script aims to solve a system of nonlinear equations using the
# Newton-Raphson method provided by the NLsolve package.

# Define the system of nonlinear equations.
# The function `system_of_equations!` modifies the input array `F` to
# represent the system of equations.
# The system to be solved is:
# 1) x[1]^2 + x[2]^2 = 2.0
# 2) x[1]^2 - x[2]^2 = 1.0
#
# Parameters:
# - F: An array that will be modified in-place to
# represent the system of equations.
# - x: An array representing the variables of the system.
function system_of_equations!(F, x)
    F[1] = x[1]^2 + x[2]^2 - 2.0  # First equation
    F[2] = x[1]^2 - x[2]^2 - 1.0  # Second equation
end

# Provide an initial guess for the solution.
# The quality of the initial guess can influence the convergence of the
# Newton-Raphson method.
# In this case, both x[1] and x[2] are initially guessed to be 1.0.
initial_guess = [1.0, 1.0]

# Use the `nlsolve` function from the NLsolve package to find the roots
# of the system.
# The function returns a result object that contains various information
# about the solution process.
result = nlsolve(system_of_equations!, initial_guess)

# Print the roots (solutions) of the system.
println("Roots found at: ", result.zero)
```

Output

```
Roots found at: [1.2247448713915894, 0.7071067811873448]
```

As we see from the code, it aims to find the roots of the system of equations,

$$x_1^2 + x_2^2 = 2,$$
$$x_1^2 - x_2^2 = 1.$$

In Julia, the “bang” notation, denoted by an exclamation mark (!) at the end of a function name, is a convention used to indicate that the function may modify its arguments in-place. This convention is not enforced by the language itself but is widely adopted in the Julia community to signal the side effects of a function.

For example, consider the function

```
function system_of_equations!(F, x)
    F[1] = x[1]^2 + x[2]^2 - 2.0
    F[2] = x[1]^2 - x[2]^2 - 1.0
end
```

The presence of the ! in the function name system_of_equations! suggests that the function might modify the contents of the arguments F and/or x directly, rather than returning a new modified value. **Modifying arguments in place can be more memory efficient**, especially for large data structures, as it avoids the need to allocate new memory for the returned values. This can lead to performance benefits in many scenarios, especially in iterative algorithms where the same operation is performed multiple times. However, it is essential to use this convention judiciously. **Modifying arguments in place can make code harder to understand and debug if not used carefully.**

You are NOT obliged to use modification in place. Here is the same example using simpler programming constructs.

```julia
using NLsolve

# Define the system of nonlinear equations WITHOUT
# using modification in place
function system_of_equations(x)
    return [
        x[1]^2 + x[2]^2 - 2.0,
        x[1]^2 - x[2]^2 - 1.0
    ]
end

# Initial guess for the solution
initial_guess = [1.0, 1.0]

# Solve the system of equations
result = nlsolve(system_of_equations, initial_guess)

# Print the solution
println("Roots found at: ", result.zero)
```

Output

```
Roots found at: [1.2247448713915894, 0.7071067811873448]
```

Here is a more impressive example of root finding, where we would NOT want to solve these equations by hand.

```julia
using NLsolve
using Random
Random.seed!(1958)

function Fun!(Fun, x)
    Fun[1] = x[1]^2 + x[2] - x[3] + x[4]^3 - 10
    Fun[2] = x[2]^2 + x[5] - x[6] + x[7]^2 - 20
    Fun[3] = x[3]^2 + x[8] - x[9] + x[1]^2 - 30
    Fun[4] = x[4]^2 + x[5]^3 - x[6] + x[9] - 40
    Fun[5] = x[7]^2 + x[8]^3 - x[9] + x[2] - 50
    Fun[6] = x[1] + x[2]^2 - x[3]^3 + x[4] - 5
```

```
    Fun[7] = x[5] + x[6]^2 - x[7]^3 + x[8] - 15
    Fun[8] = x[9] + x[1]^2 - x[2]^3 + x[3] - 25
    Fun[9] = x[4] + x[5]^2 - x[6]^3 + x[7] - 35
end

# Trying a different initial guess
x0 = randn(9) * 10

# Adjust solver settings
result = nlsolve(Fun!, x0, method = :trust_region, ftol = 1e-10,
xtol = 1e-10, iterations = 10000)

println("Roots found at: ", result.zero)
println(" ")

# After obtaining result from nlsolve
Fun_evaluated = Vector{Float64}(undef, length(x0))
Fun!(Fun_evaluated, result.zero)
println("Values of Fun at the root: ", Fun_evaluated)
```

Output

```
Roots found at: [-7.772786278382927, 4.157370417426024,
0.9040581712865199, -3.7720366206105242, -2.352593029768065,
-3.258024109655881, -1.3456746012402663, 4.301070357560675,
35.53459806404852]

Values of Fun at the root: [7.105427357601002e-15,
-3.552713678800501e-15, 3.552713678800501e-15,
7.105427357601002e-15, 2.1316282072803006e-14,
-3.552713678800501e-15, 0.0, 2.4868995751603507e-14,
-7.105427357601002e-15]
```

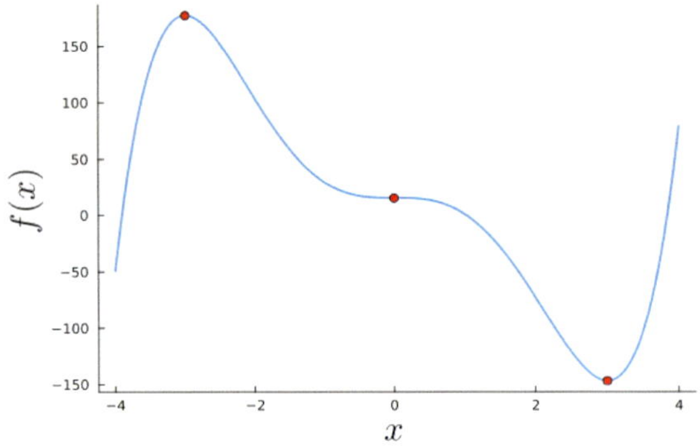

(a)

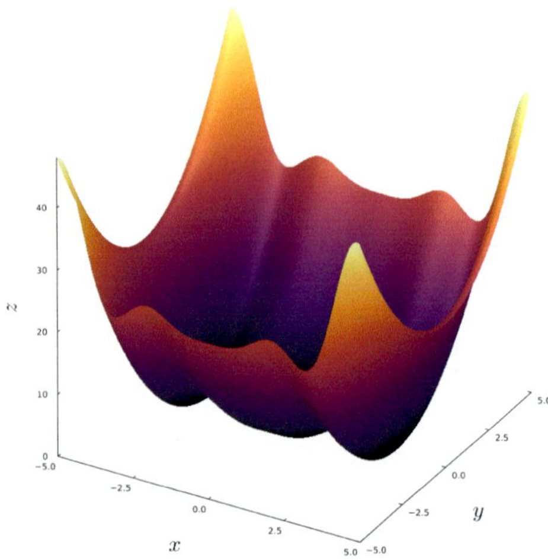

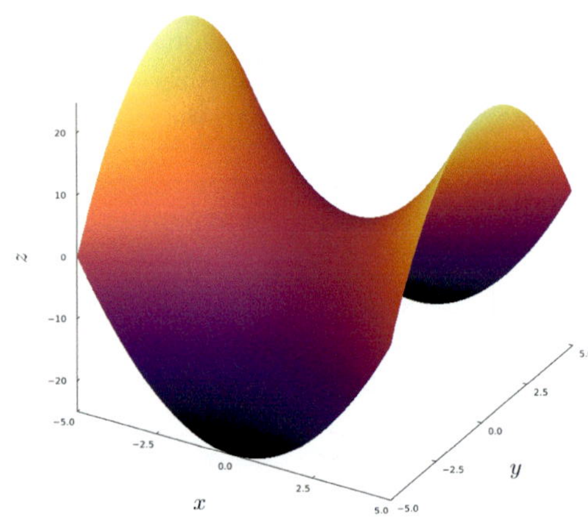

(b) (c)

Figure 6.3: Three cost functions. (a) $f(x) = x^5 - 15x^3 + 15$ has a local maximum at -3, an inflection point at the origin, and a local minimum at 3. (b) $f(x, y) = (x \cdot \sin(x))^2 + y^2$ is riddled with local minima and maxima. (c) $f(x, y) = x^2 - y^2$ has a saddle point at the origin. The name could not be more clear!

6.2.2 Minimization without Constraints

Optimization is the process of finding one or more values $x \in \mathbb{R}^n$ that **minimize a function** $f : \mathbb{R}^n \to \mathbb{R}$, where the scalar-valued function f is called the **cost function**. The cost function should be designed such that it has a minimum at a point x^* that is of interest to us, it should be "small" for values of x that are "near" x^*, and it should be "large" for values of x that are "far" from our preferred value, x^*. In other symbols, the function should be " bowl-shaped" like $f(x) = (x - x^*)^2$. In machine learning, a cost function for minimization is also called a **regret function**, in the sense that we regret being far from our preferred value, x^*, and our regret is minimum at x^*. We will abuse notation[2] and write:

$$x^* = \arg\min_{x \in \mathbb{R}^n} f(x) \tag{6.10}$$

to denote the value of x achieving the minimum, even when there may be more than one such x.

One can also perform the opposite of minimization, which is to **maximize a cost function**. In this case, the machine learning community calls the cost function a **reward**; indeed, they maximize reward and minimize

[2]Abusing notation means that one is being a bit sloppy with one's use of symbols, which is what notation is! One often abuses notation when doing the right thing is painful and would cause more of a distraction to explain the good notation than to caution about using poor notation.

regret! We will adhere to minimization throughout this section because

$$\arg\max_{x\in\mathbb{R}^n} f(x) = \arg\min_{x\in\mathbb{R}^n} (-f(x));$$

that is, by a simple sign change, we can convert a maximization problem into a minimization problem. Most of the numerical optimization packages take advantage of this property.

Fundamental Idea and Vocabulary

Because the gradient is the transpose of the Jacobian, the linearization of f via the gradient can be written as

$$f(x) \approx f(x_0) + (\nabla f(x_0))^{\top} \cdot (x - x_0) = f(x_0) + \nabla f(x_0) \bullet (x - x_0), \tag{6.11}$$

where $u \bullet v$ denotes the **dot product or inner product** of the two n-vectors, u and v. Based on this equation, we make the following observations:

(a) If the gradient evaluated at a point x_0 is the zero vector, that is, $\nabla f(x_0) = 0_{n\times 1}$, then the linear approximation of f is equal to the constant (vector) $f(x_0)$. This means that in the vicinity of x_0, irrespective of the direction[3] in which we move from x_0, the function satisfies $f(x) \approx f(x_0)$. Such points x_0 where $\nabla f(x_0) = \mathbf{0}$ are referred to as **extrema** or **stationary points**. The term "extrema" is more common in calculus courses, whereas "stationary points" seems to be the go-to term for modern optimization packages!

The essential vocabulary for discussing points where $\nabla f(x_0) = 0_{n\times 1}$ is the following:

- $x_0 \in \mathbb{R}^n$ is a **stationary point** if $\nabla f(x_0) = 0_{n\times 1}$.
- a stationary point $x_0 \in \mathbb{R}^n$ is a **local minimum** if $f(x_0) \le f(x)$ for all x "near x_0" (i.e., there exists $\delta > 0$ such that for all $||x - x_0|| < \delta$, $f(x_0) \le f(x)$).
- a stationary point $x_0 \in \mathbb{R}^n$ is a **local maximum** if $f(x_0) \ge f(x)$ for all x "near x_0" (i.e., there exists $\delta > 0$ such that for all $||x - x_0|| < \delta$, $f(x_0) \ge f(x)$).
- A stationary point x_0 that is **neither** a local minimum **nor** a local maximum is called a **saddle point**. In the special case of functions $f : \mathbb{R} \to \mathbb{R}$, saddle points are called **stationary points of inflection**.
- The crowning accomplishment in optimization is to find a **global minimum**, an x_0 that is a local minimum and $f(x_0) \le f(x)$ for all $x \in \mathbb{R}^n$, i.e., we can remove the "near x_0." The same idea applies to a **global maximum**. We will content ourselves with local minima and maxima.

Figure 6.3 illustrates the various types of stationary points.

The subject of **inflection points** can be quite subtle, as explained on Wikipedia. We will stick with stationary points of inflection.

(b) If $\nabla f(x_0) \neq 0_{n\times 1}$, then we have the possibility of choosing x near x_0 that causes f to change.

- If $\nabla f(x_0) \bullet (x - x_0) > 0$, then (6.11) implies that $f(x) > f(x_0)$, so we are making f increase.
- If $\nabla f(x_0) \bullet (x - x_0) < 0$, then (6.11) implies that $f(x) < f(x_0)$, so we are making f decrease.
- If $\nabla f(x_0) \bullet (x - x_0) = 0$, then (6.11) implies that $f(x) \approx f(x_0)$, so f is remaining approximately constant. Therefore **these directions are not fruitful to explore.**

To understand the non-fruitful directions, let us see how well we recall null spaces! When $\nabla f(x_0) \neq 0_{n\times 1}$, it has rank one, and its null space has dimension[4] $(n-1)$. Moreover, all vectors $v \in \text{null}(\nabla f(x_0))$ satisfy:

$$\nabla f(x_0) \bullet v = 0 \quad \text{(same as writing } \nabla f(x_0) \perp v\text{)},$$

[3] In $\mathbb{R}^n$, there are n possible directions from x_0. Each direction can have a positive or negative coefficient.

[4] By the Rank-Nullity Theorem, for the $n \times 1$ matrix $\nabla f(x_0)$, we have $\text{rank}(\nabla f(x_0)) + \text{nullity}(\nabla f(x_0)) = n$.

which leaves us one dimension that is orthogonal to $\text{null}(\nabla f(x_0))$ **and it is spanned by** $\nabla f(x_0)$**, the gradient of** f**!**

(c) Moving in the direction given by $-\nabla f(x_0)$ gives rise to **gradient descent**, because

$$(x - x_0) = -\nabla f(x_0) \implies \nabla f(x_0) \bullet (x - x_0) = -\nabla f(x_0) \bullet \nabla f(x_0) = -||\nabla f(x_0)||^2,$$

and hence, by (6.11),

$$f(x) \approx f(x_0) - ||\nabla f(x_0)||^2 < f(x_0).$$

When we move in the opposite direction,

$$(x - x_0) = \nabla f(x_0) \implies \nabla f(x_0) \bullet (x - x_0) = \nabla f(x_0) \bullet \nabla f(x_0) = ||\nabla f(x_0)||^2,$$

and hence

$$f(x) \approx f(x_0) + ||\nabla f(x_0)||^2 > f(x_0).$$

This is called **gradient ascent**.

6.2.3 Gradient Descent Algorithm

Method 6.7: Gradient Descent

We assume that we are given an initial vector $x_0 \in \mathbb{R}^n$. Then the update law,

$$x_{k+1} = x_k - s\left(\nabla f(x_k)\right), \tag{6.12}$$

where $s > 0$ is called the **step size**, moves x in the direction of a local minimum.

Warning: If $s > 0$ is selected too large, then (6.12) may not converge. Hence, the sophisticated implementations found in today's software packages adaptively adjust the step size. In fact, they often perform a **line search**, meaning, they evaluate (6.12) for multiple values of $s \in \{s_1, s_2, \dots, s_n\}$, and choose the value of s_i resulting in the smallest value of

$$f(x_{k+1}) = f\left(x_k - s_i\left[\nabla f(x_k)\right]\right).$$

Here is a short video, Backtracking Line Search in Gradient Descent by Deep Shallownet, discussing this process (the computer voice employed is so appropriate).

Here, the focus is on the Julia Package JuMP.

Example 6.8. ***(Minimization of the Rosenbrock Function:)*** *The Rosenbrock function, often used as a performance test for optimization algorithms, is defined as:*

$$f(x, y) = (1 - x)^2 + 100(y - x^2)^2.$$

Note that the term $(y - x^2)^2$ makes $f(x, y)$ not a simple, bowl-shaped quadratic that is familiar to us from minimum norm problems in Linear Algebra. The function appears as a narrow, curved valley, and its global minimum is inside a long, narrow, parabolic-shaped flat valley. Finding the valley is trivial, but converging to the global minimum within the valley is difficult due to the flatness of the valley, making it a challenging test for optimization algorithms.

The term $(1 - x)^2$ vanishes at $x = 1$ and the term $(y - x^2)^2$ vanishes at $y = x^2 = 1$. The function, therefore, has a global minimum of zero at:

$$(x, y) = (1, 1).$$

Solution:

```
using JuMP, Ipopt

# Initialize a new optimization model
# We're using the Ipopt solver, which is suitable for nonlinear
# optimization problems
model = Model(Ipopt.Optimizer)

# Define decision variables for the optimization problem
# We're defining two variables, x and y, both bounded between -20 and 20
# We also provide an initial guess for the solver to start from.
# This can influence the solution found.
@variable(model, -20 <= x <= 20, start = -20.0)  # Initial guess set to -20
# for x
@variable(model, -20 <= y <= 20, start = 20.0)   # Initial guess set to 20
# for y

# Define the objective function we want to minimize
# Here, we're using the Rosenbrock function, a common test problem for
# optimization algorithms
@NLobjective(model, Min, (1 - x)^2 + 100 * (y - x^2)^2)

# Adjust solver settings to improve performance and accuracy
# "max_iter" sets the maximum number of iterations the solver will take
# "tol" sets the convergence tolerance. The solver stops when
# improvements are smaller than this value.
set_optimizer_attribute(model, "max_iter", 1000)  # Allow up to 1000 iterations
set_optimizer_attribute(model, "tol", 1e-9)       # Require a very accurate
#    solution

# Start the optimization process
# The solver will try to find values of x and y that minimize the
#    Rosenbrock function
optimize!(model)

# Once optimization is complete, display the results
# "value()" retrieves the optimal value of a decision variable
# "objective_value()" retrieves the minimum value of the objective
# function
println("Optimal parameters (x, y): (", value(x), ", ", value(y), ")")
println("Minimum objective value: ", objective_value(model))
```

Summary of the Output

```
EXIT: Optimal Solution Found.
Optimal parameters (x, y): (0.9999999792197696, 0.9999999583699614)
Minimum objective value: 4.3230208956579026e-16
```

Complete Output

```
This is Ipopt version 3.14.13, running with linear solver MUMPS 5.6.1.

Number of nonzeros in equality constraint Jacobian...:        0
Number of nonzeros in inequality constraint Jacobian.:        0
Number of nonzeros in Lagrangian Hessian.............:        3

Total number of variables............................:        2
```

```
                     variables with only lower bounds:        0
                variables with lower and upper bounds:        2
                     variables with only upper bounds:        0
Total number of equality constraints.................:        0
Total number of inequality constraints...............:        0
        inequality constraints with only lower bounds:        0
   inequality constraints with lower and upper bounds:        0
        inequality constraints with only upper bounds:        0

iter    objective    inf_pr   inf_du lg(mu)  ||d||  lg(rg) alpha_du alpha_pr  ls
   0  1.3856695e+07 0.00e+00 9.70e+01  -1.0 0.00e+00    -  0.00e+00 0.00e+00   0
   1  4.1877282e+06 0.00e+00 3.94e+01  -1.0 4.81e+00    -  4.03e-02 1.00e+00f  1
   2  7.4626085e+05 0.00e+00 1.16e+01  -1.0 4.68e+00    -  9.94e-01 1.00e+00f  1
   3  1.1541790e+05 0.00e+00 3.26e+00  -1.0 2.98e+00    -  1.00e+00 1.00e+00f  1
   4  1.3175100e+04 0.00e+00 8.37e-01  -1.0 1.79e+00    -  1.00e+00 1.00e+00f  1
   5  7.1409431e+02 0.00e+00 1.60e-01  -1.7 8.86e-01    -  1.00e+00 1.00e+00f  1
   6  3.4096513e+01 0.00e+00 1.36e-02  -2.5 2.80e-01    -  1.00e+00 1.00e+00f  1
   7  2.8367005e+01 0.00e+00 2.56e-04  -3.8 1.63e-01    -  9.81e-01 1.00e+00f  1
   8  2.6050947e+01 0.00e+00 4.41e-03  -5.7 4.88e+01    -  3.33e-02 4.96e-02f  5
   9  2.3356648e+01 0.00e+00 5.03e-03  -5.7 2.43e+00    -  9.91e-01 1.00e+00f  1
iter    objective    inf_pr   inf_du lg(mu)  ||d||  lg(rg) alpha_du alpha_pr  ls
  10  2.0568742e+01 0.00e+00 2.84e-03  -5.7 1.66e+00    -  1.00e+00 1.00e+00f  1
  11  1.9006996e+01 0.00e+00 6.06e-03  -5.7 2.57e+00    -  1.00e+00 1.00e+00f  1
  12  1.5917375e+01 0.00e+00 1.05e-03  -5.7 7.47e-01    -  1.00e+00 1.00e+00f  1
  13  1.5661315e+01 0.00e+00 5.92e-03  -5.7 4.71e+00    -  1.00e+00 5.00e-01f  2
  14  1.2150777e+01 0.00e+00 5.87e-04  -5.7 3.75e-01    -  1.00e+00 1.00e+00f  1
  15  1.0995766e+01 0.00e+00 2.62e-03  -5.7 5.42e+00    -  1.00e+00 2.50e-01f  3
  16  9.2807587e+00 0.00e+00 1.09e-03  -5.7 7.27e-01    -  1.00e+00 1.00e+00f  1
  17  8.3235430e+00 0.00e+00 1.52e-03  -5.7 1.56e+00    -  1.00e+00 5.00e-01f  2
  18  7.0460250e+00 0.00e+00 1.33e-03  -5.7 8.03e-01    -  1.00e+00 1.00e+00f  1
  19  5.8858703e+00 0.00e+00 1.01e-03  -5.7 6.37e-01    -  1.00e+00 1.00e+00f  1
iter    objective    inf_pr   inf_du lg(mu)  ||d||  lg(rg) alpha_du alpha_pr  ls
  20  4.9033128e+00 0.00e+00 9.38e-04  -5.7 5.86e-01    -  1.00e+00 1.00e+00f  1
  21  3.9797479e+00 0.00e+00 5.72e-04  -5.7 3.77e-01    -  1.00e+00 1.00e+00f  1
  22  3.3193263e+00 0.00e+00 6.86e-04  -5.7 4.40e-01    -  1.00e+00 1.00e+00f  1
  23  2.5481245e+00 0.00e+00 2.37e-04  -5.7 1.31e-01    -  1.00e+00 1.00e+00f  1
  24  2.1475610e+00 0.00e+00 3.11e-04  -5.7 4.04e-01    -  1.00e+00 5.00e-01f  2
  25  1.6279972e+00 0.00e+00 1.61e-04  -5.7 1.54e-01    -  1.00e+00 1.00e+00f  1
  26  1.3005814e+00 0.00e+00 3.15e-04  -5.7 2.19e-01    -  1.00e+00 1.00e+00f  1
  27  8.8740291e-01 0.00e+00 6.30e-05  -5.7 9.78e-02    -  1.00e+00 1.00e+00f  1
  28  6.9660011e-01 0.00e+00 2.02e-04  -5.7 3.22e-01    -  1.00e+00 5.00e-01f  2
  29  4.5955435e-01 0.00e+00 7.80e-05  -5.7 1.09e-01    -  1.00e+00 1.00e+00f  1
iter    objective    inf_pr   inf_du lg(mu)  ||d||  lg(rg) alpha_du alpha_pr  ls
  30  3.7378503e-01 0.00e+00 2.58e-04  -5.7 1.98e-01    -  1.00e+00 1.00e+00f  1
  31  1.7419935e-01 0.00e+00 1.86e-05  -5.7 9.55e-02    -  1.00e+00 1.00e+00f  1
  32  1.1678685e-01 0.00e+00 1.61e-04  -5.7 3.13e-01    -  1.00e+00 5.00e-01f  2
  33  5.1675933e-02 0.00e+00 2.27e-05  -5.7 1.03e-01    -  1.00e+00 1.00e+00f  1
  34  3.9425897e-02 0.00e+00 2.04e-04  -5.7 2.09e-01    -  1.00e+00 1.00e+00f  1
  35  5.1649798e-03 0.00e+00 2.71e-06  -5.7 5.44e-02    -  1.00e+00 1.00e+00f  1
  36  1.9553865e-03 0.00e+00 5.70e-05  -8.6 1.23e-01    -  9.95e-01 1.00e+00f  1
  37  6.6464788e-06 0.00e+00 5.67e-08  -8.6 1.02e-02    -  1.00e+00 1.00e+00f  1
  38  4.4011255e-09 0.00e+00 8.67e-08  -8.6 5.14e-03    -  1.00e+00 1.00e+00f  1
  39  3.3519648e-13 0.00e+00 1.29e-13  -8.6 1.55e-05    -  1.00e+00 1.00e+00f  1
iter    objective    inf_pr   inf_du lg(mu)  ||d||  lg(rg) alpha_du alpha_pr  ls
  40  4.3230209e-16 0.00e+00 4.08e-15 -10.0 1.12e-06    -  1.00e+00 1.00e+00f  1

Number of Iterations....: 40

                                   (scaled)                 (unscaled)
Objective...............:   1.4220267008343644e-20    4.3230208956579026e-16
Dual infeasibility......:   4.0802855406541762e-15    1.2404239415581401e-10
Constraint violation....:   0.0000000000000000e+00    0.0000000000000000e+00
Variable bound violation:   0.0000000000000000e+00    0.0000000000000000e+00
Complementarity.........:   9.0909232958894218e-11    2.7636788638282267e-06
Overall NLP error.......:   9.0909232958894218e-11    2.7636788638282267e-06


Number of objective function evaluations             = 80
Number of objective gradient evaluations             = 41
Number of equality constraint evaluations            = 0
Number of inequality constraint evaluations          = 0
```

```
Number of equality constraint Jacobian evaluations   = 0
Number of inequality constraint Jacobian evaluations = 0
Number of Lagrangian Hessian evaluations             = 40
Total seconds in IPOPT                               = 0.032

EXIT: Optimal Solution Found.
Optimal parameters (x, y): (0.9999999792197696, 0.9999999583699614)
Minimum objective value: 4.3230208956579026e-16
```

■

JuMP's native output is quite verbose. We can suppress it entirely with the following parameter setting:

```
# Suppress all of the output
set_optimizer_attribute(model, "print_level", 0)

# Solve the optimization problem
optimize!(model)
```

Output

Yes, nothing comes out. One can play with the `print_level` setting to find a preferred level of output. We suggest

set_optimizer_attribute(model, "print_level", 4).

Method 6.9: Additional Explanation on Using JuMP and Ipopt for Optimization

JuMP is a modeling language used in Julia for mathematical optimization. It provides a user-friendly interface to define optimization problems and connect with various solvers. Among the powerful solvers that can be used with JuMP is **Ipopt**, which stands for Interior Point OPTimizer. It us particularly suited for nonlinear optimization problems.

(a) **Setting up the Problem:** To start, you need to define your optimization problem. This involves:

- Specifying the decision variables (i.e., the variables that can be varied in the optimization, such as $x = (x_1, x_2, \ldots, x_n)$).
- Setting the objective function (i.e., the cost function).
- Adding any constraints.

(b) **Decision Variables:** In JuMP, you can define decision variables using the `@variable` macro. For example:

```
@variable(model, -2 <= x <= 2)
```

defines a variable x that is bounded between -2 and 2. If you want no bounds on x, that is, $-\infty < x < \infty$, then you use

```
@variable(model, x)
```

and if you want $x \geq 0$, you use

```
@variable(model, x >= 0)
```

That is fairly intuitive.

(c) **Objective Function:** The objective function is what you aim to minimize (or maximize). In JuMP, you can set the objective function using the `@objective` macro. For nonlinear objectives, use `@NLobjective`. For instance:

```
@NLobjective(model, Min, (1 - x)^2 + 100 * (y - x^2)^2)
```

This sets the Rosenbrock function as the objective (aka cost function) to be minimized.

(d) **Constraints:** In Section 6.3, we will deal with minimization subject to constraints, such as $g(x) \leq c$ or $g(x) = c$. Constraints restrict the feasible region of the optimization problem. In JuMP, you can add constraints using the `@constraint` macro. For nonlinear constraints, use `@NLconstraint`. For example,

```
@NLconstraint(model, x^2 + y^2 <= 1)
```

ensures that the solution (x, y) lies inside a unit circle. **We are not using this option yet, but we will later.**

(e) **Solving the Problem:** Once the problem is defined, you can solve it using an optimization solver. Here is how to set up and solve the problem using the Ipopt solver:

```
# Create a model with Ipopt as the solver
model = Model(Ipopt.Optimizer)

# Define variables, objective, and constraints...

# Solve the problem
optimize!(model)
```

(f) **Retrieving the Solution:** After solving, you can retrieve the optimal solution and objective value using:

```
optimal_x = value(x)
optimal_y = value(y)
min_value = objective_value(model)
```

Tips for Success:

- **Initial Guess**: The starting point can influence the solution. Use the `start` attribute to set an initial guess, e.g., `@variable(model, -2 <= x <= 2, start = -1.2)`.
- **Solver Options**: Adjust solver options for better convergence. For instance, increase the maximum number of iterations or set a tighter convergence tolerance.
- **Scaling**: Ensure that the problem's scale is appropriate. If variables or constraints have vastly different magnitudes, consider rescaling.
- If your code has a bug in it, paste the full code into ChatGPT and ask it to find and fix the error.

Bottom Line: JuMP, combined with the Ipopt solver, offers a powerful and flexible environment for solving optimization problems in Julia. With its intuitive syntax and robust capabilities, engineers can tackle a wide range of optimization challenges.

Warning: There is no guarantee that a GLOBAL MINIMUM will be found with any numerical optimizer. Before adjusting the parameters in Jump, via

```
set_optimizer_attribute(model, "max_iter", 1000)  # Allow up to 1000 iterations
set_optimizer_attribute(model, "tol", 1e-9)       # Require a very accurate
  # solution
```

it reported the "optimal solution" to the Rosenbrock Problem as $(0.7864151510570021, 0.6176983075999913)$, even though the true answer is $(1.0, 1.0)$.

```
using JuMP, Ipopt

# Create a new model with Ipopt as the solver
model = Model(Ipopt.Optimizer)

# Define variables
@variable(model, -2 <= x <= 2, start = -1.2)  # Set initial guess
@variable(model, -2 <= y <= 2, start = 1.0)   # Set initial guess

# Define the Rosenbrock function
@NLobjective(model, Min, (1 - x)^2 + 100 * (y - x^2)^2)

# Solve the optimization problem
optimize!(model)

# Display results
println("Optimal parameters (x, y): (", value(x), ", ", value(y), ")")
println("Minimum objective value: ", objective_value(model))
```

Abridged Output

```
EXIT: Optimal Solution Found.
Optimal parameters (x, y): (0.7864151510570021, 0.6176983075999913)
Minimum objective value: 0.04567481005305849
```

A recommended example: Gradient descent, how neural networks learn by **3Blue1Brown**.

6.2.4 Second-Derivative Tests for Local Min and Max

We cover both single-variable and multivariable problems.

Second-Order Derivative Test for a Scalar Variable

Proposition 6.10: Scalar Variable Second-order Test for Local Max and Min

Given a function $f : \mathbb{R} \to \mathbb{R}$, with a stationary point at $x = x^*$, that is, its first derivative $\frac{df(x^*)}{dx} = 0$, the nature of the stationary point can be determined using the second derivative of f at x^*.

- If $\frac{d^2 f(x^*)}{dx^2} > 0$, then f has a local minimum at $x = x^*$.
- If $\frac{d^2 f(x^*)}{dx^2} < 0$, then f has a local maximum at $x = x^*$.

- If $\frac{d^2 f(x^*)}{dx^2} = 0$, the test is inconclusive.

Note: To remember these, take the second derivatives of $f(x) = x^2$, which has a minimum at the origin, $f(x) = -x^2$, which has a maximum at the origin, $f(x) = x^3$, which has an inflexion point at the origin, and $f(x) = x^4$, which has a minimum at the at the origin, but the derivative test is inconclusive because $\frac{d^2 f(0)}{dx^2} = 12x^2\big|_{x^*=0} = 0$. Alternatively, Example 6.13, which shows how Taylor expansions can be used to establish the result, can be used to derive the second-derivative test.

Example 6.11. *$f(x) = x^2 - x^3 + x^4$ has a stationary point at the origin. Does the second-derivative test confirm it to be a local min, a local max, or is it inconclusive?*

Solution: Ans. Local minimum because $\frac{d^2 f(x)}{dx^2} = 2 - 6x + 12x^2$ and thus, $\frac{d^2 f(0)}{dx^2} = 2 > 0$. ■

Example 6.12. *Use Jump to find a local minimum of each of the following functions,*

(a) $f(x) = x^4 \sin(1 + x^2)$

(b) $f(x) = e^x \sin(1 + x^2)$,

and then verify or refute the results using the second-derivative test.

Solutions:

(a) Jump reports a local minimum for $f(x) = x^4 \sin(1 + x^2)$ at $x^* = 0.0$ but the second-derivative test is inconclusive.

(b) Jump reports a local minimum for $f(x) = e^x \sin(1 + x^2)$ at $x^* = -1.8572587077723928$, which is confirmed by the second-derivative test.

In the following, the second-derivative test is applied using `FiniteDiff`.

```
using JuMP, Ipopt, FiniteDiff

# Define the function
if false
    f(x) = x^4*sin(1 + x^2)
else
    f(x) = exp(x)*sin(1 + x^2)
end

# Create a new model with Ipopt as the solver
model = Model(Ipopt.Optimizer)

# Manually register the function with the model to use automatic
# differentiation
register(model, :f, 1, f; autodiff = true)

# Define variables
@variable(model, x)

# Define the cost function using the registered function
@NLobjective(model, Min, f(x))
```

```
# Fine-tune a few parameters
set_optimizer_attribute(model, "max_iter", 1000)  # Allow up to 1000 iterations
set_optimizer_attribute(model, "tol", 1e-9)       # Require a very accurate
#   solution

# Suppress much of the output
set_optimizer_attribute(model, "print_level", 0)

# Solve the optimization problem
optimize!(model)

# Minimizing value
xStar = value(x)

# Display results
println("\n Optimal parameters x: ", xStar)
println("Minimum objective value: ", objective_value(model))

# Define a method to compute the first derivative
function df(x)
    return FiniteDiff.finite_difference_derivative(f, x)
end

# Compute the second derivative using finite_difference_derivative on
# the first derivative function
second_derivative = FiniteDiff.finite_difference_derivative(df, xStar)

println("The second derivative of f(x) at x = $xStar is: ", second_derivative[1, 1])
```

Output for $f(x) = x^4 \sin(1 + x^2)$

```
Optimal parameters x: 0.0

Minimum objective value: 0.0

The second derivative of f(x) at x = 0.0 is: 2.4684402317159913e-10
```

Output $f(x) = e^x \sin(1 + x^2)$

```
Optimal parameters x: -1.8572587077723928

Minimum objective value: -0.15073321671888845

The second derivative of f(x) at x = -1.8572587077723928 is:
2.149336845346529
```

■.

Example 6.13. *Use a second-order Taylor expansion to prove (or at least, plausibly argue the veracity of) Prop. 6.10.*

Solution: In the prime notation for derivatives, we are given that $f'(x^*) = 0$. Hence, the second order Taylor expansion about x^* is

$$f(x) = f(x^*) + f''(x^*) \cdot (x - x^*)^2.$$

If $f''(x^*) > 0$, then f is locally an upward pointing parabola about x^*, and thus x^* is a local minimum. If $f''(x^*) < 0$, then f is locally a downward pointing parabola about x^*, and thus x^* is a local maximum. If $f''(x^*) = 0$, then higher-order terms in the Taylor expansion would be needed to reach any conclusion. ■

(Optional Read) Second-Order Derivative Test for a Multivariable Function

For multivariable functions, stationary points are the roots of the gradient. To check if a stationary point is a local max, min, or neither, we must use the Hessian from Def. 5.43.

Proposition 6.14: Multivariable Second-order Test for Local Max and Min

Consider a function $f : \mathbb{R}^n \to \mathbb{R}$ that is twice continuously differentiable. Let $\mathbf{x}^* = (x_1^*, x_2^*, \ldots, x_n^*)$ be a stationary point of f; i.e., the gradient $\nabla f(\mathbf{x}^*) = \mathbf{0}_{n\times 1}$.

To determine whether $\mathbf{x}^*$ is a local minimum, local maximum, or neither, we examine the Hessian matrix of f at $\mathbf{x}^*$, denoted by $H_f(\mathbf{x}^*)$.

- If $H_f(\mathbf{x}^*)$ is **positive definite**, then f has a local minimum at $\mathbf{x}^*$.
- If $H_f(\mathbf{x}^*)$ is **negative definite**, then f has a local maximum at $\mathbf{x}^*$.
- If $H_f(\mathbf{x}^*)$ is **indefinite** then $\mathbf{x}^*$ is a saddle point, or the test is inconclusive.

To determine if the Hessian matrix is positive definite, negative definite, or indefinite, we can examine its eigenvalues:

- All eigenvalues of $H_f(\mathbf{x}^*)$ are positive $\implies$ $H_f(\mathbf{x}^*)$ is positive definite.
- All eigenvalues of $H_f(\mathbf{x}^*)$ are negative $\implies$ $H_f(\mathbf{x}^*)$ is negative definite.
- Eigenvalues of $H_f(\mathbf{x}^*)$ are of mixed signs or one is zero $\implies$ $H_f(\mathbf{x}^*)$ is indefinite.

A fully numerical example follows below.

Instructions for ChatGPT4: This prompt guides ChatGPT in generating a Julia solution that not only addresses an interesting optimization problem but also includes educational documentation throughout the code. It ensures that the solution will be accessible and informative for students, providing a valuable learning resource on numerical optimization, the use of specific Julia packages, and the mathematical analysis of optimization problems.

"Create a comprehensive Julia solution to analyze the Himmelblau function. The solution should include the following components, each thoroughly documented with comments in the code:

- **Function Definition:** Implement the Himmelblau function $f(x, y) = (x^2 + y - 11)^2 + (x + y^2 - 7)^2$ in Julia. Provide a brief comment on what the Himmelblau function represents in optimization theory.
- **Optimization Setup with JuMP and Ipopt:** Use the JuMP package to define and solve the optimization problem for finding critical points of the Himmelblau function. Employ Ipopt as the optimizer. Document how to set up the optimization model, define variables x and y, and specify the objective function within JuMP.
- **Moreover:** Initialize the optimization from multiple starting points: (3, 2), (-2.805118, 3.283186), (-3.779310, -3.283186), and (3.584428, -1.848126). Explain the choice of these points and their significance in exploring the function's landscape.
- **Hessian Computation with FiniteDiff:** After identifying a critical point, use the FiniteDiff.jl package to compute the Hessian matrix at that point. Include comments explaining the purpose of computing the Hessian and how it is done using FiniteDiff.
- **Eigenvalue Analysis with LinearAlgebra:** Calculate the eigenvalues of the Hessian matrix using the LinearAlgebra package. Document the process of eigenvalue calculation and its importance in classifying critical points.

- **Critical Point Classification:** Based on the eigenvalues of the Hessian, classify each critical point as a local minimum, maximum, or saddle point. Include a detailed explanation of how the classification is determined from the eigenvalues.

Ensure the Julia code is clean, well organized, and fully documented, with comments explaining each step of the process and the significance of each operation. The documentation should be detailed enough for someone unfamiliar with the process to understand the methodology, purpose, and implementation details."
You will obtain code that is more thorough than the example below!

```julia
using JuMP, Ipopt, FiniteDiff, LinearAlgebra

# Himmelblau function definition remains the same
function himmelblau(x, y)
    return (x^2 + y - 11)^2 + (x + y^2 - 7)^2
end

# Extended function to also compute Hessian and its eigenvalues
function optimize_and_analyze_himmelblau(x_init, y_init)
    model = Model(Ipopt.Optimizer)

    @variable(model, x, start = x_init)
    @variable(model, y, start = y_init)

    @NLobjective(model, Min, (x^2 + y - 11)^2 + (x + y^2 - 7)^2)

    set_optimizer_attribute(model, "max_iter", 1000)
    set_optimizer_attribute(model, "tol", 1e-9)
    set_optimizer_attribute(model, "print_level", 0)

    optimize!(model)

    xStar, yStar = value(x), value(y)
    println("\nOptimal parameters (x, y): (", xStar, ", ", yStar, ")")
    println("Minimum objective value: ", objective_value(model))

    # Define a wrapper function for FiniteDiff
    function himmelblau_wrapper(v)
        return himmelblau(v[1], v[2])
    end

    # Compute the Hessian at the optimal point
    hessian = FiniteDiff.finite_difference_hessian(himmelblau_wrapper, [xStar, yStar])

    # Compute the eigenvalues of the Hessian
    eigenvalues = eigvals(hessian)

    println("Hessian at the optimal point: \n", hessian)
    println("Eigenvalues of the Hessian: ", eigenvalues)

    # Determine the nature of the critical point based on the
    # eigenvalues
    if all(eig -> eig > 0, eigenvalues)
        println("The point is a local minimum.")
    elseif all(eig -> eig < 0, eigenvalues)
        println("The point is a local maximum.")
    else
```

```
            println("The point is a saddle point or the test is inconclusive.")
        end
end

# Starting points near the four local minima
starting_points = [(3, 2), (-2.805118, 3.283186),
    (-3.779310, -3.283186), (3.584428, -1.848126)]

# Optimize Himmelblau's function and analyze each result from the
# starting points
for (x_init, y_init) in starting_points
    println("Starting from point: (", x_init, ", ", y_init, ")")
    optimize_and_analyze_himmelblau(x_init, y_init)
end
```

Output

```
Starting from point: (3, 2)

Optimal parameters (x, y): (3.0, 2.0)
Minimum objective value: 0.0
Hessian at the optimal point:
[74.0000002682209 20.0; 20.0 34.00000011920929]
Eigenvalues of the Hessian: [25.715728893569633, 82.28427149386056]
The point is a local minimum.

Starting from point: (-2.805118, 3.283186)

Optimal parameters (x, y): (-2.805118086952745, 3.131312518250573)
Minimum objective value: 7.888609052210118e-31
Hessian at the optimal point:
[64.949500088456 1.3047777251870691; 1.3047777251870691
80.44094498775873]
Eigenvalues of the Hessian: [64.84037300554613, 80.5500720706686]
The point is a local minimum.

Starting from point: (-3.77931, -3.283186)

Optimal parameters (x, y): (-3.779310253377774, -3.2831859912861767)
Minimum objective value: 3.996369345849646e-26
Hessian at the optimal point:
[116.26548835580718 -28.24998497866954; -28.24998497866954
88.23448234835634]
Eigenvalues of the Hessian: [70.71435509450659, 133.78561560965693]
The point is a local minimum.

Starting from point: (3.584428, -1.848126)

Optimal parameters (x, y): (3.584428340330554, -1.8481265269646447)
Minimum objective value: 9.516628633969033e-25
Hessian at the optimal point:
[104.78501259863707 6.945207253476179; 6.945207253476179
29.32457337913531]
Eigenvalues of the Hessian: [28.690677260855026, 105.41890871691734]
The point is a local minimum.
```

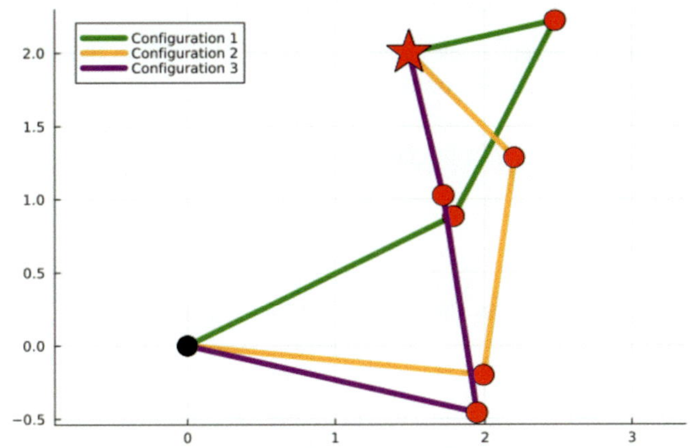

Figure 6.4: This is the same 3-link manipulator from **Fig. 1.7-(b)**. Here, the red star ($\star$) indicates the desired position of the end of the last link, meaning it is where we are trying to position the robot's end effector. The precise point in Cartesian space is $p_3 = (1.5, 2)$. We see that there are numerous configurations (i.e., different values for θ_i) of the robot that achieve the same position of the end effector (we only show three of an infinite number of possible configurations). Hence the problem is **underdetermined**. What is an aspiring Roboticist to do with this messy situation? **The solution is constrained minimization.**

Example 6.15. *Sketch out a plausible argument for Proposition 6.14.*

Solution: We use facts about symmetric matrices.

Let $x^* \in \mathbb{R}^n$ be a critical point of $f(x)$, meaning that $\nabla f(x^*) = 0$. To determine the nature of a stationary point, we will analyze the behavior of the second-order Taylor expansion of $f(x)$ around x^*.

The Hessian matrix of f at x^*, denoted $H_f(x^*)$, is symmetric. Therefore, the eigenvalues of $H_f(x^*)$, denoted $\{\lambda_1, \lambda_2, \dots, \lambda_n\}$, are real, and $H_f(x^*)$ has a full set of linearly independent eigenvectors, $\{v_1, \dots, v_n\}$. We can select these eigenvectors such that the matrix

$$Q := [v_1 \;\; v_2 \;\; \cdots \;\; v_n]$$

is orthogonal, i.e., $Q^\top \cdot Q = I_n$.

Step 1: Change of Variables

We perform a change of variables by introducing the new coordinate system $y = Q^\top(x - x^*)$. In this coordinate system, we will examine the Hessian and show that it is diagonal. Since Q is orthogonal, this transformation preserves distances, so the nature of the critical point does not change.

Step 2: Taylor Expansion

The second-order Taylor expansion of $f(x)$ around x^* is:

$$f(x) \approx f(x^*) + \frac{1}{2}(x - x^*)^\top H_f(x^*)(x - x^*).$$

Because $\nabla f(x^*) = 0$, there is no first-order term. Now, using the change of variables $y = Q^\top(x - x^*)$, we express the quadratic form in terms of y:

$$f(x) \approx f(x^*) + \frac{1}{2} y^\top Q^\top H_f(x^*) Q y$$

Step 3: Diagonalizing the Hessian

Because Q is the matrix of eigenvectors of $H_f(x^*)$, we know that:

$$Q^\top H_f(x^*)Q = \text{diag}(\lambda_1, \lambda_2, \dots, \lambda_n).$$

This is a diagonal matrix with the eigenvalues $\lambda_1, \lambda_2, \dots, \lambda_n$ of $H_f(x^*)$ on the diagonal. Therefore, in the new coordinates y, the second-order term becomes:

$$f(x) \approx f(x^*) + \frac{1}{2}\sum_{i=1}^{n} \lambda_i y_i^2.$$

Step 4: Analyzing the Eigenvalues

- If all the eigenvalues $\{\lambda_1, \dots, \lambda_n\}$ of the Hessian matrix are positive, then the quadratic form

  $$\frac{1}{2}\sum_{i=1}^{n} \lambda_i y_i^2$$

  is strictly positive for all nonzero y. This implies that the function $f(x)$ is locally bowl-shaped in a neighborhood around x^*, meaning x^* is a local minimum of $f(x)$. Similarly, if all of the eigenvalues are negative, we have the analogy of a downward pointing parabola, and hence x^* is a local maximum.
- When the eigenvalues have mixed signs, such as $y_1^2 - y_2^2$, we have the classic saddle point shown in Fig. 6.3.
- When one of the eigenvalues is zero, then the test is inconclusive, just as in the scalar case.

Step 5: Chain Rule Explanation

In this proof, we used the chain rule implicitly during the change of variables. When we transformed the variable from x to $y = Q^\top(x - x^*)$, the Hessian transforms according to the rule:

$$H_y = Q^\top H_f(x^*)Q$$

This is a standard result in multivariable calculus, showing how the Hessian matrix transforms under a change of coordinates, which diagonalizes it when expressed in the eigenvector basis. ■

6.3 Constrained Optimization

We first motivate the problem being addressed via a physical example, namely **inverse kinematics** . Then we cover the typical material that you would learn in multivariate calculus for use in an engineering course. We conclude with numerical methods for larger-scale problems.

6.3.1 Motivating Problems

As motivation, we pose two example problems based on Fig. 6.4. Recall that the end effector position of the robot as a function of $\theta = (\theta_1, \theta_2, \theta_3)$ was given in (1.10). We will first take just the x-component of the end effector and define the function

$$g_1(\theta) := L1 \cdot \cos(\theta_1) + L2 \cdot \cos(\theta_1 + \theta_2) + L3 \cdot \cos(\theta_1 + \theta_2 + \theta_3) - \mathbf{1.5}, \tag{6.13}$$

so that $g_1(\theta)$ vanishes when the end effector is at the desired x-position, namely $\mathbf{1.5}$. Given this function, we pose:

Motivating Problem 1:

$$\begin{aligned} &\text{Minimize} \quad f(\theta) := (\theta_1)^2 + (\theta_2)^2 + (\theta_3)^2, \\ &\text{subject to} \quad g_1(\theta) = 0. \end{aligned} \tag{6.14}$$

The "**subject to**" means that a solution to this problem is **constrained** to find values for θ_1, θ_2, and θ_3 such that $g(\theta_1, \theta_2, \theta_3) = 0$. **We designed g_1 so that it only vanishes for values of θ where the x-position of the end effector is at the desired value of 1.5. This is why engineers use constraints in optimization problems: we often have goals to achieve that are distinct from minimizing some quantity.**

What is the role of the cost function? Well, we already saw in **Fig. 6.4** that there are an infinite number of values of θ_1, θ_2, and θ_3 that achieve the full constraint

$$p_3 = \begin{bmatrix} 1.5 \\ 2.0 \end{bmatrix},$$

so there must be even more solutions that only constrain the x-component of p_3 and allows the y-component to be **unconstrained**. The cost function sorts through this infinite set of values and finds the one that minimizes the cost while satisfying the constraint. **This is a mainstay of modern engineering**.

Next, we define

$$g(\theta) := \begin{bmatrix} g_1(\theta) \\ g_2(\theta) \end{bmatrix} = \begin{bmatrix} L1 \cdot \cos(\theta_1) + L2 \cdot \cos(\theta_1 + \theta_2) + L3 \cdot \cos(\theta_1 + \theta_2 + \theta_3) - \mathbf{1.5} \\ L1 \cdot \sin(\theta_1) + L2 \cdot \sin(\theta_1 + \theta_2) + L3 \cdot \sin(\theta_1 + \theta_2 + \theta_3) - \mathbf{2.0} \end{bmatrix}, \tag{6.15}$$

so that $g(\theta)$ vanishes when the end effector is at the desired x- **and** y-positions. Given this function, we pose:

Motivating Problem 2:

$$\begin{aligned} \text{Minimize} \quad & f(\theta) := (\theta_1)^2 + (\theta_2)^2 + (\theta_2)^2, \\ \text{subject to} \quad & g(\theta) = 0. \end{aligned} \tag{6.16}$$

We can also write the problem as

$$\begin{aligned} \text{Minimize} \quad & f(\theta) := (\theta_1)^2 + (\theta_2)^2 + (\theta_2)^2, \\ \text{subject to} \quad & g_1(\theta) = 0 \text{ and } g_2(\theta) = 0. \end{aligned} \tag{6.17}$$

Once again, the "**subject to**" means that a solution to this problem is **constrained** (e.g., obliged or forced) to find values for θ_1, θ_2, and θ_3 such that $g(\theta_1, \theta_2, \theta_3) = 0$. And once again, we designed g so that it only vanishes when the (x, y)-position of the end effector is at the desired point in Cartesian space. The cost function takes care of the ambiguity in trying to place the end effector (that depends on three angles) at the desired location (specified by a 2-vector in the plane): we have three variables and only two constraints. **Hence, we have an underdetermined problem, and constrained optimization is a perfect method for resolving the ambiguity associated with undetermined problems.**

Note: These two problems are typical examples of inverse kinematics, where one seeks to solve the joint angles of a robot mechanism so that the end effector achieves a desired "pose", that is, a given position and orientation.

6.3.2 Vocabulary of Constrained Optimization

In the context of constrained optimization, several key terms and concepts are fundamental to understanding and solving optimization problems. Below is an overview of the primary vocabulary:

- **Cost Function**: Also known as the objective function, this is the function $f : \mathbb{R}^n \to \mathbb{R}$ that we aim to minimize or maximize within the feasible set. The cost function quantifies the "cost" associated with a particular choice of $x = (x_1, x_2, \ldots, x_n)$.

- **Decision Variables**: For a cost function, $f(x) = f(x_1, x_2, \ldots, x_n)$, the components $x_i, 1 \leq i \leq n$ are the decision variables. They are the variables we can search over to find values that extremize the cost function.

- **Constraints**: Conditions that the solution x must satisfy. Constraints can be classified into two types:

- **Equality Constraints**: These require that the constraint functions equal zero, $g(x) = 0$.
- **Inequality Constraints**: These specify that the constraint functions must be greater than or equal to zero, $h(x) \geq 0$, or less than or equal to zero, depending on the problem formulation.

- **Feasible Set**: The set of all points x that satisfy the constraints of the optimization problem. For equality constraints, this set includes all x such that $g(x) = 0$, where $g(x)$ represents the constraint functions.

- **Constraint Qualification**: A set of conditions that ensure the mathematical program has a well-defined solution. One common qualification is that the rows of the Jacobian matrix of the constraint functions $g(x)$ are linearly independent at all points in the feasible set. Equivalently, the gradients of the constraints are linearly independent as column vectors.

- **Lagrange Multipliers**: Numbers λ associated with each constraint that measure the sensitivity of the objective function's optimal value to changes in the constraint boundary. The method of Lagrange multipliers is a strategy for finding the local maxima and minima of a function subject to equality constraints.

- **KKT Conditions** (Karush-Kuhn-Tucker): Necessary conditions for a solution in nonlinear programming to be optimal, given certain regularity conditions. These conditions generalize the method of Lagrange multipliers to handle inequality constraints. (Typically covered in 400-level courses; it is not treated here.)

- **Slack Variables**: Additional variables introduced to transform inequality constraints into equality constraints, making it easier to apply certain optimization methods. (Typically covered in 400-level courses; is not treated here.)

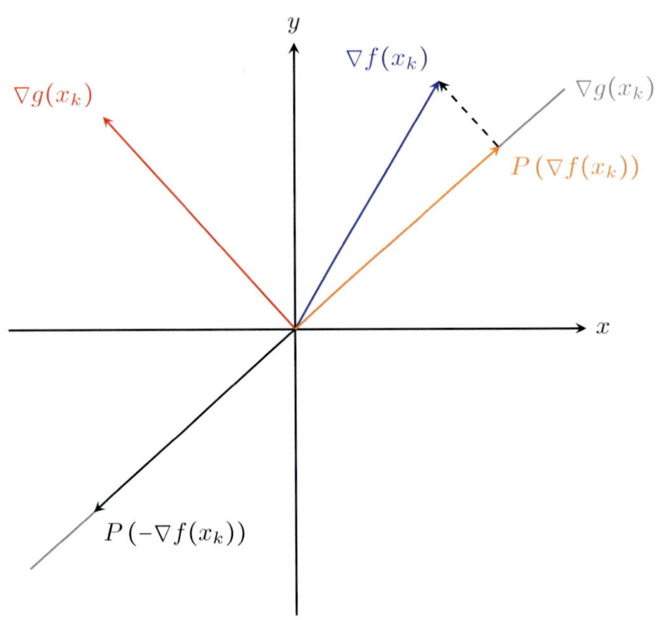

Figure 6.5: Geometry of gradient descent with an equality constraint. In the image, the line labeled $\nabla g(x_k)^\perp$ is the set of points (actually, the subspace) orthogonal to $\nabla g(x_k)$, the gradient of the cost. The angle between $\nabla f(x_k)$ and $\nabla g(x_k)$ is 72°. The dotted line shows $\alpha \cdot \nabla g(x_k)$, the proportion of $\nabla g(x_k)$ that must be removed (aka subtracted) from $\nabla f(x_k)$ so that $\nabla f(x_k) - \alpha \nabla g(x_k)$ is orthogonal to the constraint gradient. The new vector, denoted $P(\nabla f(x_k))$, is an ascent direction that respects the constraint, whereas $P(-\nabla f(x_k))$, is a descent direction that respects the constraint.

6.3.3 Gradient Descent with a Single Equality Constraint

Suppose that the functions $f : \mathbb{R}^n \to \mathbb{R}$ and $g : \mathbb{R}^n \to \mathbb{R}$ define the cost and the constraint in the constrained minimization problem:

$$\begin{aligned} \text{Minimize} \quad & f(x), \\ \text{subject to} \quad & g(x) = 0. \end{aligned} \tag{6.18}$$

We seek to modify the gradient descent algorithm to accommodate the equality constraint. Let $x_k \in \mathbb{R}^n$ be the current value of the vector of decision variables, which we assume to be feasible, that is $g(x_k) = 0$. We seek Δx such that:

$$\begin{aligned} & f(x_k + \Delta x) < f(x_k) \\ \text{and} & \\ & g(x_k + \Delta x) = 0. \end{aligned} \tag{6.19}$$

We use our gradient notation to linearize these equations about x_k,

$$\begin{aligned} & f(x_k + \Delta x) \approx f(x_k) + \nabla f(x_k) \bullet \Delta x \\ \text{and} & \\ & g(x_k + \Delta x) \approx \nabla g(x_k) \bullet \Delta x. \end{aligned} \tag{6.20}$$

From (6.19) and (6.20), we deduce that Δx must satisfy:

$$\begin{aligned} \nabla f(x_k) \bullet \Delta x &< 0, \\ \nabla g(x_k) \bullet \Delta x &= 0. \end{aligned} \tag{6.21}$$

We pose $\Delta x := -\nabla f(x_k) + \alpha \nabla g(x_k)$ and seek to choose α such that $g(x_k) \bullet \Delta x = 0$. Next, we compute

$$\begin{aligned} 0 =& \nabla g(x_k) \bullet \Delta x \\ & \Updownarrow \\ 0 =& \nabla g(x_k) \bullet (-\nabla f(x_k) + \alpha \nabla g(x_k)) \\ & \Updownarrow \\ 0 =& -\nabla g(x_k) \bullet \nabla f(x_k) + \alpha \nabla g(x_k) \bullet \nabla g(x_k) \end{aligned} \tag{6.22}$$

Hence, if $\nabla g(x_k) \neq 0_{n \times 1}$, then

$$\alpha = \frac{\nabla g(x_k) \bullet \nabla f(x_k)}{\nabla g(x_k) \bullet \nabla g(x_k)} \tag{6.23}$$

and we have

$$\Delta x = -\nabla f(x_k) + \frac{\nabla g(x_k) \bullet \nabla f(x_k)}{\nabla g(x_k) \bullet \nabla g(x_k)} \cdot \nabla g(x_k). \tag{6.24}$$

Figure 6.5 shows that this is the orthogonal projection of $-\nabla f(x_k)$ onto $[\nabla g(x_k)]^\perp := \{v \in \mathbb{R}^n \mid v \perp \nabla g(x_k)\}$. In simpler terms, we have removed from $-\nabla f(x_k)$ its component lying in the direction of $\nabla g(x_k)$. This gives the basic setup for gradient descent with a single equality constraint.

Method 6.16: Gradient Descent with a Single Equality Constraint

Consider the constrained optimization problem (6.18) for functions $f : \mathbb{R}^n \to \mathbb{R}$ and $g : \mathbb{R}^n \to \mathbb{R}$ with optimal value x^*. If the gradient vector $\nabla g(x^*) \neq 0_{n \times 1}$, then a locally optimal x^* can be found as follows:

Initialization: Find a feasible point $x_0 \in \mathbb{R}^n$, that is, a point such that $g(x_0) = 0$; this can be accomplished with `NLsolve`.

Loop: Given a point $x_k \in \mathbb{R}^n$, repeat the following steps to either find x_{k+1} or terminate.

(a) **Compute:** $\nabla f(x_k)$, the gradient of f at x_k and $\nabla g(x_k)$, the gradient of the constraint vector.

(b) **Compute a Descent Direction that Respects the Constraint:** $\Delta x = -\nabla f(x_k) + \frac{\nabla g(x_k) \bullet \nabla f(x_k)}{\nabla g(x_k) \bullet \nabla g(x_k)} \cdot \nabla g(x_k)$

(c) **Update Step:** $x_{k+1} = x_k + s \cdot \Delta x$, where $s > 0$ is the step size.

(d) **Termination Condition:** $||\Delta x|| < a_{\text{tol}}$; in other words, no more progress can be made, and thus x_k is (to numerical precision) a stationary point of the optimization problem.

Notes:

- In practice, the updates x_k will "slowly" exit the feasible set. Hence, when $|g(x_k)|$ exceeds a user-defined tolerance, a new call to `NLsolve` should be made. This is illustrated in the following code block.
- We will see shortly that **`JuMP` does all of this and a whole lot more**. It is our best option for use in professional projects. Algorithms, like the one presented here, are essential for building fundamental understanding of gradient descent, the workhorse of machine learning (ML) and inverse kinematics!

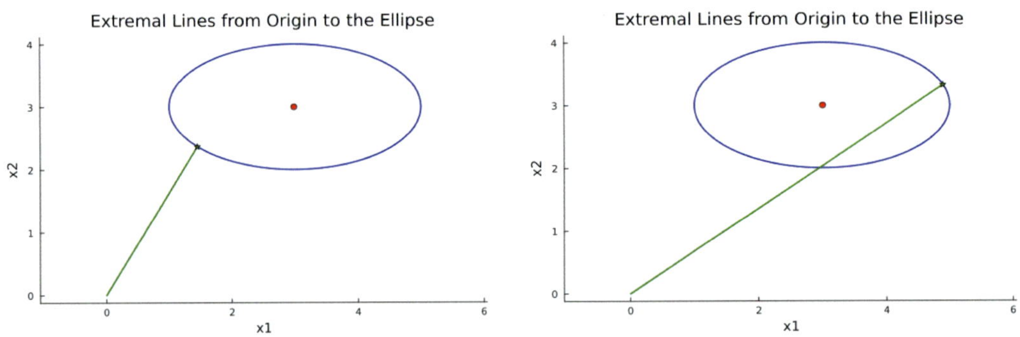

(a) (b)

Figure 6.6: Optimization with equality constraints. In these examples, the cost function is $f(x,y) = (x)^2 + (y)^2$, the distance squared from the origin to (x,y). The constraint is $g(x,y) = \frac{(x-x_c)^2}{a^2} + \frac{(y-y_c)^2}{b^2} - 1$, an ellipse. (a) Shows the minimum distance solution subject to (x,y) lying on the ellipse, while (b) shows the maximum distance solution. The first problem uses gradient **descent** with one equality constraint. The second problem uses gradient **ascent** with one equality constraint, achieved by changing the sign in the update step.

Example 6.17. *Solve the optimization problem (6.18) where the cost is $f(x,y) = x^2 + y^2$, the distance squared from the origin, and the constraint is that (x,y) lies on the ellipse defined by $g(x,y) = 0$, with $g(x,y) = \frac{(x-x_c)^2}{a^2} + \frac{(y-y_c)^2}{b^2} - 1$. Also find the maximizing solution.*

Solution: The solution is worked out in the code block below. The results are plotted in Fig. 6.6. ■

```
using ForwardDiff, LinearAlgebra, NLsolve

# Define the cost function f(x) and the equality constraints G(x)
xc = 3.0; yc = 3.0 # centers
a = 2 # x semi-major axis
b = 1.0 #y semi-major axis

# note: a = b = 1/r for a circle

```

```
# Define the cost (distance squared)
function cost(x)
    return x[1]^2 + x[2]^2
end

# Define the equality constraint: point must lie on the ellipse
function geq(x)
    # note: a = b = 1/r for a circle
    return (1/a)^2 * (x[1] - xc)^2 + (1/b)^2 * (x[2] - yc)^2 - 1
end

# Compute the gradient of the cost and constraint using ForwardDiff
grad_cost(x) = ForwardDiff.gradient(cost, x)
grad_geq(x)  = ForwardDiff.gradient(geq, x)

# Remove from the gradient of the cost any component lying along
# the gradient(s) of the constraint(s)
function modGradientCost(x)
    gradCost       = grad_cost(x)
    gradConstraint = grad_geq(x)

    if norm(gradConstraint) < 1e-6
        println("Gradient of the constraint is essentially zero and was ignored")
        alpha = 0.0
    else
        alpha = dot(gradCost, gradConstraint) / dot(gradConstraint, gradConstraint)
    end

    modGradCost = gradCost - alpha * gradConstraint
    return modGradCost
end

# Function to find a feasible starting point using NLsolve
function find_feasible_point(geq, x_guess)
    # Build function for NLsolve
    function constraint!(F, x)
        F[1] = geq(x)
    end
    xFeasible = nlsolve(constraint!, x_guess)
    return xFeasible.zero
end

# Starting guess (not feasible initially)
x_guess = [0.1; 0.1]
x0 = find_feasible_point(geq, x_guess) # Make the initial guess feasible

# Set step size, max iterations, and tolerances
s = 0.1
maxIter = 1000
gradTol = 1e-6
constraintTol = 1e-4

xk = x0
for k = 1:maxIter
    modGradCost = modGradientCost(xk)   # Use the modified gradient cost
    # function
    if norm(modGradCost) < gradTol   # Convergence check
```

```
        println("---- Iteration $k ----")
        println("Converged.")
        println("Cost: ", round.(cost(xk), digits=5))
        println("Constraints (should be near 0): ", round.(geq(xk), digits=5))
        println("------------------ ")
        break
    else
        xk = xk - s * modGradCost  # Update using the modified gradient
    end
    if norm(geq(xk)) > constraintTol
        println("Updated x to maintain g(xk)=0 at interation = $k")
        xk = find_feasible_point(geq, xk)  # Ensure feasibility
    end
end

xStar = xk
costStar = cost(xStar)
geqStar = geq(xStar)

println("\nFinal Results:")
println("-----------------")
println("Optimal x: ", round.(xStar, digits=5))
println("Optimal cost(xStar): ", round(costStar, digits=5))
println("Constraint(xStar): ", round(geqStar, digits=5))
```

Output

```
Updated x to maintain g(xk)=0 at interation = 1
Updated x to maintain g(xk)=0 at interation = 2
Updated x to maintain g(xk)=0 at interation = 3
Updated x to maintain g(xk)=0 at interation = 4
Updated x to maintain g(xk)=0 at interation = 5
Updated x to maintain g(xk)=0 at interation = 6
Updated x to maintain g(xk)=0 at interation = 7
Updated x to maintain g(xk)=0 at interation = 8
---- Iteration 25 ----
Converged.
Cost: 7.70976
Constraints (should be near 0): 9.0e-5
-----------------

Final Results:
-----------------
Optimal x: [1.45048, 2.36767]
Optimal cost(xStar): 7.70976
Constraint(xStar): 9.0e-5
```

6.3.4 Gradient Descent with a Vector of Equality Constraints

Suppose now that in addition to the cost function $f : \mathbb{R}^n \to \mathbb{R}$, there are m equality constraints, $g_i : \mathbb{R}^n \to \mathbb{R}$ in the problem

$$\begin{aligned} &\text{minimize} && f(x), \\ &\text{subject to} && g_i(x) = 0,\ 1 \le i \le m. \end{aligned} \tag{6.25}$$

We seek to modify the gradient descent algorithm to accommodate a vector of equality constraints. Let $x_k \in \mathbb{R}^n$ be the current value of the vector of decision variables, which we assume to be feasible, that is

$g_i(x_k) = 0$, $1 \leq i \leq m$. We seek Δx such that:

$$\begin{aligned} & f(x_k + \Delta x) < f(x_k) \\ \text{and} & \\ & g_i(x_k + \Delta x) = 0, \; 1 \leq i \leq m. \end{aligned} \tag{6.26}$$

We use our gradient notation to linearize these equations about x_k,

$$\begin{aligned} & f(x_k + \Delta x) \approx f(x_k) + \nabla f(x_k) \bullet \Delta x \\ \text{and} & \\ & g_i(x_k + \Delta x) \approx \nabla g_i(x_k) \bullet \Delta x, \; 1 \leq i \leq m. \end{aligned} \tag{6.27}$$

From (6.19) and (6.27), we deduce that Δx must satisfy:

$$\begin{aligned} \nabla f(x_k) \bullet \Delta x &< 0, \\ \nabla g_i(x_k) \bullet \Delta x &= 0, \; 1 \leq i \leq m. \end{aligned} \tag{6.28}$$

Analogously to the single-constraint case, we pose

$$\Delta x := -\nabla f(x_k) + \sum_{i=1}^{m} \alpha_i \nabla g_i(x_k) = -\nabla f(x_k) + \underbrace{\begin{bmatrix} \nabla g_1(x_k) & \nabla g_2(x_k) & \cdots & \nabla g_m(x_k) \end{bmatrix}}_{\nabla G(x_k)} \cdot \underbrace{\begin{bmatrix} \alpha_1 \\ \alpha_2 \\ \vdots \\ \alpha_m \end{bmatrix}}_{\alpha} \tag{6.29}$$

and seek to solve for $\{\alpha_1, \alpha_2, \ldots, \alpha_m\}$ such that

$$\nabla g_i(x_k) \bullet \Delta x = 0, \; 1 \leq i \leq m. \tag{6.30}$$

The above is equivalent to $\nabla G(x_k)^\top \cdot \Delta x = 0_{m \times 1}$, yielding

$$-\nabla G(x_k)^\top \cdot \nabla f(x_k) + \nabla G(x_k)^\top \cdot \nabla G(x_k) \cdot \alpha = 0_{m \times 1}. \tag{6.31}$$

If the columns of $\nabla G(x_k)$ are linearly independent, then the above equation has a unique solution. For small problems, we can invert the matrix $\nabla G(x_k)^\top \cdot \nabla G(x_k)$ to solve for the vector of coefficients α, but for larger problems, it is better to use matrix factorization, such as LU or QR. In Julia, the backslash command will choose the most appropriate method for the given problem. An alternative method to compute Δx in (6.29), is to apply Gram-Schmidt to the vectors

$$\{\nabla g_1(x_k), \nabla g_2(x_k), \ldots, \nabla g_m(x_k), -\nabla f(x_k)\}, \tag{6.32}$$

where it is very important to place the gradient of the cost **after** the gradients of the constraints.

Example 6.18. *Define the cost $f : \mathbb{R}^4 \to \mathbb{R}$ by $f(x_1, x_2, x_3, x_4) = \sum_{i=1}^{4} (x_i)^2$, distance squared from the origin to a point $x \in \mathbb{R}^4$, with the constraint that x must satisfy*

$$0_{2\times 1} = g(x) = \begin{bmatrix} g_1(x) \\ g_2(x) \end{bmatrix} = \begin{bmatrix} \frac{(x_1-3)^2}{2^2} + \frac{(x_2-3)^2}{1^2} - 1 \\ x_1 + x_2 + x_3 + x_4 \end{bmatrix}.$$

This is a minimum distance problem from the origin to the feasible set, that is, $\{x \in \mathbb{R}^4 \mid g(x) = 0_{2\times 1}\}$. Use gradient descent with equality constraints to solve (6.25).

Solution: Using Gram-Schmidt to remove the components of $\nabla g_1(x_k)$ and $\nabla g_2(x_k)$ from $\nabla f(x_k)$.

```
"""
    gram_schmidt(U::Matrix)
```

```

Performs the Gram-Schmidt process on the columns of the matrix `U`
(assumed to be linearly independent)
to produce an orthogonal set of vectors stored in matrix `V` and a set
of orthonormal vectors stored
in `Vnormalized`.

# Arguments
- `U::Matrix`: A matrix whose columns are linearly
    independent vectors.

# Returns
- `V::Matrix`: A matrix containing the orthogonal
    vectors corresponding to the columns of `U`.
- `Vnormalized::Matrix`: A matrix containing the
    orthonormal vectors obtained by normalizing
    the columns of `V`.

"""

using LinearAlgebra

function gram_schmidt(U::Matrix)
    m, n = size(U)  # m = number of rows,
                    # n = number of columns (vectors)
    V = Matrix{Float64}(undef, m, n)  # Matrix to store
                                      # orthogonal vectors
    Vnormalized = Matrix{Float64}(undef, m, n)
    # Matrix to store normalized (orthogonal) vectors

    for i in 1:n
        # Start with the original vector from U
        vi = U[:, i]

        # Subtract projections onto previous vectors
        for j in 1:i-1
            vj = V[:, j]
            vi -= (dot(vi, vj) / dot(vj, vj)) * vj
        end

        # Store the orthogonal vector
        V[:, i] = vi

        # Normalize the vector and store in Vnormalized
        Vnormalized[:, i] = vi / norm(vi)
    end

    return (V=V, Vnormalized=Vnormalized)
end
```

Output

```
gram_schmidt (generic function with 1 method)
```

```
# Example gradient descent with several equality constraints
# Not in the textbook

```

```
using ForwardDiff, LinearAlgebra, NLsolve, Random

# Set the seed to a specific value, for example, 1234
Random.seed!(1234)



# Define the cost function f(x) and the equality constraint G(x)
xc = 3.0; yc = 3.0 # centers
a = 2.0 # x semi-major axis
b = 1.0 # y semi-major axis

# Define the cost (distance squared)
function cost(x)
    return x[1]^2 + x[2]^2 + x[3]^2 + x[4]^2
end

# Define the equality constraint: point must lie on the ellipse
function Geq(x)
    # Equality constraints evaluated at x
    g1 = (1/a)^2 * (x[1] - xc)^2 + (1/b)^2 * (x[2] - yc)^2 - 1 # ellipse
      # in R4
    g2 = x[1] + x[2] + x[3] + x[4] # 3 dimensional plane through the
      # origin
    return [g1, g2]  # Return evaluated constraints as an array
end

# Compute the gradient of the cost and constraint using ForwardDiff
grad_cost(x) = ForwardDiff.gradient(cost, x)
Jac_constraints(x) = ForwardDiff.jacobian(Geq, x)

# Remove the component of the gradient of the cost lying along the
# gradient of the constraints
function modNegGradientCost(x)
    gradCost = grad_cost(x)
    matGradientsConstraints = Jac_constraints(x)' # transpose of Jacobian for
    # constraint gradients

    # Perform Gram-Schmidt on constraint gradients and cost gradient
    U = [matGradientsConstraints gradCost]
    F = gram_schmidt(U)
    modNegGradCost = -F.V[:,end] #it's the negative of the last vector
    # in the matrix when doing Gradient Descent
    return modNegGradCost
end

# Function to find a feasible starting point using NLsolve
function find_feasible_point(Geq, x_guess)
    function constraint!(F, x)
        F[1] = Geq(x)[1]
        F[2] = Geq(x)[2]
    end
    result = nlsolve(constraint!, x_guess)
    return result.zero
end

# Set step size, max iterations, and tolerances
```

```
s = 0.1
maxIter = 1000
gradTol = 1e-6
constraintTol = 1e-5

# Starting guess (not feasible initially)
x_guess = rand(4,1)
xk = find_feasible_point(Geq, x_guess) # Make the initial guess feasible

# Gradient Descent with equality constraints loop
for k = 1:maxIter
    modNegGradCost = modNegGradientCost(xk)  # Use the modified gradient
    # cost function
    if norm(modNegGradCost) < gradTol  # Convergence check
        println("\n---- Iteration $k ----")
        println("Converged.")
        println("Cost: ", round.(cost(xk), digits=5))
        println("Constraints (should be near 0): ", round.(Geq(xk), digits=6))
        println("----------------- ")
        break
    else
        xk = xk + s * modNegGradCost  # Update using the modified gradient
    end

    if norm(Geq(xk)) > constraintTol  # Feasibility check
        println("Updating for feasibility: G(xk) = 0_{m x 1} at iteration k = $k")
        xk = find_feasible_point(Geq, xk)  # Ensure feasibility
    end
end

# Final results
xStar = xk
costStar = cost(xStar)

println("\nFinal Results:")
println("-----------------")
println("Optimal x: ", round.(xStar, digits=5))
println("Optimal cost(xStar): ", round(costStar, digits=5))
println("Minimal Distance is sqrt(cost(xStar)): ", round(sqrt(costStar), digits=5))
```

Output

```
Updating for feasibility: G(xk) = 0_{m x 1} at iteration k = 1
Updating for feasibility: G(xk) = 0_{m x 1} at iteration k = 2

---- Iteration 63 ----
Converged.
Cost: 14.91277
Constraints (should be near 0): [0.0, -0.0]
-----------------

Final Results:
-----------------
Optimal x: [1.33912; 2.44289; -1.89101; -1.89101;;]
Optimal cost(xStar): 14.91277
Minimal Distance is sqrt(cost(xStar)): 3.86171
```

Solution: This time using (6.31) to update Δx.

```
# Example gradient descent with several equality constraints
# Not in the textbook

using ForwardDiff, LinearAlgebra, NLsolve, Random

# Set the seed to a specific value, for example, 1234
Random.seed!(1234)



# Define the cost function f(x) and the equality constraint G(x)
xc = 3.0; yc = 3.0 # centers
a = 2.0 # x semi-major axis
b = 1.0 # y semi-major axis

# Define the cost (distance squared)
function cost(x)
    return x[1]^2 + x[2]^2 + x[3]^2 + x[4]^2
end

# Define the equality constraint: point must lie on the ellipse
function Geq(x)
    # Equality constraints evaluated at x
    g1 = (1/a)^2 * (x[1] - xc)^2 + (1/b)^2 * (x[2] - yc)^2 - 1 # ellipse
    # in R4
    g2 = x[1] + x[2] + x[3] + x[4] # 3 dimensional plane through
    # the origin
    return [g1, g2]  # Return evaluated constraints as an array
end

# Compute the gradient of the cost and constraint using ForwardDiff
grad_cost(x) = ForwardDiff.gradient(cost, x)
Jac_constraints(x) = ForwardDiff.jacobian(Geq, x)

# Remove the component of the gradient of the cost lying along the
# gradient of the constraints
function modNegGradientCost(x)
    gradCost = grad_cost(x)
    matGradientsConstraints = Jac_constraints(x)' # transpose of Jacobian
            #for constraint gradients

    alpha = (matGradientsConstraints)'*matGradientsConstraints \
                matGradientsConstraints'*gradCost

    modNegGradCost = -gradCost + matGradientsConstraints *alpha

    return modNegGradCost
end

# Function to find a feasible starting point using NLsolve
function find_feasible_point(Geq, x_guess)
    function constraint!(F, x)
        F[1] = Geq(x)[1]
        F[2] = Geq(x)[2]
    end
```

```
    result = nlsolve(constraint!, x_guess)
    return result.zero
end

# Set step size, max iterations, and tolerances
s = 0.1
maxIter = 1000
gradTol = 1e-6
constraintTol = 1e-5

# Starting guess (not feasible initially)
x_guess = rand(4,1)
xk = find_feasible_point(Geq, x_guess) # Make the initial guess feasible

# Gradient Descent with equality constraints loop
for k = 1:maxIter
    modNegGradCost = modNegGradientCost(xk)  # Use the modified gradient
    # cost function
    if norm(modNegGradCost) < gradTol  # Convergence check
        println("\n---- Iteration $k ----")
        println("Converged.")
        println("Cost: ", round.(cost(xk), digits=5))
        println("Constraints (should be near 0): ", round.(Geq(xk), digits=6))
        println("----------------- ")
        break
    else
        xk = xk + s * modNegGradCost  # Update using the modified gradient
    end

    if norm(Geq(xk)) > constraintTol  # Feasibility check
        println("Updating for feasibility: G(xk) = 0_{m x 1} at iteration k = $k")
        xk = find_feasible_point(Geq, xk)  # Ensure feasibility
    end
end

# Final results
xStar = xk
costStar = cost(xStar)

println("\nFinal Results:")
println("-----------------")
println("Optimal x: ", round.(xStar, digits=5))
println("Optimal cost(xStar): ", round(costStar, digits=5))
println("Minimal Distance is sqrt(cost(xStar)): ", round(sqrt(costStar), digits=5))
```

Output

```
Updating for feasibility: G(xk) = 0_{m x 1} at iteration k = 1
Updating for feasibility: G(xk) = 0_{m x 1} at iteration k = 2

---- Iteration 63 ----
Converged.
Cost: 14.91277
Constraints (should be near 0): [0.0, -0.0]
-----------------

Final Results:
```

```
------------------
Optimal x: [1.33912; 2.44289; -1.89101; -1.89101;;]
Optimal cost(xStar): 14.91277
Minimal Distance is sqrt(cost(xStar)): 3.86171
```

6.3.5 Lagrange Multiplier for a Problem with a Single Equality Constraint

The method of Lagrange multipliers is a tool that reduces optimization problems subject to equality constraints to what looks like an unconstrained problem. The new problem has dimension $(n+1)$ instead of n, but at least we can apply the methods of Section 6.2.2 to solve it. Mathematicians are very good at taking what looks like a new, super-hard problem and reducing its solution to a problem they already know how to solve.

Proposition 6.19: Lagrange Multiplier Theorem for Constrained Optimization

Suppose that the functions $f : \mathbb{R}^n \to \mathbb{R}$ and $g : \mathbb{R}^n \to \mathbb{R}$ define the cost and the constraint in the constrained minimization problem

$$\begin{aligned} &\text{Minimize} \quad f(x), \\ &\text{subject to} \quad g(x) = 0. \end{aligned} \tag{6.33}$$

If x^* is a solution to (6.33) **and** $\frac{\partial g(x^*)}{\partial x} \neq 0$, then there exists a unique scalar $\lambda^* \in \mathbb{R}$ such that

$$\nabla f(x^*) + \lambda^* \nabla g(x^*) = 0, \tag{6.34}$$

in other words, x^* is a **stationary point of the unconstrained (cost) function**

$$L(x, \lambda^*) := f(x) = \lambda^* g(x).$$

Remark 6.20. *For reasons we will see shortly,*

*(a) the constant λ^* is called the **Lagrange Multiplier** associated with the constrained minimization problem, (6.33), and*

*(b) $L(x, \lambda) : f(x) + \lambda g(x)$ (with a general λ instead of a specific value) is the **Lagrangian** associated with (6.33).*

(c) So far, we have not stated how to find the Lagrange multiplier; that is done in Method 6.21. But we understand why it is important: it reduces a problem we do not know how to solve to one we do know how to solve.

(d) Because maximizing a function is the same as minimizing its negative, all of the above holds for constrained maximization problems as well. In fact, if we succeed in computing all of the stationary points of L, we will have found all of the minima of f, all of the maxima of f, and its saddle points. The stationary conditions are agnostic to min versus max.

(e) A similar result can be given when there are several constraints, as in our Motivating Problem 2 in (6.16). The multi-constraint result will be placed in an "Optional Read" section along with an algebraic proof of the Method of Lagrange Multipliers.

Method 6.21: Method of Lagrange Multipliers

Consider the constrained optimization problem defined by (6.33) for functions $f : \mathbb{R}^n \to \mathbb{R}$ and $g : \mathbb{R}^n \to \mathbb{R}$ with optimal value x^*. If $\frac{\partial g(x^*)}{\partial x} \neq 0$, then the optimal x^* **and** λ^* can be found as follows:

Step 1: Form the **Lagrangian** function

$$L(x, \lambda) := f(x) + \lambda g(x). \tag{6.35}$$

Step 2: Solve for the **stationary points**[a] of L, that is, find all solutions to

$$0_{(n+1)\times 1} = \nabla L(x, \lambda). \tag{6.36}$$

When the compact notation in (6.36) is expanded out, the terms needed in a calculation become more evident, namely

$$0_{n\times 1} = \begin{bmatrix} \frac{L(x,\lambda)}{\partial x_1} \\ \frac{L(x,\lambda)}{\partial x_2} \\ \vdots \\ \frac{L(x,\lambda)}{\partial x_n} \end{bmatrix} = \nabla f(x) + \lambda \nabla g(x),$$
$$0_{1\times 1} = \frac{\partial L(x, \lambda)}{\partial \lambda} = g(x). \tag{6.37}$$

For small problems, the calculations are easy to perform by hand. For "smallish to medium-sized problems," the `Symbolics` package makes short work of computing the gradient of L, and `NLSolve` can be "coaxed" into finding the roots of the gradient, that is, the stationary points.

The notation, $0_{1\times 1}$, instead of 0, emphasizes that this line is scalar-valued; in fact, it imposes the scalar-valued (feasibility) constraint, $g(x) = 0$. We note that (6.37) has $(n+1)$ equations in $(n+1)$ unknowns, (x, λ), which is why we **can now solve** for the Lagrange multiplier λ^*, in addition to x^*.

Step 3: Evaluate the cost function $f(x)$ at each of the stationary points and choose the one (or ones) yielding a minimum. At this stage, we no longer have to worry about the constraint, $g(x)$, because it was taken care of in the computation of the stationary points of the Lagrangian, L.

[a] A nice discussion on this aspect of Lagrange's Method is given in Example 5: Numerical optimization of Wikipedia.

We next solve a few examples that are on the level of difficulty found in 200- and 300-level engineering courses.

Example 6.22. *Use the Method of Lagrange Multipliers to minimize $x+y$ subject to the constraint $x^2+y^2 = 1$. Note that the problem is employing standard "non-vector" notation where one uses (x, y) to denote a point in $\mathbb{R}^2$ instead of $x = (x_1, x_2)^\top$.* **Figure 6.7** *provides a geometric illustration of the constraint set and the cost function.*

The Example is based on Wikipedia, Example 1. Another worthwhile example is Wikipedia, Example 2.

Solution: Even though it is not required, we write down the constrained optimization problem,

$$\begin{aligned} &\text{Minimize} \quad f(x, y) = x + y, \\ &\text{subject to} \quad 0 = g(x, y) = x^2 + y^2 - 1, \end{aligned}$$

so as to make the identification of $f : \mathbb{R}^2 \to \mathbb{R}$ and $g : \mathbb{R}^2 \to \mathbb{R}$ crystal clear. Following the steps of Method 6.21, we define

$$L(x, y, \lambda) := (x + y) + \lambda \left(x^2 + y^2 - 1\right).$$

We then compute the required partial derivatives

$$\begin{aligned} 0 &= \frac{\partial L(x, y, \lambda)}{\partial x} = 1 + 2\lambda \cdot x, \\ 0 &= \frac{\partial L(x, y, \lambda)}{\partial y} = 1 + 2\lambda \cdot y, \end{aligned} \tag{6.38}$$

and

$$0 = \frac{\partial L(x, y, \lambda)}{\partial \lambda} = x^2 + y^2 - 1.$$

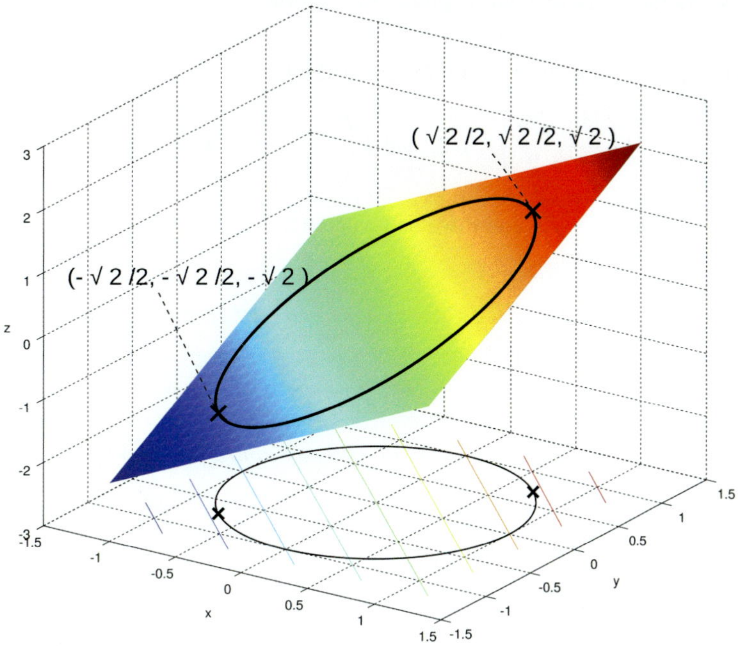

Figure 6.7: Illustration of a simple constrained optimization problem. For those of us who prefer geometric reasoning, the colored lines shown in the (x, y)-plane are the contour lines of the cost function, that is, lines along which $z = f(x, y) = x + y$ is constant. The unit circle is the constraint set, $g(x, y) = x^2 + y^2 - 1 = 0$. Hence, the set of feasible points is the unit circle. **Image Credit:** Wikipedia: Lagrange Multipliers: Example 1.

Note that the first two lines of the above equation correspond to the first 2×1 block of (6.37), and the last line of the above equation corresponds to the last line of (6.37).

Fortunately for us—**and it has to be like this for the problem to be doable by hand**—we can easily solve the stationary conditions in (6.38) for x and y in terms of λ, giving us

$$x = -\frac{1}{2\lambda}$$

and

$$y = -\frac{1}{2\lambda}.$$

Plugging these values into the constraint yields

$$0 = -1 + x^2 + y^2 = -1 + \left(-\frac{1}{2\lambda}\right)^2 + \left(-\frac{1}{2\lambda}\right)^2 = -1 + \frac{1}{2}\frac{1}{\lambda^2},$$

or

$$\lambda^2 = \frac{1}{2},$$

and therefore, $\lambda = \pm\frac{1}{\sqrt{2}} = \pm\frac{\sqrt{2}}{2}$. Putting all of this together, which is algebra and not calculus, yields the hand-computed stationary points of the Lagrangian:

$$\begin{bmatrix} x^* \\ y^* \\ \lambda^* \end{bmatrix} = \begin{bmatrix} \frac{\sqrt{2}}{2} \\ \frac{\sqrt{2}}{2} \\ -\frac{\sqrt{2}}{2} \end{bmatrix} \quad \text{and} \quad \begin{bmatrix} x^* \\ y^* \\ \lambda^* \end{bmatrix} = \begin{bmatrix} -\frac{\sqrt{2}}{2} \\ -\frac{\sqrt{2}}{2} \\ \frac{\sqrt{2}}{2} \end{bmatrix}.$$

As promised, we have found λ^*! (Exclamation point, not factorial.)

Evaluating the cost function $f(x,y) = x + y$ at these points yields

$$f\left(\frac{\sqrt{2}}{2}, \frac{\sqrt{2}}{2}\right) = \sqrt{2}, \text{ and } f\left(-\frac{\sqrt{2}}{2}, -\frac{\sqrt{2}}{2}\right) = -\sqrt{2}.$$

Thus the constrained minimum is $-\sqrt{2}$, and if we had wanted it, the constrained maximum is $\sqrt{2}$. ■.

Example 6.23. *Let us check that Prop. 6.19 really does work by re-solving Example 6.22 with the side information that*

$$\lambda^* = \frac{\sqrt{2}}{2}.$$

The goal is to check that we can find the minimum to the constrained problem by computing the stationary points of the unconstrained problem

$$L^*(x,y) := L(x,y,\lambda^*) = x + y + \underbrace{\frac{\sqrt{2}}{2}}_{\lambda^*}\left(x^2 + y^2 - 1\right).$$

Solution: We compute the gradient of L^* and set it equal to zero, namely

$$0_{2\times 1} = \nabla L^*(x,y) = \begin{bmatrix} \frac{\partial L^*(x,y)}{\partial x} \\ \frac{\partial L^*(x,y)}{\partial y} \end{bmatrix} = \begin{bmatrix} 1 + \sqrt{2}x \\ 1 + \sqrt{2}y \end{bmatrix}.$$

Solving for x and y gives the same answer we computed in Example 6.22, once we simplify via $\frac{1}{\sqrt{2}} = \frac{\sqrt{2}}{2}$.

Notes:

- If we had wanted to find the maximum via the unconstrainted problem, we would have used $\lambda^* = -\frac{\sqrt{2}}{2}$.
- All of this is just a theoretical exercise to underline that Lagrange's Method is intimately tied to the gradient of an appropriately defined cost function. If you don't get it, just realize that no 200- or 300-level Engineering course will test you on the theoretical underpinnings of Lagrange's Method. And if it is broached in a 400-level course, it will be taught (again) in the course. Second time is charm!

■

This next problem is ubiquitous in standard calculus courses: Minimize the surface area of a cylindrical can, subject to a constraint on its volume. Why minimize surface area? Because that accounts for the amount of material it takes to manufacture the can, and hence its cost. To formulate this problem, we need to recall two facts from high school geometry about cylindrical geometry:

- Volume of a right circular cylinder of height $h > 0$ and radius $r > 0$ is $V(r,h) = \pi \cdot r^2 \cdot h$.
- Surface area of the same cylinder is $S(r,h) = \underbrace{2\pi \cdot r \cdot h}_{\text{side of cylinder}} + \underbrace{2\pi \cdot r^2}_{\text{area of ends}} = 2\pi \cdot (r \cdot h + r^2)$.

Yes, a can has two ends!

Example 6.24. *Minimize the surface area of a cylindrical can of radius r and height h subject to its volume being equal to V_0.*

Solution: Even though it is not required, we write down the constrained optimization problem:

$$\begin{aligned}\text{Minimize} \quad & f(r,h) = S(r,h) = 2\pi\cdot(r\cdot h + r^2),\\ \text{subject to} \quad & 0 = g(r,h) = V(r,h) - V_0 = 2\pi\cdot r^2\cdot h - V_0,\end{aligned}$$

so as to make the identification of $f : \mathbb{R}^2 \to \mathbb{R}$ and $g : \mathbb{R}^2 \to \mathbb{R}$ crystal clear. Following the steps of Method 6.21, we define

$$L(r,h,\lambda) := 2\pi\cdot(r\cdot h + r^2) + \lambda\left(2\pi\cdot r^2\cdot h - V_0\right).$$

We then compute the required partial derivatives:

$$\begin{aligned}0 &= \frac{\partial L(r,h,\lambda)}{\partial r} = 2\pi\cdot(h+2r) + 4\pi\cdot\lambda\cdot r\cdot h = 2\pi\cdot(h + 2r + 2\lambda\cdot r\cdot h),\\ 0 &= \frac{\partial L(r,h,\lambda)}{\partial h} = 2\pi\cdot r + 2\pi\lambda\cdot r^2 = 2\pi\cdot r\cdot(1+\lambda\cdot r),\\ \text{and} \quad 0 &= \frac{\partial L(r,h,\lambda)}{\partial \lambda} = 2\pi\cdot r^2\cdot h - V_0.\end{aligned} \tag{6.39}$$

Again, note that the first two lines of the above equation correspond to the first 2×1 block of (6.37), and the last line of the above equation corresponds to the last line of (6.37).

Fortunately for us—**and it has to be like this for the problem to be doable by hand**—we can solve for r, h, and λ using elementary algebraic manipulations. Indeed, from the second stationary condition, if $r \neq 0$, which is required for a physical to have a nonzero volume, we obtain

$$\lambda\cdot r = -1.$$

Using this value in the first stationary condition, we obtain:

$$(0 = h + 2r - 2h) \iff (h = 2r).$$

Using $h = 2r$ in the last stationary condition yields

$$\left(0 = 2\pi\cdot r^2\cdot(2r) - V_0\right) \iff V_0 = 4\pi\cdot r^3.$$

Hence,

$$r = \sqrt[3]{\frac{V_0}{4\pi}}$$

and

$$h = 2\cdot\sqrt[3]{\frac{V_0}{4\pi}}.$$

We could stop here. However, for the sake of completeness, putting all of this together after using

$$(\lambda\cdot r = -1) \iff \left(\lambda = -\frac{1}{r}\right),$$

we obtain:

$$\begin{bmatrix} r^* \\ h^* \\ \lambda^* \end{bmatrix} = \begin{bmatrix} \sqrt[3]{\frac{V_0}{4\pi}} \\ 2\cdot\sqrt[3]{\frac{V_0}{4\pi}} \\ -\sqrt[3]{\frac{4\pi}{V_0}} \end{bmatrix}.$$

As promised, using Lagrange's Method, we have found λ^* and the appropriate dimensions of the can. ■

Note: In the above problem solution, we only found one extremal, the minimizing solution. We might ask ourselves, why? **Can we not maximize the surface area of a tin can?** Well, theoretically, we can do it. If we go back to our second stationary condition,

$$0 = \frac{\partial L(r,h,\lambda)}{\partial h} = 2\pi \cdot r \cdot (1 + \lambda \cdot r),$$

there was a second solution, namely, $r = 0$. Replacing that with $r = \epsilon > 0$ and solving for h such that $V_0 = 2\pi \cdot r^2 \cdot h$ gives

$$h = \frac{V_0}{2\pi\epsilon^2}.$$

Now, take the limit as $\epsilon \to 0^+$, and we get a "can" that is infinitely tall and narrow with a fixed volume. What is its surface area?

$$S = 2\pi \cdot (r \cdot h + r^2) = 2\pi \cdot \left(\epsilon \cdot \frac{V_0}{2\pi\epsilon^2} + \epsilon^2\right) = 2\pi \cdot \left(\frac{V_0}{2\pi\epsilon} + \epsilon^2\right) \to \infty \text{ as } \epsilon \to 0^+.$$

Hence, if we pose the nonsensical question of maximizing the surface area subject to a fixed volume, we can compute the nonsensical answer, as long as we believe in limits! To be clear, if we give this sort of problem to a numerical solver, it will likely diverge.

If you enjoy the idea of objects that have strange properties, such as finite volume and infinite surface area, you may enjoy the video Gabriel's Horn Paradox by Numberphile (aka The **Painter's Paradox**) or Torricelli's 2nd Paradox and its 14th-century genius monk resolution by Mathologer. The volume of the horn is finite, it is built from infinitely thin material so that the surface areas inside and outside the horn are the same, has a finite "bell" or "flare" at one end (instead of our uniformly thin cylinder), and yet it cannot be painted with a finite volume of paint. What's the paradox? Well, you can fill the horn with a finite volume of paint, and that volume is touching the inner surface of the horn, isn't it?

Example 6.25. *For $x \in \mathbb{R}^n$, minimize the norm squared of x subject to $a \bullet x = a^\top x = b_0$, where $0_{n\times 1} \neq a \in \mathbb{R}^n$ and $b \in \mathbb{R}$.*

Solution: Though it is not required, we write down the constrained optimization problem,

$$\begin{aligned} \text{Minimize} \quad & f(x) = x \bullet x \\ \text{subject to} \quad & 0 = g(x) = a \bullet x - b_0, \end{aligned}$$

so as to make the identification of $f : \mathbb{R}^n \to \mathbb{R}$ and $g : \mathbb{R}^n \to \mathbb{R}$ crystal clear. Following the steps of Method 6.21, we define

$$L(x,\lambda) := x \bullet x + \lambda\,(a \bullet x - b_0)\,.$$

We then compute the required partial derivatives in (6.37) using the vector product rule in Prop. 5.42:

$$0_{n\times 1} = \nabla f(x) + \lambda \nabla g(x) = 2x + \lambda a$$

and

$$0_{1\times 1} = \frac{\partial L(x,\lambda)}{\partial \lambda} = a \bullet x - b_0 = a^\top x - b_0.$$

Hence, the first equation gives $x = -\frac{1}{2}\lambda a$, and substituting the result into the second equation gives:

$$a^\top \left(-\frac{1}{2}\lambda a\right) - b_0 = 0,$$

which in turn gives:

$$\lambda = -2\frac{1}{a^\top \cdot a} \cdot b_0.$$

Therefore, substituting this result back into $x = -\frac{1}{2}\lambda a$ gives

$$x = \frac{b_0}{a^\top a} \cdot a.$$

■

6.3.6 (Optional Read) Lagrange Multipliers for a Problem with a Vector of Equality Constraints

The method of Lagrange multipliers has a natural extension to problems with a vector of equality constraints. Let $f : \mathbb{R}^n \to \mathbb{R}$ be the real-valued cost function as before and assume we have a vector of equality constraints $g : \mathbb{R}^n \to \mathbb{R}^m$, where

$$g(x) = \begin{bmatrix} g_1(x) \\ g_2(x) \\ \vdots \\ g_m(x) \end{bmatrix} = \begin{bmatrix} g_1(x_1, x_2, \ldots, x_n) \\ g_2(x_1, x_2, \ldots, x_n) \\ \vdots \\ g_m(x_1, x_2, \ldots, x_n) \end{bmatrix}.$$

Optimization Problem with Several Constraints:

$$\begin{aligned} &\text{Minimize} \quad f(x), \\ &\text{subject to} \quad g_i(x) = 0,\ 1 \le i \le m. \end{aligned} \tag{6.40}$$

We will skip the vector version of Prop. 6.19 and go straight to the heart of the matter, Lagrange's Method, now that you see how it works.

Method 6.26: Method of Lagrange Multipliers with a Vector of Equality Constraints

Consider the constrained optimization problem (6.40) for functions $f : \mathbb{R}^n \to \mathbb{R}$ and $g : \mathbb{R}^n \to \mathbb{R}^m$ with optimal value x^*. If the column vectors from the gradients $\{\nabla g_1(x^*), \nabla g_2(x^*), \ldots, \nabla g_m(x^*)\}$ are linearly independent (equivalently, the rows of the Jacobian $\frac{\partial g(x^*)}{\partial x}$ are linearly independent), then a locally optimal x^* **and** vector of Lagrange multipliers

$$\lambda^* = \begin{bmatrix} \lambda_1^* \\ \lambda_2^* \\ \vdots \\ \lambda_m^* \end{bmatrix}$$

can be found as follows:

Step 1: Form the **Lagrangian**

$$L(x, \lambda) := f(x) + \sum_{i=1}^{m} \lambda_i g_i(x) = f(x) + \lambda_1 g_1(x) + \lambda_2 g_2(x) + \cdots + \lambda_m g_m(x). \tag{6.41}$$

Step 2: Solve for the **stationary points** of L, that is, find all solutions to:

$$0_{(n+m)\times 1} = \nabla L(x, \lambda),$$

which, when expanded out, becomes:

$$0_{n\times 1} = \begin{bmatrix} \frac{L(x,\lambda)}{\partial x_1} \\ \frac{L(x,\lambda)}{\partial x_2} \\ \vdots \\ \frac{L(x,\lambda)}{\partial x_n} \end{bmatrix} = \nabla f(x) + \sum_{i=1}^{m} \lambda_i \nabla g_i(x)$$

and (6.42)

$$0_{m\times 1} = \begin{bmatrix} \frac{L(x,\lambda)}{\partial \lambda_1} \\ \frac{L(x,\lambda)}{\partial \lambda_2} \\ \vdots \\ \frac{L(x,\lambda)}{\partial \lambda_m} \end{bmatrix} = \begin{bmatrix} g_1(x) \\ g_2(x) \\ \vdots \\ g_m(x) \end{bmatrix}.$$

We note that the bottom $m \times 1$ block of (6.42) imposes the vector of (feasibility) constraints, $g(x) = 0$. Moreover, (6.42) has $(n+m)$ equations in $(n+m)$ unknowns, (x, λ), giving us enough information to solve for x^* and the Lagrange multipliers, λ^*. When the equations are nonlinear, there can be multiple solutions or no solutions.

Step 3: Evaluate the cost function $f(x)$ at each of the stationary points and choose the one (or ones) yielding a minimum. Good luck!

Note: The condition that the columns of the gradient $\nabla g(x^*)$ be linearly independent (equivalently, the rows of the Jacobian, $\frac{\partial g(x^*)}{\partial x}$ are linearly independent) ensures that the Lagrange multipliers are unique.

We next solve an example by "pen and paper" to illustrate the method.

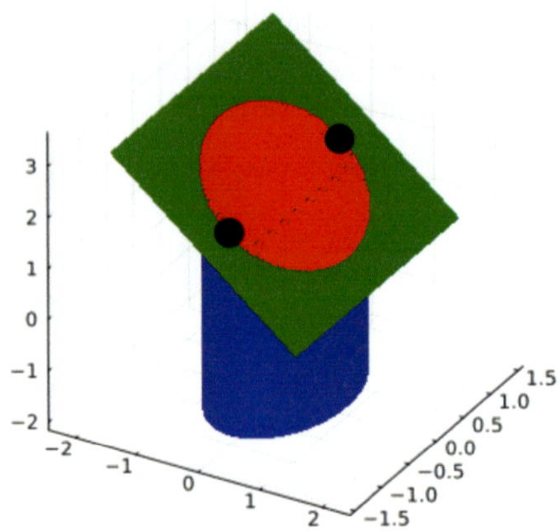

Figure 6.8: The first constraint in Example 6.27 gives a cylinder about the z-axis having radius one, while the second constraint defines a plane. The red elliptical/circular shape in the plane, excluding its interior, is the intersection of the two constraints, thereby defining the feasible set over which the cost is to be optimized. The two black points are the extrema, one corresponding to a maximum of $f(x, y, z) = x + y + z$ and the other to a minimum.

Example 6.27. *Maximize $f(x,y,z) = x+y+z$ subject to the constraints $x^2+y^2=1$ and $x+z=2$. The maximization is not a typo. We wish to illustrate our "trick" of minimizing the negative of the objective function.* **Figure 6.8** *depicts the two constraints.*

The way we are approaching the problem, namely through Lagrange Multipliers, we can find all of the extreme points, whether we include the minus sign on the cost or not. With typical numerical solvers, this is not the case.

Solution: Even though it is not required, we write down the constrained optimization problem:

$$\begin{aligned} &\text{Minimize} \quad f(x,y,z) = -(x+y+z), \\ &\text{subject to} \quad \begin{bmatrix} 0 \\ 0 \end{bmatrix} = \begin{bmatrix} g_1(x,y,z) \\ g_2(x,y,z) \end{bmatrix} = \begin{bmatrix} x^2+y^2-1 \\ x+z-2 \end{bmatrix}, \end{aligned}$$

so as to make the identification of $f : \mathbb{R}^3 \to \mathbb{R}$ and $g : \mathbb{R}^3 \to \mathbb{R}^2$ crystal clear. Following the steps of Method 6.26, we define

$$L(x,y,z,\lambda_1,\lambda_2) = -(x+y+z) + \lambda_1(x^2+y^2-1) + \lambda_2(x+z-2).$$

Taking the required partial derivatives, stacking them in a column, and setting them equal to zero, we obtain the stationary conditions:

$$\begin{aligned} 0 &= \frac{\partial L(x,y,z,\lambda_1,\lambda_2)}{\partial x} = -1 + 2\lambda_1 \cdot x + \lambda_2, \\ 0 &= \frac{\partial L(x,y,z,\lambda_1,\lambda_2)}{\partial y} = -1 + 2\lambda_1 \cdot y, \\ 0 &= \frac{\partial L(x,y,z,\lambda_1,\lambda_2)}{\partial z} = -1 + \lambda_2, \\ 0 &= \frac{\partial L(x,y,z,\lambda_1,\lambda_2)}{\partial \lambda_1} = x^2 + y^2 - 1, \\ 0 &= \frac{\partial L(x,y,z,\lambda_1,\lambda_2)}{\partial \lambda_2} = x + z - 2. \end{aligned} \tag{6.43}$$

Fortunately for us, we can solve for x, y, z, λ_1 and λ_2 using elementary algebraic manipulations. Indeed,

$$\begin{aligned} &\text{eqn \# 3:} \quad 0 = -1 + \lambda_2 \implies \lambda_2 = 1 \\ &\text{eqn \# 1:} \quad 0 = 2\lambda_1 \cdot x \implies x = 0 \ \text{(because eqn \# 2 implies neither y nor } \lambda_1 \text{ can equal zero)} \\ &\text{eqn \# 5:} \quad 0 = x + z - 2 \implies z = 2 \ \text{(because } x = 0) \\ &\text{eqn \# 4:} \quad 0 = x^2 + y^2 - 1 \implies y = \pm 1 \\ &\text{eqn \# 2:} \quad 0 = -1 + 2\lambda_1 \cdot y \implies \lambda_1 = \text{sign}(y) \cdot \frac{1}{2}, \end{aligned}$$

where eqn # i: corresponds to row i of (6.43). If there is a confusing point, it is the conclusion that eqn # 1 implies $x = 0$. This is true because if $\lambda_1 = 0$, then eqn # 2: becomes $0 = -1$, which is impossible. Hence, λ_1 cannot be equal to zero, and hence, eqn # 1 implies $x = 0$. This is basic algebra. **Calculus gave us the equations; algebra is being used to solve them!**

Putting all of this together for the "real" cost function $f(x,y,z) = x+y+z$ that we wanted to maximize, **the cost without the minus sign**, yields:

$$\begin{bmatrix} x^* \\ y^* \\ z^* \\ \lambda_1^* \\ \lambda_2^* \end{bmatrix} = \begin{bmatrix} 0 \\ 1 \\ 2 \\ \frac{1}{2} \\ 1 \end{bmatrix} \text{ giving a maximum of } f^* = 3, \text{ and } \begin{bmatrix} x^* \\ y^* \\ z^* \\ \lambda_1^* \\ \lambda_2^* \end{bmatrix} = \begin{bmatrix} 0 \\ -1 \\ 2 \\ -\frac{1}{2} \\ 1 \end{bmatrix} \text{ giving a minimum of } f^* = 1.$$

■

Let us now see how to apply Lagrange Multipliers in code! The solution combines our ability to compute gradients symbolically, understand that stationary points correspond to roots of the gradient, and numerically solve root-finding problems. The following code puts it all together.

```
using Symbolics, NLsolve, Random

# -------------------
# PROBLEM DEFINITION
# -------------------

# Define symbolic variables for the optimization
@variables var[1:3]  # Actual problem variables
@variables lam[1:2]   # Lagrange multipliers

# Objective function to be minimized
f = -(var[1] + var[2] + var[3])

# Constraint equations
g1 = var[1]^2 + var[2]^2 - 1
g2 = var[1] + var[3] - 2

# Formulate the Lagrangian: objective function + lagrange multipliers
# for constraints
L = f + lam[1]*g1 + lam[2]*g2

# Compute the gradient of the Lagrangian with respect to both variables
# and Lagrange multipliers
gradL = Symbolics.gradient(L, [var; lam])

# For debugging: display the gradient or convert it to LaTeX for better
# visualization
if false
    display(gradL)
else
    latexify(gradL) |> println
end

# ---------------------------
# NUMERICAL SOLUTION APPROACH
# ---------------------------

# Convert the symbolic gradient expression to a numerical function
# for computational efficiency
gradL_num = build_function(gradL, [var; lam])
gradL_num = eval(gradL_num[1])

# Initialize an array to store solutions
f_solutions = Float64[]

# Solve the optimization problem for multiple initial guesses
# This is done to explore different solutions since NLsolve can get
# trapped in local minima
for k = 1:100
    # Generate a random initial guess for the optimization variables
```

```
    # and Lagrange multipliers
    # randn is the normal or gaussian distribution, which includes
    # positive and negative values
    initial_guess = randn(5)

    # Use nlsolve to find where the gradient is zero, which corresponds
    # to an extremum
    result = nlsolve(gradL_num, initial_guess, xtol=1e-6, ftol=1e-7)


# Process the solution only if nlsolve successfully converged
if converged(result)
    # Extract the values of the problem variables from the solution
    var_val = result.zero[1:3]
    lam_val = result.zero[4:5]

    # Compute the value of the objective function at the found extremum
    f_extremePoint = -sum(var_val)

    # Add the found solution only if it is sufficiently different from
    # the previously found solutions
    if all(abs(f_extremePoint - sol) > 1e-2 for sol in f_solutions)
        push!(f_solutions, f_extremePoint)
            @show [var_val; lam_val]
    end
end
end

# Sort solutions for easier analysis
sorted_solutions = sort(f_solutions)
```

Output

$$\begin{bmatrix} -1.0000 + lam2 + 2.0000 lam1 var1 \\ -1.0000 + 2.0000 lam1 var2 \\ -1.0000 + lam2 \\ -1.0000 + var1^2 + var2^2 \\ -2.0000 + var1 + var3 \end{bmatrix} \tag{6.44}$$

```
[var_val; lam_val] = [2.5557001655682698e-9, -1.0000000021988782,
1.9999999974442995, -0.4999999970071819, 1.0]
[var_val; lam_val] = [2.672589761676145e-12, 1.000000000001114,
1.9999999999973277, 0.49999999999892936, 1.0]

2-element Vector{Float64}:
 -3.000000000001114
 -0.9999999978011215
```

Remark 6.28. *The above problem has two distinct roots. To which one will* `NLsolve` *converge? It depends on the initial condition. That is why the solver is run 100 times with random initial conditions, giving us a chance to find both of them.*

Here is a solution using `Jump`. As a professional, this is preferred over the above solution using Lagrange Multipliers. Why? **Jump's code has been optimized for speed and tested by enough users that it is essen-**

tially bug-free. For doing HW in engineering courses, being able to run the symbolic version of Lagrange Multipliers will help you tremendously. Moreover, you now know several ways to check your HW and learn from your mistakes.

```julia
using JuMP, Ipopt

# Create a new model with Ipopt as the solver
model = Model(Ipopt.Optimizer)

# Define the variables
@variable(model, var[1:3])

# Set the objective function
@NLobjective(model, Min, -(var[1] + var[2] + var[3]))

# Add the constraint equations
@NLconstraint(model, var[1]^2 + var[2]^2 == 1)
@NLconstraint(model, var[1] + var[3] == 2)

# Fine tune a few params
set_optimizer_attribute(model, "max_iter", 1000)  # Allow up to 1000 iterations
set_optimizer_attribute(model, "tol", 1e-9)        # Require a very accurate
                                                   # solution

# Suppress much of the output
set_optimizer_attribute(model, "print_level", 4)

# Solve the problem
optimize!(model)

# Display the results
println("Objective value: ", objective_value(model))
println("Solution: ")
println("var[1] = ", value(var[1]))
println("var[2] = ", value(var[2]))
println("var[3] = ", value(var[3]))
```

Output

```
Total number of variables............................:        3
                     variables with only lower bounds:        0
                variables with lower and upper bounds:        0
                     variables with only upper bounds:        0
Total number of equality constraints.................:        2
Total number of inequality constraints...............:        0
        inequality constraints with only lower bounds:        0
   inequality constraints with lower and upper bounds:        0
        inequality constraints with only upper bounds:        0


Number of Iterations....: 15

                                   (scaled)                 (unscaled)
Objective...............:  -3.0000000000000000e+00   -3.0000000000000000e+00
Dual infeasibility......:   0.0000000000000000e+00    0.0000000000000000e+00
Constraint violation....:   0.0000000000000000e+00    0.0000000000000000e+00
Variable bound violation:   0.0000000000000000e+00    0.0000000000000000e+00
Complementarity.........:   0.0000000000000000e+00    0.0000000000000000e+00
```

```
Overall NLP error.......:   0.0000000000000000e+00    0.0000000000000000e+00


Number of objective function evaluations             = 16
Number of objective gradient evaluations             = 16
Number of equality constraint evaluations            = 16
Number of inequality constraint evaluations          = 0
Number of equality constraint Jacobian evaluations   = 16
Number of inequality constraint Jacobian evaluations = 0
Number of Lagrangian Hessian evaluations             = 15
Total seconds in IPOPT                               = 0.012

EXIT: Optimal Solution Found.
Objective value: -3.0
Solution:
var[1] = 4.579141502905379e-17
var[2] = 1.0
var[3] = 2.0
```

Example 6.29. *Consider an underdetermined system of linear equations $Ax = b$, where A is $m \times n$. When the rows of A are linearly independent (equivalently, the columns of $A^\top$ are linearly independent), determine a minimum norm squared solution to $Ax = b$.*

Solutions: Even though it is not required, we write down the constrained optimization problem as follows:

$$\begin{aligned}&\text{Minimize} \quad f(x) = x \bullet x = x^\top \cdot x,\\ &\text{subject to} \quad 0_{m\times 1} = g(x) = Ax - b,\end{aligned}$$

so as to make the identification of $f : \mathbb{R}^n \to \mathbb{R}$ and $g : \mathbb{R}^n \to \mathbb{R}^m$ crystal clear. Following the steps of Method 6.21, we define

$$L(x, \lambda) := x \bullet x + \lambda_1(a_1 x - b_1) + \lambda_2(a_2 x - b_2) + \cdots + \lambda_m(a_m x - b_m),$$

where a_i are the rows of A and b_i are the entries of b. We then compute the required derivatives:

$$\begin{aligned}0_{n\times 1} &= \nabla(x \bullet x) + \lambda_1 \nabla(a_1 x - b_1) + \lambda_2 \nabla(a_2 x - b_2) + \cdots + \lambda_m \nabla(a_m x - b_m)\\ &= 2x + \lambda_1 \cdot (a_1)^\top + \lambda_2 \cdot (a_2)^\top + \cdots + \lambda_m \cdot (a_m)^\top\\ &= 2x + A^\top \cdot \lambda,\\ 0_{m\times 1} &= Ax - b.\end{aligned}$$

From the first n equations, we obtain $x = -\frac{1}{2} A^\top \cdot \lambda$, and from the last m equations we obtain the constraint $Ax = b$. Substituting the first into the second and doing a bit of algebra yields:

$$\left(A \cdot A^\top\right) \cdot \lambda = -2b.$$

Solving for λ and substituting into $x = -\frac{1}{2} A^\top \cdot \lambda$ yields:

$$x = A^\top \cdot \left(A \cdot A^\top\right)^{-1} \cdot b.$$

■

6.3.7 3-Link Manipulator Meets Inequality Constraints

We return to the three-link manipulator. Our goal is to place the end effector in a box, which will use **equality constraints**, and we want to avoid contacting an obstacle, which will use **inequality constraints**. We first set the problem up and then solve it using `Jump`.

A standard form for a constrained optimization problem includes both **equality and inequality constraints**. Many solvers require you to write the constraints in the form

$$g_j^{\text{in}}(x) \le 0,$$

as we show in (6.45). Some solvers, such as `JuMP` accept constraints in the form

- $g_j^{\text{in}}(x) \le c_j$
- $g_j^{\text{in}}(x) < c_j$
- $g_j^{\text{in}}(x) > c_j$
- $g_j^{\text{in}}(x) \ge c_j$.

Standard form for many optimization problems with inequality and equality constraints:

$$\begin{aligned} \text{Minimize} \quad & f(x), \\ \text{subject to} \quad & g_i^{\text{eq}}(x) = 0, \; 1 \le i \le m, \\ & g_j^{\text{in}}(x) \le 0, \; 1 \le j \le p, \end{aligned} \tag{6.45}$$

For the 3-link manipulator, we will define the cost function as

$$f(\theta) = (\theta_1)^2 + (\theta_2)^2 + (\theta_3)^2 .$$

We need two equality constraints to place the end effector in the box located at $(3.5, 0)$, namely

$$g^{\text{eq}}(\theta) = \begin{bmatrix} g_1^{\text{eq}}(\theta) \\ g_2^{\text{eq}}(\theta) \end{bmatrix} = \begin{bmatrix} L1 \cdot \cos(\theta_1) + L2 \cdot \cos(\theta_1 + \theta_2) + L3 \cdot \cos(\theta_1 + \theta_2 + \theta_3) - 3.5 \\ L1 \cdot \sin(\theta_1) + L2 \cdot \sin(\theta_1 + \theta_2) + L3 \cdot \sin(\theta_1 + \theta_2 + \theta_3) - 0.0 \end{bmatrix},$$

and we will impose four inequality constraints on the vertical components of p_1 and p_2 of the robot, namely

$$g^{\text{in}}(\theta) = \begin{bmatrix} g_1^{\text{in}}(\theta) \\ g_2^{\text{in}}(\theta) \\ g_3^{\text{in}}(\theta) \\ g_2^{\text{in}}(\theta) \end{bmatrix} = \begin{bmatrix} L1 \cdot \sin(\theta_1) \ge 0.1 \\ L1 \cdot \sin(\theta_1) + L2 \cdot \sin(\theta_1 + \theta_2) \ge 0.1 \\ L1 \cdot \sin(\theta_1) \le 0.95 \\ L1 \cdot \sin(\theta_1) + L2 \cdot \sin(\theta_1 + \theta_2) \le 0.95 \end{bmatrix}.$$

We do not impose constraints on their horizontal positions because we do not care where the joints end up. We simply want the links to be above the obstacle.

Here is what it looks like in `JuMP`

```julia
# JuMP solves the problem
using Plots, JuMP, Ipopt

# Define the model parameters for the robot manipulator
function modelParameters()
    # Lengths of the three links
    L1, L2, L3 = [2, 1.5, 1]
    return (L1=L1, L2=L2, L3=L3)
end

# Calculate the positions of the robot's links based on the joint angles
```

```
function linkPostions(th1,th2,th3)
    params = modelParameters()
    # Base position
    p0 = [0.0; 0.0]
    # Calculate positions of the other joints
    p1 = p0 + [params.L1 * cos(th1); params.L1 * sin(th1)]
    p2 = p1 + [params.L2 * cos(th1+th2); params.L2 * sin(th1+th2)]
    p3 = p2 + [params.L3 * cos(th1+th2+th3); params.L3 * sin(th1+th2+th3)]
    return (p0=p0, p1=p1, p2=p2, p3=p3)
end

# Plot a rectangle on the current plot
function plot_rectangle(xc, yc, w, h; color="brown")
    # Calculate the corners of the rectangle
    x1 = xc - w/2
    x2 = xc + w/2
    y1 = yc
    y2 = yc + h
    # Plot the rectangle
    plot!([x1, x2, x2, x1, x1], [y1, y1, y2, y2, y1], fill=true, fillalpha=1.0,
    fillcolor=color, linecolor=color, legend=false)
end

# Plot the robot's configuration
function plot_points(positions, label; line_thickness=5, ball_size=10)
    # Extract joint positions
    p0 = positions.p0
    p1 = positions.p1
    p2 = positions.p2
    p3 = positions.p3
    # Plot obstacles
    plot_rectangle(1.75, 1, 4, .1)
    plot_rectangle(1.75, -.1, 4, .1)
    # Plot the robot's configuration
    scatter!([p3[1]], [p3[2]], color=:green, markersize=2*ball_size,
    marker=:square, label=nothing)
    plot!([p0[1], p1[1], p2[1], p3[1]], [p0[2], p1[2], p2[2], p3[2]],
        linewidth=line_thickness, color=:blue, label=label)
    scatter!([p0[1], p1[1], p2[1], p3[1]], [p0[2], p1[2], p2[2], p3[2]],
    color=:red, markersize=ball_size, label=nothing)
end

# Solve the optimization problem and plot the solution
function solve_and_plot(use_inequality_constraints)
    params = modelParameters()
    model = Model(Ipopt.Optimizer)
    # Solver settings
    set_optimizer_attribute(model, "max_iter", 1000)
    set_optimizer_attribute(model, "tol", 1e-9)
    set_optimizer_attribute(model, "print_level", 1)
    # Define optimization variables and constraints
    @variable(model, -π<= th1 <= π)
    @variable(model, -π <= th2 <= π)
    @variable(model, -π <= th3 <= π)
    @constraint(model, g1, params.L1*cos(th1) + params.L2*cos(th1 + th2) +
        params.L3*cos(th1 + th2 + th3) - 2.5 == 0)
    @constraint(model, g2, params.L1*sin(th1) + params.L2*sin(th1 + th2) +
```

```
            params.L3*sin(th1 + th2 + th3) - 0.75 == 0)
    if use_inequality_constraints
        @constraint(model, g3, params.L1*sin(th1) <= 0.95)
        @constraint(model, g4, params.L1*sin(th1) +
            params.L2*sin(th1 + th2) <= 0.95)
        @constraint(model, g5, params.L1*sin(th1) >=0.1)
        @constraint(model, g6, params.L1*sin(th1) +
            params.L2*sin(th1 + th2) >=0.1)
    end
    # Define the objective function
    @objective(model, Min, th1^2 + th2^2 +  th3^2)
    # Solve the optimization problem
    optimize!(model)
    # Display the results
    label = use_inequality_constraints ?
        "With Inequality Constraints" : "No Inequality Constraints"
    println(label)
    println("Objective value: ", objective_value(model))
    println("Solution: ")
    println("th1Star = ", value(th1))
    println("th2Star = ", value(th2))
    println("th3Star = ", value(th3))
    println(" ")
    positions = linkPostions(value(th1), value(th2), value(th3))
    plot_points(positions, label)
end

# Initialize the plot
fig = plot(; aspect_ratio=:equal, legend=false)
# Solve and plot without and with inequality constraints
solve_and_plot(false)
solve_and_plot(true)
# Display and save the plot
display(fig)
png(fig, "constrainedInequalityOptimizationFor3LinkRobotKinematicChain")
```

Output (Angles in radians)

```
No Inequality Constraints
Objective value: 3.4536798144131877
Solution:
th1Star = -0.6590530918483652
th2Star = 1.3848362097307525
th3Star = 1.0495510986878442

With Inequality Constraints
Objective value: 6.812490744951165
Solution:
th1Star = 0.05002085181111906
th2Star = 0.18815229215108328
th3Star = 2.6028037525509964
```

The configuration of the robot is shown in Fig. 6.9.

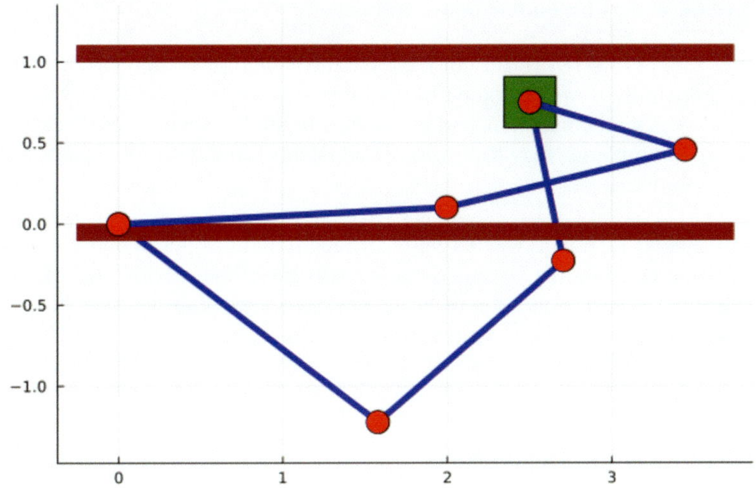

Figure 6.9: A pseudo-realistic manipulation task that requires inequality constraints in addition to equality constraints. The brown objects represent obstacles we must avoid while reaching into the green box with the robot's end effector. A solution with only equality constraints does reach into the green box, but "passes through" the obstacles. By adding inequality constraints, Jump finds the desired solution. The cost of the tightly constrained problem is approximately double the cost of the other.

6.4 Dynamics à la Lagrange

You know for sure that $F = m \cdot a$ for a point mass, as shown in **Fig. 6.10-(a)**, and you have likely seen the equation for the pendulum of **Fig. 6.10-(b)**,

$$\ddot{\theta}(t) = -\frac{g}{L} \cdot \sin(\theta). \tag{6.46}$$

But could you derive the equations for a person riding a Hoverboard? Or the 3-link manipulator in **Fig. 1.7-(b)**? At the end of this section, you will be able to do all of this and more. You will be able to derive by hand the equations for simple models such as the pendulum, and, via the `Symbolics` package, handle more complicated models.

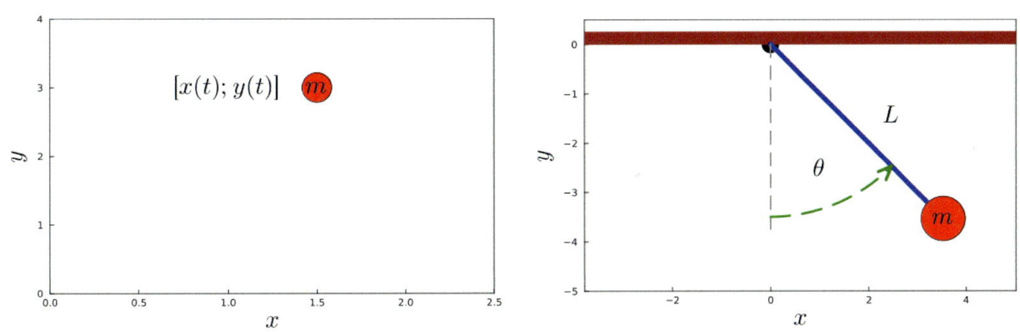

(a) (b)

Figure 6.10: (a) A point mass in the plane with its Cartesian coordinates. The origin is in the bottom left corner. (b) The classic pendulum formed by a massless rod with a point mass attached at its end. The angle θ is measured CCW from the vertical. The origin is placed at the pivot point attached to the "ceiling".

Secrets of the Arcane 6.30: Fun Facts about Lagrange

Joseph-Louis Lagrange (1736–1813) did not have the advantage of computers or modern software to assist with computations, as we do today. In his time, mathematics was done by hand using pen and paper. Though some mathematicians had colleagues or students with whom they might discuss ideas or collaborate on problems, they didn't have "assistants" in the sense of someone performing computational tasks for them.

Oh yeah, the Fun Facts: Lagrange was largely self-taught and did not obtain a university degree. Fascinated by maxima and minima of functions, he was the principal founder of the calculus of variations. In a far-reaching reformulation of Isaac Newton's laws, Lagrange created a brilliant new vision of mechanics. We are working toward that now in this section!

6.4.1 The Dot Notation for Derivatives

In (6.46), $\ddot{\theta}(t)$ denotes the **angular acceleration** of the pendulum. The dot notation for derivatives is a very popular quasi-standard notation for denoting derivatives with respect to time. For example, if the vector

$$p(t) := \begin{bmatrix} x(t) \\ y(t) \end{bmatrix} \tag{6.47}$$

is the Cartesian position of the particle in the plane, then its Cartesian **velocity** is the first derivative with respect to time of the position vector, namely,

$$v(t) := \begin{bmatrix} \dot{x}(t) \\ \dot{y}(t) \end{bmatrix} := \frac{d}{dt}\begin{bmatrix} x(t) \\ y(t) \end{bmatrix} = \begin{bmatrix} \frac{dx(t)}{dt} \\ \frac{dy(t)}{dt} \end{bmatrix}, \tag{6.48}$$

which is read as x and y dot. In a similar manner, the derivative of the velocity vector with respect to time is the Cartesian **acceleration vector**,

$$a(t) := \begin{bmatrix} \ddot{x}(t) \\ \ddot{y}(t) \end{bmatrix} := \frac{d}{dt}\begin{bmatrix} \dot{x}(t) \\ \dot{y}(t) \end{bmatrix} = \begin{bmatrix} \frac{d^2x(t)}{dt^2} \\ \frac{d^2y(t)}{dt^2} \end{bmatrix}, \tag{6.49}$$

which is read as "x and y double dot."

For a point mass in the plane, as in Fig. 6.10-(a), with its Cartesian position, velocity, and acceleration given in (6.47) to (6.49), you may recall from a physics course that when the only force acting on the mass is gravity, $g = 9.81\frac{\mathrm{m}}{\mathrm{s}^2}$, Newton's laws give:

$$\begin{bmatrix} m \cdot \ddot{x}(t) \\ m \cdot \ddot{y}(t) \end{bmatrix} = \begin{bmatrix} 0.0 \\ -m \cdot g \end{bmatrix}. \tag{6.50}$$

These equations are called the **equations of motion**, or EoM for short. The equations of motion for the Cassie bipedal robot are "somewhat" similar to the above, except they would fill 100 pages!

Though we will not need the third derivative with respect to time in this section, we note that it is called the jerk. Car and elevator manufacturers are very careful about the "smoothness of acceleration" (keeping its rate of change continuous and not too large), so they use bounds on **jerk**, the third derivative of position, in the design of their products. When you see clothing[5] with

"Don't be a $\frac{d^3x(t)}{dt^3}$",

you are now in the know.

[5]Easily available on the web. Here, three dots would be too subtle.

6.4.2 Kinetic and Potential Energy of a Point Mass in the Plane

- **Potential Energy of a Point Mass** is equal to "mass times gravity times height", or in symbols:

$$V := m \cdot g \cdot y,$$

where we have deliberately dropped the explicit dependence on time; we should keep in mind, however, that the particle's height does vary with time.

- **Kinetic Energy of a Point Mass** by "one half mass times the magnitude (or norm) of the velocity squared", or in symbols,

$$K := \frac{1}{2} m \cdot \left(\dot{x}^2 + \dot{y}^2 \right), \tag{6.51}$$

where, again, we have deliberately dropped the explicit dependence on time, but we keep in mind that the particle's velocity does vary with time.

For later use, we note that both the potential and kinetic energies are scalar quantities. Their units are Joules. **We do want to observe that the kinetic energy is quadratic in the velocity components of the particle.** In fact, the Cartesian velocity, $v(t)$, is a 2-vector (because we are working in the plane) and its kinetic energy is

$$K := \frac{1}{2} m \cdot ||v(t)||^2.$$

We note this relation so that in 3-space, we know how to compute the kinetic energy of a point mass! It is always the same formula: one-half the mass times the norm squared of the velocity.

This all looks very innocent and perhaps a touch boring. For an unconstrained particle in the plane, computing its potential and kinetic energy is easy, so let's consider, instead, the pendulum, which has a bob of nonzero mass attached to a massless rod. We emphasize that the rod is massless because it then has no potential or kinetic energy. The potential and kinetic energies of the pendulum reside solely in the bob of mass $m > 0$.

Using the geometry given in **Fig. 6.10-(b)**, the Cartesian position of the bob is given by:

$$p = \begin{bmatrix} x \\ y \end{bmatrix} = \begin{bmatrix} L \cdot \sin(\theta) \\ -L \cdot \cos(\theta) \end{bmatrix}. \tag{6.52}$$

The sines and cosines show up in different places than the 3-link example in Section 1.6.3 because the angle θ for the pendulum is now referenced with respect to the vertical, with $\theta = 0$ corresponding to straight down, i.e., the pendulum at rest. We can now compute the potential energy for the pendulum,

$$V_{\text{pend}} = m \cdot g \cdot y = -m \cdot g \cdot L \cdot \cos(\theta). \tag{6.53}$$

The potential energy is at its lowest when the pendulum is hanging downwards and at its peak when the pendulum has rotated to $\theta = \pi$, straight up. Moreover, the potential energy is a nonlinear function of θ.

To determine the kinetic energy of the pendulum's bob, we differentiate the position, namely,

$$v = \begin{bmatrix} \dot{x} \\ \dot{y} \end{bmatrix} = \begin{bmatrix} \frac{d}{dt}\left(L \cdot \sin(\theta)\right) \\ \frac{d}{dt}\left(-L \cdot \cos(\theta)\right) \end{bmatrix} = \begin{bmatrix} L \cdot \cos(\theta) \cdot \dot{\theta} \\ L \cdot \sin(\theta) \cdot \dot{\theta} \end{bmatrix}, \tag{6.54}$$

where we used the chain rule for $x(\theta)$ with $\theta = \theta(t)$ and $\frac{dx}{dt} = \frac{dx}{d\theta} \cdot \frac{d\theta}{dt}$, and similarly for $y(\theta)$. The corresponding kinetic energy is given by:

$$\begin{aligned} K_{\text{pend}} &= \frac{1}{2} m \cdot \left((\dot{x})^2 + (\dot{y})^2) \right) \\ &= \frac{1}{2} m \cdot \left((L \cdot \cos(\theta) \cdot \dot{\theta})^2 + (L \cdot \sin(\theta) \cdot \dot{\theta})^2) \right) \\ &= \frac{1}{2} m \cdot L^2 \cdot \left(\cos^2(\theta) + \sin^2(\theta) \right) \cdot \dot{\theta}^2 \\ &= \frac{1}{2} m \cdot L^2 \cdot \dot{\theta}^2. \end{aligned} \tag{6.55}$$

The kinetic energy is observed to be a quadratic function of the **angular velocity**, $\dot{\theta}$. The kinematics of the pendulum are specified by a single quantity, its angle θ, and the time rate of change of that quantity shows up in the kinetic energy.

The Chain Rule: We come back to computing $\dot{x}$ and $\dot{y}$ in (6.54).

- We need to differentiate $x(t) = L \cdot \sin(\theta(t))$ with respect to time, t.

  Define $f : \mathbb{R} \to \mathbb{R}$, by $f(z) = \sin(z)$ and $g : \mathbb{R} \to \mathbb{R}$ by $g(t) = \theta(t)$. Then $x(t) = f(g(t))$; yes, z in $f(z)$ is substituted in by $g(t) = \theta(t)$. By the Chain Rule, using the prime notation for differentiation,

$$\dot{x}(t) = x'(t) = L \cdot (f(g(t)))' = L \cdot f'(g(t)) \cdot g'(t).$$

  We compute the various terms:

$$\begin{aligned} f'(z) &= (\sin(z))' = \cos(z) \\ g'(t) &= (\theta(t))' = \frac{d}{dt}\theta(t) = \dot{\theta}(t). \end{aligned}$$

  Putting the pieces together yields:

$$\begin{aligned} \dot{x}(t) &= L \cdot f'(g(t)) \cdot g'(t) \\ &= L \cdot \sin(g(t)) \cdot g'(t) \\ &= L \cdot \sin(\theta(t)) \cdot \dot{\theta}(t) \\ &= L \cdot \sin(\theta) \cdot \dot{\theta} \text{ after dropping the } t\text{'s}. \end{aligned}$$

  Professionals drop the t's because they know they are there and carrying them along for the ride in the derivations is "heavy."

  You may not need the next one, but a minus sign does disappear, and it's good to see how that happens.

- We need to differentiate $y(t) = -L \cdot \cos(t)$ with respect to time, t.

  Define $f : \mathbb{R} \to \mathbb{R}$, by $f(z) = \cos(z)$ and $g : \mathbb{R} \to \mathbb{R}$ by $g(t) = \theta(t)$. Then $y(t) = f(g(t))$; again, z in $f(z)$ is substituted in by $g(t) = \theta(t)$. By the Chain Rule, using the prime notation for differentiation,

$$\dot{y}(t) = y'(t) = -L \cdot (f(g(t)))' = -L \cdot f'(g(t)) \cdot g'(t).$$

  We compute the various terms,

$$\begin{aligned} f'(z) &= (\cos(z))' = -\sin(z) \\ g'(t) &= (\theta(t))' = \frac{d}{dt}\theta(t) = \dot{\theta}(t). \end{aligned}$$

  Putting the pieces togeter yields,

$$\begin{aligned} \dot{y}(t) &= -L \cdot f'(g(t)) \cdot g'(t) \\ &= -L \cdot (-\sin(g(t))) \cdot g'(t) \\ &= L \cdot \sin(\theta(t)) \cdot \dot{\theta}(t) \text{ (double minus sign canceled)} \\ &= L \cdot \sin(\theta) \cdot \dot{\theta} \text{ after dropping the t's}. \end{aligned}$$

In the next subsection, we will compute the potential and kinetic energies of the 3-link robot.

6.4.3 Kinetic and Potential Energy of the 3-Link Manipulator

Just as with the pendulum, we assume the links of the robot are massless[6] except for point masses $m_i > 0$ at the ends of each link, namely, the red balls in **Fig. 1.7-(b)** and in **Fig. 6.4**.

[6]A bad assumption in practice, but a good one for learning.

The kinetic energy functions become quite complex quite quickly! Hence, as a warm-up exercise, we leave off the third link in the manipulator, giving us a 2-link robot. To set us up for **Lagrange's Method** in the next section, we use what are called **generalized coordinates**, where we replace:

$$\begin{aligned} \theta_1 &\to q_1 \\ \theta_2 &\to q_2 \\ \theta_3 &\to q_3, \end{aligned}$$

and define

$$q = \begin{bmatrix} q_1 \\ q_2 \\ q_3 \end{bmatrix}.$$

Even for the 2-link version of the robot, the kinetic energy will not fit on one line! Fortunately for us engineers, kinetic energy has some beautiful structure, which we highlight in the following result.

Proposition 6.31: Kinetic Energy Is a Quadratic Form

The kinetic energy of any robot comprised of rigid links (e.g., our 3-link manipulator) can be expressed in the following quadratic form:

$$K(q,\dot{q}) = \frac{1}{2}\,\dot{q}^\top\, D(q)\,\dot{q}, \tag{6.56}$$

where

$$q := \begin{bmatrix} q_1 \\ q_2 \\ \vdots \\ q_n \end{bmatrix} \quad \text{and} \quad \dot{q} := \begin{bmatrix} \dot{q}_1 \\ \dot{q}_2 \\ \vdots \\ \dot{q}_n \end{bmatrix}$$

are the generalized position and velocity coordinates, respectively, and

$$D(q) = \begin{bmatrix} d_{11}(q) & d_{12}(q) & \cdots & d_{1n}(q) \\ d_{21}(q) & d_{22}(q) & \cdots & d_{2n}(q) \\ \vdots & \vdots & \ddots & \vdots \\ d_{n1}(q) & d_{n2}(q) & \cdots & d_{nn}(q) \end{bmatrix} \tag{6.57}$$

is a **positive-definite**[a] **symmetric matrix.**

$D(q)$ is called the **mass-inertia matrix. Its determinant is always positive (meaning greater than zero), and hence $D(q)$ is always invertible.**

[a]All eigenvalues are positive real numbers.

Example 6.32. *Determine the mass-inertia matrices for the systems in* Fig. 6.10*, that is, a point mass in the plane moving under the influence of gravity and a pendulum consisting of a point mass at the end of a massless bar.*

Solutions:

For Fig. 6.10-(a), we defined the kinetic energy in (6.51). All we have to do is write it in the form (6.56).

Doing so yields

$$K = \frac{1}{2} m \cdot \left(\dot{x}^2 + \dot{y}^2 \right)$$
$$= \frac{1}{2} \begin{bmatrix} \dot{x} \\ \dot{y} \end{bmatrix}^{\top} \cdot \begin{bmatrix} m & 0 \\ 0 & m \end{bmatrix} \cdot \begin{bmatrix} \dot{x} \\ \dot{y} \end{bmatrix}.$$

Hence,

$$D(x, y) = \begin{bmatrix} m & 0 \\ 0 & m \end{bmatrix},$$

a diagonal matrix with the mass on the diagonal. Not so interesting, except it is clearly symmetric and positive definite as long as $m > 0$!

For **Fig. 6.10-(b)**, we computed in (6.55) that

$$\mathrm{KE}_{\mathrm{pend}} = \frac{1}{2} m \cdot L^2 \cdot \dot{\theta}^2.$$

Hence,

$$D(\theta) = m \cdot L^2,$$

a 1×1 matrix depending on mass and length. Not so interesting, except it is clearly symmetric and positive definite as long as $m > 0$ and $L > 0$.

Will the multi-link robots be more interesting? In fact, they will be so interesting that writing KE by hand in the form of (6.56) will be highly error-prone.

Method 6.33: Recovering the Mass-Inertia Matrix from the Kinetic Energy

Given an arbitrarily complicated (correct) expression $KE(q, \dot{q})$ for the Kinetic Energy of a rigid n-link mechanism, the mass-intertia matrix can be extracted via a double Jacobian, that is,

$$D(q) = \frac{\partial}{\partial \dot{q}} \left[\left(\frac{\partial}{\partial \dot{q}} KE(q, \dot{q}) \right)^{\top} \right].$$

Note: This follows from part (f) of Prop. 5.44 in Chapter 5.7.5. We will now provide the symbolic code needed to compute $D(q)$.

■

2-Link Manipulator

Here is the Symbolic code for computing the energies of a two-link robot.

```
using Symbolics, SymPy, Latexify
set_default(fmt = "%.1f", convert_unicode = false)
#latexify(b) |> println

# This is for a two-link manipulator

# For the angles, see the schematic of the 3-link manipulator
# and remove the last link

@Symbolics.variables q1 q2 dq1 dq2 g m1 m2 L1 L2

```

```
# We have substituted theta1 and theta2 by q1 and q2
# Not only is this more compact, but it's also
# standard notation for Lagrange's Method for computing
# dynamical equations of motion

# Form vectors for use in Jacobians
q = [q1, q2]
dq = [dq1, dq2]

# Kinematics
p1 = L1*[cos(q1), sin(q1)]
p2 = p1 + L2*[cos(q1+q2), sin(q1+q2)]

# Potential Energy
PE = g*m1*p1[2] + g*m2*p2[2]
PE = simple(PE)
println("PE = ", PE)

# Velocity vectors via the Jacobian!
v1 = Symbolics.jacobian(p1, q) * dq
v2 = Symbolics.jacobian(p2, q) * dq

# Kinetic Energy of each mass on the Robot
KE1 = 0.5 * m1 * sum(v1.*v1)
KE2 = 0.5 * m2 * sum(v2.*v2)

# Sum the individual KEs to get the total Kinetic Energy
KE = KE1 + KE2

# Extract the mass-inertia matrix from KE
if false
    D = Symbolics.jacobian(Symbolics.jacobian([KE], dq), dq)
else
    D = Symbolics.hessian(KE, dq)
end

D = simple(D)
# Choose display method for the mass-inertia matrix
if false
    print("D = ")
    display(D)
else
    n= size(D,1)
    for i = 1:n
        for j = i:n
            print("\n \n D[$i,$j] = ", D[i,j])
        end
    end
end
```

Output After Simplication Using a Custom Function Developed in Collaboration with ChatGPT4, 09 to 14 Oct 2023. More on this later.

```
PE = g*(L1*m1*sin(q1) + m2*(L1*sin(q1) + L2*sin(q1 + q2)))


 D[1,1] = L1^2*m1 + m2*(L1^2 + 2*L1*L2*cos(q2) + L2^2)
```

```
D[1,2] = L2*m2*(L1*cos(q2) + L2)

D[2,2] = L2^2*m2
D[2,2] = L2^2*m2
```

Typeset Mass-Inertia Matrix

$$D(q) = \left[\begin{array}{cc} L_1^2 \cdot m_1 + m_2 \cdot \left(L_1^2 + 2 \cdot L_1 \cdot L_2 \cdot \cos\left(q_2\right) + L_2^2\right) & L_2 \cdot m_2 \cdot \left(L_1 \cdot \cos\left(q_2\right) + L_2\right) \\ L_2 \cdot m_2 \cdot \left(L_1 \cdot \cos\left(q_2\right) + L_2\right) & L_2^2 \cdot m_2 \end{array}\right]; \tag{6.58}$$

this is a more interesting mass-inertia matrix! We are done with the warmup.

3-Link Manipulator: The Real Deal

```
using Symbolics, SymPy, Latexify
set_default(fmt = "%.1f", convert_unicode = false)
#latexify(b) |> println

# This is for a three-link manipulator

# For the angles, see the schematic of the 3-link manipulator
# and remove the last link

@Symbolics.variables q1 q2 q3 dq1 dq2 dq2 dq3 g m1 m2 m3 L1 L2 L3

# We have substituted theta_i by q_i
# Not only is this more compact, but it's also
# standard notation for Lagrange's Method for computing
# dynamical equations of motion

# Form vectors for use in Jacobians
q = [q1, q2, q3]
dq = [dq1, dq2, dq3]

# Kinematics
p1 = L1*[cos(q1), sin(q1)]
p2 = p1 + L2*[cos(q1+q2), sin(q1+q2)]
p3 = p2 + L3*[cos(q1+q2+q3), sin(q1+q2+q3)]

# Potential Energy
PE = g*m1*p1[2] + g*m2*p2[2] + g*m3*p3[2]
PE = simple(PE)
println("PE = ", PE)

# Velocity vectors via the Jacobian!
v1 = Symbolics.jacobian(p1, q) * dq
v2 = Symbolics.jacobian(p2, q) * dq
v3 = Symbolics.jacobian(p3, q) * dq

# Kinetic Energy of each mass on the Robot
KE1 = 0.5 * m1 * sum(v1.*v1)
KE2 = 0.5 * m2 * sum(v2.*v2)
KE3 = 0.5 * m3 * sum(v3.*v3)

# Sum the individual KEs to get the total Kinetic Energy
```

```
KE = KE1 + KE2 + KE3
KE = Symbolics.simplify(KE)

# Extract the mass-inertia matrix from KE
if false
    D = Symbolics.jacobian(Symbolics.jacobian([KE], dq), dq)
else
    D = Symbolics.hessian(KE, dq)
end

# Do simplification on D with a custom function
D = simple(D)

# Choose display method for the mass-inertia matrix
if false
    print("D = ")
    display(D)
else
    n= size(D,1)
    for i = 1:n
        for j = i:n
            println("\n D[$i,$j] = ", D[i,j])
            #display(D[1,j])
        end
    end
end
```

Output

```
PE = g*(L1*m1*sin(q1) + m2*(L1*sin(q1) + L2*sin(q1 + q2))
+ m3*(L1*sin(q1) + L2*sin(q1 + q2) + L3*sin(q1 + q2 + q3)))

 D[1,1] = L1^2*m1 + m2*(L1^2 + 2*L1*L2*cos(q2) + L2^2) + m3*(L1^2
 + 2*L1*L2*cos(q2) + 2*L1*L3*cos(q2 + q3) + L2^2 + 2*L2*L3*cos(q3)
 + L3^2)

 D[1,2] = L2*m2*(L1*cos(q2) + L2) + m3*(L1*L2*cos(q2)
 + L1*L3*cos(q2 + q3) + L2^2 + 2*L2*L3*cos(q3) + L3^2)

 D[1,3] = L3*m3*(L1*cos(q2 + q3) + L2*cos(q3) + L3)

 D[2,2] = L2^2*m2 + m3*(L2^2 + 2*L2*L3*cos(q3) + L3^2)

 D[2,3] = L3*m3*(L2*cos(q3) + L3)

 D[3,3] = L3^2*m3
```

6.4.4 Remarks on Simplifying Symbolic Equations

We have been using the `Symbolics` package in Julia. It works quite well and has native Julia syntax. Its main weakness lies in its simplification of trigonometric expressions: if you really want to combine all the trig terms, it comes up short. The competing package, `SymPy`, is based in Python and has a "Julia wrapper" so that it can be used in Julia. Its syntax is Python-centric, which is good for some users and awkward for others. One of its strengths is that its simplification command is more powerful: if there is a way to combine some gnarly trig expressions to produce a shorter output string, it will likely find it.

Currently, it is hard to go back and forth between the two packages. ChatGPT helped your author build a simplification tool that translates a `Symbolic` expression to a `Sympy` expression for simplification. We give it below.

```julia
# New Function created 12 March 2026 using ChatGPT5

using Symbolics
using SymPy

# ------------------------------------------------------------------
# simple(expr; SymbolicsOut=false, SymPyOut=false, expand_first=true)
#
# Purpose:
#   Simplify a symbolic expression, vector, or matrix.
#
# Output modes:
#
#   1. simple(expr)
#      Returns a String (or an array of Strings).
#      This is the default mode and preserves the old workflow.
#
#   2. simple(expr; SymbolicsOut=true)
#      Returns a Symbolics expression (or array of Symbolics expressions).
#      but without promising the same strength as SymPy.
#      Use this when you want to continue symbolic manipulation in Julia.
#
#   3. simple(expr; SymPyOut=true)
#      Returns a SymPy expression (or array of SymPy expressions).
#      Use this when you want SymPy-native output.
#
# Notes:
#   - At most one of SymbolicsOut=true and SymPyOut=true may be used.
#   - If both are false, the output is a display-friendly String.
#   - Arrays are handled entry-by-entry.
# ------------------------------------------------------------------

function simple(expr; SymbolicsOut=false, SymPyOut=false,
      expand_first=true)

      if SymbolicsOut && SymPyOut
            error("Choose at most one of SymbolicsOut=true and SymPyOut=true.")
      end

      if SymbolicsOut
            return _simple_symbolics(expr; expand_first=expand_first)
      elseif SymPyOut
            return _simple_sympy(expr; expand_first=expand_first)
      else
            return _simple_string(expr; expand_first=expand_first)
      end
end

# Optional positional backward-compatible form:
function simple(expr, flag::Bool)
      return simple(expr; SymbolicsOut=flag)
```

```
end

# ------------------------------
# Symbolics-native output
# ------------------------------

function _simple_symbolics(A::AbstractArray; expand_first=true)
    return map(x -> _simple_symbolics(x; expand_first=expand_first), A)
end

function _simple_symbolics(expr; expand_first=true)
    out = expr
    if expand_first
        out = Symbolics.expand_derivatives(out)
    end
    out = Symbolics.simplify(out)
    return out
end

# ------------------------------
# SymPy-native output
# ------------------------------

function _simple_sympy(A::AbstractArray; expand_first=true)
    return map(x -> _simple_sympy(x; expand_first=expand_first), A)
end

function _simple_sympy(expr; expand_first=true)
    sym_expr = _simple_symbolics(expr; expand_first=expand_first)
    return _to_sympy(sym_expr)
end

# ------------------------------
# String output (default)
# ------------------------------

function _simple_string(A::AbstractArray; expand_first=true)
    return map(x -> _simple_string(x; expand_first=expand_first), A)
end

function _simple_string(expr; expand_first=true)
    py = _simple_sympy(expr; expand_first=expand_first)
    return _pretty_string(py)
end

# ------------------------------
# Conversion helpers
# ------------------------------

function _to_sympy(expr)
    raw = expr isa Symbolics.Num ? Symbolics.value(expr) : expr
    s = string(raw)

    # Convert Julia-style syntax to SymPy-readable syntax
    s = replace(s, "//" => "/")
    s = replace(s, "^" => "**")

```

```
    # Insert explicit multiplication where needed
    s = replace(s, r"(\d)([A-Za-z\(])" => s"\1*\2")
    s = replace(s, r"(\))([A-Za-z0-9])" => s"\1*\2")

    py = SymPy.sympify(s)
    py = SymPy.expand(py)
    py = SymPy.simplify(py)

    return py
end

function _pretty_string(x)
    str = string(x)
    str = replace(str, "**" => "^")
    str = replace(str, "*" => "")
    str = replace(str, r"(?<![\d\.])1\.0\*(?![\d\.])" => "")
    return str
end
```

Output

```
_pretty_string (generic function with 1 method)
```

Warning:

- **In its default mode, the function `simple(expr)` converts the result into a `String`. Once this happens, the expression is no longer symbolic and you cannot perform further symbolic operations on it.** Hence, only use the default form when you are "printing out" your final expressions.
- If you want to continue symbolic manipulation after simplification, use
  - `simple(expr; SymbolicsOut=true)` or
  - `simple(expr; SymPyOut=true)`,

  which return symbolic objects rather than strings.
- For intermediate simplification of expressions while remaining entirely within the `Symbolics` framework, you may also use `Symbolics.simplify`.

6.4.5 Lagrange's Equations

Lagrange's equation is a profound and elegant reformulation of classical mechanics. Instead of navigating the intricacies of vector forces and torques in Newtonian mechanics, Lagrange's equation offers a scalar approach, focusing on the potential and kinetic energy of moving bodies. Working with the concept of generalized coordinates, it provides a systematic method to derive equations of motion for complex mechanical systems, even those with constraints. This approach not only simplifies the mathematical treatment of complicated systems but also offers deeper insights into the conservation laws and symmetries inherent in physical systems. In essence, Lagrange's equation bridges the gap between the microscopic laws of forces as encapsulated in Newton's "$F = ma$" and the macroscopic behavior of systems, making it an indispensable tool for physicists and engineers alike.

Consider a mechanical system with generalized position coordinates, $q = (q_1, q_2, \ldots, q_n)$, and generalized velocities, $\dot{q} = (\dot{q}_1, \dot{q}_2, \ldots, \dot{q}_n)$; both q and $\dot{q}$ are column vectors. For the problems we will study, there is one generalized coordinate for each independent link or moving part in our robots.

Lagrange's Equations of Motion

Let $V(q)$ and $K(q,\dot{q})$ be the potential and kinetic energies, respectively, of a mechanical system.

Definition 6.34. *The* ***Lagrangian*** *is defined as*

$$\mathcal{L}(q,\dot{q}) := K(q,\dot{q}) - V(q). \tag{6.59}$$

Lagrange's Equations of Motion (EoM) *are given by:*

$$\frac{d}{dt}\frac{\partial \mathcal{L}(q,\dot{q})}{\partial \dot{q}} - \frac{\partial \mathcal{L}(q,\dot{q})}{\partial q} = \Gamma, \tag{6.60}$$

where the partial derivatives are column vectors and Γ is a column vector of external forces acting on the system. For us, they will primarily be motor torques.

Note that we are using the traditional convention (meaning nearly universal) in Lagrange's equations of motion that the partial derivatives are organized into a column vector.

Note: Written in the compact form (6.60), Lagrange's equations of motion are at once awe-inspiring due to their elegance and terrifying due to the many complicated-looking symbols. We will unpack (6.60) little by little to make it manageable, and then we'll turn loose the `Symbolics` package on it and compute some real equations of motion! After that, we'll deal with Γ. One thing at a time!

As a first step, we unpack the partial derivatives, just as we did with the Method of Lagrange Multipliers, giving us

$$\begin{bmatrix} \frac{d}{dt}\frac{\partial \mathcal{L}(q,\dot{q})}{\partial \dot{q}_1} \\ \frac{d}{dt}\frac{\partial \mathcal{L}(q,\dot{q})}{\partial \dot{q}_2} \\ \vdots \\ \frac{d}{dt}\frac{\partial \mathcal{L}(q,\dot{q})}{\partial \dot{q}_n} \end{bmatrix} - \begin{bmatrix} \frac{\partial \mathcal{L}(q,\dot{q})}{\partial q_1} \\ \frac{\partial \mathcal{L}(q,\dot{q})}{\partial q_2} \\ \vdots \\ \frac{\partial \mathcal{L}(q,\dot{q})}{\partial q_n} \end{bmatrix} = \begin{bmatrix} \Gamma_1 \\ \Gamma_2 \\ \vdots \\ \Gamma_n \end{bmatrix}. \tag{6.61}$$

While that $\frac{d}{dt}$ looks like it could cause complications, it is simply the total derivative covered in Chapter 5.8. You may recall that it means we need to use a vector version of the Chain Rule.

Example 6.35. *Apply Lagrange's equations of motion to the unconstrained point mass in the plane, shown in Fig. 6.10-(a).*

Solution: From Section 6.4.2, we have

$$V(x,y) = m \cdot g \cdot y$$

and

$$K(x,y,\dot{x},\dot{y}) = \frac{1}{2}m\left(\dot{x}^2 + \dot{y}^2\right).$$

Therefore, the Lagrangian is

$$\mathcal{L}(x,y,\dot{x},\dot{y}) = \frac{1}{2}m\left(\dot{x}^2 + \dot{y}^2\right) - m \cdot g \cdot y$$

In Lagrange's equations, we have the two column vectors $q = (x,y)$ and $\dot{q} = (\dot{x},\dot{y})$. Let us compute some partial derivatives:

$$\frac{\partial \mathcal{L}(q,\dot{q})}{\partial q_1} = \frac{\partial}{\partial x}\left(\frac{1}{2}m \cdot \left(\dot{x}^2 + \dot{y}^2\right) - m \cdot g \cdot y\right) = 0 \;\; (x \text{ does not appear in the Lagrangian}),$$

$$\frac{\partial \mathcal{L}(q,\dot{q})}{\partial q_2} = \frac{\partial}{\partial y}\left(\frac{1}{2}m \cdot \left(\dot{x}^2 + \dot{y}^2\right) - m \cdot g \cdot y\right) = -m \cdot g \;\; (y \text{ appears linearly in the Lagrangian}).$$

Continuing with the partial derivatives,

$$\frac{\partial \mathcal{L}(q,\dot{q})}{\partial \dot{q}_1} = \frac{\partial \mathcal{L}(x,y,\dot{x},\dot{y})}{\partial \dot{x}} = m \cdot \dot{x} \text{ ($\dot{x}$ appears quadratically in the Lagrangian, and there's a one-half)},$$
$$\frac{\partial \mathcal{L}(q,\dot{q})}{\partial \dot{q}_2} = \frac{\partial \mathcal{L}(x,y,\dot{x},\dot{y})}{\partial \dot{y}} = m \cdot \dot{y} \text{ ($\dot{y}$ appears quadratically in the Lagrangian, and there's a one-half)}.$$

Next, we apply the time derivative:

$$\frac{d}{dt}\frac{\partial \mathcal{L}(q,\dot{q})}{\partial \dot{q}_1} = \frac{d}{dt}(m \cdot \dot{x}) = m \cdot \ddot{x}$$

and

$$\frac{d}{dt}\frac{\partial \mathcal{L}(q,\dot{q})}{\partial \dot{q}_2} = \frac{d}{dt}(m \cdot \dot{y}) = m \cdot \ddot{y}.$$

Combining these results, Lagrange's equations give:

$$\underbrace{\begin{bmatrix} m \cdot \ddot{x} \\ m \cdot \ddot{y} \end{bmatrix}}_{\frac{d}{dt}\frac{\partial \mathcal{L}(q,\dot{q})}{\partial \dot{q}}} - \underbrace{\begin{bmatrix} 0 \\ -mg \end{bmatrix}}_{\frac{\partial \mathcal{L}(q,\dot{q})}{\partial q}} = \underbrace{\begin{bmatrix} 0 \\ 0 \end{bmatrix}}_{\Gamma}, \tag{6.62}$$

where $\Gamma = 0_{2\times 1}$ because there are no external forces acting on the particle. While gravity is in the picture, it and springs are (almost) always included in the potential energy term. If we were modeling an airplane, we would have an external thrusting force from its propeller or jet engine. For our simple particle, there is no such force.

Bottom Line: For this simple system, Lagrange's Equations of Motion (EoM) may seem to be overkill, but we certainly recovered the same model we had previously. ■.

Example 6.36. *Apply Lagrange's equations of motion to the constrained point mass in the plane, shown in Fig. 6.10-(b), otherwise known as a pendulum.*

Solution: From Section 6.4.2, we have

$$V(\theta) = -m \cdot g \cdot L \cdot \cos(\theta)$$

and

$$K(\theta,\dot{\theta}) = \frac{1}{2} m \cdot L^2 \cdot \dot{\theta}^2.$$

Therefore, the Lagrangian given by (6.59) is:

$$\mathcal{L}(\theta,\dot{\theta}) = \frac{1}{2} m \cdot L^2 \cdot \dot{\theta}^2 + m \cdot g \cdot L \cdot \cos(\theta),$$

where we replaced the double minus signs with a single plus sign.

In Lagrange's equations, we have scalar generalized coordinates $q = \theta$ and $\dot{q} = \dot{\theta}$. Let us compute some partial derivatives:

$$\frac{\partial \mathcal{L}(q,\dot{q})}{\partial q} = \frac{\partial}{\partial \theta}\left(\frac{1}{2} m \cdot L^2 \cdot \dot{\theta}^2 + m \cdot g \cdot L \cdot \cos(\theta)\right) = -m \cdot g \cdot L \cdot \sin(\theta).$$

Continuing with the partial derivatives,

$$\frac{\partial \mathcal{L}(q,\dot{q})}{\partial \dot{q}} = \frac{\partial}{\partial \dot{\theta}}\left(\frac{1}{2} m \cdot L^2 \cdot \dot{\theta}^2 + m \cdot g \cdot L \cdot \cos(\theta)\right) = m \cdot L^2 \cdot \dot{\theta}.$$

Applying $\frac{d}{dt}$ leads to:

$$\frac{d}{dt}\frac{\partial \mathcal{L}(q,\dot{q})}{\partial \dot{q}_1} = \frac{d}{dt}(m \cdot L^2 \cdot \dot{\theta}) = m \cdot L^2 \cdot \ddot{\theta}$$

because the mass and length are constant. If the bob were a bucket of water, with water splashing over the sides as it oscillated, we would have needed to compute $\dot{m}$, but not in our case.

Integrating everything together, Lagrange's equations give:

$$\underbrace{m \cdot L^2 \cdot \ddot{\theta}}_{\frac{d}{dt}\frac{\partial \mathcal{L}(q,\dot{q})}{\partial \dot{q}}} - \underbrace{(-m \cdot g \cdot L \cdot \sin(\theta))}_{\frac{\partial \mathcal{L}(q,\dot{q})}{\partial q}} = \underbrace{0}_{\Gamma}, \tag{6.63}$$

where $\Gamma = 0_{1\times 1}$ because we have not yet installed a motor at the pivot. After a tiny bit of algebra, we recover the equations we had earlier in (6.46). ■

The following result will help us to interpret the output of Lagrange's EoM for the 3-link manipulator and, in fact, for all models of connected rigid bodies.

Proposition 6.37: The Robot Equations

When Lagrange's EoM in (6.60) are expanded out for a standard mechanical system (i.e., most robots), they take the form

$$D(q) \cdot \ddot{q} + C(q,\dot{q}) \cdot \dot{q} + G(q) = \Gamma \qquad \textbf{(``the Robot Equations'')}, \tag{6.64}$$

where

- $D(q)$ is the mass-inertia matrix from the kinetic energy and is therefore invertible,
- $G(q) := \nabla V(q)$, is the gradient of the potential energy,
- and $C(q,\dot{q}) \cdot \dot{q}$ contains all the remaining terms in (6.60).

Example 6.38. *Identify the terms in the robot equations for the unconstrained point mass in the plane and the pendulum.*

Solution:

The unconstrained point mass in the plane has equations of motion,

$$\begin{bmatrix} \ddot{x} \\ \ddot{y} \end{bmatrix} + \begin{bmatrix} 0 \\ mg \end{bmatrix} = \begin{bmatrix} 0 \\ 0 \end{bmatrix},$$

after removing the double minus sign. Hence,

$$q = \begin{bmatrix} x \\ y \end{bmatrix}, \;\; D(q) = I_2, \;\; C(q,\dot{q}) = 0_{2\times 2}, \;\; G(q) = \begin{bmatrix} 0 \\ mg \end{bmatrix}, \text{ and } \Gamma = \begin{bmatrix} 0 \\ 0 \end{bmatrix}.$$

The pendulum has equations of motion,

$$m \cdot L^2 \cdot \ddot{\theta} + (m \cdot g \cdot L \cdot \sin(\theta)) = 0.$$

Hence,

$$q = \theta, \;\; D(q) = m L^2, \;\; C(q,\dot{q}) = 0_{1\times 1}, \;\; G(q) = m \cdot g \cdot L \cdot \sin(\theta), \text{ and } \Gamma = 0.$$

■

Remark 6.39. *If you find the Symbolic computations of the next section impressive, please read Secrets of the Arcane 6.40. Today, robot models can be extracted directly from the CAD files used to design the robot in a* ***Unified Robot Description Format****, or URDF for short. You can think of a URDF as a written blueprint for a robot: "Link-1 is connected to Link-2 via a revolute joint. Link-2 is connected to Link-3 via a prismatic joint." Moreover, the URDF contains all of the mechanical parameters of each link, such as length, mass, center of mass, moments of inertia about various points on the link, etc. It is all coming together so that you can spend your time on creative electromechanical design, perception systems and signal processing, planning and mapping, feedback control design, human-robot interaction (HRI), etc. It's a great time to be a Roboticist!*

Videos on the Calculus of Variations and Lagrange's Equation: While this is definitely advanced material, it may inspire future course selections in mathematics and physics.

- Introduction to Variational Calculus - Deriving the Euler-Lagrange Equation by Good Vibrations with Freeball.
- Derivation of the Euler-Lagrange Equation | Calculus of Variations by Faculty of Khan.
- Derivation of the Euler-Lagrange Equation by The Caribbean Bookworm.

6.4.6 Lagrange's Equations of Motion for the 3-Link Manipulator

```
using Symbolics, SymPy, Latexify
set_default(fmt = "%.1f", convert_unicode = false)
#latexify(b) |> println

# This is for a three-link manipulator

@Symbolics.variables q1 q2 q3 dq1 dq2 dq2 dq3 g m1 m2 m3 L1 L2 L3

# We have substituted theta_i by q_i
# Not only is this more compact, but it's also
# standard notation for Lagrange's Method for computing
# dynamical equations of motion

# Form vectors for use in Jacobians
q = [q1, q2, q3]
dq = [dq1, dq2, dq3]

# Kinematics
p1 = L1*[cos(q1), sin(q1)]
p2 = p1 + L2*[cos(q1+q2), sin(q1+q2)]
p3 = p2 + L3*[cos(q1+q2+q3), sin(q1+q2+q3)]

# Potential Energy
PE = g*m1*p1[2] + g*m2*p2[2] + g*m3*p3[2]

# Use native Symbolics simplification
PE = Symbolics.simplify(PE)

# Velocity vectors via the Jacobian!
v1 = Symbolics.jacobian(p1, q) * dq
v2 = Symbolics.jacobian(p2, q) * dq
v3 = Symbolics.jacobian(p3, q) * dq

# Kinetic Energy of each mass on the Robot
```

```
KE1 = 0.5 * m1 * sum(v1.*v1)
KE2 = 0.5 * m2 * sum(v2.*v2)
KE3 = 0.5 * m3 * sum(v3.*v3)

# Sum the individual KEs to get the total Kinetic Energy
KE = KE1 + KE2 + KE3

# Use native Symbolics simplification
KE = Symbolics.simplify(KE)

# Form the Lagrangian
L = KE - PE

# Compute terms in the Robot Equations
G = Symbolics.gradient(PE, q)
D = Symbolics.hessian(KE, dq) # Strange name explained below

# Compute C (advanced, uses Christoffel symbols)
n = length(q)
C = Array{Num}(undef, n, n) # Num is Symbolic Type
                            # Strange name explained below
for k in 1:n
    for j in 1:n
        C[k, j] = 0
        for i in 1:n
            C[k, j] += 0.5 * (Symbolics.gradient(D[k, j], q)[i] +
            Symbolics.gradient(D[k, i], q)[j] -
            Symbolics.gradient(D[i, j], q)[k]) * dq[i]
        end
    end
end

# Place an actuator at the hip on
# the relative angle between the torso and
# each leg
# Use D'Alembert's Principle
E = [q1;q2;q3]
B = Symbolics.jacobian(E, q)
B = B'
JacG = Symbolics.jacobian(G, q) # Used when we discuss linearization of
     # NL models


# Only run simple after all Symbolic manipulation is done and you
# are ready to create a function
if true
    # Simplify as much as possible at the very end
    D = simple(D)
    C = simple(C)
end

# Create a function that computes all terms
# in the Robot Equations
fcn_name = "dyn_mod_3LinkManipulator"
modelParamString = "g, L1, L2, L3, m1, m2, m3 = modelParameters()"
# Need to include line break command \n
variableNamesString = "q1, q2, q3 = q \ndq1, dq2, dq3 = dq"
```

```
writeEOM(fcn_name, D, C, G, B, JacG,modelParamString,variableNamesString)

# bring the model into the workspace for later use
include("dyn_mod_3LinkManipulator.jl")

# If the line above fails, it is because Julia is trying to read the
# file before it has
# been made available to it by the file system. Simply copy the command
# into a new cell and
# run it. The time it takes you to do that will allow the file system to
# catch up! :-)
```

Output

```
[creating DYN_MOD_3LINKMANIPULATOR.jl]
File dyn_mod_3LinkManipulator.jl created successfully!

dyn_mod_3LinkManipulator (generic function with 1 method)
```

When you open the file, you have a function that computes all the terms in Lagrange's EOM, formatted in a more-or-less human-readable form.

```
function dyn_mod_3LinkManipulator(q, dq)
# DYN_MOD_3LINKMANIPULATOR
# 2023-12-03 08:48:06
#
# Author: Grizzle
#
# Model NOTATION: D(q)ddq + C(q,dq)*dq + G(q) = B*tau
# The Robot Equations: From Lagrange's Equations of Motion
#
g, L1, L2, L3, m1, m2, m3 = modelParameters()
#
# Variable names for the model
q1, q2, q3 = q
dq1, dq2, dq3 = dq
#
D = zeros(3, 3)
  D[1, 1] = L1^2*m1 + m2*(L1^2 + 2*L1*L2*cos(q2) + L2^2) +
            m3*(L1^2 + 2*L1*L2*cos(q2) + 2*L1*L3*cos(q2 + q3) + L2^2 +
            2*L2*L3*cos(q3) + L3^2)
  D[1, 2] = L2*m2*(L1*cos(q2) + L2) + m3*(L1*L2*cos(q2) + L1*L3*cos(q2 +
             q3) + L2^2 + 2*L2*L3*cos(q3) + L3^2)
  D[1, 3] = L3*m3*(L1*cos(q2 + q3) + L2*cos(q3) + L3)
  D[2, 1] = L2*m2*(L1*cos(q2) + L2) + m3*(L1*L2*cos(q2) + L1*L3*cos(q2 +
             q3) + L2^2 + 2*L2*L3*cos(q3) + L3^2)
  D[2, 2] = L2^2*m2 + m3*(L2^2 + 2*L2*L3*cos(q3) + L3^2)
  D[2, 3] = L3*m3*(L2*cos(q3) + L3)
  D[3, 1] = L3*m3*(L1*cos(q2 + q3) + L2*cos(q3) + L3)
  D[3, 2] = L3*m3*(L2*cos(q3) + L3)
  D[3, 3] = L3^2*m3
#
C = zeros(3, 3)
  C[1, 1] = -L1*L2*dq2*m2*sin(q2) - L1*L2*dq2*m3*sin(q2) - L1*L3*dq2*m3*
            sin(q2 + q3) - L1*L3*dq3*m3*sin(q2 + q3) - L2*L3*dq3*m3*sin(q3)
  C[1, 2] = -L1*L2*dq1*m2*sin(q2) - L1*L2*dq1*m3*sin(q2) - L1*L2*dq2*m2*
```

```
                 sin(q2) - L1*L2*dq2*m3*sin(q2) - L1*L3*dq1*m3*sin(q2 + q3) - L1*
                 L3*dq2*m3*sin(q2 + q3) - L1*L3*dq3*m3*sin(q2 + q3) - L2*L3*dq3*
                 m3*sin(q3)
    C[1, 3] = -L3*m3*(L1*sin(q2 + q3) + L2*sin(q3))*(dq1 + dq2 + dq3)
    C[2, 1] = L1*L2*dq1*m2*sin(q2) + L1*L2*dq1*m3*sin(q2) + L1*L3*dq1*m3*
                 sin(q2 + q3) - L2*L3*dq3*m3*sin(q3)
    C[2, 2] = -L2*L3*dq3*m3*sin(q3)
    C[2, 3] = -L2*L3*m3*(dq1 + dq2 + dq3)*sin(q3)
    C[3, 1] = L3*m3*(L1*dq1*sin(q2 + q3) + L2*dq1*sin(q3) + L2*dq2*sin(q3))
    C[3, 2] = L2*L3*m3*(dq1 + dq2)*sin(q3)
    C[3, 3] = 0.0
#
G = zeros(3)
    G[1] = g*m2*(L1*cos(q1) + L2*cos(q1 + q2)) + g*m3*(L1*cos(q1) + L2*
                 cos(q1 + q2) + L3*cos(q1 + q2 + q3)) + L1*g*m1*cos(q1)
    G[2] = g*m3*(L2*cos(q1 + q2) + L3*cos(q1 + q2 + q3)) + L2*g*m2*cos(q1 +
                  q2)
    G[3] = L3*g*m3*cos(q1 + q2 + q3)
#
B = zeros(3, 3)
    B[1, 1] = 1
    B[2, 2] = 1
    B[3, 3] = 1
#
JacG = zeros(3, 3)
    JacG[1, 1] = g*m2*(-L1*sin(q1) - L2*sin(q1 + q2)) +
                   g*m3*(-L1*sin(q1) - L2*sin(q1 + q2) -
                   L3*sin(q1 + q2 + q3)) - L1*g*m1*sin(q1)
    JacG[1, 2] = g*m3*(-L2*sin(q1 + q2) - L3*sin(q1 + q2 + q3)) -
                   L2*g*m2*sin(q1 + q2)
    JacG[1, 3] = -L3*g*m3*sin(q1 + q2 + q3)
    JacG[2, 1] = g*m3*(-L2*sin(q1 + q2) - L3*sin(q1 + q2 + q3)) -
                   L2*g*m2*sin(q1 + q2)
    JacG[2, 2] = g*m3*(-L2*sin(q1 + q2) - L3*sin(q1 + q2 + q3)) -
                   L2*g*m2*sin(q1 + q2)
    JacG[2, 3] = -L3*g*m3*sin(q1 + q2 + q3)
    JacG[3, 1] = -L3*g*m3*sin(q1 + q2 + q3)
    JacG[3, 2] = -L3*g*m3*sin(q1 + q2 + q3)
    JacG[3, 3] = -L3*g*m3*sin(q1 + q2 + q3)
#
    return (D=D, C=C, G=G, B=B, JacG=JacG)
end
```

This brings the model into the workspace.

```
include("dyn_mod_3LinkManipulator.jl") # bring the model into the workspace
```

Output

```
dyn_mod_3LinkManipulator (generic function with 1 method)
```

6.5 (Optional Read) More on Lagrangian Dynamics and Mechanical Properties of Planar Bodies

6.5.1 A Crude Numerical Method to Solve the Robot Equations

This section describes a crude but straightforward approach to "solving" the **robot dynamic equations** via "numerical integration." Since ordinary differential equations (ODEs) have not yet been introduced in the course, we use a **basic integration method** combining the **Riemann sum** and the **trapezoidal rule**. The included code stub assumes a 2-link manipulator.

Problem Setup

We consider the robot equations without input, namely

$$D(q)\ddot{q} + C(q,\dot{q})\dot{q} + G(q) = 0_{n\times 1},$$

where, as before,

- q typically represents joint angles,
- $\dot{q}$ typically represents angular velocities, and
- $\ddot{q}$ represents the angular accelerations.

Moreover,

- $D(q)$ is the $n \times n$ **inertia matrix**,
- $C(q,\dot{q})$ represents the $n \times n$ **Coriolis and centrifugal effects**,
- $G(q)$ is the $n \times 1$ **gravity vector**.

To be extra clear, we are not including motor torques at the joints.

Angular Acceleration

For later use, the angular acceleration can be solved for and used to update the velocity term,

$$\ddot{q} = D(q)\backslash\left(-C(q,\dot{q})\dot{q} - G(q)\right).$$

Note that we always choose to avoid matrix inverses whenever possible.

Numerical Integration Approach

We solve this system over a finite time interval T using a **discrete time step**:

$$t = \{0, \Delta t, 2\Delta t, \ldots, T\}$$

where $\Delta t = \frac{T}{N}$ is the time step, and N is the total number of steps.
The numerical integration proceeds as follows:

1. **Notation for the code**
   - $q[:,i]$ is q (angular position) at time $t[i]$.
   - $dq[:,i]$ is $\dot{q}$ (angular velocity) at time $t[i]$.
   - $ddq[:,i]$ is $\ddot{q}$ (angular acceleration) at time $t[i]$.
2. **Compute angular acceleration** $\ddot{q}$ at the current time step using:

$$ddq[:,i] = D(q[:,i])\backslash(-C(q[:,i], dq[:,i])\cdot dq[:,i] - G(q[:,i])).$$

3. **Update velocity** $\dot{q}$ using a (left) **Riemann Sum**:

$$dq[:, i+1] = dq[:, i] + \Delta t \cdot ddq[:, i]$$

4. **Update position** q using the **Trapezoidal Rule**:

$$q[:, i+1] = q[:, i] + \Delta t \cdot \left(\frac{dq[:, i] + dq[:, i+1]}{2} \right)$$

```
# bring the model into the workspace. We assume a 2-link manipulator
include("dyn_mod_2LinkManipulator.jl")

# Define parameters for the model, all in one place
# 2-Link Manipulator Model
function modelParameters()
    g = 9.81
    L1 = 1.0
    L2 = 0.5
    m1 = 3.0
    m2 = 1.0
    return g, L1, L2, m1, m2
end

# Initial values for the angles and the angular velocities
q0  = [-pi/4; 0.0] # iniital angles of the 2-links
dq0 = [0.0; 0.0]   # iniital angular velocities of the 2-links



# set up a time vector
 T = 5.0 #seconds
 N = 200
 t = range(0.0, T, N+1)
dt = T/N

# Create vectors to store the angles q, and the angular velocities, dq
q  = zeros(2,N+1)
dq = zeros(2,N+1)

# Initialize everything
q[:,1]  = q0
dq[:,1] = dq0

# Set up a for loop to run a combination of a Riemann sum and TrapZ
println("\nThe computed solution is only accurate over a short time
    interval due to our crude integration method\n")
println("Numerical errors will excite the pendulum and cause an increase
    in amplitude. This is not physical behavior.\n \n")

for i = 1:N
    # Call the model and solve for the angular acceleration of the
    # links at time t[i]
    D, C, G, ~, ~ = dyn_mod_2LinkManipulator(q[:,i], dq[:,i]) # model terms
    # at time t[i]
    ddq_i = D\( -C*dq[:,i] - G ) # acceleration at time t[i]
    dq[:, i+1] = dq[:, i] + dt * ddq_i # integrate acceleration to compute
```

```
    # velocity using Riemann Sum
    q[:, i+1] =  q[:, i] + dt * ( dq[:, i] + dq[:,i+1] )/2 # integrate
    # velocity to compute position using TrapZ
end
```

Observations and Limitations

- This method is **only accurate over a short time interval** due to its numerical simplicity.
- The **Riemann sum** for velocity and the **trapezoidal rule** for position introduce small integration errors that accumulate over time.
- These numerical errors **excite the pendulum**, leading to unrealistic, unstable behavior over long simulation times.
- More sophisticated integration methods (e.g., **Runge-Kutta**, implicit solvers) are needed for **long-term accuracy**.
- Despite its simplicity, this approach provides a **basic understanding of numerical integration** without requiring any significant knowledge of **ODE solvers**. Later in the course, we will introduce **professional solvers** to improve stability and accuracy.

6.5.2 How the Robot Equations Arise from Lagrange's Equations of Motion

The Robot Equations in (6.64) arise from a careful evaluation of the terms in

$$\frac{d}{dt}\frac{\partial \mathcal{L}(q,\dot{q})}{\partial \dot{q}} - \frac{\partial \mathcal{L}(q,\dot{q})}{\partial q} = \Gamma$$

for

$$\mathcal{L}(q,\dot{q}) := K(q,\dot{q}) - V(q).$$

From Prop. 6.31, and in particular, (6.56), the kinetic energy has the quadratic form

$$K(q,\dot{q}) = \frac{1}{2}\dot{q}^{\top} D(q)\dot{q}.$$

Moreover, the potential energy $V(q)$ does not depend on the velocity terms. These observations yield

$$\begin{aligned}
\frac{\partial}{\partial \dot{q}}\mathcal{L}(q,\dot{q}) &= \frac{\partial}{\partial \dot{q}}\left(\frac{1}{2}\dot{q}_1^{\top} D(q)\dot{q} - V(q)\right) \quad \text{(derivative of a sum)} \\
&= \frac{\partial}{\partial \dot{q}}\left(\frac{1}{2}\dot{q}_1^{\top} D(q)\dot{q}\right) - \frac{\partial}{\partial \dot{q}}\left(V(q)\right) \quad \text{(equals sum of derivatives))} \\
&= \frac{\partial}{\partial \dot{q}}\left(\frac{1}{2}\dot{q}_1^{\top} D(q)\dot{q}\right) \quad (V(q) \text{ independent of } \dot{q}) \\
&= D(q)\cdot\dot{q} \quad \text{(column convention for the partial derivative).}
\end{aligned}$$

The term $\frac{d}{dt}\frac{\partial \mathcal{L}(q,\dot{q})}{\partial \dot{q}} = \frac{d}{dt}\left(D(q)\dot{q}\right)$ is the total derivative from Section 5.8. Inspired by the Product Rule of single-variable differentiation, one is tempted to write

$$\frac{d}{dt}\frac{\partial \mathcal{L}(q,\dot{q})}{\partial \dot{q}} = \frac{d}{dt}\left(D(q)\cdot\dot{q}\right) = \dot{D}(q,\dot{q})\cdot\dot{q} + D(q)\cdot\ddot{q},$$

and, in fact, it is possible to make sense of

$$\dot{D}(q,\dot{q}) := \frac{d}{dt}D(q).$$

Indeed, as in (6.57), let $d_{ij}(q)$ denote the ij-term of the mass-inertia matrix, $D(q)$. Then, by Def. 5.45, the Chain Rule applied to the ij-term of $\frac{d}{dt}D(q)$ is

$$\frac{d}{dt}d_{ij}(q) = \underbrace{\frac{\partial d_{ij}(q)}{\partial q}}_{1\times n \text{ Jacobian}} \cdot \underbrace{\dot{q}}_{n\times 1},$$

which is a scalar quantity. In other symbols,

$$\left[\frac{d}{dt}D(q)\right]_{ij} = \frac{\partial d_{ij}(q)}{\partial q} \cdot \dot{q} =: \left[\dot{D}(q,\dot{q})\right]_{ij}.$$

Next,

$$\begin{aligned}\frac{\partial \mathcal{L}(q,\dot{q})}{\partial q} &= \frac{\partial}{\partial q}\left(K(q,\dot{q}) - V(q)\right) \\ &= \frac{\partial}{\partial q}\left(K(q,\dot{q})\right) - \frac{\partial}{\partial q}V(q) \\ &= \frac{\partial}{\partial q}\left(K(q,\dot{q})\right) - \nabla V(q).\end{aligned}$$

Hence, the process leads to:

$$\frac{d}{dt}\frac{\partial \mathcal{L}(q,\dot{q})}{\partial \dot{q}} - \frac{\partial \mathcal{L}(q,\dot{q})}{\partial q} = D(q)\cdot\ddot{q} + \left(\dot{D}(q,\dot{q})\cdot\dot{q} - \frac{\partial}{\partial q}\left(K(q,\dot{q})\right)\right) + \nabla V(q).$$

We will not expand more upon the middle term in parentheses, except to note that it provides the term traditionally denoted as $C(q,\dot{q})\cdot\dot{q}$, that is,

$$C(q,\dot{q})\cdot\dot{q} := \dot{D}(q,\dot{q})\cdot\dot{q} - \frac{\partial K(q,\dot{q})}{\partial q}.$$

For all but the simplest of mechanical systems, computing Lagrange's EoM by hand is extremely tedious and error-prone. Today, however, Symbolic packages/toolboxes allow the calculations to be done accurately and quickly. Moreover, as we saw with the 3-link manipulator, we can easily turn the computed quantities into executable code. **This underlines, once again, why a calculus course like this one is needed: without seeing with our own eyes the software tools of today in action, it is all too easy to dismiss an amazing body of work as completely beyond our abilities.**

We can even move one step further, where the robot model is extracted directly from the CAD files used to design the robot! The **Unified Robot Description Format (URDF)** stands out as the leading programming-language-agnostic medium for this purpose at the present time.

Secrets of the Arcane 6.40: URDF Files

Here are some resources that can be useful for undergraduates looking to learn URDF (Unified Robot Description Format) files for robots:

1. **ROS Wiki Tutorials**:
   - The ROS (Robot Operating System) Wiki provides a series of tutorials on URDF. It covers building a visual robot model with URDF from scratch, defining movable joints in URDF, and adding physical and collision properties to a URDF model. These tutorials provide a step-by-step approach which could be very beneficial for learning URDF. (source)
2. **Medium Article on Creating URDF files**:
   - A Medium article discusses creating URDF files for robots. It seems to cover forward kinematics and mentions that it will later cover inverse kinematics as well. The author has uploaded some files on GitHub for reference. This article could provide practical insights and hands-on examples for learning URDF. (source)

3. **Learning ROS for Robotics Programming**:

   - Another book that provides a section on creating the first URDF file, which is part of building a mobile robot with four wheels and an arm with a gripper. This resource might provide a good hands-on approach to learning URDF through a practical project. (source)

6.5.3 Including Actuators in Lagrange's Equations

In the 3-link manipulator, we included the actuation term without explaining how it is done. This is not a course on Lagrangian Dynamics, so it is necessary to take some shortcuts. Just in case you are curious, this is how to compute Γ in the robot equations for planar robots with n-revolute joints when the motors, if any, are mounted either

- on the ground (aka to the world frame), with link-1 attached to the output shaft of the motor, or for $i > 1$,
- at the end of link-i with link-$i+1$ attached to the output shaft of the motor.

Not all links have to be actuated; humanoids are famously **underactuated**, which is the term for robots that do not have a motor on every joint (Why not include a lot or motors? They are heavy and use a lot of energy!) We also assume that the generalized coordinates have been assigned as we have done with the 3-link manipulator, with the base link associated to an absolute angle q_1 and subsequent links assigned relative angles q_i.

Let q_a be an $m \times 1$ column vector containing the list of actuated joint angles. For the 3-link manipulator, we assumed that all of the joints were actuated, so we set $m = 3$ and

$$q_a := \begin{bmatrix} q_1 \\ q_2 \\ q_3 \end{bmatrix}.$$

The recipe for computing Γ is then

$$\Gamma := B \cdot \tau,$$

where

$$B := \left(\frac{\partial q_a}{\partial q}\right)^\top,$$

an $n \times m$ matrix, and τ is the $m \times 1$ vector of motor torques

$$\tau := \begin{bmatrix} \tau_1 \\ \vdots \\ \tau_m \end{bmatrix}.$$

Usually, there is gearing between a motor and a joint. Moreover, the spinning of the motor's rotor contributes significantly to the kinetic energy of the overall system. All of these effects can be taken into account when building the Lagrangian model of your robot. To learn from a simplified presentation, see the bottom of page 433 of Feedback Control of Dynamic Bipedal Locomotion by Westervelt et al., **Accounting for motors and rigid gear trains**. You might also watch the video Actuation, Gearing, and Friction by Prof. Kevin Lynch of Northwestern University.

6.5.4 Moment of Inertia

From Section 3.4.5, the **moment of inertia** (I) of an object about a specific axis is a measure of how difficult it is to change the rotational motion of that object. It depends on the mass distribution of the object relative to the axis of rotation. The moment of inertia is given by the integral

$$I = \int r^2 \, dm,$$

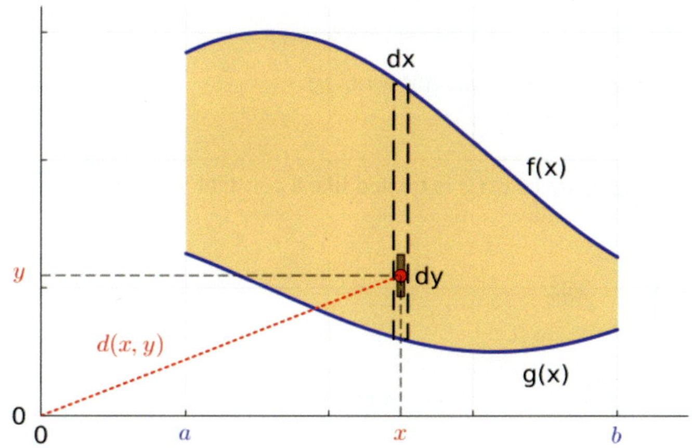

Figure 6.11: (Moment of Inertia about the z-Axis): This figure illustrates how to use a rectangle of width dx, upper bounded by $f(x)$ and lower bounded by $g(x)$, to compute the moment of inertia about the z-axis, that is, the axis perpendicular to the (x, y)-plane and passing through the origin. The body is assumed to be of constant density. The computation requires two integrals. The first one, along the y-direction, is a very simple computation of the moment of inertia of an infinitesimally thin rectangle of width dx. The second one, along the x-direction, adds up the contribution of each thin rectangle. The final result is given in (6.65).

where r is the distance of a small mass element (dm) from the axis of rotation. This formula highlights that the further the mass is distributed from the axis, the higher is the moment of inertia, thus making it harder to rotate the object.

For practical applications, common shapes have standard formulas for their moments of inertia. For example, for a solid cylinder of radius R rotating about its central axis, the moment of inertia is

$$I = \frac{1}{2} m R^2,$$

where m is the mass of the cylinder. This expression helps in understanding the rotational characteristics of the cylinder and plays a crucial role in dynamics and engineering.

Proposition 6.41: Moment of Inertia about the z-Axis for Planar Bodies of Constant Density

Given the planar body in Fig. 6.11 bounded by $y = f(x)$ and $y = g(x)$, with constant density ρ, the moment of inertia about the z-axis can be computed by a single integral from $x = a$ to $x = b$,

$$I_z = \int_a^b \rho \left(x^2 (f(x) - g(x)) + \frac{1}{3} (f^3(x) - g^3(x)) \right) dx. \tag{6.65}$$

Proof: The proof is another example of the power of the rectangle in solving calculus problems. We refer to Fig. 6.11 throughout the derivation.

Consider the vertical rectangle of width dx, and then within the rectangle, a small rectangle of height dy. The differential area of the infinitesimal rectangle is the product of height times width, that is, $dA := dy\, dx$ and the differential mass is the product of the density ρ and the differential area, giving us

$$dm = \rho\, dA = \rho\, dy\, dx. \tag{6.66}$$

The differential moment of inertia about the z-axis is defined as the distance squared times the differential mass, where

$$r^2 = d^2(x, y) = x^2 + y^2.$$

We need to add up the contribution of each differential mass along the y-axis of the thin rectangle, which results in the integral

$$\begin{aligned}
dI_z(x) &= \left[\int_{g(x)}^{f(x)} d^2(x,y)\rho\, dy\right] dx \quad \text{(integral w.r.t. } y)\\
&= \left[\int_{g(x)}^{f(x)} (x^2+y^2)\rho\, dy\right] dx \quad (x \text{ is treated like a constant when integrating w.r.t. } y)\\
&= \left[x^2\, y + \frac{y^3}{3}\right]_{g(x)}^{f(x)} \rho\, dx\\
&= \left(x^2(f(x)-g(x)) + \frac{1}{3}(f^3(x)-g^3(x))\right) \rho\, dx.
\end{aligned} \tag{6.67}$$

Next, we need to integrate with respect to x so as to add up the contribution of each thin rectangle:

$$\begin{aligned}
I_z &= \int_a^b dI_z(x)\\
&= \int_a^b \left(x^2(f(x)-g(x)) + \frac{1}{3}(f^3(x)-g^3(x))\right) \rho\, dx,
\end{aligned} \tag{6.68}$$

which establishes (6.65). ■

Suggested Videos:

- More on moment of inertia | Moments, torque, and angular momentum by Khan Academy.
- Demonstrating Rotational Inertia (or Moment of Inertia) by Flipping Physics. It's a bit goofy but full of information if you can stick through it. It's also a bit advanced.
- Moment of Inertia by DoctorPhys. This is Lesson 1 of many! Next, we call out Lesson 9 because it pertains to the BallBot.
- Moment of Inertia for a Spherical Shell by DoctorPhys. This is the moment of inertia of a basketball rolling on a floor.

6.5.5 A Simple Planar Biped

Most of the terms in the model are defined in Fig. 6.12-(a). Missing is the length of the torso (line from O_H to O_T) is L and the four line segments in the legs each have length r.

```
# Julia code
using Symbolics

# 3-link Walker Model is from 1999.
# All angles are ABSOLUTE and are
# Positive in the CW Direction (not CCW)

# Variable declarations
@variables th1 th2 th3 dth1 dth2 dth3 m Mh Mt g L r

# Generalized coordinates
q = [th1, th2, th3]

```

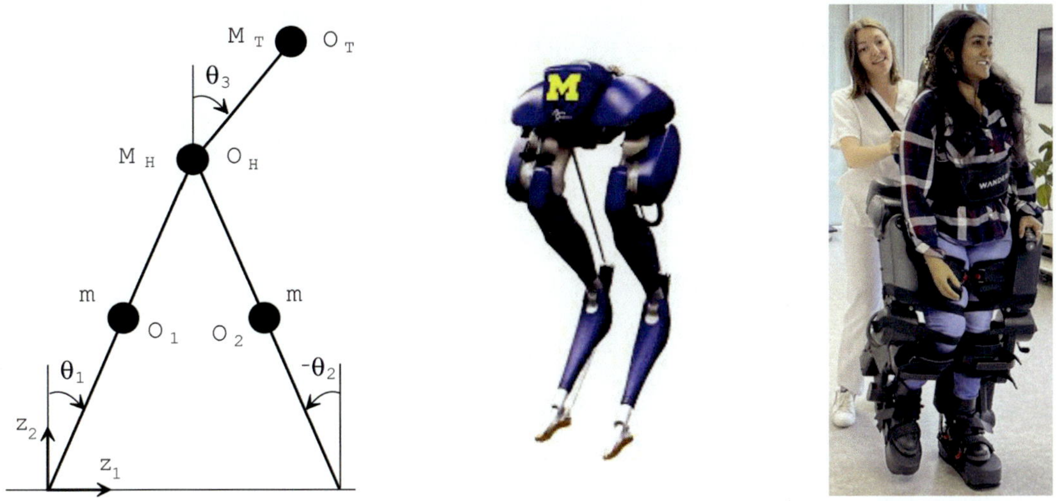

(a) (b) (c)

Figure 6.12: The University of Michigan's foray into bipedal locomotion began with the study of the 3-link walker shown in (a), and over the years progressed to Cassie in (b), and collaborating with Wandercraft on the exoskeleton, Atalante, shown in (c). Atalante allows patients with paraplegia to walk again. In addition to powering the patient's legs, it provides the overall balance for the patient and itself. It is truly a wearable robot.

```
# First derivative of generalized coordinates
dq = [dth1, dth2, dth3]


# Position of masses in the system
p_m1 = [r/2*sin(th1), r/2*cos(th1)]
p_Mh = [r*sin(th1), r*cos(th1)]
p_Mt = p_Mh + [L*sin(th3), L*cos(th3)]
p_m2 = p_Mh - [r/2*sin(th2), r/2*cos(th2)]

# Potential energy of masses in the system
PE_m1 = p_m1[2] * m * g
PE_Mh = p_Mh[2] * Mh * g
PE_Mt = p_Mt[2] * Mt * g
PE_m2 = p_m2[2] * m * g

# Total potential energy of the system
PE = PE_m1 + PE_Mh + PE_Mt + PE_m2
PE = Symbolics.simplify(PE)

# Velocities of masses in the system
v_m1 = Symbolics.jacobian(p_m1, q) * dq
v_Mh = Symbolics.jacobian(p_Mh, q) * dq
v_Mt = Symbolics.jacobian(p_Mt, q) * dq
v_m2 = Symbolics.jacobian(p_m2, q) * dq

# Kinetic Energy of each mass on the Robot
KE_m1 = 0.5 * m * sum(v_m1.*v_m1)
KE_Mh = 0.5 * Mh * sum(v_Mh.*v_Mh)
KE_Mt = 0.5 * Mt * sum(v_Mt.*v_Mt)
KE_m2 = 0.5 * m * sum(v_m2.*v_m2)

```

```
# Total kinetic energy of the system
KE = KE_m1 + KE_Mh + KE_Mt + KE_m2
KE = Symbolics.simplify(KE)

# Form the Lagrangian
L = KE - PE

# Compute terms in the Robot Equations
G = Symbolics.gradient(PE, q)
D = Symbolics.hessian(KE, dq) # Strange name explained below

# Compute C (advanced, uses Christoffel symbols)
n = length(q)
C = Array{Num}(undef, n, n) # Num is Symbolic Type
                             # Strange name explained below
for k in 1:n
    for j in 1:n
        C[k, j] = 0
        for i in 1:n
            C[k, j] += 0.5 * (Symbolics.gradient(D[k, j], q)[i] +
            Symbolics.gradient(D[k, i], q)[j] -
            Symbolics.gradient(D[i, j], q)[k]) * dq[i]
        end
    end
end

# Place an actuator at the hip on
# the relative angle between the torso and
# each leg
# Use D'Alembert's Principle
E = [th3-th1; th3-th2]
B = Symbolics.jacobian(E, q)
B = B'
JacG = Symbolics.jacobian(G, q) # Used when we discuss linearization of
    # NL models



# Only run simple after all Symbolic manipulation is done and you
# are ready to create a function
if true
    # Simplify as much as possible at the very end
    D = simple(D)
    C = simple(C)
end


# Create a function that computes all terms
# in the Robot Equations
fcn_name = "dyn_mod_3LinkWalker"
modelParamString = "g, r, L, m, Mh, Mt = modelParameters3LinkWalker() "
# Need to include line break command \n
variableNamesString = "th1, th2, th3 = q \ndth1, dth2, dth3 = dq"
writeEOM(fcn_name, D, C, G, B, JacG,modelParamString,variableNamesString)

# bring the model into the workspace for later use
include("dyn_mod_3LinkWalker.jl")
```

```
# If the line above fails, it is because Julia is trying to read the
# file before it has
# been made available to it by the file system. Simply copy the command
# into a new cell and
# run it. The time it takes you to do that will allow the file system to
# catch up! :-)
```

Output

```
[creating DYN_MOD_3LINKWALKER.jl]
File dyn_mod_3LinkWalker.jl created successfully!
dyn_mod_3LinkWalker (generic function with 1 method)
```

Here is the Lagrangian Model.

```
function modelParameters3LinkWalker()
#
     r=1;    # length of each leg
     m=5;     # mass of a leg
     Mh=15; # mass of hips
     Mt=10; # mass of torso
     L=0.5; # distance between hips and torso mass
     g=9.8; # acceleration due to gravity
#
     return  g, r, L, m, Mh, Mt
end


function dyn_mod_3LinkWalker(q, dq)
# DYN_MOD_3LINKWALKER
# 2024-02-21 17:48:08
#
# Author: Grizzle
#
# Model NOTATION: D(q)ddq + C(q,dq)*dq + G(q) = B*tau
# The Robot Equations: From Lagrange's Equations of Motion
#
g, r, L, m, Mh, Mt = modelParameters3LinkWalker()
#
# Variable names for the model
th1, th2, th3 = q
dth1, dth2, dth3 = dq
#
D = zeros(3, 3)
  D[1, 1] = r^2*(Mh + Mt + 5*m/4)
  D[1, 2] = -0.5*m*r^2*cos(th1 - th2)
  D[1, 3] = L*Mt*r*cos(th1 - th3)
  D[2, 1] = -0.5*m*r^2*cos(th1 - th2)
  D[2, 2] = 0.25*m*r^2
  D[2, 3] = 0
  D[3, 1] = L*Mt*r*cos(th1 - th3)
  D[3, 2] = 0
  D[3, 3] = L^2*Mt
```

```
    #
    C = zeros(3, 3)
      C[1, 1] = 0.0
      C[1, 2] = -0.5*dth2*m*r^2*sin(th1 - th2)
      C[1, 3] = L*Mt*dth3*r*sin(th1 - th3)
      C[2, 1] = 0.5*dth1*m*r^2*sin(th1 - th2)
      C[2, 2] = 0.0
      C[2, 3] = 0.0
      C[3, 1] = -L*Mt*dth1*r*sin(th1 - th3)
      C[3, 2] = 0.0
      C[3, 3] = 0.0
    #
    G = zeros(3)
      G[1] = -Mh*g*r*sin(th1) - Mt*g*r*sin(th1) - (3/2)*g*m*r*sin(th1)
      G[2] = (1/2)*g*m*r*sin(th2)
      G[3] = -L*Mt*g*sin(th3)
    #
    B = zeros(3, 2)
      B[1, 1] = -1
      B[2, 2] = -1
      B[3, 1] = 1
      B[3, 2] = 1
    #
    JacG = zeros(3, 3)
      JacG[1, 1] = -Mh*g*r*cos(th1) - Mt*g*r*cos(th1) - (3/2)*g*m*r*cos(th1)
      JacG[2, 2] = (1/2)*g*m*r*cos(th2)
      JacG[3, 3] = -L*Mt*g*cos(th3)
    #
      return (D=D, C=C, G=G, B=B, JacG=JacG)
    end
```

6.6 (Optional Read) Proof behind Lagrange Multipliers for Equality Constraints

Let $f : \mathbb{R}^n \to \mathbb{R}$ be a differentiable cost function, $g_i : \mathbb{R}^n \to \mathbb{R}$ be differentiable equality constraints for $1 \le i \le m$, and let $x_0 \in \mathbb{R}^n$ be an arbitrary point. We seek necessary conditions for x_0 to be a local extremum of the constrained optimization problem (6.40), which we repeat here for convenience:

$$\begin{aligned} &\text{Minimize} \quad f(x), \\ &\text{subject to} \quad g_i(x) = 0,\ 1 \le i \le m. \end{aligned} \tag{6.69}$$

We use the Jacobian to define $(n+1)$ constant row vectors

$$\begin{aligned} J_f(x_0) &:= \frac{\partial f(x_0)}{\partial x} = \left[\begin{array}{cccc} \frac{\partial f(x_0)}{\partial x_1} & \frac{\partial f(x_0)}{\partial x_2} & \cdots & \frac{\partial f(x_0)}{\partial x_n} \end{array}\right], \\ J_{g_i}(x_0) &:= \frac{\partial g_i(x_0)}{\partial x} = \left[\begin{array}{cccc} \frac{\partial g_i(x_0)}{\partial x_1} & \frac{\partial g_i(x_0)}{\partial x_2} & \cdots & \frac{\partial g_j(x_0)}{\partial x_n} \end{array}\right],\ 1 \le i \le m. \end{aligned} \tag{6.70}$$

By definition,[7] we note that $J_f(x_0)$ is a linear combination of the set of row vectors

$$\{J_{g_1}(x_0), J_{g_2}(x_0), \ldots, J_{g_m}(x_0)\}$$

if, and only if, there exist scalars $\lambda_1, \lambda_2, \ldots, \lambda_m \in \mathbb{R}$ such that

$$J_f(x_0) + \lambda_1 J_{g_1}(x_0) + \lambda_2 J_{g_2}(x_0) + \cdots + \lambda_m J_{g_m}(x_0) = 0. \tag{6.71}$$

[7]Write $J_f(x_0) = \alpha_1 J_{g_1}(x_0) + \alpha_2 J_{g_2}(x_0) + \cdots + \alpha_m J_{g_m}(x_0)$, and then define $\lambda_i := -\alpha_i$. Moving all the terms to the left side of the equation gives (6.71)

In other words,

- $J_f(x_0)$ being a linear combination of $\{J_{g_1}(x_0), J_{g_2}(x_0), \ldots, J_{g_m}(x_0)\}$ will be intimately related to the existence of Lagrange multipliers transforming the constrained optimization problem (6.69) into the unconstrained problem of finding stationary points for the Lagrangian

$$L(x,\lambda) := f(x) + \lambda_1 g_1(x) + \lambda_2 g_2(x) + \cdots + \lambda_m g_m(x). \tag{6.72}$$

- We also note that the set of row vectors $\{J_{g_1}(x_0), J_{g_2}(x_0), \ldots, J_{g_m}(x_0)\}$ is **linearly independent** if, and only if, the constraint qualification condition holds at x_0,

$$\text{rank}\left(\frac{\partial g(x_0)}{\partial x}\right) = \text{rank}\left(\begin{bmatrix} J_{g_1}(x_0) \\ J_{g_2}(x_0) \\ \vdots \\ J_{g_m}(x_0) \end{bmatrix}\right) = m. \tag{6.73}$$

- The coefficients in (6.71) are unique if, and only if, the set of row vectors

$$\{J_{g_1}(x_0), J_{g_2}(x_0), \ldots, J_{g_m}(x_0)\}$$

  is linearly independent.

We next state a result that makes it easy to relate conditions for x_0 being a local extremum of (6.69) to x_0 being a stationary point of the Lagrangian (6.72).

Proposition 6.42: When Can a Row Vector be Expressed as a Linear Combination of Other Row Vectors?

The row vector $J_f(x_0)$ is a linear combination of the set of row vectors

$$\{J_{g_1}(x_0), J_{g_2}(x_0), \ldots, J_{g_m}(x_0)\}$$

if, and only if,

$$\bigcap_{i=1}^{m} \text{null}\left(J_{g_i}(x_0)\right) \subset \text{null}\left(J_f(x_0)\right). \tag{6.74}$$

We delay the proof of Prop. 6.42 until later.

Assumptions:

- x_0 is feasible; that is $g_i(x_0) = 0$, $1 \le i \le m$.
- x_0 satisfies the constraint qualification condition, that is, (6.73) holds.

Given x_0 as above, we call $v \in \mathbb{R}^n$ a **feasible direction** if $v \in \bigcap_{i=1}^{m} \text{null}\left(J_{g_i}(x_0)\right)$. Why this name? One way to understand it is that by the linear approximation of $g_i : \mathbb{R}^n \to \mathbb{R}$ at a feasible point x_0,

$$\begin{aligned} g_i(x_0 + v) &\approx g_i(x_0) + J_{g_i}(x_0) \cdot v \\ &\approx J_{g_i}(x_0) \cdot v \\ &\approx 0. \end{aligned}$$

Hence, if x_0 is feasible, then so is $x_0 + v$ to first order.[8]

Let $v \in \bigcap_{i=1}^{m} \text{null}\left(J_{g_i}(x_0)\right)$ be arbitrary, and consider $f(x_0 + v)$ through the lens of its linear approximation,

$$f(x_0 + v) \approx f(x_0) + J_f(x_0) \cdot v.$$

From the linearization of f at x_0, we make the following conclusions:

[8] A more formal formulation uses the concept of a **directional derivative**, defined as, for v a unit vector,

$$\frac{d}{dt} g_i(x_0 + t \cdot v)\bigg|_{t=0} = \frac{\partial g_i(x_0)}{\partial x} \cdot v = J_{g_i}(x_0) \cdot v.$$

Hence, the directional derivative vanishes for $v \in \text{null}\left(J_{g_i}(x_0)\right)$.

- If $v \notin \operatorname{null}(J_f(x_0))$, meaning $J_f(x_0) \cdot v \neq 0$, then we can move along v to make $f(x_0 + v)$ smaller or larger than $f(x_0)$. In this case, x_0 is not an extreme point of (6.69).
- Therefore, the only way for x_0 to be an extreme point is for $J_f(x_0) \cdot v = 0$, that is $v \in \operatorname{null}(J_f(x_0))$.

Because $v \in \bigcap_{i=1}^{m} \operatorname{null}(J_{g_i}(x_0))$ was arbitrary, we conclude that the only way for x_0 to be an extreme point of (6.69) is for (6.74) to hold, and thus $J_f(x_0)$ can be expressed as the linear combination

$$J_f(x_0) + \lambda_1 J_{g_1}(x_0) + \lambda_2 J_{g_2}(x_0) + \cdots + \lambda_m J_{g_m}(x_0) = 0. \tag{6.75}$$

We are now essentially done, with only some bookkeeping to complete the proof. Using the coefficients in (6.75), we define the Lagrangian as in (6.72). Then, by (6.75), x_0 is a stationary point of L. ■

Proof of Prop. 6.42 Without loss of generality, we can assume that the set of row vectors

$$\{J_{g_1}(x_0), J_{g_2}(x_0), \ldots, J_{g_m}(x_0)\}$$

is linearly independent because any dependent vectors do not change the intersection of their null spaces. If there exist $\lambda_1, \ldots, \lambda_m$ such that (6.71) holds, then v satisfying $J_{g_i}(x_0) \cdot v = 0$, $1 \leq i \leq m$, implies $J_f(x_0) \cdot v = 0$. For the other direction, we assume that $\{J_{g_1}(x_0), J_{g_2}(x_0), \ldots, J_{g_m}(x_0)\}$ is linearly independent and that $J_f(x_0)$ is not a linear combination of these vectors. Hence, the $(m+1) \times n$ matrix

$$A := \begin{bmatrix} J_f(x_0) \\ J_{g_1}(x_0) \\ J_{g_2}(x_0) \\ \vdots \\ J_{g_m}(x_0) \end{bmatrix}$$

has full row rank. Therefore, there exist $v \in \mathbb{R}^n$ such that:

$$A \cdot v = \begin{bmatrix} 1 \\ 0 \\ 0 \\ \vdots \\ 0 \end{bmatrix}.$$

Hence, there exists $v \in \bigcap_{i=1}^{m} \operatorname{null}(J_{g_i}(x_0))$ such that $v \notin \operatorname{null}(J_f(x_0))$. This completes the proof. ■

6.7 (Optional Read) Proof Underlying Gradient Descent with Equality Constraints

It is a bit ironic, but the above proof of the Method of Lagrange Multipliers shows how to accomplish constrained optimization without explicitly computing Lagrange multipliers! In fact, it shows how to update gradient descent to work in the presence of equality constraints. We retain all of the assumptions made for the proof of the Method of Lagrange Multipliers with Equality Constraints.

Method 6.43: Gradient Descent with Equality Constraints

Consider the constrained optimization problem (6.40) for functions $f : \mathbb{R}^n \to \mathbb{R}$ and $g : \mathbb{R}^n \to \mathbb{R}^m$ with optimal value x^*. If the columns vectors from the gradients $\{\nabla g_1(x^*), \nabla g_2(x^*), \ldots, \nabla g_m(x^*)\}$ are linearly independent (equivalently, the rows of the Jacobian $\frac{\partial g(x^*)}{\partial x}$ are linearly independent), then a locally optimal x^* can be found as follows:

Initialization: Find a feasible point $x_0 \in \mathbb{R}^n$, that is, a point such that $g_i(x_0) = 0$ for $1 \leq i \leq m$. This can be accomplished with `NLsolve`.

Loop: Given a point $x_k \in \mathbb{R}^n$, repeat the following steps to either find x_{k+1} or terminate.

(a) **Compute:** $\nabla f(x_k)$, the gradient of f at x_k and $\frac{\partial g(x_k)}{\partial x}$, the Jacobian of the constraint vector.

From the proof of the Method of Lagrange Multipliers, the set of feasible directions is

$$V := \bigcap_{i=1}^{m} \text{null}\left(J_{g_i}(x_k)\right) = \text{null}\left(\frac{\partial g(x_k)}{\partial x}\right),$$

where $g : \mathbb{R}^n \to \mathbb{R}^m$ is the vector of equality constraints.

(b) **Compute the orthognal projection of $\nabla f(x_k)$ onto V, the set of feasible directions:** This can be accomplished by applying Gram-Schmidt. Indeed, let $\{b_1, b_2, \ldots, b_m\}$ be an orthonormal basis for the columns of $\left(\frac{\partial g(x_k)}{\partial x}\right)^\top$. Then

$$P^V_{\nabla f(x_k)} := \nabla f(x_k) - \sum_{i=1}^{m} (\nabla f(x_k) \bullet b_i) \cdot b_i$$

is the desired projection.

Why: Each term $(\nabla f(x_k) \bullet b_i) \cdot b_i$ is a component of $\nabla f(x_k)$ that is orthogonal to V, the feasible directions. From linear algebra, once these components are subtracted off, only the part of $\nabla f(x_k)$ lying in V is left.

(c) **Update or Terminate, That Is the Question:** If the termination condition is not met, define

$$x_{k+1} = x_k - s \cdot P^V_{\nabla f(x_k)},$$

where $s > 0$ is the step size. Loop back to (a).

(d) **Termination Condition:** $\|P^V_{\nabla f(x_k)}\| < a_{\text{tol}}$; in other words, $J_f(x_k)$ is to numerical precision a linear combination of $\{J_{g_1}(x_k), J_{g_2}(x_k), \ldots, J_{g_m}(x_k)\}$, and thus x_k is (to numerical precision) a stationary point of the Lagrangian.

Notes:

- In practice, the updates x_k will "slowly" exit the feasible set. Hence, when $\|g(x_k)\|$ exceeds a user-defined tolerance, a new call to `NLsolve` should be made.
- Lagrange likely understood that an algorithm like this was possible, but given that his "computer" consisted of pen and paper, it was not feasible to pursue this version of the method.
- If there are no constraints, then the null space of $\left(\frac{\partial g(x_k)}{\partial x}\right)$ is all of $\mathbb{R}^n$. The projected gradient is then the regular gradient.

- An algorithm like this one cannot be presented in most Calc III courses **because Linear Algebra traditionally comes after Calc I through Calc IV**. The University of Michigan's Robotics Department is advocating reversing this sequence and placing linear algebra before calculus.

- **`JuMP` does all of this and a whole lot more**. It is your best option for use in professional projects. For learning about gradient descent, the workhorse of machine learning (ML), algorithms like the one presented here are helpful.

Here is an instantiation of the above version of Lagrange's Method for Optimization with Equality Constraints.

```julia
using ForwardDiff, LinearAlgebra, NLsolve

# Define the cost function f(x) and the equality constraints G(x)
function f(x)
    return x[1]^2 + x[2]^2 + x[3]^2 + x[4]^2
end

function G(x)
    return [x[1] + x[2] - 3, sin(pi*x[1]) - x[3]^2 - 1 + x[4]^3]
end

# Compute the gradient of f and the Jacobian of G using ForwardDiff
grad_f(x) = ForwardDiff.gradient(f, x)
JacG(x) = ForwardDiff.jacobian(G, x)

# Define functions used for computing the projection of grad_f onto
# the nullspace of JacG, that is, onto the feasible directions

function gram_schmidt(jacobian_G)
    n, m = size(jacobian_G)
    B = zeros(n, m)  # Basis vectors

    for j = 1:m
        b = jacobian_G[:, j]
        for i = 1:j-1
            b -= dot(B[:, i], jacobian_G[:, j]) * B[:, i]
        end
        B[:, j] = b / norm(b)
    end

    return B
end

# Project the gradient onto the feasible directions,
# that is, nullspace of JacG
function project_grad_f(grad_f, B)
    # Iterate over columns of B, subtracting their projection from
    # grad_f
    projected_grad_f = grad_f
    for j = 1:size(B, 2)
        projected_grad_f = projected_grad_f - dot(grad_f, B[:, j]) * B[:, j]
    end
    return projected_grad_f
end

```

```
# Function to find a feasible starting point using NLsolve
# or to push xk back onto the feasible set when it
# drifts off
function find_feasible_start (G, x_guess)
    # build a function for NLSolve
    function constraint! (F, x)
        F[1:2] = G(x)
    end
    xFeasible = nlsolve(constraint!, x_guess)
    return xFeasible.zero
end
#
## End define functions for projected gradient descent

## Actual Gradient Descent Algorithm Starts Here

x_guess = [0.0, 0.0, 0.0, 0.0] # Not feasible
x0 = find_feasible_start(G, x_guess) # Make feasible


# set step size, maxIter, tolerances
s=0.1
maxIter=10000
gradTol = 1e-6
constraintTol = 1e-4

xk = x0
for k=1:maxIter
    jacobian_G_transposed = transpose(JacG(xk))
    B = gram_schmidt(jacobian_G_transposed)
    projected_grad_f = project_grad_f(grad_f(xk), B)
    if norm(projected_grad_f) < gradTol
        println("---- Iteration $k ----")
        println("Met tolerance.")
        println("Jacf applied to null(JacG) (should be near 0): ",
            round.((grad_f(xk)')*nullspace(JacG(xk)), digits=5))
            println("------------------ ")
        break
    else
        # Update
        xk = xk - s*projected_grad_f
    end
    if norm(G(xk)) > constraintTol
        xk = find_feasible_start(G, xk) # Again Feasible
    end
end
xStar = xk
fStar = f(xStar)

println("\nFinal Results:")
println("------------------")
println("Optimal x: ", round.(xStar, digits=5))
#
println("Optimal Cost f(xStar): ", round(fStar, digits=5))
```

Output

```
---- Iteration 184 ----
Met tolerances.
Jacf applied to null(JacG) (should be near 0): [0.0 -0.0]
-----------------

Final Results:
-----------------
Optimal x: [0.95451, 2.04549, 0.0, 0.95008]
Optimal Cost f(xStar): 5.99778
```

6.8 Words of Wisdom from One of the Planet's Greatest Living Mathematicians

You may enjoy the video, **Math isn't actually Sorcery**, **by Prof. Terence (Terry) Tao**, an Australian-American mathematician who is widely regarded as one of the greatest living mathematicians (if you hear an accent in the video, it's Aussie). He discusses the perception that Mathematics is a set of arcane symbols and magic spells, and that if you don't follow the incantations faithfully, it will all collapse. He argues, instead, that Mathematics is a way of thinking, with powerful means for breaking hard problems down into simpler ones.

Relevance to Calculus for the Modern Engineer: At each step of our adventure, we've broken a new topic into smaller building blocks from which we could assemble an approach to solving a problem. Often, the building blocks, put end-to-end, formed an algorithm. Other times, they provided an aha moment of insight into how a complicated-looking formula was a direct consequence of understanding rectangles (e.g., theory of integration and its applications) or linear approximations of functions (e.g., derivation of differentiation rules and the applications of derivatives to gradient descent and Lagrange's equations of motion). Limits of one kind or another have been the cement holding the blocks together.

Summarized from Wikipedia: Born in 1975 to Chinese immigrant parents in Adelaide, Tao was a child prodigy who exhibited extraordinary mathematical abilities from a very young age. **He skipped six grades and attended university-level mathematics courses at age 9.** At just eight years old, he scored 760 on the SAT math section, a feat achieved by only three children in the history of the Johns Hopkins Study of Exceptional Talent program.

Tao's prodigious talent was further demonstrated through his participation in the International Mathematical Olympiad. He became the youngest participant in the competition's history, winning bronze, silver, and gold medals at ages 10, 11, and 13, respectively. He remains the youngest winner of each medal type in the Olympiad's history.

Currently, Tao is a professor of mathematics at the University of California, Los Angeles (UCLA), where he holds the James and Carol Collins Chair in the College of Letters and Sciences. His research spans various areas of mathematics, including harmonic analysis, partial differential equations, combinatorics, probability theory, and analytic number theory.

Tao's contributions to mathematics have earned him numerous prestigious awards, including the Fields Medal in 2006, the Royal Medal and Breakthrough Prize in Mathematics in 2014, and a MacArthur Fellowship in 2006. He has authored or co-authored over three hundred research papers, solidifying his reputation as one of the most influential mathematicians of his generation.

Chapter 7

Antiderivatives and the Fundamental Theorems of Calculus

Learning Objectives

By the end of this chapter, students should be able to:

- Understand that antidifferentiation is essentially the inverse operation of differentiation.
- Comprehend the two Fundamental Theorems of Calculus and explain the precise sense in which differentiation and integration are inverse operations on functions.
- Understand that while antidifferentiation and definite integration are technically distinct operations, they are often perceived as identical in the context of Calculus education. This perception is partially justified by their closely related concepts. However, it is crucial to acknowledge the potential drawbacks of this viewpoint, particularly the confusion and discouragement that can arise from the complex rules associated with manually computing antiderivatives.
- Appreciate the historical and practical context of computing antiderivatives by hand, and recognize the modern approaches that render hand computations almost obsolete for all but the simplest of cases.

Outcomes

Upon successful completion of this chapter, students will:

- Be introduced to the concept of antiderivatives and familiarize themselves with elementary techniques for finding them.
- Gain a thorough understanding of the Fundamental Theorems of Calculus, including their implications in both geometric and analytic contexts.
- Develop an insight into the interplay between differentiation, antiderivatives, and definite integration, enhancing their comprehension of calculus as a whole.
- Cultivate an appreciation for the traditional methods of finding antiderivatives while also understanding the value and efficiency of using computational tools for real-world applications.

Antiderivative $\int f(x)\,dx$	Function $f(x)$	Derivative $f'(x)$
C	0	0
$kx + C$	k	0
$\frac{1}{3}x^3 + C$	x^2	$2x$
$e^x + C$	e^x	e^x
$-\cos(x) + C$	$\sin(x)$	$\cos(x)$
$x\ln(x) - x + C$	$\ln(x)$	$\frac{1}{x}$
$\ln\lvert x\rvert + C$	$\frac{1}{x}$	$-\frac{1}{x^2}$
$-\ln\lvert\cos(x)\rvert + C$	$\tan(x)$	$1 + \tan^2(x)$

Table 7.1: C is an arbitrary constant, later called a **constant of integration**. $F(x)$ is an antiderivative of $f(x)$ if $F'(x) = f(x)$. This table illustrates the relationships between an antiderivative, a function, and a derivative. The functions in the middle are the derivatives of the functions in the left column, and the functions in the right column are the derivatives of the functions in the middle column. Going the other direction, the functions in the middle column are antiderivatives of the functions in the right column. As labeled, the functions in the left column are antiderivatives of the functions in the middle column. Bottom line: $F(x) + C$, for C a constant, is always an antiderivative for $f(x) := \frac{dF(x)}{dx}$.

7.1 Introduction

This chapter marks a major waypoint in our journey to discover and understand the essence of Calculus. The **concept of an antiderivative** is a bridge connecting the micro (differential calculus, focused on the infinitesimal, df, dx, and their ratio) to the macro (integral calculus, focused on adding up an infinite number of vanishingly small areas, $dA = f(x)\,dx$). As illustrated in Table 7.1, an antiderivative of a function $f(x)$ is another function $F(x)$ such that $F'(x) = f(x)$. The Fundamental Theorems of Calculus elegantly tie together differentiation and integration, showing that these two operations are, in fact, inverse processes. As we unveil the theorems, we will discover the profound connection between the definite integral and antiderivatives, which will further lead us to study indefinite integration and, eventually, improper integrals and differential equations in subsequent chapters.

Because of the Fundamental Theorems of Calculus, the operation of determining antiderivatives shares the same notation as integration, namely

$$\int f(x)\,dx = F(x) + C,$$

where C is a **constant of integration** and $F'(x) = f(x)$. **Later, we may regard this elegant formula as the proverbial "wolf in sheep's clothing." Intrigued?**

Antiderivative of a Function

Definition 7.1. *Let $f : I \to \mathbb{R}$ be a function defined on an interval I. An* ***antiderivative*** *of f on the interval I is a function $F : I \to \mathbb{R}$ such that*

$$F'(x) = f(x)$$

for all $x \in I$.

Note: Because $\frac{dC}{dx} = 0$ for all constants C, if F is an antiderivative of f, then so is $F(x) + C$. In other words, if a function has one antiderivative, it has an infinite number of antiderivatives, and moreover, they all differ by a constant.

Example 7.2. *In the list of functions (a) - (e) below, identify those functions that are antiderivatives of other functions in the same list.*

(a) $f_a(x) = x \cdot \ln(x) - x$

(b) $f_b(x) = \ln(x)$

(c) $f_c(x) = \ln(x^2)$

(d) $f_d(x) = \cos(x^2)$

(e) $f_e(x) = -2x \cdot \sin(x^2)$

Solution: This problem is acquainting us with the process of checking whether a given function $F(x)$ is an antiderivative for another function $f(x)$ or not. We apply the definition of what it means to be an antiderivative and check whether $F'(x) = f(x)$ or $F'(x) \neq f(x)$. It is that simple.

(a)

$$\begin{aligned}\frac{df_a(x)}{dx} &= \frac{dx}{dx}\ln(x) + x\,\frac{d}{dx}\,(\ln(x)) - \frac{dx}{dx}\\ &= \ln(x) + x \cdot \frac{1}{x} - 1\\ &= \ln(x) = f_b(x),\end{aligned}$$

and hence $f_a(x)$ is an antiderivative of $f_b(x)$.

(b) $\frac{df_b(x)}{dx} = \frac{d}{dx}\,(\ln(x)) = \frac{1}{x}$, and hence $f_b(x)$ is NOT an antiderivative of any function in the list.

(c) $\frac{df_c(x)}{dx} = \frac{2}{x}$, and hence $f_c(x)$ is NOT an antiderivative of any function in the list.

(d) $\frac{df_d(x)}{dx} = -2x\sin\left(x^2\right) = f_e(x)$, and hence $f_d(x)$ is an antiderivative of $f_e(x)$.

(e) $\frac{df_e(x)}{dx} = -2\sin\left(x^2\right) - 4x^2\cos\left(x^2\right)$, and hence $f_e(x)$ is NOT an antiderivative of any function in the list.

```
using Symbolics

# Create a custom function to make differentiation convenient
function deriv(f, x)
    return expand_derivatives.(Differential(x)(f))
end

# Sample Usage
@variables x # Makes x a symbolic variable
f = x^3*cos(x) + x^2*sin(x)^2 + log(3x)
dfdx = deriv(f, x)

# Note we do not say f(x) =, just f =

fa = x * log(x) - x
fb = log(x)
fc = log(x^2)
fd = cos(x^2)
fe = -2x * sin(x^2);
```

Output

```
(nothing)
```

```
using Symbolics

@show dfadx = deriv(fa, x)

@show dfbdx = deriv(fb, x)

@show dfcdx = deriv(fc, x)

@show dfddx = deriv(fd, x)

@show dfedx = deriv(fe, x);
```

Output

```
dfadx = deriv(fa, x) = log(x)
dfbdx = deriv(fb, x) = 1 / x
dfcdx = deriv(fc, x) = 2 / x
dfddx = deriv(fd, x) = -2x*sin(x^2)
dfedx = deriv(fe, x) = -2sin(x^2) - 4(x^2)*cos(x^2)
```

■

> **How to verify that $F(x)$ is an antiderivative of $f(x)$?** Simply apply the definition: differentiate $F(x)$ and compare the result to $f(x)$. Hence, given a candidate antiderivative for $f(x)$, verifying or refuting the candidate antiderivative is straightforward. We will see that finding candidate antiderivatives is, for the most part, anything but straightforward.

Example 7.3. *ChatGPT4 (on its own, without an important plugin) has proposed the following antiderivatives for the functions in Example 7.2. Determine if they are correct or not.*

(a) *For* $f_a(x) = x\log(x) - x$*: the antiderivative is* $\int f_a(x)\,dx = \frac{1}{2}x^2\log(x) - \frac{1}{4}x^2 + C$.

(b) *For* $f_b(x) = \log(x)$*: the antiderivative is* $\int f_b(x)\,dx = x\log(x) - x + C$.

(c) *For* $f_c(x) = \log(x^2)$*: the antiderivative is* $\int f_c(x)\,dx = x\log(x^2) - x + C$.

(d) *For* $f_d(x) = \cos(x^2)$*: the antiderivative is* $\int f_d(x)\,dx = \frac{1}{2}\sqrt{\pi}\,\mathit{FresnelS}\left(\sqrt{\frac{2}{\pi}}x\right) + C$.

(e) *For* $f_e(x) = -2x\sin(x^2)$*: the antiderivative is* $\int f_e(x)\,dx = \cos(x^2) + C$.

Solution: The antiderivative for $f_d(x)$, as discussed in Wikipedia, is a special function that we will talk about later.

```
using Symbolics

# Define the symbolic variable
@variables x

# Define the functions as per their original definitions
fa(x) = x * log(x) - x
```

```
fb(x) = log(x)
fc(x) = log(x^2)  # Which is equivalent to 2*log(x)
fd(x) = cos(x^2)
fe(x) = -2x * sin(x^2)


# Antiderivatives proposed by ChatGPT on 30 March 2025.
# The constants of integration have been removed.

AntiDeriv_fa = (x^2 * log(x))/2 - (3x^2)/4  # Updated to match fa(x)
AntiDeriv_fb = x * log(x) - x
AntiDeriv_fc = 2x * log(x) - 2x  # Correct antiderivative for fc(x)
AntiDeriv_fe = cos(x^2)

# Perform differentiation and compare with the original functions
result_fa = simplify(deriv(AntiDeriv_fa, x))
result_fb = simplify(deriv(AntiDeriv_fb, x))
result_fc = simplify(deriv(AntiDeriv_fc, x))
result_fe = simplify(deriv(AntiDeriv_fe, x))

println("Derivative of AntiDeriv_fa compared to fa(x): ", result_fa, " == ", fa(x))
println("Derivative of AntiDeriv_fb compared to fb(x): ", result_fb, " == ", fb(x))
println("Derivative of AntiDeriv_fc compared to fc(x): ", result_fc, " == ", fc(x))
println("Derivative of AntiDeriv_fe compared to fe(x): ", result_fe, " == ", fe(x))
```

Output

```
Derivative of AntiDeriv_fa compared to fa(x):
  x*log(x) - x == x*log(x) - x
Derivative of AntiDeriv_fb compared to fb(x): log(x) == log(x)
Derivative of AntiDeriv_fc compared to fc(x): 2log(x) == log(x^2)
Derivative of AntiDeriv_fe compared to fe(x):
  -2x*sin(x^2) == -2x*sin(x^2)
```

■

Using logarithm rules, $\ln(x^2) = 2\ln(x)$, and hence the third one is correct.

Secrets of the Arcane 7.4: Pro Tip for Determining Antiderivatives

The `Symbolics.jl` package can compute simple antiderivatives. The competing package, `SymPy.jl`, is somewhat better. However, the hands-down winners are MapleSoft and the tools from Wolfram, namely Wolfram Alpha Pro and Mathematica. Because of its ease of use, ChatGPT can be used as a first recourse, but we should always check the result!.

Example 7.5. *Compute the antiderivative for*

$$f(x) := \frac{2x^3 \cos(x) \sin(x) e^{2+\sin^2(x)}}{1 + \left(e^{2+\sin^2(x)}\right)^2} + 3x^2 \arctan\left(e^{2+\sin^2(x)}\right).$$

Solution: ChatGPT4 failed. Both Wolfram Alpha Pro and ChatGPT5 produced the following result:

$$x^3 \arctan\left(e^{\frac{1}{2}(5-\cos(2x))}\right) + \text{C}.$$

The answer is correct because the problem was created by differentiating $F(x) := x^3 \cdot \text{atan}(\exp(\sin(x)^2 + 2))$ with respect to x, which leads to the expression for $f(x)$. The trig identity $\sin^2(x) = \frac{1-\cos(2x)}{2}$ allows one to show that $\sin(x)^2 + 2 = \frac{(5-\cos(2x))}{2}$. Few humans would want to take on this "integration problem" using the tools taught in standard calculus courses. **What ChatGPT accomplished is impressive.**

■

Example 7.6. *Compute the antiderivative for:*

$$f(x) = 3x^2\sqrt{2 + (\log(1+x^2))^2}\arcsin\left(e^{2+\sin^2(x)}\right) + \frac{2x^4\log(1+x^2)\arcsin\left(e^{2+\sin^2(x)}\right)}{\sqrt{2+(\log(1+x^2))^2}\,(1+x^2)} + \frac{2x^3\cos(x)\sin(x)\,e^{2+\sin^2(x)}\sqrt{2+(\log(1+x^2))^2}}{\sqrt{1-\left(e^{2+\sin^2(x)}\right)^2}}. \tag{7.1}$$

Solution: **ChatGPT4 failed. ChatGPT5 computed** the result:

$$f(x) = \frac{d}{dx}\underbrace{\left(x^3\cdot\arcsin\left(e^{2.0+\sin^2(x)}\right)\cdot\sqrt{2+(\log(1+x^2))^2}\right)}_{F(x)}, \tag{7.2}$$

and hence, a "relatively simple" antiderivative $F(x)$ exists. ■

7.2 Uniting Integration and Differentiation through the Fundamental Theorems of Calculus

The **First Fundamental Theorem of Calculus** states that every continuous function has an antiderivative. The **Second Fundamental Theorem of Calculus** states that all definite integrals can be evaluated through antiderivatives. Together, these two results seamlessly tie together integration and differentiation as "inverse operations".

7.2.1 Fundamental Theorems

Proposition 7.7: Fundamental Theorems of Calculus

First Fundamental Theorem of Calculus: Let $f : [a,b] \to \mathbb{R}$ be a continuous real-valued function defined on a closed interval $[a,b]$. Then, the function $F : [a,b] \to \mathbb{R}$ defined by

$$F(x) = \int_a^x f(t)\,dt \tag{7.3}$$

for all $x \in [a,b]$, is continuous on $[a,b]$, differentiable on the open interval (a,b), and $F'(x) = f(x)$ for all $x \in (a,b)$. It is also written as:

$$\frac{d}{dx}F(x) = \frac{d}{dx}\left[\int_a^x f(t)\,dt\right] = f(x), \tag{7.4}$$

to emphasize that differentiation undoes, or inverses, integration.

Second Fundamental Theorem of Calculus: Let $f : [a,b] \to \mathbb{R}$ be a real-valued function defined on a closed interval $[a,b]$ that admits an antiderivative F on $[a,b]$, that is, $F : [a,b] \to \mathbb{R}$ is differentiable on (a,b) and for all $x \in (a,b)$, $F'(x) = f(x)$. Then, f is integrable on $[a,b]$ and

$$\int_a^b f(x)\,dx = F(x)\Big|_a^b := F(b) - F(a), \tag{7.5}$$

which also is written as:

$$\int_a^b \frac{dF(x)}{dx}\, dx = F(x)\Big|_a^b := F(b) - F(a), \tag{7.6}$$

to underline that integration undoes, or inverses, differentiation.

Notes: For a nuanced presentation of the First Fundamental Theorem of Calculus, we recommend 3 Levels of Proving the Fundamental Theorem of Calculus by Trivial. The first part of the video The Essence of (Multivariable) Calculus by Foolish Chemist provides an intuitive treatment of the Second Fundamental Theorem of Calculus. If you exceed the 6:20 mark, you will be into multivariable versions of the Fundamental Theorems, which is beyond what is needed for this chapter.

7.2.2 Using the Second Fundamental Theorem of Calculus for Definite Integration

Let us roll up our sleeves and get to work using the Second Fundamental Theorem of Calculus to evaluate some definite integrals. It looks super easy, right? In Table 5.1 we had a list of functions and their derivatives. Based on the First Fundamental Theorem of Calculus, we can repurpose our table to give us a list of antiderivatives, for example:

Function	**Derivative**
C	0
$\frac{x^{k+1}}{k+1} + C$	x^k
$e^x + C$	e^x
$\ln\lvert x\rvert + C$	$\frac{1}{x}$
$\sin(x) + C$	$\cos(x)$
$\cos(x) + C$	$-\sin(x)$
$\tan(x) + C$	$1 + \tan^2(x)$
$\text{atan}(x) + C$	$\frac{1}{1+x^2}$
Antiderivative	**Function**

Table 7.2: Common Elementary Functions and Their Antiderivatives (worth committing to memory to showcase your expertise). When doing antiderivatives, it is more convenient to have the monomials presented as given here than how we did the monomial entry in Table 5.1. In addition, we have extended the natural logarithm to negative numbers via an absolute value. Otherwise, the two tables are the same.

Why is there an absolute value on the natural logarithm? Note that $\frac{1}{x}$ is defined for all $x \neq 0$. Hence, its domain can be $(0, \infty)$, the positive reals, or $(-\infty, 0)$, the negative reals. Logarithms, on the other hand, are only defined for the positive reals. When $x > 0$, $|x| = x$, and hence $\frac{d}{dx}\ln|x| = \frac{d}{dx}\ln(x) = \frac{1}{x}$. When $x < 0$, $|x| = -x$, and then, by the Chain Rule,

$$\frac{d}{dx}\ln|x| = \frac{d}{dx}\ln(-x) = \frac{1}{-x}\cdot(-1) = \frac{1}{x},$$

where the (-1) comes from differentiating $-x$ with respect to x. While it is never wrong to include the absolute value sign in the antiderivative of $\frac{1}{x}$, when the context makes it clear that $x > 0$, it is standard practice to drop it because, then, $|x| = x$.

We can also think about the derivative of the natural logarithm as follows. From the Chain Rule,

$$\frac{d}{dx}\ln(g(x)) = \frac{1}{g(x)} \cdot \frac{dg(x)}{dx},$$

for $g(x) > 0$. Substituting in $g(x) = |x|$ and $x \neq 0$ yields $\frac{d}{dx}\ln(|x|) = \frac{1}{|x|} \cdot \frac{d}{dx}|x|$. When $x > 0$, $\frac{d}{dx}|x| = 1$ and when $x < 0$, $\frac{d}{dx}|x| = -1$. Hence,

$$\frac{1}{|x|} \cdot \frac{d}{dx}|x| = \begin{cases} \frac{1}{x} \cdot 1 & x > 0 \\ \frac{1}{-x} \cdot (-1) & x < 0 \end{cases} = \frac{1}{x} \text{ for } x \neq 0.$$

Example 7.8. *On the basis of Table 7.2 and the Fundamental Theorems of Calculus, compute the following definite integrals.*

(a) $\int_1^5 \left(2x^3 + e^x\right) dx$

(b) $\int_1^e \left(\frac{1}{x} + \frac{1}{1+x^2}\right) dx$

(c) $\int_0^\pi (\sin(x) + 3\cos(x))\, dx$

(d) $\int_0^1 \left(\frac{x}{1+x^2}\right) dx$

Solution:

(a) The antiderivative of x^3 is $\frac{x^4}{4}$ and the antiderivative of e^x is itself, e^x. Hence,

$$\int_1^5 \left(2x^3 + e^x\right) dx = \underbrace{\left(2 \cdot \frac{x^4}{4} + e^x\right)}_{\text{antideriv. } F(x)} \underbrace{\Bigg|_{a=1}^{b=5}}_{\text{limits}} = \underbrace{\left(\frac{5^4}{2} + e^5\right)}_{F(b)} - \underbrace{\left(\frac{1^4}{2} + e^1\right)}_{F(a)} = 312 + e^5 - e \approx 457.69.$$

(b) The antiderivative of $\frac{1}{x}$ is $\ln(x)$ and the antiderivative of $\frac{1}{1+x^2}$ is $\text{atan}(x)$. Hence,

$$\int_1^e \left(\frac{1}{x} + \frac{1}{1+x^2}\right) dx = \underbrace{(\ln(x) + \text{atan}(x))}_{\text{antideriv. } F(x)} \underbrace{\Bigg|_{a=1}^{b=e}}_{\text{limits}}$$

$$= \underbrace{(\ln(e) + \text{atan}(e))}_{F(b)} - \underbrace{(\ln(1) + \text{atan}(1))}_{F(a)} = 1 + \text{atan}(e) - \text{atan}(1) \approx 1.43.$$

(c) The antiderivative of $\sin(x)$ is $-\cos(x)$ and the antiderivative of $\cos(x)$ is $\sin(x)$. Hence,

$$\int_0^\pi (\sin(x) + 3\cos(x))\, dx = \underbrace{(-\cos(x) + 3\sin(x))}_{\text{antideriv. } F(x)} \underbrace{\Bigg|_{a=0}^{b=\pi}}_{\text{limits}}$$

$$= \underbrace{(-\cos(\pi) + 3\sin(\pi))}_{F(b)} - \underbrace{(-\cos(0) + 3\sin(0))}_{F(a)} = 2.$$

(d) The antiderivative of $\frac{x}{1+x^2}$ is NOT in Table 7.2. Hence, we either have to compute it via ChatGPT or Wolfram or learn a more powerful means for generating antiderivatives. From ChatGPT, we have the antiderivative of $\frac{x}{1+x^2}$ is $0.5\ln(x^2+1)$. Hence,

$$\int_0^1 \left(\frac{x}{1+x^2}\right) dx = 0.5\ln(x^2+1)\Big|_0^1 = 0.5\ln(2) - 0.5\ln(1) = 0.5\ln(2) \approx 0.34657.$$

```
# Our prior means of computing definite integrals as a double check
using QuadGK
ansA, = quadgk(x ->2x^3 + exp(x), 1, 5)
ansB, = quadgk(x ->(1/x) + (1/(1+x^2)), 1, exp(1))
ansC, = quadgk(x ->sin(x) + 3cos(x), 0, pi)
ansD, = quadgk(x ->x/(1+x^2), 0, 1);
[ansA, ansB, ansC, ansD]
```

Output

```
4-element Vector{Float64}:
 457.69487727411763
   1.4328847416198296
   2.000000000000001
   0.3465735902799727
```

■

Taking Stock: For simple "everyday" functions, it is relatively easy to use the Second Fundamental Theorem of Calculus for evaluating definite integrals. It can be a bit tedious because we have to find the antiderivative for each term in the integral, and then we have to evaluate the antiderivatives at the limits of integration, but conceptually, it is straightforward. Obtaining the antiderivatives is clearly the most difficult part of this process. For some super-common functions, we have the antiderivatives available in a standard table, or even memorized. For less common functions, we have to determine the antiderivatives ourselves, one way or another. In Julia, the process of determining the antiderivatives and plugging in values is certainly harder than just running `quadgk` on the original functions. **So, when should we compute antiderivatives to evaluate integrals?**

- When we seek to establish a general result about an integral, having the closed-form solution provided by an antiderivative is highly useful. What is the volume of a sphere of radius $r > 0$? QuadGK can only compute the result for a specific value of r, and not for $r > 0$ a variable.
- In real-time coding for robotics, our available computing resources may be too slow to run `quadgk` over and over. Being able to compute the answer to an integral in closed form may provide for ultra-fast function evaluation. In this case, knowledge (i.e., Fundamental Theorems of Calculus) is power!

7.2.3 Conflating Integration and Antiderivatives Is a Pedagogical Pitfall in Calculus

In this book, we have endeavored to maintain a clear distinction between definite integrals and integration via antiderivatives. For starters, we treat them in separate chapters. However, in broader discourse (aka the big bad world outside the course), the distinction might not always be as clear-cut, and the sense of the action verb " to integrate" must be extracted from the surrounding context. The notation of calculus, while often elegant and succinct, is not helpful in this regard.

In the context of definite integrals, the integral sign $\int$ is accompanied by limits of integration, as in $\int_a^b f(x)\,dx$, indicating the interval over which the function f is to be integrated. However, when we venture into the domain of antiderivatives or indefinite integrals, the integral sign stands alone,

$$\int f(x)\,dx,$$

and represents a family of functions whose derivative is f. An alternative notation for denoting antiderivatives, albeit less common, is to use the differentiation symbol with a negative exponent, as in

$$D^{-1}f \quad \text{or} \quad \frac{d^{-1}}{dx^{-1}}f(x).$$

This notation explicitly mirrors the inverse operation being performed relative to differentiation. Yet it is not widely adopted, perhaps due to its unwieldiness or the potential confusion with reciprocal functions. Hence we do not use it.

The transition from $\int_a^b f(x)\,dx$ to $\int f(x)\,dx$ marks the evolution from a concrete computation to a more abstract concept, mirroring the increase in abstraction from arithmetic to algebra. On one hand, antidifferentiation does amplify the power of Calculus. On the other hand, all too often, the numerous methods for computing antiderivatives become the "be all and end all" of a standard Calc II course, leading many students to throw up their hands and walk away.

How could that be? As we will explain in the next Section, the methods for computing antiderivatives are based on inverting the differentiation rules in Prop. 5.11. Becoming proficient in them takes a lot of effort and is typically accompanied by deep frustration on the part of the learner. The frustration arises because (a) knowing when to use one method over another requires significant experience; (b) each method typically involves multiple steps where a simple minus sign error or confusing a multiply by a constant with a divide by the same constant destroys the entire calculation, and finally, (c), once you have computed the antiderivative, beyond the satisfaction of having solved a cool puzzle, you rarely get to use it for something meaningful . In real engineering, you are unlikely to use the hand methods of computing antiderivatives with sufficient frequency to maintain fluency, supposing you master them while in college.

You were once taught how to extract a square root by hand. How often do you do that now? **This textbook takes the attitude that while there was a time when it made sense to learn how to execute by hand all of the methods for computing antiderivatives, that time has passed.**

The next section will work through enough methods and examples so that when you are required to compute antiderivatives by hand in an engineering course, you are prepared.

7.3 The Art of the Antiderivative: Inverting Differentiation Rules to Find Antiderivatives

We go through, roughly in order of importance, various strategies for finding antiderivatives of functions.

7.3.1 The Power Rule of Integration: Inverts the Power Rule of Differentiation

From the power rule for differentiation, we have that $\frac{d}{dx}x^{k+1} = (k+1)x^k$. Hence, by the Second Fundamental Theorem of Calculus,

$$\int x^k\,dx = \frac{x^{k+1}}{k+1} + C. \tag{7.7}$$

Example 7.9. *Compute an antiderivative for* $4x^2$.

Solution: By the Power Rule, $\int 4x^2\,dx = 4\left(\frac{x^3}{3} + C_1\right) = \frac{4}{3}x^3 + C$, where $C := 4C_1$. ■

7.3.2 The Fundamental Rule: Integrating the Differential

When computing an integral (by hand) that is harder than a monomial, our first thought should be: Can we recognize the integrand (aka the function being integrated) as being the derivative of a commonly known function? The most fundamental relationship between differentiation and integration is given by the total differential of a function,

$$df(x) = \frac{df(x)}{dx}\,dx = f'(x)\,dx.$$

The Second Fundamental Theorem of Calculus says that integrating both sides with respect to x gives us back the original function (up to a constant of integration):

$$\int df(x) = \int \frac{df(x)}{dx}\,dx = \int f'(x)\,dx = f(x) + C. \tag{7.8}$$

Example 7.10. *Compute an antiderivative for* $3x^2$.

Solution: Here, we recognize that $x^2 = \frac{d}{dx}\left(\frac{x^3}{3}\right) = \left(\frac{x^3}{3}\right)'$. Therefore, the Second Fundamental Theorem of Calculus gives us

$$\int 3x^2\,dx = \int 3\left(\frac{x^3}{3}\right)'\,dx = 3\frac{x^3}{3} + C = x^3 + C.$$

■

Example 7.11. *Determine* $\int 13\,(1 + \tan^2(x))\,dx$.

Solution: This is an antiderivative problem. From Table 7.2, we recognize that $1 + \tan^2(x) = \frac{d}{dx}(\tan(x)) = (\tan(x))'$. Therefore, the Second Fundamental Theorem of Calculus gives us

$$\int 13\,(1 + \tan^2(x))\,dx = \int 13\,(\tan(x))'\,dx = 13\tan(x) + C.$$

■

Admittedly, these are not difficult integrals, but they do offer good examples of the process.

7.3.3 Inverting the Chain Rule: Integration by Substitution, aka u-Substitution

The chain rule for differentiation states that if both $f'(x) = \frac{df(x)}{dx}$ and $g'(x) = \frac{dg(x)}{dx}$ do exist, then

$$\frac{d}{dx}(f(g(x))) = (f(g(x)))' = f'(g(x))\cdot g'(x).$$

To invert this, we use **integration by substitution**. If we define $u = g(x)$, then $du = \frac{dg(x)}{dx}\,dx = g'(x)\,dx$, and

$$\int \underbrace{f'(g(x))}_{f'(u)}\cdot\underbrace{g'(x)\,dx}_{du} = \int f'(u)\,du = f(u) + C = f(g(x)) + C, \tag{7.9}$$

after substituting in $u = g(x)$. In other words, by a clever substitution, we transformed a function that did not look like a total differential into one that is a recognizable total differential so that the Fundamental Rule in (7.8) can be applied. Computing $\int f'(u)\,du$ is easy, while $\int f'(g(x))\cdot g'(x)\,dx$ looks hard before making the substitution. This is an example of useful pattern matching: looking for instances of the chain rule appearing as the integrand (aka under the integral sign).

Remark 7.12. ***(u-Substitutions)*** *It is common practice to use $u(x)$ instead of $g(x)$ AND to call this the Method of u-Substitution. Why u? When you find a satisfying answer, let us all know.*

Example 7.13. *Use integration to find* $\int 3x\cos(x^2 + 1)\,dx$.

Solution: Here, the term "integrate" means to find an antiderivative. Table 7.2 does not contain an antiderivative for $3x\cos(x^2 + 1)$. However, we see that the cosine function has been composed with $x^2 + 1$, and lucky for us, its derivative, $2x$, is ALMOST sitting in front of the function. Inspired by this, we let $f'(x) = \cos(x)$ and $u(x) = x^2 + 1$ so that $du = 2x\,dx$. Then

$$\begin{aligned}\int 3x\cos(x^2+1)\,dx &= \frac{3}{2}\cdot\int 2x\cos(x^2+1)\,dx \\ &= \frac{3}{2}\cdot\int \cos(u)\,du = \frac{3}{2}\cdot\sin(u) + C = \frac{3}{2}\cdot\sin(x^2+1) + C,\end{aligned}$$

after substituting in $u = x^2 + 1$. ■

Example 7.14. *Compute the definite integral* $\int_0^\pi 2x\cos(x^2+1)\,dx$.

Solution: We give two solutions, one that obliges us to complete the last step of substituting $u = u(x)$ back into $f(u)$ and one that allows us to avoid that step.

Approach One: From Example 7.13, we know the antiderivative of $2x\cos(x^2+1)$ is $F(x) := f(u(x)) = \sin(x^2+1) + C$. By the Second Fundamental Theorem of Calculus,

$$\int_0^\pi 2x\cos(x^2+1)\,dx = F(\pi) - F(0) = \left(\sin(\pi^2+1) + C\right) - \left(\sin(0^2+1) + C\right) = \sin(\pi^2+1) - \sin(1).$$

Because the constant of integration C always cancels out, it is common not even to write it down when using antiderivatives for definite integrals.

Approach Two: The limits of integration $a = 0$ and $b = \pi$ really mean the x-lower limit of integration is 0 and the x-upper limit of integration is π. When we perform a change of variables to $u(x)$, we should also change the limits of integration to present the u-lower limit of integration and the u-upper limit of integration, which are $u(a) = u(0) = \sin(1)$ and $u(b) = u(\pi) = \sin(\pi^2+1)$. Hence, we compute the integral as:

$$\int_0^\pi 2x\cos(x^2+1)\,dx = \int_{u(0)}^{u(\pi)} \cos(u)\,du = \sin(u)\Big|_1^{\pi^2+1} = \sin(\pi^2+1) - \sin(1).$$

When the antiderivative is more complicated, the second approach becomes more valuable. In addition, when the first substitution only gets us halfway there, and a second substitution is required, this second approach really shines. It is also common to add extra notation to help us keep track of the limits carefully when doing a change of variable, such as

$$\int_{x=0}^{x=\pi} 2x\cos(x^2+1)\,dx = \int_{u=u(0)}^{u=u(\pi)} \cos(u)\,du = \sin(u)\Big|_{u=1}^{u=\pi^2+1} = \sin(\pi^2+1) - \sin(1);$$

this helps avoid a really bad error, where the original limits are carried forward as in

$$\int_0^\pi 2x\cos(x^2+1)\,dx = \int_0^\pi \cos(u)\,du = \sin(u)\Big|_0^\pi = \sin(\pi) - \sin(0) = 0,$$

and everything in red is wrong! ■

Example 7.15. *Suppose that $f(x)$ is the antiderivative of $f'(x)$. Find the antiderivative of $f'(\alpha x - x_0)$, where $\alpha \neq 0$ and $x_0 \in \mathbb{R}$ are constants.*

Solution: We define $u(x) := \alpha x - x_0$ and note that $f'(u(x)) = f'(\alpha x - x_0)$. Moreover, $du(x) = \alpha\,dx \implies dx = \frac{1}{\alpha}\,du$. Hence,

$$\int f'(\alpha x - x_0)\,dx = \int f'(u)\frac{1}{\alpha}\,du = \frac{1}{\alpha}f(u) + C = \frac{1}{\alpha}f(\alpha x - x_0) + C. \tag{7.10}$$

■

Using the result of Example 7.15, we can immediately expand our table of known antiderivatives to include scaling and shifting.

Antiderivative	Function
$\frac{1}{\alpha}\frac{(\alpha x - x_0)^{k+1}}{k+1}$	$(\alpha x - x_0)^k$
$\frac{1}{\alpha}e^{(\alpha x - x_0)}$	$e^{(\alpha x - x_0)}$
$\frac{1}{\alpha}\ln\lvert\alpha x - x_0\rvert$	$\frac{1}{(\alpha x - x_0)}$
$\frac{1}{\alpha}\sin(\alpha x - x_0)$	$\cos(\alpha x - x_0)$
$\frac{1}{\alpha}\cos(\alpha x - x_0)$	$-\sin(\alpha x - x_0)$
$\frac{1}{\alpha}\tan(\alpha x - x_0)$	$1 + \tan^2(\alpha x - x_0)$
$\frac{1}{\alpha}\text{atan}(\alpha x - x_0)$	$\frac{1}{1+(\alpha x - x_0)^2}$

Table 7.3: Common Elementary Functions and Their Antiderivatives. The constants of integration, C, have been dropped to save space.

Example 7.16. *Compute the indefinite integral* $\int \frac{(\ln|x|))^2}{x}\,dx$.

Solution: Ans. $\int \frac{(\ln|x|)^2}{x}\,dx = \frac{(\ln(x))^3}{3} + C$ for any domain $[a, b]$ that excludes $x = 0$.
The problem uses a different vocabulary for the antiderivative so that we get used to the idea that integration and antidifferentiation are one and the same, two peas in the same pod, identical twins. Once again, we cannot find the integrand in Table 7.2. If we set $f'(x) = x^2$ and $u(x) = \ln|x|$, then $u'(x) = \frac{1}{x}$ and $du = \frac{1}{x}\,dx$. Hence, we have:

$$\frac{(\ln|x|)^2}{x}\,dx = u^2\,du = f'(u)\,du$$

Because we know that the antiderivative of $f'(u) = u^2$ is $f(u) = \frac{u^3}{3}$, we have

$$\int \frac{(\ln|x|)^2}{x}\,dx = \int u^2\,du = \frac{u^3}{3} + C = \frac{(\ln|x|)^3}{3} + C.$$

■

Secrets of the Arcane 7.17: "Danger, Will Robinson!" is 1960's Slang for "Red Flag!"

Because **Danger, Will Robinson!** is tied to a robot, we use it in this textbook.

The phrase "Danger, Will Robinson" is a classic line from the 1960s TV show **"Lost in Space"** used by the Robot to alert young Will Robinson to imminent danger. This iconic phrase has become synonymous with signaling a warning or drawing attention to a potential problem. It's a part of pop culture lexicon that transcends generations, often used humorously or dramatically to indicate caution in various contexts.

In contemporary terms, expressions like "Red Flag!" have gained popularity, especially on social media and in daily communication, to signal caution or alert someone to potential issues, dangers, or problematic situations. The term "Red Flag" is widely used to describe warning signs in relationships, behaviors, or situations that may be considered suspicious, dangerous, or indicative of deeper issues.

While "Danger, Will Robinson" harks back to a specific context of science fiction and the protective instincts of a robotic character, "Red Flag" is more universally applied across various scenarios, including personal interactions, professional environments, and online discourse. Both serve a similar purpose in signaling caution but reflect different cultural contexts and eras.

Fact 7.18: Danger, Danger Will Robinson!

When computing antiderivatives, it is easy to forget about the domains on which the integrand and the antiderivative are defined. The integrand $\frac{(\ln|x|)^2}{x}$ is defined and continuous on domains $[a,b]$ that do not include $x = 0$. Hence, its Riemann integral exists and will be finite for $[a,b] \subset (-\infty, 0)$ or $[a,b] \subset (0,\infty)$. What about the antiderivative for $\frac{(\ln|x|)^2}{x}$? It should exist and be finite in this same domain, and in fact, we observe that $\frac{(\ln|x|)^3}{3}$ exists and is finite for $[a,b] \subset (-\infty, 0)$ and $[a,b] \subset (0,\infty)$.

Where we can get into trouble: Suppose we want to evaluate the definite integral $\int_{-e}^{e} \frac{(\ln|x|)^2}{x}\,dx$. From our analysis above, we know the problem does not make sense because $0 \in [-e, e]$. However, it is ever so tempting to write

$$\int_{-e}^{e} \frac{(\ln|x|)^2}{x}\,dx = \frac{(\ln|x|)^3}{3}\Big|_{-e}^{e} = \frac{(\ln|e|)^3}{3} - \frac{(\ln|-e|)^3}{3} = 0 - 0 = 0,$$

which is wrong. **The correct answer is: the definite integral is not defined.** When computing definite integrals, it is easy to make "innocent" mistakes like this one.

In case everything has seemed totally straightforward so far, this next example will recalibrate expectations when it comes to finding antiderivatives by hand.

Example 7.19. *(A Tricky Antiderivative) Find an antiderivative for* $\sec(x) := \frac{1}{\cos(x)}$.

Solution: Ans. $\int \sec(x)\,dx = \ln|\sec(x) + \tan(x)| + C$, with its domain analyzed at the end of the solution.

To find the antiderivative of $\sec(x)$, we multiply and divide by $\sec(x) + \tan(x)$ to create a convenient form for u-substitution,

$$\begin{aligned}\int \sec(x)\,dx &= \int \sec(x)\frac{\sec(x)+\tan(x)}{\sec(x)+\tan(x)}\,dx\\ &= \int \frac{\sec^2(x)+\sec(x)\tan(x)}{\sec(x)+\tan(x)}\,dx.\end{aligned}$$

Now, let $u = \sec(x) + \tan(x)$. Then, $du = (\sec(x)\tan(x) + \sec^2(x))\,dx$, which is the numerator of the integrand. Thus, we can rewrite the integral in terms of u,

$$\begin{aligned}\int \sec(x)\,dx &= \int \frac{1}{u}\,du\\ &= \ln|u| + C\\ &= \ln|\sec(x)+\tan(x)| + C,\end{aligned}$$

where C is the constant of integration.

Where did the idea to multiply and divide by $\sec(x) + \tan(x)$ come from? Probably from knowing way too many trig identities! Finding antiderivatives by hand is definitely an art, and as with any form of art, some students will love it and others will hate it.

We need to ask ourselves, in what domain does the integration problem make sense, and the same for the proposed answer:

- We cannot compute an integral over an interval $[a,b]$ where $\sec(x) := \frac{1}{\cos(x)}$ blows up. Hence, we need to avoid intervals $[a,b]$ that contain $\frac{\pi}{2} + k\pi$, for some $k \in \mathbb{Z}$.
- Does the antiderivative blow up anywhere? The natural logarithm blows up at $x = 0$. Now, $\sec(x) + \tan(x) = \frac{1}{\cos(x)} + \frac{\sin(x)}{\cos(x)} = \frac{1+\sin(x)}{\cos(x)}$, and we note that $1 + \sin(x) = 0$ for $x = -\frac{\pi}{2} + 2k\pi$, $k \in \mathbb{Z}$, which is a subset of where $\cos(x)$ vanishes. What about the ratio? Applying L'Hôpital's Rule, we have

$$\lim_{x\to -\frac{\pi}{2}} \frac{1+\sin(x)}{\cos(x)} = \lim_{x\to -\frac{\pi}{2}} \frac{\cos(x)}{-\sin(x)} = 0.$$

  Hence, the antiderivative makes sense everywhere that the integrand makes sense, which is how it should be.

■

This example and the next one come from Advanced Integration by U-Substitution - Calculus II by the Calculus God.

Example 7.20. *Determine* $\int \cos^3(x)\,dx$.

Solution: We will offer two solutions.

Option 1: Sine and cosine raised to a power can always be expressed as a sum of sine and cosine without any powers; see Prop. 1.61. By the identity for the cosine of a triple angle,

$$\cos(3x) = 4\cos^3(x) - 3\cos(x).$$

Solving for $\cos^3(x)$ gives

$$\cos^3(x) = \frac{1}{4}\cos(3x) + \frac{3}{4}\cos(x).$$

Hence, $\int \cos^3(x)\,dx = \frac{1}{12}\sin(3x) + \frac{3}{4}\sin(x)$.

Option 2 (by the Calculus God):

$$\begin{aligned}\int \cos^3(x)\,dx &= \int \cos(x)\cdot\cos^2(x)\,dx\\ &= \int \cos(x)\cdot\left(1-\sin^2(x)\right)\,dx\\ &= \int \cos(x)\,dx - \int \cos(x)\sin^2(x)\,dx.\end{aligned}$$

The first antiderivative is easy; the second one looks harder.

Let $u = \sin(x)$ so that $du = \cos(x)\,dx$. Then,

$$\begin{aligned}\int \cos^3(x)\,dx &= \sin(x) - \int u^2\,du\\ &= \sin(x) - \frac{u^3}{3}\\ &= \sin(x) - \frac{1}{3}\cdot\sin^3(x).\end{aligned}$$

The two proposed antiderivatives appear to be different, but the trig substitution $\sin^3(x) = \frac{3}{4}\sin(x) - \frac{1}{4}\sin(3x)$, leads to $\sin(x) - \frac{1}{3}\cdot\sin^3(x) = \frac{1}{12}\sin(3x) + \frac{3}{4}\sin(x)$, establishing agreement with the previous answer. ■

Example 7.21. *Determine* $\int \frac{x}{\sqrt{x+1}}\, dx$.

Solution: We will offer two solutions. We note that the integrand, $\frac{x}{\sqrt{x+1}}$, is a well defined function for $x+1>0$, which is equivalent to $x>-1$.

Option 1: For the u-substitution, we will set $u = x+1$, which we note is strictly greater than zero. In addition, we have $du = dx$ and $x = u-1$. Applying the substitutions gives us

$$\begin{aligned}\int \frac{x}{\sqrt{x+1}}\, dx &= \int \frac{u-1}{\sqrt{u}}\, du \\ &= \int \frac{u-1}{u^{1/2}}\, du \\ &= \int u^{1/2} - u^{-1/2}\, du.\end{aligned}$$

Applying the generalized power rule gives

$$\begin{aligned}\int \frac{x}{\sqrt{x+1}}\, dx &= \int u^{1/2} - u^{-1/2}\, du \\ &= \frac{u^{3/2}}{3/2} - \frac{u^{1/2}}{1/2} \\ &= \frac{2}{3}u^{3/2} - 2u^{1/2}.\end{aligned}$$

Substituting in $u = x+1$ gives

$$\int \frac{x}{\sqrt{x+1}}\, dx = \frac{2}{3}\sqrt{(x+1)^3} - 2\sqrt{x+1}.$$

Option 2 (by the Calculus God): For the u-substitution, we set $u = \sqrt{x+1}$, and for later use, we note that $u>0$. From $u = \sqrt{x+1}$, we obtain $u^2 = x+1$ and $2u\, du = dx$. Applying the substitutions gives us

$$\begin{aligned}\int \frac{x}{\sqrt{x+1}}\, dx &= \int \frac{u^2-1}{u} \cdot 2u\, du \quad \text{(note that the } u \text{ in the denominator)} \\ &= \int 2\left(u^2-1\right) du \quad \text{(cancels with the } u \text{ in } 2u\, du\text{)} \\ &= 2\left(\frac{u^3}{3} - u\right).\end{aligned}$$

Plugging back in for x, we have

$$\int \frac{x}{\sqrt{x+1}}\, dx = \frac{2}{3}\left(\sqrt{(x+1)}\right)^3 - 2\sqrt{x+1},$$

which agrees with the answer in Option 1, because $\left(\sqrt{(x+1)}\right)^3 = \sqrt{(x+1)^3}$ by our power rules. ■

Hopefully it is now abundantly clear that the construction of antiderivatives can be approached in many ways, and even for a given problem there is not a single best way to work it. ChatGPT is still a good approach.

Recommended Videos for Additional Practice

- How to Integrate Using U-Substitution by Nancy Pi.
- Understand U-substitution, the Idea! by **BlackPenRedPen**.
- How To Integrate Using U-Substitution by the Organic Chemistry Tutor.
- U-substitution With Definite Integrals by the Organic Chemistry Tutor.

Danger, Will Robinson! When using any antiderivative, you must be careful to avoid making the "innocent" error highlighted in Fact. 7.18. It does not matter if you used ChatGPT or did the computation yourself by hand. You should always ask yourself: "What is an allowed interval of integration for the problem (aka what is an allowed domain for the integrand)?" and "Does the answer make sense in that domain?" On YouTube, this is rarely, if ever, discussed. It is always a good idea to either differentiate the antiderivative to make sure you recover the integrand or to compare to `quadgk` for a set of (definite) integration bounds.

7.3.4 Inverting the Product Rule: Integration by Parts

The product rule states that $\frac{d}{dx}(u \cdot v) = u \cdot \frac{dv}{dx} + v \cdot \frac{du}{dx}$. Or, in terms of the total differential, we have

$$d(u \cdot v) = u \cdot dv + v \cdot du.$$

Integration by parts is the inverse of the product rule, based on rearranging the above equation

$$u \cdot dv = d(u \cdot v) - v \cdot du.$$

Integrating both sides gives

$$\int u\, dv = u \cdot v - \int v\, du, \tag{7.11}$$

where $\int d(u \cdot v) = u \cdot v$, because $d(u \cdot v)$ is a total differential. This "trick" is useful when the integral we need to evaluate, $\int u\, dv$, is hard, but its other half coming from the product rule of differentiation, $\int v\, du$, is easier. The method is called **Integration by Parts** because we have broken the integral into two parts, one that is trivial $\int d(u \cdot v) = u \cdot v$, and one, $\int v\, du$, that we "hope" is easier than the original integral, $\int u\, dv$. We will return to this method when we examine Laplace Transforms in Chapter 10.

Example 7.22. *Solve for* $\int x \sin(x)\, dx$.

Solution: For this first example, we do not have a lot of choices to make when defining $u(x)$ and $dv(x)$. One of them has to be associated with x and the other has to be associated with $\sin(x)$. We explore both options.

Option 1: Let $u = x$ and $dv = \sin(x)\, dx$. Then $du = dx$ and $v = -\cos(x)$. It was very important to choose dv such that it has a known antiderivative. Armed with this information, we have

$$\int \underbrace{x \sin(x)\, dx}_{u\, dv} = \underbrace{-x\cos(x)}_{u \cdot v} + \int \underbrace{\cos(x)\, dx}_{v\, du} = -x\cos(x) + \sin(x) + C.$$

Option 2: Let $u = \sin(x)$ and $dv = x\, dx$. Then $du = \cos(x)\, dx$ and $v = \frac{x^2}{2}$. For this problem, it looks like both choices are yielding easy antiderivatives. Armed with this information, we proceed and find that

$$\int \underbrace{x \sin(x)\, dx}_{u\, dv} = \underbrace{\frac{x^2}{2}\sin(x)}_{u \cdot v} - \int \underbrace{\frac{x^2}{2}\cos(x)\, dx}_{v\, du}.$$

At this point, we go Ooops! We've created an even more challenging antiderivative to compute, namely, the antiderivative for $\frac{x^2}{2}\cos(x)$ has a higher power of x than we started with.

The second option illustrates another aspect of the **Art of the Antiderivative**; we have to make good choices for $u(x)$ and $v(x)$. In this problem, the x term is causing the problem. We know how to integrate $\sin(x)$. It is smart to choose $u(x) = x$ because when we differentiate $u(x)$, we get a constant, which puts us back to dealing with an integral with a simple antiderivative. ■

Example 7.23. *Find an antiderivative for* $\ln(x)$.

Solution: Without an integral sign in sight, it is hard to even think about applying integration by parts, so we rewrite the problem as evaluate $\int \ln(x)\,dx$. We do not know how to integrate $\ln(x)$, but we sure know how to differentiate it. Hence, we set $u(x) = \ln(x)$ and $dv = dx$. With these choices, $du = \frac{1}{x}\,dx$ and $v(x) = x$. Hence, our integration by parts formula becomes

$$\int \underbrace{\ln(x)\,dx}_{u\,dv} = \underbrace{\ln(x)\cdot x}_{u\cdot v} - \int \underbrace{x\cdot\frac{1}{x}\,dx}_{v\,du} = \ln(x)\cdot x - \int 1\,dx = x\cdot\ln(x) - x + C.$$

■

Example 7.24. *Evaluate* $\int x^3\cdot e^x\,dx$.

Solution: This problem requires three successive applications of the integration by parts formula,

$$\int u\,dv = uv - \int v\,du.$$

For the first application, we choose:

$$\begin{aligned} u_1 &= x^3 & &\Longrightarrow & du_1 &= 3x^2\,dx, \\ dv_1 &= e^x\,dx & &\Longrightarrow & v_1 &= e^x. \end{aligned}$$

Applying the integration by parts formula:

$$\begin{aligned} \int x^3e^x\,dx &= u_1v_1 - \int v_1\,du_1 \\ &= x^3e^x - \int 3x^2e^x\,dx. \end{aligned}$$

For the second application, we choose:

$$\begin{aligned} u_2 &= x^2 & &\Longrightarrow & du_2 &= 2x\,dx, \\ dv_2 &= e^x\,dx & &\Longrightarrow & v_2 &= e^x. \end{aligned}$$

Applying the integration by parts formula again:

$$\begin{aligned} \int 3x^2e^x\,dx &= 3(u_2v_2 - \int v_2\,du_2) \\ &= 3(x^2e^x - \int 2xe^x\,dx). \end{aligned}$$

For the third application, we choose:

$$\begin{aligned} u_3 &= x & &\Longrightarrow & du_3 &= dx, \\ dv_3 &= e^x\,dx & &\Longrightarrow & v_3 &= e^x. \end{aligned}$$

Applying the integration by parts formula a third time:

$$\begin{aligned} \int 2xe^x\,dx &= 2(u_3v_3 - \int v_3\,du_3) \\ &= 2(xe^x - \int e^x\,dx) \\ &= 2xe^x - 2e^x. \end{aligned}$$

Combining all the applications, we get:

$$\begin{aligned} \int x^3e^x\,dx &= x^3e^x - 3(x^2e^x - 2(xe^x - e^x)) \\ &= x^3e^x - 3x^2e^x + 6xe^x - 6e^x + C. \end{aligned}$$

Thus, the antiderivative of x^3e^x is

$$x^3e^x - 3x^2e^x + 6xe^x - 6e^x + C,$$

where C is the constant of integration. ■

Method 7.25: (Optional Read) DI Method or Integration by Parts on Steroids

For those who wish to master problems requiring multiple applications of integration by parts, please see Integration by Parts, DI Method, VERY EASY by **BlackPenRedPen**. The DI Method is a tabular form of integration by parts that makes it easy to keep track of the signs and understand how the initial choices of u and dv propagate through successive steps of integration by parts. The D-column contains u, the function that is differentiated when applying integration by parts, while the I-column contains dv, the function that is to be integrated. After watching the video by **BlackPenRedPen**, you will be able to build a table like the one below and extract from it the antiderivative of $x^3 \cdot e^x$. Not only is the table very compact, but the alternating signs in the left column make it easy to avoid sign errors.

	D	I
$+$	x^3	e^x
$-$	$3x^2$	e^x
$+$	$6x$	e^x
$-$	6	e^x
$+$	0	e^x

$$\int x^3 e^x \, dx = x^3 e^x - 3x^2 e^x + 6x e^x - 6e^x + C.$$

You can perfect your technique by watching a few more DI Method Videos by **BlackPenRedPen**.

Recommended Videos for Additional Practice:

- Integration by Parts... How? by Nancy Pi.
- Integration by Parts by the Organic Chemistry Tutor.
- Integration by Parts, Calculus 2, AP Calculus BC by **BlackPenRedPen**.

7.3.5 Antiderivatives of Rational Functions by Partial Fraction Expansion (PFE)

Consider a rational function expanded as follows,

$$R(x) = \frac{k_1}{x - r_1} + \frac{k_2}{x - r_2} + \cdots + \frac{k_n}{x - r_n}, \tag{7.12}$$

for real "roots" $r_i \in \mathbb{R}$ and coefficients $k_i \in \mathbb{R}$. On the basis of Table 7.3, computing its antiderivative is straightforward:

$$\int R(x) \, dx = k_1 \ln|x - r_1| + k_2 \ln|x - r_2| + \cdots + k_n \ln|x - r_n|,$$

because $\frac{d}{dx} \ln|x - x_0| = \frac{1}{x - x_0}$.

Note: The antiderivative "blows up" when the argument of the natural logarithm is zero. Hence, the domain of the antiderivative is equal to $\{x \in \mathbb{R} \mid x \notin \{r_1, r_2, \ldots, r_n\}\}$, which is also the domain of $R(x)$. Everything checks out.

Partial Fraction Expansion for Real Roots

If we were given (7.12) in the form $R(x) = \frac{P(x)}{Q(x)}$, then it would be far less obvious how to compute an antiderivative. As an illustration, if we start out with a few terms similar to (7.12), and put them over a common denominator,

$$
\begin{aligned}
R(x) &= \frac{1}{x-1} + \frac{2}{x+2} + \frac{3}{x-3} \\
&= \frac{(x+2)(x-3)}{(x-1)(x+2)(x-3)} + \frac{2(x-1)(x-3)}{(x-1)(x+2)(x-3)} + \frac{3(x-1)(x+2)}{(x-1)(x+2)(x-3)} \\
&= \frac{6x^2-6x-6}{(x-1)(x+2)(x-3)} \qquad (7.13)\\
&= \frac{6x^2-6x-6}{x^3-2x^2-5x+6} \\
&=: \frac{P(x)}{Q(x)}.
\end{aligned}
$$

Finding the antiderivative of $\frac{6x^2-6x-6}{x^3-2x^2-5x+6}$ looks much more difficult than finding the antiderivative of $\frac{1}{x-1} + \frac{2}{x+2} + \frac{3}{x-3}$.

The process called **Partial Fraction Expansion (PFE)** was invented to take a rational function like (7.13) and turn it into a sum of first-order terms, as in (7.12). We will begin with a simplified version of partial fraction expansion that, like (7.13), works for rational functions satisfying the following two properties:

- $\deg(P(x)) < \deg(Q(x))$, the numerator polynomial has a strictly lower degree than the denominator polynomial.
- the roots of the denominator polynomial, $Q(x)$, are real and distinct (which means $Q(x)$ has no complex roots and no repeated roots).

For (7.13), we note that $2 = \deg(P(x)) < \deg(Q(x)) = 3$, and that the roots of $Q(x) = x^3 - 2x^2 - 5x + 6 = (x-1)(x+2)(x-3)$, being $r_1 = 1$, $r_2 = -2$, and $r_3 = 3$, are real and distinct.

Proposition 7.26: Antiderivative with a Simplified Partial Fraction Expansion (PFE)

Consider a rational function $R(x) = \frac{P(x)}{Q(x)}$ where

- the degree of $P(x)$ is strictly lower than the degree of $Q(x)$,
- $Q(x)$ is monic (aka leading coefficient is one),
- the roots of $Q(x)$, $\{r_1, r_2, \dots, r_m\}$, are real and distinct, so that $Q(x) = \prod_{i=1}^{m}(x - r_i)$.

Then $R(x)$ always has a partial fraction expansion

$$R(x) = \frac{k_1}{x-r_1} + \frac{k_2}{x-r_2} + \dots + \frac{k_m}{x-r_m}, \qquad (7.14)$$

where the coefficients $k_1, k_2, \dots, k_m$ satisfy

$$k_i = \lim_{x \to r_i} \frac{(x-r_i)P(x)}{Q(x)} = \frac{P(r_i)}{(r_i-r_1)\cdots(r_i-r_{i-1})(r_i-r_{i+1})\cdots(r_i-r_m)}. \qquad (7.15)$$

Note: Because we assumed the roots to be distinct, once we canceled the term $(x-r_i)$ in its numerator and denominator,

$$\frac{(\boldsymbol{x}-\boldsymbol{r_i})P(x)}{Q(x)} = \frac{(\boldsymbol{x}-\boldsymbol{r_i})P(x)}{(x-r_1)\cdots(x-r_{i-1})\cdot(\boldsymbol{x}-\boldsymbol{r_i})\cdot(x-r_{i+1})\cdots(x-r_m)},$$

the limit is simply the evaluation of the reduced polynomial at $x = r_i$.

Note: There is an alternative method for determining the coefficients k_i that involves setting up a system of linear equations. In the end, it is better to use software tools for anything more than a quadratic.

Example 7.27. *Determine* $\int \frac{1}{x^2-1}\, dx$.

Solution: We check to see if our simplified method of partial fraction expansion applies to the problem. The degree of the numerator is zero, which is strictly less than two, the degree of the denominator. The denominator factors as $x^2 - 1 = (x+1)\cdot(x-1)$, giving us distinct real roots, $r_1 = -1$ and $r_2 = 1$. Hence, we can decompose the integrand into partial fractions:

$$\frac{1}{x^2-1} = \frac{k_1}{(x-1)} + \frac{k_2}{(x+1)}.$$

To compute the coefficients in the partial fraction expansion,

- $k_1 = \lim\limits_{x\to r_1} \dfrac{(x+1)}{(x+1)(x-1)} = \lim\limits_{x\to -1} \dfrac{1}{(x-1)} = \dfrac{1}{(-1-1)} = -\dfrac{1}{2}.$
- $k_2 = \lim\limits_{x\to r_2} \dfrac{(x-1)}{(x+1)(x-1)} = \lim\limits_{x\to +1} \dfrac{1}{x+1} = \dfrac{1}{1+1} = \dfrac{1}{2}.$

Armed with the coefficients, we decompose the integrand as the sum:

$$\frac{1}{x^2-1} = \frac{1}{2}\frac{1}{x-1} - \frac{1}{2}\frac{1}{x+1}.$$

Thus,

$$\int \frac{1}{x^2-1}\, dx = \frac{1}{2}\ln|x-1| - \frac{1}{2}\ln|x+1| + C.$$

Note: The antiderivative "blows up" when the argument of the natural logarithm is zero. Hence, the domain of the antiderivative is equal to $\{x \in \mathbb{R} \mid x \notin \{-1,+1\}\}$, the same points where the original rational function is well defined. ■

Example 7.28. *Compute the partial fraction expansion of* $R(x) = \frac{6x^2-6x-6}{(x-1)(x-(-2))(x-3)}$ *and then use it to evaluate the antiderivative.*

Solution: The degree of the numerator is one less than the denominator, and we have $r_1 = 1$, $r_2 = -2$, and $r_3 = 3$. The roots are therefore real and distinct, and so a partial fraction expansion exists.

- $k_1 = \lim\limits_{x\to r_1} \dfrac{(x-1)(6x^2-6x-6)}{(x-1)(x-(-2))(x-3)} = \dfrac{(6(1)^2-6(1)-6)}{(1-(-2))(1-3)} = \dfrac{-6}{-6} = 1$
- $k_2 = \lim\limits_{x\to r_2} \dfrac{(x-(-2))(6x^2-6x-6)}{(x-1)(x-(-2))(x-3)} = \dfrac{(6(-2)^2-6(-2)-6)}{((-2)-1)((-2)-3)} = \dfrac{30}{15} = 2$
- $k_3 = \lim\limits_{x\to r_3} \dfrac{(x-3)(6x^2-6x-6)}{(x-1)(x-(-2))(x-3)} = \dfrac{(6(3)^2-6(3)-6)}{(3-1)(3-(-2)} = \dfrac{30}{10} = 3$

Hence, we have

$$\int \frac{6x^2-6x-6}{(x-1)(x-(-2))(x-3)}\, dx = \ln|x-1| + 2\ln|x+2| + 3\ln|x-3|.$$

■

Antiderivatives for a Pair of Complex Conjugate Roots

The denominator of the real rational function

$$R(x) = \frac{ax + b}{(x - \alpha)^2 + \omega^2}$$

has complex roots, $z = \alpha + \mathbf{i}\omega$ and $z^* = \alpha - \mathbf{i}\omega$, where z^* is the **complex conjugate** of z. While it is true that $R(x)$ can be expanded as

$$R(x) = \frac{\kappa}{x - z} + \frac{\kappa^*}{x - z^*},$$

trying to find the antiderivative in this form would take us into the realm of the Calculus of Complex Variables, which is beyond the scope of the book. Instead, we work directly with $R(x)$ as a ratio of a linear numerator and a quadratic denominator.

Proposition 7.29: Antiderivative for a Pair of Complex Conjugate Roots

Consider a real rational function $R(x) = \frac{ax+b}{(x-\alpha)^2+\omega^2}$. Its antiderivative is:

$$\int \frac{ax + b}{(x - \alpha)^2 + \omega^2}\, dx = \frac{a}{2} \ln\left((x - \alpha)^2 + \omega^2\right) + \frac{a\alpha + b}{\omega} \arctan\left(\frac{x - \alpha}{\omega}\right) + C, \qquad (7.16)$$

where C is a constant of integration.

Proof: The antiderivative of the function $\frac{ax+b}{(x-\alpha)^2+\omega^2}$ is found by combining u-substitution and the recognition of a total derivative. The integral is broken down into two parts based on the numerator,

$$\int \frac{ax + b}{(x - \alpha)^2 + \omega^2}\, dx = \int \frac{a(x - \alpha) + (a\alpha + b)}{(x - \alpha)^2 + \omega^2}\, dx,$$

where splitting the integral gives,

$$= a \int \frac{x - \alpha}{(x - \alpha)^2 + \omega^2}\, dx + (a\alpha + b) \int \frac{1}{(x - \alpha)^2 + \omega^2}\, dx.$$

Step 1: Using u-Substitution:
For the first part,

$$a \int \frac{x - \alpha}{(x - \alpha)^2 + \omega^2}\, dx,$$

we let $u = (x-\alpha)^2 + \omega^2$, then $du = 2(x-\alpha)dx$, and $\frac{1}{2}du = (x-\alpha)dx$. Thus, this part of the integral becomes,

$$\frac{a}{2} \int \frac{1}{u} du = \frac{a}{2} \ln|u| + C_1 = \frac{a}{2} \ln\left((x - \alpha)^2 + \omega^2\right) + C_1$$

after substituting back in $u = (x - \alpha)^2 + \omega^2$. Because $(x - \alpha)^2 + \omega^2 > 0$, the absolute value can be removed; indeed, if $\omega = 0$, then the root is not complex.

Step 2: Recognizing the Derivative of Arctangent
For the second part,

$$(a\alpha + b) \int \frac{1}{(x - \alpha)^2 + \omega^2}\, dx,$$

is a total differential of the arctangent function, namely,

$$\frac{a\alpha + b}{\omega} \arctan\left(\frac{x - \alpha}{\omega}\right) + C_2,$$

where once again, we used the fact that because the root is complex, ω cannot equal zero.

Final Step:
Combining both parts, we get the full antiderivative,

$$\int \frac{ax+b}{(x-\alpha)^2+\omega^2}\,dx = \frac{a}{2}\ln\left((x-\alpha)^2+\omega^2\right) + \frac{a\alpha+b}{\omega}\arctan\left(\frac{x-\alpha}{\omega}\right) + C,$$

where $C := C_1 + C_2$ is the constant of integration. ■

Example 7.30. *Find an antiderivative for* $R(x) = \frac{2x+1}{3x^2-6x+15}$.

Solution: **Ans.** $\frac{1}{3}\cdot\log\left(x^2-2\cdot x+5\right)+\frac{1}{2}\cdot\arctan\left(\frac{x}{2}-\frac{1}{2}\right)+C.$

We identify $P(x) = 2x+1$ and $Q(x) = 3x^2-6x+15$, which clearly satisfy the degree of the numerator being less than the degree of the denominator. To see if the denominator has real or complex poles, we check the discriminant

$$b^2-4ac = (-6)^2-4(3)(15) = -144 < 0,$$

and hence, the roots are complex. Indeed, applying the quadratic formula yields $z = 1 \pm 2\,\mathbf{i}$, and thus

$$\alpha = 1 \quad \text{and} \quad \omega = 2;$$

it is fine to choose $\omega = -2$, as it will not change the answer.

The final step for applying Proposition 7.29 is to make the denominator monic by factoring out the coefficient of x^2,

$$R(x) = \frac{1}{3}\cdot\frac{2x+1}{x^2-2x+5} = \frac{1}{3}\cdot\frac{2x+1}{(x-1)^2+2^2}.$$

Substituting into (7.16) yields

$$\int R(x)\,dx = \frac{1}{3}\cdot\left(\log\left(x^2-2\cdot x+5\right)+\frac{3\cdot\arctan\left(\frac{x}{2}-\frac{1}{2}\right)}{2}\right)+C.$$

■

Antiderivatives for a Mixture of Real and Complex Conjugate Roots

Real polynomials often have a mixture of real roots and complex roots, with the complex roots occurring in complex conjugate pairs. Pairs of complex conjugate roots can be recombined to give a quadratic polynomial with real coefficients, which can then be integrated as in Prop. 7.29. We will first show how the corresponding partial fraction expansions works in software, and then we will state for the record the corresponding theoretical result.

Example 7.31. *Use software to find first the partial fraction expansion and then the antiderivative of*

$$R(x) = \frac{2\cdot x^3-x+4}{x^6-4\cdot x^4+10\cdot x^3-11\cdot x^2+10\cdot x-6}.$$

Clearly, doing this by hand would be a challenge. You already need an algorithm to find the roots of the denominator, so why not use another algorithm to find the coefficients in the PFE?

```
using SymPy

# Define the variable
x = symbols("x")

P = 2*x^3 - x + 4
Q = x^6 - 4*x^4 + 10*x^3 - 11*x^2 + 10*x - 6
```

```
R = P/Q

# Compute the PFE
PFE = apart(R)

println(PFE)
```

Output

```
-(4*x + 3)/(10*(x^2 + 1)) - (5*x - 8)/(17*(x^2 - 2*x + 2))
+ 47/(680*(x + 3)) + 5/(8*(x - 1))
```

or, nicely typeset,

$$\mathrm{PFE}(x) = \frac{-(4\cdot x+3)}{10\cdot(x^2+1)} - \frac{5\cdot x-8}{17\cdot(x^2-2\cdot x+2)} + \frac{47}{680\cdot(x+3)} + \frac{5}{8\cdot(x-1)}.$$

The `apart` function in `SymPy` has conveniently expanded the rational function as a sum of first-order terms and second-order terms. We could not ask for more, because now we can easily compute the antiderivative for each term! But instead of doing them by hand, we continue with software.

Continuing with the antiderivative:

```
antiderivative = integrate(PFE, x)
println(antiderivative)
```

Output

```
5*log(x - 1)/8 + 47*log(x + 3)/680 - log(x^2 + 1)/5
- 5*log(x^2 - 2*x + 2)/34
- 3*atan(x)/10 + 3*atan(x - 1)/17
```

or nicely typeset,

$$\int R(x)\,dx = \frac{5\cdot\log(x-1)}{8} + \frac{47\cdot\log(x+3)}{680} - \frac{\log\left(x^2+1\right)}{5} - \frac{5\cdot\log\left(x^2-2\cdot x+2\right)}{34} - \frac{3\cdot\arctan(x)}{10} + \frac{3\cdot\arctan(x-1)}{17}.$$

As you might guess, with software, the antiderivative can be computed in one step, in other words, without first doing a PFE:

```
# Find the antiderivative of the rational function
antiderivative = integrate(R, x)
println(antiderivative)
```

Output

```
5*log(x - 1)/8 + 47*log(x + 3)/680 - log(x^2 + 1)/5 - 5*log(x^2
- 2*x + 2)/34 - 3*atan(x)/10 + 3*atan(x - 1)/17
```

or nicely typeset,

$$\int R(x)\,dx = \frac{5\cdot\log(x-1)}{8} + \frac{47\cdot\log(x+3)}{680} - \frac{\log\left(x^2+1\right)}{5} - \frac{5\cdot\log\left(x^2-2\cdot x+2\right)}{34} - \frac{3\cdot\arctan(x)}{10} + \frac{3\cdot\arctan(x-1)}{17}.$$

■

Below is how the subject is taught (if at all) in a traditional Calculus course.

Proposition 7.32: (Optional Read) Antiderivative for a Mixture of Real and Complex Roots

Consider a real rational function

$$R(x) = \frac{P(x)}{Q(x)},$$

where $\deg(P) < \deg(Q)$, $Q(x)$ is monic, and all roots of $Q(x)$ are distinct. Because $Q(x)$ has real coefficients, its nonreal roots occur in complex conjugate pairs. List the roots as

- **real roots:** $r_1, r_2, \ldots, r_m$,
- **nonreal roots:** $z_1, z_2, \ldots, z_p$ together with their complex conjugates $z_1^*, z_2^*, \ldots, z_p^*$,

and write

$$\text{Real}\{z_i\} =: \alpha_i, \qquad \text{Imag}\{z_i\} =: \omega_i, \qquad z_i = \alpha_i + \mathbf{i}\omega_i, \qquad \omega_i > 0.$$

Under these assumptions,

$$Q(x) = \left(\prod_{i=1}^{m}(x - r_i)\right)\left(\prod_{i=1}^{p}\left((x-\alpha_i)^2 + \omega_i^2\right)\right).$$

Moreover, $R(x)$ can be expressed as a partial fraction expansion (PFE)

$$R(x) = \sum_{i=1}^{m}\frac{k_i}{x - r_i} + \sum_{i=1}^{p}\frac{A_i(x-\alpha_i) + B_i}{(x-\alpha_i)^2 + \omega_i^2},$$

for unique real numbers k_i, A_i, B_i.

How the coefficients are found (overview):

- For a real root r_i, the coefficient k_i is given by the "cover-up" formula

$$k_i = \lim_{x \to r_i}\frac{(x - r_i)P(x)}{Q(x)}.$$

- The remaining coefficients A_i and B_i are found by multiplying the PFE by $Q(x)$ and equating coefficients of the resulting polynomials.

Antiderivative of a Linear-Factor Term:

$$\int \frac{k}{x - r}\,dx = k\ln|x - r| + C.$$

Antiderivative of an Irreducible-Quadratic Term: Let $\omega > 0$. Then

$$\int \frac{A(x-\alpha) + B}{(x-\alpha)^2 + \omega^2}\,dx = \frac{A}{2}\ln\left((x-\alpha)^2 + \omega^2\right) + \frac{B}{\omega}\arctan\left(\frac{x-\alpha}{\omega}\right) + C.$$

Therefore: Once a PFE is obtained in the form above, an antiderivative of $R(x)$ is found by integrating term-by-term using the two generic formulas.

7.3.6 Trigonometric Substitutions for Radicals

Trigonometric substitution is a technique for computing antiderivatives (and integrals), and it is particularly useful when the integrand involves expressions like $\sqrt{a^2 - x^2}$, $\sqrt{a^2 + x^2}$, or $\sqrt{x^2 - a^2}$. Here, we carefully work out examples where trigonometric substitution is commonly used, highlighting common pitfalls along the way.

Heads up: Trig substitution is the most technically and mentally taxing of all the methods for finding antiderivatives by hand. Hence, if you find it difficult, you are not alone. You might see it a few times in an engineering course, though it arises less frequently than u-substitution and integration by parts. We will include images that help you to understand how the transformations were "discovered".

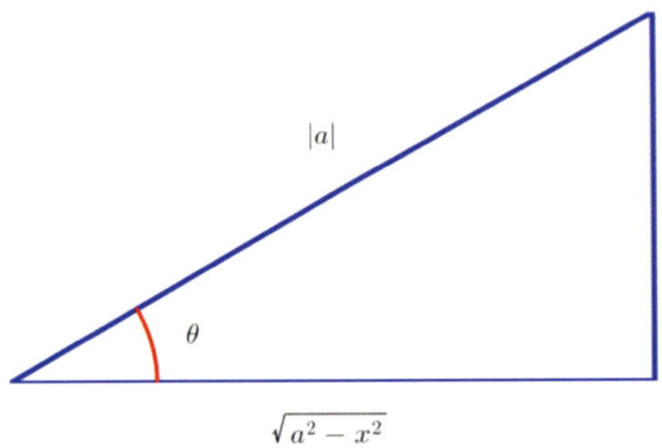

The above triangle satisfies the Pythagorean Theorem, and hence it is a right triangle. From basic trigonometry it follows that the opposite side of the triangle is given by $x = |a|\sin(\theta)$, which motivates the change of variable we use in the next problem.

Example 7.33. *Evaluate* $\int \sqrt{a^2 - x^2}\, dx$.

Solution: Ans. For $|x| \le |a|$, $\int \sqrt{a^2 - x^2}\, dx = \frac{a^2}{2} \cdot \arcsin\left(\frac{x}{|a|}\right) + \frac{x}{2} \cdot \sqrt{a^2 - x^2} + C$.

Because a is squared in the integrand, we can either replace it with $|a|$ or simply assume that $a > 0$. We make fewer errors if we use $|a|$, as in the diagram.

The integrand exists and is continuous for $(a^2 - x^2 \ge 0) \iff (-|a| \le x \le |a|)$. The standard trig substitution for this integral is $x = |a|\sin(\theta)$ and $dx = |a|\cos(\theta)\, d\theta$. We note that with $-\frac{\pi}{2} \le \theta \le \frac{\pi}{2}$, it follows that $-|a| \le x \le |a|$ and that $0 \le \cos(\theta) \le 1$. Moreover, later, when we need to inverse the transformation, we are working with a principal domain for defining $\arcsin(x)$. Hence, everything checks out!

Using this substitution and recalling that $1 - \sin^2(x) = \cos^2(x)$, the integral becomes

$$\begin{aligned}\int \left(\sqrt{a^2 - a^2\sin^2(\theta)}\right) \cdot |a|\cos(\theta)\, d\theta &= \int \left(\sqrt{a^2 \cdot \cos^2(\theta)}\right) \cdot |a|\cos(\theta)\, d\theta \\ &= \int \left(|a| \cdot |\cos(\theta)|\right) \cdot |a|\cos(\theta)\, d\theta \\ &= a^2 \int \cos^2(\theta)\, d\theta,\end{aligned}$$

where the last equality follows because we (carefully) noted earlier that $\cos(\theta) \ge 0$ for $-\frac{\pi}{2} \le \theta \le \frac{\pi}{2}$, the domain where the substitution was defined.

A visual derivation based on the triangle: From $x = |a|\sin(\theta)$, we deduce that $dx = |a|\cos(\theta)\,d\theta$. We also have the adjacent side of the triangle is $|a|\cos(\theta) = \sqrt{a^2 - x^2}$, which can be used to simplify the integrand. Substituting these into the integral, we have

$$\int \sqrt{a^2 - x^2}\,dx = \int \underbrace{|a|\cos(\theta)}_{\sqrt{a^2-x^2}} \cdot \underbrace{|a|\cos(\theta)\,d\theta}_{dx} = a^2 \int \cos^2(\theta)\,d\theta.$$

Using the half-angle identity, $\cos^2(\theta) = \frac{1+\cos(2\theta)}{2}$, we have

$$a^2 \int \cos^2(\theta)\,d\theta = \frac{a^2}{2} \int (1 + \cos(2\theta))\,d\theta = \frac{a^2}{2}\left(\theta + \frac{\sin(2\theta)}{2}\right) + C.$$

To return to x, we use $\sin(\theta) = \frac{x}{|a|}$ and $\cos(\theta) = \sqrt{1 - \sin^2(\theta)} = \sqrt{1 - \left(\frac{x}{|a|}\right)^2}$. Importantly, as noted earlier, we are working with a principal domain of the function arcsin. Thus

$$\theta = \arcsin\left(\frac{x}{a}\right), \text{ and } \sin(2\theta) = 2\sin(\theta)\cos(\theta) = \frac{2x}{|a|}\sqrt{1 - \left(\frac{x}{|a|}\right)^2}.$$

The final antiderivative in terms of x is

$$\int \sqrt{a^2 - x^2}\,dx = \frac{a^2}{2}\left(\arcsin\left(\frac{x}{|a|}\right) + \frac{x}{|a|}\sqrt{1 - \left(\frac{x}{|a|}\right)^2}\right) + C.$$

Because $-1 \le \frac{x}{|a|} \le 1$ for all x in the domain of the integrand, the terms $\arcsin\left(\frac{x}{|a|}\right)$ and $\sqrt{1 - \left(\frac{x}{|a|}\right)^2}$ in the antiderivative are well defined (i.e., make mathematical sense[1]); hence, the computed antiderivative is defined for all x in the domain of the integrand.

A more aesthetically pleasing form of the answer can be obtained with a small bit of algebra:

For $|x| \le |a|$, $\displaystyle\int \sqrt{a^2 - x^2}\,dx = \frac{a^2}{2} \cdot \arcsin\left(\frac{x}{|a|}\right) + \frac{x}{2} \cdot \sqrt{a^2 - x^2} + C.$

Note: From the right triangle associated with this problem, we have $\theta = \arcsin\left(\frac{x}{|a|}\right)$ and $\theta = \operatorname{atan}\left(\frac{x}{\sqrt{a^2-x^2}}\right)$, and thus

$$\arcsin\left(\frac{x}{|a|}\right) = \operatorname{atan}\left(\frac{x}{\sqrt{a^2 - x^2}}\right).$$

It follows that the answer can also be written as

For $|x| \le |a|$, $\displaystyle\int \sqrt{a^2 - x^2}\,dx = \frac{a^2}{2} \cdot \operatorname{atan}\left(\frac{x}{\sqrt{a^2 - x^2}}\right) + \frac{x}{2} \cdot \sqrt{a^2 - x^2} + C.$

■

[1]As an example, if $\frac{x}{|a|}$ were greater than one, then neither the arcsine nor the square root would make sense.

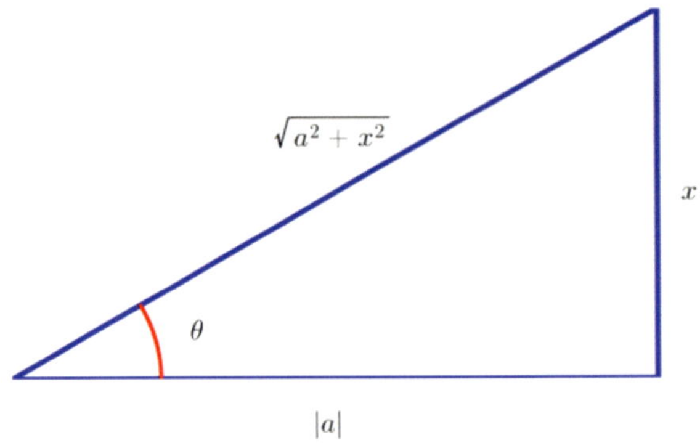

Note that the above triangle satisfies the Pythagorean Theorem, and hence it is a right triangle. It follows that $\tan(\theta) = \frac{x}{|a|}$, or, $x = |a|\tan(\theta)$, which motivates the change of variable we use in the next problem.

Example 7.34. *Evaluate the integral*

$$\int \sqrt{a^2+x^2}\, dx.$$

Solution: Ans. For $x \in \mathbb{R}$, $\int \sqrt{a^2+x^2}\, dx = \frac{a^2}{2}\ln\left(x + \sqrt{a^2+x^2}\right) + \frac{x}{2}\cdot\sqrt{a^2+x^2} + C.$

The integrand $\sqrt{a^2+x^2}\, dx$ is defined and continuous in $(-\infty, \infty)$. Because a is squared, we replace it with $|a|$ to avoid sign errors later on.

The substitution $x = |a|\tan(\theta)$ is defined for $-\frac{\pi}{2} < \theta < \frac{\pi}{2}$, and gives $dx = |a|\cdot\sec^2(\theta)\, d\theta$. The integral becomes

$$\int \sqrt{a^2 + a^2\tan^2(\theta)}\cdot|a|\cdot\sec^2(\theta)\, d\theta = \int \sqrt{a^2\left(1+\tan^2(\theta)\right)}\cdot|a|\cdot\sec^2(\theta)\, d\theta.$$

Next, we use the trig identity, $1+\tan^2(\theta) = \sec^2(\theta)$, giving us

$$\begin{aligned}\int \sqrt{a^2\left(1+\tan^2(\theta)\right)}\cdot|a|\cdot\sec^2(\theta)\, d\theta &= \int \sqrt{a^2\cdot\sec^2(\theta)}\cdot|a|\cdot\sec^2(\theta)\, d\theta\\ &= \int |a||\sec(\theta)|\cdot|a|\cdot\sec^2(\theta)\, d\theta\\ &= a^2\int \sec^3(\theta)\, d\theta,\end{aligned}$$

where in the last line we used the fact that $\sec(\theta) := \frac{1}{\cos(\theta)}$ is positive for $-\frac{\pi}{2} < \theta < \frac{\pi}{2}$, the domain of the change of variable.

Computing the antiderivative of $\int \sec^3(\theta)\, d\theta$ requires a new technique that is useful to know when performing integration by parts. Define

$$\begin{aligned}I &:= \int \sec^3(\theta)\, d\theta\\ &= \int \sec(\theta)\cdot\left(1+\tan^2(x)\right)\, d\theta \ \text{(trig identity, } 1+\tan^2(\theta) = \sec^2(\theta)\text{)}\\ &= \int \sec(\theta)\, d\theta + \int \sec(\theta)\cdot\tan^2(x)\, d\theta \ \text{(algebra)}\\ &= \ln|\sec(\theta)+\tan(\theta)| + \int \sec(\theta)\cdot\tan^2(x)\, d\theta \ \text{(first integral is from Example 7.19).}\end{aligned} \tag{7.17}$$

The remaining integral in (7.17) requires integration by parts. Define

$$
\begin{aligned}
u &= \tan(\theta) \implies du = \sec^2(\theta)\, d\theta, \\
v &= \sec(\theta) \implies dv = \sec(\theta) \cdot \tan(\theta)\, d\theta,
\end{aligned}
$$

yielding

$$
\begin{aligned}
\int \underbrace{\sec(\theta)\tan^2(\theta)\, d\theta}_{u\,dv} &= \underbrace{\tan(\theta)\sec(\theta)}_{u\cdot v} - \int \underbrace{\sec(\theta)\sec^2(\theta)\, d\theta}_{v\,du} \\
&= \tan(\theta)\sec(\theta) - \int \sec^3(\theta)\, d\theta \\
&= \tan(\theta)\sec(\theta) - I.
\end{aligned}
\tag{7.18}
$$

Substituting back into (7.17), we have:

$$
\begin{aligned}
I := \int \sec^3(\theta)\, d\theta &= \ln|\sec(\theta) + \tan(\theta)| + \int \sec(\theta) \cdot \tan^2(x)\, d\theta \\
&= \ln|\sec(\theta) + \tan(\theta)| + \tan(\theta)\sec(\theta) - I.
\end{aligned}
\tag{7.19}
$$

Solving for I yields

$$
\begin{aligned}
2I &= \ln|\sec(\theta) + \tan(\theta)| + \tan(\theta)\sec(\theta) \\
&\Downarrow \\
I &= \frac{1}{2}\ln|\sec(\theta) + \tan(\theta)| + \frac{1}{2}\tan(\theta)\sec(\theta).
\end{aligned}
\tag{7.20}
$$

Putting the (numerous) steps together yields

$$
\int \sqrt{a^2 + x^2}\, dx = a^2 \int \sec^3(\theta)\, d\theta = a^2 I = \frac{a^2}{2}\ln|\sec(\theta) + \tan(\theta)| + \frac{a^2}{2}\tan(\theta)\sec(\theta).
$$

To return to an answer in x , we use the trigonometric substitutions

$$
\begin{aligned}
\tan(\theta) &= \frac{x}{|a|}\,, \\
\sec(\theta) &= \sqrt{1 + \tan^2(\theta)} = \sqrt{1 + \left(\frac{x}{|a|}\right)^2}.
\end{aligned}
$$

The final antiderivative in terms of x is

$$
\begin{aligned}
\int \sqrt{a^2 + x^2}\, dx &= \frac{a^2}{2}\ln\left|\sqrt{1 + \left(\frac{x}{|a|}\right)^2} + \frac{x}{|a|}\right| + \frac{a^2}{2}\frac{x}{|a|} \cdot \sqrt{1 + \left(\frac{x}{|a|}\right)^2} + C \\
&= \frac{a^2}{2}\ln\left(\sqrt{1 + \left(\frac{x}{|a|}\right)^2} + \frac{x}{|a|}\right) + \frac{x}{2} \cdot \sqrt{a^2 + x^2} + C,
\end{aligned}
$$

where C is the constant of integration. The absolute value signs in the natural logarithm were removed because

$$
\sqrt{1 + \left(\frac{x}{|a|}\right)^2} + \frac{x}{|a|} > 0
$$

for all $x \in \mathbb{R}$.

It is traditional to simplify further the answer by taking advantage of the arbitrary constant of integration, C.

$$\ln\left(\sqrt{1+\left(\frac{x}{|a|}\right)^2}+\frac{x}{|a|}\right)=\ln\left(\frac{1}{|a|}\cdot\left(\sqrt{a^2+x^2}+x\right)\right)$$

$$=\ln\left(\frac{1}{|a|}\right)+\ln\left(\sqrt{a^2+x^2}+x\right).$$

Absorbing the constant $\ln\left(\frac{1}{|a|}\right)$ into C gives a much nicer final answer,

For $x \in \mathbb{R}$, $\int \sqrt{a^2+x^2}\,dx = \frac{a^2}{2}\ln\left(x+\sqrt{a^2+x^2}\right)+\frac{x}{2}\cdot\sqrt{a^2+x^2}+C.$

■

Note: The answer looks so innocent. Why was it so hard to compute?

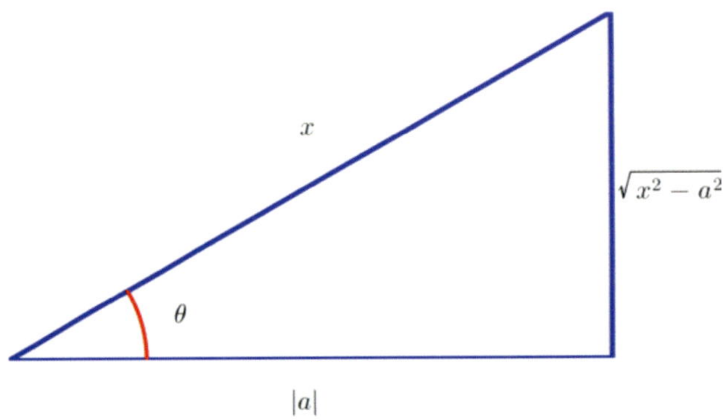

Note that the above triangle satisfies the Pythagorean Theorem, and hence it is a right triangle. It follows that $\cos(\theta)=\frac{|a|}{x}$, or, $x=\frac{|a|}{\cos(\theta)}=|a|\sec(\theta)$, which motivates a common change of variables for the next problem. There is also another viable change of variable that we wish to illustrate. **Hyperbolic Trig Functions** are often used in integrals that have terms of the form x^2-a^2. This is because of identities such as, $\cosh^2(x)-\sinh^2(x)=1$, which can also be written as $\cosh^2(x)-1=\sinh^2(x)$. This next example follows a similar pattern to the previous examples, ensuring that the domain of the integrand is carefully delineated and each change of variables respects the domain in which the integrand is defined.

Example 7.35. *Find an antiderivative for* $\sqrt{x^2-a^2}$.

Solution: Ans. For $|x|\geq|a|$, $\int\sqrt{x^2-a^2}\,dx=\frac{x}{2}\sqrt{x^2-a^2}-\text{sign}(x)\cdot\frac{a^2}{2}\ln\left(|x|+\sqrt{x^2-a^2}\right)+C$, where

$$\text{sign}(x)=\begin{cases}+1 & x\geq 0,\\ -1 & x<0.\end{cases}$$

First Approach: Standard Trig Substitution: The integrand of $\int\sqrt{x^2-a^2}\,dx$ is defined and continuous wherever $x^2-a^2\geq 0$; that is, $|x|\geq|a|$. We can therefore develop an antiderivative for $x\geq|a|$, and a second one for $x\leq-|a|$. Here, we will develop a solution for $x\leq-|a|$ using standard trig functions, and with the hyperbolic trig functions, we will develop the solution for $x\geq|a|$.

For $x\leq-|a|$, we choose the change of variable $x=-|a|\sec(\theta)$, which gives $dx=-|a|\sec(\theta)\cdot\tan(\theta)\,d\theta$. For all $0\leq\theta<\frac{\pi}{2}$, $1\leq\sec(\theta)<\infty$, and hence the change of variables respects the domain of definition of the integrand. Moreover, for later use, we note that for all $0 \le \theta < \frac{\pi}{2}$, $\tan(\theta) \ge 0$.

Using the change of variable, the integral becomes

$$\begin{aligned}
\int \sqrt{x^2 - a^2}\, dx &= \int \sqrt{a^2 \sec^2(\theta) - a^2} \cdot (-1) \cdot |a| \sec(\theta) \cdot \tan(\theta)\, d\theta \\
&= -a^2 \int \sqrt{\tan^2(\theta)} \cdot \sec(\theta) \cdot \tan(\theta)\, d\theta \quad (\sec^2(\theta) - 1 = \tan^2(\theta)) \\
&= -a^2 \int |\tan(\theta)| \cdot \sec(\theta) \cdot \tan(\theta)\, d\theta \quad (\tan(\theta) \ge 0) \\
&= -a^2 \int \tan^2(\theta) \cdot \sec(\theta)\, d\theta \quad (\tan^2(\theta) = \sec^2(\theta) - 1) \\
&= -a^2 \int \left(\sec^2(\theta) - 1\right) \cdot \sec(\theta)\, d\theta \\
&= -a^2 \int \sec^3(\theta)\, d\theta + a^2 \int \sec(\theta)\, d\theta.
\end{aligned}$$

Fortunately for us, we have already computed antiderivatives for both of these integrands,

$$\begin{aligned}
\int \sec(\theta)\, d\theta &= \ln|\sec(\theta) + \tan(\theta)| + C, \\
\int \sec^3(\theta)\, d\theta &= \frac{1}{2} \ln|\sec(\theta) + \tan(\theta)| + \frac{1}{2} \tan(\theta) \sec(\theta) + C.
\end{aligned}$$

Hence,

$$\begin{aligned}
\int \sqrt{x^2 - a^2}\, dx &= -\frac{a^2}{2} \ln|\sec(\theta) + \tan(\theta)| - \frac{a^2}{2} \tan(\theta) \sec(\theta) + a^2 \ln|\sec(\theta) + \tan(\theta)| + C \\
&= \frac{a^2}{2} \ln|\sec(\theta) + \tan(\theta)| - \frac{a^2}{2} \tan(\theta) \sec(\theta) + C.
\end{aligned}$$

From $x = -|a| \sec(\theta)$, we have that $\sec(\theta) = -\frac{x}{|a|}$. From the trig diagram for the problem, we have

$$\tan(\theta) = \frac{\sqrt{x^2 - a^2}}{|a|}.$$

Substituting in, we have

$$\begin{aligned}
\int \sqrt{x^2 - a^2}\, dx &= \frac{a^2}{2} \ln\left| -\frac{x}{|a|} + \frac{\sqrt{x^2 - a^2}}{|a|} \right| - \frac{a^2}{2} \frac{\sqrt{x^2 - a^2}}{|a|} \cdot \left(\frac{-x}{|a|}\right) + C \\
&= \frac{a^2}{2} \ln|-x + \sqrt{x^2 - a^2}| - \ln(|a|) + \frac{x}{2}\sqrt{x^2 - a^2} + C \\
&= \frac{a^2}{2} \ln|-x + \sqrt{x^2 - a^2}| + \frac{x}{2}\sqrt{x^2 - a^2} + C,
\end{aligned}$$

where in the last line we absorbed $\ln(|a|)$ into the constant of integration, C.

Second Approach: Hyperbolic Trig Functions: The integrand of $\int \sqrt{x^2 - a^2}\, dx$ is defined and continuous wherever $x^2 - a^2 \ge 0$; that is, $|x| \ge |a|$. We can therefore develop an antiderivative for $x \ge |a|$, and a second one for $x \le -|a|$.

For $x \ge |a|$, we choose the hyperbolic substitution $x = |a| \cosh(t)$, which gives $dx = |a| \sinh(t)\, dt$. For all $t \ge 0$, $\cosh(t) \ge 1$, and hence the change of variable respects the domain of definition of the integrand. Moreover, for later use, we note that for all $t \ge 0$, $\sinh(t) \ge 0$.

Using the change of variable, the integral becomes

$$\begin{aligned}\int \sqrt{x^2 - a^2}\, dx &= \int \sqrt{a^2 \cosh^2(t) - a^2} \cdot |a| \sinh(t)\, dt \\ &= a^2 \int \sqrt{\cosh^2(t) - 1} \cdot \sinh(t)\, dt \\ &= a^2 \int \sqrt{\sinh^2(t)} \cdot \sinh(t)\, dt \\ &= a^2 \int |\sinh(t)| \cdot \sinh(t)\, dt \\ &= a^2 \int \sinh^2(t)\, dt,\end{aligned}$$

where we used $\sinh(t) \geq 0$.
Using the identity $\sinh^2(t) = \frac{\cosh(2t)-1}{2}$, the integral simplifies to

$$\begin{aligned}a^2 \int \sinh^2(t)\, dt &= \frac{a^2}{2} \int (\cosh(2t) - 1)\, dt \\ &= \frac{a^2}{2}\left(\frac{\sinh(2t)}{2} - t\right) + C \\ &= \frac{a^2}{4}\sinh(2t) - \frac{a^2}{2} t + C.\end{aligned}$$

To return to x, we use $\cosh(t) = \frac{x}{|a|}$ and $\sinh(t) = \sqrt{\cosh^2(t) - 1}$. Thus

$$\sinh(2t) = 2\cosh(t)\sinh(t) = 2\frac{x}{|a|} \cdot \sqrt{\left(\frac{x}{|a|}\right)^2 - 1},$$

$$t = \operatorname{acosh}\left(\frac{x}{|a|}\right).$$

The final antiderivative in terms of x is

$$\frac{a^2}{4} \cdot 2\frac{x}{|a|}\sqrt{\left(\frac{x}{|a|}\right)^2 - 1} - \frac{a^2}{2}\operatorname{acosh}\left(\frac{x}{|a|}\right) + C,$$

which simplifies to

$$\frac{x}{2}\sqrt{x^2 - a^2} - \frac{a^2}{2}\operatorname{acosh}\left(\frac{x}{|a|}\right) + C.$$

From Table 1.9 on the inverse hyperbolic trig functions,

$$\begin{aligned}\operatorname{acosh}\left(\frac{x}{|a|}\right) &= \ln\left(\frac{x}{|a|} + \sqrt{\frac{x^2}{a^2} - 1}\right) \\ &= \ln\left(\frac{1}{|a|}\left(x + \sqrt{x^2 - a^2}\right)\right) \\ &= \ln\left(\frac{1}{|a|}\right) + \ln\left(x + \sqrt{x^2 - a^2}\right).\end{aligned}$$

For $\geq |a|$, absorbing the term $\ln\left(\frac{1}{|a|}\right)$ into the constant of integration gives:

$$\int \sqrt{x^2 - a^2}\, dx = \frac{x}{2}\sqrt{x^2 - a^2} - \frac{a^2}{2}\ln\left(x + \sqrt{x^2 - a^2}\right) + C.$$

The answers for $x \geq |a|$ and $x \leq -|a|$ can be combined into a single formula:

$$\text{For } |x| \geq |a|, \int \sqrt{x^2 - a^2}\, dx = \frac{x}{2}\sqrt{x^2 - a^2} - \text{sign}(x) \cdot \frac{a^2}{2} \ln\left(|x| + \sqrt{x^2 - a^2}\right) + C.$$

■

Recommended Videos for Additional Practice

- Trig Substitution... How? by Nancy Pi.
- Trig substitution - How to solve? by Krista King.
- How to solve EVERY trigonometric substitution problem ever! by Krista King.
- Trigonometric Substitution by the Organic Chemistry Tutor.
- Calculus 2: Integration - Trig Substitution (2 of 28) Applicable Integrals by Prof. Michel van Biezen. It's highly recommended that you take a peek to see an impressive list of integrals that can be solved with the methods we just developed.
- Playlist for Trig Substitutions by **BlackPenRedPen**. You'll have to find your favorite one.
- Introduction to trigonometric substitution by Khan Academy.

We repeat: Trig substitutions are the "hardest to master" of all the methods for finding antiderivatives; they require great care if you want to avoid nonsense, even when using "tables." Moreover, when using ChatGPT or Wolfram AlphaPro, it is still on you to understand the domain on which the antiderivative is defined. You cannot just throw in a function, get an answer, snap your fingers, and say you are done! To understand the domain of the antiderivative, analyze that of the integrand (aka the function you are integrating).

7.4 Coming Up for Air and Taking Stock

Now is a good time to re-read Section 7.2.3, **Conflating Integration and Antiderivatives Is a Pedagogical Pitfall in Calculus**. It is rare in engineering practice that a function to be integrated is simple enough for you to find an antiderivative by hand, even after spending another year learning methods more advanced than those we have presented. And in the rare instances when you can find an antiderivative by hand, ChatGPT or Wolfram Alpha Pro can almost surely do it too, and not make a sign error in the process.

In regards to integration, we recommend that you:

1. Appreciate that integration and differentiation are inverse operations (up to a constant of integration). The Fundamental Theorems of Calculus are amazing.
2. Use numerical integration tools whenever possible. At the very least, check your analytically derived integrals using computer-based methods by either symbolically differentiating the antiderivative to make sure it gives the original function, or by using numerical integration to check the integral.
3. Use ChatGPT or Wolfram Alpha Pro when you absolutely must determine the antiderivative of a function that is neither in a standard integral table nor among the six or eight that are worth memorizing.
4. Read Secrets of the Arcane 7.36 to learn why antiderivatives limit the ability to compute definite integrals.

Secrets of the Arcane 7.36: Elementary Functions and Antiderivatives

In mathematics, when we talk about **elementary functions**, we are referring to functions that are constructed using basic arithmetic operations (addition, subtraction, multiplication, and division) and compositions of:

1. **Polynomials**: Functions like $f(x) = x^2$, $g(x) = 3x^3 - 2x + 1$.
2. **Rational Functions**: Ratios of two polynomials, e.g., $R(x) = \frac{x^2+1}{x-2}$.
3. **Root Functions**: Functions involving square roots, cube roots, etc., like $h(x) = \sqrt{x}$.
4. **Trigonometric Functions**: The standard sine, cosine, tangent, etc., and their inverses.
5. **Exponential and Logarithmic Functions**: Functions like e^x, $\ln(x)$.

These functions are termed "elementary" because they are among the first functions introduced in mathematics education and are foundational in calculus and analysis. They are well-understood and have known properties, such as **the derivative of an elementary function is also elementary**. When you combine elementary functions using operations like addition, multiplication, or composition, you get a wide variety of functions that can model many different phenomena, but they still retain the "elementary" label because they are built from these basic components.

Most elementary functions do not have antiderivatives that are also elementary functions.[a] Even the very common function, e^{-x^2}, used in Radial Basis Functions and Gaussian Probability Distributions, does not have an antiderivative that can be expressed in terms of anything we would recognize. The same is true for the antiderivative of $\cos(x^2)$ in Example 7.3-(d). In these cases, the antiderivative is literally defined through the First Fundamental Theorem of Calculus; that is,

$$\text{the antiderivative of } f(x) \text{ is } F(x) = \int_a^x f(t)\,dt + C,$$

which is the integral you were trying to compute in the first place. Fortunately, for functions that arise sufficiently often, numerical analysts seek fast and accurate means to compute the new antiderivative. If you are lucky, your antiderivative may be one of the non-elementary antiderivatives that have been tabulated. Are you ready to roll the dice?

The *coup de grâce* in all of this: The hand methods for determining antiderivatives allow you to find antiderivatives that are already available in what used to be called **Integral Tables**. You can simply look up your antiderivative. Back in the day, engineering students purchased books with thousands of antiderivatives in them. Today, no one needs to do that. We have software tools that provide access to more integrals than all of the books we students of yore used to lug around.

Further Reading: Non-elementary Integrals; **Proving the Nonexistence of Elementary Antiderivatives for Certain Elementary Functions**; **Impossibility Theorems for Elementary Integration**; How WolframAlpha defines these nonelementary integrals; and **Closure of the Set of Elementary Functions**.

[a] In plain words, seeking an antiderivative that you can express in understandable terms can be a fool's errand.

7.5 Software Tools for Finding Antiderivatives

7.5.1 SymPy in Julia

In the opinion of your author, `SymPy` beats `Symbolics` when it comes to finding antiderivatives, also known as "integration". We illustrate several use cases here.

u-Substitution: Example 7.13

```
using SymPy

# Define symbolic variable
x = symbols("x")

# Define your integrand here
integrand = 3cos(x^2+1)
antiderivative = integrate(integrand, x)

# Print the problem and results
println("Integrand = ", integrand, "\n")
println("Antiderivative = ",antiderivative, "\n")
```

Output

```
Integrand = 3*cos(x^2 + 1)

Antiderivative = 3*sqrt(2)*sqrt(pi)*(cos(1)*fresnelc(sqrt(2)*x/sqrt(pi))
- sin(1)*fresnels(sqrt(2)*x/sqrt(pi)))/2
```

Integration by Parts: The "Tricky" Secant Integrand from Example 7.19

```
using SymPy
# Define symbolic variable
x = symbols("x")

# Define your integrand here
integrand = 1/cos(x)
antiderivative = integrate(integrand, x)

# Print the problem and results
println("Integrand = ", integrand, "\n")
println("Antiderivative = ",antiderivative, "\n")
```

Output

```
Integrand = 1/cos(x)

Antiderivative = -log(sin(x) - 1)/2 + log(sin(x) + 1)/2
```

Integration by Parts Example 7.22

```
using SymPy
# Define symbolic variable
x = symbols("x")

# Define your integrand here
integrand = x*sin(x)
antiderivative = integrate(integrand, x)
```

```

# Print the problem and results
println("Integrand = ", integrand, "\n")
println("Antiderivative = ",antiderivative, "\n")
```

Output

```
Integrand = x*sin(x)

Antiderivative = -x*cos(x) + sin(x)
```

Rational Functions and PFEs Example 7.31

```
using SymPy

# Define the variable
x = symbols("x")

P = 2*x^3 - x + 4
Q = x^6 - 4*x^4 + 10*x^3 - 11*x^2 + 10*x - 6

R = P/Q

# Define your integrand here
integrand = R
antiderivative = integrate(integrand, x)

# Print the problem and results
println("Integrand = ", integrand, "\n")
println("Antiderivative = ",antiderivative, "\n")
```

Output

```
Integrand = (2*x^3 - x + 4)/(x^6 - 4*x^4 + 10*x^3 - 11*x^2 + 10*x - 6)

Antiderivative = 5*log(x - 1)/8 + 47*log(x + 3)/680 - log(x^2 + 1)/5
- 5*log(x^2 - 2*x + 2)/34 - 3*atan(x)/10 + 3*atan(x - 1)/17
```

Trig Substituions for Radicals Example 7.33

```
using SymPy

# Define the variable
x= symbols("x", real=true)

# Define your integrand here
a=pi
integrand =sqrt(a^2 - x^2)
antiderivative = integrate(integrand, x)

# Print the problem and results
println("Integrand = ", integrand, "\n")
```

```
println ("Antiderivative = ", antiderivative, "\n")
```

Output

```
Integrand = 3.14159265358979*sqrt(1 - 0.101321183642338*x^2)

Antiderivative =
3.14159265358979*Piecewise((0.5*I*x*sqrt(0.101321183642338*x^2 - 1)
- 1.5707963267949*I*x*acosh(0.318309886183791*Abs(x))/Abs(x)
+ 0.785398163397448*pi*x/Abs(x), 0.101321183642338*x^2 > 1),
(-0.0506605918211689*x^3/sqrt(1 - 0.101321183642338*x^2)
+ 1.5707963267949*x*asin(0.318309886183791*Abs(x))/Abs(x)
+ 0.5*x/sqrt(1 - 0.101321183642338*x^2), True))
```

Or "semi-nicely" typeset,

$$3.14159 \cdot \begin{cases} 0.5 \cdot \mathbf{i} \cdot x \cdot \sqrt{0.101321 \cdot x^2 - 1} - \frac{1.570796 \cdot \mathbf{i} \cdot x \cdot \operatorname{arccosh}(0.318310 \cdot \operatorname{Abs}(x))}{\operatorname{Abs}(x)} \\ \quad + \frac{0.785398 \cdot \pi \cdot x}{\operatorname{Abs}(x)} & x^2 > 9.869604401089337, \\ \frac{-0.050661 \cdot x^3}{\sqrt{1 - 0.101321 \cdot x^2}} + \frac{1.570796 \cdot x \cdot \arcsin(0.318310 \cdot \operatorname{Abs}(x))}{\operatorname{Abs}(x))} \\ \quad + \frac{0.5 \cdot x}{\sqrt{1.000000 - 0.101321 \cdot x^2}} & \text{otherwise.} \end{cases}$$

`SymPy` has given us the solution for $x^2 > \pi^2$, which makes the integrand complex-valued, as well as the answer we computed for a real integrand where $\pi^2 - x^2 \geq 0$. We will let you dig into the intricacies of `SymPy` to avoid computing the complex answer in addition to the real answer you likely would want, because, as you will see in the next subsection, Wolfram Alpha solves the problem with supreme elegance.

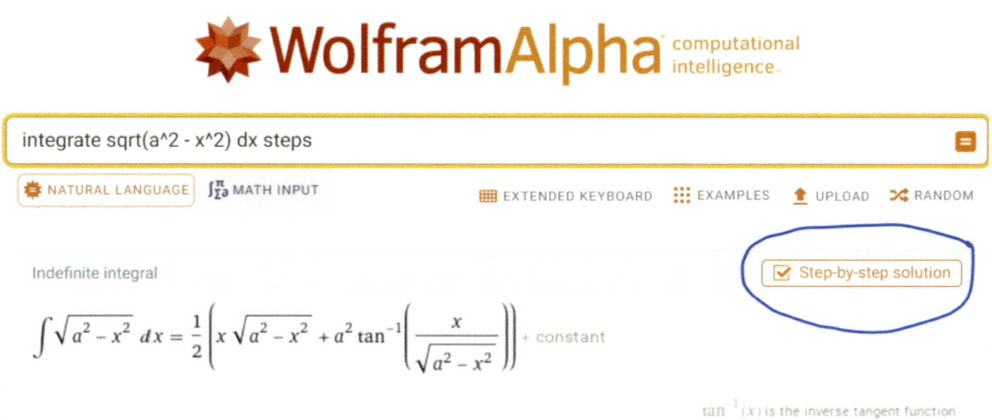

Figure 7.1: The Wolfram Alpha website provides more information than ChatGPT. See the circled box, "Step-by-step solution".

7.5.2 More on Wolfram Alpha

Additional features can be accessed through the Wolfram Alpha Website, such as step-by-step solutions. At the present time, the step-by-step solution is a paid feature, with student pricing available. Here is an example of the level of detail provided by Wolfram Alpha's Step-by-Step Solutions feature.

Indefinite Integral of $\sqrt{a^2 - x^2}$

STEP 1
Take the integral:

$$\int \sqrt{a^2 - x^2}\, dx$$

STEP 2
For the integrand $\sqrt{a^2 - x^2}$, (assuming all variables are positive) substitute $x = a\sin(u)$ and $dx = a\cos(u)\, du$. Then $\sqrt{a^2 - x^2} = \sqrt{a^2 - a^2\sin^2(u)} = a\cos(u)$ and $u = \sin^{-1}\left(\frac{x}{a}\right)$:

$$= a\int a\cos^2(u)\, du$$

STEP 3
Factor out constants:

$$= a^2\int \cos^2(u)\, du$$

STEP 4
Write $\cos^2(u)$ as $\frac{1}{2}\cos(2u) + \frac{1}{2}$:

$$= a^2\int \left(\frac{1}{2}\cos(2u) + \frac{1}{2}\right) du$$

STEP 5
Integrate the sum term by term and factor out constants:

$$= \frac{a^2}{2}\int \cos(2u)\, du + \frac{a^2}{2}\int 1\, du$$

STEP 6
For the integrand $\cos(2u)$, substitute $s = 2u$ and $ds = 2\, du$:

$$= \frac{a^2}{4}\int \cos(s)\, ds + \frac{a^2}{2}\int 1\, du$$

STEP 7
The integral of $\cos(s)$ is $\sin(s)$:

$$= \frac{a^2}{4}\sin(s) + \frac{a^2}{2}\int 1\, du$$

STEP 8
The integral of 1 is u:

$$= \frac{a^2}{4}\sin(s) + \frac{a^2 u}{2} + \text{constant}$$

STEP 9
Substitute back for $s = 2u$:

$$= \frac{a^2 u}{2} + \frac{a^2}{4}\sin(2u) + \text{constant}$$

STEP 10
Apply the double angle formula $\sin(2u) = 2\sin(u)\cos(u)$:

$$= \frac{a^2 u}{2} + \frac{a^2}{2}\sin(u)\cos(u) + \text{constant}$$

STEP 11
Express $\cos(u)$ in terms of $\sin(u)$ using $\cos^2(u) = 1 - \sin^2(u)$:

$$= \frac{a^2 u}{2} + \frac{a^2}{2}\sin(u)\sqrt{1 - \sin^2(u)} + \text{constant}$$

STEP 12
Substitute back for $u = \sin^{-1}\left(\frac{x}{a}\right)$:

$$= \frac{1}{2}x\sqrt{a^2 - x^2} + \frac{a^2}{2}\sin^{-1}\left(\frac{x}{a}\right) + \text{constant}$$

STEP 13
Factor the answer a different way:

$$= \frac{1}{2}\left(x\sqrt{a^2 - x^2} + a^2\sin^{-1}\left(\frac{x}{a}\right)\right) + \text{constant}$$

STEP 14
Which is equivalent for restricted x and a values to:

$$\text{Answer:} \quad \frac{1}{2}\left(x\sqrt{a^2 - x^2} + a^2\tan^{-1}\left(\frac{x}{\sqrt{a^2 - x^2}}\right)\right) + \text{constant}$$

Indefinite Integral of $x^2\sin^3(x)$

STEP 1
Take the integral:

$$\int x^2\sin^3(x)\,dx$$

STEP 2
For the integrand $x^2\sin^3(x)$, use the trigonometric identity $\sin^2(x) = \frac{1}{2}(1 - \cos(2x))$:

$$= \frac{1}{2}\int x^2\sin(x)(1 - \cos(2x))\,dx$$

STEP 3
Expanding the integrand $x^2\sin(x)(1 - \cos(2x))$ gives $x^2\sin(x) - x^2\sin(x)\cos(2x)$:

$$= \frac{1}{2}\int (x^2\sin(x) - x^2\sin(x)\cos(2x))\,dx$$

STEP 4
Integrate the sum term by term and factor out constants:

$$= -\frac{1}{2}\int x^2\sin(x)\cos(2x)\,dx + \frac{1}{2}\int x^2\sin(x)\,dx$$

STEP 5
Use the trigonometric identity $\sin(\alpha)\cos(\beta) = \frac{1}{2}(\sin(\alpha - \beta) + \sin(\alpha + \beta))$, where $\alpha = x$ and $\beta = 2x$:

$$= -\frac{1}{4}\int x^2(\sin(3x) - \sin(x))\,dx + \frac{1}{2}\int x^2\sin(x)\,dx$$

STEP 6
Expanding the integrand $x^2(\sin(3x) - \sin(x))$ gives $x^2\sin(3x) - x^2\sin(x)$:

$$= -\frac{1}{4}\int (x^2\sin(3x) - x^2\sin(x))\,dx + \frac{1}{2}\int x^2\sin(x)\,dx$$

STEP 7
Integrate the sum term by term and factor out constants:

$$= -\frac{1}{4}\int x^2\sin(3x)\,dx + \frac{3}{4}\int x^2\sin(x)\,dx$$

STEP 8
For the integrand $x^2\sin(3x)$, integrate by parts, $\int f\,dg = fg - \int g\,df$, where $f = x^2$, $dg = \sin(3x)\,dx$, $df = 2x\,dx$, $g = -\frac{1}{3}\cos(3x)$:

$$= \frac{1}{12}x^2\cos(3x) - \frac{1}{6}\int x\cos(3x)\,dx + \frac{3}{4}\int x^2\sin(x)\,dx$$

STEP 9
For the integrand $x\cos(3x)$, integrate by parts, $\int f\,dg = fg - \int g\,df$, where $f = x$, $dg = \cos(3x)\,dx$, $df = dx$, $g = \frac{1}{3}\sin(3x)$:

$$= \frac{1}{12}x^2\cos(3x) - \frac{1}{18}x\sin(3x) + \frac{1}{18}\int \sin(3x)\,dx + \frac{3}{4}\int x^2\sin(x)\,dx$$

STEP 10
For the integrand $\sin(3x)$, substitute $u = 3x$ and $du = 3dx$:

$$= \frac{1}{12}x^2\cos(3x) - \frac{1}{18}x\sin(3x) + \frac{1}{54}\int \sin(u)\,du + \frac{3}{4}\int x^2\sin(x)\,dx$$

STEP 11
The integral of $\sin(u)$ is $-\cos(u)$:

$$= -\frac{\cos(u)}{54} + \frac{1}{12}x^2\cos(3x) - \frac{1}{18}x\sin(3x) + \frac{3}{4}\int x^2\sin(x)\,dx$$

STEP 12
For the integrand $x^2\sin(x)$, integrate by parts, $\int f\,dg = fg - \int g\,df$, where $f = x^2$, $dg = \sin(x)\,dx$, $df = 2x\,dx$, $g = -\cos(x)$:

$$= -\frac{\cos(u)}{54} - \frac{3}{4}x^2\cos(x) + \frac{1}{12}x^2\cos(3x) - \frac{1}{18}x\sin(3x) + \frac{3}{2}\int x\cos(x)\,dx$$

STEP 13
For the integrand $x\cos(x)$, integrate by parts, $\int f\,dg = fg - \int g\,df$, where $f = x$, $dg = \cos(x)\,dx$, $df = dx$, $g = \sin(x)$:

$$= -\frac{\cos(u)}{54} - \frac{3}{4}x^2\cos(x) + \frac{1}{12}x^2\cos(3x) + \frac{3}{2}x\sin(x) - \frac{1}{18}x\sin(3x) - \frac{3}{2}\int \sin(x)\,dx$$

STEP 14
The integral of $\sin(x)$ is $-\cos(x)$:

$$= -\frac{\cos(u)}{54} - \frac{3}{4}x^2\cos(x) + \frac{1}{12}x^2\cos(3x) + \frac{3}{2}x\sin(x) - \frac{1}{18}x\sin(3x) + \frac{3\cos(x)}{2} + \text{constant}$$

STEP 15
Substitute back for $u = 3x$:

$$= -\frac{3}{4}x^2\cos(x) + \frac{1}{12}x^2\cos(3x) + \frac{3}{2}x\sin(x) - \frac{1}{18}x\sin(3x) + \frac{3\cos(x)}{2} - \frac{1}{54}\cos(3x) + \text{constant}$$

STEP 16
Which is equal to:

Answer: $$\frac{1}{108}(-81(x^2-2)\cos(x) + (9x^2-2)\cos(3x) - 6x(\sin(3x) - 27\sin(x))) + \text{constant}$$

7.6 (Optional Read) Proofs Associated with the Chapter

Proof of Prop. 7.7 (First Fundamental Theorem of Calculus)

Let $f : [a,b] \to \mathbb{R}$ be a continuous real-valued function defined on a closed interval $[a,b]$. Then, the function $F : [a,b] \to \mathbb{R}$ defined by

$$F(x) = \int_a^x f(t)\,dt \tag{7.21}$$

for all $x \in [a,b]$, is continuous on $[a,b]$, differentiable on the open interval (a,b), and $F'(x) = f(x)$ for all $x \in (a,b)$. You will also see it written as

$$\frac{d}{dx}F(x) = \frac{d}{dx}\left[\int_a^x f(t)\,dt\right] = f(x), \tag{7.22}$$

to emphasize that differentiation undoes, or inverses, integration.

Proof: We give a proof that is straightforward and fully rigorous for functions $f : [a,b] \to \mathbb{R}$ that are **nondecreasing**. The general case is proven in a standard course on Real Analysis.

Claim 1: Continuity of F: Because f is nondecreasing,[2]

$$M := \max_{x\in[a,b]} |f(x)| = \max\{|f(a)|, f(b)\} < \infty.$$

Let $\epsilon > 0$ and define $\delta := \frac{\epsilon}{M+1} > 0$. We will show for all $x, x_0 \in [a,b]$, that $|x - x_0| < \delta$ implies

$$|F(x) - F(x_0)| < \epsilon.$$

Proof: Consider the change in F as x varies from x_0,

$$F(x) - F(x_0) = \int_a^x f(t)\,dt - \int_a^{x_0} f(t)\,dt = \int_{x_0}^x f(t)\,dt.$$

This equation holds by the properties of the definite integral, specifically the additivity over adjacent intervals. The absolute value of the integral is bounded by

$$|F(x) - F(x_0)| = \left|\int_{x_0}^x f(t)\,dt\right| \le \int_{x_0}^x |f(t)|\,dt.$$

Given that $|f(t)| \le M$ for all t in $[a,b]$, the integral simplifies further,

$$|F(x) - F(x_0)| \le \int_{x_0}^x M\,dt = M|x - x_0|.$$

The integral $\int_{x_0}^x M\,dt$ evaluates to $M \cdot |x - x_0|$ because the integrand M is constant, and the length of the interval from x_0 to x is $|x - x_0|$.

Now, by the choice of δ, if $|x - x_0| < \delta$ then:

$$|F(x) - F(x_0)| \le M|x - x_0| < M\delta = M\left(\frac{\epsilon}{M+1}\right) < \epsilon.$$

Thus, we have shown that:

$$|F(x) - F(x_0)| < \epsilon,$$

for all x such that $|x - x_0| < \delta$. Therefore, F is continuous at x_0, and since x_0 was arbitrary, F is continuous on $[a,b]$.

[2]If $f(b) \ge 0$, then no absolute value sign is needed on $f(b)$. If $f(b) < 0$, then $f(a) \le f(b) \implies |f(a)| \ge |f(b)| \ge f(b)$. Hence, once again, no absolute value sign is needed.

Claim 2: $F'(x) = f(x)$ for all $a < x < b$:

Proof: Consider the difference quotient for F at some point x in (a,b),

$$\begin{aligned}\frac{F(x+h)-F(x)}{h} &= \frac{1}{h}\left(\int_a^{x+h} f(t)\,dt - \int_a^x f(t)\,dt\right)\\ &= \frac{1}{h}\int_x^{x+h} f(t)\,dt.\end{aligned}$$

Assume first that $h > 0$. Then, using once again that f is nondecreasing,

$$f(x)\cdot h \le \int_x^{x+h} f(t)\,dt \le f(x+h)\cdot h.$$

Plugging into the previous equation and canceling the h's, we have

$$f(x) \le \frac{1}{h}\int_x^{x+h} f(t)\,dt \le f(x+h).$$

Because f is continuous, $\lim_{h\to 0^+} f(x+h) = f(x)$. Therefore, by the Squeeze Theorem,

$$\lim_{h\to 0^+} \frac{1}{h}\int_x^{x+h} f(t)\,dt = f(x).$$

The same argument can be repeated with minor variations to show that

$$\lim_{h\to 0^-} \frac{1}{h}\int_x^{x+h} f(t)\,dt = f(x).$$

This completes the proof that $F'(x) = f(x)$, which is a fundamental connection between the differential and integral calculus for continuous functions.

■

Proof of Prop. 7.7 (Second Fundamental Theorem of Calculus)

Let $f : [a,b] \to \mathbb{R}$ be a real-valued function defined on a closed interval $[a,b]$ that admits an antiderivative F on $[a,b]$, that is, $F : [a,b] \to \mathbb{R}$ is differentiable on (a,b) and for all $x \in (a,b)$, $F'(x) = f(x)$. Then, f is integrable on $[a,b]$ and

$$\int_a^b f(x)\,dx = F(x)\Big|_a^b := F(b) - F(a). \tag{7.23}$$

We rewrite this as

$$\int_a^b F'(x)\,dx = F(x)\Big|_a^b := F(b) - F(a), \tag{7.24}$$

to underline that integration undoes, or inverses, differentiation.

Proof: We use a method due to D. Joyce, with one modification: we ASSUME that $f : [a,b] \to \mathbb{R}$ is continuous, thereby guaranteeing that $f(x) = F'(x)$ is integrable.

Define

$$G(x) := \int_a^x F'(t)\,dt, \text{ for } x \in [a,b].$$

Then, by the First Fundamental Theorem of Calculus, $G'(x) = F'(x)$. Therefore, G and F differ by a constant C, that is, $G(x) - F(x) = C$ for all $x \in [a,b]$. But

$$G(a) = \int_a^a F'(t)\,dt = 0,$$

and, therefore, $G(a) - F(a) = C$, so $\boxed{C = -F(a)}$.

It follows that $\boxed{G(x) - F(x) = -F(a)}$ for all $x \in [a,b]$. In particular, $G(b) - F(b) = -F(a)$, so $G(b) = F(b) - F(a)$; that is,

$$\int_a^b F'(t)\,dt = F(b) - F(a).$$

This completes the proof that the integral of a function over an interval is equal to the difference in the values of any of its antiderivatives evaluated at the endpoints of the interval, affirming that integration can be seen as the inverse operation to differentiation. ■

We next explore how restrictive it was to assume that a function is nondecreasing.

1) Property of max and min functions: For any $y \in \mathbb{R}$, the functions $\max(y,0)$ and $\min(y,0)$ have the property that:

$$\max(y,0) + \min(y,0) = y.$$

Proof: Consider two cases based on the value of y:

Case 1: $y \geq 0$. In this case, we have

$$\max(y,0) = y \quad \text{and} \quad \min(y,0) = 0.$$

Thus,

$$\max(y,0) + \min(y,0) = y + 0 = y.$$

Case 2: $y < 0$. In this case, we have

$$\max(y,0) = 0 \quad \text{and} \quad \min(y,0) = y.$$

Thus,

$$\max(y,0) + \min(y,0) = 0 + y = y.$$

In both cases, $\max(y,0) + \min(y,0)$ simplifies to y, proving the statement. ■

2) Decomposition of continuously differentiable functions: If $f : [a,b] \to \mathbb{R}$ is a continuously differentiable function, then f can be expressed as the difference of two continuous, nondecreasing functions.

Construction: Define two functions g and k over $[a,b]$ by:

$$g(x) = f(a) + \int_a^x \max(f'(t),0)\,dt$$

and

$$k(x) = -\int_a^x \min(f'(t),0)\,dt.$$

Properties: Since f is continuously differentiable, f' exists and is continuous. The function g is nondecreasing because the integrand, $\max(f'(t),0)$, is nonnegative and strictly positive whenever $f'(t) > 0$.

The function k is nondecreasing because the integrand $-\min(f'(t),0)$ is nonnegative, being zero when $f'(t) \geq 0$ and positive when $f'(t) < 0$.

Expressing f as a difference:

$$\begin{aligned} g(x) - k(x) &= f(a) + \left(\int_a^x \max(f'(t), 0)\, dt\right) - \left(-\int_a^x \min(f'(t), 0)\, dt\right) \\ &= f(a) + \int_a^x (\max(f'(t), 0) + \min(f'(t), 0))\, dt \\ &= f(a) + \int_a^x f'(t)\, dt \\ &= f(a) + (f(x) - f(a)) \\ &= f(x). \end{aligned}$$

This construction expresses f as the difference of two continuous, nondecreasing functions over the interval $[a, b]$. Hence, the proof of Prop. 7.7 given for nondecreasing functions can be applied to functions that are continuously differentiable. ■

Chapter 8

Improper Integrals

Learning Objectives

By the end of this chapter, students should be able to:

- Define improper integrals and understand their significance within the context of calculus.
- Identify integrals that may pose difficulties due to their unbounded nature or the behavior of the function being integrated.
- Answer the pressing question: can an integral be doubly improper?
- Explore the applications of improper integrals, emphasizing their practical importance in statistics and probability.

Outcomes

Upon successful completion of this chapter, students will:

- Master the technique of integrating over unbounded domains using limits.
- Learn to manage and integrate functions with singularities or discontinuities by applying limits to circumvent infinite behavior at finite points, thus handling vertical asymptotes effectively.
- Gain a comprehensive understanding of the analytical methods for solving improper integrals, employing antiderivatives to facilitate calculation.
- Acquire skills in applying numerical methods, specifically Julia's `QuadGK`, for evaluating improper integrals, particularly when analytical solutions are challenging to obtain.
- Delve into two probability theory examples that utilize improper integrals, reinforcing the concept's application in real-world scenarios and theoretical studies.

8.1 Introduction

Improper? The term "improper integral" serves as a descriptive label to indicate that the integral in question deviates from the standard conditions that apply to definite integrals, and hence, extra caution is advised. The word "improper" in this context does not imply that there is something wrong or less valid about the integrals we examine in this Chapter. Rather, it indicates that they do not fit within the framework of standard definite integrals, and therefore they require special techniques for their evaluation. These techniques involve computing limits, as the interval of integration or the function itself is unbounded. The terminology likely originated to help mathematicians categorize and study different types of integrals, especially as the field of real analysis began to formalize concepts related to integration. **The term helps to signal that additional care or "alternative methods" may be needed to evaluate the integral.**

8.2 Type-I Improper Integrals: Unbounded Limits of Integration

We first look at cases where the integrand, f, is piecewise continuous, and hence, for any closed and bounded interval $[a,b]$, the definite integral $\int_a^b f(x)\,dx$ exists and is finite. We are interested in what happens as one or more of the limits of integration is unbounded, that is, $\int_a^\infty f(x)\,dx$, $\int_{-\infty}^b f(x)\,dx$, and $\int_{-\infty}^\infty f(x)\,dx$. To help us understand what the issues are, let us look at a few examples first.

- Let $f:[0,\infty)\to\mathbb{R}$ by $f(x)=e^{-x}$. Then, $\int_0^b f(x)\,dx = -e^{-x}\,|_0^b = 1-e^{-b}$. Hence, $\lim\limits_{b\to\infty}\displaystyle\int_0^b f(x)\,dx = \lim\limits_{b\to\infty}\left(1-e^{-b}\right) = 1.0$. Because the value of the integral is finite, we say it is **convergent**.

- Let $f:[0,\infty)\to\mathbb{R}$ by $f(x)=1$. Then, $\int_0^b f(x)\,dx = b$. Hence, $\lim\limits_{b\to\infty}\displaystyle\int_0^b f(x)\,dx=\infty$. Because the value of the integral is unbounded, we say it is **divergent**.

- Let $f:(-\infty,0]\to\mathbb{R}$ by $f(x)=\cos(x)$. Then, $\int_a^0 f(x)\,dx = \sin(x)\Big|_a^0 = -\sin(a)$. Hence,

  $$\lim_{a\to-\infty}\int_a^0 f(x)\,dx = \lim_{a\to-\infty} -\sin(a)$$

  is **undefined**.

The following definitions are based on the previous three examples.

Unbounded Upper or Lower Limit: Take 1

Definition 8.1. *Suppose that $f:[a,\infty)\to\mathbb{R}$ is a real-valued function, and for all $b>a$ (finite), $\int_a^b f(x)\,dx$ exists and is finite.*[a] *Then,*

$$\int_a^\infty f(x)\,dx := \lim_{b\to\infty}\int_a^b f(x)\,dx, \tag{8.1}$$

if the limit exists. The integral is ***undefined*** *otherwise. If the limit exists but is unbounded (i.e., equals $\pm\infty$), then we say the integral is* ***divergent****, while if the limit exists and is finite, we say the integral is* ***convergent****.*

Similarly, suppose that $f:(-\infty,b]\to\mathbb{R}$ is a real-valued function, and for all $a<b$ (finite), $\int_a^b f(x)\,dx$ exists and is finite.[b] *Then,*

$$\int_{-\infty}^b f(x)\,dx := \lim_{a\to-\infty}\int_a^b f(x)\,dx, \tag{8.2}$$

if the limit exists. The integral is undefined otherwise. If the limit exists but is unbounded, (i.e., equals $\pm\infty$), then we say the integral is ***divergent****, while if the limit exists and is finite, we say the integral is*

convergent.

[a]If $f : [a, \infty) \to \mathbb{R}$ is piecewise continuous, then this condition is met.
[b]If $f : (-\infty, b] \to \mathbb{R}$ is piecewise continuous, then this condition is met.

Example 8.2. *Determine values, if possible, for the following improper integrals. You may wish to use the antiderivatives in Table. 7.2 and Table. 7.3 because the integrands are very standard functions.*

(a) $\int_0^\infty x\,dx$.

(b) $\int_0^\infty e^{-x}\,dx$.

(c) $\int_{-\infty}^0 e^{-x}\,dx$.

(d) $\int_{-\infty}^0 \sin(x)\,dx$.

Solutions:

(a) **Ans.** $\int_0^\infty x\,dx = \infty$.

From Table. 7.2, $\int_0^b x\,dx = \frac{1}{2}x^2\Big|_0^b = \frac{b^2}{2}$. Hence, $\int_0^\infty x\,dx := \lim_{b\to\infty}\int_0^b x\,dx = \lim_{b\to\infty}\frac{1}{2}b^2 = \infty$.

(b) **Ans.** $\int_0^\infty e^{-x}\,dx = 1$.

From Table. 7.3, $\int_0^b e^{-x}\,dx = -e^{-x}\Big|_0^b = -\left(e^{-b}-1\right) = 1 - e^{-b}$. Hence,

$$\int_0^\infty e^{-x}\,dx := \lim_{b\to\infty}\int_0^b e^{-x}\,dx = \lim_{b\to\infty} 1 - e^{-b} = 1.$$

We can also use quadGK:

```
using QuadGK

f(x) = exp(-x)
Data = Array{Float64}(undef, 0, 3)
#
for b=0:2:30
    value, error = quadgk(f, 0, b)
    Data = [Data; b value error]
end
Data
```

Output

```
16×3 Matrix{Float64}:
  0.0  0.0       0.0
  2.0  0.864665  7.77156e-16
  4.0  0.981684  1.00392e-11
  6.0  0.997521  1.74719e-9
  8.0  0.999665  1.02231e-11
 10.0  0.999955  1.804e-10
 12.0  0.999994  1.75152e-9
 14.0  0.999999  1.12599e-8
```

```
16.0   1.0        2.81922e-11
18.0   1.0        7.20663e-11
20.0   1.0        2.09573e-10
22.0   1.0        6.26783e-10
24.0   1.0        1.77795e-9
26.0   1.0        4.65932e-9
28.0   1.0        1.1276e-8
30.0   1.0        2.97104e-11
```

The improper integral converges very quickly to one.

(c) **Ans.** $\int_{-\infty}^{0} e^{-x}\,dx = \infty$.

From Table. 7.3, $\int_a^0 e^{-x}\,dx = -e^{-x}\Big|_a^0 = -(1-e^{-a}) = e^{-a} - 1$. Hence,

$$\int_{-\infty}^{0} e^{-x}\,dx := \lim_{a\to-\infty}\int_a^0 e^{-x}\,dx = \lim_{a\to-\infty} e^{-a} - 1 = \infty.$$

(d) **Ans.** $\int_{-\infty}^{0} \sin(x)\,dx$ is undefined because the limit does not exist.

Intuition: We know that $\sin(x)$ oscillates between plus and minus one. Moreover, $\sin(x+\pi) = -\sin(x)$, and hence, the integral over any 2π-interval $[k\pi, (k+2)\pi]$ is equal to zero. Therefore, we do not expect the integral to converge. But we check it out anyway, just to confirm our intuition.

We can use antiderivatives for an analytical approach: $\int_a^0 \sin(x)\,dx = -\cos(x)\Big|_a^0 = \cos(a) - 1$. Hence, $\int_{-\infty}^{0} \sin(x)\,dx = -1 + \lim_{a\to-\infty}\cos(a)$, which does not exist.

We can also see via numerical integration that something is not right!

```julia
using QuadGK, Plots, LaTeXStrings

f(x) = sin(x)
Data = Array{Float64}(undef, 0, 3)
#
a0=-30*pi
for a=a0:-.01:(a0-6pi)
    value, error = quadgk(f, a, 0)
    Data = [Data; a value error]
end
p1 = plot(Data[:,1], Data[:,2], linewidth=4, color=:blue, label=false)
plot!(xlabel=L"$a$", ylabel=L"$y= \int_a^0 \sin(x) \, dx$")

display(p1)
```

Output

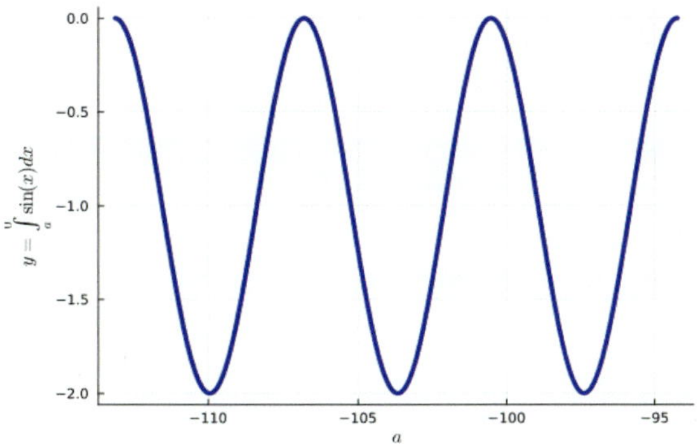

■

Functions defined on $(-\infty,\infty)$ pose an extra layer of difficulty. To see why, consider, $f:\mathbb{R}\to\mathbb{R}$ by $f(x)=4x^3$. Then, $\int_a^b f(x)\,dx = x^4\Big|_a^b = b^4-a^4$. Hence, $\lim\limits_{a\to-\infty}\int_a^b f(x)\,dx = \lim\limits_{a\to-\infty} b^4-a^4=-\infty$. But if we evaluate $\lim\limits_{b\to\infty}\int_a^b f(x)\,dx = \lim\limits_{b\to\infty} b^4-a^4=\infty$. So what value should we assign to $\int_{-\infty}^{\infty} f(x)\,dx$?

Secrets of the Arcane 8.3: A Word of Caution for Doubly-Infinite Integrals

In general, $\int_{-\infty}^{\infty} f(x)\,dx \neq \lim\limits_{c\to\infty}\int_{-c}^{c} f(x)\,dx$. **You must compute the two limits,** $\lim\limits_{c\to\infty}\int_{-c}^{0} f(x)\,dx$ **and** $\lim\limits_{c\to\infty}\int_{0}^{c} f(x)\,dx$, **SEPARATELY!!!**. Why? Consider

$$\int_{-c}^{c} x\,dx.$$

Then, as illustrated in the figure below, the magnitude of the negative area is exactly equal to the positive area, and hence for any interval $[-c,c]$ that is symmetric about the origin, $\int_{-c}^{c} x\,dx = 0.0$, i.e., the positive and negative areas cancel out, exactly. However, $\int_{-\infty}^{0} x\,dx=-\infty$ and $\int_0^{\infty} x\,dx=+\infty$, which do NOT cancel out; their sum is undefined. Moreover, if we computed $\lim\limits_{c\to\infty}\int_{-c}^{c+1} x\,dx$, we would obtain positive infinity, and if we computed $\lim\limits_{c\to\infty}\int_{-c^3}^{c^2} x\,dx$, we would obtain negative infinity. In other words, the limit changes if we advance more quickly in one direction than the other. **We avoid this problem by assessing the two limits separately and then analyzing whether their sum makes sense (or not). The sum will make sense except when both limits are unbounded and have opposite signs, giving rise to the notorious, "$\infty-\infty$".**

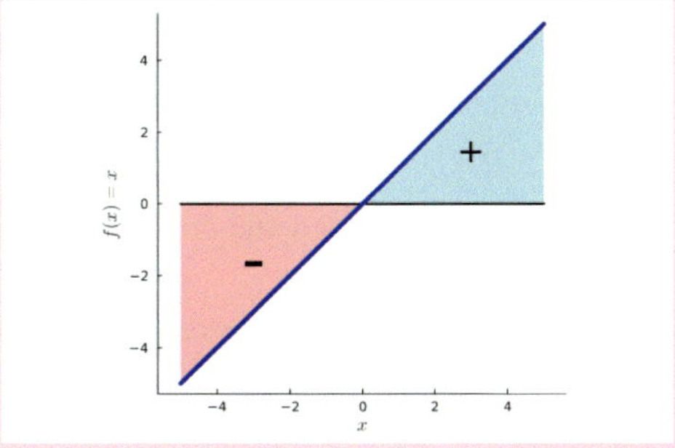

Note: The symmetric limit, $\lim_{c\to\infty}\int_{-c}^{c} f(x)\,dx$, has a name: the **Cauchy Principal Value**; see Wikipedia for more information.

Unbounded Upper and Lower Limit: Take 2

Definition 8.4. *Suppose that $f : (-\infty, \infty) \to \mathbb{R}$ is a real-valued function, and for all $a < b$ (both finite), $\int_a^b f(x)\,dx$ exists and is finite.*[a] *Then,*

$$\int_{-\infty}^{\infty} f(x)\,dx := \lim_{a\to-\infty}\int_a^0 f(x)\,dx + \lim_{b\to\infty}\int_0^b f(x)\,dx, \tag{8.3}$$

if both limits exist and are finite (i.e., both integrals are convergent). If one of the limits is undefined, then the integral is undefined. If both limits exist but at least one is unbounded (i.e, $\pm\infty$), more care is required to avoid $\infty - \infty$. Hence,

- *if one limit is unbounded while the other is finite, then the value of the unbounded limit is the value of the integral;*
- *if both limits are unbounded and have the same sign, then the common value of the unbounded limits is the value of the integral; and as was said above,*
- *if one limit equals $+\infty$ and the other $-\infty$, then the integral is* ***undefined.***

[a] If $f : (-\infty, \infty) \to \mathbb{R}$ is piecewise continuous, then this condition is met.

Example 8.5. *Determine values, if possible, for the following improper integrals. You may wish to use the antiderivatives in Table. 7.2 and Table. 7.3 because the integrands are very standard functions.*

(a) $\int_{-\infty}^{\infty} x\,dx$.

(b) $\int_{-\infty}^{\infty} x^2\,dx$.

(c) $\int_{-\infty}^{\infty} e^{-x}\,dx$.

Solutions:

(a) **Ans.** $\int_{-\infty}^{\infty} x\,dx$ is undefined because we have "$-\infty + \infty$".

From Table. 7.2, $\int_a^0 x\,dx = -\frac{1}{2}a^2$. Hence, $\int_0^{\infty} x\,dx := \lim_{a\to-\infty}\int_a^0 x\,dx = \lim_{a\to-\infty} -\frac{1}{2}a^2 = -\infty$. From part (a) of Example 8.2, $\int_0^{\infty} x\,dx = \infty$. Hence, $\int_{-\infty}^{\infty} x\,dx = -\infty + \infty =$ undefined.

(b) **Ans.** $\int_{-\infty}^{\infty} x^2\,dx = \infty$ because the two limits share the same sign and at least one is unbounded.

From Table. 7.2, $\int_a^0 x^2\,dx = -\frac{1}{3}a^3$. Hence, $\int_{-\infty}^{0} x^2\,dx := \lim_{a\to-\infty}\int_a^0 x^2\,dx = \lim_{a\to-\infty} -\frac{1}{3}a^3 = +\infty$. Similarly, $\int_0^b x^2\,dx = \frac{1}{3}b^3$. Hence, $\int_0^{\infty} x^2\,dx := \lim_{b\to\infty}\int_0^b x^2\,dx = \lim_{b\to\infty}\frac{1}{3}b^3 = \infty$. Therefore, $\int_{-\infty}^{\infty} x^2\,dx = \infty + \infty = \infty$.

(c) **Ans.** $\int_{-\infty}^{\infty} e^{-x}\,dx = \infty$ because one limit is bounded and the other is unbounded.

In Example 8.2, we computed each limit separately. The rest is left to the reader. ■

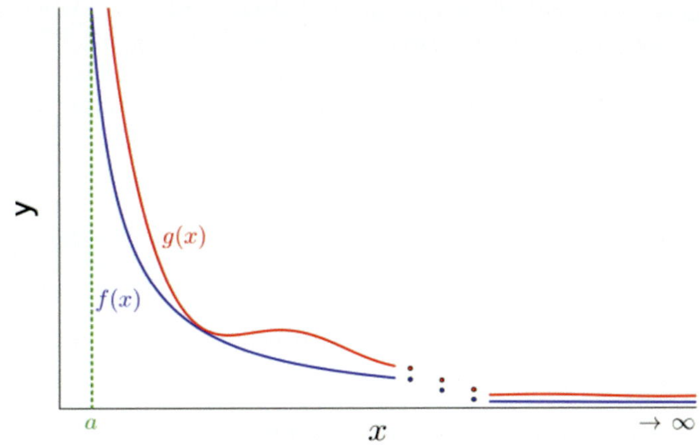

Figure 8.1: **Comparison Test:** Suppose that $f : [a, \infty) \to [0, \infty)$ and $g : [a, \infty) \to [0, \infty)$ are piecewise continuous functions so that their Riemann Integrals exist and are finite for any bounded interval $[a, b]$. When are they finite on $[a, \infty)$? In the plot, we observe that $f(x) \le g(x)$ for all $x \in [a, \infty)$, and hence, on one hand, if the area under g is finite, so must be the area under f, and on the other hand, if the area under f is unbounded, so must be the area under g. **These simple observations are the heart of the comparison test.**

8.2.1 Comparison Test

It would be nice to have some rules of thumb for determining when an improper integral should exist and be finite and when it is likely to be unbounded. Figure 8.1 shows that by comparing one function against another function's KNOWN behavior, we can make conclusions about an integral of the form $\int_a^\infty f(x)\,dx$ by comparing it to another integral $\int_a^\infty g(x)\,dx$. Indeed, if $0 \le f(x) \le g(x)$, then if

- $\int_a^\infty g(x)\,dx < \infty \implies \int_a^\infty f(x)\,dx < \infty$, and if
- $\int_a^\infty f(x)\,dx = \infty \implies \int_a^\infty g(x)\,dx = \infty$.

While this is interesting, what are some relevant comparison functions to use? We dig into our antiderivative bag of easy integrals and make some observations.

(a) For all $p > 1$, $\int_1^b \frac{1}{x^p}\,dx = \int_1^b x^{-p}\,dx = \frac{x^{1-p}}{1-p}\Big|_1^b = \frac{b^{1-p}}{1-p} - \frac{1}{1-p} = \frac{1}{1-p} \cdot \left(\frac{1}{b^{p-1}} - 1\right)$. For $p > 1$, it follows that $p - 1 > 0$, and hence $\lim_{b\to\infty} \frac{1}{b^{p-1}} = 0$. Therefore, $\int_1^\infty \frac{1}{x^p}\,dx = \frac{-1}{1-p} = \frac{1}{p-1} < \infty$.

(b) $\int_1^b \frac{1}{x}\,dx = \ln(x)\Big|_1^b = \ln(b)$ and hence $\int_1^\infty \frac{1}{x}\,dx = \lim_{b\to\infty} \ln(b) = \infty$. This makes sense because as $p \to 1^+$ in part (a), we surrmise that the value of the integral, $\frac{1}{p-1}$, diverges to infinity.

(c) For all $0 \le p < 1$, $\int_1^b \frac{1}{x^p}\,dx = \int_1^b x^{-p}\,dx = \frac{x^{1-p}}{1-p}\Bigg|_1^b = \frac{b^{1-p}}{1-p} - \frac{1}{1-p} = \frac{1}{1-p} \cdot \left(b^{1-p} - 1\right)$. For $p < 1$, it follows that $1 - p > 0$, and hence $\lim_{b\to\infty} b^{1-p} = \infty$. Therefore, $\int_1^\infty \frac{1}{x^p}\,dx = \infty$.

The above analysis says that $\frac{1}{x}$ is a "line of demarcation" between an improper integral of $\frac{1}{x^p}$ being unbounded (for $0 \le p \le 1$) and being finite (for $p > 1$). Hence, we can use $\frac{1}{x^p}$ in the **Comparison Test** illustrated in Fig. 8.1 and stated more formally below.

Suppose f is piecewise continuous on $[a, \infty)$, and let $c \ge a$. Then

$$\int_a^\infty f(x)\,dx = \int_a^c f(x)\,dx + \int_c^\infty f(x)\,dx.$$

Because f is piecewise continuous on the finite interval $[a, c]$, the integral $\int_a^c f(x)\,dx$ is automatically finite. Therefore, the convergence of $\int_a^\infty f(x)\,dx$ depends entirely on the "tail integral", $\int_c^\infty f(x)\,dx$. In particular, $\int_a^\infty f(x)\,dx$ is finite if and only if, for some $c \geq a$, $\int_c^\infty f(x)\,dx$ is finite. Why is this a helpful observation? We will demonstrate through examples that by choosing c sufficiently large, we can sometimes use "asymptotic reasoning" more easily to form an upper bound or a lower bound on $f : [c, \infty) \to \mathbb{R}$ than we can for $f : [a, \infty) \to \mathbb{R}$.

Proposition 8.6: Comparision Test

Consider a **nonnegative piecewise continuous** function $f : [0, \infty) \to [0, \infty)$ and a second non-negative piecewise continuous function, $g : [c, \infty) \to [0, \infty)$, where $c \geq 0$. Then the following are true:

(a) **Test for a convergent (aka finite) improper integral:**

- Suppose that for all $x \geq c$, $f(x) \leq g(x)$ (one says that f is upper bounded by g on $[c, \infty)$, or f is eventually upper bounded by g), and
- $\int_c^\infty g(x)\,dx$ exists and is finite.

Then $\int_0^\infty f(x)\,dx$ exists and is finite. (In other words, f being eventually upper bounded by a function with a finite improper integral ensures that its improper integral is also finite).

(b) **Test for a divergent (aka unbounded) improper integral:**

- Suppose that for all $x \geq c$, $g(x) \leq f(x)$ (one says that f is lower bounded by g on $[c, \infty)$, or f is eventually lower bounded by g), and
- $\int_c^\infty g(x)\,dx$ exists and is infinite.

Then $\int_0^\infty f(x)\,dx$ exists and is infinite. (In other words, f being lower bounded by a function with an unbounded improper integral implies that its improper integral is also unbounded.)

Note: We leave the cases for functions defined on $(-\infty, 0]$ to the reader.

Example 8.7. *Use the Comparison Test to determine which of the following improper integrals are convergent or divergent.*

(a) $\int_0^\infty \frac{1}{x^2+1}\,dx$, *where we note that the denominator has no real roots, and hence* $\frac{1}{x^2+1}$ *is finite for all* $x \in \mathbb{R}$.

(b) $\int_0^\infty \frac{\sin(x)}{x^2+1}\,dx$.

(c) $\int_0^\infty \frac{x}{x^2+1}\,dx$.

(d) $\int_0^\infty \frac{x^3+4}{x^5-3x+21}\,dx$, *where the denominator has one real root at* $x \approx -1.93017$ *and four complex roots.*

(e) $\int_0^\infty \frac{x+e^{-x}}{x^2+1}\,dx$.

(f) $\int_0^\infty e^{-x} \cdot \frac{x^4+111}{x^2+1}\,dx$.

Solutions:

(a) **Ans.** $\int_0^\infty \frac{1}{x^2+1}\,dx$ is convergent.

Let $f(x) := \frac{1}{x^2+1} = \frac{\frac{1}{x^2}}{1+\frac{1}{x^2}}$, where we divided the numerator and denominator by x^2. Next, we note that $1 + \frac{1}{x^2} > 1$ for all $x \geq 0$. Hence, $f(x) \leq g(x) := \frac{1}{x^2}$ for all $x \geq 0$. We know that $\int_0^\infty g(x)\,dx$ is convergent, and therefore, so is $\int_0^\infty f(x)\,dx$.

Note: For this integrand, we know that $\text{atan}(x)$ is an antiderivative of $\frac{1}{x^2+1}$. Hence, $\int_0^\infty \frac{1}{x^2+1}\,dx = \text{atan}(x)\Big|_0^\infty = \frac{\pi}{2}$.

(b) **Ans.** $\int_0^\infty \frac{\sin(x)}{x^2+1}\,dx$ is convergent.

Because $|\sin(x)| \leq 1$, we have that $\left|\frac{\sin(x)}{x^2+1}\right| \leq \frac{1}{x^2+1}$. From (a), we know that $\int_0^\infty \frac{1}{x^2+1}\,dx$ is convergent. Hence, $\int_0^\infty \frac{\sin(x)}{x^2+1}\,dx$ is also convergent.

(c) **Ans.** $\int_0^\infty \frac{x}{x^2+1}\,dx$ is divergent.

Let $f(x) := \frac{x}{x^2+1} = \frac{\frac{1}{x}}{1+\frac{1}{x^2}}$, where we divided the numerator and denominator by x^2. Next, we note that $1 + \frac{1}{x^2} \leq 2$ for all $x \geq 0$. Hence, $f(x) \geq g(x) := \frac{1}{2} \cdot \frac{1}{x}$ for all $x \geq 0$. We know that $\int_0^\infty g(x)\,dx$ is divergent, and therefore, so is $\int_0^\infty f(x)\,dx$.

(d) **Ans.** $\int_0^\infty \frac{x^3+4}{x^5-3x+21}\,dx$ is convergent.

First of all, why was it important to note that the denominator has one real root at $x \approx -1.93017$ and four complex roots? Because there are no positive real roots, the denominator does not vanish for any $x \geq 0$, and hence, we are never dividing by zero. Therefore,

$$f(x) := \frac{x^3+4}{x^5-3x+21}$$

is a continuous function on $[0, \infty)$.

Now, we turn to checking if the integral is convergent or divergent. We divide the numerator and denominator by x^5, giving us

$$f(x) = \frac{\frac{1}{x^2} + \frac{4}{x^5}}{1 - \frac{3}{x^4} + \frac{21}{x^5}}.$$

For all $x \geq 10$, $\text{den}(x) := 1 - \frac{3}{x^4} + \frac{21}{x^5} \geq 1 - \frac{3}{10^4} + \frac{21}{10^5} \geq 1 - \frac{3}{10^4} \geq 0.99$. Using this bound, we have that for all $x \geq 10$:

$$f(x) = \frac{\frac{1}{x^2} + \frac{4}{x^5}}{1 - \frac{3}{x^4} + \frac{21}{x^5}} \leq \frac{\frac{1}{x^2} + \frac{4}{x^5}}{0.99}.$$

Because $\frac{1}{x^2}$ and $\frac{1}{x^5}$ are both convergent, we deduce that $\int_{10}^\infty \frac{x^3+4}{x^5-3x+21}\,dx$ is convergent, which then implies that $\int_0^\infty \frac{x^3+4}{x^5-3x+21}\,dx$ is convergent.

(e) **Ans.** $\int_0^\infty \frac{x+e^{-x}}{x^2+1}\,dx$ is divergent.

We write the integral as

$$\int_0^\infty \frac{x+e^{-x}}{x^2+1}\,dx = \int_0^\infty \frac{x}{x^2+1}\,dx + \int_0^\infty \frac{e^{-x}}{x^2+1}\,dx.$$

Because $e^{-x} \leq 1$ for all $x \geq 0$,

$$\int_0^\infty \frac{e^{-x}}{x^2+1}\,dx \leq \int_0^\infty \frac{1}{x^2+1}\,dx,$$

which we know to be convergent. Moreover, we already know $\int_0^\infty \frac{x}{x^2+1}\,dx$ is divergent. Because the sum of something unbounded and something bounded is unbounded, we conclude that the overall integral is divergent. If both had been divergent, we would have had to check that we are not committing the crime of $\infty - \infty$! Clear enough?

(f) **Ans.** $\int_0^\infty e^{-x} \cdot \frac{x^4+111}{x^2+1}\,dx$ is convergent.

Ignoring the exponential for a moment and focusing on the rational function,

$$\frac{x^4+111}{x^2+1} = \frac{x^4}{x^2}\frac{1+\frac{111}{x^4}}{1+\frac{1}{x^2}} \quad \text{(factor our highest power from numerator and denominator)}$$

$$\le x^2 \frac{1+\frac{111}{x^4}}{1} \quad \text{(lower bound on the denominator holds for all } x \ge 0)$$

$$\le 2x^2 \quad (\text{for all } x > \sqrt[4]{111} \approx 3.247).$$

This yields

$$\int_4^\infty e^{-x} \cdot \frac{x^4+111}{x^2+1}\,dx \le \int_4^\infty 2x^2 e^{-x}\,dx = -2x^2e^{-x} + 4xe^{-x} - 4e^{-x}\Big|_4^\infty \approx 0.3663,$$

and because

$$\int_0^4 e^{-x} \cdot \frac{x^4+111}{x^2+1}\,dx$$

is finite, the overall integral is finite; in other words, it converges.

■

Way back in Proposition 2.30 and Corollary 2.31, we learned that exponentials dominate polynomials and rational functions when taking limits as $x \to \infty$. You may suspect that the presence of a decaying or exploding exponential in an integrand affects whether the integral is convergent or not, and you would be right!

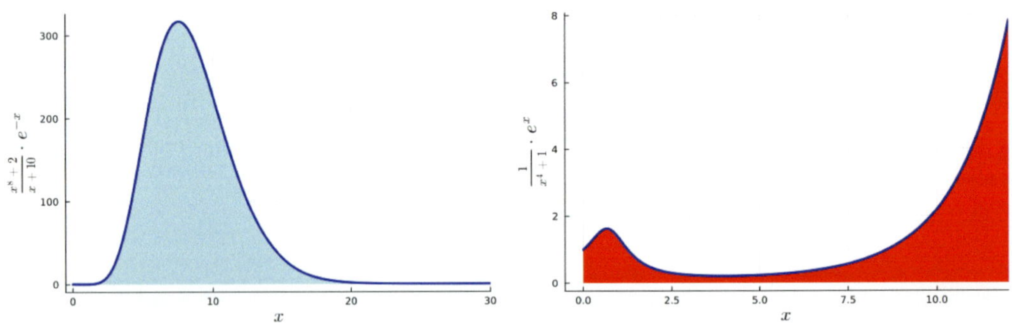

(a) (b)

Figure 8.2: When Present, Exponentials Determine Convergence of Improper Integrals: (a) Shows an exploding rational function, $\frac{x^8+2}{x+10}$, times a decaying exponential, e^{-x}. The area under the function is finite. (b) Shows a decaying rational function, $\frac{1}{x^4+1}$, times an exploding exponential, e^x. The area under the function is infinite.

Proposition 8.8: Families of Convergent and Divergent Improper Integrals: Take 1

For all real numbers $\alpha > 0$ and integers $k \ge 0$, the following hold:

(a) $\int_0^\infty x^k \cdot e^{-\alpha x} dx$ is convergent.

(b) $\int_0^\infty \frac{1}{x^k} \cdot e^{\alpha x}\, dx$ is divergent.

Proof: Part (a): For $k \geq 0$, define

$$\gamma(k) := \int_0^\infty x^k \cdot e^{-\alpha x} dx.$$

We prove by induction that $\gamma(k) < \infty$ for all $k \geq 0$.

The base case is

$$\gamma(0) = \int_0^\infty e^{-\alpha x} dx = \frac{1}{\alpha}.$$

Next, assume $\gamma(k) < \infty$ for some $k \geq 0$ and show $\gamma(k+1) < \infty$. Apply integration by parts to $x^{k+1}e^{-\alpha x}$ with $u := x^{k+1}$ and $dv := e^{-\alpha x}dx$. Then $du = (k+1)x^k dx$ and $v = -\frac{1}{\alpha}e^{-\alpha x}$, so for any $0 < a < \infty$,

$$\begin{aligned}\int_0^a x^{k+1} \cdot e^{-\alpha x} dx &= x^{k+1} \cdot \left(\frac{-1}{\alpha} e^{-\alpha x}\right)\Big|_0^a - \int_0^a \left(\frac{-1}{\alpha} e^{-\alpha x}\right) \cdot \left((k+1)x^k\right) dx \\ &= \frac{-1}{\alpha} \cdot \left(x^{k+1} \cdot e^{-\alpha x}\right)\Big|_0^a + \frac{k+1}{\alpha} \underbrace{\int_0^a x^k \cdot e^{-\alpha x} dx}_{\longrightarrow \gamma(k) \text{ as } a \to \infty}.\end{aligned}$$

Taking limits as $a \to \infty$, the upper limit of the first term satisfies

$$\lim_{a \to \infty} a^{k+1} e^{-\alpha a} = 0,$$

and the lower limit at $x = 0$ is 0. Hence,

$$\gamma(k+1) = \frac{k+1}{\alpha} \gamma(k).$$

Therefore, if $\gamma(k)$ is finite, then $\gamma(k+1)$ is finite as well.

Part (b): From Prop. 2.30,

$$\lim_{x \to \infty} \frac{1}{x^k} \cdot e^{\alpha x} = \infty.$$

Hence, there exists $M < \infty$ such that $\frac{1}{x^k} \cdot e^{\alpha x} > 1$ for all $x > M$. It follows that

$$\int_0^\infty \frac{1}{x^k} \cdot e^{\alpha x} dx \geq \int_M^\infty 1\, dx = \infty,$$

so the integral is divergent. ■.

Next, we recall our "trick" from Chapter 2.5.1, namely

$$\frac{b_m x^m + b_{m-1}x^{m-1} + \cdots + b_0}{a_n x^n + a_{n-1}x^{n-1} + \cdots + a_0} = \boxed{\frac{x^m}{x^n}} \cdot \frac{b_m + b_{m-1}\frac{1}{x} + \cdots + b_0 \frac{1}{x^m}}{a_n + a_{n-1}\frac{1}{x} + \cdots + a_0 \frac{1}{x^n}}.$$

Because

$$\lim_{x \to \infty} \frac{b_m + b_{m-1}\frac{1}{x} + \cdots + b_0 \frac{1}{x^m}}{a_n + a_{n-1}\frac{1}{x} + \cdots + a_0 \frac{1}{x^n}} = \frac{b_m}{a_n},$$

we can immediately extend Prop. 8.8 to rational functions. The proof is left as an exercise.

Proposition 8.9: Families of Convergent and Divergent Improper Integrals: Take 2

Consider a rational function

$$f(x) = \frac{b_m x^m + b_{m-1}x^{m-1} + \cdots + b_0}{a_n x^n + a_{n-1}x^{n-1} + \cdots + a_0}$$

and assume it is not the zero function. If the denominator $a_n x^n + a_{n-1}x^{n-1} + \cdots + a_0$ has any real roots, choose $c \geq 0$ strictly greater than the largest root; otherwise, set $c = 0$. Then the function $f : [c, \infty) \to \mathbb{R}$ is continuous and the following hold:

(a) For all $\alpha > 0$,

$$\int_c^\infty e^{-\alpha x} \cdot \frac{b_m x^m + b_{m-1}x^{m-1} + \cdots + b_0}{a_n x^n + a_{n-1}x^{n-1} + \cdots + a_0}\, dx$$

is convergent.

(b) For all $\beta > 0$,

$$\int_c^\infty e^{\beta x} \cdot \frac{b_m x^m + b_{m-1}x^{m-1} + \cdots + b_0}{a_n x^n + a_{n-1}x^{n-1} + \cdots + a_0}\, dx$$

is divergent.

(c) When there is no exponential term,

$$\int_c^\infty \frac{b_m x^m + b_{m-1}x^{m-1} + \cdots + b_0}{a_n x^n + a_{n-1}x^{n-1} + \cdots + a_0}\, dx \quad \begin{cases} \text{is convergent} & n \geq m+2, \\ \text{is divergent} & \text{otherwise.} \end{cases}$$

Note: When there is no exponential term present, the comparison function $\frac{1}{x^p}$ applies, as we essentially demonstrated in the solution to Example 8.7-(d). When the degree of the denominator is at least two greater than the degree of the numerator, the integrand behaves as $\frac{1}{x^p}$ for $p \geq 2$, which is convergent; otherwise, the integrand behaves as $\frac{1}{x^p}$ for $p \leq 1$, which is divergent.

When an exponential term is present, the numerator and denominator degrees can be anything; for example, $\int_1^\infty x^{1000} \cdot e^{-0.001x}\, dx$ is convergent and $\int_1^\infty \frac{1}{x^{1000}} \cdot e^{0.001x}\, dx$ is divergent. If you try these cases with a numerical integrator, your results may vary due to numerical inaccuracy. The theory is useful.

Example 8.10. *Use Prop. 8.9 to determine which of the following improper integrals are convergent or divergent.*

(a) $\int_0^\infty e^{-x} \cdot \frac{x^{24}+2}{x^2+1}\, dx.$

(b) $\int_0^\infty e^{x} \cdot \frac{x^1+1}{x^{24}+1}\, dx.$

(c) $\int_0^\infty e^{-x} \cdot \frac{x^{24}+2}{x^2-1}\, dx$

(d) $\int_0^\infty \frac{x^1+1}{x^{24}+1}\, dx.$

(e) $\int_0^\infty \frac{x^{24}+2}{x^2+1}\, dx.$

(f) $\int_0^\infty \frac{x^{24}+x^7-\pi x^4+2}{x^{26}+1}\, dx.$

(g) $\int_0^\infty \frac{x^{25}+x^7-\pi x^4+2}{x^{26}+1}\, dx.$

(h) $\int_0^\infty \frac{x^{25}+x^7-\pi x^4+2}{(x^{24}+1)\cdot(x^2-1)}\, dx.$

Solutions:

(a) **Ans.** $\int_0^\infty e^{-x} \cdot \frac{x^{24}+2}{x^2+1}\, dx$ is convergent due to the decaying exponential out front and x^2+1 has no real roots.

(b) **Ans.** $\int_0^\infty e^{x} \cdot \frac{x^1+1}{x^{24}+1}\, dx$ is divergent due to the exploding exponential out front.

(c) **Ans.** $\int_0^\infty e^{-x} \cdot \frac{x^{24}+2}{x^2-1}\, dx$ **cannot be addressed** with Prop. 8.9 because the denominator, $x^2-1 = (x+1)\cdot(x-1)$, has a real root $r_1 = 1$ within the domain of integration, $[0, \infty)$. If we changed the integral to $\int_c^\infty e^{-x} \cdot \frac{x^{24}+2}{x^2-1}\,dx$, for $c > 1$, then it would be convergent. **When applying a Theorem, Proposition, Corollary, etc., it is crucial to check the conditions under which it holds.**

(d) **Ans.** $\int_0^\infty \frac{x^1+1}{x^{24}+1}\,dx$ is convergent because (i) $x^{24}+1$ has no real roots, and (ii) the degree of the denominator exceeds the degree of the numerator by 23, which is greater or equal to two.

(e) **Ans.** $\int_0^\infty \frac{x^{24}+2}{x^2+1}\,dx$ is divergent because (i) x^2+1 has no real roots, and (ii) the degree of the numerator exceeds the degree of the numerator.

(f) **Ans.** $\int_0^\infty \frac{x^{24}+x^7-\pi x^4+2}{x^{26}+1}\,dx$ is convergent because (i) $x^{26}+1$ has no real roots, and (ii) the degree of the denominator exceeds the degree of the numerator by two.

(g) **Ans.** $\int_0^\infty \frac{x^{25}+x^7-\pi x^4+2}{x^{26}+1}\,dx$ is divergent because (i) $x^{26}+1$ has no real roots, and (ii) the degree of the denominator DOES NOT EXCEED the degree of the numerator by two (it only exceeds it by one).

(h) **Ans.** $\int_0^\infty \frac{x^{25}+x^7-\pi x^4+2}{(x^{24}+1)\cdot(x^2-1)}\,dx$ **cannot be addressed** with Prop. 8.9 because the denominator, $(x^{24}+1)\cdot(x^2-1) = (x^{24}+1)\cdot(x+1)\cdot(x-1)$ has a real root $r_1 = 1$ within the domain of integration, $[0,\infty)$. If we changed the integral to

$$\int_c^\infty \frac{x^{25}+x^7-\pi x^4+2}{(x^{24}+1)\cdot(x^2-1)}\,dx,$$

for $c > 1$, then it would be divergent because the degree of the numerator is only one less than the degree of the denominator. **When applying a Theorem, Proposition, Corollary, etc., it is crucial to check the conditions under which it holds.**

■

Recommended Videos for Additional Practice

- Improper Integrals - Convergence and Divergence - Calculus 2 by the Organic Chemistry Tutor.
- Type 1 improper integrals, calculus 2 by **BlackPenRedPen**.
- Improper Integrals - Convergent or Divergent (Made Easy) by Quoc Dat Phung.
- Comparison test for convergence and divergence of improper integrals by **BlackPenRedPen**.
- Comparison Test for Improper Integrals by Dr. Trefor Bazett.

8.2.2 Absolute Integrability

Consider a function $f : [a,\infty) \to \mathbb{R}$ that takes on both positive and negative values. As illustrated in Fig. 8.3, an easy way to make sense of its improper integral is to examine the function's absolute value. Because the absolute value is nonnegative, the comparison test can be applied.

Absolute Integrability

Definition 8.11. *A piecewise continuous function $f : [a,\infty) \to \mathbb{R}$ is **absolutely integrable** if*

$$\int_a^\infty |f(x)|\,dx < \infty. \tag{8.4}$$

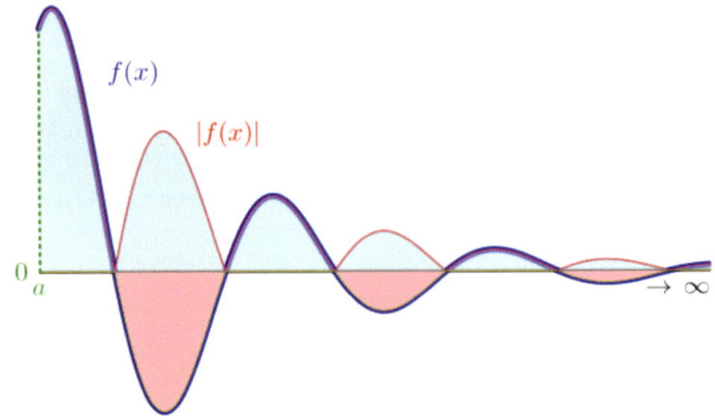

Figure 8.3: Absolute Integrability: Suppose that $f : [a, \infty) \to \mathbb{R}$ is a piecewise continuous function so that its Riemann Integral exists and is finite for any bounded interval $[a, b]$. We observe that the area under $|f|$ is the sum of the area where $f \geq 0$ and THE NEGATIVE OF the area where $f \leq 0$. Hence, if the area under $|f|$ is finite, then each of the positive (shaded blue) and negative (shaded pink) areas of f must be finite, and therefore, the total area under f is well defined (aka does not involve $\infty - \infty$) and finite.

Proposition 8.12: Usefulness of Absolute Integrability

Suppose that $f : [a, \infty) \to \mathbb{R}$ is piecewise continuous. Then the following hold:

(a) If f is absolutely integrable, then $\int_a^\infty f(t)\,dt$ (without absolute value signs) exists and is finite.

(b) If $\lim\limits_{t\to\infty} f(t)$ exists and is nonzero, then f is not absolutely integrable.

(c) If f is differentiable on (a, ∞) and f' (its derivative) is absolutely integrable, then $\lim\limits_{t\to\infty} f(t)$ exists and is finite.

(d) If both f and f' (its derivative) are absolutely integrable on (a, ∞), then $\lim\limits_{t\to\infty} f(t) = 0$.

Notes: Each of these properties has its uses: (a) allows the comparison test to be used on functions that take on both positive and negative values; (b) provides a sufficient condition for a function NOT TO BE absolutely integrable; (c) provides a sufficient condition for a function to have a limit as t goes to infinity; and (d) provides an extra condition for that limit to be zero.

Example 8.13. *Functions of the form* $t^k\,e^{at}\,\sin(\omega t)$ *and* $t^k\,e^{at}\,\cos(\omega t)$ *show up very often in engineering and STEM fields. Show that for all integers* $k \geq 0$ *and* $\omega \in (0, \infty)$*, these functions are absolutely integrable on* $[0, \infty)$ *if, and only if,* $a < 0$.

Solution: The proofs for sine and cosine are nearly identical, so we consider only one of them. For $\sin(\omega t)$, it is important to exclude $\omega = 0$, while for $\cos(\omega t)$, it is not. The essence of the proofs can be deduced from Fig. 8.4.

Part 1: $a < 0 \implies \int_0^\infty |t^k\,e^{at}\,\sin(\omega t)|\,dt < \infty$. We break the proof into discrete steps to make it easier to follow.

- $|t^k\,e^{at}\,\sin(\omega t)| \leq t^k\,e^{at}$ for all $t \geq 0$ because $-1 \leq \sin(\omega t) \leq 1$ and $t^k\,e^{at}$ is nonnegative.
- Define $b := \frac{|a|}{2} > 0$ so that $a + b = \frac{a}{2} < 0$.

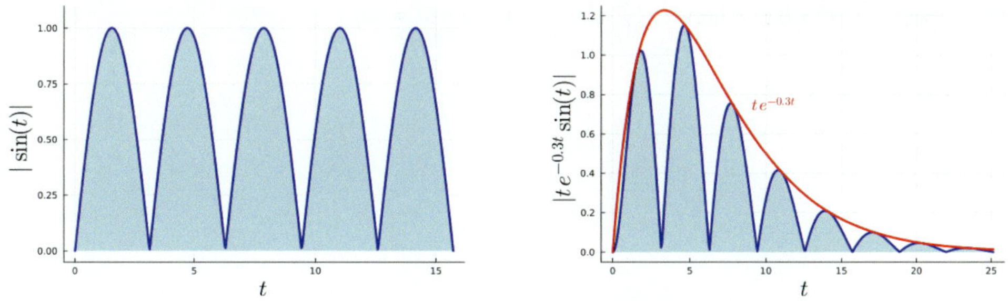

(a) (b)

Figure 8.4: Common Functions and Their Absolute Integrability: (a) shows that $\sin(t)$ is not absolutely integrable over an unbounded interval, while (b) shows that a negative exponential damps out $t\sin(t)$ so that it becomes absolutely integrable; moreover, (b) also illustrates, by the red curve, a decaying exponential eventually overtakes an exploding monomial. The same holds for higher powers of t.

- Because $b > 0$, and an exploding exponential dominates any monomial, there exists $0 < T < \infty$ such that $t^k \leq e^{bt}$ for all $t \geq T$.
- Combining these inequalities, we obtain

$$\begin{aligned}\int_0^\infty |t^k\, e^{at}\, \sin(\omega t)|\, dt &\leq \int_0^\infty t^k\, e^{at}\, dt \\ &\leq \int_0^T t^k\, e^{at}\, dt + \int_T^\infty e^{bt}\, e^{at}\, dt \\ &\leq \int_0^T t^k\, e^{at}\, dt + \int_T^\infty e^{\frac{a}{2}t}\, dt \\ &\leq \int_0^T t^k\, e^{at}\, dt + \int_0^\infty e^{\frac{a}{2}t}\, dt,\end{aligned} \tag{8.5}$$

where the last line is true because integrating a positive function over a larger interval cannot decrease the value of the integral.

The first integral is finite because we have a continuous function over a bounded interval. The second integral evaluates to $-\frac{2}{a} = \frac{2}{|a|}$, which is finite. Hence, the sum of the two integrals is finite, completing the proof.

Part 2: $a \geq 0 \implies \int_0^\infty |t^k\, e^{at}\, \sin(\omega t)|\, dt = \int_0^\infty t^k\, e^{at}\, |\sin(\omega t)|\, dt = \infty$. Again, we break the proof into discrete steps to make it easier to follow.

- Let $T := \frac{2\pi}{\omega} > 0$ be the period of $\sin(\omega t)$.
- $\displaystyle\int_0^\infty t^k\, e^{at}\, |\sin(\omega t)|\, dt := \lim_{N\to\infty} \int_0^{NT} t^k\, e^{at}\, |\sin(\omega t)|\, dt.$
- $\displaystyle\int_0^{NT} t^k\, e^{at}\, |\sin(\omega t)|\, dt = \sum_{n=0}^{N-1} \int_{nT}^{(n+1)T} t^k\, e^{at}\, |\sin(\omega t)|\, dt.$
- Because $t^k\, e^{at}$ is non-decreasing,[1] for all $n \geq 0$, $\displaystyle\int_{nT}^{(n+1)T} t^k\, e^{at}\, |\sin(\omega t)|\, dt \geq \int_0^T t^k\, e^{at}\, |\sin(\omega t)|\, dt.$
- Hence, $\displaystyle\sum_{n=0}^{N-1} \int_{nT}^{(n+1)T} t^k\, e^{at}\, |\sin(\omega t)|\, dt \geq \sum_{n=0}^{N-1} \int_0^T t^k\, e^{at}\, |\sin(\omega t)|\, dt = N \int_0^T t^k\, e^{at}\, |\sin(\omega t)|\, dt.$

Combining these results together,

$$\int_0^\infty t^k\, e^{at}\, |\sin(\omega t)|\, dt \geq \lim_{N\to\infty} N \int_0^T t^k\, e^{at}\, |\sin(\omega t)|\, dt. \tag{8.6}$$

[1] For $k > 0$ or $a > 0$, it is strictly increasing, but for $0 = k = a$, it is constant.

But, $\int_0^T t^k e^{at} |\sin(\omega t)| dt > 0$ and hence, $\int_0^\infty t^k e^{at} |\sin(\omega t)| dt = \infty$.

Parts 1 and 2 together prove the result. ■

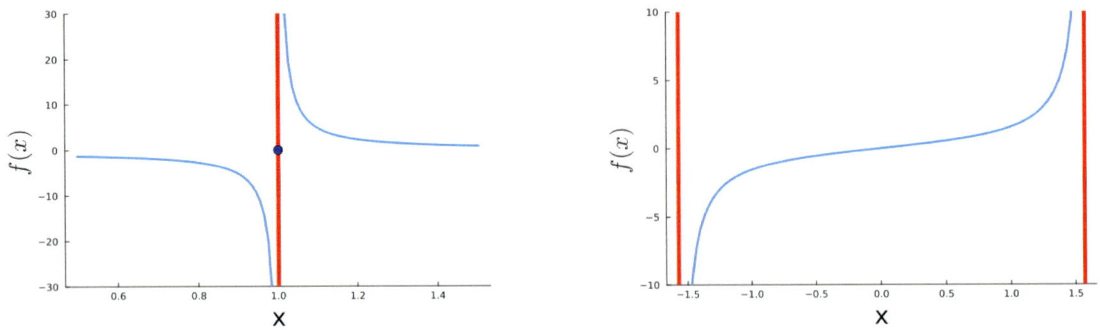

Figure 8.5: Type-II Improper Integrals: As shown in (a), the function $\frac{1}{x^2-1}$ has a vertical asymptote at $x = 1$. Can we make sense of the integral $\int_0^1 \frac{1}{x^2-1} dx$ or $\int_0^2 \frac{1}{x^2-1} dx$? Similarly, as shown in (b), the function $\tan(x)$ has vertical asymptotes at $x = \pm\frac{\pi}{2}$. Can we make sense of the integral $\int_0^{\frac{\pi}{2}} \tan(x)\, dx$?

8.3 Type-II Improper Integrals: Vertical Asymptotes

A function that has a vertical asymptote, as shown in Fig. 8.5, diverges to $\pm\infty$ at some finite point $x_0 \in \mathbb{R}$. The unbounded nature of a function near a finite point demands extra attention when evaluating its integral, akin to situations where the limits of integration are unbounded. This section is dedicated to exploring such integrals.

Before getting into the details of the process, let us consider a few basic examples.

Consider $f : (0,2) \to \mathbb{R}$ by $f(x) = \frac{1}{x^3}$. The function is continuous on the set $(0,2)$, and for all $0 < a < 2$, $\int_a^2 \frac{1}{x^3} dx$ exists and is finite. Indeed,

$$\int_a^2 \frac{1}{x^3} dx = \int_a^2 x^{-3} dx = \frac{1}{-2} x^{-2} \Big|_a^2 = \frac{1}{2x^2} \Big|_2^a = \frac{1}{2a^2} - \frac{1}{8}.$$

We next take the limit as a approaches zero from the right,

$$\int_0^2 \frac{1}{x^3} dx := \lim_{a \to 0^+} \int_a^2 \frac{1}{x^3} dx = \lim_{a \to 0^+} \left(\frac{1}{2a^2} - \frac{1}{8} \right) = \infty,$$

and hence the integral is **divergent**.

Consider $f : (0,2) \to \mathbb{R}$ by $f(x) = \frac{1}{\sqrt{x}}$. The function is continuous on the set $(0,2)$, and for all $0 < a < 2$, $\int_a^2 \frac{1}{\sqrt{x}} dx$ exists and is finite. Indeed,

$$\int_a^2 \frac{1}{\sqrt{x}} dx = \int_a^2 x^{-\frac{1}{2}} dx = \frac{1}{\frac{1}{2}} x^{\frac{1}{2}} \Big|_a^2 = 2\left(\sqrt{2} - \sqrt{a}\right).$$

We next take the limit as a approaches zero from the right,

$$\int_0^2 \frac{1}{\sqrt{x}} dx := \lim_{a \to 0^+} \int_a^2 \frac{1}{\sqrt{x}} dx = \lim_{a \to 0^+} 2\left(\sqrt{2} - \sqrt{a}\right) = 2 \cdot \sqrt{2},$$

and hence the integral is **convergent**.

Consider $f : [0,1) \to \mathbb{R}$ by $f(x) = \frac{x}{\sqrt[4]{1-x^2}}$, which has a vertical asymptote at $x_0 = 1$. The function is continuous on the set $(0,1)$, and for all $0 < b < 1$, $\int_0^b \frac{x}{\sqrt[4]{1-x^2}}\,dx$ exists and is finite. Next, we change the variable to $u = 1 - x^2 \implies du = -2x$, and compute

$$\int_{x=0}^{x=b} \frac{x}{\sqrt[4]{1-x^2}}\,dx = \int_{u=1}^{u=1-b^2} -\frac{1}{2}\frac{1}{\sqrt[4]{u}}\,du = \frac{1}{2}\int_{u=1-b^2}^{u=1} u^{-\frac{1}{4}}\,du = \frac{1}{2}\frac{1}{\frac{3}{4}}u^{\frac{3}{4}}\Bigg|_{u=1-b^2}^{u=1} = \frac{2}{3}\left(1-\left(1-b^2\right)^{\frac{3}{4}}\right).$$

Taking the limit as b approaches 1 from the left leads to:

$$\int_0^1 \frac{x}{\sqrt[4]{1-x^2}}\,dx := \lim_{b\to 1^-}\int_0^b \frac{x}{\sqrt[4]{1-x^2}}\,dx = \lim_{b\to 1^-}\frac{2}{3}\left(1-\left(1-b^2\right)^{\frac{3}{4}}\right) = \frac{2}{3},$$

and hence the integral is **convergent**.

From the above, we see that the process is essentially identical to what we did for improper integrals of Type I: we set up the integral for a domain where the integrand is well behaved (continuous and bounded), and then use a single-sided limit to approach the "problematic point" from the left or the right; for Type-I Integrals, the "problematic point" was $\pm\infty$, while for Type-II Integrals, it is ${x_0}^+$ or x_0^-, the right and left sides of a vertical asymptote. Next, we will consider one more example to illustrate what happens when a vertical asymptote is strictly between the lower and upper bounds of integration, instead of being one of the endpoints of integration.

Consider[2] $f : [0,2] \to \mathbb{R}$ by

$$f(x) = \begin{cases} 0 & x = 1, \\ \frac{x}{\sqrt[4]{|1-x^2|}} & \text{otherwise,} \end{cases}$$

where an absolute value was added to define the square root for $x > 1$. The function is well behaved at each of the endpoints (does not blow up), so the relevant question is: does the integral $\int_0^2 f(x)\,dx$ make sense? Especially when, as is often the case, it is written as:

$$\int_0^2 \frac{x}{\sqrt[4]{|1-x^2|}}\,dx?$$

The answer is: to make sense of it, we need to break the integral into two pieces,

$$\underbrace{\int_0^{1^-} \frac{x}{\sqrt[4]{|1-x^2|}}\,dx}_{\text{left of the asymptote}} + \underbrace{\int_{1^+}^2 \frac{x}{\sqrt[4]{|1-x^2|}}\,dx}_{\text{right of the asymptote}},$$

and then analyze each piece separately, just as we did with an integral having unbounded lower and upper limits. Then, as long as we can avoid the dreaded "$\infty - \infty$", we can sum the two answers and arrive at a value for the overall integral. If you think of $+\infty$ as approaching infinity from the left and $-\infty$ as approaching infinity from the right, then dealing with a Type-I Improper Integral with doubly infinite limits and a Type-II Improper Integral with an asymptote between the limits of integration, are really one and the same.

Fact 8.14: Universal Approach for all Improper Integrals

To address an improper integral, the universal approach involves initially setting up a definite integral,

$$\int_a^b f(x)\,dx,$$

[2]Often, the person posing the problem will not be so careful as to define the function at the vertical asymptote, but they should be that careful.

where $f : [a,b] \to \mathbb{R}$ is piecewise continuous and bounded. Subsequently, evaluate the limit as a or b approaches either $\pm\infty$ or converges to the left or right of a vertical asymptote. This process is realized one limit at a time if both bounds of integration involve limits or if the original integral involves a vertical asymptote occurring at $a < x_0 < b$ (strictly between the limits).

Example 8.15. *Evaluate the following integrals.*

(a) $\int_0^1 \frac{1}{x}\,dx$.

(b) $\int_0^1 \frac{1}{\sqrt{x}}\,dx$.

(c) $\int_0^2 \frac{1}{\sqrt[4]{|x-1|}}\,dx$.

(d) $\int_0^{\frac{\pi}{2}} \tan(x)\,dx$.

(e) $\int_{-1}^1 \frac{x}{\sqrt{1-x^2}}\,dx$.

Solutions:

(a) **Ans.** $\int_0^1 \frac{1}{x}\,dx = \infty$.

There is a vertical asymptote at $x_0 = 0$. Hence,

$$\int_0^1 \frac{1}{x}\,dx := \lim_{a\to 0^+}\int_a^1 \frac{1}{x}\,dx = \lim_{a\to 0^+} \ln(x)\Big|_a^1 = \lim_{a\to 0^+}(\ln(1)-\ln(a)) = \infty.$$

(b) **Ans.** $\int_0^1 \frac{1}{\sqrt{x}}\,dx = 1$.

There is a vertical asymptote at $x_0 = 0$. Hence,

$$\int_0^1 \frac{1}{\sqrt{x}}\,dx := \lim_{a\to 0^+}\int_a^1 x^{-\frac{1}{2}}\,dx = \lim_{a\to 0^+}\frac{1}{2}\sqrt{x}\Big|_a^1 = \lim_{a\to 0^+}\left(1-\sqrt{a}\right) = 1.$$

(c) **Ans.** $\int_0^2 \frac{1}{\sqrt[4]{|x-1|}}\,dx = \frac{8}{3}$.

There is a vertical asymptote at $x_0 = 1$. Hence, $\int_0^2 \frac{1}{\sqrt[4]{|x-1|}}\,dx := \int_0^{1^-} \frac{1}{\sqrt[4]{|x-1|}}\,dx + \int_{1^+}^2 \frac{1}{\sqrt[4]{|x-1|}}\,dx$. Taking them one at a time gives:

$$\begin{aligned}
\int_0^{1^-} \frac{1}{\sqrt[4]{|x-1|}}\,dx &= \lim_{b\to 1^-} \int_0^b \frac{1}{\sqrt[4]{|x-1|}}\,dx \\
&= \lim_{b\to 1^-} \int_0^b \frac{1}{\sqrt[4]{1-x}}\,dx \quad \text{(because } x-1 \le 0\text{)} \\
&= \lim_{b\to 1^-} \int_0^b (1-x)^{-\frac{1}{4}}\,dx \quad \text{(prepare to apply generalized power rule)} \\
&= \lim_{b\to 1^-} \frac{1}{\frac{3}{4}} \cdot (-1) \cdot (1-x)^{\frac{3}{4}} \Big|_0^b \quad \text{(}-1 \text{ is from the chain rule)} \\
&= \lim_{b\to 1^-} -\frac{4}{3}\left((1-b)^{\frac{3}{4}} - 1\right) \\
&= -\frac{4}{3}\left((1-1)^{\frac{3}{4}} - 1\right) \\
&= \frac{4}{3}
\end{aligned}$$

and, next,

$$\begin{aligned}
\int_{1^+}^{2} \frac{1}{\sqrt[4]{|x-1|}}\,dx &= \lim_{a\to 1^+} \int_a^2 \frac{1}{\sqrt[4]{|x-1|}}\,dx \\
&= \lim_{a\to 1^+} \int_a^2 \frac{1}{\sqrt[4]{x-1}}\,dx \quad \text{(because } x-1 \ge 0\text{)} \\
&= \lim_{a\to 1^+} \int_a^2 (1-x)^{-\frac{1}{4}}\,dx \quad \text{(prepare to apply generalized power rule)} \\
&= \lim_{a\to 1^+} \frac{1}{\frac{3}{4}} \cdot (1-x)^{\frac{3}{4}} \Big|_a^2 \\
&= \lim_{a\to 1^+} \frac{4}{3}\left((2-1)^{\frac{3}{4}} - (a-1)^{\frac{3}{4}}\right) \\
&= \frac{4}{3}\left((2-1)^{\frac{3}{4}} - (1-1)^{\frac{3}{4}}\right) \\
&= \frac{4}{3}.
\end{aligned}$$

Hence, $\int_0^2 \frac{1}{\sqrt[4]{|x-1|}}\,dx = \frac{4}{3} + \frac{4}{3} = \frac{8}{3}$.

Note: In the first integral, we can use a u-substitution via $u = 1 - x \implies du = -dx$:

$$\begin{aligned}
\int_{x=0}^{x=b} (1-x)^{-\frac{1}{4}}\,dx &= \int_{u=1}^{u=1-b} u^{-\frac{1}{4}}\,(-1)\,du \\
&= \int_{u=1-b}^{u=1} u^{-\frac{1}{4}}\,du \quad \text{(swap bounds of integration to remove minus sign)} \\
&= \frac{4}{3} u^{\frac{3}{4}} \Big|_{1-b}^{1}.
\end{aligned}$$

(d) **Ans.** $\int_0^{\frac{\pi}{2}} \tan(x)\,dx = \infty.$

There is a vertical asymptote at $x_0 = \frac{\pi}{2}$. Before dealing with that, let us obtain an antiderivative for $\tan(x) = \frac{\sin(x)}{\cos(x)}$ via the u-substitution, $u = \cos(x) \implies du = -\sin(x)\,dx$. Hence,

$$\int \frac{\sin(x)}{\cos(x)}\,dx = \int -\frac{1}{u}du = -\ln(u) + C = -\ln(\cos(x)) + C.$$

Knowing the antiderivative, we compute

$$\begin{aligned}
\int_0^{\frac{\pi}{2}} \tan(x)\,dx &= \lim_{b\to\frac{\pi}{2}^-} \int_0^b \tan(x)\,dx \\
&= \lim_{b\to\frac{\pi}{2}^-} -\ln(\cos(x))\Big|_0^b \\
&= \lim_{b\to\frac{\pi}{2}^-} \ln(\cos(x))\Big|_b^0 \quad \text{(swap bounds of integration to remove minus sign)} \\
&= \lim_{b\to\frac{\pi}{2}^-} (\ln(\cos(0)) - \ln(\cos(b))) \\
&= 0 - (-\infty) \\
&= \infty.
\end{aligned}$$

Note: If we were to compute $\displaystyle\int_{-\frac{\pi}{2}}^{\frac{\pi}{2}} \tan(x)\,dx$, we would have two vertical asymptotes, one at $-\frac{\pi}{2}$ and one at $\frac{\pi}{2}$, and we would arrive at the integral being undefined due to $-\infty + \infty$. If this point is not clear, see **Fig. 8.5-(b)**.

(e) **Ans.** $\int_{-1}^1 \frac{x}{\sqrt{1-x^2}}\,dx = 0.$

There are vertical asymptotes at ±1. Before dealing with that, let us find an antiderivative for $\frac{x}{\sqrt{1-x^2}}$ via the u-substitution $u = 1 - x^2 \implies du = -2x\,dx$. Hence,

$$\begin{aligned}
\int \frac{x}{\sqrt{1-x^2}}\,dx &= \int -\frac{1}{2}\frac{1}{\sqrt{1-x^2}}(-2x)\,dx \\
&= \int -\frac{1}{2}\frac{1}{\sqrt{u}}\,du \\
&= \int -\frac{1}{2}u^{-\frac{1}{2}}\,du \\
&= -\frac{1}{2}\cdot\frac{1}{\frac{1}{2}}u^{\frac{1}{2}} + C \\
&= -\sqrt{u} + C \\
&= -\sqrt{1-x^2} + C
\end{aligned}$$

Now that we have an antiderivative, $\int_{-1}^1 \frac{x}{\sqrt{1-x^2}}\,dx = \displaystyle\lim_{a\to-1^+}\int_a^0 \frac{x}{\sqrt{1-x^2}}\,dx + \lim_{b\to1^-}\int_0^b \frac{x}{\sqrt{1-x^2}}\,dx.$

Taking the integrals one at a time, we have, for the first integral,

$$\begin{aligned}
\lim_{a \to -1^+} \int_a^0 \frac{x}{\sqrt{1-x^2}}\, dx &= \lim_{a \to -1^+} -\sqrt{1-x^2}\Bigg|_a^0 \\
&= \lim_{a \to -1^+} \sqrt{1-x^2}\Bigg|_0^a \quad \text{(swap bounds of integration to remove minus sign)} \\
&= \lim_{a \to -1^+} \left(\sqrt{1-a^2} - \sqrt{1-0}\right) \\
&= \frac{1}{4}\left(\sqrt{1-1} - \sqrt{1}\right) \\
&= -1.
\end{aligned}$$

Evaluating the second integral,

$$\begin{aligned}
\lim_{b \to 1^-} \int_0^b \frac{x}{\sqrt{1-x^2}}\, dx &= \lim_{b \to 1^-} -\sqrt{1-x^2}\Bigg|_0^b \\
&= \lim_{b \to 1^-} \sqrt{1-x^2}\Bigg|_b^0 \quad \text{(swap bounds of integration to remove minus sign)} \\
&= \lim_{b \to 1^-} \left(\sqrt{1-0} - \sqrt{1-b^2}\right) \\
&= \lim_{b \to 1^-} \left(1 - \sqrt{1-1}\right) \\
&= 1.
\end{aligned}$$

Hence, $\int_{-1}^{1} \frac{x}{\sqrt{1-x^2}}\, dx = \lim_{a \to -1^+} \int_a^0 \frac{x}{\sqrt{1-x^2}}\, dx + \lim_{b \to 1^-} \int_0^b \frac{x}{\sqrt{1-x^2}}\, dx = -1 + 1 = 0.$

■

Example 8.16. *Use* `quadgk` *to evaluate the following integrals. The primary goal is to determine whether it detects that there is a "problem" due to its integrand blowing up over the domain of integration or not. A secondary goal would be to estimate the value of the integral using* `quadgk`*.*

(a) $\int_0^1 \frac{1}{x}\, dx.$

(b) $\int_{-2}^{2} \frac{x}{\sqrt{|1-x^2|}}\, dx$*, has two vertical asymptotes within the domain of integration.*

Solutions:

a) We apply numerical techniques to $\int_0^1 \frac{1}{x}\, dx.$

```
using QuadGK
fa(x)  = 1/x
sol,   = quadgk(fa, 0, 1)
```

Output

```
DomainError with 3.5601181736115222e-307:
integrand produced Inf in the interval (0.0, 7.120236347223045e-307)
```

That is reassuring: `quadgk` detected there is an issue. So, we ask next, can we use `quadgk` to determine if the integral is convergent or divergent?

```
using QuadGK
k=collect(1:2:10)
limitRange = 10.0.^(-k)
for a in limitRange
    sol,  = quadgk(fa, a, 1)
    display([a sol])
end
```

Output

```
1×2 Matrix{Float64}:
 0.1   2.30259
1×2 Matrix{Float64}:
 0.001   6.90776
1×2 Matrix{Float64}:
 1.0e-5   11.5129
1×2 Matrix{Float64}:
 1.0e-7   16.1181
1×2 Matrix{Float64}:
 1.0e-9   20.7233
```

Once again, this is reassuring: as $a \to 0^+$, the integral is not converging.

b) We apply numerical techniques to $\int_{-2}^{2} \frac{x}{\sqrt{|1-x^2|}}\, dx$, which has vertical asymptotes at ± 1.

```
using QuadGK
fe(x) = x/sqrt(abs(1 - x^2))
# There are two vertical asymptotes within the domain of integration
sol,  = quadgk(fe, -2, 2)
```

Output

```
DomainError with -1.0:
integrand produced NaN in the interval (-2.0, 0.0)
```

Sweet! It caught the first vertical asymptote at $x_0 = -1$, even though it was not on the boundary. We push our luck and continue.

```
using QuadGK
fe(x) = x/sqrt(abs(1 - x^2))
# There remains one vertical asymptote within the domain of integration
sol,  = quadgk(fe, -0.99, 2) # -0.99 approx = -1^+ = -1 + a tiny bit
```

Output

```
(1.8731177232031915, 2.1226263027166235e-8)
```

It does not complain about the vertical asymptote at $+1$.

```
using QuadGK
fe(x) = x/sqrt(abs(1 - x^2))
# There remains one vertical asymptote within the domain of integration
sol,  = quadgk(fe, -1.0+1e-4, 2) # -0.9999 approx = -1^+ = -1 + a tiny bit
```

Output

```
(1.746192377358709, 2.5638760207248433e-8)
```

As we will see later, that is a very good approximation of the integral. QuadGK has integrated through the singularity. You can play with the code and try `sol, = quadgk(fe, -1.0+1e-6, 2)` to see what happens.

We know where the vertical asymptotes are. Let us see how `quadgk` does if we use that information.

$\int_{-2}^{-1^-} \frac{x}{\sqrt{|1-x^2|}}\,dx$

```
# Integrate from -2 to -1
using QuadGK
k=collect(1:3:15)
epsilonRange = 10.0.^(-k)
b=-1
for epsilon in epsilonRange
    sol, = quadgk(fe, -2, b-epsilon)
    display([b-epsilon sol])
end
```

Output

```
1×2 Matrix{Float64}:
 -1.1  -1.27379
1×2 Matrix{Float64}:
 -1.0001  -1.71791
1×2 Matrix{Float64}:
 -1.0  -1.7316
1×2 Matrix{Float64}:
 -1.0  -1.73204
1×2 Matrix{Float64}:
 -1.0  -1.73205
```

Awesome! We get that

$$\int_{-2}^{1^-} \frac{x}{\sqrt{|1-x^2|}}\,dx \approx -1.73205.$$

Next, we are going to "cheat" a bit and go for $\int_{-1^+}^{1^-} \frac{x}{\sqrt{|1-x^2|}}\,dx$, even though we should break it into two integrals, over $[-1^+, 0]$ and $[0, 1^-]$.

```
# Integrate from -1 to +1
using QuadGK
k=collect(1:3:15)
epsilonRange = 10.0.^(-k)
a = -1
b = 1
for  epsilon in epsilonRange
    sol, = quadgk(fe, a + epsilon, b - epsilon)
    display([a + epsilon b - epsilon sol])
end
```

Output

```
1×3 Matrix{Float64}:
```

```
 -0.9  0.9  4.66156e-17
1×3 Matrix{Float64}:
 -0.9999  0.9999  1.93258e-16
1×3 Matrix{Float64}:
 -1.0  1.0  1.19294e-15
1×3 Matrix{Float64}:
 -1.0  1.0  6.37856e-16
1×3 Matrix{Float64}:
 -1.0  1.0  -1.35308e-16
```

Sweet, we get zero. Finally, we go for $\int_{1^+}^{2} \frac{x}{\sqrt{|1-x^2|}}\, dx$:

```
# Integrate from +1 to +2
using QuadGK
k=collect(1:3:15)
epsilonRange = 10.0.^(-k)
a = 1
b = 2
for  epsilon in epsilonRange
     sol,  = quadgk(fe, a + epsilon, b - epsilon)
     display([a + epsilon b - epsilon sol])
end
```

Output

```
1×3 Matrix{Float64}:
 1.1  1.9  1.15729
1×3 Matrix{Float64}:
 1.0001  1.9999  1.71779
1×3 Matrix{Float64}:
 1.0  2.0  1.7316
1×3 Matrix{Float64}:
 1.0  2.0  1.73204
1×3 Matrix{Float64}:
 1.0  2.0  1.73205
```

And we obtain $\int_{1^+}^{2} \frac{x}{\sqrt{|1-x^2|}}\, dx \approx 1.73205$. When we add up all the pieces, we obtain zero. ■

Bottom line: QuadGK is a valid option for evaluating improper integrals, though additional care is required.

Recommended Videos for Additional Practice

- 8 Examples of Type II Improper Integrals, Calculus 2 by **BlackPenRedPen**.
- Type II Improper Integrals - Discontinuous Integrands - Convergence and Divergence by K.O. Math has very nice graphics to help with intuition.
- Type I vs Type II Improper Integrals (10 examples) by **BlackPenRedPen**.
- Improper Integrals (6 of 16) Type 2 Example by Prof. Michel van Biezen.

8.4 Improper Integrals through the Lens of Probability

Improper integrals are common in probability theory, the "science of randomness."

Secrets of the Arcane 8.17: Understanding Randomness through Exam Scores

Imagine you're about to take an important exam. You might think that your score will depend solely on how well you've studied. However, even if you're fully prepared, there are countless unforeseen factors that could influence your performance. These factors introduce an element of **randomness** into what your final score might be, where **randomness** refers to the outcome of events that are not certain and the complex ways in which these uncertain events influence outcomes. It's a concept found everywhere around us, from the roll of dice to the weather each day.

Exam Score Factors: Consider your score on an exam as a random variable. This does not mean your score is entirely left to chance; rather, it's affected by a combination of predictable elements (like how much you studied) and unpredictable ones (randomness), such as the factors listed below:

- **Did You Sleep Well?** A good night's sleep is crucial for memory and concentration. Poor sleep could unexpectedly lower your performance.
- **Transportation Troubles**: Being late because the bus was delayed can cause panic and rush, impacting your focus and, consequently, your score.
- **Unexpected News**: Receiving distressing news from a friend right before the exam can unsettle you, or on the flip side, hearing good news (like winning the lottery) might overly excite you, both affecting your concentration.
- **Personal Well-Being**: Sudden headaches, feeling unwell, or even the room being too cold or too hot can distract you during the exam.
- **Antsy People Near You**: A nearby student who twiddles their pencil or keeps moving around in their chair can cause a loss of concentration.
- **The Order the Questions Are Placed on the Exam**: Nailing the first question can give you a sense of calm for proceeding to the next questions, while flailing away on the first question can undermine your confidence.
- **Adequate Time to Study:** Maybe you have a job and had to work overtime, or another course scheduled its exam on the same day with very little notice.

You should agree that most of the factors listed above are not in your direct control and that you cannot predict ahead of time with any degree of certainty whether or not they will occur for you on any given exam. That is what randomness is all about. It's a concept that stretches far beyond the simple outcome of flipping a coin or rolling dice. Randomness infiltrates every aspect of our lives, influencing events in ways that are often subtle, sometimes profound, and always unavoidable.

The subjects of **probability** and **statistics** are the mathematical siblings born out of the need to understand and quantify this randomness. Through these disciplines, humanity has developed a toolkit not just for gauging the likelihood of events but also for making sense of the random variability that surrounds us. You have likely heard the terms **mean**, **variance**, and **standard deviation**, which are critical in this endeavor. Let us delve a bit deeper into how these quantities help us grapple with randomness:

- The **mean**, often referred to as the average, gives us a central value for a set of data. When considering exam scores across many students, the mean score provides a quick snapshot of the overall performance level.
- **Variance** measures how spread out a set of data is. If we're looking at the variance in exam scores, we're essentially asking, "How much do students' scores differ from each other and from the mean?" High variance means a wide dispersion of scores, indicating that factors of randomness may have had varied impacts on the students.

- The **standard deviation**, the square root of the variance, offers a digestible measure of this spread in the same units as the data itself. It tells us, on average, how far each student's exam score deviates from the mean score. Understanding standard deviation allows students to appreciate the diversity of outcomes produced by both controllable factors (like study habits) and uncontrollable factors (like those random events we discussed).

Through probability and statistics, engineers learn to model, analyze, and interpret the effects of randomness. We use these mathematical methods to study random events, enabling us to predict probabilities, assess risks, and make informed decisions—even in the face of uncertainty. So, while we might not be able to control or predict every factor that affects our lives, these disciplines give us a powerful way to understand and manage the complexity of randomness.

8.4.1 What Is a Probability Density?

Formal courses on probability and statistics explore different ways to model (aka mathematically represent) random quantities, including probability measures, distributions, and densities. This section highlights **probability densities**—functions $f : \mathbb{R} \to [0, \infty)$ that are nonnegative, sum up (integrate) to one, and where the integral $\int_a^b f(x)\,dx$ signifies the probability of a random quantity falling within the range $[a, b]$, such as your exam score being between 86% and 93%. Some probability densities are defined on bounded sets, and others on unbounded sets. Those defined on unbounded sets require improper integrals.

8.4.2 The Role of Improper Integrals

Consider a web server where events correspond to users requesting information. The time between one user request and the next is a random variable. The time between these events is often modeled using the exponential distribution, a choice motivated by its so-called "memoryless property": the probability of the next event occurring does not depend on how much time has elapsed since the previous event occurred.

Example 8.18. *The probability density function (PDF) for the* ***exponential distribution*** *is given by*

$$f(x) = \lambda e^{-\lambda x},$$

where $\lambda > 0$ is the rate parameter, and $x \geq 0$ represents the time between events in clock cycles. Given the high clock frequencies of modern processors, the time x can stretch towards infinity, making this an example of an improper integral. For the exponential distribution, perform the following:

(a) Show that the PDF integrates to 1.

(b) Compute the mean value.

(c) Compute the variance.

Solution:

(a) We need to show that $\displaystyle\int_0^\infty \lambda e^{-\lambda x}\,dx = 1$. Let us compute this integral by considering it as an improper integral due to the infinite upper bound:

$$\begin{aligned}\int_0^\infty \lambda e^{-\lambda x}\,dx &= \lim_{t\to\infty} \int_0^t \lambda e^{-\lambda x}\,dx \\ &= \lim_{t\to\infty} \left[-e^{-\lambda x}\right]_0^t \\ &= \lim_{t\to\infty} \left(-e^{-\lambda t} + e^{-\lambda \cdot 0}\right) \\ &= \lim_{t\to\infty} \left(-e^{-\lambda t} + 1\right) \\ &= 1.\end{aligned}$$

(b) In a course on probability and statistics, you will learn that the mean value, often denoted by μ, satisfies:

$$\mu = \int_0^\infty x f(x)\, dx.$$

Specializing this formula to the exponential distribution gives

$$\mu = \int_0^\infty x\lambda e^{-\lambda x}\, dx.$$

To find the mean value μ, we use integration by parts, where $u = x$ and $dv = \lambda e^{-\lambda x} dx$, giving $du = dx$ and $v = -e^{-\lambda x}$. Thus,

$$\begin{aligned}\mu &= -xe^{-\lambda x}\Big|_0^\infty + \int_0^\infty e^{-\lambda x}\, dx \\ &= \lim_{t\to\infty}\left(-te^{-\lambda t}\right) + \left[-\frac{1}{\lambda}e^{-\lambda x}\right]_0^\infty \\ &= 0 + \left(0 - \left(-\frac{1}{\lambda}\right)\right) \\ &= \frac{1}{\lambda}.\end{aligned}$$

(c) In a course on probability and statistics, you will learn that the variance, often denoted by σ^2, satisfies

$$\sigma^2 = \int_0^\infty (x-\mu)^2 f(x)\, dx.$$

Specializing this formula to the exponential distribution gives

$$\sigma^2 = \int_0^\infty (x-\mu)^2 \lambda e^{-\lambda x}\, dx.$$

To find the variance σ^2, we first substitute in $\mu = \frac{1}{\lambda}$ and then expand the quadratic, giving,

$$\sigma^2 = \int_0^\infty \left(x - \frac{1}{\lambda}\right)^2 \lambda e^{-\lambda x}\, dx = \int_0^\infty x^2\lambda e^{-\lambda x}\, dx - 2\frac{1}{\lambda}\int_0^\infty x\lambda e^{-\lambda x}\, dx + \left(\frac{1}{\lambda}\right)^2 \int_0^\infty \lambda e^{-\lambda x}\, dx.$$

The second and third terms have been computed in part (b) and part (a) respectively. For the first term, using integration by parts twice (details left as an exercise), we find:

$$\int_0^\infty x^2 \lambda e^{-\lambda x}\, dx = \frac{2}{\lambda^2}.$$

Thus,

$$\sigma^2 = \frac{2}{\lambda^2} - 2\left(\frac{1}{\lambda}\right)\left(\frac{1}{\lambda}\right) + \left(\frac{1}{\lambda}\right)^2 = \frac{1}{\lambda^2}.$$

■

We focus next on the normal distribution, which is one of the most fundamental distributions in probability and statistics due to the Central Limit Theorem. Unlike the exponential distribution example, the integration to show that the normal distribution density integrates to one is complex and typically beyond the scope of a first calculus course. Hence, we will concentrate on verifying the mean.

Example 8.19. *The probability density function (PDF) for the normal distribution is given by:*

$$f(x) = \frac{1}{\sigma\sqrt{2\pi}} e^{-\frac{1}{2}\left(\frac{x-\mu}{\sigma}\right)^2},$$

where μ is (supposed to represent) the mean, σ is (supposed to be) the standard deviation, and $x \in \mathbb{R}$ represents the value of the random variable. Verify that the mean value really is given by μ.

Solution:
The integral to find the mean μ is

$$\mu = \int_{-\infty}^{\infty} x \cdot \frac{1}{\sigma\sqrt{2\pi}}\, e^{-\frac{1}{2}\left(\frac{x-\mu}{\sigma}\right)^2}\, dx.$$

This is a doubly improper integral. We break it into two parts

$$\int_{\mu}^{\infty} x \cdot \frac{1}{\sigma\sqrt{2\pi}}\, e^{-\frac{1}{2}\left(\frac{x-\mu}{\sigma}\right)^2}\, dx := \lim_{b\to\infty} \int_{\mu}^{b} x \cdot \frac{1}{\sigma\sqrt{2\pi}}\, e^{-\frac{1}{2}\left(\frac{x-\mu}{\sigma}\right)^2}\, dx,$$
$$\int_{-\infty}^{\mu} x \cdot \frac{1}{\sigma\sqrt{2\pi}}\, e^{-\frac{1}{2}\left(\frac{x-\mu}{\sigma}\right)^2}\, dx := \lim_{a\to-\infty} \int_{a}^{\mu} x \cdot \frac{1}{\sigma\sqrt{2\pi}}\, e^{-\frac{1}{2}\left(\frac{x-\mu}{\sigma}\right)^2}\, dx.$$

Focusing on the first integral, substitute $z = \frac{x-\mu}{\sigma}$, thus $dx = \sigma dz$, and $x = \mu + \sigma z$, yielding

$$\lim_{b\to\infty} \int_{0}^{\frac{b-\mu}{\sigma}} (\mu + \sigma z) \frac{1}{\sqrt{2\pi}}\, e^{-\frac{1}{2}z^2}\, \sigma\, dz.$$

Splitting the integral gives

$$\lim_{b\to\infty} \mu \int_{0}^{\frac{b-\mu}{\sigma}} \frac{1}{\sqrt{2\pi}}\, e^{-\frac{1}{2}z^2}\, \sigma\, dz + \lim_{b\to\infty} \sigma \int_{0}^{\frac{b-\mu}{\sigma}} z \frac{1}{\sqrt{2\pi}}\, e^{-\frac{1}{2}z^2}\, \sigma\, dz.$$

To evaluate the first integral, you have to know that the integral of the Gaussian density is 1, and half of the Gaussian density gives $\frac{1}{2}$. The second integral can be done by u-substiution and yields one. Hence,

$$\lim_{b\to\infty} \int_{0}^{\frac{b-\mu}{\sigma}} (\mu + \sigma z) \frac{1}{\sqrt{2\pi}}\, e^{-\frac{1}{2}z^2}\, \sigma\, dz = \frac{\mu}{2} + \sigma.$$

The second integral is similar and yields,

$$\lim_{a\to-\infty} \int_{a}^{\mu} x \cdot \frac{1}{\sigma\sqrt{2\pi}}\, e^{-\frac{1}{2}\left(\frac{x-\mu}{\sigma}\right)^2}\, dx = \frac{\mu}{2} - \sigma.$$

Adding them up completes the verification. ■

Example 8.20. *While proving that the Gaussian density really does integrate to 1 is a bit advanced for a first calculus course, showing that the integral exists and is bounded is much easier. So that you can focus on the essential elements, prove that $\int_0^{\infty} e^{-x^2}\, dx$ exists and is finite.*

Solution: We use the Comparison Test. For $x \geq 1$, $e^{-x^2} \leq x\, e^{-x^2}$. Hence,

$$\begin{aligned}
\int_1^{\infty} e^{-x^2}\, dx &\leq \int_1^{\infty} x\, e^{-x^2}\, dx \\
&\leq \int_0^{\infty} x\, e^{-x^2}\, dx \quad \text{(because the integrand is nonnegative)} \\
&\leq \frac{1}{2} \int_0^{\infty} e^{-u}\, du \quad \text{(u-substitution } u = x^2, du = 2x dx\text{)} \\
&\leq \frac{1}{2} e^{-u}\Big|_0^{\infty} \\
&\leq \frac{1}{2}.
\end{aligned}$$

Because e^{-x^2} is continuous, $\int_0^1 e^{-x^2}\, dx$ exists and is finite. Hence, by combining this with the previous bound, we obtain the desired result.

Alternatively, one can apply the Comparison Test like this: for $x \geq 1$, $e^{-x^2} \leq e^{-x}$. We ask the reader to adapt the above proof to this new bound. ■.

Functions of the form $c\frac{1}{x^k}$, $x \geq x_{\text{min}} > 0$ are called **power-law densities** or **Pareto densities**. They occur in Psychology to model forgetting, Climate Science to model energy dissipation in cyclones, in Computer Engineering to model the relationship between a CPU's cache size and the number of cache misses, and in Economics to model relationships varying from the distribution of artists by the average price of their artworks to income distribution in a market economy.

Example 8.21. *Assume a power-law density $f(x) = c\frac{1}{x^3}$, $x \in [1, \infty)$. Perform the following:*

(a) Find c such that f is a density.

(b) Compute the mean value.

(c) Compute the variance.

Solution:

(a) To find c such that f is a density, we integrate $f(x)$ over its domain and set the integral equal to 1, namely

$$1 = \int_1^\infty c\frac{1}{x^3}\,dx = c\frac{x^{-2}}{-2}\bigg|_1^\infty = \frac{c}{2}.$$

Using this result gives $c = 2$.

(b) The mean value μ is computed as,

$$\mu = \int_1^\infty x f(x)\,dx = \int_1^\infty x\frac{2}{x^3}\,dx = 2\frac{x^{-1}}{-1}\bigg|_1^\infty = 2.$$

(c) The variance σ^2 is theoretically obtained via,

$$\sigma^2 = \int_1^\infty (x-\mu)^2 f(x)\,dx.$$

However, the improper integral is divergent. In a course on probability, one would say that the infinite variance is due to the power-law's "heavy tail" (meaning, the tail of the density does not drop off to zero rapidly enough to yield a finite integral).

To see qualitatively that the integral is divergent, note that x^2 times $\frac{1}{x^3}$ behaves like $\frac{1}{x}$, which is the line of demarcation for convergent and divergent integrals, as shown in the preliminaries to Section 8.2.1. To show divergence analytically, note that

$$(x-\mu)^2 f(x) = \left(x^2 - 2\mu x + \mu^2\right)\frac{2}{x^3},$$

and that

$$\int_1^\infty x^2\frac{2}{x^3}\,dx = \int_1^\infty \frac{2}{x}\,dx = 2\ln(x)\Big|_1^\infty = \infty,$$

while

$$\int_1^\infty (-2\mu x)\frac{2}{x^3}\,dx = -4\int_1^\infty x\frac{2}{x^3}\,dx = -8$$

and

$$\int_1^\infty \mu^2\frac{2}{x^3}\,dx = 4\int_1^\infty x\frac{2}{x^3}\,dx = 8.$$

We computed the last two integrals just to assure ourselves that we did not land in $\infty - \infty$, which would have placed us in the "undefined" category, instead of divergent.

■

8.5 (Optional Read) Proofs Associated with the Chapter

Proof of Prop. 8.12 (Usefulness of Absolute Integrability)

Suppose that $f : [a, \infty) \to \mathbb{R}$ is piecewise continuous.

(a) If f is absolutely integrable, then $\int_a^\infty f(t)\,dt$ (without absolute value signs) exists and is finite.

(b) If $\lim_{t\to\infty} f(t)$ exists and is nonzero, then f is not absolutely integrable.

(c) If f is differentiable on (a, ∞) and f' (its derivative) is absolutely integrable, then $\lim_{t\to\infty} f(t)$ exists and is finite.

(d) If both f and f' (its derivative) are absolutely integrable on (a, ∞), then $\lim_{t\to\infty} f(t) = 0$.

Proof:

(a) Decompose f into $f^+ := \max\{f, 0\}$ and $f^- := \min\{f, 0\}$. Both f^+ and f^- are piecewise continuous since f is. The absolute integrability of f means $\int_a^\infty |f(t)|\,dt < \infty$.

We note that by construction, $f = f^+ + f^-$, and $|f| = f^+ - f^-$. Moreover, for all $t \in [a, \infty)$, $0 \le f^+(t) \le |f(t)|$ and $0 \le -f^-(t) \le |f(t)|$. Hence, by the Comparison Test, both f^+ and $-f^-$ are integrable[3] over $[a, \infty)$. Because $-f^-$ is integrable, so is its negative, $-(-f^-) = f^-$. Then, because the sum of two integrable functions over $[a, \infty)$ is also integrable, the proof is complete.

The last step can also be written as: Because

$$\lim_{T\to\infty} \int_a^T f^+(t)\,dt \ \text{ exists and is finite, and because}$$

$$\lim_{T\to\infty} \int_a^T f^-(t)\,dt \ \text{ exists and is finite, it follows that their sum,}$$

$$\lim_{T\to\infty} \int_a^T f^+(t)\,dt + \lim_{T\to\infty} \int_a^T f^-(t)\,dt = \lim_{T\to\infty} \int_a^T \underbrace{\left(f^+(t) + f^-(t)\right)}_{f(t)}\,dt, \ \text{ exists and is finite.}$$

(b) We assume that $L := \lim_{t\to\infty} f(t)$ exists and is nonzero. Moreover, we'll assume that $L > 0$ as the proof for $L < 0$ follows the same reasoning. Because the limit exists, for $\epsilon = \frac{L}{2} > 0$, there exists $T < \infty$ such that $|f(t) - L| \le \frac{L}{2}$ for all $t \ge T$, which in turn implies that $f(t) \ge \frac{L}{2}$. Because a positive constant $\frac{L}{2}$ is not absolutely integrable[4], by the Comparison Test, neither is $f(t)$.

(c) By the Fundamental Theorem of Calculus, $f(T) = f(a) + \int_a^T f'(t)\,dt$. Then, by part (a), because $f'(t)$ is absolutely integrable,

$$\lim_{T\to\infty} f(t) = \lim_{T\to\infty} \left(f(a) + \int_a^T f'(t)\,dt \right) = f(a) + \lim_{T\to\infty} \int_a^T f'(t)\,dt$$

exists and is finite.

(d) By part (b), if we assume in addition that f is absolutely integrable, then it cannot be the case that the limit exists and is nonzero. But by (c), the limit does exist, and hence, it must be zero.

■

[3]Means, $\lim_{b\to\infty} \int_a^b f^+(t)\,dt$ exists and is finite; same for $-f^-$.

[4] $\lim_{T\to\infty} \int_a^T \frac{L}{2}\,dt = \lim_{T\to\infty} (T-a)\,\frac{L}{2} = \infty$ when $\frac{L}{2} > 0$.

Chapter 9

Ordinary Differential Equations

Learning Objectives

By completing this chapter, students will:

- Gain an introduction to differential equations, focusing on their significance and the various classifications.
- Learn the foundational theories and techniques for solving one-dimensional first-order ordinary differential equations (ODEs).
- Acquire knowledge on the fundamentals and solution methods for vector first-order ODEs.
- Explore the application of numerical methods in solving vector first-order ODEs.
- Delve into the characteristics of linear systems of first-order ODEs and their implications for mathematical modeling.

Outcomes

Upon mastering the content of this chapter, students will be equipped to:

- Understand the definition and classification of ODEs based on their order, linearity, and whether they are homogeneous or non-homogeneous.
- Engage with various strategies for tackling first-order ODEs, covering separable, linear, and select nonlinear equations.
- Dive into the theory and application of solving systems of linear differential equations, highlighting their importance across diverse applications.
- Comprehend the utilization of ODEs in modeling and solving real-world problems across various domains.
- Grasp the fundamentals of numerical methods designed for solving ODEs in scenarios where analytic solutions are challenging or unattainable.
- Deeply investigate linear systems of ODEs, focusing on:
  - The structure and solution of linear vector ODEs, symbolized as $\dot{x} = Ax, x(t_0) = x_0$.
  - The concept and significance of the matrix exponential, denoted as e^{At}.
  - Essential attributes of the matrix exponential, such as $\frac{d}{dt}e^{At} = Ae^{At} = e^{At}A$.
  - Methods for solving linear vector ODEs, exemplified by $x(t) = e^{A(t-t_0)}x_0$.

- The role of complex scalar exponentials, $e^{(a+i\omega)t}$, in the context of ODE solutions.
- The interplay between eigenvectors, eigenvalues, and the matrix exponential, such as $Av = \lambda v \implies e^{A(t-t_0)}v = e^{\lambda(t-t_0)}v$.
- The criteria for stability in linear systems, $\text{real}(\lambda) < 0 \implies \lim_{t\to\infty} e^{A(t-t_0)}v = 0_{n\times 1}$, underlining how the real parts of eigenvalues influence the system's long-term behavior.
- The application of these principles to general initial conditions when A has a complete set of eigenvectors.

(a) (b) (c)

Figure 9.1: What do we have in common? (a) Illustration of the solar system. (b) Space-X landing a booster. (c) Wandercraft's exoskeleton allows patients with paraplegia (paralysis from the waist down) to walk again. **Credits:** (a) NASA and rawpixel, and (c) TF1 Info: Wandercraft fait remarcher les paraplégiques grâce à un exosquelette.

9.1 Introduction

Figure 9.1 illustrates three "objects" where **ordinary differential equations (ODEs)** are required to understand how they function. A differential equation is an equation that includes one or more derivatives, such as

$$\dot{x}(t) + x(t) - \sin(t) = 0, \quad \text{or}$$
$$\ddot{x}(t) + \dot{x}(t) + (x(t) + 1)^3 = 0;$$

if there were no derivatives, we could simply solve the equations for the unknown $x(t)$ and be done with it. Then what makes an equation with derivatives "ordinary"? It is an "ordinary" differential equation if the included derivatives are all with respect to the same scalar variable. If there are derivatives with respect to several variables (or a vector of variables), then they are called partial derivatives, and we would have a partial differential equation, or PDE for short. We've encountered a few ODEs already, in Example 5.48 dealing with the total derivative, and in Section 6.4 on Lagrangian Dynamics, where we derived the equations of motion for a point mass in $\mathbb{R}^2$, a pendulum, a three-link manipulator, and a three-link bipedal walker.

In the present chapter, we explore what it means for a function (eventually, a vector of functions) to be a solution to an ODE. As with antiderivatives, we will learn that, in some cases, a closed-form solution to an ODE can be found, whereas in other cases its solution cannot be expressed in terms of elementary functions, in which case we rely on numerical methods to determine a solution.

Secrets of the Arcane 9.1: Understanding the Solar System Has Been a Major Driver of Scientific Inquiry

Watch: We Did This by Neil deGrasse Tyson. Differential equations are amazing!

The understanding of the solar system has evolved significantly over the centuries[a]. In the beginning, we humans postulated paths or orbits for the planets, such as perfect circles around the earth. Eventually, data showed that the circular model was wrong, and it was replaced with ellipses around the sun. These early models were **descriptive**, meaning they described the observational data but did not attempt to explain the "how" or the "why" of the planets' orbits.

Newton proposed the first **mechanistic** model, with his famous law of gravitational attraction. It was eventually realized that Newton's laws were highly accurate for all of the planets except Mercury, which presented a significant puzzle to astronomers for many years. The issue was with its perihelion, the point in its orbit closest to the Sun. Over time, the perihelion of Mercury was observed to advance or shift in a way that couldn't be explained by Newtonian mechanics. *Mercury's perihelion advanced by an additional 43 arcseconds per century more than what was predicted by Newton's laws.*

Mercury's orbital discrepancy was resolved by Albert Einstein's general theory of relativity in 1915, which is grounded in an advanced form of calculus on surfaces. General relativity describes gravity not as a force, but as a curvature of spacetime caused by mass. According to this theory, the path of a planet is a geodesic (a path of shortest distance, such as a great circle route on the earth) in the curved spacetime caused by the Sun.

Timeline

1. **Geocentric Model (Ancient Times - 16th Century)**: *Ancient Greece (circa 4th century BCE)*: The geocentric model, with Earth at the center of the universe, was developed by Greek philosophers like Aristotle and later refined by Ptolemy (Egypt) in the 2nd century ACE.
2. **Nicolaus Copernicus (1543)**: *Copernicus* proposed the heliocentric model in *De revolutionibus orbium coelestium,* suggesting that Earth and other planets revolve around the Sun.
3. **Tycho Brahe (late 16th century)**: *Brahe* proposed a hybrid model combining aspects of both geocentric and heliocentric theories.
4. **Kepler's Laws of Planetary Motion (Early 17th Century)**: *Johannes Kepler (1609 – 1619)* formulated his three laws of planetary motion, describing elliptical orbits of planets around the Sun.
5. **Newton's Law of Universal Gravitation (Late 17th Century)**: *Isaac Newton (1687)*, in his *Philosophiae Naturalis Principia Mathematica*, formulated the law of universal gravitation, explaining planetary motions and laying the groundwork for Newtonian physics.
6. **Einstein's Theory of General Relativity (Early 20th Century)**: *Albert Einstein (1915)*: Einstein's theory of general relativity further refined the understanding of gravitational forces and introduced the concept of spacetime curvature.

[a] The day the universe changed by James Burke. The video starts off quite slow, but gets very interesting at 33 minutes, where you see calculus thinking being employed 100 years before calculus was formally invented by Newton and Leibniz.

9.2 Let's Start Simple: One Equation, One Unknown, One Derivative

Our mission is to understand how ODEs can be used to model physical phenomena and how to "solve" some of the simplest possible ODEs. In fact, part of our mission is to understand what it even means to be a solution to an ODE.

Fact 9.2: Four Models of Chutes

We use Newton's Law, $F = ma$, written as $m\frac{dv(t)}{dt} = F$ to model an object braking under air resistance[a] while rolling on flat ground and an object falling vertically under the combined influence of gravity and air resistance. In the following, v (m/s) is the velocity of the object, m (kg) is the mass, κ (units depend on the model) is a drag coefficient, and g (m/s^2) is the gravitational constant.

(a) **Model 1: Horizontal Motion, Linear Drag:** $m\frac{dv(t)}{dt} = -\kappa v(t), \;\; v(t_0) = v_0.$

(b) **Model 2: Horizontal Motion, Quadratic Drag:** $m\frac{dv(t)}{dt} = -\kappa v^2(t), \;\; v(t_0) = v_0.$

(c) **Model 3: Vertical (Falling) Motion, Linear Drag:** $m\frac{dv(t)}{dt} = -\kappa v(t) - mg, \;\; v(t_0) = v_0.$

(d) **Model 4: Vertical (Falling) Motion, Quadratic Drag:** $m\frac{dv(t)}{dt} = \kappa v^2(t) - mg, \;\; v(t_0) = v_0.$

The term "$v(t_0) = v_0$" is called the **initial condition** of the differential equation. We understand intuitively that the velocity of a ball at some future time depends on its velocity when it leaves our hand. For example, if you drop a ball at rest versus tossing it in the air at one meter per second, you know that its evolution will be different. Yet, the moment the ball leaves your hand, the physics is the same: the ball evolves under the action of gravity and air resistance.

Regarding the signs in Models 2 and 4, the correct model for drag is $-\kappa|v(t)| \cdot v(t)$, so that the **drag force acts opposite to the direction of motion.** Figure 9.2 shows two examples of chutes being used as brakes. For the Space Shuttle orbiter, we can assume that it is moving from left to right at touchdown, and hence $v(t) > 0$. Therefore, the model for drag simplifies to $-\kappa v^2(t)$. For a parachutist, however, $v(t) < 0$, and therefore, the model for drag simplifies to $+\kappa v^2(t)$, where the plus sign is included to emphasize that the drag force is acting "opposite to the downward force of gravity" for the parachutist.

Note: The **order of an ordinary differential equation** is the degree of the highest derivative appearing in the equation. A **homogeneous ordinary differential equation (ODE)** is a type of differential equation in which every term is a function of the dependent variable and its derivatives. In particular, there are no terms that are just constants or functions of the independent variable alone. Hence, in the above, Model 1 is a first-order linear homogeneous ODE and Model 2 is a first-order nonlinear homogeneous ODE. Models 3 and 4 are first-order ODEs, but they are not homogeneous due to the inclusion of gravity in the model. Models that are not homogeneous are **inhomogeneous**, but if you say, **non-homogeneous**, everyone will understand you just fine.

[a] The video Physics of a Parachute by EA Reach provides an explanation of drag, aka air resistance. The only significant error in the video is the claim that the opening of a parachute "initially causes a parachutist to move upward until gravity gradually takes over." While it surely feels that way to a parachutist, our senses often lie to us. Can you explain the feeling? The key is in Fig. 9.4-(b).

Informal Definition of a Solution: A differentiable function of time, $\varphi(t)$, is a solution to the ODE $\frac{dx(t)}{dt} = f(x(t))$, with $x(t_0) = x_0$, if (i) it satisfies the initial condition, meaning, $\varphi(t_0) = x_0$ and (ii) it and its derivative together satisfy $\frac{d\varphi(t)}{dt} = f(\varphi(t))$; i.e., the left side equals the right side, as in any equation!

Note: For $\frac{dx(t)}{dt} = f(x(t))$ with $x(t_0) = x_0$, it is common practice to denote the solution by $x(t)$, the variable used in writing down the ODE, instead of using a different variable name, such as φ. For the drag chute and parachute models in Fact 9.2, we will consequently denote their solutions as $v(t)$.

(a) (b)

Figure 9.2: Using a chute as a brake. (a) A drag chute used for braking the Space Shuttle orbiter Endeavour ; and (b) a parachutist. **Credits:** (a) NASA and (b) pixabay.

9.2.1 Analytical Solutions via Antiderivatives

Example 9.3. *Compute and plot a solution to the first-order linear homogeneous ordinary differential equation*

$$m\frac{dv(t)}{dt} = -\kappa v(t), \quad v(t_0) = v_0. \tag{9.1}$$

Solution: **Ans.** $v(t) = v_0 \cdot e^{-\frac{\kappa}{m}\cdot(t-t_0)}$.

First, we write v in place of $v(t)$ and replace dt with $d\tau$. Why? The next steps will be more clear when we use something other than t as the dummy variable of integration. Proceeding,

$$\begin{aligned}
\frac{dv}{d\tau} &= -\frac{\kappa}{m}v \\
&\Updownarrow \quad (\text{multiply both sides by } \frac{d\tau}{v}) \\
\frac{dv}{d\tau}\cdot\frac{d\tau}{v} &= -\frac{\kappa}{m}v\cdot\frac{d\tau}{v} \\
&\Updownarrow \\
\frac{dv}{v} &= -\frac{\kappa}{m}\,d\tau.
\end{aligned}$$

We now have v on one side of the equation and τ on the other side. It is for this reason that the method is called **separation of variables**. We next integrate both sides to obtain:

$$\int_{v_0}^{v(t)} \frac{dv}{v} = -\frac{\kappa}{m}\int_{t_0}^{t} d\tau \quad (\text{this is why we wanted } \tau \text{ instead of } t \text{ as the independent variable}).$$

In the above, t_0 is the time the orbiter *Endeavour* touches the runway, and $v_0 = v(t_0)$ is the initial velocity of *Endeavour* at touchdown; often, we take $t_0 = 0$, but we could also use a different time reference, such as the time since the orbiter reentered the atmosphere.

We know antiderivatives for both integrands, giving us

$$\begin{aligned}
\ln(v)\Big|_{v_0}^{v(t)} &= -\frac{\kappa}{m}\cdot\tau\Big|_{t_0}^{t} \\
&\Updownarrow \text{ (using } \ln(v(t)) - \ln(v_0) = \ln\left(\frac{v(t)}{v_0}\right) \\
\ln\left(\frac{v(t)}{v_0}\right) &= -\frac{\kappa}{m}\cdot(t-t_0) \\
&\Updownarrow \text{ (take exponential of both sides)} \\
\frac{v(t)}{v_0} &= e^{-\frac{\kappa}{m}\cdot(t-t_0)} \\
&\Updownarrow \text{ (multiply both sides by } v_0) \\
v(t) &= v_0\cdot e^{-\frac{\kappa}{m}\cdot(t-t_0)}.
\end{aligned}$$

This leads to the general solution to the ODE as

$$\boxed{m\frac{dv(t)}{dt} = -\kappa v(t),\;\; v(t_0) = v_0 \iff v(t) = v_0\cdot e^{-\frac{\kappa}{m}\cdot(t-t_0)},} \tag{9.2}$$

which was the main point of the exercise.

Bonus: Let us explore the result further because we are engineers. What should the value of κ be so that in 3,000 m the speed of the orbiter is reduced to 10% of its landing speed of 120 m/s (over 400 km/hr), given that it weighs 78,000 kg? The assumption is that friction brakes could take over at that point.

Well, distance is the integral of speed. If we set $t_0 = 0$, we have

$$x(t) = \int_0^T v_0\cdot e^{-\frac{\kappa}{m}\cdot\tau}\,d\tau = v_0\frac{m}{\kappa}\cdot\left(1-e^{-\frac{\kappa}{m}\cdot T}\right),$$

where T is the time it takes to reduce the velocity to $0.1v_0$. Using our expression for velocity, we compute that T satisfies

$$0.1v_0 = v_0\cdot e^{-\frac{\kappa}{m}\cdot T} \iff e^{-\frac{\kappa}{m}\cdot T} = 0.1$$

Substituting this into our formula for distance, we have

$$3000 = v_0\frac{m}{\kappa}\cdot\left(1-e^{-\frac{\kappa}{m}\cdot T}\right) = v_0\frac{m}{\kappa}\cdot(1-0.1) \implies \boxed{\kappa = \frac{0.9\cdot v_0\cdot m}{3000} = 2808}.$$

A plot of $v(t)$ vs. t is given in Fig. 9.3-(a). The red dot marks the time at which the orbiter has traveled 3 km and reduced its speed to 12 m/s or approx 27 mph. The runway is 4 km long. (See **Table 9.1** for examples of everyday speeds in both mph and m/s.) ■

Videos on the Method of Separation of Variables

- Separable differential equations introduction: First order differential equations by Khan Academy.
- Separable First Order Differential Equations - Basic Introduction by The Organic Chemistry Tutor.
- Separation of Variables // Differential Equations by Dr. Trefor Bazett.

Example 9.4. *Compute and plot a solution to the first-order nonlinear, homogeneous ordinary differential equation,*

$$m\frac{dv(t)}{dt} = -\kappa v^2(t),\;\; v(t_0) = v_0. \tag{9.3}$$

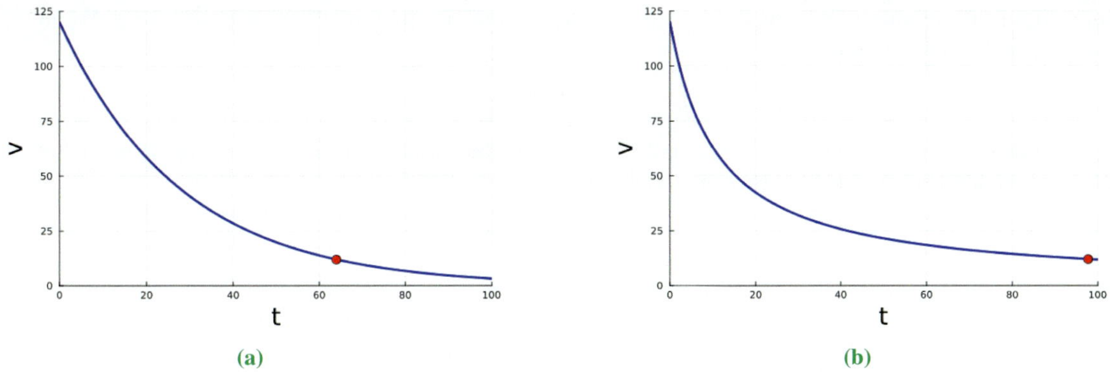

Figure 9.3: Velocity profiles of the Space Shuttle orbiter *Endeavour* braking with a chute in meters per second (m/s). (a) Linear drag model (Example 9.3). (b) Nonlinear drag model (Example 9.5). The linear model optimistically predicts that *Endeavour*'s speed is reduced to 12 m/s approximately 30 seconds sooner than the (probably more accurate) nonlinear drag model.

Table 9.1: Everyday speed examples in mph and m/s. For a quick conversion, using a factor of two is not bad, while 2.2 and 0.45 get you even closer.

Example	**Speed (mph)**	**Speed (m/s)**
Average human walking	3.1	1.4
Cyclist (moderate speed)	15	6.7
Car (city driving)	30	13.4
Car (highway speed)	60	26.8
Propeller airplane	150	67.1
Commercial jet	550	245.9
High-speed train	200	89.4
Cheetah	75	33.5
Sound speed at sea level	767	343

Solution: **Ans.** $v(t) = \dfrac{v_0}{1 + v_0 \cdot \frac{\kappa}{m} \cdot (t - t_0)}$.

Things will be a bit clearer in the following steps if we write v in place of $v(t)$ and replace dt with $d\tau$. Doing so yields:

$$\frac{dv}{d\tau} = -\frac{\kappa}{m}v^2$$

$$\Updownarrow \quad \text{(multiply both sides by } \frac{d\tau}{v^2}\text{)}$$

$$\frac{dv}{d\tau} \cdot \frac{d\tau}{v^2} = -\frac{\kappa}{m}v^2 \cdot \frac{d\tau}{v^2}$$

$$\Updownarrow$$

$$\frac{dv}{v^2} = -\frac{\kappa}{m}\,d\tau.$$

Separation of variables: We now have v on one side of the equation and τ on the other side. We next integrate both sides to obtain

$$\int_{v_0}^{v(t)} \frac{dv}{v^2} = -\frac{\kappa}{m}\int_{t_0}^{t} d\tau.$$

In the above, t_0 is the time of touchdown, and $v_0 = v(t_0)$ is the initial velocity of *Endeavour* at touchdown. We know antiderivatives for these two integrands, giving us

$$\begin{aligned}
-\frac{1}{v}\Bigg|_{v_0}^{v(t)} &= -\frac{\kappa}{m}\cdot\tau\Bigg|_{t_0}^{t} \\
&\Updownarrow \\
-\left(\frac{1}{v(t)} - \frac{1}{v_0}\right) &= -\frac{\kappa}{m}\cdot(t - t_0) \\
&\Updownarrow \\
-\frac{1}{v(t)} &= -\frac{1}{v_0} - \frac{\kappa}{m}\cdot(t - t_0) \\
&\Updownarrow \ (\text{solve for } v(t)) \\
v(t) &= \frac{1}{\frac{1}{v_0} + \frac{\kappa}{m}\cdot(t - t_0)} \\
&\Updownarrow \\
v(t) &= \frac{v_0}{1 + v_0\cdot\frac{\kappa}{m}\cdot(t - t_0)}.
\end{aligned}$$

This leads to the general solution to the ODE as:

$$\boxed{m\frac{dv(t)}{dt} = -\kappa v^2(t),\ v(t_0) = v_0 \Longleftrightarrow v(t) = \frac{v_0}{1 + v_0\cdot\frac{\kappa}{m}\cdot(t - t_0)},} \tag{9.4}$$

which was the main point of the exercise.

Bonus: With this nonlinear model of drag, what should the value of κ be so that in 3,000 m, the speed of the orbiter is reduced to 10% of its landing speed of 120 m/s (over 400 km/hr), given that it weighs 78,000 kg? The assumption, again, is that friction brakes could take over at that point.

The time it takes to slow down to $0.1v_0$ is

$$0.1v_0 = \frac{v_0}{1 + v_0\cdot\frac{\kappa}{m}\cdot T} \Longrightarrow 10 = 1 + v_0\cdot\frac{\kappa}{m}\cdot T \Longrightarrow T = \frac{9m}{\kappa v_0}.$$

With this nonlinear model, integrating velocity to obtain distance will require a u-substitution, $u = 1 + v_0\cdot\frac{\kappa}{m}t$, $du = v_0\cdot\frac{\kappa}{m}\,dt$, resulting in

$$\begin{aligned}
x(T) &= \int_{t=0}^{t=T} \frac{v_0}{1 + v_0\cdot\frac{\kappa}{m}\cdot t}\,dt \\
&= \int_{u=1}^{u=10} \frac{m}{\kappa}\frac{1}{u}\,du \\
&= \frac{m}{\kappa}\cdot\ln(10).
\end{aligned}$$

Imposing $3000 = \frac{m}{\kappa}\cdot\ln(10) \Longrightarrow \boxed{\kappa = \frac{m}{3000}\cdot\ln(10) = 59.8}$. A plot of $v(t)$ vs. t is given in **Fig. 9.3-(b)**.

Note: The required drag coefficient predicted by the nonlinear model is very different from the value predicted by the linear model, $\kappa = 59.8$ versus $\kappa = 2808$. This is because of the extra factor of v in the nonlinear drag model, $\kappa \cdot v \cdot v$, versus the linear drag model $\kappa \cdot v$. Dividing the linear drag coefficient by 120 gives 23.4, which is in the same ballpark. When you understand that a linear approximation of $\kappa \cdot v \cdot v$ about the landing speed is $2\kappa \cdot v_0 \cdot (v - v_0)$, the values match up even better, $\kappa = 59.8$ versus $\kappa = 46.8$. ■

We now turn to analyzing the models for a parachutist,[1] where gravity has an effect. The differential equations are now inhomogeneous. For the linear model, we need a new idea when seeking to compute a solution.

Fact 9.5: Integrating Factors and Integration by Parts

The concept of an **integrating factor** in solving differential equations and the technique of **integration by parts** in integral calculus are both based on the **product rule of differentiation**.

1. **Integration by Parts in Integral Calculus:**
   - Integration by parts is a technique derived from the product rule of differentiation. It is used to solve integrals involving the product of two functions and is expressed as $\int u\,dv = uv - \int v\,du$.
   - It allows you to exchange the task of finding an antiderivative for $\int u\,dv$ for the "hopefully" easier task of finding an antiderivative for $\int v\,du$.
   - This formula is a direct consequence of the product rule. If you take the derivative of the product of two functions $u \cdot v$ using the product rule and then integrate both sides, you arrive at the integration by parts formula.
2. **Integrating Factor in Differential Equations:**
   - In the context of first-order linear differential equations, the integrating factor method involves finding a function (the integrating factor) that, when multiplied with the original differential equation, transforms the original ODE into a form where the left side becomes a total differential, as in $d(u \cdot v)$, making it trivial to integrate. The "hope" is that an antiderivative can be found for the right side of the equation.
   - For a differential equation of the form $\frac{dy(t)}{dt} + P(t)y(t) = Q(t)$, an integrating factor is $e^{\int_a^t P(\tau)\,d\tau}$, where a is an arbitrary constant, because:
     - Multiplying both sides of the ODE by $e^{\int_a^t P(\tau)\,d\tau}$ gives
$$\left(e^{\int_a^t P(\tau)\,d\tau}\right) \cdot \left(\frac{dy(t)}{dt} + P(t)y(t)\right) = \left(e^{\int_a^t P(\tau)\,d\tau}\right) \cdot Q(t).$$
     - By the Chain Rule and the First Fundamental Theorem of Calculus,
$$\frac{d}{dt}\left(e^{\int_a^t P(\tau)\,d\tau}\right) = \left(e^{\int_a^t P(\tau)\,d\tau}\right) \cdot \frac{d}{dt}\int_{t_0}^{t} P(\tau)\,d\tau = e^{\int_a^t P(\tau)\,d\tau} \cdot P(t).$$
     - By the Product Rule,
$$\begin{aligned}\frac{d}{dt}\left(y(t) \cdot e^{\int_a^t P(\tau)\,d\tau}\right) &= \frac{dy(t)}{dt} \cdot \left(e^{\int_a^t P(\tau)\,d\tau}\right) + y(t) \cdot \frac{d}{dt}\left(e^{\int_a^t P(\tau)\,d\tau}\right) \\ &= \frac{dy(t)}{dt} \cdot e^{\int_a^t P(\tau)\,d\tau} + y(t) \cdot e^{\int_a^t P(\tau)\,d\tau} \cdot P(t).\end{aligned}$$
     - Hence, $e^{\int_a^t P(\tau)\,d\tau} \cdot \left(\frac{dy(t)}{dt} + P(t)y(t)\right) = \frac{d}{dt}\left(y(t) \cdot e^{\int_a^t P(\tau)\,d\tau}\right)$ is a total differential, making it trivial to integrate (aka anti-differentiate).
     - Then, as in Integration by Parts, we "hope" that we can find an antiderivative for the right side of the equation,
$$Q(t) \cdot e^{\int_a^t P(\tau)\,d\tau}.$$

Note: When computing the integrating factor, $e^{\int_a^t P(\tau)\,d\tau}$, we are free to choose $\boldsymbol{a}$ to simplify the integral, $\int_{\boldsymbol{a}}^t P(\tau)\,d\tau$. Zero and minus infinity are often convenient values.

[1] A "parachutist" is someone who is trained and experienced in the art of skydiving. They have likely completed many jumps and have received formal training and certification. On the other hand, a "parachuter" is someone who simply jumps out of a plane with a parachute, without necessarily having any formal training or experience.

Example 9.6. *Compute and plot a solution to the first-order linear inhomogeneous ODE*

$$m\frac{dv(t)}{dt} = -\kappa v(t) - mg, \quad v(t_0) = v_0. \tag{9.5}$$

Solution: Ans. $v(t) = v_0 e^{-\frac{\kappa}{m}(t-t_0)} - \frac{mg}{\kappa}\cdot\left(1 - e^{-\frac{\kappa}{m}(t-t_0)}\right)$.

Given the differential equation $m\frac{dv(t)}{dt} = -\kappa v(t) - mg$, we can rewrite it as:

$$\frac{dv(t)}{dt} + \frac{\kappa}{m}v(t) = -g.$$

The solution to this equation can be found using an **integrating factor**, as explained in Fact 9.5. In this case, $P(t) = \frac{\kappa}{m}$, which has antiderivative $\frac{\kappa}{m}\cdot t + a$. Setting $a = 0$ gives the integrating factor as:

$$e^{\frac{\kappa}{m}t}.$$

As developed in Fact 9.5, the integrating factor has the "magical property" that when we multiply both sides of the ODE by $e^{\frac{\kappa}{m}t}$, we have:

$$e^{\frac{\kappa}{m}t}\frac{dv(t)}{dt} + \frac{\kappa}{m}e^{\frac{\kappa}{m}t}v(t) = -ge^{\frac{\kappa}{m}t},$$

and the left side of the equation is equal to $\frac{d}{dt}\left(e^{\frac{\kappa}{m}t}v(t)\right)$. Therefore, integrating both sides with respect to t gives:

$$\int \frac{d}{dt}\left(e^{\frac{\kappa}{m}t}v(t)\right)dt = -\int ge^{\frac{\kappa}{m}t}dt.$$

Determining an antiderivative for the left-hand side, which was the job of the integrating factor, is now trivial. Determining an antiderivative for the right-hand side is straightforward because it involves an exponential. Integrating both sides, we therefore obtain:

$$e^{\frac{\kappa}{m}t}v(t) = -\frac{mg}{\kappa}\cdot e^{\frac{\kappa}{m}t} + C,$$

where C is the constant of integration. To find C, we use the initial condition $v(t_0) = v_0$, per

$$e^{\frac{\kappa}{m}t_0}v_0 = -\frac{mg}{\kappa}\cdot e^{\frac{\kappa}{m}t_0} + C.$$

Solving for C and then substituting back into the equation for $v(t)$ yields the final solution:

$$v(t) = v_0 e^{-\frac{\kappa}{m}(t-t_0)} - \frac{mg}{\kappa}\left(1 - e^{-\frac{\kappa}{m}(t-t_0)}\right).$$

Thus, the general solution to the ODE is:

$$\boxed{m\frac{dv(t)}{dt} = mg - \kappa v(t), \; v(t_0) = v_0 \iff v(t) = v_0 e^{-\frac{\kappa}{m}(t-t_0)} - \frac{mg}{\kappa}\left(1 - e^{-\frac{\kappa}{m}(t-t_0)}\right).} \tag{9.6}$$

The limit as $t \to \infty$ is called the "terminal velocity." No, it's not the speed that terminates a beginner parachuter, it's the steady-state speed reached by the parachutist,

$$\lim_{t\to\infty} v(t) = \lim_{t\to\infty} v_0 e^{-\frac{\kappa}{m}(t-t_0)} - \frac{mg}{\kappa}\left(1 - e^{-\frac{\kappa}{m}(t-t_0)}\right) = -\frac{mg}{\kappa}.$$

From this, we see that if a 75 kg parachutist wants to land with a speed of −2m/s, they need a parachute that provides a drag coefficient of $\boxed{\kappa = 367.88}$. A plot is given in Fig. 9.4-(a). ■

Example 9.7. *Compute and plot a solution to the first-order nonlinear, inhomogeneous ordinary differential equation*

$$m\frac{dv(t)}{dt} = \kappa v^2(t) - mg, \quad v(t_0) = v_0. \tag{9.7}$$

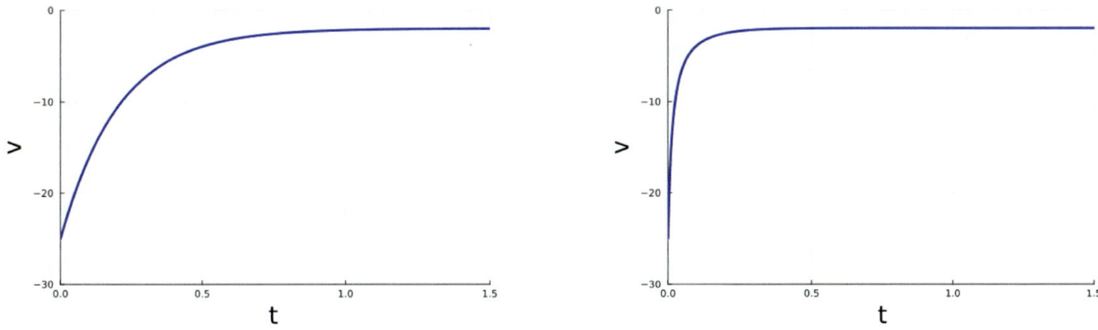

(a) (b)

Figure 9.4: Velocity profiles of parachutists, with an emphasis on when each model "achieves a terminal velocity". (a) Linear drag model. (b) Nonlinear drag model. The linear model predicts a more gentle deceleration to the terminal velocity. In practice, a chute does not open instantaneously, so the braking effect of a real parachute (with nonlinear drag) may not be as harsh as we are predicting!

Solution: A closed-form solution cannot be given in terms of elementary functions, as defined in Secrets of the Arcane 7.36. To prove this fact requires Differential Galois Theory, which few non-mathematicians acquire in their mathematical studies. Even though a closed-form solution cannot be found, we can still compute numerical solutions, as shown in **Fig. 9.4-(b)**. Moreover, qualitative analysis of the equation can be performed as in the video Parachute Physics by LabRat Scientific.

So yes, we can still compute a terminal velocity, $v_{\rm ss}$. At the terminal velocity, $\frac{dv(t)}{dt} \equiv 0$. Using this as a definition, we have

$$\frac{dv(t)}{dt} \equiv 0 \implies \kappa v_{\rm ss}^2 = mg \implies v_{\rm ss} = \pm\sqrt{\frac{mg}{\kappa}}.$$

We can also turn it around to "size the parachute" because larger parachutes provide more air resistance. For a parachutist to land with a velocity of $v_{\rm ss}$ m/s implies

$$\kappa = \frac{mg}{v_{\rm ss}^2}.$$

For a parachutist with a mass of 75 kg and desiring a terminal velocity of -2 m/s, $\boxed{\kappa = 183.94}$. A plot is given in **Fig. 9.4-(b)**. ■

> It is remarkable that, just by adding a constant term to the right side of an ODE with one derivative, one unknown, and one equation, we went from something solvable by an elementary method, namely separation of variables, to an ODE where it is provably impossible to write down an analytical solution. Now, go back and look at the ODEs we derived via Lagrange's method and ask yourself: What's the chance we can find an analytical solution to those equations? Ah, what is one in a googol (one in 10^{100})?

9.2.2 Numerical Solutions

Solving ODEs in Julia Using DifferentialEquations.jl

Among the many powerful tools in Julia for solving Ordinary Differential Equations (ODEs) is the `DifferentialEquations.jl` package. This package provides a comprehensive suite of solvers and utilities for dealing with various types of differential equations. To begin, you need to define the differential equation in a function. For example, consider an ODE of the form $\frac{dy}{dt} = f(y,t)$, where f is some function of y and t. In Julia, you would define this as `function f(y, t)`. Next, you set the initial condition and the time span for which you want to solve the ODE. This is done by creating an `ODEProblem` object with your function, initial condition, and time span. Finally, you solve the ODE using the `solve` function from the package. This function takes the problem object and optionally a solver algorithm and other parameters. The result is a solution object that can be analyzed or plotted to understand the behavior of the system described by the ODE. For more information and examples, visit the DifferentialEquations.jl documentation.

Endeavour with a Linear Drag Model

```
# Import useful packages
using DifferentialEquations
using Plots
gr()

# Space Shuttle orbiter Endeavour with Linear Drag Model
struct Params
    m::Float64          # Mass in kilograms
    kappa::Float64    # Drag coefficient
    g::Float64         # Acceleration due to gravity in m/s^2
end

# Initialize the Params structure with specific values
m = 78e3 # kg  78,000 kg or 39 metric tons
g = 9.81 # m/s^2
v0 = 120 # m/s
kappa = (0.9*v0*m)/3000; @show kappa
#
params = Params(m, kappa, g)

# define the ODE dv/dt = f(v,params,t)
function f(v, params, t)
    dv = -(params.kappa / params.m) * v
    return dv
end

# Set the initial condition as a vector!
v0 = 120 # m/s
v0= [v0]

# Set the time interval
T = (0.0, 100) # 100 second simulation period?

# Setup the ODE problem with out-of-place function
problem = ODEProblem{false}(f, v0, T, params)

# solve the ODE problem using the Runge Kutta Tsitouras 5/4 Integrator
sol = solve(problem, Tsit5())

# plot the solution
p1 = plot(sol, lw=3, guidefont = 20,  xlabel="t", ylabel="v",
                    legend=false, color=:blue)
ylims!(p1, (0, 125))

# e^{ -\frac{\kappa}{m} \cdot Tf}  = 0.1 <==> Tf = -(m/kappa)*log(0.1)
Tf = -(m/kappa)*log(0.1); @show Tf
p1 = scatter!(p1, [Tf], [0.1*v0], color=:red, markersize=6)
```

Output

```
kappa = 2808
```

```
Tf = 63.960697027612376
```

***Endeavour* with a Nonlinear Drag Model**

```
# Import useful packages
using DifferentialEquations
using Plots
gr()

# Space Shuttle orbiter Endeavour with Nonlinear Drag Model
struct Params
    m::Float64          # Mass in kilograms
    kappa::Float64   # Drag coefficient
    g::Float64          # Acceleration due to gravity in m/s^2
end

# Initialize the Params structure with specific values
m = 78e3 # 78,000 kg
g = 9.81
kappa = (m/3000)*log(10); @show kappa
params = Params(m, kappa, g)

# define the ODE dv/dt = f(v,params,t)
function f(v, params, t)
    dv = -(params.kappa / params.m) * v.^2
    return dv
end

# Set the initial condition as a vector!
v0 = 120 # m/s
v0= [v0]

# Set the time interval
T = (0.0, 100)

# Setup the ODE problem with out-of-place function
problem = ODEProblem{false}(f, v0, T, params)

# solve the ODE problem using the Runge Kutta Tsitouras 5/4 Integrator
sol = solve(problem, Tsit5())

# plot the solution
p1 = plot(sol, lw=3, guidefont = 20,  xlabel="t", ylabel="v",
                    legend=false, color=:blue)

# e^{ -\frac{\kappa}{m} \cdot Tf}  = 0.1 <==> Tf = -(m/kappa)*log(0.1)
Tf = 9*m/(kappa*v0[1]); @show Tf
p1 = scatter!(p1, [Tf], [0.1*v0], color=:red, markersize=6)
ylims!(p1, (0, 125))
```

Output

```
kappa = 59.867212417845195
Tf = 97.71625842823164
```

Parachutist with a Linear Drag Model

```
# Import useful packages
using DifferentialEquations
using Plots
gr()

# Parachutist with a linear drag model
struct Params
    m::Float64        # Mass in kilograms
    kappa::Float64   # Drag coefficient
    g::Float64       # Acceleration due to gravity in m/s^2
end

#vTerminal =  -\frac{m g}{\kappa}

# Initialize the Params structure with specific values
m = 75 # kg
g = 9.81 # m/s^2

vTerminal = -2 # m/s
kappa = -m*g/vTerminal; @show kappa
params = Params(m, kappa, g)

# define the ODE dv/dt = f(v,params,t)
function f(v, params, t)
    dv = -params.g .-(params.kappa / params.m) * v
    return dv
end

# Set the initial condition as a vector
v0 = -25 # m/s speed when chute is opened
v0= [v0]

# Set the time interval
T = (0.0, 1.5) #

# Setup the ODE problem with out-of-place function
problem = ODEProblem{false}(f, v0, T, params)

# solve the ODE problem using the Runge Kutta Tsitouras 5/4 Integrator
sol = solve(problem, Tsit5());

# plot the solution
p1 = plot(sol, lw=3, guidefont = 20, xlabel="t", ylabel="v",
                     legend=false, color=:blue)
ylims!(p1, (-30, 0))
```

Output

```
kappa = 367.875
```

Parachutist with a Nonlinear Drag Model

```
# Import useful packages
using DifferentialEquations
```

```
using Plots
gr()

# Parachutist with a nonlinear drag model
struct Params
    m::Float64        # Mass in kilograms
    kappa::Float64    # Drag coefficient
    g::Float64        # Acceleration due to gravity in m/s^2
end

# Initialize the Params structure with specific values
m = 75 # kg
g = 9.81 # m/s^2


vTerminal = -2 # m/s
kappa = m*g/(vTerminal^2); @show kappa
params = Params(m, kappa, g)

# define the ODE dv/dt = f(v,params,t)
function f(v, params, t)
    dv = dv = -params.g .-((params.kappa / params.m) .* abs.(v).*v)
    return dv
end

# Set the initial condition as a vector
v0 = -25 # m/s speed when chute is opened
v0= [v0]

# Set the time interval
T = (0.0, 1.5) #

# Setup the ODE problem with out-of-place function
problem = ODEProblem{false}(f, v0, T, params)

# solve the ODE problem using the Runge Kutta Tsitouras 5/4 Integrator
sol = solve(problem, Tsit5());

# plot the solution
p1 = plot(sol, lw=3, guidefont = 20,  xlabel="t", ylabel="v",
                     legend=false, color=:blue)
ylims!(p1, (-30, 0))
```

Output

```
kappa = 183.9375
```

9.2.3 Finite Escape Time

The ODEs we have investigated so far have solutions that exist for all $t \geq 0$. This is NOT always the case. Consider

$$\frac{dx(t)}{dt} = 1 + x^2(t), \tag{9.8}$$

with initial condition $x_0 = x(0) = 0$. Then, using the method of separation of variables, we arrive at

$$\frac{1}{1+x^2}\,dx = d\tau$$
$$\Updownarrow$$
$$\int_0^{x(t)} \frac{1}{1+x^2}\,dx = \int_0^t d\tau$$
$$\Updownarrow$$
$$\mathrm{atan}(x(t)) = t$$
$$\Updownarrow$$
$$x(t) = \tan(t).$$

We know the function $\tan(t)$ has a vertical asymptote at $\frac{\pi}{2}$, that is,

$$\lim_{t\to\frac{\pi}{2}^-} \tan(t) = \infty.$$

When the solution of the ODE has a vertical asymptote, we say that the ODE itself has a **finite escape time**: in finite time, the solution of the ODE ceases to exist; said another way, the solution explodes to $\pm\infty$ at a finite time called t_{escape}. As you might imagine, finite escape times drive a numerical solver crazy.

```
# Import useful packages
using DifferentialEquations
using Plots
gr()

# Finite Escape Time

# define the ODE dv/dt = f(v,params,t)
function f(x,~, t) # If there are no paramters to pass, use a tilde
    dx = 1 .+ x.^2 # Function is vectorized
    return dx
end

# Set the initial condition as a vector
x0 = 0.0 # m/s
x0= [x0]

# Set the time interval
T = (0.0, 3)

# Setup the ODE problem with out-of-place function
problem = ODEProblem{false}(f, x0, T)

# solve the ODE problem using the Runge Kutta Tsitouras 5/4 Integrator
sol = solve(problem, Tsit5());

# plot the solution
plot(sol)
```

Output

```
Warning: dt(4.440892098500626e-16) <= dtmin(4.440892098500626e-16) at
t=1.5707114767768096. Aborting. There is either an error in your model
specification or the true solution is unstable.
```

By "unstable", the solver means "numerically unstable". For you, it means the solutions blew up, and you need to decide if that is a "feature" or a "bug"! In our case, it is not a bug in the code. It is a feature of the model we are trying to simulate. While it may not be a desirable feature, the finite escape time is inherent in the ODE itself and has nothing to do with the capabilities of the numerical solver. If we had set the time interval to any value $0 < T < \frac{\pi}{2}$, the solver would have found a very good approximation of the solution. In the output's first line, the solver tells us it encountered a problem at $t = 1.5707114767768096$, which is very close to $\frac{\pi}{2}$.

9.2.4 One-Dimensional ODE with Multiple Solutions

So far, we have associated a single solution to each ODE we have solved. Some nonlinear ODEs have more than one solution as highlighted in this fun video The Dome Paradox: A Loophole in Newton's Laws by Up and Atom. Here, we present a simple example of a one-dimensional ordinary differential equation (ODE) that exhibits multiple solutions. **Spoiler alert: it solves the dome paradox!**

Consider the differential equation

$$\frac{dx(t)}{dt} = (x(t))^{2/3},\ x(0) = 0. \tag{9.9}$$

The right side of this ODE, $x^{2/3} = \sqrt[3]{x^2}$, is notable because it is not differentiable at $x = 0$. Indeed,

$$\frac{d}{dx}\left(x^{2/3}\right) = \frac{2}{3}x^{-\frac{1}{3}} = \frac{2}{3}\frac{1}{\sqrt[3]{x}},$$

which blows up at the origin.

In terms of solutions, the trivial solution

$$x(t) \equiv 0 \tag{9.10}$$

satisfies the ODE and the initial condition. Surprisingly, there are other (nontrivial) solutions as well. For instance, the function $\varphi : [0, \infty) \to \mathbb{R}$ given by

$$\varphi(t) = \frac{t^3}{27}, \tag{9.11}$$

also satisfies the initial condition and the differential equation. Indeed,

- $\frac{d}{dt}\varphi(t) = \frac{d}{dt}\frac{t^3}{27} = \frac{t^2}{9}$, and
- $(\varphi(t))^{2/3} = \left(\left(\frac{t}{3}\right)^3\right)^{2/3} = \left(\frac{t}{3}\right)^2 = \frac{t^2}{9}$.

Perhaps even more mind-blowing,[2] for all constants $c > 0$,

$$\varphi_c(t) = \begin{cases} 0 & \text{if } t < c, \\ \frac{(t-c)^3}{27} & \text{if } t \geq c, \end{cases} \tag{9.12}$$

is also a solution. This example illustrates that some differential equations, particularly those where the right side is not everywhere continuously differentiable, can have multiple solutions. Understanding the criteria under which a unique solution exists is crucial in the study of ODEs. We address this topic in Section 9.4.3.

A related phenomenon is that the trajectories of some ODEs can be highly sensitive to their initial conditions. In such cases, "small" changes in the initial condition can lead to "large" variations in the resulting time trajectory. In both mathematics and popular science, this behavior is often referred to as **"Chaos"**. If you would like an accessible introduction to this topic, we recommend the video Chaos: The Science of the Butterfly Effect by Veritasium.

[2]Why is this mind-blowing? Suppose you are building a gadget and through much labor, you show that one of its variables satisfies (9.9). Then the variable behaves almost like something "random" because, at any moment in time, the "universe" rolls a die, chooses $c > 0$, and spookily produces a nonzero solution. Wild!

9.2.5 (Optional Read) Is It Possible for an ODE to Have No Solution?

You see this question a lot on the internet, so we thought to address it by giving various examples. If they are all a bit wacky, that's a sign that most ODEs do have a solution.

(a) $\left(\frac{dx(t)}{dt}\right)^2 + 1 = 0$ does not have a real solution for any initial condition for the same reason that $y^2 + 1 = 0$ has no real solutions.

(b) $\frac{dx(t)}{dt} = \frac{1}{x(t)}$ does not have a solution for the initial condition $x(0) = 0$ because the right side of the ODE blows up at $x(t) = 0$.

(c) $\frac{dx(t)}{dt} = \begin{cases} 1 & x \in \mathbb{Q} \\ 0 & \text{otherwise,} \end{cases}$ does not have a solution for any initial condition because the function on the right side of the ODE is not Riemann-integrable. Why is that a problem? By the First Fundamental Theorem of Calculus, if the equation has a solution $x(t)$, then it must be an antiderivative of $x'(t)$, meaning that $x(t) = x(a) + \int_a^t x'(\tau)\,d\tau$. But for that to hold, we need to make sense of the integral. Because we only know Riemann-integration, we are stuck. Now, if you learn the Lebesgue Integral, you have a way out of this corner.

The following result comes from functional analysis and topology, branches of mathematics that math majors take their senior year or in grad school. It is only included here to underline that any ODE that may be of interest to an engineer can be proven to have a solution, justifying the wackiness of our "counterexamples."

[Peano's Existence Theorem] Let G be an open set in $\mathbb{R} \times \mathbb{R}^n$ and let $f : G \to \mathbb{R}^n$ be a continuous function. If (t_0, x_0) is a point in G, then there exists an interval I containing t_0 and a function $x : I \to \mathbb{R}^n$ that is a solution to the initial value problem

$$\frac{dx}{dt} = f(t,x), \quad x(t_0) = x_0. \tag{9.13}$$

Note: The proof of Peano's Existence Theorem is based on the application of the Arzelà–Ascoli theorem and Schauder's fixed point theorem. The idea is to construct a sequence of approximate solutions using successive approximations and show that this sequence has a convergent subsequence that converges to an actual solution of the differential equation. The continuity of f ensures that the sequence of approximate solutions is equicontinuous and uniformly bounded, which allows the application of the Arzelà–Ascoli theorem. ■

9.2.6 (Optional Read) The Independent Variable in an ODE Can Be Any Strictly Increasing Quantity

The study of heat conduction is a fundamental aspect of thermodynamics and continuum mechanics. In particular, one-dimensional steady-state heat conduction can be modeled using a **second-order** ordinary differential equation. This model provides insights into how temperature varies along a rod or a similar object when the system has reached a steady state. If time were also considered, we would have a PDE.

Consider a long, thin rod of length L, with thermal conductivity k, and wrapped in thermal insulation along its length, except at the ends. The rod is heated at one end and cooled at the other, creating a temperature gradient along its length. Let x be a position along the rod and $T(x)$ the temperature at the point x. The temperature distribution $T(x)$ along the rod is described by the steady-state heat equation,

$$-k\frac{d^2T}{dx^2} = 0. \tag{9.14}$$

The boundary conditions might be, for example, $T(0) = T_0$ (temperature at the heated end) and $T(L) = T_L$ (temperature at the cooled end).

The general solution to this differential equation is an affine (aka linear plus a constant) temperature profile,

$$T(x) = C_1 x + C_2, \tag{9.15}$$

where C_1 and C_2 are constants determined by the boundary conditions. For the given boundary conditions, the solution becomes,

$$T(x) = \frac{T_L - T_0}{L} x + T_0. \tag{9.16}$$

This equation describes a linear temperature gradient from the heated end to the cooled end of the rod. While the examples we study will mostly have time as the independent variable, you will encounter various independent variables in other engineering courses.

9.3 Higher-Order ODEs and a Direct Current (DC) Motor Model

Most models of physical systems involve more than one derivative, or more than one variable, or both! Fig. 9.5-(b) shows the pendulum we examined in Section 6.4. Its differential equation is:

$$m_p \cdot L_p^2 \cdot \ddot{\theta} + m_p \cdot g \cdot L_p \cdot \sin(\theta) = \tau_p, \tag{9.17}$$

where τ_p is the torque applied to the pendulum at the revolute joint. This is an example of a **second-order ODE in a single variable**, θ. It is second-order because the second derivative, $\ddot{\theta}$, appears in the model.

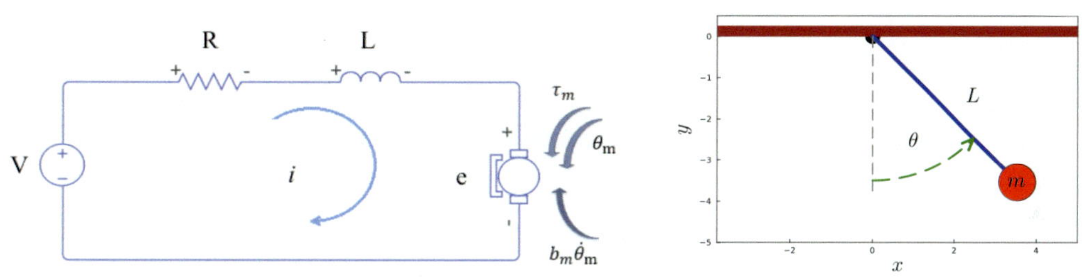

(a) (b)

Figure 9.5: Connecting two (sub)systems to form a single system. (a) A circuit model of DC motor by Research Gate. (b) The classic pendulum formed by a massless rod with a point mass attached at its end from Fig. 6.10-(b). The torque from the motor can be used to achieve a desired behavior, such as balancing the pendulum in the upright position. **Notice the clash in notation, where ME-types use L for length and EE-types use L for inductance. This is very common and you have to learn how to navigate it.** In the text, we add a subscript p to the parameters of the pendulum and a subscript m to the parameters of the motor.

This section aims to show that real differential equations typically have orders higher than one and contain multiple variables. In other words, higher-order ODEs are not just a mathematical curiosity invented to torture young learners.

Fig. 9.5-(a) shows a standard circuit representation of a DC motor; more comprehensive treatments can be found in DC Motor Speed: System Modeling by the MathWorks (creator of MATLAB) and How does an Electric Motor work? DC Motor explained by The Engineering Mindset. Using Kirchoff's Current and Voltage Laws, its model can be written as

$$L_m \frac{di}{dt} + R_m i = V - b_m \dot{\theta}_m, \tag{9.18}$$

where θ_m is the angle of the motor's shaft, L_m is the inductance of the motor's windings and R_m is their resistance, V is the DC source voltage applied to the motor, and $b_m\dot{\theta}$ represents the "back emf" (Back Electromotive Force[3]). The output shaft of the motor is typically connected to the revolute joint of the pendulum through some sort of gearing that imposes

$$\theta_m = N\theta,$$

[3]The back (or counter) electromotive force (emf) is the voltage generated by a running motor that acts to counter the supplied voltage. In the simple case of a permanent-magnet DC motor, it is proportional to the rotational speed of the motor.

where N is the gear ratio, which could be in the range of 100. Gearing is needed because typical motors are efficient when spinning fast and producing low torque, while most robotics applications require high torque and low speed! **The current of the motor is now a second variable in the differential equations modeling the overall system.** The torque produced by a DC motor is proportional to the current, $\tau_m = K_m \cdot i$, where K_m is called the motor torque constant.

We have presented these details for two reasons:

- The differential equations you will encounter in engineering come from physical systems. Each discipline has its collection of important differential equation models, and students typically do not learn how to develop the models in a mathematics course. Back in the day we learned how to model through courses in physics, chemistry, EE, and ME (when robotics did not yet exist as an academic discipline), and through consulting work in industry. In your courses, pay attention to the models. Even if the models are simplified, they open doors to the more realistic models you will encounter in practice.

- The ODE model of the Cassie bipedal robot has 40 variables in it and fills up at least 100 pages in 10-point font. Through the application of appropriate software tools, the students who work with Cassie manage the robot's ODE model[4] just fine!

Fact 9.8: A Third-Order ODE Model for a Motorized Pendulum

The overall system comprising a DC motor connected to a pendulum, as shown in Fig. 9.5, obeys the following third-order ODE:

$$m_p \cdot L_m \cdot L_p^2 \cdot \dddot{\theta} + \left(R_m \cdot m_p \cdot L_p^2\right) \cdot \ddot{\theta} + \left(m_p \cdot g \cdot L_m \cdot L_p \cdot \cos(\theta) + N \cdot K_m \cdot b_m\right) \cdot \dot{\theta} \\ + R_m \cdot m_p \cdot g \cdot L_p \cdot \sin(\theta) = N \cdot K_m \cdot V, \tag{9.19}$$

where $V(t)$ is the voltage supplied to the motor. The derivation is given below as an optional read.

High-Order ODEs of a Single Variable

Definition 9.9. *An ODE of the form,*

$$y^{(n)} = f(y^{(n-1)}, y^{(n-2)}, \ldots, \dot{y}, y, t) \tag{9.20}$$

*is called an **n-th order (nonlinear) ODE**. While an ODE of the form,*

$$y^{(n)} + a_{n-1}y^{(n-1)} + a_{n-2}y^{(n-2)} + \cdots + a_1\dot{y} + a_0 y = 0 \tag{9.21}$$

*is called an **n-th order linear time-invariant ODE** when $a_k \in \mathbb{R}$ are constants, $0 \le k \le n-1$.*

(a) Simple algebra is required to express (9.19) in the form of (9.20).

(b) After which, replacing $\cos(\theta)$ with 1.0 and $\sin(\theta)$ with θ yields a linear time-invariant ODE model.

(c) It is OK to have a term multiplying $y^{(n)}$ as long as it never vanishes.

Example 9.10. *(Optional Read) From the following equations for the pendulum driven by a DC motor, derive Fact 9.8.*

[4] Y. Gong was a second-year MS student when designing the gait controller for Cassie. You do not see the model in the paper because 100 pages of printed formulas would serve no useful purpose.

$$
\begin{aligned}
m_p \cdot L_p^2 \cdot \ddot{\theta} + m_p \cdot g \cdot L_p \cdot \sin(\theta) &= \tau_p && (a)\\
L_m \frac{di}{dt} + R_m i &= V - b_m \dot{\theta}_m && (b)\\
\theta_m &= N\theta && (c)\\
\tau_m &= K_m i && (d)\\
\tau_p &= N\tau_m && (e)
\end{aligned}
\tag{9.22}
$$

Solution: Substituting (9.22)-(d) and -(e) into (9.22)-(a) gives

$$m_p \cdot L_p^2 \cdot \ddot{\theta} + m_p \cdot g \cdot L_p \cdot \sin(\theta) = N \cdot K_m \cdot i.$$

Differentiating the above equation with respect to time gives

$$m_p \cdot L_p^2 \cdot \dddot{\theta} + m_p \cdot g \cdot L_p \cdot \cos(\theta) \cdot \dot{\theta} = N \cdot K_m \cdot \frac{di}{dt}, \tag{9.23}$$

where the dot notation for the derivative was not used on the motor current i, for obvious reasons. To arrive at (9.19), we need to eliminate $\frac{di}{dt}$ from the above equation. To accomplish this goal, we solve (9.22)-(b) for $\frac{di}{dt}$, giving us

$$\frac{di}{dt} = -\frac{R_m}{L_m} i + \frac{1}{L_m} V - \frac{b_m}{L_m} \dot{\theta}_m.$$

Next, we replace the current i with

$$
\begin{aligned}
i &= \frac{1}{K_m} \tau_m \\
&= \frac{1}{N \cdot K_m} \tau_p \\
&= \frac{1}{N \cdot K_m} \left(m_p \cdot L_p^2 \cdot \ddot{\theta} + m_p \cdot g \cdot L_p \cdot \sin(\theta) \right) \\
&= \frac{m_p \cdot L_p^2}{N \cdot K_m} \cdot \ddot{\theta} + \frac{m_p \cdot g \cdot L_p}{N \cdot K_m} \cdot \sin(\theta),
\end{aligned}
$$

resulting in

$$
\begin{aligned}
\frac{di}{dt} &= -\frac{R_m}{L_m} \left(\frac{m_p \cdot L_p^2}{N \cdot K_m} \cdot \ddot{\theta} + \frac{m_p \cdot g \cdot L_p}{N \cdot K_m} \cdot \sin(\theta) \right) + \frac{1}{L_m} V - \frac{b_m}{L_m} \dot{\theta}_m \\
&= -\frac{R_m \cdot m_p \cdot L_p^2}{L_m \cdot N \cdot K_m} \cdot \ddot{\theta} - \frac{R_m \cdot m_p \cdot g \cdot L_p}{L_m \cdot N \cdot K_m} \cdot \sin(\theta) + \frac{1}{L_m} V - \frac{b_m}{L_m} \dot{\theta}_m.
\end{aligned}
\tag{9.24}
$$

The final step is to substitute (9.24) into (9.23), giving

$$m_p \cdot L_p^2 \cdot \dddot{\theta} + m_p \cdot g \cdot L_p \cdot \cos(\theta) \cdot \dot{\theta} = N \cdot K_m \cdot \left(-\frac{R_m \cdot m_p \cdot L_p^2}{L_m \cdot N \cdot K_m} \cdot \ddot{\theta} - \frac{R_m \cdot m_p \cdot g \cdot L_p}{L_m \cdot N \cdot K_m} \cdot \sin(\theta) + \frac{1}{L_m} V - \frac{b_m}{L_m} \dot{\theta}_m \right). \tag{9.25}$$

Simplifying then gives (9.19).

Note: For a real project, one would use a Symbolic Package to combine and simplify the above equations. Hand calculations are error-prone and do not lend themselves to good documentation.

The following is inelegant and contains some hand calculations. However, the substitutions and simplifications are done symbolically, which avoids common errors.

```
using SymPy

# Define the symbolic variables
@syms th dth ddth mp Lp g taup Lm Rm i V bm N Km
```

```

# Define the expressions
#i = taup / (N * Km)
di = (V - bm * dth - Rm * i) / Lm

# Perform the substitution
di = di.subs(i, taup / (N * Km))
#
# (mp*Lp^2) * ddth = taup - mp*g*Lp*sin(th) and hence,
# taup = (mp*Lp^2) * ddth + mp*g*Lp*sin(th)
di = di.subs(taup, (mp*Lp^2)*ddth + mp*g*Lp*sin(th))
di = simplify(di)

# Display the substituted expression
println(di)
```

Output

```
(-Km*N*(-V + bm*dth) - Lp*Rm*mp*(Lp*ddth + g*sin(th)))/(Km*Lm*N)
```

```
#ddth = (N * Km * i - mp*g*Lp*sin(th))/(mp*Lp^2)
dddth = (N * Km * di - mp*g*Lp*cos(th)*dth)/(mp*Lp^2)
dddth = simplify(dddth)

# Extracting numerator and denominator
num, den = dddth.as_numer_denom()

# print numerator and denominator of dddth, theta triple dot
println(num); println(" ")
println(den)
dendddth = simplify(dddth * den)

println("")
println("ODE is: $den * dddth = ", dendddth)
```

Output

```
-Km*N*(-V + bm*dth) - Lm*Lp*dth*g*mp*cos(th) - Lp*Rm*mp*(Lp*ddth
+ g*sin(th))

Lm*Lp^2*mp

ODE is: Lm*Lp^2*mp * dddth = Km*N*(V - bm*dth) - Lm*Lp*dth*g*mp*cos(th)
        - Lp*Rm*mp*(Lp*ddth + g*sin(th))
```

All of the terms match up with those in the hand calculations. ■

9.4 Vector ODEs: Multidimensional Dynamics, Single Independent Variable

We first discuss an important source of "vector ODEs" with the goal of understanding what that even means! Then, we can discuss what a solution of the ODE means, when solutions are unique, and when solutions exist for an unbounded time interval (i.e., no finite escape time).

9.4.1 The Return of the Robot Equations

In Prop. 6.37, we introduced the Robot Equations,

$$D(q) \cdot \ddot{q} + C(q, \dot{q}) \cdot \dot{q} + G(q) = B \cdot \tau, \tag{9.26}$$

where

- $q \in \mathbb{R}^n$ is a vector of generalized position coordinates,
- $\dot{q} \in \mathbb{R}^n$ is the corresponding vector of velocities,
- $\ddot{q} \in \mathbb{R}^n$ is the corresponding vector of accelerations,
- and $\tau \in \mathbb{R}^m$ is a vector of motor torques.

Moreover,

- $D(q)$ is the $n \times n$ mass-inertia matrix and is invertible,
- $G(q) := \nabla V(q)$ is the $n \times 1$ gradient of the potential energy,
- the $n \times 1$ vector $C(q, \dot{q}) \cdot \dot{q}$ contains all the remaining terms in (6.60),
- and B is the $n \times m$ torque distribution matrix.

Due to the appearance of $\ddot{q} = \frac{d^2}{dt^2} q$ in (9.26), the Robot Equations are a **vector of second-order ODEs in the independent variable t**. They look totally intimidating. Our first goal is to rewrite them as a vector of first-order ODEs,

$$\dot{x} = f(x, \tau),$$

which, beyond appearing "more approachable", will greatly help us with both our analytical work and our numerical solvers. **It is important to note that because so few of the Robot Equations have closed-form solutions, we routinely use numerical algorithms to compute solutions.**

> **Method 9.11: From Second-Order ODEs to First-Order ODEs**
>
> Consider the Robot Equations in (9.26) and let $x_1 := q$, $x_2 := \dot{q}$. In terms of these variables, we have
>
> $$\begin{aligned} \dot{x}_1 &= \dot{q} = x_2, \\ \dot{x}_2 &= \ddot{q} = -D^{-1}(q) \cdot [C(q, \dot{q}) \cdot \dot{q} + G(q)] + D^{-1}(q) \cdot B \cdot \tau \\ &= -D^{-1}(x_1) \cdot [C(x_1, x_2) \cdot x_2 + G(x_1)] + D^{-1}(x_1) \cdot B \cdot \tau. \end{aligned} \tag{9.27}$$
>
> Once we stack x_1 and x_2 to form the $2n \times 1$ vector x, (9.26) becomes equivalent to
>
> $$\underbrace{\begin{bmatrix} \dot{x}_1 \\ \dot{x}_2 \end{bmatrix}}_{\dot{x}} = \underbrace{\begin{bmatrix} x_2 \\ -D^{-1}(x_1) \cdot [C(x_1, x_2) \cdot x_2 + G(x_1)] + D^{-1}(x_1) \cdot B \cdot \tau \end{bmatrix}}_{f(x, \tau)}. \tag{9.28}$$
>
> An equation of the form
>
> $$\dot{x} = f(x, \tau) \tag{9.29}$$
>
> is called in various situations:
>
> - a vector of first-order ODEs,
> - a vector of controlled first-order ODEs (due to the presence of the motor torques τ, which are used to control the robot),

- a **system of first-order ODEs**,
- or a **state-variable model**.

The components of the vector x are called state variables. They are important because we need to provide initial conditions for each state variable in order to uniquely specify a solution to the ODE.

Note: Instead of computing inverses of matrices, we recognize that $\dot{x}_2$ is the unique solution to the linear system of equations

$$D(x_1) \cdot \dot{x}_2 = [C(x_1, x_2) \cdot x_2 + G(x_1)] + B \cdot \tau,$$

which we would solve via LU Factorization or QR Factorization. In Julia, we have the backslash operator, which would select either LU or QR as it deems most appropriate for solving the equations, giving us

$$\dot{x}_2 = D(x_1) \setminus (-[C(x_1, x_2) \cdot x_2 + G(x_1)] + B \cdot \tau)$$

as the better way to implement in code the second row of (9.28).

Example 9.12. *Here is the pendulum model from Example 6.36,*

$$\underbrace{m \cdot L^2 \cdot \ddot{\theta}}_{\frac{d}{dt}\frac{\partial \mathcal{L}(q,\dot{q})}{\partial \dot{q}}} - \underbrace{(-m \cdot g \cdot L \cdot \sin(\theta))}_{\frac{\partial \mathcal{L}(q,\dot{q})}{\partial q}} = \underbrace{0}_{\Gamma}, \tag{9.30}$$

where $\Gamma = 0_{1\times 1}$ because a motor has not been installed at the pivot and there is no friction. Express the model as a vector first-order ODE and provide a simulation with initial conditions $\theta_0 = \frac{3\pi}{8}$ and $\dot{\theta}_0 = 0$. Recall that the pendulum is pointing straight down at rest with $\theta_0 = 0$ and $\dot{\theta}_0 = 0$.

Solution: Following the template given in Method 9.11, we define $x_1 := \theta$ and $x_2 := \dot{\theta}$. After a bit of algebra, this gives

$$\begin{aligned} \dot{x}_1 &= x_2, \\ \dot{x}_2 &= -\frac{g}{L} \cdot \sin(x_1), \end{aligned}$$

because $\frac{m \cdot g \cdot L}{m \cdot L^2} = \frac{g}{L}$. It is perfectly fine to leave the equations in this form. It is equally fine to express them as

$$\underbrace{\begin{bmatrix} \dot{x}_1 \\ \dot{x}_2 \end{bmatrix}}_{\dot{x}} = \underbrace{\begin{bmatrix} x_2 \\ -\frac{g}{L} \cdot \sin(x_1) \end{bmatrix}}_{f(x)}.$$

The initial condition becomes

$$x(0) = \begin{bmatrix} \theta_0 \\ \dot{\theta}_0 \end{bmatrix} = \begin{bmatrix} \frac{3\pi}{8} \\ 0 \end{bmatrix}.$$

```
# Import useful packages
using DifferentialEquations
using Plots
using LaTeXStrings
gr()

```

```
# Pendulum Model

struct Params
    m::Float64          # Mass in kilograms
    L::Float64          # Length in meters
    g::Float64          # Acceleration due to gravity in m/s^2
end


# Initialize the Params structure with specific values
m = 2
L = 1
g = 9.81
params = Params(m, L, g)


# define the ODE dx/dt = f(x,params,t)
function f(x, params, t)
    dx = [x[2], -(params.g/params.L)*sin.(x[1])]
    return dx
end

# Set the initial condition as a vector
x0= [3*pi/8; 0.0]

# Set the time interval
T = (0.0, 10)

# Setup the ODE problem with out-of-place function
problem = ODEProblem{false}(f, x0, T, params)

# solve the ODE problem using the Runge Kutta Tsitouras 5/4 Integrator
sol = solve(problem, Tsit5());

# plot the solution
p1 = plot(sol, lw=3, guidefont = 20,  xlabel=L"t",
    ylabel=L"(\theta, \dot{\theta})", legend=false)
```

Output The blue trace is the angle of the pendulum, while the orange trace is its angular velocity. The pendulum oscillates as expected.

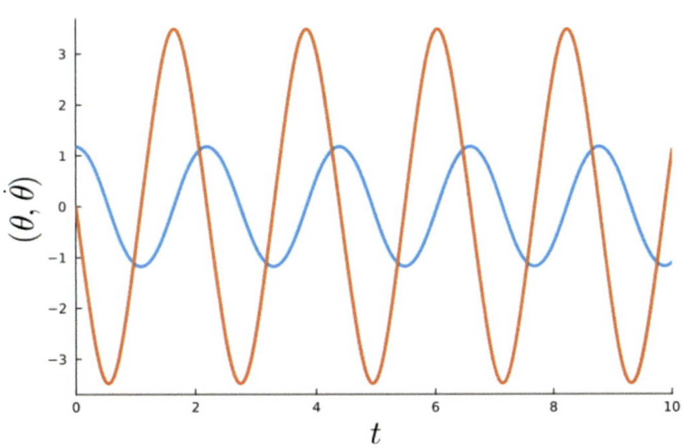

■

Fact 9.13: 2-Link Manipulator

The base angle q_1 is an absolute angle, with its zero position aligned with the x-axis. The angle of the second link, q_2, is relative to the base (first) link. Hence, $q_2 = 0$ has the second link extending straight out from the base link. The angular velocities are denoted dq_1 and dq_2 in the code. Actuators have been placed at the base link and the second link, though we will not be using them in this chapter.

```
function dyn_mod_2LinkManipulator(q, dq)
# DYN_MOD_2LINKMANIPULATOR
# 2023-11-20 17:26:11
#
# Author: Grizzle
#
# Model NOTATION: D(q)ddq + C(q,dq)*dq + G(q) = B*tau
# The Robot Equations: From Lagrange's Equations of Motion
#
g, L1, L2, m1, m2 = modelParameters()
#
# Variable names for the model
q1, q2 = q
dq1, dq2 = dq
#
D = zeros(2, 2)
  D[1, 1] = L1^2*m1 + m2*(L1^2 + 2*L1*L2*cos(q2) + L2^2)
  D[1, 2] = L2*m2*(L1*cos(q2) + L2)
  D[2, 1] = L2*m2*(L1*cos(q2) + L2)
  D[2, 2] = L2^2*m2
#
C = zeros(2, 2)
  C[1, 1] = -L1*L2*dq2*m2*sin(q2)
  C[1, 2] = -L1*L2*m2*(dq1 + dq2)*sin(q2)
  C[2, 1] = L1*L2*dq1*m2*sin(q2)
#
G = zeros(2)
  G[1] = g*m2*(L1*cos(q1) + L2*cos(q1 + q2)) + L1*g*m1*cos(q1)
  G[2] = L2*g*m2*cos(q1 + q2)
#
B = zeros(2, 2)
  B[1, 1] = 1
  B[2, 2] = 1
#
JacG = zeros(2, 2)
  JacG[1, 1] = g*m2*(-L1*sin(q1) - L2*sin(q1 + q2)) - L1*g*m1*sin(q1)
  JacG[1, 2] = -L2*g*m2*sin(q1 + q2)
  JacG[2, 1] = -L2*g*m2*sin(q1 + q2)
  JacG[2, 2] = -L2*g*m2*sin(q1 + q2)
#
  return (D=D, C=C, G=G, B=B, JacG=JacG)
end
```

To bring the model into the workspace, use

```
# bring the model into the workspace for later use
include("dyn_mod_2LinkManipulator.jl")
```

Example 9.14. *A two-link manipulator model is given in Fact 9.13, following the same angle convention as in* Fig. 1.7-(b). *Express the model as a vector first-order ODE and provide a simulation with initial conditions*

$$\begin{aligned} q_1(0) &= -\frac{3\pi}{8}, \\ q_2(0) &= 0, \\ \dot{q}_1(0) &= 0, \\ \dot{q}_2(0) &= 0. \end{aligned}$$

Solution: Following the template given in Method 9.11, we define

$$x_1 := \begin{bmatrix} q_1 \\ q_2 \end{bmatrix} \text{ and } x_2 := \begin{bmatrix} \dot{q}_1 \\ \dot{q}_2 \end{bmatrix}.$$

We do not symbolically compute the inverse of the mass-inertia matrix because, in larger and more realistic models, it would make our simulation very slow. Hence, we write the model as:

$$\dot{x}_1 = x_2$$

$$D(x_1)\dot{x}_2 = -C(x_1, x_2) \cdot x_2 - G(x_1),$$

and use the "backslash" operator to solve for $\dot{x}_2$. In code, it looks quite clean.

```
# Import useful packages
using DifferentialEquations
using Plots
using LaTeXStrings
using LinearAlgebra # for the backslash command
gr()

# 2-Link Manipulator Model
function modelParameters()
    g = 9.81
    L1 = 1.0
    L2 = 0.5
    m1 = 1.0
    m2 = 3.0
    return g, L1, L2, m1, m2
end

# bring the model into the workspace for later use
include("dyn_mod_2LinkManipulator.jl")

# define the ODE dx/dt = f(x,t)
#
# the parameters are passed via modelParameters() in the
# function dyn_mod_2LinkManipulator(q, dq)
#
function f(x,~,t)
    n = floor(Int, length(x) / 2) # In Julia n/2 is a Float64
    q = x[1:n]
    dq = x[n+1:end]
    model = dyn_mod_2LinkManipulator(q, dq)
    dx1 = dq
    dx2 = (model.D) \ (-model.C*dq-model.G) # note the use of backslash
    dx=[dx1;dx2]
```

```
        return dx
end

# Set the initial condition as a vector
x0= [-3pi/8; 0.0; 0; 0]

# Set the time interval
T = (0.0, 5)

# Setup the ODE problem with out-of-place function
# params removed
problem = ODEProblem{false}(f, x0, T)

# solve the ODE problem using the Runge Kutta Tsitouras 5/4 Integrator
sol = solve(problem, Tsit5());

# plot the solution
p1 = plot(sol, lw=3, guidefont=20, xlabel=L"t",
     ylabel=L"(q_1, q_2, \dot{q}_1, \dot{q}_2)",
     label=["q1" "q2" "dq1" "dq2"], legend=:topright)
```

Output The initial condition has the manipulator arm hanging downward like a pendulum ($q_1(0) = -\frac{\pi}{2}$ would be straight down). Consequently, q_1 oscillates similar to a single-link downward-pointing pendulum, with one important change: some of the "energy" is transferred to the second link, which begins oscillating as well, even though it started in perfect alignment with the base link. Robot models with even two planar links have very complicated dynamics. Can you imagine the dynamics of a 3D bipedal robot, such as Cassie, with 14 links? Your answer should be an emphatic **NO!**

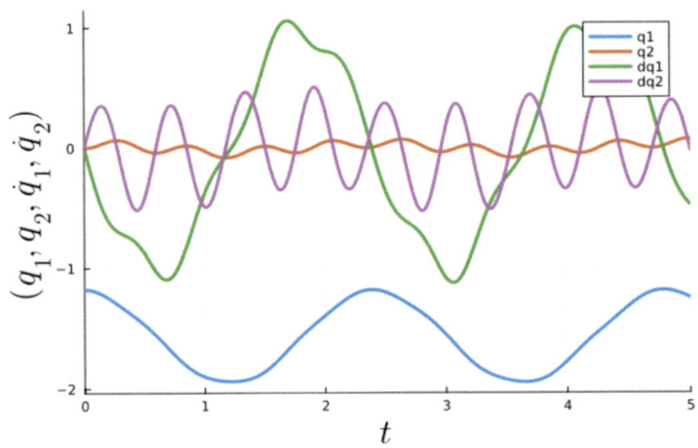

■.

Interesting Video: The Oxford Professor and his Pendulums by OxfordUni. Oxford Mathematics Prof. Jon Chapman explains why he loves his double pendulum. The clip is part of the "Oxford Mathematicians and their Toys" series, which is part of the Oxford Mathematics Instagram page. Warning: The term "stability" is used differently here. It does not mean asymptotic stability; its meaning is closer to "not unstable": two initial conditions that are close to one another give rise to solutions that remain close to one another.

9.4.2 Solutions of First-Order Vector ODEs

What Is a Solution?

Definition 9.15. *A function,* $\varphi : [t_0, T) \to \mathbb{R}^n$, $T > t_0$, *is a* **solution** *to the Ordinary Differential Equation (ODE)*

$$\dot{x}(t) = f(x(t)), \ \ x(t_0) = x_0, \tag{9.31}$$

if

(a) $\varphi : [t_0, T) \to \mathbb{R}^n$ *is continuous (meaning, each of its components is continuous);*

(b) $\varphi(t_0) = x_0$ *(satisfies the initial condition); and*

(c) for all $t \in (t_0, T)$*, the derivative* $\dot{\varphi}(t) := \frac{d}{dt}\varphi(t)$ *exists and* $\dot{\varphi}(t) - f(\varphi(t)) = 0_{n\times 1}$ *(satisfies the differential equation).*

Note: It is very common to suppress the time variable and write the ODE as $\dot{x} = f(x), \ x(t_0) = x_0$. The "dot" notation for the derivative indicates the dependence on time.

Example 9.16. *Show that for all constants* $c > 0$,

$$\varphi_c(t) = \begin{cases} 0 & \text{if } t < c, \\ \frac{(t-c)^3}{27} & \text{if } t \geq c, \end{cases}$$

is a solution to the differential equation $\dot{x} = \sqrt[3]{x^2}$, $x(0) = 0$.

Solution We need to check the three conditions in Definition 9.15.

(a) **Continuity:** The only point where continuity of $\varphi_c(t)$ may be in doubt is $t = c$. We check, therefore, the two one-sided limits and compare them to the value of the function at c, viz,

$$\lim_{t\to c^-} \varphi_c(t) = \lim_{t\to c^-} 0 = 0 \ \ \text{(the function is identically zero for } t < c\text{)}$$

$$\lim_{t\to c^+} \varphi_c(t) = \lim_{t\to c^+} \frac{(t-c)^3}{27} = 0, \ \ \text{(using the function's definition for } t > c\text{)}$$

$$\varphi_c(t)\bigg|_{t=c} = \frac{(t-c)^3}{27}\bigg|_{t=c} = 0.$$

Hence, the function is continuous at c and therefore, at all points in its domain.

(b) **Satisfies the Initial Condition:** $0 = \frac{(t-c)^3}{27}\Big|_{t=c} = x_0$.

(c) **Satisfies the Differential Equation:**

- For $t < c$, $\varphi_c(t)$ is identically zero, and hence its derivative is also zero. Also, $(0)^{2/3} = 0$. Hence, $\dot{\varphi}_c(t) - f(\varphi_c(t)) = 0$ for all $t < c$.
- For $t > c$, $\varphi_c(t) = \frac{(t-c)^3}{27}$, and hence by the Power Rule,

$$\frac{d}{dt}\varphi_c(t) = 3\frac{(t-c)^2}{27} = \frac{(t-c)^2}{3^2} = \left(\frac{(t-c)}{3}\right)^2.$$

This takes care of the left side of the differential equation. Turning to the right side,

$$(\varphi_c(t))^{2/3} = \left(\frac{(t-c)^3}{27}\right)^{2/3} = \left(\left(\frac{(t-c)}{3}\right)^3\right)^{2/3} = \left(\frac{(t-c)}{3}\right)^2.$$

Therefore, for all $t \neq c$, $\frac{d}{dt}\varphi_c(t) - (\varphi_c(t))^{2/3} = 0$. **What about $t = c$**, the point where the solution transitions from being identically zero to being a quadratic function of t? We need to evaluate the derivative at that point using left and right limits. It all works out, but we skip the details because there is a more general notion of a solution given in Section 9.4.4 that obviates the need for the calculation!

■

9.4.3 Existence and Uniqueness of Solutions

The **Euclidean norm** of a vector $x = \begin{bmatrix} x_1 & x_2 & \cdots & x_n \end{bmatrix}$ is defined as

$$||x||_2 := \sqrt{(x_1)^2 + (x_2)^2 + \cdots + (x_n)^2} = \sqrt{\sum_{i=1}^{n} (x_i)^2}. \tag{9.32}$$

There are also other norms, such

$$||x||_p := \sqrt[p]{\sum_{i=1}^{n} (x_i)^p}, \quad 1 \leq p < \infty \quad \text{and} \quad ||x||_{\max} := \max_{1 \leq i \leq n} |x_i|.$$

Norm of a Matrix

Definition 9.17. *Let A be an $n \times m$ matrix with entries a_{ij}. The* ***Euclidean norm*** *of A is defined as*

$$||A||_2 := \sqrt{\sum_{i=1}^{n} \sum_{j=1}^{m} (a_{ij})^2}, \tag{9.33}$$

which is the square root of the sum of all squared entries of A. This is the same as stacking the columns of A into a large vector and applying the Euclidean norm to the resulting vector.

Note: The Euclidean or 2-norm is the default norm in Julia obtained with `norm(A)`. There are many other ways to define norms on matrices. It is often easiest to use the following matrix norm:

$$||A||_\infty := \max_{i,j} |a_{ij}|, 1 \leq i, j \leq n, \tag{9.34}$$

which goes by the name **max norm** or the **infinity norm**.

Open Ball of Radius r

Open balls generalize to $\mathbb{R}^n$ the notion of open intervals in $\mathbb{R}$.

Definition 9.18. *Let x_0 be a point in $\mathbb{R}^n$ and $r > 0$ a constant. The* ***open ball of radius r centered about x_0*** *is defined as*

$$B_r(x_0) := \{x \in \mathbb{R}^n \mid ||x - x_0|| < r\}. \tag{9.35}$$

The ball is called "open" because it does not include the boundary points where $||x - x_0|| = r$.

Note: Open balls are used to specify when some property holds **locally** (meaning near a given point x_0) versus **globally** (meaning it holds for all $x \in \mathbb{R}^n$).

Example 9.19. *Sketch the open balls about the origin for the 2-norm, 1-norm, and max-norm.*

Solution:

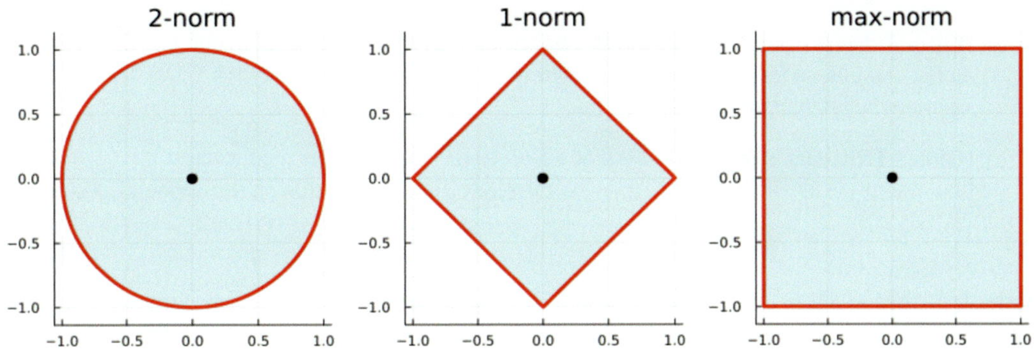


```
using Plots

# Function to plot the unit ball for a given norm
function plot_unit_ball(norm_type::String)
    theta = range(0, stop=2π, length=1000) # parameter for the angle

    if norm_type == "2-norm"
        # Circle: x^2 + y^2 = 1
        x = cos.(theta)
        y = sin.(theta
    elseif norm_type == "1-norm"
        # Diamond: |x| + |y| = 1, simplified representation
        x = [0, 1, 0, -1, 0]
        y = [1, 0, -1, 0, 1]
    elseif norm_type == "max-norm"
        # Square: max(|x|, |y|) = 1, simplified representation
        x = [1, 1, -1, -1, 1]
        y = [1, -1, -1, 1, 1]
    else
        error("Unknown norm type")
    end

    plot(x, y, label=" ", aspect_ratio=:equal, title="$(norm_type)",
    fill=(0, 0.5, :lightblue), lw=3, color=:red)

end

# Plot unit balls for each norm
p1 = plot_unit_ball("2-norm")
p2 = plot_unit_ball("1-norm")
p3 = plot_unit_ball("max-norm")

# Display all plots in a single figure
pall= plot(p1, p2, p3, layout=(1, 3), size=(900, 300))
```

Output See the above plot. ■

Proposition 9.20: Existence and Uniqueness of Solutions

Consider a vector-valued ODE,

$$\dot{x} = f(x), \quad x(t_0) = x_0 \in \mathbb{R}^n. \tag{9.36}$$

(a) **(Local Existence and Uniqueness of a Solution)** If for some $r > 0$, the Jacobian, $\frac{\partial f(x)}{\partial x}$, exists and is continuous for all $x \in B_r(x_0)$, then there exists $\delta > 0$ such that the ODE has a unique solution on the interval $[t_0, t_0 + \delta]$.

(b) **(Global Existence and Uniqueness of a Solution)** If there exists a constant $0 \le L < \infty$ such that, for all $x \in \mathbb{R}^n$, $||\frac{\partial f(x)}{\partial x}|| \le L$, then the ODE has a unique solution on the unbounded interval $[t_0, \infty)$.

Note: It is often easier to check the condition for global existence and uniqueness with the max (aka infinity norm) because it can be computed element by element; indeed, $||\frac{\partial f(x)}{\partial x}||_\infty \le L$ for all $x \in \mathbb{R}^n$ if, and only if, $\left|\frac{\partial f_i(x)}{\partial x_j}\right| \le L$ for all $i, j \le n$ and $x \in \mathbb{R}^n$.

Note 2: Most of the ODEs that appear in engineering practice satisfy the condition for local existence and uniqueness of a solution. However, the vector ODEs that model robots are typically quadratic in the angular velocities. Hence, their partial derivatives will be linear in the angular velocities, which means the global bound $||\frac{\partial f(x)}{\partial x}|| \le L$ fails.

Example 9.21. *Apply Prop. 9.20 to the ODEs given in Fact 9.2, Four Models of Chutes, to check for the local or global existence and uniqueness of their solutions. Comment on whether your conclusions match those of our previous calculations (of the solutions), or not.*

For ease of reference, we repeat the ODEs here, rewritten slightly so that the left-hand side only contains a derivative.

(a) ***Model 1: Horizontal Motion, Linear Drag:*** $\frac{dv(t)}{dt} = -\frac{\kappa}{m}v(t), \quad v(t_0) = v_0.$

(b) ***Model 2: Horizontal Motion, Quadratic Drag:*** $\frac{dv(t)}{dt} = -\frac{\kappa}{m}v^2(t), \quad v(t_0) = v_0.$

(c) ***Model 3: Vertical Motion, Linear Drag:*** $\frac{dv(t)}{dt} = -\frac{\kappa}{m}v(t) - g, \quad v(t_0) = v_0.$

(d) ***Model 4: Vertical Motion, Quadratic Drag:*** $\frac{dv(t)}{dt} = \frac{\kappa}{m}v^2(t) - g, \quad v(t_0) = v_0.$

Solutions:

We use the generic notation, $\dot{v} = f(v)$, for each of the models.

(a) $\dot{v} = -\frac{\kappa}{m}v, \quad v(t_0) = v_0.$ **Ans. Prop. 9.20 implies global existence and uniqueness of solutions.**

We have $f(v) = -\frac{\kappa}{m}v$. Therefore, $\frac{df(v)}{dv} = -\frac{\kappa}{m}$ and $|\frac{df(v)}{dv}| = |-\frac{\kappa}{m}| = \frac{\kappa}{m}$ is finite for all $v \in \mathbb{R}$. Therefore, Prop. 9.20 implies a solution exists and is unique for all $t > t_0$ and all $v_0 \in \mathbb{R}$.

Comparison with our calculations in Example 9.3: we obtained a globally defined solution but had no means to check uniqueness.

(b) $\dot{v} = -\frac{\kappa}{m}v^2, \quad v(t_0) = v_0.$ **Ans. Prop. 9.20 implies local existence and uniqueness of solutions.**

We have $f(v) = -\frac{\kappa}{m}v^2$. Therefore, $\frac{df(v)}{dv} = -2\frac{\kappa}{m}v$. The magnitude of the Jacobian, $|\frac{df(v)}{dv}|$, is finite on any open ball $B_r(v_0)$, $0 < r < \infty$, but grows unbounded as $|v| \to \infty$. Therefore, Prop. 9.20 implies the existence of $\delta > 0$ such that a solution exists and is unique for all $t \in [t_0, t_0 + \delta]$ and all $v_0 \in \mathbb{R}$, but does not imply global existence and uniqueness.

Comparison with our calculations in Example 9.4: We obtained a globally defined solution, even though the sufficient conditions of the Proposition only guaranteed a local solution. When we solved this problem earlier, we had no means to check uniqueness.

Note: If we "flip the sign" on the ODE to $\frac{dv}{dt} = \frac{\kappa}{m}v^2$, then $|\frac{df(v)}{dv}| = 2\frac{\kappa}{m}|v|$ as before, again implying only local existence and uniqueness of solutions. This time, the method of separation of variables gives

$$v(t) = \frac{v_0}{1 - v_0 \frac{\kappa}{m}(t - t_0)},$$

which blows up at $t = t_0 + \frac{m}{\kappa \cdot v_0}$ because the denominator vanishes. By flipping the sign in front of the equation, we go from one with a globally defined solution to one that blows up in finite time. The test given in Prop. 9.20 is based on the norm of the Jacobian of the right side of the equation, and norms are invariant under sign changes.

(c) $\dot{v} = -\frac{\kappa}{m}v(t) - g, \;\; v(t_0) = v_0$. **Ans. Prop. 9.20 implies global existence and uniqueness of solutions.**

We have $f(v) = -\frac{\kappa}{m}v - g$. Therefore, $\frac{df(v)}{dv} = -\frac{\kappa}{m}$ and $|\frac{df(v)}{dv}| = |-\frac{\kappa}{m}| = \frac{\kappa}{m} < \infty$. Therefore, Prop. 9.20 implies a solution exists and is unique for all $t > t_0$ and all $v_0 \in \mathbb{R}$.

Comparison with our calculations in Example 9.6: We obtained a globally defined solution but had no means to check uniqueness.

(d) $\dot{v} = \frac{\kappa}{m}v^2(t) - g, \;\; v(t_0) = v_0$. **Ans. Prop. 9.20 implies local existence and uniqueness of solutions.**

We have $f(v) = -\frac{\kappa}{m}v^2 - g$. Therefore, $\frac{df(v)}{dv} = -2\frac{\kappa}{m}v$ and the magnitude of the Jacobian, $|\frac{df(v)}{dv}|$, is finite on any open ball $B_r(v_0)$, $0 < r < \infty$, but grows unbounded as $|v| \to \infty$. Therefore, Prop. 9.20 implies the existence of $\delta > 0$ such that a solution exists and is unique for all $t \in [t_0, t_0 + \delta]$ and all $v_0 \in \mathbb{R}$.

Comparison with our calculations in Example 9.7: We were unable to obtain an analytical solution. Our numerical calculations showed the existence of a solution over an adequately large time domain. Our numerical work provided no means to check uniqueness.

■

Example 9.22. *Revisit the ODE in (9.9), where we produced an infinite number of solutions (non-uniqueness on steroids),*

$$\dot{x} = x^{2/3}, \; x(0) = 0.$$

In what manner does it fail the conditions in Prop. 9.20?

Solution: The discussion in Chapter 9.2.4 gave us a heads-up on the issue, but we will continue as if that had not happened! Define $f(x) := x^{2/3}$. Prop. 9.20 tells us to evaluate

$$\frac{df(x)}{dx} = \frac{d}{dx}\left(x^{2/3}\right) = \frac{2}{3}x^{-\frac{1}{3}} = \frac{2}{3}\frac{1}{\sqrt[3]{x}},$$

which blows up at the initial condition, $x_0 = 0$. Hence, none of the conditions of Prop. 9.20 are met, so it has nothing to say about the solutions of the ODE. ■

Example 9.23. *Consider the epidemic model*

$$\frac{dx_1}{dt} = -\beta x_1 x_2 - u, \tag{9.37}$$

$$\frac{dx_2}{dt} = \beta x_1 x_2 - \gamma x_2, \tag{9.38}$$

where $x_1(t)$ and $x_2(t)$ denote the number of susceptible and infectious individuals, respectively. The term $\beta x_1 x_2$ models transmission through contact between susceptible and infectious individuals, where $\beta > 0$ is the infection rate. Infectious individuals recover (or are removed) at rate γx_2, where $\gamma > 0$ is the recovery rate. Vaccination is represented by a constant rate u, which reduces the susceptible population. This model, by delineating interaction dynamics, aids in predicting and mitigating disease spread, underscoring the value of mathematical approaches in public health.

The epidemic model depends on the parameters $\beta > 0$, $\gamma > 0$, and the states x_1, x_2. Assuming u is constant, determine:

(a) whether local existence and uniqueness of solutions can be guaranteed for all $x_1 \geq 0$, $x_2 \geq 0$;

(b) whether global existence and uniqueness of solutions can be guaranteed.

Solution:

Define $f(x) := \begin{bmatrix} -\beta x_1 x_2 - u \\ \beta x_1 x_2 - \gamma x_2 \end{bmatrix}$.

(a) Local existence and uniqueness of solutions: **Ans.** Yes.

The Jacobian,

$$\frac{\partial f(x)}{\partial x} = \begin{bmatrix} -\beta x_2 & -\beta x_1 \\ \beta x_2 & \beta x_1 - \gamma \end{bmatrix},$$

is continuous everywhere, and hence, the condition for local existence and uniqueness is met for all $x_1 \geq 0$, $x_2 \geq 0$.

(b) Global existence and uniqueness of solutions: **Ans.** No.

We use the max norm and note immediately that the $(1,1)$ element of the Jacobian satisfies

$$\sup_{x_1 \geq 0, x_2 \geq 0} \left| \frac{f_1(x)}{\partial x_1} \right| = \sup_{x_2 \geq 0} | -\beta x_2 | = \infty,$$

and hence

$$\| \frac{\partial f(x)}{\partial x} \|_\infty > L$$

for all finite L.

Note: Does this mean that solutions definitely blow up? No! Our calculations mean that we cannot prove global existence and uniqueness with the elementary means treated in the textbook. More powerful methods must be employed. Consider $V(x_1, x_2) := x_1 + x_2$, the total population of individuals. Then an easy calculation shows that $\frac{d}{dt}V(x_1, x_2) = \dot{x}_1 + \dot{x}_2 = -u - \gamma x_2 \leq 0$ when $u \geq 0$. Hence, the total population remains bounded, ruling out a finite escape time, and ensuring global existence and uniqueness of solutions.

■

ODE Examples from a Plethora of Engineering Domains: The (free) textbook, State Models and Dynamical Systems by Prof. Emeritus N. Harris McClamroch, aims to enhance undergraduate students' understanding of dynamic physical processes through applied mathematics and systems theory. It is written assuming that readers have a background in calculus, differential equations, Laplace transforms, physics, mechanics, and circuits. The book introduces mathematical systems theory, using case studies to clarify complex concepts.

Example 9.24. *Show that the condition for global existence and uniqueness fails for the 2-link manipulator model from Fact 9.13 and Example 9.14.*

Solution: Let $\dot{x} = f(x)$ be the state variable model as expressed in Example 9.14, after inverting $D(q)$. Using symbolic computations, we compute

$$J_{33} := \frac{\partial f_3(q_1, q_2, \dot{q}_1, \dot{q}_2)}{\partial \dot{q}_1} = \alpha(q_1, q_2)\dot{q}_1 + \beta(q_1, q_2)\dot{q}_2,$$

where

$$\alpha(q_1, q_2) = \frac{-4 \cdot m_2 \cdot (L_2 + L_1 \cdot \cos(q_2)) \cdot \sin(q_2)}{L_1 \cdot (-2 \cdot m_1 - m_2 + m_2 \cdot \cos(2 \cdot q2))}$$

and

$$\beta(q_1, q_2) = \frac{-4 \cdot L_2 \cdot m_2 \cdot \sin(q_2)}{-2 \cdot L_1 \cdot m_1 - L_1 \cdot m_2 + L_1 \cdot m_2 \cdot \cos(2 \cdot q_2)}.$$

Hence, for $q_2 \neq \pm\frac{\pi}{2}$ (modulo π), $\alpha(q_1, q_2) \neq 0$ and then

$$\sup_{\dot{q}_1} J_{33} = \infty,$$

and therefore $||\frac{\partial f}{\partial x}||$ cannot be uniformly bounded. More generally, if a robot model has two or more revolute joints, then it is pretty much guaranteed to fail the test we have given for global existence and uniqueness of solutions for the same reason underlined here: at least one component of the Jacobian will be linear (affine) in at least one component of the generalized velocity coordinates. ■

9.4.4 (Optional Read) More General Notion of a Solution

Consider a rather general vector ODE that may even depend explicitly upon time,

$$\dot{x}(t) = f(x(t), t), \quad x(t_0) = x_0. \tag{9.39}$$

An ODE is an equation relating $x(t)$ and its derivative, $\dot{x}(t)$. We know that $x(t)$ is always an antiderivative of $\dot{x}(t)$. The First Fundamental Theorem of Calculus[5] states that if $x(t)$ is an antiderivative of $x'(t)$ ($x'(t) = \dot{x}(t) = \frac{d}{dt}x(t)$ depending on the notation of the moment), the two functions must be related by

$$x(t) = x(t_0) + \int_{t_0}^{t} x'(\tau)\, d\tau \quad \textbf{and} \quad \frac{d}{dt}\left(x(t_0) + \int_{t_0}^{t} x'(\tau)\, d\tau\right) = x'(t). \tag{9.40}$$

We can then use (9.39) in (9.40) and obtain that $x(t)$, $f(x(t), t)$, $x'(t)$, and $\dot{x}(t)$ are related by

$$x(t) = x(t_0) + \int_{t_0}^{t} \underbrace{f(x(\tau), \tau)}_{x'(\tau)}\, d\tau \quad \textbf{and} \quad \frac{d}{dt}\left(x(t_0) + \int_{t_0}^{t} \underbrace{f(x(\tau), \tau)}_{x'(\tau)}\, d\tau\right) = \underbrace{\dot{x}(t)}_{x'(t)}. \tag{9.41}$$

Therefore, the equation

$$x(t) = x_0 + \int_{t_0}^{t} f(x(\tau), \tau)\, d\tau, \tag{9.42}$$

[5]While the Fundamental Theorems of Calculus apply to scalar-valued functions of a single variable, we can apply them to each component of a vector-valued function of a single variable, t.

relating $x(t)$ and an indefinite integral depending upon $x(t)$, is equivalent to the ODE (9.39), relating $x(t)$ and its derivative $\dot{x}(t)$. The interesting property of (9.42) is that the derivative of $x(t)$ does not show up explicitly, and hence functions that are only "piecewise differentiable", instead of differentiable everywhere, can also be solutions of the ODE (9.39). **We care about this as engineers, because when we design feedback controllers for a system (e.g., a robot), we often send command signals to the closed-loop system that are only piecewise continuous; that is, the command signals have jumps in them. Engineer to Robot: "Move forward at 1 m/s." The engineer sees an obstacle. Engineer screams to Robot: "Stop! Stop immediately..." as they slam the joystick in full reverse! As engineers, we would like to know that a solution to the differential equations that govern our robot does not cease to make sense as soon as we send a very abrupt (nondifferentiable) change in the command signal.**

9.5 Linear Systems of ODEs

When the right side of the ODE $\dot{x} = f(x)$ is linear in x, one would hope that it is simpler to compute analytical solutions and/or understand various properties of the solution. This is indeed the case and motivates the present section.

Homogeneous Linear Systems

Definition 9.25. *For an $n \times n$ real matrix A, the equation*

$$\dot{x} = Ax, \; x(t_0) = x_0 \tag{9.43}$$

is called a ***homogeneous linear system of differential equations****.*

Notes: By Prop. 5.44-(a), $\frac{\partial}{\partial x}(Ax) = A$, which is independent of x. Hence, Prop. 9.20-(b) implies the global existence and uniqueness of solutions. In other words, for all $x_0 \in \mathbb{R}^n$, $t_0 \in \mathbb{R}$, there exists a solution on $[t_0, \infty)$ and the solution is unique.

9.5.1 Examples from Circuits and Mechanics

Example 9.26. *Using Kirchoff's laws for linear RLC circuits, derive a differential equation model for the given circuit. The basics of Kirchoff's laws will be given in the solution. For more details, you can see Kirchhoff's Voltage Law - KVL Circuits, Loop Rule & Ohm's Law - Series Circuits, Physics and Kirchhoff's Law, Junction & Loop Rule, Ohm's Law - KCL & KVL Circuit Analysis - Physics, both by the Organic Chemistry Tutor.*

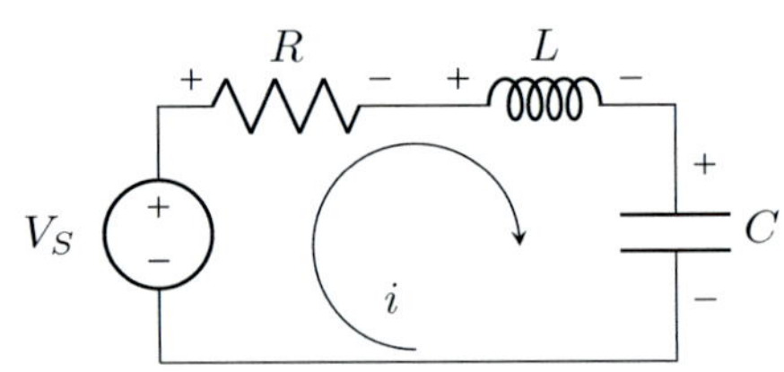

Solution: V_s is a voltage source, which could be a battery or something more capable, such as a power supply. R denotes a resistor, L an inductor, and C a capacitor. The loop denotes the reference direction for the current flowing through the circuit, $i(t)$. **Kirchoff's Voltage Law** says that the sum of the voltages around any loop must equal zero, which leads to

$$V_S + V_R + V_L + V_C = 0. \tag{9.44}$$

So far, there are no explicit derivatives in the equation, but the **constitutive relations** for the three circuit elements are

$$\begin{aligned} V_R &= R \cdot i, \\ V_L &= L\frac{di}{dt}, \\ i &= C\frac{dV_C}{dt}. \end{aligned} \tag{9.45}$$

The voltage across a resistor is proportional to the current flowing through it, the voltage across the inductor is proportional to the time derivative of the current, and the current through the capacitor is proportional to the time derivative of the voltage across the capacitor, denoted here as $\frac{dV_c}{dt}$. Now that we have derivatives, we can build an ODE model. The rest is "algebra".

If we assign state variables by

$$\begin{aligned} x_1 &:= V_C \\ x_2 &:= i, \end{aligned} \tag{9.46}$$

then, from line three of (9.45), we have

$$\dot{x}_1 = \frac{1}{C}x_2. \tag{9.47}$$

The equation for $\dot{x}_2$ comes from Kirchoff's Voltage Law as given by (9.44) in tandem with lines one and two of (9.45). Indeed, from (9.44), we obtain

$$V_L = -V_s - V_R - V_C,$$

which from (9.45) gives,

$$L\frac{di}{dt} = -V_s - r \cdot i - V_C.$$

From the definition of the state variables in (9.46), we obtain

$$\begin{aligned} L\dot{x}_2 &= -Cx_1 - Rx_2 - V_s \\ &\Updownarrow \\ \dot{x}_2 &= -\frac{C}{L}x_1 - \frac{R}{L}x_2 - \frac{1}{L}V_S. \end{aligned} \tag{9.48}$$

Combining (9.47) and (9.48) gives

$$\begin{bmatrix} \dot{x}_1 \\ \dot{x}_2 \end{bmatrix} = \begin{bmatrix} 0 & \frac{1}{C} \\ -\frac{C}{L} & -\frac{R}{L} \end{bmatrix} \begin{bmatrix} x_1 \\ x_2 \end{bmatrix} + \begin{bmatrix} 0 \\ \frac{1}{L} \end{bmatrix} V_s. \tag{9.49}$$

Equation (9.49) is an inhomogeneous system of linear equations due to the voltage source, V_s. If we assume $V_s \equiv 0$, we obtain

$$\begin{bmatrix} \dot{x}_1 \\ \dot{x}_2 \end{bmatrix} = \begin{bmatrix} 0 & \frac{1}{C} \\ -\frac{C}{L} & -\frac{R}{L} \end{bmatrix} \begin{bmatrix} x_1 \\ x_2 \end{bmatrix}. \tag{9.50}$$

The initial conditions for (9.50) come from the initial voltage across the capacitor and the initial current through the inductor. ■

Hence, RLC circuits give rise to linear systems of ODEs. In a similar manner, mechanical systems without "rotating elements" (i.e., only have elements that move in a straight line) "mostly" give rise also to linear systems of ODEs when stiction and dead zones are ignored.

Example 9.27. *Using either Newton's Laws or Lagrange's Equations of Motion, derive a linear differential equation model for the mass-spring-damper system below.*

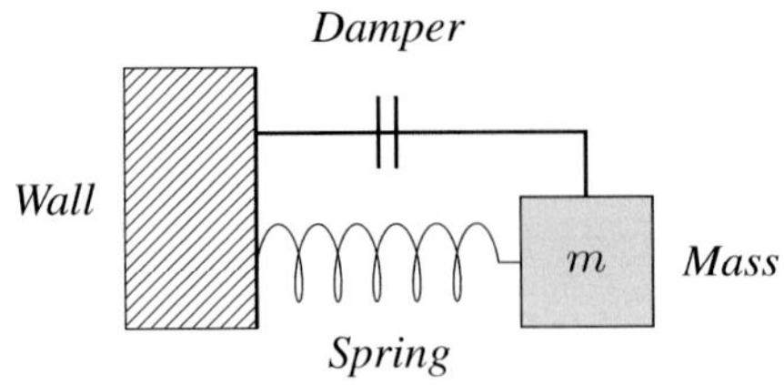

Solution: We assume the wall is immovable and let x be the distance of the mass from the wall, with positions to the right of the wall being positive. The velocity of the mass is denoted $\dot{x}$, and its Kinetic Energy is $\frac{1}{2}m\dot{x}^2$. Let the rest position of the spring be x_s (the position of the spring when it is neither compressed nor stretched). The Potential Energy of the spring is then $\frac{1}{2}k\left(x - x_s\right)^2$. The effect of the linear damper, $-b\dot{x}$, must be included in the right-hand side of (9.26). With these assumptions, Lagrange's equations become

$$\begin{aligned} L(x,\dot{x}) &= \frac{1}{2}m\dot{x}^2 - \frac{1}{2}k\left(x - x_s\right)^2 \\ \frac{d}{dt}\frac{\partial L}{\partial \dot{x}} - \frac{\partial L}{\partial x} &= \Gamma, \end{aligned} \tag{9.51}$$

where k is the spring constant, b is the damping coefficient, and $\Gamma = -b\dot{x}$. Computing the indicated derivatives gives

$$\begin{gathered} \frac{d}{dt}\left(m\dot{x}\right) + k\left(x - x_s\right) = -b\dot{x} \\ \Updownarrow \\ m\ddot{x} + b\dot{x} + k\left(x - x_s\right) = 0. \end{gathered} \tag{9.52}$$

Assigning state variables as

$$\begin{aligned} x_1 &:= x - x_s \\ x_2 &:= \dot{x}, \end{aligned} \tag{9.53}$$

gives the relationships

$$\begin{aligned} \dot{x}_1 &= x_2 \\ m\dot{x}_2 &= -kx_1 - bx_2. \end{aligned} \tag{9.54}$$

Therefore, the linear homogeneous model is

$$\begin{bmatrix} \dot{x}_1 \\ \dot{x}_2 \end{bmatrix} = \begin{bmatrix} 0 & 1 \\ -\frac{k}{m} & -\frac{b}{m} \end{bmatrix} \begin{bmatrix} x_1 \\ x_2 \end{bmatrix}. \tag{9.55}$$

The initial conditions for (9.55) come from the initial position of the mass (relative to the rest position of the spring) and the initial velocity of the mass.

■

9.5.2 Expressing Higher-Order Linear ODEs in First-Order Form

In Definition 9.9, we introduced the notion of "higher order ODES of a single variable", where the linear model takes the form

$$y^{(n)} + a_{n-1}y^{(n-1)} + a_{n-2}y^{(n-2)} + \cdots + a_1\dot{y} + a_0 y = 0.$$

Such models can be written as a vector first-order system of linear ODEs by assigning state variables as follows

$$\begin{bmatrix} x_1 \\ x_2 \\ \vdots \\ x_{n-1} \\ x_n \end{bmatrix} = \begin{bmatrix} y \\ \dot{y} \\ \vdots \\ y^{(n-2)} \\ y^{(n-1)} \end{bmatrix}, \tag{9.56}$$

so that

$$\frac{d}{dt}\begin{bmatrix} x_1 \\ x_2 \\ \vdots \\ x_{n-1} \\ x_n \end{bmatrix} = \begin{bmatrix} \dot{y} \\ \ddot{y} \\ \vdots \\ y^{(n-1)} \\ y^{(n)} \end{bmatrix} = \begin{bmatrix} x_2 \\ x_3 \\ \vdots \\ x_n \\ -a_0 x_1 - a_1 x_2 - \cdots - a_{n-1} x_n \end{bmatrix}. \tag{9.57}$$

In matrix form, the system of ODEs becomes

$$\dot{x} = \underbrace{\begin{bmatrix} 0 & 1 & 0 & 0 & \cdots & 0 & 0 \\ 0 & 0 & 1 & 0 & \cdots & 0 & 0 \\ 0 & 0 & 0 & 1 & \cdots & 0 & 0 \\ \vdots & \vdots & \vdots & \vdots & \ddots & 0 & 0 \\ 0 & 0 & 0 & 0 & \ddots & 1 & 0 \\ 0 & 0 & 0 & 0 & \ddots & 0 & 1 \\ -a_0 & -a_1 & -a_2 & -a_3 & \cdots & -a_{n-2} & -a_{n-1} \end{bmatrix}}_{A} \cdot x. \tag{9.58}$$

Example 9.28. *Rewrite the second-order linear mass-spring-damper model,*

$$m\ddot{x} + b\dot{x} + k\left(x - x_s\right) = 0,$$

as a linear state-variable model.

Solutions: We first solve for $\ddot{x}$, yielding

$$\ddot{x} = -\frac{k}{m}\left(x - x_s\right) - \frac{b}{m}\dot{x}.$$

Assigning state variables as

$$\begin{aligned} x_1 &:= x - x_s \\ x_2 &:= \dot{x}, \end{aligned}$$

gives the relationships

$$\begin{aligned} \dot{x}_1 &= x_2 \\ \dot{x}_2 &= -\frac{k}{m}x_1 - \frac{b}{m}x_2. \end{aligned}$$

Therefore, the linear homogeneous model is

$$\begin{bmatrix} \dot{x}_1 \\ \dot{x}_2 \end{bmatrix} = \begin{bmatrix} 0 & 1 \\ -\frac{k}{m} & -\frac{b}{m} \end{bmatrix} \begin{bmatrix} x_1 \\ x_2 \end{bmatrix},$$

just as we derived earlier in Example 9.27. ■

9.5.3 Linearization of Nonlinear ODE Models

In engineering practice, linear models predominantly arise from local approximations of nonlinear models around equilibrium points, that is, initial conditions where the system remains constant unless disturbed. The utility of linear approximations might seem counterintuitive, given our ability to compute solutions for complex nonlinear models numerically. However, often, the real challenge lies not in simulating systems but in modifying their behavior. Consider the example of a Segway: while creating and simulating its nonlinear dynamical model is straightforward, designing a feedback control system to maintain its balance is far more complex. Here, linearization becomes invaluable. By simplifying a nonlinear system to its linear counterpart around an equilibrium point, we gain a more manageable framework. This linear approximation facilitates the use of well-developed linear control techniques like PID control and state-space analysis, offering crucial insights into the system's stability and response near the equilibrium. Such simplifications are essential for effective control system design and analysis despite being local approximations that may not fully capture the system's behavior in more extreme conditions.

Equilibrium Points

Definition 9.29. *A vector $x_e \in \mathbb{R}^n$ is an* **equilibrium point** *of the differential equation $\dot{x} = f(x)$ if $f(x_e) = 0_{n\times 1}$. Equivalently, x_e is an equilibrium point if the constant function $\varphi(t, t_0, x_e) := x_e$ is a solution of the ODE.*

Notes:

- In simple terms, equilibrium points are initial conditions from which the solution of the ODE does not change; it remains constant and equal to the initial condition. For example, if a pendulum is initialized at rest in the downward hanging position, it stays there!
- Determining points x_e such that $f(x_e) = 0$ is a root-finding problem; hence, Newton-Raphson is an effective means for finding equilibrium points in cases where the physics of a problem does not make them obvious.
- For a linear ODE, $\dot{x} = Ax$, x_e is an equilibrium point if, and only if, $x_e \in \text{null}(A)$. In particular, for a linear ODE, the origin is always an equilibrium point.

Example 9.30. *Find equilibrium points for the following ordinary differential equations:*

(a) $\frac{dv(t)}{dt} = \frac{\kappa}{m}v^2(t) - g$ *(parachutist with nonlinear drag, from Example 9.7).*

(b) $\begin{bmatrix} \dot{x}_1 \\ \dot{x}_2 \end{bmatrix} = \begin{bmatrix} x_2 \\ -\frac{g}{L}\cdot\sin(x_1) \end{bmatrix}$ *(pendulum model from Examples 6.36 and 9.12).*

(c) $\begin{bmatrix} \dot{x}_1 \\ \dot{x}_2 \end{bmatrix} = \begin{bmatrix} x_2 \\ -D^{-1}(x_1)\cdot[C(x_1,x_2)\cdot x_2 + G(x_1)] \end{bmatrix}$ *(the robot equations from Method 9.11).*

Solutions:

(a) $\dot{v} = \frac{\kappa}{m}v^2 - g$ **Ans.** $v_e = -\sqrt{\frac{mg}{\kappa}}$.

We set the right side of the ODE to zero and solve for v, which gives $v^2 = \frac{mg}{\kappa}$. Solving for v gives $v_e = \pm\sqrt{\frac{mg}{\kappa}}$. However, only the negative velocity makes sense because the parachutist is not falling upward. This is where "engineering" judgment comes into play. If we knew nothing about the model, we would have reported both solutions as potential equilibria.

Note: The equilibrium point is what we called the "terminal velocity" earlier when examining this problem. It corresponds to the speed at which the downward acceleration of gravity is in equilibrium with the upward drag force generated by the parachute.

(b) $\begin{bmatrix} \dot{x}_1 \\ \dot{x}_2 \end{bmatrix} = \begin{bmatrix} x_2 \\ -\frac{g}{L}\cdot\sin(x_1) \end{bmatrix}$ **Ans.** $x_e = \begin{bmatrix} 0 \\ 0 \end{bmatrix}$ and $x_e = \begin{bmatrix} \pi \\ 0 \end{bmatrix}$.

We set the right side to zero and solve for x, which gives $x_2 = 0$ for the angular velocity of the pendulum, and $\sin(x_1) = 0$, which implies $x_1 = k\pi$, $k \in \mathbb{Z}$. $k = 0$ corresponds to the downward equilibrium and $k = 1$ gives the upward equilibrium.

(c) $\begin{bmatrix} \dot{x}_1 \\ \dot{x}_2 \end{bmatrix} = \begin{bmatrix} x_2 \\ -D^{-1}(x_1)\cdot[C(x_1,x_2)\cdot x_2 + G(x_1)] \end{bmatrix}$ **Ans.** $x_e = \left\{\begin{bmatrix} q_e \\ 0 \end{bmatrix} \in \mathbb{R}^{2n} \mid G(q_e) = 0_{n\times1}\right\}$

We set the right side to zero and solve for x. The top row gives $x_2 = 0$, the generalized velocity. Substituting this into the bottom row gives $-D^{-1}(x_1)\cdot G(x_1) = 0_{n\times1} \iff G(x_1) = 0_{n\times1}$. Because x_1 is the generalized position of the system, we'll use q_e to denote the roots of the potential energy gradient vector, $G(q_e) = 0_{n\times1}$. There could be no roots, one root, or multiple roots, as with any nonlinear function.

■

Example 9.31. *We consider the 2-link manipulator augmented with springs on each of the revolute joints. The springs are installed so that their rest positions correspond to the origin, namely,* $(\theta_1,\theta_2) = (0,0)$. *With this modification, and after setting the model parameters to*

$$\begin{bmatrix} g \\ L_1 \\ L_2 \\ m_1 \\ m_2 \\ k_1 \\ k_2 \end{bmatrix} = \begin{bmatrix} 9.81\ m/s^2 \\ 1.0\ m \\ 0.5\ m \\ 1.0\ kg \\ 3.0\ kg \\ 24\ N/rad \\ 24\ N/rad \end{bmatrix},$$

we have

$$G(q_1,q_2) = \begin{bmatrix} 14.715\cdot\cos(q_1+q_2) + 24q_1 + 39.24\cdot\cos(q_1) \\ 14.715\cdot\cos(q_1+q_2) + 24q_2 \end{bmatrix}$$

and

$$Jac_G(q_1, q_2) = \begin{bmatrix} 24.0 - 14.715 \cdot \sin(q_1 + q_2) - 39.24 \cdot \sin(q_1) & -14.715 \cdot \sin(q_1 + q_2) \\ -14.715 \cdot \sin(q_1 + q_2) & 24.0 - 14.715 \cdot \sin(q_1 + q_2) \end{bmatrix}.$$

Find an equilibrium point.

Solution: Ans. $q^* = \begin{bmatrix} -1.03536 \\ -0.20116 \end{bmatrix}$ radians.

The rest position of the springs corresponds to both links at $(\theta_1, \theta_2) = (0, 0)$, which corresponds to $p_1 = [0; 0]$ and $p_2 = [0; 0]$, that is, the links are fully extended along the positive x-axis. With nothing to support the robot arm, gravity seeks to rotate its links downward, thereby engaging the springs. In other words, when hanging partially downward, the springs provide restorative (supporting) forces at the two joints. The stiffer the springs, the less motion is required by the links to be in equilibrium with gravity.

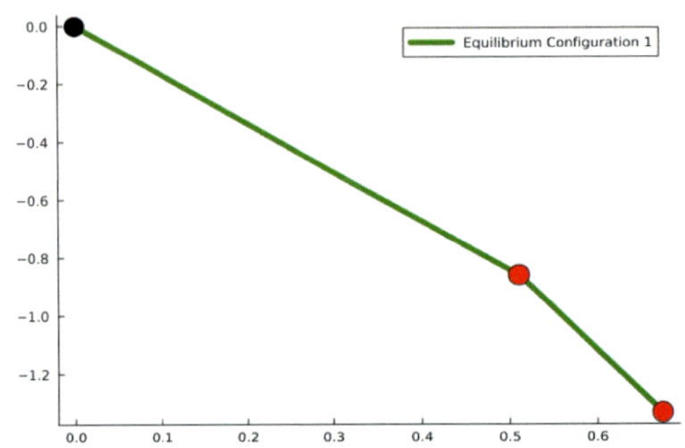

The following code implements a Newton-Raphson routine to find $G(q_1^*, q_2^*) = [0; 0]$.

```
using LinearAlgebra
include("dyn_mod_2LinkManipulatorWithSprings.jl")
x0 = [pi/2, pi/4]
Model = dyn_mod_2LinkManipulatorWithSprings(x0, 0*x0)
G = Model.G
JacG = Model.JacG
k = 0
aTol = 1e-5
s = 0.1
xk = x0
while (k<1e5) && (norm(G)>aTol)
    k = k+1
    Delta_x = -JacG\G
    xk = xk + s*Delta_x
    Model = dyn_mod_2LinkManipulatorWithSprings(xk, 0*xk)
    G = Model.G
    JacG = Model.JacG
end
Model = dyn_mod_2LinkManipulatorWithSprings(xk, 0*xk)
G = Model.G
JacG = Model.JacG

display(G)
[xk x0]
```

Output The (first) vector is $G(q_1^*, q_2^*)$ and the (second) matrix compares the starting point of the Newton-Raphson Algorithm to its final value.

```
2-element Vector{Float64}:
 -2.959527982682175e-6
  9.381671145725079e-6

2×2 Matrix{Float64}:
 -1.03536   1.5708
 -0.201158  0.785398
```

■

Linearization about an Equilibrium Point

Let $x_e \in \mathbb{R}^n$ be an **equilibrium point** of the differential equation $\dot{x} = f(x)$ and suppose that f is differentiable at x_e. Then from (5.34), the linearization of f about x_e is

$$f(x) = f(x_e) + \left.\frac{\partial f(x)}{\partial x}\right|_{x=x_e} \cdot (x - x_e) = \frac{\partial f(x_e)}{\partial x} \cdot (x - x_e),$$

where $f(x_e) = 0_{n\times 1}$ follows from x_e being an equilibrium point.

Definition 9.32. *For a nonlinear vector ODE $\dot{x} = f(x)$, the* ***linearization about an equilibrium point*** *x_e is the linear vector ODE*

$$\delta\dot{x} = \frac{\partial f(x_e)}{\partial x} \cdot \delta x, \tag{9.59}$$

where $\delta x := x - x_e$.

Notes:

- If $\delta x(t)$ is a solution to the linearized model (9.59), then $x_e + \delta x(t)$ is an **approximate solution** to the original nonlinear model.
- When $x_e = 0_{n\times 1}$, it is common to write the linearized model as $\dot{x} = Ax$, for $A := \frac{\partial f(x_e)}{\partial x}$, thereby ignoring the δ.
- Even when $x_e \neq 0_{n\times 1}$, it is still rather uncommon to use δx as the name of the state variables in the linear model. Some engineers abuse notation and reuse the variable x, others opt for a new variable name, such as z or χ.

Example 9.33. *Linearize the following nonlinear models about the given equilibrium points.*

(a) $\frac{dv(t)}{dt} = \frac{\kappa}{m}v^2(t) - g$ *with* $v_e = -\sqrt{\frac{m\,g}{\kappa}}$ *(parachutist with nonlinear drag, from Example 9.7).*

(b) $\begin{bmatrix} \dot{x}_1 \\ \dot{x}_2 \end{bmatrix} = \begin{bmatrix} x_2 \\ -\frac{g}{L}\cdot\sin(x_1) \end{bmatrix}$ *with* $x_e = \begin{bmatrix} 0 \\ 0 \end{bmatrix}$ *and* $x_e = \begin{bmatrix} \pi \\ 0 \end{bmatrix}$ *(pendulum model from Examples 6.36 and 9.12).*

Solutions:

(a) $\dot{v} = \frac{\kappa}{m}v^2 - g$ **Ans.** $\delta\dot{v} = -\left(2\sqrt{\frac{\kappa\,g}{m}}\right)\cdot\delta v$, where $\delta v = v - v_e$.

We identify $f(v) = \frac{\kappa}{m}v^2 - g$ and compute $\frac{df(v)}{dv} = \frac{d}{dv}(\frac{\kappa}{m}v^2 - g) = 2\frac{\kappa}{m}v$. Evaluating the "Jacobian" at $v_e = -\sqrt{\frac{mg}{\kappa}}$ gives $\frac{df(v_e)}{dv} = -2\sqrt{\frac{\kappa g}{m}}$. Hence, the linearized model is

$$\delta\dot{v} = -\left(2\sqrt{\frac{\kappa g}{m}}\right) \cdot \delta v, \quad \delta v = v - v_e.$$

Note: The scalar coefficient in front of δv is negative. Later, we will learn that this negative sign imparts a "self-stabilizing action" to the model, meaning that if the velocity is initially greater than the equilibrium velocity, then it will decrease toward the equilibrium velocity, and conversely, if the velocity is initially lower than the equilibrium velocity, then it will increase toward the equilibrium velocity. If this were not the case, the parachutist would have to continually adjust the surface area of the parachute to regulate their speed, considerably augmenting the difficulty of the jump!

(b) $\begin{bmatrix} \dot{x}_1 \\ \dot{x}_2 \end{bmatrix} = \begin{bmatrix} x_2 \\ -\frac{g}{L} \cdot \sin(x_1) \end{bmatrix}$ **Ans.** $\begin{bmatrix} \dot{x}_1 \\ \dot{x}_2 \end{bmatrix} = \begin{bmatrix} 0 & 1 \\ -\frac{g}{L} & 0 \end{bmatrix} \begin{bmatrix} x_1 \\ x_2 \end{bmatrix}$ and $\begin{bmatrix} \delta\dot{x}_1 \\ \delta\dot{x}_2 \end{bmatrix} = \begin{bmatrix} 0 & 1 \\ \frac{g}{L} & 0 \end{bmatrix} \begin{bmatrix} \delta x_1 \\ \delta x_2 \end{bmatrix}$.

We identify $f(x) = f(x_1, x_2) = \begin{bmatrix} x_2 \\ -\frac{g}{L} \cdot \sin(x_1) \end{bmatrix}$ and compute $\frac{\partial f(x_1, x_2)}{\partial x} = \begin{bmatrix} 0 & 1 \\ -\frac{g}{L} \cdot \cos(x_1) & 0 \end{bmatrix}$. Evaluating the Jacobian at the two equilibria gives

$$\frac{\partial f(0,0)}{\partial x} = \begin{bmatrix} 0 & 1 \\ -\frac{g}{L} & 0 \end{bmatrix} \quad \text{and} \quad \frac{\partial f(\pi,0)}{\partial x} = \begin{bmatrix} 0 & 1 \\ \frac{g}{L} & 0 \end{bmatrix}.$$

Hence, the two linear models are

$$\begin{bmatrix} \dot{x}_1 \\ \dot{x}_2 \end{bmatrix} = \begin{bmatrix} 0 & 1 \\ -\frac{g}{L} & 0 \end{bmatrix} \begin{bmatrix} x_1 \\ x_2 \end{bmatrix} \quad \text{and} \quad \begin{bmatrix} \delta\dot{x}_1 \\ \delta\dot{x}_2 \end{bmatrix} = \begin{bmatrix} 0 & 1 \\ \frac{g}{L} & 0 \end{bmatrix} \begin{bmatrix} \delta x_1 \\ \delta x_2 \end{bmatrix},$$

where in the first model, because the equilibrium is at the origin, we kept x as the state variable, and in the second model, we are using $\begin{bmatrix} \delta x_1 \\ \delta x_2 \end{bmatrix} := \begin{bmatrix} x_1 - \pi \\ x_2 \end{bmatrix}$. There are no hard and fast rules governing these choices. You have a lot of freedom to do what you prefer.

Note: The two linear models differ by a minus sign in the (2,1) position of the "A-matrix". Later, we will learn that the "A-matrix" of the first model has purely imaginary eigenvalues $\lambda = \pm\mathbf{i}\sqrt{\frac{g}{L}}$, indicating oscillatory behavior, while the second model has two real eigenvalues $\lambda = \pm\sqrt{\frac{g}{L}}$, with the positive eigenvalue indicating "instability". This comports with what we know about an inverted (aka upward-pointing) pendulum: if we perturb it ever so slightly, it will fall over and not return to the upright position, while if we perturb a normal pendulum, it will oscillate. Segways, BallBots, bipedal robots, and humans each contain an (unstable) inverted pendulum that must be balanced! If this were not the case, babies would have a "self-righting" mechanism built in at birth.

■

Proposition 9.34: Linearization of the Robot Equations

We initially encountered the robot equations expressed as a set of nonlinear second-order differential equations:

$$D(q)\cdot\ddot{q}+C(q,\dot{q})\cdot\dot{q}+G(q)=0_{n\times1},$$

where we are assuming here that there are no external forces acting on the robot. In Example 9.30, we determined that their equilibrium points correspond to $G(q_e)=0_{n\times1}$ and $\dot{q}_e=0_{n\times1}$. Their linearization about an equilibrium can be expressed in both second-order and first-order forms, namely,

$$D(q_e)\cdot\delta\ddot{q}+\frac{\partial G(q_e)}{\partial q}\cdot\delta q=0_{n\times1}, \tag{9.60}$$

where $\delta q=q-q_e$, and

$$\begin{bmatrix}\delta\dot{x}_1\\ \delta\dot{x}_2\end{bmatrix}=\begin{bmatrix}0_n & I_n\\ A_{21} & 0_n\end{bmatrix}\begin{bmatrix}\delta x_1\\ \delta x_2\end{bmatrix}, \tag{9.61}$$

where $A_{21}=-D(q_e)\backslash\frac{\partial G(q_e)}{\partial q}$, I_n is the $n\times n$ identity matrix, $\delta x_1=q-q_e$, and $\delta x_2=\dot{q}-0_{n\times1}$.

Note: At first, it is surprising that the term $C(q,\dot{q})\cdot\dot{q}$ does not show up in the linearization. The reason is that it consists of quadratic terms of the form $c_{ij}(q)\cdot\dot{q}_i\cdot\dot{q}_j$, which when linearized about $\dot{q}_e=0_{n\times1}$, $1\le i\le n$, all vanish. By taking advantage of this property, the linearization of the robot equations is trivial to compute: we only have to compute one Jacobian, $\frac{\partial G(q_e)}{\partial q}$, and evaluate the mass inertia matrix at q_e.

Example 9.35. *In Section 6.4.6, we derived Lagrange's equations for a 3-link manipulator. Using Julia, compute its linearized model about the downward equilibrium* $q_e^d=\begin{bmatrix}-\frac{\pi}{2}\\ 0\\ 0\end{bmatrix}$ *and the upward equilibrium*

$$q_e^u=\begin{bmatrix}\frac{\pi}{2}\\ 0\\ 0\end{bmatrix}.$$

Use the following function to define physical values for the parameters.

```julia
function modelParameters()
    g = 9.81 # m/s^2
    L1 = 1 # m
    L2 = 0.7
    L3 = 0.5
    m1 = 15 # kg
    m2 = 10
    m3 = 5
    return g, L1, L2, L3, m1, m2, m3
end
```

Solutions:

Ans. Downward Equilibrium

$$\begin{bmatrix} \dot{x}_1 \\ \dot{x}_2 \\ \dot{x}_3 \\ \dot{x}_4 \end{bmatrix} = \begin{bmatrix} 0.0000 & 0.0000 & 0.0000 & 1.0000 & 0.0000 & 0.0000 \\ 0.0000 & 0.0000 & 0.0000 & 0.0000 & 1.0000 & 0.0000 \\ 0.0000 & 0.0000 & 0.0000 & 0.0000 & 0.0000 & 1.0000 \\ -9.8100 & 9.8100 & 0.0000 & 0.0000 & 0.0000 & 0.0000 \\ 9.8100 & -37.8386 & 7.0071 & 0.0000 & 0.0000 & 0.0000 \\ 0.0000 & 28.0286 & -36.4371 & 0.0000 & 0.0000 & 0.0000 \end{bmatrix} \cdot \begin{bmatrix} x_1 \\ x_2 \\ x_3 \\ x_4 \\ x_4 \\ x_6 \end{bmatrix}$$

Ans. Upward Equilibrium

$$\begin{bmatrix} \dot{x}_1 \\ \dot{x}_2 \\ \dot{x}_3 \\ \dot{x}_4 \end{bmatrix} = \begin{bmatrix} 0.0000 & 0.0000 & 0.0000 & 1.0000 & 0.0000 & 0.0000 \\ 0.0000 & 0.0000 & 0.0000 & 0.0000 & 1.0000 & 0.0000 \\ 0.0000 & 0.0000 & 0.0000 & 0.0000 & 0.0000 & 1.0000 \\ 9.8100 & -9.8100 & 0.0000 & 0.0000 & 0.0000 & 0.0000 \\ -9.8100 & 37.8386 & -7.0071 & 0.0000 & 0.0000 & 0.0000 \\ 0.0000 & -28.0286 & 36.4371 & 0.0000 & 0.0000 & 0.0000 \end{bmatrix} \cdot \begin{bmatrix} x_1 \\ x_2 \\ x_3 \\ x_4 \\ x_4 \\ x_6 \end{bmatrix}$$

Note: The A_{21} block (in blue) for the upward equilibrium is the negative of the A_{21} block (in red) for the downward equilibrium. This generalizes the observation we made for the single-link pendulum to the three-link configuration. In Example 9.36, we explore what this means for the eigenvalues of the respective linearized models.

The function `dyn_mod_3LinkManipulator(q, dq)` (see file `dyn_mod_3LinkManipulator.jl`), **returns (D=D, C=C, G=G, B=B, JacG=JacG)**, where JacG = $\frac{\partial G}{\partial q}$. We will use D and $JacG$.

We only show the code for the downward equilibrium.

```
# Linear Model for the 3-link Manipulator
#
# bring the model into the workspace for later use
include("dyn_mod_3LinkManipulator.jl")

#Build a function to define the parameters of the model

function modelParameters()
    g = 9.81 # m/s^2
    L1 = 1 # m
    L2 = 0.7
    L3 = 0.5
    m1 = 15 # kg
    m2 = 10
    m3 = 5
    return g, L1, L2, L3, m1, m2, m3
```

```
end

# Set equilibirum Point
if true
    qe = [-pi/2, 0, 0]
else
    qe = [pi/2, 0, 0]
end

F = dyn_mod_3LinkManipulator(qe, 0*qe)
D = F.D; display(D)
JacG = F.JacG; display(JacG)

# build the linearized model about the given equilibrium point
A21 = -D\JacG; A21 = cleanUp(A21); display(A21)
n, ~ = size(A21)
A = [zeros(n,n) I(n); A21 zeros(n,n)]
```

Output

```
3×3 Matrix{Float64}:
 68.1   25.1   5.5
 25.1   12.1   3.0
  5.5    3.0   1.25

3×3 Matrix{Float64}:
 421.83    127.53    24.525
 127.53    127.53    24.525
  24.525    24.525   24.525

3×3 Matrix{Float64}:
 -9.81     9.81       0.0
  9.81   -37.8386     7.00714
  0.0     28.0286   -36.4371

6×6 Matrix{Float64}:
  0.0      0.0        0.0       1.0  0.0  0.0
  0.0      0.0        0.0       0.0  1.0  0.0
  0.0      0.0        0.0       0.0  0.0  1.0
 -9.81     9.81       0.0       0.0  0.0  0.0
  9.81   -37.8386     7.00714   0.0  0.0  0.0
  0.0     28.0286   -36.4371    0.0  0.0  0.0
```

■

Example 9.36. *Use Julia to compute the eigenvalues for both linearized models of Example 9.35.*

Solutions: **Ans.** $\begin{bmatrix} \lambda_1 \\ \lambda_2 \\ \lambda_3 \\ \lambda_4 \\ \lambda_5 \\ \lambda_6 \end{bmatrix}_d = \begin{bmatrix} -7.2389\,\mathrm{i} \\ 7.2389\,\mathrm{i} \\ -2.4528\,\mathrm{i} \\ 2.4528\,\mathrm{i} \\ -5.0664\,\mathrm{i} \\ 5.0664\,\mathrm{i} \end{bmatrix}$ and **Ans.** $\begin{bmatrix} \lambda_1 \\ \lambda_2 \\ \lambda_3 \\ \lambda_4 \\ \lambda_5 \\ \lambda_6 \end{bmatrix}_u = \begin{bmatrix} -7.2389 \\ -5.0664 \\ -2.4528 \\ 2.4528 \\ 5.0664 \\ 7.2389 \end{bmatrix}.$

Note: For the downward equilibrium, the eigenvalues are distinct and purely imaginary, which, we will later show, implies sustained periodic behavior. For the upward equilibrium, on the other hand, the eigenvalues are real and occur in positive-negative pairs, $\pm\lambda$. We will observe later that the positive real eigenvalues imply "instability," meaning solutions have norms that tend to infinity.

(Optional Read) Is there a logical explanation for why the eigenvalues change so drastically from one model to the other? Yes. There are two explanations. The first explanation is based on our physical experience with inverted pendula, as we discussed in Example 9.30(b); once the system is disturbed, it topples over. The second and more mathematical explanation is that for the robot equations, we have

$$A = \begin{bmatrix} 0_n & I_n \\ A_{21} & 0_n \end{bmatrix}.$$

When A_{21} is an invertible $n \times n$ matrix, we can show that

$$\det(\lambda I_{2n} - A) = \det \begin{bmatrix} \lambda I_n & -I_n \\ -A_{21} & \lambda I_n \end{bmatrix} = \det(\lambda^2 I_n - A_{21}).$$

Therefore, the eigenvalues of A are the plus and minus square roots of the eigenvalues of A_{21}. The sign change on the entries of A_{21} for upward and downward equilibria changes purely imaginary eigenvalues into purely real eigenvalues. The reason behind the sign change of the entries of A_{21} is more technical. It arises because in the upward equilibrium, the potential energy is a maximum, while at the lower equilibrium, it is a minimum. Indeed, because $G(q) = \nabla PE(q)$, $\frac{\partial G(q)}{\partial q}$ is the Hessian of the potential energy. The Hessian is a positive definite matrix at a local minimum and a negative definite matrix at a local maximum. Linear algebra is very powerful!

Code for the Downward Equilibrium

```
E=eigen(A)
evals = cleanUp(E.values); display(evals)
```

Output

```
6-element Vector{ComplexF64}:
 0.0 - 7.23885032225449im
 0.0 + 7.23885032225449im
 0.0 - 2.452784229773624im
 0.0 + 2.452784229773624im
 0.0 - 5.066419822703628im
 0.0 + 5.066419822703628im
```

CleanUp Function that handles complex numbers

```
function cleanUp(A, tol=1e-10)
    B = copy(A)
    for i in eachindex(B)
        if isa(B[i], Complex)
            # Clean up the real part
            real_part = abs(real(B[i])) < tol ? 0.0 : real(B[i])
            # Clean up the imaginary part
            imag_part = abs(imag(B[i])) < tol ? 0.0 : imag(B[i])
            # Reconstruct the complex number
            B[i] = real_part + imag_part * im
        else
```

```
            # Original cleanup for non-complex numbers
            if abs(B[i]) < tol
                B[i] = 0.0
            end
        end
    end
    return B
end
```

Output

```
cleanUp (generic function with 2 methods)
```

9.5.4 The Matrix Exponential

We recall that for $a \in \mathbb{R}$, the Maclaurin/Taylor Expansions for e^{at} and $e^{a(t-t_0)}$ are:

$$\begin{aligned} e^{at} &= \sum_{k=0}^{\infty} a^k \cdot \frac{t^k}{k!} = 1 + a \cdot t + a^2 \cdot \frac{t^2}{2!} + \cdots + a^k \cdot \frac{t^k}{k!} + \cdots \text{ and} \\ e^{a(t-t_0)} &= \sum_{k=0}^{\infty} a^k \cdot \frac{(t-t_0)^k}{k!} = 1 + a \cdot (t-t_0) + a^2 \cdot \frac{(t-t_0)^2}{2!} + \cdots + a^k \cdot \frac{(t-t_0)^k}{k!} + \cdots, \end{aligned} \tag{9.62}$$

respectively.[6] In the following, we initially set $t_0 = 0$, so that the Maclaurin and Taylor expansions are the same.

Matrix Exponential and Exponential Function

In the following, I_n denotes the $n \times n$ identity matrix.

Definition 9.37. *For an $n \times n$ real matrix A,*

$$e^A := \sum_{k=0}^{\infty} A^k \cdot \frac{1}{k!} = I_n + A + A^2 \cdot \frac{1}{2!} + \cdots + A^k \cdot \frac{1}{k!} + \cdots \tag{9.63}$$

*is called the **matrix exponential**. It is also denoted as $\exp(A)$.*

***Note:** If $A = 0_n$, the $n \times n$ zero matrix, then $\exp(0_n) := I_n$.* **Warning: This shows that the matrix exponential is not the exponential of each term of the matrix.** *In other symbols, $\left[e^A\right]_{ij} \neq \exp(a_{ij})$. If it were, then $\exp(0_n)$ would equal an $n \times n$ matrix filled with ones, instead of being an identity matrix.*

Definition 9.38. *For an $n \times n$ real matrix A and $t \in \mathbb{R}$,*

$$e^{At} := \sum_{k=0}^{\infty} A^k \cdot \frac{t^k}{k!} = I_n + A \cdot t + A^2 \cdot \frac{t^2}{2!} + \cdots + A^k \cdot \frac{t^k}{k!} + \cdots \tag{9.64}$$

*is called the **matrix exponential function**. It is also denoted as $\exp(At)$.*

Notes:

- The matrix exponential function is simply the matrix exponential evaluated at the matrix $A \cdot t = t \cdot A$, which depends upon time.

[6] The definition should be written as $e^{at} = 1 + \lim_{N \to \infty} \sum_{k=1}^{N} a^k \cdot \frac{t^k}{k!}$ so that the need for a limit is clear, and so that we do not have to use $0^0 = 1$. Normally, these "niceties" are assumed to be understood by the reader.

- Sometimes e^{At} is inaccurately referred to as the matrix exponential instead of the matrix exponential "function" because context usually makes the true meaning very clear.
- A more formal definition of the matrix exponential is $e^A := I_n + \lim_{N\to\infty} \sum_{k=1}^{N} A^k \cdot \frac{1}{k!}$, making the need to evaluate a limit clear. Using techniques from a branch of mathematics called real analysis, the existence of the limit can be established for ALL $n \times n$ matrices with entries in $\mathbb{R}$ or $\mathbb{C}$.

Example 9.39. *Using the series definition, compute* e^{At} *for*

(a) $A_1 = \begin{bmatrix} a_{11} & 0 \\ 0 & a_{22} \end{bmatrix}$,

(b) $A_2 = \begin{bmatrix} \lambda_1 & 0 & 0 \\ 0 & \ddots & 0 \\ 0 & 0 & \lambda_n \end{bmatrix}$, *and*

(c) $A_3 = \begin{bmatrix} 0 & 1 \\ -1 & 0 \end{bmatrix}$.

Solutions: In each case, we need to find a pattern that makes evaluating the infinite sum feasible. The first two problems are remarkably easy.

(a) **Ans.** $A_1 = \begin{bmatrix} a_{11} & 0 \\ 0 & a_{22} \end{bmatrix} \Longrightarrow e^{A_1 t} = \begin{bmatrix} e^{a_{11}t} & 0 \\ 0 & e^{a_{22}t} \end{bmatrix}$.

Before applying the definition, we note that raising a diagonal matrix to a power gives a diagonal matrix with entries raised to the same power, that is,

$$\begin{bmatrix} a_{11} & 0 \\ 0 & a_{22} \end{bmatrix}^k = \underbrace{\begin{bmatrix} a_{11} & 0 \\ 0 & a_{22} \end{bmatrix} \times \cdots \times \begin{bmatrix} a_{11} & 0 \\ 0 & a_{22} \end{bmatrix}}_{k-\text{times}} = \begin{bmatrix} (a_{11})^k & 0 \\ 0 & (a_{22})^k \end{bmatrix},$$

as is easily checked for $k = 2$ and $k = 3$. The general case follows by induction (which would be good practice). Using this result, we have,

$$\begin{aligned} e^{A_1 t} &= I_2 + \begin{bmatrix} a_{11} & 0 \\ 0 & a_{22} \end{bmatrix} \cdot t + \begin{bmatrix} (a_{11})^2 & 0 \\ 0 & (a_{22})^2 \end{bmatrix} \cdot \frac{t^2}{2!} + \cdots + \begin{bmatrix} (a_{11})^k & 0 \\ 0 & (a_{22})^k \end{bmatrix} \cdot \frac{t^k}{k!} + \cdots \\ &= \begin{bmatrix} 1 & 0 \\ 0 & 1 \end{bmatrix} + \begin{bmatrix} a_{11}t & 0 \\ 0 & a_{22}t \end{bmatrix} + \begin{bmatrix} (a_{11})^2 \frac{t^2}{2!} & 0 \\ 0 & (a_{22})^2 \frac{t^2}{2!} \end{bmatrix} + \cdots + \begin{bmatrix} (a_{11})^k \frac{t^k}{k!} & 0 \\ 0 & (a_{22})^k \frac{t^k}{k!} \end{bmatrix} + \cdots \\ &= \begin{bmatrix} e^{a_{11}t} & 0 \\ 0 & e^{a_{22}t} \end{bmatrix}. \end{aligned}$$

(b) **Ans.** $A_2 = \begin{bmatrix} \lambda_1 & 0 & 0 \\ 0 & \ddots & 0 \\ 0 & 0 & \lambda_n \end{bmatrix} \implies e^{A_2 t} = \begin{bmatrix} e^{\lambda_1 t} & 0 & 0 \\ 0 & \ddots & 0 \\ 0 & 0 & e^{\lambda_n t} \end{bmatrix}$

The calculations follow exactly the same steps as part (a).

(c) **Ans.** $A_3 = \begin{bmatrix} 0 & 1 \\ -1 & 0 \end{bmatrix} \implies e^{A_3 t} = \begin{bmatrix} \cos(t) & \sin(t) \\ -\sin(t) & \cos(t) \end{bmatrix}.$

Let us pause for just a moment and appreciate that a matrix exponential of a real-valued matrix can oscillate, while scalar exponentials of real variables are monotonic. Later, in Definition 9.51 and Prop. 9.52, we will discover that this remarkable feature comes from Euler's formula, $e^{\mathbf{i}\theta} = \cos(\theta) + \mathbf{i}\sin(\theta)$, or, if we replace θ with ωt, $\boxed{e^{\mathbf{i}\omega t} = \cos(\omega t) + \mathbf{i}\sin(\omega t).}$

To reduce clutter, we replace A_3 with A in the following.

The pattern occurring in the powers of A is more "subtle" this time. The first few powers are straightforward: by brute force multiplication, we obtain:

$$A^0 = I_2, \quad A = \begin{bmatrix} 0 & 1 \\ -1 & 0 \end{bmatrix}, \quad A^2 = \begin{bmatrix} -1 & 0 \\ 0 & -1 \end{bmatrix}, \quad A^3 = \begin{bmatrix} 0 & -1 \\ 1 & 0 \end{bmatrix}, \quad A^4 = \begin{bmatrix} 1 & 0 \\ 0 & 1 \end{bmatrix} = I_2,$$

where we notice that A^4 brings us back to where we started, the identity matrix. That is exactly the pattern we are seeking!

We note immediately that:

$$A^{4k} = \underbrace{A^4 \times \cdots \times A^4}_{\text{k times}} = I_2 \times \cdots \times I_2 = I_2.$$

Next,

$$A^{4k+0} = I_2, \quad A^{4k+1} = A = \begin{bmatrix} 0 & 1 \\ -1 & 0 \end{bmatrix}, \quad A^{4k+2} = A^2 = \begin{bmatrix} -1 & 0 \\ 0 & -1 \end{bmatrix}, \quad A^{4k+3} = A^3 = \begin{bmatrix} 0 & -1 \\ 1 & 0 \end{bmatrix},$$

which is kind of amazing the first time you see it.

To compute e^{At}, we need to substitute the above results into our power series for the matrix exponential,

$$e^{At} = I_2 + At + A^2\frac{t^2}{2!} + A^3\frac{t^3}{3!} + A^4\frac{t^4}{4!} + A^5\frac{t^5}{5!}\cdots.$$

By doing so, we obtain:

$$\left[e^{At}\right]_{11} = 1 + 0 - \frac{t^2}{2} + 0 + \frac{t^4}{4!} + \cdots = \cos(t),$$

$$\left[e^{At}\right]_{12} = 0 + t + 0 - \frac{t^3}{3!} + 0 + \frac{t^5}{5!} + \cdots = \sin(t),$$

$$\left[e^{At}\right]_{21} = 0 - t + 0 + \frac{t^3}{3!} + 0 - \frac{t^5}{5!} + \cdots = -\sin(t),$$

$$\left[e^{At}\right]_{22} = 1 + 0 - \frac{t^2}{2} + 0 + \frac{t^4}{4!} + \cdots = \cos(t),$$

after recognizing the Maclaurin/Taylor expansions for cosine and sine. Therefore,

$$e^{At} = \begin{bmatrix} \cos(t) & \sin(t) \\ -\sin(t) & \cos(t) \end{bmatrix}.$$

■

9.5.5 Software Tools

Fortunately, there are software tools to help us with computing both e^{A} (the matrix exponential) and e^{At} (the matrix exponential function). We illustrate these next.

```
using LinearAlgebra
# Mass spring damper system
m = 5 # kg
k = 2 # N/m
b = 3 # Ns/m (Newton seconds per meter)
A = [0 1; -k/m -b/m]
expA = exp(A)
```

Output

```
2×2 Matrix{Float64}:
  0.839867  0.703132
 -0.281253  0.417988
```

Below is a comparison between the built-in function and a highly truncated implementation of the power series. We can see that the power series converges rapidly. This rapid convergence means that a finite number of terms is sufficient to provide a good approximation of the matrix exponential. We will differentiate the matrix exponential **function** with respect to time in the next subsection, using an argument based on the matrix exponential as being well approximated by a finite number of terms. Its formal proof relies on analytically proving the absolute convergence of the Maclaurin/Taylor expansions for the matrix exponential, a step we will not take in this textbook.

```
# Compare to the builtin exp function
N=9
expA2 = I(2)
AkOverKfactorial = I(2)
for k = 1:N
    AkOverKfactorial = A*AkOverKfactorial/k
    expA2 = expA2 + AkOverKfactorial
end
[expA expA2]
```

Output

```
2×4 Matrix{Float64}:
  0.839867  0.703132   0.839867  0.703132
 -0.281253  0.417988  -0.281253  0.417988
```

Note: In MATLAB, we use `expA = expm(A)`, where the "m" appended to "exp" tells MATLAB to compute the matrix exponential. Continuing with MATLAB, the command, `expA = exp(A)` would compute the exponential of each term of the matrix, which is very much not the matrix exponential. Because Julia uses **broadcasting** to apply a scalar function to a vector- or matrix-valued object, it does not need the `expm( )` command.

```
# Comparison of exp(A) and exp.(A)
[exp(A) exp.(A)] # Not the same
```

Output

```
2×4 Matrix{Float64}:
  0.839867  0.703132  1.0       2.71828
 -0.281253  0.417988  0.67032   0.548812
```

For large square matrices, the matrix exponential can be computed numerically. **The symbolic computation of the matrix exponential function, on the other hand, is limited to "smallish matrices with simple structure." In practice, this tends not to be an issue, because of eigenvalues and eigenvectors, as we will highlight in the next section.**

We continue with "modest" results on symbolic calculations applied to Example 9.39.

```
using SymPy
#
@vars a11 a22 t
A1 = [a11 0; 0 a22]
A1t = A1*t # Define a 2x2 square matrix
expA1t =exp(A1t)   # Compute the matrix exponential function
```

Output

$$\begin{bmatrix} e^{a11\cdot t} & 0.0 \\ 0.0 & e^{a22\cdot t} \end{bmatrix}$$

```
using SymPy
#
@vars lam1 lam2  lam3 lam4 t
A2 = Diagonal([lam1, lam2,  lam3, lam4])
A2t = A2*t # Define a 2x2 square matrix
expA2t =exp(A2t)   # Compute the matrix exponential function
```

Output

$$\begin{bmatrix} e^{lam1\cdot t} & 0.0 & 0.0 & 0.0 \\ 0.0 & e^{lam2\cdot t} & 0.0 & 0.0 \\ 0.0 & 0.0 & e^{lam3\cdot t} & 0.0 \\ 0.0 & 0.0 & 0.0 & e^{lam4\cdot t} \end{bmatrix}$$

```
using SymPy
#
@vars t
A3 = [0 1; -1 0]
A3t = A3*t  # Define a 2x2 square matrix
expA3t =exp(A3t)  # Compute the matrix exponential function
```

Output

$$\begin{bmatrix} \frac{e^{\mathrm{i}\cdot t}}{2.0} + \frac{e^{(-\mathrm{i})\cdot t}}{2.0} & \frac{(-\mathrm{i})\cdot e^{\mathrm{i}\cdot t}}{2.0} + \frac{\mathrm{i}\cdot e^{(-\mathrm{i})\cdot t}}{2.0} \\ \frac{\mathrm{i}\cdot e^{\mathrm{i}\cdot t}}{2.0} - \frac{\mathrm{i}\cdot e^{(-\mathrm{i})\cdot t}}{2.0} & \frac{e^{\mathrm{i}\cdot t}}{2.0} + \frac{e^{(-\mathrm{i})\cdot t}}{2.0} \end{bmatrix}.$$

We use Euler's Formula from Chapter 1.11 to understand the answer! Moreover, for $\sin(t)$, we recognize that

$$\frac{(-\mathrm{i})\cdot e^{\mathrm{i}\cdot t}}{2.0} + \frac{\mathrm{i}\cdot e^{(-\mathrm{i})\cdot t}}{2.0} = \frac{e^{\mathrm{i}\cdot t} - e^{-\mathrm{i}\cdot t}}{2\,\mathrm{i}}$$

because $\frac{1}{\mathrm{i}} = \frac{1}{\mathrm{i}} \cdot \frac{\mathrm{i}}{\mathrm{i}} = -\mathrm{i}$. Currently, applying `SymPy.simplify` to the answer does not help.

9.5.6 Solutions to Linear ODEs

Proposition 9.40: Derivative of the Matrix Exponential Function

For all $t \in \mathbb{R}$ and all $n \times n$ (constant) matrices,

$$\frac{d}{dt} e^{A\cdot t} = A \cdot e^{A\cdot t} = e^{A\cdot t} \cdot A. \tag{9.65}$$

Example 9.41. *(Optional Read) Provide a non-rigorous, but mostly persuasive, argument that (9.65) holds.*

Solution: The obstruction to being rigorous is that we do not have enough math to understand when the derivative of an infinite sum is the infinite sum of the derivatives. Hence, we approximate the matrix exponential by a finite sum,

$$e^{At} \approx e_N^{A\cdot t} := I_n + \sum_{k=1}^{N} A^k \cdot \frac{t^k}{k!} = I_n + A\cdot t + A^2 \cdot \frac{t^2}{2!} + A^3 \cdot \frac{t^3}{3!} + \cdots + A^{N-1} \cdot \frac{t^{N-1}}{(N-1)!} + A^N \cdot \frac{t^N}{N!},$$

where N sets the number of terms in the approximation. We can now use the fact that the derivative of a finite sum is the finite sum of the derivatives.

Computing derivatives term by term gives

$$\frac{d}{dt} I_n = 0_n \text{ and}$$

$$\begin{aligned} \frac{d}{dt}\left(A^k \cdot \frac{t^k}{k!}\right) &= A^k \cdot k \cdot \frac{t^{k-1}}{k!} \\ &= A^k \cdot \frac{t^{k-1}}{(k-1)!} \\ &= A\left(A^{k-1} \cdot \frac{t^{k-1}}{(k-1)!}\right) = \left(A^{k-1} \cdot \frac{t^{k-1}}{(k-1)!}\right) A, \end{aligned}$$

where we note that A can be factored to either side.

Using the above result,

$$\frac{d}{dt}\underbrace{\left(I_n + \sum_{k=1}^{N} A^k \cdot \frac{t^k}{k!}\right)}_{\approx e^{At}} = 0_n + \sum_{k=1}^{N} A\left(A^{k-1} \cdot \frac{t^{k-1}}{(k-1)!}\right)$$

$$= A\sum_{k=1}^{N}\left(A^{k-1} \cdot \frac{t^{k-1}}{(k-1)!}\right)$$

$$= A \cdot \underbrace{\left(I_n + \sum_{k=1}^{N-1} \frac{t^k}{k!}\right)}_{\approx e^{At}}.$$

Once again, we note that when A was factored out, it could have gone to either side. This completes our "mostly" persuasive argument. ■

Proposition 9.42: Explicit Solution to a Linear System of ODEs

The unique solution to $\dot{x} = A \cdot x,\ x(t_0) = x_0$ is

$$\varphi(t, t_0, x_0) := e^{A(t-t_0)} \cdot x_0. \tag{9.66}$$

Note: It is perfectly acceptable to denote the solution by $\varphi(t) = e^{A(t-t_0)} \cdot x_0$, dropping the explicit dependence on the initial time, t_0, and the initial state, x_0, or to use $\varphi(t, x_0) = e^{At} \cdot x_0$, when $t_0 = 0$.

Proof: All the conditions of Def. 9.15 are met; indeed:

- Prop. 9.40 states that the matrix exponential is differentiable; it follows that it must be continuous (meaning that each of the entries of the matrix exponential function is continuous in t).
- $\varphi(t_0, t_0, x_0) := e^{A(t_0-t_0)} \cdot x_0 = I_n \cdot x_0 = x_0$ (satisfies the initial condition).
- $\frac{\partial}{\partial t}\varphi(t, t_0, x_0) := \left(\frac{\partial}{\partial t} e^{A(t-t_0)}\right) \cdot x_0 = \left(A \cdot e^{A(t-t_0)}\right) \cdot x_0 = A \cdot e^{A(t-t_0)} \cdot x_0 = A \cdot \varphi(t, t_0, x_0)$ (satisfies the differential equation).

We already noted in Def. 9.15 why the solution is unique. Indeed, by Prop. 5.44-(a), $\frac{\partial}{\partial x}(Ax) = A$ is independent of x. Hence, Prop. 9.20-(b) implies that for all $x_0 \in \mathbb{R}^n$, $t_0 \in \mathbb{R}$, there exists a solution on $[t_0, \infty)$ and the solution is unique. ■

We rarely use (9.66) to compute the solution to a real engineering problem. Instead, we use solvers, such as "DifferentialEquations.jl". Equation (9.66) is used to deduce GENERAL PROPERTIES of SOLUTIONS to linear ODES, while a solver is used to COMPUTE SOLUTIONS for SPECIFIC INITIAL CONDITIONS.

Example 9.43. *Compute the solution to the mass-spring-damper system in Example 9.27, using two methods:*

- *A numerical solver, such as* `DifferentialEquations.jl`
- *Prop. 9.42.*

Use the parameters as mass $m = 5$ Kg, spring constant $k = 2$ N/m (newtons per meter), and damping coefficient $b = 3$ Ns/m (newton-seconds per meter).

Solution:

```
# Import useful packages
using DifferentialEquations
using Plots
using LaTeXStrings
using LinearAlgebra # for the backslash command
gr()

# Mass spring damper system
m = 5 # kg
k = 2 # N/m
b = 3 # Ns/m (newton-seconds per meter)
A = [0 1; -k/m -b/m]

# Define the ODE
function MassSpringDamper(du, x, A, t)
    du .= A*x
end


# Set the initial condition as a vector
x0=[1; 0.0]

# Set the time interval
T = (0.0, 20)

# Setup the ODE problem
params = A
problem = ODEProblem(MassSpringDamper, x0, T, params)

# solve the ODE problem using the Runge Kutta Tsitouras 5/4 Integrator
sol = solve(problem, Tsit5())

# plot the continuous solution
p1 = plot(sol, lw=3, guidefont=20, xlabel=L"t", ylabel=L"(x, v)",
    label=["x" "v"], legend=:topright)


# Discrete time vector
t=0:dt:T[2]

# use a for loop to compute the matrix exponential for A*t[k] and
# then multiply by the initial condition to get the solution at t[k]
N = length(t)
x = zeros(2, N)
for k = 1:length(t)
    x[:,k] = exp(A*t[k])*x0
end

# Overlay the discrete-time solution with the continuous-time solution
scatter!(t, x[1,:], label = false, color=:blue)
scatter!(t, x[2,:], label = false, color=:red)
```

Output The solid lines are the solutions from the numerical solver, while the dots are the points computed by evaluating the matrix exponential function at $t_k = k \cdot 0.4$ seconds, for $0 \leq k \leq 50$. They are indistinguishable.

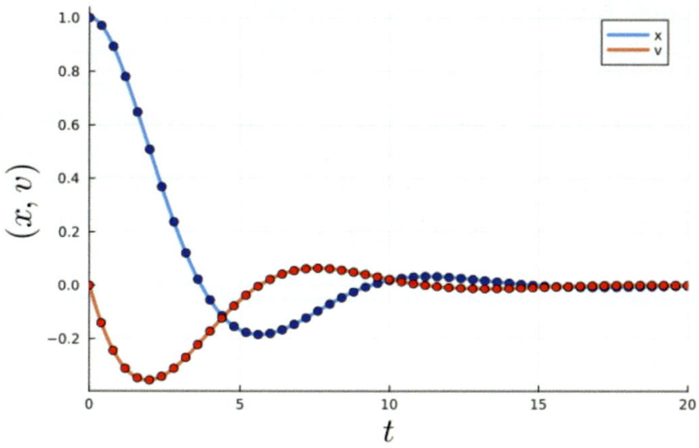

■

Proposition 9.44: (Optional Read) Useful Properties of the Matrix Exponential

Here, we highlight the similarities and differences between the matrix exponential function and the scalar exponential function. For an $n \times n$ matrix A,

(a) $\det\left(e^{A}\right) = e^{\mathrm{trace}(A)}$, where $\mathrm{trace}(A)$ is the sum of its diagonal elements, and

(b) $e^{A(s+t)} = e^{As} \cdot e^{At}$ for all real numbers s and t.

The first property says that, like the scalar exponential function, the matrix exponential is always invertible. The second property is similar to $e^{as} \cdot e^{at} = e^{a(s+t)}$, though it has a more profound interpretation in terms of solutions of linear ODEs. In particular, if we define $\varphi(t, t_0, x_0) := e^{A(t-t_0)} \cdot x_0$, then

$$\begin{aligned} \varphi(s+t, t_0, x_0) &:= e^{A(s+t-t_0)} \cdot x_0 \\ &= e^{As} \cdot e^{A(t-t_0)} \cdot x_0 \\ &= e^{As} \cdot \varphi(t, t_0, x_0). \end{aligned}$$

In other words, multiplying on the left by the matrix exponential $e^{A(t-t_0)}$ transitions the solution of $\dot{x} = Ax, x(t_0) = x_0$ from the value of the solution at the initial time t_0, to the value of the solution at time t, and multiplying on the left again by the matrix exponential e^{As} transitions the solution from time t to time to $s + t$.

Property (b) implies the following two additional properties,

(c) $\left(e^{At}\right)^{-1} = e^{-At}$ for all real numbers t, and

(d) $\left(e^{At_d}\right)^{k} = e^{A \cdot k \cdot t_d}$ for all real numbers t_d and integers k.

A key difference between the scalar exponential function and the matrix exponential function involves $e^{(A+B)t}$, for B an $n \times n$ matrix.

(e) $e^{(A+B)t} = e^{At} \cdot e^{Bt} \iff A \cdot B - B \cdot A = 0_n$, which, in words, states that the matrix exponential of a sum of matrices equals the product of the matrix exponentials if, and only if, A and B commute. Since most matrices do not commute, it is generally false that the matrix exponential of a sum is equal to the product of the matrix exponentials. Of course, if the matrices are 1×1, meaning they are scalars, then they also commute.

9.5.7 Eigenvalues and Eigenvectors to the Rescue

Punchline for This Section: We have given several hints that eigenvalues are important. In this Section, we confirm it! From the eigenvalues of A, we can determine whether all solutions of $\dot{x} = Ax$ remain bounded and converge to zero as t goes to infinity, there are sustained oscillations, or some solutions tend to infinity. The reason behind this phenomenal fact? **If λ is an eigenvalue of A with eigenvector v, then $e^{\lambda t}$ is an eigenvalue of e^{At} with corresponding eigenvector v.** When λ is real, it is easy to understand whether $e^{\lambda t}$ converges to zero or explodes. Thanks to Euler's Formula, we can also handle the case of complex eigenvalues.

Review of Complex Numbers

For $\alpha, \beta \in \mathbb{R}$, and $\mathbf{i}^2 = -1$,

$$z := \alpha + \mathbf{i}\,\beta$$

is a **complex number**. Its **complex conjugate** is

$$z^* := \alpha - \mathbf{i}\,\beta,$$

where the sign is flipped on the imaginary part. A **complex vector** is simply a vector with entries that are complex numbers. Hence, it can be written as

$$v := v_R + \mathbf{i}\,v_I,$$

where $v_R, v_I \in \mathbb{R}^n$; v_R is called the **real part** of the vector and v_I is called the **imaginary part** of the vector. It is important to understand that the imaginary part DOES NOT include **i**, the imaginary unit number, and thus v_I is a real vector. The complex conjugate of the vector is

$$v^* := v_R - \mathbf{i}\,v_I,$$

where the sign is flipped on the imaginary part. Finally, the **magnitude** of the complex number $z^* := \alpha + \mathbf{i}\,\beta$ is defined to be

$$|z| := \sqrt{\alpha^2 + \beta^2} = \sqrt{z \cdot z^*}.$$

The **(Euclidean) norm of a complex vector** is

$$||v|| := \sqrt{(v^*)^\top \cdot v} = \sqrt{\sum_{i=1}^{n} |v_i|^2},$$

where v_i are the components of v. Just as with norms of real vectors,

$$||\alpha v|| = |\alpha| \cdot ||v||,$$

for all $\alpha \in \mathbb{C}$ and $v \in \mathbb{C}^n$.

Eigenvalues and Eigenvectors of Matrices

Definition of Eigenvalues and Eigenvectors

Definition 9.45. *Let A be an $n \times n$ matrix with real or complex coefficients. A scalar $\lambda \in \mathbb{C}$ is an **eigenvalue** of A, if there exists a nonzero vector $v \in \mathbb{C}^n$ such that $Av = \lambda v$. Any such vector v is called an **eigenvector** associated with λ.*

Notes:

- Eigenvectors are not unique. If v is an eigenvector, then so is $\alpha \cdot v$ for any nonzero constant α.

- Most linear algebra packages normalize the eigenvector to have a norm of one.
- If A is a real matrix and $\lambda \in \mathbb{R}$, then we take $v \in \mathbb{R}^n$.
- If A is a real matrix and $\lambda \in \mathbb{C}$ satisfies $Av = \lambda v$, then λ^*, the complex conjugate of λ, is also an eigenvalue of A with corresponding eigenvector, v^*. In particular, $Av^* = \lambda^* v^*$. We say the eigenvalues and eigenvectors occur in **complex conjugate pairs**.

This next example helps us to understand why eigenvalues tell us so much about solutions to linear systems of ODEs,

$$\dot{x} = A \cdot x, \quad x(t_0 = 0) = x_0,$$

in particular, when the solutions converge to the origin, oscillate, or blow up.

Example 9.46. **(Must Read)** *The* (2×2) *matrix* $A = \begin{bmatrix} a & \omega \\ -\omega & a \end{bmatrix}$ *has eigenvalues* $\lambda_1 = a+\mathbf{i}\omega$ *and* $\lambda_2 = a-\mathbf{i}\omega$.

Use `SymPy` *to compute* e^{At}. **On the basis of e^{At}, interpret the qualitative behavior of the solutions to the linear ODE $\dot{x} = Ax$ when the real parts of the eigenvalues** ***are negative, zero, and positive.***

Solution:

```
using SymPy

@vars a w t
A = [a w; -w a]*t  # Define a 2x2 square matrix
expA = A.exp()  # Another way to compute the matrix exponential
```

Output

$$\begin{bmatrix} \dfrac{e^{a\cdot t-\mathbf{i}\cdot t\cdot w}}{2.0} + \dfrac{e^{a\cdot t+\mathbf{i}\cdot t\cdot w}}{2.0} & \dfrac{\mathbf{i}\cdot e^{a\cdot t-\mathbf{i}\cdot t\cdot w}}{2.0} - \dfrac{\mathbf{i}\cdot e^{a\cdot t+\mathbf{i}\cdot t\cdot w}}{2.0} \\ \dfrac{(-\mathbf{i})\cdot e^{a\cdot t-\mathbf{i}\cdot t\cdot w}}{2.0} + \dfrac{\mathbf{i}\cdot e^{a\cdot t+\mathbf{i}\cdot t\cdot w}}{2.0} & \dfrac{e^{a\cdot t-\mathbf{i}\cdot t\cdot w}}{2.0} + \dfrac{e^{a\cdot t+\mathbf{i}\cdot t\cdot w}}{2.0} \end{bmatrix} = \begin{bmatrix} e^{at}\cos(\omega t) & e^{at}\sin(\omega t) \\ -e^{at}\sin(\omega t) & e^{at}\cos\omega t) \end{bmatrix}.$$

The simplification on the right-hand side was performed by hand, based on

$$\begin{aligned} \frac{e^{a\cdot t-\mathbf{i}\cdot t\cdot\omega}}{2.0} + \frac{e^{a\cdot t+\mathbf{i}\cdot t\cdot\omega}}{2.0} &= e^{at}\left(\frac{e^{-\mathbf{i}\cdot t\cdot\omega} + e^{\mathbf{i}\cdot t\cdot\omega}}{2}\right) \\ &= e^{at}\left(\frac{\cos(t\cdot\omega) - \mathbf{i}\sin(t\cdot\omega) + \cos(t\cdot\omega) + \mathbf{i}\sin(t\cdot\omega)}{2}\right) \\ &= e^{at}\cos(\omega\, t) \end{aligned}$$

and

$$\begin{aligned} \frac{\mathbf{i}\, e^{a\cdot t-\mathbf{i}\cdot t\cdot\omega}}{2.0} - \frac{\mathbf{i}\, e^{a\cdot t+\mathbf{i}\cdot t\cdot\omega}}{2.0} &= e^{at}\left(\frac{\mathbf{i}\, e^{-\mathbf{i}\cdot t\cdot\omega} - \mathbf{i}\, e^{\mathbf{i}\cdot t\cdot\omega}}{2}\right) \\ &= e^{at}\left(\frac{\mathbf{i}\cos(t\cdot\omega) + \sin(t\cdot\omega) - \mathbf{i}\cos(t\cdot\omega) + \sin(t\cdot\omega)}{2}\right) \\ &= e^{at}\sin(\omega\, t). \end{aligned}$$

Thanks to eigenvalues and eigenvectors, **complex exponentials**, aka exponentials of the form, $e^{(a+\mathrm{i}\omega)t}$, play a fundamental role in understanding the solutions of linear systems of ODEs.

For the present problem, we note that for all $x_0 \neq 0_{2\times 1}$, the solution of $\dot{x} = Ax$, $x(t_0 = 0) = x_0$,

$$\varphi(t, t_0, x_0) := \begin{bmatrix} e^{at}\cos(\omega t) & e^{at}\sin(\omega t) \\ -e^{at}\sin(\omega t) & e^{at}\cos(\omega t) \end{bmatrix} \cdot \begin{bmatrix} x_{01} \\ x_{02} \end{bmatrix},$$

- converges to zero if $a < 0$ (decaying exponentially) when eigenvalues have a negative real part;
- oscillates forever if $a = 0$ (no exponential term) when eigenvalues are purely imaginary; and
- blows up if $a > 0$ (exploding exponentially) when eigenvalues have a positive real part.

■

Example 9.47. *Use Julia to compute the eigenvalues and eigenvectors of the* (3×3) *matrix*

$$A := \begin{bmatrix} 1.3916 & -0.4173 & -1.0494 \\ -0.4823 & -0.4498 & -0.2974 \\ 1.1709 & -0.2278 & 1.9801 \end{bmatrix}.$$

In addition, verify the following properties:

(a) The eigenvalues and eigenvectors occur in complex conjugate pairs.

(b) $Av = \lambda v$.

(c) The eigenvectors are linearly independent and hence form a basis for $\mathbb{C}^3$.

(d) If $v = v_R + \mathrm{i}\, v_I$ *is a complex eigenvector, then* v *and its complex conjugate* v^* *can be replaced with* v_R *and* v_I*, the real and imaginary parts of* v*, to build a basis for* $\mathbb{R}^3$.

Solutions:

(a) The eigenvalues and eigenvectors occur in complex conjugate pairs.

```julia
using LinearAlgebra

A = [1.39164    -0.417337   -1.04938
     -0.482258   -0.449806   -0.297397
      1.1709     -0.2278      1.98006]

E = eigen(A)
```

Output

```
Eigen{ComplexF64, ComplexF64, Matrix{ComplexF64},
Vector{ComplexF64}} values:
3-element Vector{ComplexF64}:
 -0.5492714681705643 + 0.0im
  1.7355827340852823 - 1.0454627840167563im
  1.7355827340852823 + 1.0454627840167563im
vectors:
```

```
3×3 Matrix{ComplexF64}:
    0.206377+0.0im     0.17411+0.639401im      0.17411-0.639401im
    0.978444+0.0im   0.105157-0.0907931im   0.105157+0.0907931im
 -0.00741578+0.0im  -0.735901-0.0im          -0.735901+0.0im
```

Hence, the eigenvalues are

$$\lambda_1 = -0.5493 + 0.0000\,\mathbf{i}, \quad \lambda_2 = 1.7356 - 1.0455\,\mathbf{i}, \quad \lambda_3 = 1.7356 + 1.0455\,\mathbf{i},$$

and the corresponding eigenvectors are

$$v_1 = \begin{bmatrix} 0.2064 + 0.0000\,\mathbf{i} \\ 0.9784 + 0.0000\,\mathbf{i} \\ -0.0074 + 0.0000\,\mathbf{i} \end{bmatrix}, \quad v_2 = \begin{bmatrix} 0.1741 + 0.6394\,\mathbf{i} \\ 0.1052 - 0.0908\,\mathbf{i} \\ -0.7359 + 0.0000\,\mathbf{i} \end{bmatrix}, \quad v_3 = \begin{bmatrix} 0.1741 - 0.6394\,\mathbf{i} \\ 0.1052 + 0.0908\,\mathbf{i} \\ -0.7359 + 0.0000\,\mathbf{i} \end{bmatrix}.$$

λ_1 is real, with real eigenvector v_1, while λ_2 and λ_3 are complex, with complex eigenvectors, v_2 and v_3. Moreover, $\lambda_3 = \lambda_2^*$ and $v_3 = v_2^*$.

(b) $Av = \lambda v$.

```
# verify A*v = lambda*v
for k = 1:3
    v=E.vectors[:,k]
    lam = E.values[k]
    display([A*v lam*v])
end
```

The columns of $[Av \quad \lambda v]$ are identical.

Output

```
3×2 Matrix{ComplexF64}:
  -0.113357+0.0im    -0.113357+0.0im
  -0.537432+0.0im    -0.537432+0.0im
 0.00407327+0.0im   0.00407327-0.0im

3×2 Matrix{ComplexF64}:
  0.970653+0.927707im    0.970653+0.927707im
 0.0875883-0.267517im   0.0875883-0.267517im
  -1.27722+0.769357im    -1.27722+0.769357im

3×2 Matrix{ComplexF64}:
  0.970653-0.927707im    0.970653-0.927707im
 0.0875883+0.267517im   0.0875883+0.267517im
  -1.27722-0.769357im    -1.27722-0.769357im
```

(c) The eigenvectors are linearly independent and hence form a basis for $\mathbb{C}^3$.

```
# Check for a basis
V=E.vectors

# Note V' computes the complex conjugate transpose of the complex
# matrix V
```

```
# V'*V is theoretically guaranteed to have a real determinant,
# though numerically, it may have a small imaginary part

# Check linear independence of the columns of V
det(V'*V)
```

Output

```
0.897060699129927 + 1.6064687371449844e-18im
```

We note that the determinant equals $0.8971+\mathbf{i}\,1.6065\times10^{-18}$; that is, the imaginary part is the numerical equivalent of zero. Because the determinant is nonzero, the columns of V are linearly independent.

(Optional Read) Note: $V'\cdot V := (V^*)^\top\cdot V$ is a **Hermitian Matrix**, which is a square matrix equal to the transpose of its (complex) conjugate matrix. The diagonal elements of a Hermitian matrix are all real numbers, and the element of the (i, j) position is equal to the (complex) conjugate of the element in the (j, i) position. In our case,

$$V'\cdot V := (V^*)^\top\cdot V = \begin{bmatrix} 1.0000+0.0000\,\mathbf{i} & 0.1443+0.0431\,\mathbf{i} & 0.1443-0.0431\,\mathbf{i} \\ 0.1443-0.0431\,\mathbf{i} & 1.0000+0.0000\,\mathbf{i} & 0.1658-0.2036\,\mathbf{i} \\ 0.1443+0.0431\,\mathbf{i} & 0.1658+0.2036\,\mathbf{i} & 1.0000+0.0000\,\mathbf{i} \end{bmatrix},$$

satisfies all of these properties.

(d) If $v = v_R + \mathbf{i}\,v_I$ is a complex eigenvector, then v and its complex conjugate v^* can be replaced with v_R and v_I, the real and imaginary parts of v, to build a basis for $\mathbb{R}^3$.

The vectors $V_{\text{new}} := [v_1, \text{real}(v_2), \text{imag}(v_2)]$ are linearly independent because the determinant of $V_{\text{new}}^\top\cdot V_{\text{new}}$ equals 0.224 and hence is nonzero.

```
# real(E.vectors[:,1]) removes the zero imaginary part and makes it
# a Float64 instead a ComplexF64

# real(E.vectors[:,2]) computes the real part of v2
# imag(E.vectors[:,2]) computes the imaginary part of v2

# Stack candidate basis vectors as columns of V_new
V_new = [real(E.vectors[:,1]) real(E.vectors[:,2]) imag(E.vectors[:,2])]

display(V_new)

# Check linear independence
det(V_new'*V_new)
```

Output

```
3×3 Matrix{Float64}:
  0.206377     0.17411     0.639401
  0.978444     0.105157   -0.0907931
 -0.00741578  -0.735901   -0.0
```

```
0.22426517478248179
```

■

Eigenvalues and Eigenvectors of the Matrix Exponential

Remarkably, eigenvectors of A are also eigenvectors of e^{At}, for all $t \in \mathbb{R}$, while the eigenvalues are transformed under the exponential map in a simple manner.

Proposition 9.48: Eigenvalues and Eigenvectors of the Matrix Exponential

Suppose that A is an $n \times n$ real matrix and that $Av = \lambda v$, with $v \neq 0_{n\times 1}$. Then, the following properties are true,

(a) $A^k \cdot v = \lambda^k v$ for all integers $k \geq 1$,

(b) $e^A \cdot v = e^\lambda v$, and

(c) $e^{At} \cdot v = e^{\lambda t} v$ for all $t \in \mathbb{R}$.

Hence, if $v \neq 0_{n\times 1}$ is an eigenvector of A with corresponding eigenvalue λ, then v is an eigenvector of the matrix exponential function e^{At}, with corresponding eigenvalue $e^{\lambda t}$. The surprising part is that only the eigenvalue is a function of time, and it (aka $e^{\lambda t}$) is either a standard (real) exponential function of time or a complex exponential function of time. The eigenvectors of e^{At} are constant and identical to the eigenvectors of A.

Proof: (a) is proved by induction. We know the base case holds because we assumed that $Av = \lambda v$. Hence, assume $A^k \cdot v = \lambda^k v$ holds for some $k \geq 1$; we need to show it holds for $k+1$.

$$\begin{aligned} A^{k+1} \cdot v &= A \cdot \left(A^k \cdot v\right) \quad \text{(algebra)} \\ &= A \cdot \left(\lambda^k \cdot v\right) \quad \text{(by the induction hypothesis)} \\ &= \lambda^k \cdot (A \cdot v) \quad \text{(algebra, swapped the order of } A \text{ and } \lambda^k) \\ &= \lambda^k \cdot (\lambda v) \quad \text{(because } Av = \lambda v) \\ &= \lambda^{k+1} v \quad \text{(algebra).} \end{aligned}$$

To prove (c), we use (a) with the power series for the matrix exponential, namely,

$$\begin{aligned} e^A \cdot v &= \left(I_n + \sum_{k=1}^{\infty} \frac{t^k}{k!} A^k\right) \cdot v \quad \text{(definition of } e^A) \\ &= v + \sum_{k=1}^{\infty} \frac{t^k}{k!} A^k \cdot v \quad \text{(multiply through by } v) \\ &= v + \sum_{k=1}^{\infty} \frac{t^k}{k!} \lambda^k \cdot v \quad \text{(from part (a))} \\ &= \left(1 + \sum_{k=1}^{\infty} \frac{t^k}{k!} \lambda^k\right) \cdot v \quad \text{(factor out } v) \\ &= e^{\lambda t} v \quad \text{(apply the definition of } e^{\lambda t}). \end{aligned}$$

Part (b) is obtained from (c) by substituting $t = 1$. ■

Example 9.49. *Use the eigenvectors of* $A = \begin{bmatrix} -2.7690 & -4.1472 \\ 0.6061 & 1.2690 \end{bmatrix}$ *as initial conditions for* $\dot{x} = Ax$*. Observe the "qualitative properties" of the corresponding solutions.*

Solution: Let us find the eigenvalues and eigenvectors of A.

```
using LinearAlgebra

A = [-2.76899   -4.14723; 0.606141   1.26899]

E = eigen(A)
```

Output

```
Eigen{Float64, Float64, Matrix{Float64}, Vector{Float64}}
values:
2-element Vector{Float64}:
 -2.0000057922545804
  0.50000579225458
vectors:
2×2 Matrix{Float64}:
 -0.98324    0.785355
  0.182314  -0.619045
```

Hence, the eigenvalues

$$\lambda_1 = -2, \quad \lambda_2 = 0.5$$

are both real and distinct, and the corresponding eigenvectors

$$v_1 = \begin{bmatrix} -0.983240 \\ 0.182314 \end{bmatrix}, \quad v_2 = \begin{bmatrix} 0.785355 \\ -0.619045 \end{bmatrix}$$

are also real and linearly independent.

We note that $\lambda_1 < 0$ while $\lambda_2 > 0$. We next compute and plot solutions to $\dot{x} = Ax$, with $x_0 = v_1$ and $x_0 = v_2$. In fact, we are going to plot the norm of each solution, $||e^{At}x_0||$. We show only the code for the first eigenvector as the initial condition.

```
# Compute and plot ||x(t) = exp(At)*x0|| for x0 an eigenvector.
using Plots
using LinearAlgebra
using LaTeXStrings

function normSolution(t; A=A, x0=E.vectors[:,1])
    eAt = exp(A*t)
    xt = eAt * x0
    y = norm(xt)
    return y
end

# time values where we compute the solution
t = 0:.1:8

y = normSolution.(t)

p1 = plot(t,y, guidefont = 20, legend = false, xlabel=L"t",
    ylabel=L"||e^{At}\cdot x_0 ||", lw=3, color=:blue)
```

Output

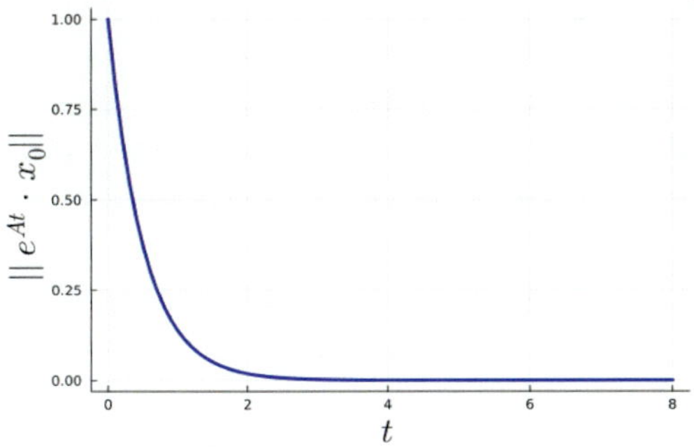

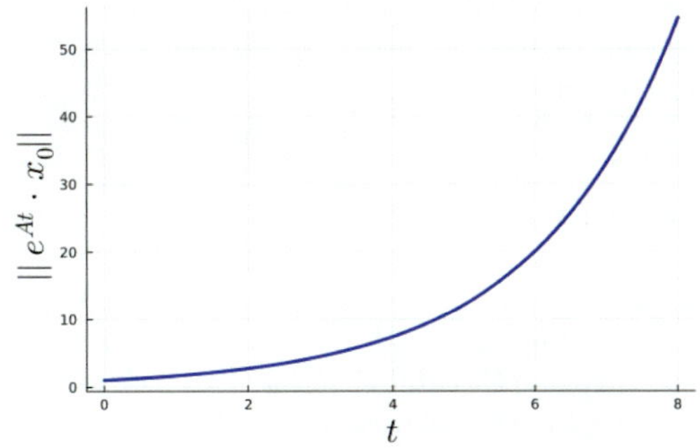

The image on the left shows $||x(t)||$ decaying like $e^{\lambda_1 t} = e^{-2t}$ and the image on the right shows $||x(t)||$ exploding like $e^{\lambda_2 t} = e^{0.5t}$. This is because, when $x_0 = v$, an eigenvector, then $e^{At} \cdot v = e^{\lambda t} v$, and hence

$$||e^{At} \cdot v|| = ||e^{\lambda t} \cdot v|| = |e^{\lambda t}| \, ||v||,$$

which implies for real eigenvalues,

$$\lim_{t \to \infty} ||e^{At} \cdot v|| = \begin{cases} 0 & \lambda < 0 \\ ||v|| & \lambda = 0 \\ \infty & \lambda > 0. \end{cases}$$

■

Our next challenge is to understand what happens in the case of complex eigenvalues and eigenvectors. Suppose that A is real with complex eigenvalue $\lambda = a + \mathbf{i}\omega$ and corresponding eigenvector, $v = v_R + \mathbf{i} v_I$. From Prop. 9.48,

$$e^{At} v = e^{\lambda t} v = e^{(a + \mathbf{i}\omega)t} v.$$

Hence,

$$||e^{At} v|| = ||e^{(a+\mathbf{i}\omega)t} v|| = |e^{(a+\mathbf{i}\omega)t}| \cdot ||v|| = |e^{at} \cdot e^{\mathbf{i}\omega t}| \cdot ||v|| = |e^{at}| \cdot |e^{\mathbf{i}\omega t}| \cdot ||v|| = e^{at} \cdot |e^{\mathbf{i}\omega t}| \cdot ||v||.$$

By Euler's Formula,

$$e^{\mathbf{i}\omega t} = \cos(\omega t) + \mathbf{i} \sin(\omega t),$$

and therefore,

$$|e^{\mathbf{i}\omega t}| = |\cos(\omega t) + \mathbf{i} \sin(\omega t)| = \sqrt{\cos^2(\omega t) + \sin^2(\omega t)} = 1.$$

After all of this, we obtain that

$$||e^{At} v|| = e^{at} ||v||,$$

where $a = \text{real}(\lambda)$, the real part of the eigenvalue λ, allowing our analysis of the real case to be applied here.

■

Example 9.50. *Use the eigenvectors of* $A = \begin{bmatrix} -1.8931 & 11.4911 & 2.8358 & -21.4883 \\ 1.9318 & -6.8883 & -0.6626 & 9.1683 \\ -1.4704 & 2.0461 & 0.1305 & -8.2359 \\ 1.5956 & -4.4190 & -0.9576 & 5.6509 \end{bmatrix}$ *as initial conditions for* $\dot{x} = Ax$. *Observe the "qualitative properties" of the corresponding solutions.*

Solution: Let us find the eigenvalues and eigenvectors of A.

```
# Compute eigenvalues and eigenvectors
A = [-1.89311  11.4911    2.83579   -21.4883
  1.93182  -6.88831  -0.662625    9.16832
 -1.47045   2.04612   0.130518   -8.23588
  1.59557  -4.41903  -0.95763     5.65091]

E=eigen(A)

for k = 1:2
    println("Pair of eigenvalues and vectors")
    display(E.values[(2*k-1):2*k])
    display(E.vectors[:,(2*k-1):2*k])
    println(" ")
end
```

Output

```
Pair of eigenvalues and vectors
2-element Vector{ComplexF64}:
 -2.000000213236534 - 1.9999961984135264im
 -2.000000213236534 + 1.9999961984135264im
4×2 Matrix{ComplexF64}:
  -0.314334+0.172478im   -0.314334-0.172478im
 -0.0323151-0.563835im  -0.0323151+0.563835im
  -0.654925-0.0im        -0.654925+0.0im
  -0.121327-0.329915im   -0.121327+0.329915im

Pair of eigenvalues and vectors
2-element Vector{ComplexF64}:
 0.5000042132365348 - 2.9999956474834977im
 0.5000042132365348 + 2.9999956474834977im
4×2 Matrix{ComplexF64}:
    0.77073-0.0im          0.77073+0.0im
 -0.0107431+0.313272im  -0.0107431-0.313272im
   0.385818-0.320985im    0.385818+0.320985im
 -0.0406637+0.232767im  -0.0406637-0.232767im
```

Hence, the eigenvalues

$$\lambda_1 = -2 - \mathbf{i}\,2, \quad \lambda_2 = -2 + \mathbf{i}\,2, \quad \lambda_3 = 0.5 - \mathbf{i}\,3, \quad \lambda_4 = 0.5 + \mathbf{i}\,3$$

are complex and respect complex conjugate symmetry, namely, $\lambda_2 = \lambda_1^*$ and $\lambda_4 = \lambda_3^*$. The corresponding eigenvectors

$$v_1 = \begin{bmatrix} -0.314334 + 0.172478\,\mathbf{i} \\ -0.0323151 - 0.563835\,\mathbf{i} \\ -0.654925 - 0.000000\,\mathbf{i} \\ -0.121327 - 0.329915\,\mathbf{i} \end{bmatrix}, \quad v_2 = \begin{bmatrix} -0.314334 - 0.172478\,\mathbf{i} \\ -0.0323151 + 0.563835\,\mathbf{i} \\ -0.654925 + 0.000000\,\mathbf{i} \\ -0.121327 + 0.329915\,\mathbf{i} \end{bmatrix},$$

$$v_3 = \begin{bmatrix} 0.77073 - 0.000000\,\mathbf{i} \\ -0.0107431 + 0.313272\,\mathbf{i} \\ 0.385818 - 0.320985\,\mathbf{i} \\ -0.0406637 + 0.232767\,\mathbf{i} \end{bmatrix}, \quad v_4 = \begin{bmatrix} 0.77073 + 0.000000\,\mathbf{i} \\ -0.0107431 - 0.313272\,\mathbf{i} \\ 0.385818 + 0.320985\,\mathbf{i} \\ -0.0406637 - 0.232767\,\mathbf{i} \end{bmatrix},$$

are also complex and respect complex conjugate symmetry, $v_2 = v_1^*$ and $v_4 = v_3^*$.

We note that $\text{real}(\lambda_1) = \text{real}(\lambda_2) = -2 < 0$ while $\text{real}(\lambda_3) = \text{real}(\lambda_4) = 0.5 > 0$. We next compute and plot solutions to $\dot{x} = Ax$, with $x_0 = v_1$ and $x_0 = v_3$. In fact, we are going to plot the norm of each solution, $||e^{At}x_0||$. We show only the code for the first eigenvector as the initial condition.

```
# Compute and plot ||x(t) = exp(At)*x0|| for x0 an eigenvector.
using Plots
using LinearAlgebra
using LaTeXStrings

function normSolution(t; A=A, x0=E.vectors[:,1])
    eAt = exp(A*t)
    xt = eAt * x0
    y = norm(xt)
    return y
end

# time values where we compute the solution
t = 0:.1:8

y = normSolution.(t)

p1 = plot(t,y, guidefont = 20, legend = false, xlabel=L"t",
         ylabel=L"||e^{At}\cdot x_0 ||", lw=3, color=:blue )
```

Output

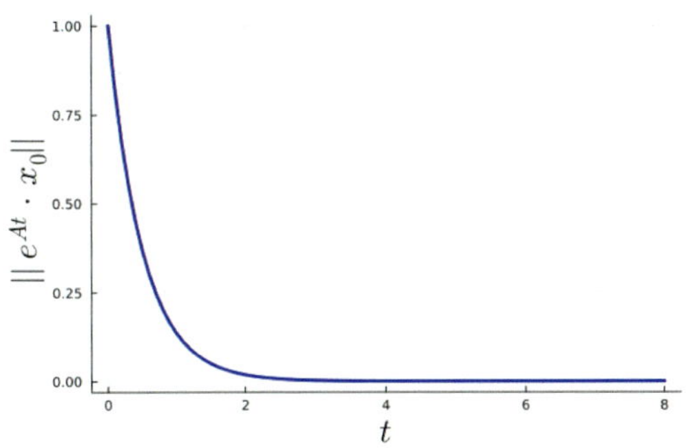

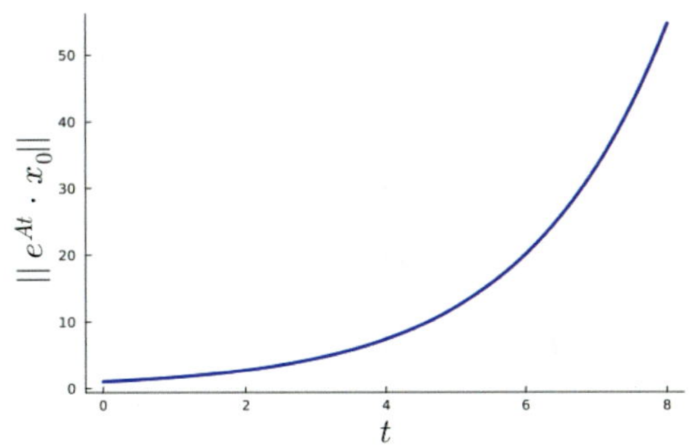

The image on the left shows $||x(t)||$ decaying like $e^{\text{real}(\lambda_1)t} = e^{-2t}$ and the image on the right shows $||x(t)||$ exploding like $e^{\text{real}(\lambda_3)t} = e^{0.5t}$. This is because, when $x_0 = v$, an eigenvector, then $e^{At} \cdot v = e^{\lambda t} v$, and hence,

$$||e^{At} \cdot v|| = ||e^{\lambda t} \cdot v|| = |e^{\lambda t}| \cdot ||v|| = e^{\text{real}(\lambda)t} \cdot ||v||,$$

which implies that for complex eigenvalues,

$$\lim_{t \to \infty} ||e^{At} \cdot v|| = \begin{cases} 0 & \text{real}(\lambda) < 0 \\ ||v|| & \text{real}(\lambda) = 0 \\ \infty & \text{real}(\lambda) > 0. \end{cases}$$

What does it mean when the initial condition is a complex vector? That seems very strange! You make a good point. We've made great progress in understanding the solutions of $\dot{x} = Ax$ for special initial conditions, namely, x_0 is an eigenvector of A. But yeah, on a real system, we cannot set x_0 to a complex value. So, did we just do a bunch of work for nothing?

Hardly. However, it does take a bit more linear algebra to make sense of our calculations.

Complex (Scalar) Exponential Functions

Complex exponentials are a small yet very powerful generalization of Euler's Formula. If you like to learn from videos, here is The MAGIC of Complex Exponentials! by FiguR3 iT ouT (sic).

Complex Exponetials

Definition 9.51. *For* $a, \omega \in \mathbb{R}$,

$$e^{(a+\mathbf{i}\omega)t} := e^{at} \cdot (\cos(\omega t) + \mathbf{i}\sin(\omega t)) \tag{9.67}$$

is called a ***complex exponential.***

Equivalently, let $z := a + \mathbf{i}\omega \in \mathbb{C}$ be a complex number, then e^{zt} is a **complex exponential.**

Note: We first encountered a complex exponential in the context of Example 9.49. Complex exponentials arise in the solution of linear ODEs when there are complex eigenvalues.

Proposition 9.52: Properties of Complex Exponentials

Complex exponentials are fundamental in various fields of mathematics, engineering, and physics. Below are some of the key properties of complex exponentials, for $a, \omega \in \mathbb{R}$ and $z := a + \mathbf{i}\omega \in \mathbb{C}$:

(a) **Periodicity:** The function $e^{\mathbf{i}\omega t}$ is periodic with period $T = \frac{2\pi}{|\omega|}$, for $\omega \neq 0$.

(b) **Multiplicative Property:** For any two complex numbers $z_1 = a_1 + \mathbf{i}\omega_1$ and $z_2 = a_2 + \mathbf{i}\omega_2$, the exponential function satisfies $e^{z_1 t} \cdot e^{z_2 t} = e^{(z_1+z_2)t}$.

(c) **Inverse:** The inverse of e^{zt} is given by e^{-zt}.

(d) **Differentiation and Integration:** The derivative and antiderivative of e^{zt} with respect to t are:

- $\frac{d}{dt}e^{zt} = ze^{zt}$ and
- for $z \neq 0$, $\int e^{zt}dt = \frac{1}{z}e^{zt} + C$, where C is the constant of integration.

(e) **Conjugation:** The complex conjugate of $e^{(a+\mathbf{i}\omega)t}$ is $e^{(a-\mathbf{i}\omega)t}$.

(f) **Magnitude:** The magnitude of $e^{(a+\mathbf{i}\omega)t}$ is $|e^{(a+\mathbf{i}\omega)t}| = e^{at}$, which exhibits exponential growth or decay based on the sign of the real part, a.

Note: These properties make complex exponentials an indispensable tool in the analysis of oscillatory and wave phenomena, signal processing, control theory, and many other disciplines.

Understanding the Roles Played by the Real and Imaginary Parts of Complex Eigenvalues and Eigenvectors

Proposition 9.53: Real and Imaginary Parts of a Complex Eigenvector as Initial Conditions

Suppose that A is a real $n \times n$ matrix with a complex eigenvalue $\lambda = a + \mathbf{i}\omega$ and corresponding eigenvector $v = v_R + \mathbf{i}v_I$. Then

$$\begin{aligned} e^{At}v_R &= e^{at}\cos(\omega t)v_R - e^{at}\sin(\omega t)v_I \\ e^{At}v_I &= e^{at}\sin(\omega t)v_R + e^{at}\cos(\omega t)v_I. \end{aligned} \tag{9.68}$$

Note: Equation (9.68) helps us to observe the oscillatory aspect of the solutions of linear ODEs when there are complex eigenvalues. The imaginary part of the eigenvalue, $\omega := \text{imag}(\lambda)$ sets the (angular) frequency of the oscillation, while the real part of the eigenvalue, $a := \text{real}(\lambda)$, determines the "amplification" or "attenuation" factor, depending on whether a is positive, negative, or zero.

When the initial condition is a real eigenvector, the solution of $\dot{x} = Ax$ evolves entirely along the eigenvector via $e^{\lambda t}v$. It is "living" in a one-dimensional subspace given by $\text{span}\{v\}$. For a complex eigenvector, when the ODE is initialized at v_R or v_I, the solution "lives" in a two-dimensional subspace given by $\text{span}\{v_R, v_I\}$. Within that subspace, the solution spirals to the origin if $a < 0$, spirals to infinity if $a > 0$, and forms a circle if $a = 0$. To be able to deduce such detailed information through linear algebra is pretty awesome, don't you agree?

Example 9.54. *Using our new insights, re-analyze the solutions to $\dot{x} = Ax$ for the matrix in Example 9.50. The key data are:*

$$A = \begin{bmatrix} -1.8931 & 11.4911 & 2.8358 & -21.4883 \\ 1.9318 & -6.8883 & -0.6626 & 9.1683 \\ -1.4704 & 2.0461 & 0.1305 & -8.2359 \\ 1.5956 & -4.4190 & -0.9576 & 5.6509 \end{bmatrix},$$

which has eigenvalues

$$\lambda_1 = -2 - \mathbf{i}\,2, \quad \lambda_2 = \lambda_1^* = -2 + \mathbf{i}\,2, \quad \lambda_3 = 0.5 - \mathbf{i}\,3, \quad \lambda_4 = \lambda_3^* = 0.5 + \mathbf{i}\,3$$

and corresponding eigenvectors

$$v_1 = \begin{bmatrix} -0.3143 \\ -0.0323 \\ -0.6549 \\ -0.1213 \end{bmatrix} + \mathbf{i} \begin{bmatrix} 0.1725 \\ -0.5638 \\ -0.0000 \\ -0.3299 \end{bmatrix}, \quad v_2 = v_1^* = \begin{bmatrix} -0.3143 \\ -0.0323 \\ -0.6549 \\ -0.1213 \end{bmatrix} - \mathbf{i} \begin{bmatrix} 0.1725 \\ -0.5638 \\ -0.0000 \\ -0.3299 \end{bmatrix},$$

$$v_3 = \begin{bmatrix} 0.7707 \\ -0.0107 \\ 0.3858 \\ -0.0407 \end{bmatrix} + \mathbf{i} \begin{bmatrix} -0.0000 \\ 0.3133 \\ -0.3210 \\ 0.2328 \end{bmatrix}, \quad v_4 = v_3^* = \begin{bmatrix} 0.7707 \\ -0.0107 \\ 0.3858 \\ -0.0407 \end{bmatrix} - \mathbf{i} \begin{bmatrix} -0.0000 \\ 0.3133 \\ -0.3210 \\ 0.2328 \end{bmatrix}.$$

Solutions: Compute the solution of $e^{At} \cdot v_{R,1}$ and plot its "projection" onto $\text{span}\{v_{R,1}, v_{I,1}\}$.

```
using LinearAlgebra
using Plots
using LaTeXStrings

A = [-1.89311  11.4911    2.83579   -21.4883
  1.93182  -6.88831  -0.662625    9.16832
 -1.47045   2.04612   0.130518   -8.23588
  1.59557  -4.41903  -0.95763     5.65091]

E=eigen(A)
```

```
(lam1, lam2, lam3, lam4) = E.values
(v1, v2, v3, v4) = E.vectors

t = 0.0:0.02:100

# For x0 = real(v1)
x =  exp.(real(lam1)*t).*cos.(imag(lam1)*t)
y = -exp.(real(lam1)*t).*sin.(imag(lam1)*t)

p1 = plot(x, y, guidefont = 15, lw=3, color=:blue, legend=false,
          aspect_ratio=1, xlabel=L"e^{at}\cos(\omega t) v_R",
          ylabel=L"-e^{at}\sin(\omega t) v_I")

# mark the initial condition
scatter!([1], [0], markersize=8, color=:red)


display(p1)
```

Compute the solution of $e^{At} \cdot v_{I,3}$ and plot its "projection" onto $\text{span}\{v_{R,3}, v_{I,3}\}$.

```
using LinearAlgebra
using Plots
using LaTeXStrings

A = [-1.89311  11.4911    2.83579   -21.4883
      1.93182  -6.88831  -0.662625    9.16832
     -1.47045   2.04612   0.130518   -8.23588
      1.59557  -4.41903  -0.95763     5.65091]

E=eigen(A)
(lam1, lam2, lam3, lam4) = E.values
(v1, v2, v3, v4) = E.vectors

t = 0.0:0.02:8

# For x0 = imag(v3)
x = exp.(real(lam3)*t).*sin.(imag(lam3)*t)
y = exp.(real(lam3)*t).*cos.(imag(lam3)*t)

p1 = plot(x, y, guidefont = 15, lw=3, color=:red, legend=false,
          aspect_ratio=1, xlabel=L"e^{at}\sin(\omega t) v_R",
          ylabel=L"e^{at}\cos(\omega t) v_I")

# mark the initial condition
scatter!([0], [1], markersize=6, color=:blue)

display(p1)
```

Output

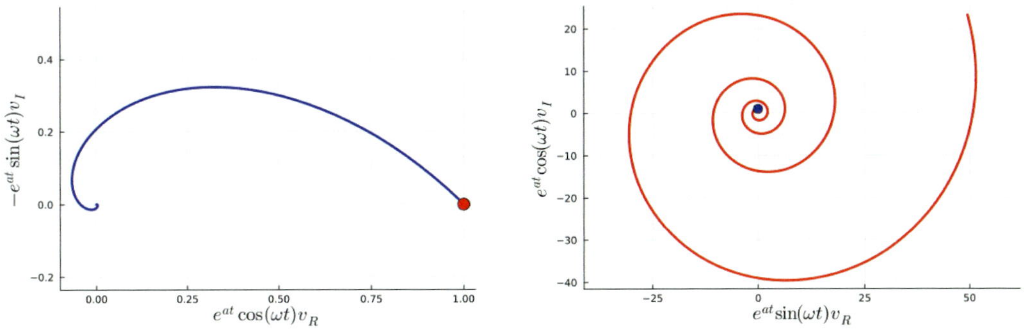

On the left is the solution of $e^{At}v_{R,1}$, projected onto $\mathrm{span}\{v_{R,1}, v_{I,1}\}$, with the red dot marking the initial condition. The function e^{-2t} decays so quickly that we cannot see with our eyes very much of the spiral into the origin. On the right is the solution of $e^{At}v_{I,3}$, projected onto $\mathrm{span}\{v_{R,3}, v_{I,3}\}$, with the blue dot marking the initial condition. The function $e^{0.5t}$ expands slowly enough that we can see it spiral outward to infinity. ■

Based on the results in this subsection, we understand how the solutions to $\dot{x} = Ax$ evolve for initial conditions tied to eigenvectors. The next subsection treats general initial conditions.

9.5.8 Exponential Stability of Linear Systems of ODEs with Implications for Nonlinear ODEs

Equilibrium Points and Stability Properties

Let A be a real $n \times n$ matrix and consider the linear system of ODEs $\dot{x} = Ax$. We specialize Def. 9.29 to the case of linear systems.

Definition 9.55. *$x_e \in \mathbb{R}^n$ is an **equilibrium point** of $\dot{x} = Ax$ if $Ax_e = 0$. Equivalently, the constant vector $\varphi(t, t_0, x_e) := x_e$ is a solution of the ODE.*

Note: For a linear ODE, the origin is always an equilibrium point.

Definition 9.56. *$x_e \in \mathbb{R}^n$ is a **globally asymptotically stable equilibrium point** of $\dot{x} = Ax$ if for all $x_0 \in \mathbb{R}^n$,*

$$\lim_{t\to\infty} e^{At}x_0 = x_e.$$

Note: For a linear system, if x_e is a globally asymptotically stable equilibrium point, then it necessarily equals the origin. This is because, taking $x_0 = 2x_e$ results in $\lim_{t\to\infty} e^{At}(2x_e) = x_e \iff 2\lim_{t\to\infty} e^{At}x_e = x_e \iff 2\lim_{t\to\infty} x_e = x_e \iff 2x_e = x_e \iff x_e = 0_{n\times 1}$, where we used the definition of an equilibrium point to simplify $e^{At}x_e = x_e$.

Definition 9.57. *The origin is a **globally exponentially stable equilibrium point** of $\dot{x} = Ax$ if there exist $\kappa > 0$ and $\gamma < \infty$ such that, for all $x_0 \in \mathbb{R}^n$,*

$$||e^{At}x_0|| \le \gamma \cdot e^{-\kappa t} \cdot ||x_0||.$$

Note: For a linear system of ODEs, Global **exponential** stability of an equilibrium point and global **asymptotic** stability of an equilibrium point are equivalent concepts. For a nonlinear system of ODEs, they are different notions of stability. In this textbook, understanding that each notion implies that solutions converge to zero as time marches to infinity is sufficient. The rate at which the convergence takes place is less relevant.

We laid the seeds for guessing that the eigenvalues provide all the information needed to check for asymptotic stability. The eigenvectors were present because, when used as an initial condition, the solution to a linear ODE is very simple, $e^{At} \cdot v = e^{\lambda t} v$.

Proposition 9.58: Exponential Stability of Linear Systems of ODEs

Let A be a real $n \times n$ matrix with eigenvalues $\lambda_1, \lambda_2, \ldots, \lambda_n$, and consider the linear system of ODEs,

$$\dot{x} = Ax.$$

If $\text{real}(\lambda_i) < 0$ for $1 \leq i \leq n$ (aka the real parts of all eigenvalues are negative), then the **origin** is globally exponentially stable.

Note: The proof is given in Section 9.9 for the case of distinct eigenvalues.

Note: The converse of the proposition is also true: if the origin is globally exponentially stable, then all of the eigenvalues have negative real parts.

Example 9.59. *Determine if the origin is a globally exponentially stable equilibrium point for the following linear systems of ODEs.*

(a) *Consider the mass-spring-damper system in Example 9.27. When $m = 5$ kg, $k = 2$ N/m, and $b = 3$ Ns/m (Newton seconds per meter), the model becomes*

$$\dot{x} = \begin{bmatrix} 0.0000 & 1.0000 \\ -0.4000 & -0.6000 \end{bmatrix} \cdot x.$$

(b) *Consider the RLC network in Example 9.26. For a consumer-grade RLC circuit in an older WiFi radio receiver, where discrete circuit elements were still used, typical component values might be:*

- *(R) 10 kilo-ohms (kΩ)*
- *(L) 20 nanohenries (nH)*
- *(C) 10 nanofarads (nF),*

yielding $\dot{x} = Ax$, with

$$A = \begin{bmatrix} 0.0000e+00 & 1.0000e+08 \\ -5.0000e-01 & -5.0000e+11 \end{bmatrix}.$$

Solutions:

(a) Mass-spring-damper system: **Ans.** The origin is globally exponentially stable because the eigenvalues all have negative real parts.

```
using LinearAlgebra
# Mass spring damper system
m = 5 # kg
k = 2 # N/m
b = 3 # Ns/m (Newton seconds per meter)
A = [0 1; -k/m -b/m]
E = eigen(A)
```

Output

```
Eigen{ComplexF64, ComplexF64, Matrix{ComplexF64},
Vector{ComplexF64}} values:
2-element Vector{ComplexF64}:
 -0.29999999999999993 - 0.5567764362830022im
 -0.29999999999999993 + 0.5567764362830022im

vectors:
2×2 Matrix{ComplexF64}:
  0.845154-0.0im        0.845154+0.0im
 -0.253546-0.470562im  -0.253546+0.470562im
```

(b) RLC network: **Ans.** The origin is globally exponentially stable because all of the eigenvalues have negative real parts.

```
using LinearAlgebra
# RLC Circuit
R = 10e3 # ohms
L = 20e-9 # Henries
C = 10e-9 # farads
A = [0 1/C; -C/L -R/L]
E = eigen(A)
```

Output

```
Eigen{Float64, Float64, Matrix{Float64}, Vector{Float64}}
values:
2-element Vector{Float64}:
 -4.999999999999999e11
 -0.0001220703125

vectors:
2×2 Matrix{Float64}:
 -0.0002   1.0
  1.0     -1.0e-12
```

■

Example 9.60. *Take a random initial condition for the mass-spring-damper system and plot the norm of its solution.*

Solution:

```
# Take a random initial condition for the mass-spring-damper
# system and plot the norm of its solution.
using Plots
using LinearAlgebra
using Random
using LaTeXStrings

# Mass spring damper system
m = 5 # kg
k = 2 # N/m
b = 3 # Ns/m (Newton seconds per meter)
A = [0 1; -k/m -b/m]

```

```
Random.seed!(543210)
x0 = randn(2,1); display(x0)
function normSolution(t; A=A, x0=x0)
    eAt = exp(A*t)
    xt = eAt * x0
    y = norm(xt)
    return y
end

# time values where we compute the solution
t = 0:0.01:20

y = normSolution.(t)

p1 = plot(t,y, guidefont = 20, legend = false, xlabel=L"t", ylabel=L"||e^{At}\cdot x_0 ||", lw=3,
    color=:blue )
```

Output

```
2×1 Matrix{Float64}:
 -0.9072986899852817
  0.7151631134658601
```

Because the eigenvalues are complex, we see oscillation in the norm as the solution decays to the origin, $x_e = [0; 0]$.

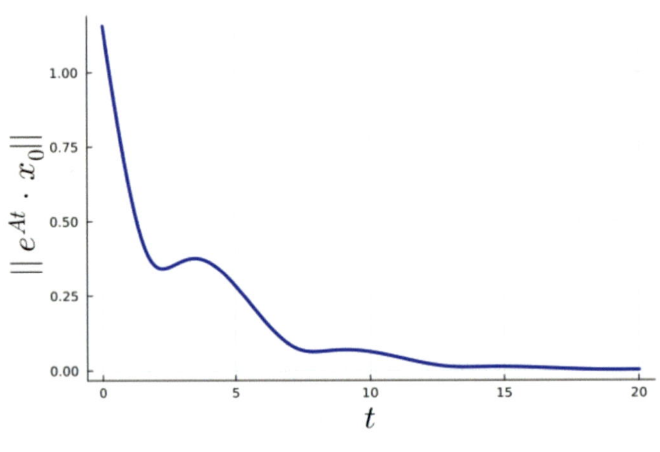

■

Example 9.61. *Determine if the origin is a globally exponentially stable equilibrium point for the linearized model of the 2-link robot arm, for (a) the upward equilibrium case and (b) the downward equilibrium case. We chose the 2-link arm, so that we need to do everything from scratch:*

- *build the dynamic model,*
- *compute the linearization for each of the equilibria, and*
- *evaluate the eigenvalues.*

In addition, for a random initial condition, compute solutions to each of the linearized models.

Solutions:

(a) Upward equilibrium. **Ans.** The origin is not exponentially stable. There are four real eigenvalues, of which two are positive.

(b) Downward equilibrium. **Ans.** The origin is not exponentially stable. There are four purely imaginary eigenvalues, meaning their real part is identically zero.

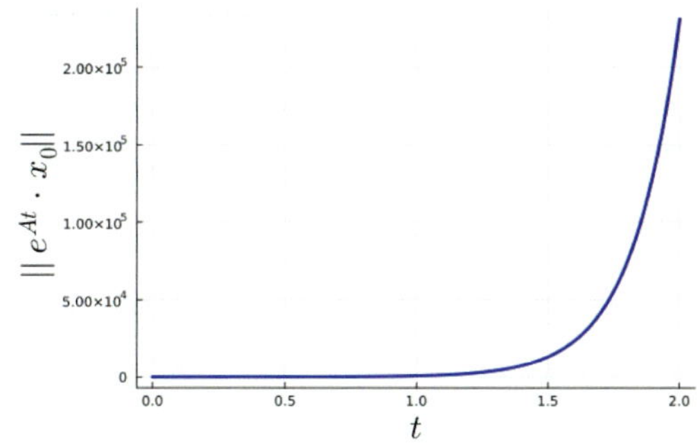

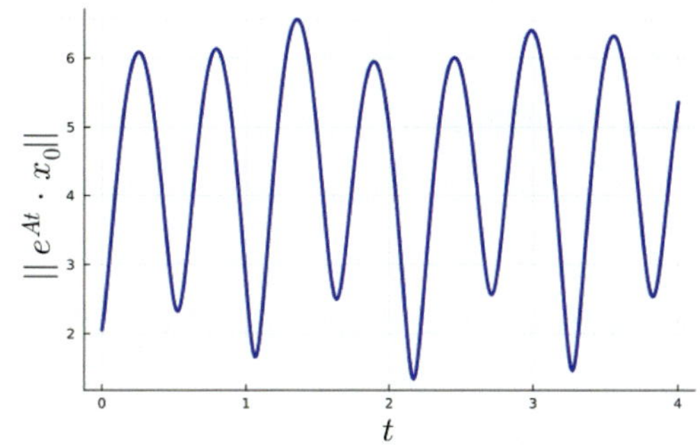

Using Lagrange's Method, we obtain:

```
function dyn_mod_2LinkManipulator(q, dq)
# DYN_MOD_2LINKMANIPULATOR
# 2023-12-05 14:45:06
#
# Author: Grizzle
#
# Model NOTATION: D(q)ddq + C(q,dq)*dq + G(q) = B*tau
# The Robot Equations: From Lagrange's Equations of Motion
#
g, L1, L2, m1, m2 = modelParameters()
#
# Variable names for the model
q1, q2= q
dq1, dq2 = dq
#
D = zeros(2, 2)
    D[1, 1] = L1^2*m1 + m2*(L1^2 + 2*L1*L2*cos(q2) + L2^2)
    D[1, 2] = L2*m2*(L1*cos(q2) + L2)
    D[2, 1] = L2*m2*(L1*cos(q2) + L2)
    D[2, 2] = L2^2*m2
#
C = zeros(2, 2)
    C[1, 1] = -L1*L2*dq2*m2*sin(q2)
    C[1, 2] = -L1*L2*m2*(dq1 + dq2)*sin(q2)
    C[2, 1] = L1*L2*dq1*m2*sin(q2)
    C[2, 2] = 0.0
#
G = zeros(2)
    G[1] = g*m2*(L1*cos(q1) + L2*cos(q1 + q2)) + L1*g*m1*cos(q1)
    G[2] = L2*g*m2*cos(q1 + q2)
#
B = zeros(2, 2)
    B[1, 1] = 1
```

```
    B[2, 2] = 1
#
JacG = zeros(2, 2)
    JacG[1, 1] = g*m2*(-L1*sin(q1) - L2*sin(q1 + q2)) - L1*g*m1*sin(q1)
    JacG[1, 2] = -L2*g*m2*sin(q1 + q2)
    JacG[2, 1] = -L2*g*m2*sin(q1 + q2)
    JacG[2, 2] = -L2*g*m2*sin(q1 + q2)
#
    return (D=D, C=C, G=G, B=B, JacG=JacG)
end
```

Upward Equilibrium

```
# Linear Model for the 2-link Manipulator
#
# bring the model into the workspace for later use
include("dyn_mod_2LinkManipulator.jl")

# Build a function to define the parameters of the model

function modelParameters()
      g = 9.81 # m/s^2
      L1 = 1 # m
      L2 = 0.7
      m1 = 15 # kg
      m2 = 10
      return g, L1, L2, m1, m2
end

# Set equilibirum Point
if false
      qe = [-pi/2, 0] #downward
else
      qe = [pi/2, 0] # upward
end

F = dyn_mod_2LinkManipulator(qe, 0*qe)
D = F.D;
println("D matrix"); display(D)
JacG = F.JacG;
println("JacG matrix"); display(JacG)

# build the linearized model about the give equilibrium point
A21 = -D\JacG; A21 = cleanUp(A21);
println("A21 matrix"); display(A21)
n, ~ = size(A21)
A = [zeros(n,n) I(n); A21 zeros(n,n)];
println("A matrix"); display(A)

# Compute eigenvalues
E = eigen(A)
```

Output

```
D matrix
2×2 Matrix{Float64}:
```

```
 43.9  11.9
 11.9   4.9

JacG matrix
2×2 Matrix{Float64}:
 -313.92  -68.67
  -68.67  -68.67

A21 matrix
2×2 Matrix{Float64}:
  9.81  -6.54
 -9.81  29.8971

A matrix
4×4 Matrix{Float64}:
  0.0    0.0     1.0  0.0
  0.0    0.0     0.0  1.0
  9.81  -6.54    0.0  0.0
 -9.81  29.8971  0.0  0.0

Eigen{Float64, Float64, Matrix{Float64}, Vector{Float64}}
values:
4-element Vector{Float64}:
 -5.718391382198321
 -2.647100840002673
  2.647100840002673
  5.71839138219832

vectors:
4×4 Matrix{Float64}:
 -0.0473235  -0.324822  0.324822   0.0473235
  0.165632   -0.139209  0.139209  -0.165632
  0.270614    0.859836  0.859836   0.270614
 -0.947151    0.368501  0.368501  -0.947151
```

Downward Equilibrium

```
# Linear Model for the 2-link Manipulator
#
# bring the model into the workspace for later use
include("dyn_mod_2LinkManipulator.jl")

#Build a function to define the parameters of the model

function modelParameters()
    g = 9.81 # m/s^2
    L1 = 1 # m
    L2 = 0.7
    m1 = 15 # kg
    m2 = 10
    return g, L1, L2, m1, m2
end

# Set equilibrium Point
if true
    qe = [-pi/2, 0] #downward
else
```

```
    qe = [pi/2, 0] # upward
end

F = dyn_mod_2LinkManipulator(qe, 0*qe)
D = F.D;
println("D matrix"); display(D)
JacG = F.JacG;
println("JacG matrix"); display(JacG)

# build the linearized model about the given equilibrium point
A21 = -D\JacG; A21 = cleanUp(A21);
println("A21 matrix"); display(A21)
n, ~ = size(A21)
A = [zeros(n,n) I(n); A21 zeros(n,n)];
println("A matrix"); display(A)

# Compute eigenvalues
E = eigen(A)
```

Output

```
D matrix
2×2 Matrix{Float64}:
 43.9  11.9
 11.9   4.9

JacG matrix
2×2 Matrix{Float64}:
 313.92  68.67
  68.67  68.67

A21 matrix
2×2 Matrix{Float64}:
 -9.81    6.54
  9.81  -29.8971

A matrix
4×4 Matrix{Float64}:
  0.0     0.0      1.0  0.0
  0.0     0.0      0.0  1.0
 -9.81    6.54     0.0  0.0
  9.81  -29.8971   0.0  0.0

Eigen{ComplexF64, ComplexF64, Matrix{ComplexF64},
Vector{ComplexF64}} values:
4-element Vector{ComplexF64}:
 2.220446049250297e-16 - 2.6471008400026728im
 2.220446049250297e-16 + 2.6471008400026728im
  3.945433097643709e-16 - 5.718391382198318im
  3.945433097643709e-16 + 5.718391382198318im

vectors:
4×4 Matrix{ComplexF64}:
  6.06496e-17-0.324822im     ...  1.65597e-17-0.0473235im
 -2.13366e-17-0.139209im         2.56542e-17+0.165632im
    -0.859836-0.0im                 0.270614+2.991e-17im
    -0.368501+2.66706e-16im        -0.947151+0.0im
```

Solutions are computed using the same code as in Example 9.60. ■

This next example illustrates the power of feedback control. The particular "type" of controller employed is linear state-variable feedback. The example uses linear systems theory to design the feedback controller and illustrates two important points:

- **an unstable equilibrium point can often be made into an exponentially stable equilibrium point with a properly designed feedback controller; and**
- **global exponential stability of a linearized model about an equilibrium point implies local exponential stability of the nonlinear model about the same equilibrium point.**

Many feedback controllers in engineering practice are designed on the basis of these two facts.

Example 9.62. *Consider the following linearized model of the 2-link robot arm about the upward equilibrium, where motors (aka actuators) have been added to the two joints,*

$$\delta\dot{x} = \underbrace{\begin{bmatrix} 0.000 & 0.000 & 1.000 & 0.000 \\ 0.000 & 0.000 & 0.000 & 1.000 \\ 9.810 & -6.540 & 0.000 & 0.000 \\ -9.810 & 29.897 & 0.000 & 0.000 \end{bmatrix}}_{A} \delta x + \underbrace{\begin{bmatrix} 0.000 & 0.000 \\ 0.000 & 0.000 \\ 0.067 & -0.162 \\ -0.162 & 0.597 \end{bmatrix}}_{B} \delta\tau, \tag{9.69}$$

$$\text{where } \delta x := \underbrace{\begin{bmatrix} x_1 \\ x_2 \\ x_3 \\ x_4 \end{bmatrix}}_{x} - \underbrace{\begin{bmatrix} 1.571 \\ 0.000 \\ 0.000 \\ 0.000 \end{bmatrix}}_{x_e} \text{ and } \delta\tau := \underbrace{\begin{bmatrix} \tau_1 \\ \tau_2 \end{bmatrix}}_{\tau} - \underbrace{\begin{bmatrix} 0 \\ 0 \end{bmatrix}}_{\tau_e}.$$

τ_e is the value of the motor torque at the equilibrium point. In this case, it equals $0_{2\times1}$, because no torque is required at the equilibrium.

*Because motors have been attached to both joints (aka the joints are now actuated), we have the possibility to apply motor torques (τ_1, τ_2) to render the upward equilibrium exponentially stable. When the applied motor torques are made a linear function of the states of the model, this is called **linear state-variable feedback control**: $\tau = -K \cdot (x - x_e) = -K\delta x$, where, in our case, K is a 2×4 matrix of **controller gains**. Consider a linear state-variable feedback controller given by:*

$$\delta\tau = -\underbrace{\begin{bmatrix} 357.820 & 80.570 & 62.084 & 16.829 \\ 80.570 & 73.570 & 16.829 & 6.930 \end{bmatrix}}_{K} \delta x. \tag{9.70}$$

Your tasks are:

(a) Determine the eigenvalues of the closed-loop linearized system.

(b) Apply the linear state-variable feedback to the nonlinear model of the 2-link robot arm and produce a simulation for $x_0 = \begin{bmatrix} 0.785 \\ -0.785 \\ 0.000 \\ 0.000 \end{bmatrix}$*, which has the base link at* $\frac{\pi}{4}$ *from the horizontal and the second link at* $-\frac{\pi}{4}$ *from the base link.*

Will it topple over, or will it stand up and balance?

Solution: Ans. Here is a GIF generated from a simulation of the nonlinear 2-link robot arm, using the linear feedback controller. See the **Caveat** below.

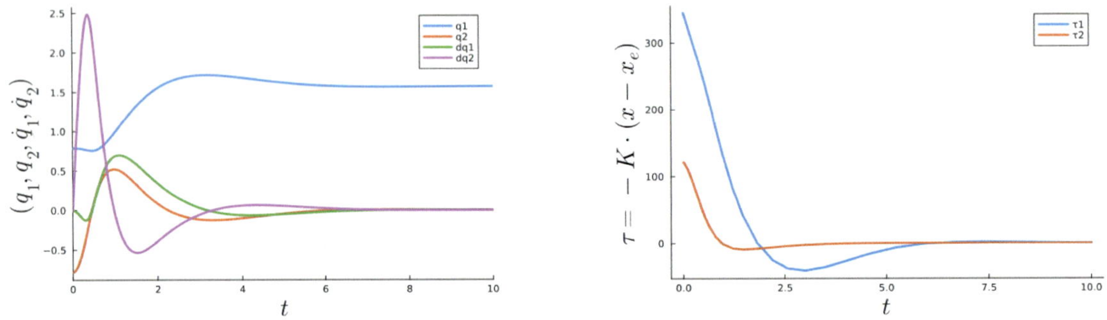

Caveat: The model we have chosen is particularly simple because (1) each joint of the robot arm is actuated, and (2), we have assumed the motors can produce as much torque as the controller demands of them. Not all robots have as many actuators as joints and real motors have limitations. Chapter 10 will examine these issues. For now, we ignore them and simply marvel at the fact that we were able to balance the arm in the inverted position.

Build the linearized model:

```
# Linear Model for the 2-link Manipulator
#
# bring the model into the workspace for later use
include("dyn_mod_2LinkManipulator.jl")

#Build a function to define the parameters of the model

function modelParameters()
    g = 9.81 # m/s^2
    L1 = 1 # m
    L2 = 0.7
    m1 = 15 # kg
    m2 = 10
    return g, L1, L2, m1, m2
end

# Set equilibirum Point
if false
    qe = [-pi/2, 0] #downward
else
```

```
    qe = [pi/2, 0] # upward
end

F = dyn_mod_2LinkManipulator(qe, 0*qe)
D = F.D;
JacG = F.JacG;
B = F.B;

# build the linearized model about the given equilibrium point
A21 = -D\JacG; A21 = cleanUp(A21);
n, ~ = size(A21)
A = [zeros(n,n) I(n); A21 zeros(n,n)];
B = [zeros(n,n); F.D \ B];
display(A)
display(B)
```

Output

```
4×4 Matrix{Float64}:
  0.0     0.0      1.0  0.0
  0.0     0.0      0.0  1.0
  9.81   -6.54     0.0  0.0
 -9.81   29.8971   0.0  0.0

4×2 Matrix{Float64}:
  0.0          0.0
  0.0          0.0
  0.0666667   -0.161905
 -0.161905     0.597279
```

Using the given feedback gain matrix K, compute the closed-loop eigenvalues.

Consider a linear model $\dot{x} = Ax + Bu$, where u is the traditional notation for the control input in the Feedback Control Community. Suppose that a linear feedback $u = -Kx$ is applied to the system. Then, we have:

$$\begin{aligned} \dot{x} &= Ax + Bu \\ &= Ax + B \cdot (-Kx) \quad \text{(substitute in the feedback)} \\ &= (A - B \cdot K)\, x \quad \text{(algebra)}. \end{aligned}$$

Hence, the eigenvalues of $A - B \cdot K$ determine the asymptotic stability of the closed-loop system.

```
K=[
 357.82  80.57  62.084   16.8291
  80.57  73.57  16.8291   6.92965
]
eigen(A-B*K)
```

Output

```
Eigen{ComplexF64, ComplexF64, Matrix{ComplexF64}, Vector{ComplexF64}}
values:
4-element Vector{ComplexF64}:
 -0.7071159639815681 - 0.7070975982722726im
 -0.7071159639815681 + 0.7070975982722726im
 -0.7071061788755746 - 0.7071073834970084im
 -0.7071061788755746 + 0.7071073834970084im
```

```
vectors:
4×4 Matrix{ComplexF64}:
 -0.223141-2.18102e-11im  -0.223141+2.18102e-11im  …  -0.233087-5.00046e-11im
  0.670976-0.0im           0.670976+0.0im              -0.667586+0.0im
  0.157786+0.157782im      0.157786-0.157782im          0.164817-0.164818im
 -0.474458-0.474445im     -0.474458+0.474445im          0.472054-0.472055im
```

The closed-loop eigenvalues have been placed at $-\frac{\sqrt{2}}{2} \pm i\,\frac{\sqrt{2}}{2}$. Hence, the upright equilibrium is now globally exponentially stable for the linearized model.

Simulate the Nonlinear Closed-Loop System

```
# Import useful packages
using DifferentialEquations
using Plots
using LaTeXStrings
using LinearAlgebra # for the backslash command
gr()


# bring the model into the workspace for later use
include("dyn_mod_2LinkManipulator.jl")

# Parameters for the Feedback controller

struct Params
    qe::Vector{Float64}        # equilibrium point
    K ::Matrix{Float64}         # LQR Feedback gain matrix
end
# Initialize the Params structure with specific values
params = Params(qe, K)


# define the ODE dx/dt = f(x,t)
#
# the parameters are passed via modelParameters() in the
# function dyn_mod_2LinkManipulator(q, dq)
#
function f(x,params,t)
    n = floor(Int, length(x) / 2) # In Julia n/2 is a Float64
    q = x[1:n]
    dq = x[n+1:end]
    K = params.K
    qe = params.qe
    xe = [qe; 0*qe]
    tau = -K*(x-xe)
    model = dyn_mod_2LinkManipulator(q, dq)
    dx1 = dq
    dx2 = (model.D) \ ( -model.C*dq-model.G+model.B*tau) # note the use of
         # backslash
    dx=[dx1;dx2]
    return dx
end

# Set the initial condition as a vector
x0= [qe;0*qe] - [pi/4; pi/4; 0; 0]

# Set the time interval
T = (0.0, 10)

```

```
# Setup the ODE problem with out-of-place function
problem = ODEProblem{false}(f, x0, T, params)

# solve the ODE problem using the Runge Kutta Tsitouras 5/4 Integrator
 sol = solve(problem, Tsit5());
#sol = solve(problem, Vern9());


# plot the solution
p1 = plot(sol, lw=3, guidefont=20, xlabel=L"t", ylabel=L"(q_1, q_2, \dot{q}_1, \dot{q}_2)",
    label=["q1" "q2" "dq1" "dq2"], legend=:topright)
display(p1)


# Extract state variables from the solution
x_sol = Array(sol)

# Preallocate the array for control signals
tau_sol = zeros(2,size(x_sol,2))

# Define terms in the feedback controller
K = params.K
qe = params.qe
xe = [qe; 0*qe]

# Compute the control signal at each time step
for i in 1:size(x_sol, 2)
    x = x_sol[:, i]
    tau_sol[:, i] = -K * (x - xe)
end

# Plot the control signal
p2 = plot(sol.t, tau_sol', lw=3, guidefont=20, xlabel=L"t", ylabel=L"\tau = -K \cdot(x-x_e)",
          label=["tau_1" "tau_2"], legend=:topright)
display(p2)
```

Output The plots were moved to the beginning of the solution.

Make a GIF

```
using Plots

# Robot parameters
g, L1, L2, m1, m2 = modelParameters()

# Function to calculate forwardKinematics
function forwardKinematics(q1, q2)
    p1 = L1 * [cos(q1), sin(q1)]
    p2 = p1 + L2 * [cos(q1 + q2), sin(q1 + q2)]
    return p1, p2
end

# Extract q1 and q2 from the solution
q1_vals = x_sol[1, :]
q2_vals = x_sol[2, :]

# Create a plot for each time step and save as a GIF
anim = @animate for i in 1:length(q1_vals)
    p1, p2 = forwardKinematics(q1_vals[i], q2_vals[i])
    plot([0, p1[1], p2[1]], [0, p1[2], p2[2]], legend=false, lw=3, color=:blue,
        xlims=(-(L1+L2), L1+L2), ylims=(-(L1+L2)/10, 1.1*(L1+L2)), aspect_ratio=1)
```

```
    scatter!([p1[1]], [p1[2]], color=:red, markersize = m1)
    scatter!([p2[1]], [p2[2]], color=:red, markersize = m2)
    scatter!([0.0], [0.0], color=:black, markersize = m1+m2)
end

gif(anim, "inverted2LinkRobotArm.gif", fps = 10)
```

Output Here is the GIF; it is the same as the one provided at the beginning of the solution. **Just Kidding!** The GIF shows what happens if one messes up the SIGN on the feedback controller gain. Instead of implementing $u = -Kx$, the sign was flipped and the feedback implemented it as $u = Kx$. The resulting eigenvalues are:

$$\lambda_{1,\ldots,4} = \begin{bmatrix} -7.472 \\ -3.232 \\ 4.646 \\ 8.886 \end{bmatrix}.$$

In the linearized model, the states head off to infinity (in norm). In the nonlinear model, there are rotational joints, where arithmetic modulo 2π comes into play ... at least in theory. In most real robots, the limbs cannot make a full rotation without having a self-collision with another part of the robot, or, hitting a hard stop that has been installed for safety reasons. Hitting the hard stop makes a loud noise and usually breaks something. If you are fortunate enough to be working on a modern commercial robot, the system likely has a piece of software supervising the commands to the robot joints and it will shut down the robot before anything dangerous has happened.

> **In a feedback loop, if the sign on the gains is wrong, the closed-loop system is usually toast. Many other things can cause a feedback controller to fail, but getting the sign wrong is the most egregious error.** It seems like a silly mistake, one nearly impossible to make, but all it takes is one miscommunication with a team member for it to happen, as your author experienced early in his career. And it was on me as the lead control system designer. Talk about egg on your face ... it's like putting your hand on a hot stove ... you do not repeat the error.

■

Julia Packages for Feedback Control Design

- **ControlSystems.jl:** A comprehensive control systems toolbox in Julia for system modeling, analysis, controller design, and simulation. Supports state-space and frequency domain methods.
- **RobustAndOptimalControl.jl:** Focuses on robust and optimal control, extending the capabilities of ControlSystems.jl. Includes advanced techniques like H-infinity and robust control.
- **Systems.jl:** Offers tools for modeling, simulation, and analysis of dynamical systems, essential in control design.
- **LCMCore.jl:** Useful for control system applications, particularly in communication between system components.

9.6 When in a Bind, Euler's Method Is Always There for You

Yes, him again, the Einstein of his era! Euler's Method is one of the most fundamental and historically significant algorithms for computing numerical solutions to ODEs. Named after the uber-prolific Swiss mathematician Leonhard Euler, this method provides a straightforward yet powerful approach for approximating solutions to ODEs, a cornerstone in mathematical modeling of natural and engineered processes. Euler's Method is particularly renowned for its simplicity and ease of implementation, making it an excellent starting point for understanding numerical techniques in differential equations. At its core, the method transforms the often daunting task of solving complex ODEs into a `for loop`. This is achieved by approximating the slope of the solution curve at a series of points, thereby constructing a piecewise linear approximation to the solution. While Euler's Method may not boast the precision of more advanced techniques, such as Runge-Kutta methods, its conceptual clarity offers invaluable insights into the behavior of differential equations. It serves as a stepping stone to more sophisticated algorithms, providing a foundational understanding of how numerical integration can bridge the gap between analytical mathematics and practical computation.

Secrets of the Arcane 9.63: Euler's Method and Katherine Johnson, an American Hero!

A super interesting video: Katherine Johnson and Euler's Method by the UCLA Modeling Class.

In the bustling halls of NASA during the 1950s through 1980s, Katherine Johnson, an African-American mathematician, emerged as a pivotal figure. Amidst the Space Race, Johnson's brilliance shone, especially in her application of Euler's method, a key mathematical technique. Euler's method, akin to a mathematical magic wand, simplifies complex equations into smaller, manageable steps. Johnson wielded this tool with precision, transforming the trajectories of spacecraft into a series of calculations that even non-mathematicians could fathom. Her most notable use of Euler's method was in calculating the trajectory for John Glenn's 1962 orbital mission. Johnson's calculations were so accurate that Glenn famously insisted on her verification before his flight, saying, "If she says they're good, then I'm ready to go." By breaking down the arcane into the understandable, Johnson not only propelled rockets into space but also carved a path for future mathematicians. Her legacy, intertwined with Euler's enduring method, reminds us that complex problems often have elegant, step-by-step solutions.

Katherine Johnson graduated with highest honors in Mathematics and French from West Virginia State College in 1937. Her exceptional talent in mathematics was evident from a young age and was the foundation of her illustrious career at NASA.

Key Dates:

- **1953:** Katherine Johnson joins NASA.
- **1961:** Calculates trajectory for Mercury mission.
- **1962:** Verifies calculations for John Glenn's orbit.
- **1969:** Leads team computing orbit for Apollo 11's lunar landing.

Katherine Johnson, a mathematician who made the Moon reachable, and Euler's method, her trusty guide, together remind us of the power of simplicity in solving some of the most complex problems faced by humankind. **Related Videos:** NASA Trailblazer: Katherine Johnson by National Geographic and I found Katherine Johnson's actual calculations at NASA by Tibees.

Method 9.64: Euler's Method Is There When You Do Not Have Access to Packages

Consider a vector ordinary differential equation, $\dot{x} = f(x,t)$, with initial condition $x(t_0) = x_0$. If the ODE has a solution, $x(t, t_0, x_0)$ defined on $[t_0, T)$, then it can be numerically approximated as follows:

- Choose a step time $h > 0$.
- In a for loop, compute
$$x_{n+1} = x_n + h \cdot f(x_n, t_n), \tag{9.71}$$
from $n = 0$ to $n = \texttt{floor}\left(\frac{T}{h}\right)$.

The process starts at t_0 with x_0 and proceeds in steps of h to approximate the solution at subsequent points. The sequence of vectors x_n is an approximation of $x(t, t_0, x_0)$ at time t_n for $t_{n+1} = t_n + h$.

Note: The global error in an ODE Solver is the error in the numerical solution after a finite number of steps. For Euler's method, the global error is proportional to the step size, making it a first-order method. Specifically, the global error is of the order $O(h)$. This means that if you halve the step size, the total accumulated error over a fixed interval is also roughly halved.

The rate at which error accumulates in Euler's method can be problematic, especially for stiff equations or equations where the solution exhibits rapid changes. The method's simplicity and ease of implementation come at the cost of accuracy and stability, particularly for larger step sizes or over longer intervals. For more accurate results, especially over longer intervals or with more complex ODEs, higher-order methods like the Runge-Kutta methods are often preferred, as they offer better accuracy for a given step size, albeit with increased computational complexity.

Example 9.65. *Use Euler's Method to solve the following ODEs:*

(a) The single-link pendulum from Examples 6.36 and 9.12.

(b) The 2-link robot arm stabilized in the upward equilibrium from Example 9.62.

Solutions:

(a) The single-link pendulum from Examples 6.36 and 9.12.

```julia
# Import useful packages
using Plots
using LaTeXStrings
gr()

# Single-link Pendulum Model

struct ParamsPendulum
        m::Float64          # Mass in kilograms
        L::Float64          # Length in meters
        g::Float64          # Acceleration due to gravity in m/s^2
end


# Initialize the Params structure with specific values
m = 2
L = 1
g = 9.81
params = ParamsPendulum(m, L, g)


# define the ODE dx/dt = f(x,params,t)
function f(x, params, t)
        dx = [x[2], -(params.g/params.L)*sin.(x[1])]
        return dx
end

# Set the initial condition as a vector
x0= [3*pi/8; 0.0]

# Set the time interval
T = (0.0, 10)

# Euler's method
dt = 1e-2
N = floor(Int64,T[2]/dt)
x=zeros(2,N+1)
x[:,1]=x0
t=zeros(N+1)
t[1] = T[1]
for n=1:N
        t[n+1] = t[n]+dt
        x[:,n+1] = x[:,n] + dt*f(x[:,n],params,t[n])
end

p1 = plot(t, x', lw=3, guidefont = 20,  xlabel=L"t",
            ylabel=L"(\theta, \dot{\theta})", legend=false)
display(p1)
png(p1, "PendulumSimulationChap09Eulerdt1eminus2")
```

Output In the left-side plot for $dt = 0.01$, the simulation does not faithfully represent a pendulum; we can see the velocity growing with each swing of the pendulum. In the right-side plot, already for $dt = 0.001$, the simulation more faithfully represents a pendulum. If we look closely, we can still see the velocity growing each swing of the pendulum, but much slower. This is a drawback of Euler for "marginally stable" models; that is, models whose linearization has purely imagininary eigenvalues.

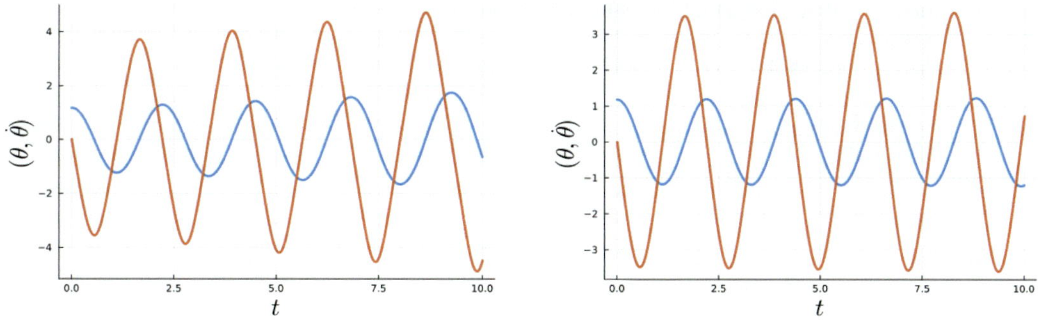

(b) The 2-link robot arm stabilized in the upward equilibrium from Example 9.62.

```
# Import useful packages
using Plots
using LaTeXStrings
using LinearAlgebra # for the backslash command
gr()


# bring the model into the workspace for later use
include("dyn_mod_2LinkManipulator.jl")

#Build a function to define the parameters of the model
function modelParameters()
    g = 9.81 # m/s^2
    L1 = 1 # m
    L2 = 0.7
    m1 = 15 # kg
    m2 = 10
    return g, L1, L2, m1, m2
end

# Parameters for the Feedback controller

struct Params
    qe::Vector{Float64}        # equilibrium point
    K ::Matrix{Float64}         # LQR Feedback gain matrix
end
# Initialize the Params structure with specific values

K=[ 357.82  80.57  62.084   16.8291
  80.57  73.57  16.8291    6.92965]
qe = [pi/2, 0] # upward
params = Params(qe, K)


# define the ODE dx/dt = f(x,t)
#
# the parameters are passed via modelParameters() in the
# function dyn_mod_2LinkManipulator(q, dq)
#
function f(x,params,t)
    n = floor(Int, length(x) / 2) # In Julia n/2 is a Float64
    q = x[1:n]
    dq = x[n+1:end]
    K = params.K
    qe = params.qe
    xe = [qe; 0*qe]
    tau = -K*(x-xe)
```

```
    model = dyn_mod_2LinkManipulator(q, dq)
    dx1 = dq
    dx2 = (model.D) \ ( -model.C*dq-model.G+model.B*tau) # note the
        # use of backslash
    dx=[dx1;dx2]
    return dx
end

# Set the initial condition as a vector
x0= [qe;0*qe] - [pi/4; pi/4; 0; 0]

# Set the time interval
T = (0.0, 8)

# Euler's method
dt = 1e-2
N = floor(Int64,T[2]/dt)
x=zeros(4,N+1)
x[:,1]=x0
t=zeros(N+1)
t[1] = T[1]
for n=1:N
    t[n+1] = t[n]+dt
    x[:,n+1] = x[:,n] + dt*f(x[:,n],params,t[n])
end
# p1 = plot(t, x[1:2,:]', lw=3, guidefont = 20,  xlabel=L"t",
# ylabel=L"(q_1, q_2)", legend = ("q_1", "q_2"))
p1 = plot(t, x[1, :], lw=3, label="q1", guidefont=20, xlabel=L"t",
                    ylabel=L"(q_1, q_2)")
plot!(p1, t, x[2, :], lw=3, label="q2")
display(p1)
```

Output The feedback controller in this system was designed so that the eigenvalues of the closed-loop linearized model have negative real parts. In such cases, Euler tends to perform quite well. Here, we see the first link converging to the upright position, $q_1 = \frac{\pi}{2}$, and the second link aligning itself with the first link, that is, $q_2 = 0$.

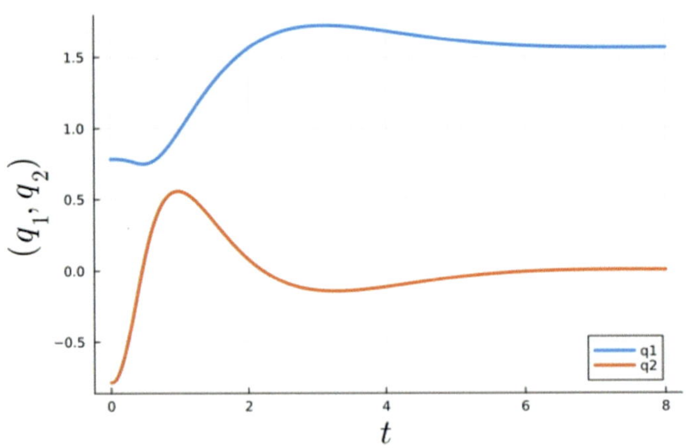

Fact 9.66: Derivation of Euler's Method, or How to Never Forget It

Consider an ODE $\dot{x} = f(x,t)$. Using a forward difference to approximate the derivative of $x(t)$ gives:

$$\frac{x(t+h)-x(t)}{h} = f(x(t),t)$$

$$\Updownarrow$$

$$x(t+h)-x(t) = h\cdot f(x(t),t)$$

$$\Updownarrow$$

$$x(t+h) = x(t) + h\cdot f(x(t),t)$$

$$\Updownarrow \quad \text{(substituting in } t_{n+1} := t_n + h\text{)}$$

$$x(t_{n+1}) = x(t_n) + h\cdot f(x(t_n),t_n).$$

The derivation is amazingly simple and easy to re-derive at a moment's notice when you need it most.

9.7 Resonance in ODEs

The collapse of the Tacoma Narrows Bridge (1940) (see also the video Tacoma Narrows Bridge Collapse "Gallopin' Gertie") in Washington State in 1940 serves as a stark reminder of the importance of considering dynamic effects in structural design. This event is often associated with the concept of resonance; however, the actual cause of the collapse is more accurately attributed to a phenomenon known as *aeroelastic flutter*, which differs from resonance.

Resonance occurs when a system oscillates at its natural frequency due to external periodic driving forces at that same frequency. Initially, it was believed that the Tacoma Narrows Bridge collapsed due to resonance caused by wind. Further analysis, however, has shown that the situation was more complex.

Aeroelastic Flutter is a self-exciting and self-sustaining oscillation caused by the interaction between aerodynamic forces, elastic properties, and inertial forces of a structure. For the Tacoma Narrows Bridge, wind provided an external force that initiated oscillations in the bridge. The bridge's design—particularly its narrow, flexible, and lightweight construction—made it exceptionally vulnerable to these aeroelastic oscillations.

Below we consider a mass-spring system, with no damping. It has a natural frequency of oscillation, $\omega_n = \sqrt{k/m}$, where k is the spring constant and m is the mass. If one gives the system a tap, it will simply oscillate at its natural frequency. In the code below, we force the system with $\cos(\omega_0\, t)$, where ω_0 is chosen as the natural frequency of the system. This creates a **resonance response** that grows like $t\,\cos(\omega_0\, t + \theta_0)$. The increasing amplitude with time will eventually cause any real-world system to reach its physical limits and break, similar to what aeroelastic flutter did to the Tacoma Narrows Bridge.

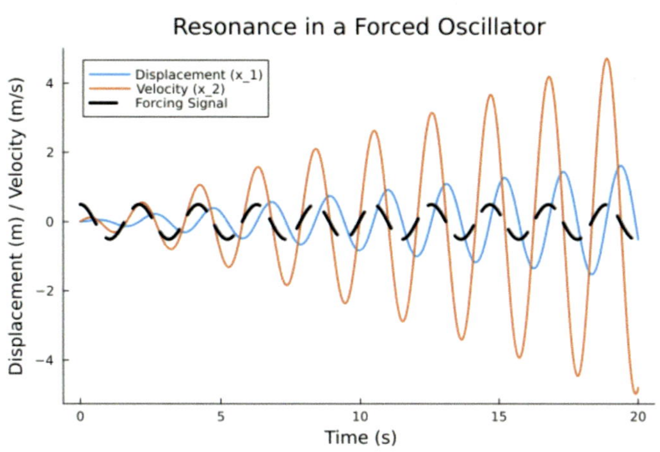


```
using DifferentialEquations, Plots

# Constants
```

```
m = 1.0    # Mass (kg)
k = 9.0    # Spring constant (N/m)
F0 = 0.5   # Amplitude of the external force (N)
w_n = sqrt(k / m)   # Natural frequency (rad/s)
w = w_n    # Forcing frequency = Natural frequency for resonance

# ODE Definition
function forced_oscillator!(du, u, p, t)
    x, v = u
    du[1] = v
    du[2] = (F0*cos(w*t) - k*x) / m
end

# Initial conditions
u0 = [0.0, 0.0]   # Initial displacement and velocity
tspan = (0.0, 20.0)   # Time span

# Solve the ODE
prob = ODEProblem(forced_oscillator!, u0, tspan)
sol = solve(prob, Tsit5(), reltol=1e-8, abstol=1e-8)

# Create a finer time grid for interpolation
fine_t = range(tspan[1], tspan[2], length=10000) # Increased number of points
                                                 # for finer interpolation

# Interpolate the solution at the finer time grid
fine_x = [sol(t)[1] for t in fine_t]
fine_v = [sol(t)[2] for t in fine_t]

# Forcing signal
forcing_signal = F0*cos.(w.*fine_t)

# Plotting
p1 = plot(fine_t, fine_x, label="Displacement (x_1)", xlabel="Time (s)",
          ylabel="Displacement (m) / Velocity (m/s)",
          title="Resonance in a Forced Oscillator", linewidth=2)
plot!(fine_t, fine_v, label="Velocity (x_2)", linewidth=2) # Smooth velocity curve
# Forcing signal
plot!(fine_t, forcing_signal, label="Forcing Signal", linestyle=:dash, color=:black, lw=3)

# Display the plot
plot!(legend=:topleft)
```

Output See the above plot.

(Optional Read) Videos for Enhancing Your Understanding and Intuition:

- A better description of resonance by Steve Mould. He illustrates wave patterns via the heights of flames! If you believe that fire makes anything better, then this video is for you. It's a bit slow initially, so you may want to jump ahead and then return to the beginning after you've seen the punchline.
- Resonance and the Sounds of Music by Walter Lewin. Very old-school setup by an absolute master with 1.64 M subscribers.
- Resonance Introduction using 9 Demonstrations by Flipping Physics (so nerdy that the humor enhances understanding).

9.8 (Optional Read) More on Julia Packages for Solving ODEs and PDEs

In Julia, the ecosystem for solving differential equations, including ordinary differential equations (ODEs) and partial differential equations (PDEs), is extensive and under continuous development. The primary effort that began under the umbrella of "JuliaDiffE" has since been integrated into a larger, more comprehensive initiative known as the SciML (Scientific Machine Learning) ecosystem. This initiative aims to provide a cohesive suite of tools for scientific computing, which includes solving differential equations, machine learning in the context of scientific problems, and much more.

- **For ODEs:**
  - **DifferentialEquations.jl**: A flagship package within the SciML ecosystem offering a unified interface for solving a broad spectrum of differential equations, including ODEs, PDEs, stochastic differential equations (SDEs), and more. It is highly flexible and features many algorithms for solving ODEs.
  - **OrdinaryDiffEq.jl**: A part of the DifferentialEquations.jl suite focused specifically on ODEs, providing a comprehensive array of solvers and analysis tools.
- **For PDEs:**
  - **DifferentialEquations.jl**: It also acts as the go-to tool for PDEs, offering methods to discretize and solve PDEs either directly or through conversion into ODEs or SDEs.
  - **FEniCS.jl**: A Julia wrapper for the FEniCS project, an acclaimed open-source computing platform for solving PDEs using finite element methods, enabling Julia users to utilize FEniCS's potent PDE-solving capabilities.
  - **Gridap.jl**: Provides a comprehensive toolkit for the simulation of PDEs using the finite element method (FEM), offering a powerful and adaptable framework for complex multiphysics simulations.
  - **JuAFEM.jl**: Aimed at solving finite element method problems, focusing on a simple and intuitive interface for Julia users.
- **Other Notable Packages:**
  - **ModelingToolkit.jl**: Part of the SciML ecosystem designed to ease the modeling of scientific problems, including differential equations creation and manipulation, often used alongside DifferentialEquations.jl.
  - **ParameterizedFunctions.jl**: Facilitates defining systems of differential equations with additional features like symbolic derivatives and parameter estimation, enhancing DifferentialEquations.jl's functionality.
  - **PDEOperators.jl**: Offers tools for defining differential operators for PDE discretization, simplifying work with complex differential operators.
  - **ApproxFun.jl**: Though not exclusively for differential equations, this package enables function approximation and can be used for solving differential equations.

These packages are integral to a broader, highly integrated ecosystem that includes tools for optimization, machine learning, and beyond, allowing for advanced problem-solving approaches across various scientific fields. The Julia community actively develops and maintains these packages, ensuring regular updates and expansions to the ecosystem.

9.9 (Optional Read) Proofs Associated with the Chapter

Proposition 9.67: Real Basis Vectors from Eigenvectors

Suppose that A is a real $n \times n$ matrix. Then we know its eigenvalues and eigenvectors are either real or they occur in complex conjugate pairs. Let $\lambda_1, \dots, \lambda_p, \lambda_1^*, \dots, \lambda_p^*$ be the complex eigenvalues and let $\lambda_{2p+1}, \dots, \lambda_n$ be the real eigenvalues, with a similar enumeration for the corresponding eigenvectors $\{v_1, \dots, v_p, v_1^*, \dots, v_p^*, v_{2p+1}, \dots, v_n\}$. If the eigenvalues are distinct, that is $\lambda_i \neq \lambda_j$ for $i \neq j$, then the set of n real vectors

$$\{v_{R1}, \dots, v_{Rp}, v_{I1}, \dots, v_{Ip}, v_{2p+1}, \dots, v_n\}$$

is linearly independent and forms a basis of $\mathbb{R}^n$.

Moreover, if $\lambda = \lambda_R + \mathbf{i}\,\lambda_I$ and $v = v_R + \mathbf{i}\,v_I$ are a complex eigenvalue and eigenvector, then

$$\begin{aligned} Av_R &= \lambda_R v_R - \lambda_I v_I \\ Av_I &= \lambda_I v_R + \lambda_R v_I. \end{aligned} \tag{9.72}$$

Proof: The eigenvalues being distinct implies that the set of eigenvectors

$$\{v_1, \dots, v_p, v_1^*, \dots, v_p^*, v_{2p+1}, \dots, v_n\}$$

forms a basis of $(\mathbb{C}^n, \mathbb{C})$. Hence,

$$\dim\ \mathrm{span}\{v_1, \ldots, v_p, v_1^*, \ldots, v_p^*, v_{2p+1}, \ldots, v_n\} = n,$$

the dimension of $(\mathbb{C}^n, \mathbb{C})$.

By the definition of the real and imaginary parts of vectors, for each $1 \le j \le p$, $v_j = v_{Rj} + \mathbf{i}\, v_{Ij}$ and $v_j^* = v_{Rj} - \mathbf{i}\, v_{Ij}$, and moreover, $v_{Rj} = \frac{v_j + v_j^*}{2}$ and $v_{Ij} = \frac{v_j - v_j^*}{2\mathbf{i}}$. Hence,

$$\mathrm{span}\{v_j, v_j^*\} = \mathrm{span}\{v_{Rj}, v_{Ij}\},$$

and

$$\mathrm{span}\{v_1, \ldots, v_p, v_1^*, \ldots, v_p^*, v_{2p+1}, \ldots, v_n\} = \{v_{R1}, \ldots, v_{Rp}, v_{I1}, \ldots, v_{Ip}, v_{2p+1}, \ldots, v_n\}.$$

It follows that

$$\dim\ \mathrm{span}\{v_{R1}, \ldots, v_{Rp}, v_{I1}, \ldots, v_{Ip}, v_{2p+1}, \ldots, v_n\} = n,$$

establishing that the vectors are linearly independent in $(\mathbb{C}^n, \mathbb{C})$, and hence also in $(\mathbb{R}^n, \mathbb{R})$.

To establish (9.72), we use

$$\begin{aligned} Av &= A(v_R + \mathbf{i}\, v_I) = (\lambda_R + \mathbf{i}\,\lambda_I) \cdot (v_R + \mathbf{i}\, v_I) = \lambda v \\ Av^* &= A(v_R - \mathbf{i}\, v_I) = (\lambda_R - \mathbf{i}\,\lambda_I) \cdot (v_R - \mathbf{i}\, v_I) = \lambda^* v^*. \end{aligned}$$

This gives

$$\begin{aligned} Av_R &= A\frac{v + v^*}{2} = \lambda_R v_R - \lambda_I v_I \\ Av_I &= A\frac{v - v^*}{2\mathbf{i}} = \lambda_I v_R + \lambda_R v_I. \end{aligned}$$

An alternative proof is

$$\begin{aligned} Av_R &= \mathrm{real}(Av) = \mathrm{real}\left((\lambda_R + \mathbf{i}\,\lambda_I) \cdot (v_R + \mathbf{i}\, v_I)\right) = \lambda_R v_R - \lambda_I v_I, \\ Av_I &= \mathrm{imag}(Av) = \mathrm{imag}\left((\lambda_R + \mathbf{i}\,\lambda_I) \cdot (v_R + \mathbf{i}\, v_I)\right) = \lambda_I v_R + \lambda_R v_I. \end{aligned}$$

■

Proposition 9.68: $e^{At} v_R$ and $e^{At} v_I$

$$\begin{aligned} e^{At}\, v_R &= e^{\lambda_R t}\, \cos(\lambda_I t)\, v_R - e^{\lambda_R t}\, \sin(\lambda_I t)\, v_I \\ e^{At}\, v_I &= e^{\lambda_R t}\, \sin(\lambda_I t)\, v_R + e^{\lambda_R t}\, \cos(\lambda_I t)\, v_I. \end{aligned} \tag{9.73}$$

Proof: As in Prop. 9.67, we write $\lambda = \lambda_R + \mathbf{i}\,\lambda_I$ and $v = v_R + \mathbf{i}\, v_I$. Then,

$$\begin{aligned} e^{At}\, v_R &= \mathrm{real}\left(e^{At}\, v\right) \\ &= \mathrm{real}\left(e^{\lambda t}\, v\right) \\ &= \mathrm{real}\left(e^{\lambda_R t}\, (\cos(\lambda_I t) + \mathbf{i}\sin(\lambda_I t))\, (v_R + \mathbf{i}\, v_I)\right) \\ &= e^{\lambda_R t}\, \cos(\lambda_I t)\, v_R - e^{\lambda_R t}\, \sin(\lambda_I t)\, v_I. \end{aligned}$$

Similarly,

$$\begin{aligned} e^{At}\, v_I &= \mathrm{imag}\left(e^{At}\, v\right) \\ &= \mathrm{imag}\left(e^{\lambda t}\, v\right) \\ &= \mathrm{imag}\left(e^{\lambda_R t}\, (\cos(\lambda_I t) + \mathbf{i}\sin(\lambda_I t))\, (v_R + \mathbf{i}\, v_I)\right) \\ &= e^{\lambda_R t}\, \sin(\lambda_I t)\, v_R + e^{\lambda_R t}\, \cos(\lambda_I t)\, v_I. \end{aligned}$$

■

From Prop. 9.34 Linearization of the Robot Equations, we have that the linearized model about any equilibrium point has the form

$$\begin{bmatrix} \delta\dot{x}_1 \\ \delta\dot{x}_2 \end{bmatrix} = \begin{bmatrix} 0_n & I_n \\ A_{21} & 0_n \end{bmatrix} \begin{bmatrix} \delta x_1 \\ \delta x_2 \end{bmatrix}. \tag{9.74}$$

Proposition 9.69: Eigen-Facts about the Robot Equations

Suppose the matrices A and B are invertible and that matrix C is defined as

$$C = \begin{bmatrix} 0 & B \\ A & 0 \end{bmatrix}.$$

Then,

$$\det(C - \lambda I) = \det \begin{bmatrix} -\lambda I & B \\ A & -\lambda I \end{bmatrix} = \det(\lambda^2 I - AB).$$

Therefore, the eigenvalues of C are the square roots of the eigenvalues of AB. So, they depend on the eigenvalues of AB rather than the individual eigenvalues of the matrices A and B.

Note: If AB has a negative real eigenvalue, then C will have a pair of purely imaginary eigenvalues. If AB has a positive real eigenvalue, then C will have both a positive and negative eigenvalue. If AB has a pair of complex conjugate eigenvalues, then C will have two complex conjugate pairs of eigenvalues.

Proof: Based on Eigenvalues of block matrices with zero diagonal blocks. Given two vectors $u, v \in \mathbb{R}^n$, let

$$w = \begin{bmatrix} u \\ v \end{bmatrix} \in \mathbb{R}^{2n}$$

be the vector whose first n components are those of u and whose last n are those of v. The matrix-vector product $C \bullet w$ gives

$$\begin{bmatrix} 0 & B \\ A & 0 \end{bmatrix} \begin{bmatrix} u \\ v \end{bmatrix} = \begin{bmatrix} Bv \\ Au \end{bmatrix}.$$

Therefore, solving $Cw = \lambda w$ reduces to solving the following two matrix equations,

$$\begin{cases} Bv = \lambda u \\ Au = \lambda v. \end{cases}$$

Applying A to the first equation and B to the second gives

$$\begin{cases} ABv = \lambda Au \\ BAu = \lambda Bv. \end{cases}$$

Using again the relations $Au = \lambda v$ and $Bv = \lambda u$, we obtain

$$\begin{cases} ABv = \lambda^2 v \\ BAu = \lambda^2 u. \end{cases}$$

Hence, if λ is an eigenvalue of C, then λ^2 is an eigenvalue of AB and also of BA.
Conversely, if μ is an eigenvalue of AB with eigenvector $v \neq 0$, and if $\lambda \in \mathbb{C}$ satisfies $\lambda^2 = \mu$, then setting

$$u = \frac{1}{\lambda} Bv$$

gives

$$Bv = \lambda u \quad \text{and} \quad Au = \frac{1}{\lambda} ABv = \frac{1}{\lambda} \mu v = \lambda v.$$

Therefore,

$$C\begin{bmatrix} u \\ v \end{bmatrix} = \begin{bmatrix} Bv \\ Au \end{bmatrix} = \begin{bmatrix} \lambda u \\ \lambda v \end{bmatrix} = \lambda \begin{bmatrix} u \\ v \end{bmatrix},$$

showing that λ is an eigenvalue of C.

We conclude that $\lambda \in \mathbb{C}$ is an eigenvalue of C if, and only if, λ^2 is an eigenvalue of AB (or BA, since they have the same eigenvalues). ■

Proof of Prop. 9.52 (Properties of Complex Exponentials)

Complex exponentials are fundamental in various fields of mathematics, engineering, and physics. Below are some of the key properties of complex exponentials, for $a, \omega \in \mathbb{R}$ and $z := a + \mathbf{i}\omega \in \mathbb{C}$:

(a) **Periodicity:** The function $e^{\mathbf{i}\omega t}$ is periodic with period $T = \frac{2\pi}{|\omega|}$, for $\omega \neq 0$.

(b) **Multiplicative Property:** For any two complex numbers $z_1 = a_1 + \mathbf{i}\omega_1$ and $z_2 = a_2 + \mathbf{i}\omega_2$, the exponential function satisfies $e^{z_1 t} \cdot e^{z_2 t} = e^{(z_1+z_2)t}$.

(c) **Inverse:** The inverse of e^{zt} is given by e^{-zt}.

(d) **Differentiation and Integration:** The derivative and antiderivative of e^{zt} with respect to t are:

- $\frac{d}{dt}e^{zt} = ze^{zt}$ and
- for $z \neq 0$, $\int e^{zt}dt = \frac{1}{z}e^{zt} + C$, where C is the constant of integration.

(e) **Conjugation:** The complex conjugate of $e^{(a+\mathbf{i}\omega)t}$ is $e^{(a-\mathbf{i}\omega)t}$.

(f) **Magnitude:** The magnitude of $e^{(a+\mathbf{i}\omega)t}$ is $|e^{(a+\mathbf{i}\omega)t}| = e^{at}$, which exhibits exponential growth or decay based on the sign of the real part, a.

Proof:

(a) **Periodicity:** By Euler's formula, $e^{\mathbf{i}\omega t} = \cos(\omega t) + \mathbf{i}\sin(\omega t)$. Because $\cos(\omega(t + \frac{2\pi}{|\omega|})) = \cos(\omega t + \operatorname{sign}(\omega)2\pi) = \cos(\omega t)$ and $\sin(\omega(t + \frac{2\pi}{|\omega|})) = \sin(\omega t + \operatorname{sign}(\omega)2\pi) = \sin(\omega t)$, it follows that $e^{\mathbf{i}\omega(t+\frac{2\pi}{|\omega|})} = e^{\mathbf{i}\omega t}$, proving periodicity with period $\frac{2\pi}{|\omega|}$.

(b) **Multiplicative Property:** Let $z_1 = a_1 + \mathbf{i}\omega_1$ and $z_2 = a_2 + \mathbf{i}\omega_2$. Then

$$\begin{aligned} e^{z_1 t} &= e^{a_1 t}\cos(\omega_1 t) + \mathbf{i}\, e^{a_1 t}\sin(\omega_1 t) \\ e^{z_2 t} &= e^{a_2 t}\cos(\omega_2 t) + \mathbf{i}\, e^{a_2 t}\sin(\omega_2 t). \end{aligned}$$

Therefore,

$$\begin{aligned} e^{z_1 t} \cdot e^{z_2 t} &= \left(e^{a_1 t}\cos(\omega_1 t) + \mathbf{i}\, e^{a_1 t}\sin(\omega_1 t)\right) \cdot \left(e^{a_2 t}\cos(\omega_2 t) + \mathbf{i}\, e^{a_2 t}\sin(\omega_2 t)\right) \\ &= e^{a_1 t} \cdot e^{a_2 t}\left(\cos(\omega_1 t) + \mathbf{i}\sin(\omega_1 t)\right) \cdot \left(\cos(\omega_2 t) + \mathbf{i}\sin(\omega_2 t)\right) \\ &= e^{a_1 t} \cdot e^{a_2 t}\left(\cos(\omega_1 t)\cos(\omega_2 t) - \sin(\omega_1 t)\sin(\omega_2 t)\right) \\ &\quad + \mathbf{i}\, e^{a_1 t} \cdot e^{a_2 t}\left(\cos(\omega_1 t)\sin(\omega_2 t) + \sin(\omega_1 t)\cos(\omega_2 t)\right) \\ &= e^{a_1 t} \cdot e^{a_2 t}\cos((\omega_1 + \omega_2)t) + \mathbf{i}\, e^{a_1 t} \cdot e^{a_2 t}\sin((\omega_1 + \omega_2)t) \\ &= e^{(a_1+a_2)t + (\omega_1+\omega_2)t} \\ &= e^{(z_1+z_2)t}. \end{aligned}$$

(c) **Inverse:** From (b), $e^{zt} \cdot e^{-zt} = e^{(z-z)t} = e^0 = 1$, and hence, e^{-zt} is the multiplicative inverse of e^{zt}.

(d) **Differentiation and Integration:** The most important observation is that by using Euler's formula, we are only differentiating or "anti-differentiating" real functions.

Suppose $z = a + \mathbf{i}\omega$, and write $e^{zt} = e^{at}\left(\cos(\omega t) + \mathbf{i}\sin(\omega t)\right) = e^{at}\cos(\omega t) + \mathbf{i}\,e^{at}\sin(\omega t)$.

- By the product rule and chain rule,

$$\begin{aligned}\frac{d}{dt}e^{zt} &= \frac{d}{dt}\left(e^{at}\cos(\omega t)\right) + \mathbf{i}\frac{d}{dt}\left(e^{at}\sin(\omega t)\right)\\ &= \left(a\,e^{at}\cos(\omega t) - \omega e^{at}\sin(\omega t)\right) + \mathbf{i}\left(a\,e^{at}\sin(\omega t) + \omega e^{at}\cos(\omega t)\right)\\ &= a\,e^{at}\left(\cos(\omega t) + \mathbf{i}\sin(\omega t)\right) + \mathbf{i}\omega\,e^{at}\left(\cos(\omega t) + \mathbf{i}\sin(\omega t)\right)\\ &= (a + \mathbf{i}\omega)\cdot e^{at}\left(\cos(\omega t) + \mathbf{i}\sin(\omega t)\right)\\ &= ze^{zt}.\end{aligned}$$

- We assume $z \neq 0$. Then, by the previous result, $\frac{d}{dt}\left(\frac{1}{z}e^{zt}\right) = \frac{z}{z}e^{zt} = e^{zt}$, proving that $\frac{1}{z}e^{zt}$ is an antiderivative of e^{zt}.

(e) **Conjugation:** Because $e^{(a+\mathbf{i}\omega)t} = e^{at}\left(\cos(\omega t) + \mathbf{i}\sin(\omega t)\right)$, the complex conjugate of $e^{(a+\mathbf{i}\omega)t}$ is $e^{at}\left(\cos(\omega t) - \mathbf{i}\sin(\omega t)\right) = e^{at}\left(\cos(-\omega t) + \mathbf{i}\sin(-\omega t)\right) = e^{(a-\mathbf{i}\omega)t}$.

(f) **Magnitude:** The magnitude of $e^{(a+\mathbf{i}\omega)t} = e^{at}\left(\cos(\omega t) + \mathbf{i}\sin(\omega t)\right)$ is

$$\begin{aligned}\left|e^{at}\cos(\omega t) + \mathbf{i}\,e^{at}\sin(\omega t)\right| &= \sqrt{\left(e^{at}\cos(\omega t)\right)^2 + \left(e^{at}\sin(\omega t)\right)^2}\\ &= \sqrt{\left(e^{at}\right)^2\cdot\left(\cos^2(\omega t) + \sin^2(\omega t)\right)} = e^{at}.\end{aligned}$$

■

Proof of Prop. 9.58 (Exponential Stability of Linear Systems of ODEs)

Let A be a real $n \times n$ matrix with eigenvalues $\lambda_1, \lambda_2, \ldots, \lambda_n$, and consider the linear system of ODEs,

$$\dot{x} = Ax.$$

If $\mathrm{real}(\lambda_i) < 0$ for $1 \leq i \leq n$ (aka the real parts of all eigenvalues are negative), then the **origin** is globally exponentially stable.

Proof: We give the proof for the case of distinct eigenvalues. By Prop. 9.67, set of n real vectors

$$\{v_{R1}, \ldots, v_{Rp}, v_{I1}, \ldots, v_{Ip}, v_{2p+1}, \ldots, v_n\}$$

is linearly independent and forms a basis of $\mathbb{R}^n$. Hence, an arbitrary $x_0 \in \mathbb{R}^n$ can be written as a linear combination

$$x_0 = \sum_{j=1}^{p}\left(\alpha_j v_{Rj} + \beta_j v_{Ij}\right) + \sum_{j=2p+1}^{n}\gamma_j v_j.$$

Next, applying Prop. 9.68, we have

$$
\begin{aligned}
e^{At}x_0 &= \sum_{j=1}^{p} e^{At}\left(\alpha_j v_{Rj} + \beta_j v_{Ij}\right) + \sum_{j=2p+1}^{n} \gamma_j e^{At} v_j \\
&= \sum_{j=1}^{p} \left(\alpha_j e^{At} v_{Rj} + \beta_j e^{At} v_{Ij}\right) + \sum_{j=2p+1}^{n} \gamma_j e^{At} v_j \\
&= \sum_{j=1}^{p} \Big(\alpha_j \left(e^{\lambda_{Rj}t}\cos(\lambda_{Ij}t)\, v_{Rj} - e^{\lambda_{Rj}t}\sin(\lambda_{Ij}t)\, v_{Ij}\right) \\
&\quad + \beta_j \left(e^{\lambda_{Rj}t}\sin(\lambda_{Ij}t)\, v_{Rj} + e^{\lambda_{Rj}t}\cos(\lambda_{Ij}t)\, v_{Ij}\right)\Big) + \sum_{j=2p+1}^{n} \gamma_j e^{\lambda_j t} v_j \\
&= \sum_{j=1}^{p} e^{\lambda_{Rj}t}\Big(\alpha_j \left(\cos(\lambda_{Ij}t)\, v_{Rj} - \sin(\lambda_{Ij}t)\, v_{Ij}\right) \\
&\quad + \beta_j \left(\sin(\lambda_{Ij}t)\, v_{Rj} + \cos(\lambda_{Ij}t)\, v_{Ij}\right)\Big) + \sum_{j=2p+1}^{n} \gamma_j e^{\lambda_j t} v_j.
\end{aligned}
$$

From this result, we see that if all eigenvalues have negative real parts, then $x(t) := e^{At}x_0 \to 0_{n\times 1}$ as $t \to \infty$, completing the proof.

Note: Suppose one of the eigenvalues has real part that is non-negative. Then by taking x_0 to be the (real part of) the corresponding eigenvector, we then have a solution that does not converge to zero. It will either oscillate, stay constant, or explode. ■

Chapter 10

Laplace Transforms through the Lens of Feedback Control

Learning Objectives

- To introduce the concept and mathematical theory of Laplace Transforms.
- To become familiar with common Laplace transforms and their properties.
- To understand the process and methods of finding inverse Laplace transforms.
- To apply Laplace transforms for solving linear ordinary differential equations (ODEs).
- To define and derive transfer functions for linear time-invariant systems.
- To understand the importance and application of transfer functions in the design of Singe-Input Single-Output (SISO) feedback loops.
- To use appropriate software tools for each of the above topics.

Outcomes

- Gaining a foundational understanding of Laplace transforms and their application in engineering problems.
- Mastery of standard Laplace transforms of common functions and their key properties.
- Ability to understand inverse Laplace transforms using techniques such as partial fraction decomposition and tools in Julia.
- Proficiency in applying Laplace transforms to simplify and solve linear ODEs with initial conditions.
- Comprehensive understanding of transfer functions, their derivation, and significance in system analysis.
- Applying knowledge of transfer functions to design and analyze SISO feedback loops in engineering systems, particularly in robotics.

The Laplace Transform Magically Turns Linear Differential Equations into Algebraic Equations. This metamorphosis is not merely a theoretical curiosity but an immensely practical tool that unlocks a new perspective on ODEs. **By translating the differential equations into algebraic ones, Laplace transforms simplify the process of understanding solutions of ODEs, showing how to modify the equations to achieve asymptotic stability. This technique allows for a more straightforward analysis and design of systems governed by ODEs, thereby bridging the gap between complex mathematical theory and real-world application.**

10.1 Introduction

Thanks to Chapter 9, we now understand what an ODE is, how to compute its solution for a given set of initial conditions, and whether its solutions asymptotically converge to an equilibrium point or not. That is already a big accomplishment. We now make a radical change in focus from analyzing solutions of ODEs to actively modifying their solutions so that they exhibit properties that we deem desirable. To achieve this, we introduce controlled inputs into our ODE models of physical systems. These inputs serve as a mechanism to influence and steer the evolution of the system's state. By strategically manipulating these inputs, we can guide the system's trajectory toward our objectives, such as achieving exponential stability, driving an important signal to a desired value (e.g., setting a robot's speed), or attenuating the effects of perturbations. This approach marks a significant shift from passive analysis to active control, opening the door to designing systems that respond effectively to commanded inputs and achieve specific, desired outcomes.

The fascinating possibilities of feedback control are introduced within the context of linear Single-Input Single-Output (SISO) systems, utilizing Laplace transforms and transfer functions. We employ the Segway model as a practical example to demonstrate the foundational principles of feedback control, while also touching upon concepts like PD controllers and precompensators. Through our exploration of Laplace transforms and transfer functions, we aim to provide an initial framework for understanding how these tools can be used to design basic, yet effective, feedback control strategies. The chapter builds upon the concepts of stability and feedback introduced in Chapter 9 and Example 9.62, illustrating the impact of these tools in engineering.

Our principal goal is to provide the concepts and methods required for stabilizing an underactuated mobile robot. It is important to note, however, that the material in this chapter, while illustrative and intriguing, remains an introductory overview and is not a substitute for more advanced courses in feedback control systems.

Figure 10.1: An Input-Output System. The signal $u(t)$ represents the action of an actuator, such as a DC motor, a rocket engine, or a voltage supply. The output $y(t)$ is a signal that is important to the system designer, such as the speed of a robot or the angle of a joint. If $u(t)$ and $y(t)$ are scalar-valued, then we have a Single-Input, Single-Output (SISO) System. If $u(t)$ and/or $y(t)$ are vectors, then we have a Multi-Input Multi-Output (MIMO) System.

10.2 Single-Input Single-Output (SISO) Linear Systems

A system is characterized by an ODE with both inputs and outputs. A drone (aka quadrotor) is a system. A system's inputs are signals that cause changes in the motion of the entire system or parts of the system. A quadrotor-drone has four DC motors with propellers attached to them. These are the actuators. The output of the drone could be its 3D position and its 3D orientation in $\mathbb{R}^3$. A simpler output would be its speed in a single direction, say along the x-axis. It would be up to you (or your team) to define the important quantities to be regulated in the system. Here, we define linear systems with a single input and output. It is THE PLACE to start. The planar Segway model that will guide our discussion has one input (a motor at the wheels) and two outputs (the angle of the Segway's body and its speed) that we will learn to treat successively. Controlling two outputs with one input is more challenging than controlling one output. That's Robotics in a nutshell.

10.2.1 Input-Output and State-Variable Models

Every domain has its "secret codes and rituals." In **Feedback Control**, the generic notation for an input is $u(t)$, and for an output, it is $y(t)$. We will use this notation throughout this chapter.

SISO Linear Systems

Definition 10.1.

$$y^{(n)} + a_{n-1}y^{(n-1)} + a_{n-2}y^{(n-2)} + \cdots + a_1\dot{y} + a_0 y = b_0 u + b_1\dot{u} + \cdots + b_m u^{(m)} \tag{10.1}$$

*is a **linear input-output system**. It is called an input-output system because the only variables present are the input, the output, and their derivatives. Because we will always assume that both $u(t)$ and $y(t)$ are scalars, it is a **SISO linear system**.*

Note: We are assuming the coefficients a_k are constants. In this case, the model is called a **Linear Time-Invariant (LTI)** input-output system. If the coefficients depended on time, then we would have a **Linear Time-Varying (LTV)** input-output system.

Definition 10.2. *Consider a linear system of ODEs, $\dot{x} = Ax$, with $x \in \mathbb{R}^n$. It becomes a **control system** if we add an input and an output to the model. In particular, for b, an $n \times 1$ column vector, and for c, a $1 \times n$ row vector,*

$$\begin{aligned} \dot{x} &= Ax + bu, \\ y &= cx, \end{aligned} \tag{10.2}$$

*is a **SISO linear state-variable model**. While studying MIMO linear state-variable models is super interesting, we will not do that.*

Note: In a course solely dedicated to feedback control, we learn that the models (10.1) and (10.2) are two sides of the same coin. So why do we give both forms? Because the first one makes it very easy to motivate **transfer functions**, and the second one is how we will obtain linear control models for the Segway.

Note: An input-output system is **causal** if its output at time t does not depend upon future values of the input signal. It is a nontrivial fact that the system in (10.1) is causal if, and only if, $m \le n$, that is, the highest derivative appearing in the input is less than or equal to the highest derivative appearing in the output. The state-variable model (10.2) is always causal.

We illustrate how these models arise in circuits and mechanics.

10.2.2 Input-Output Models from Circuits

Example 10.3. *For the RLC circuit in Example 9.26, derive a SISO input-output model instead of a state-variable model, using $y := V_C$ and $u := V_S$.*

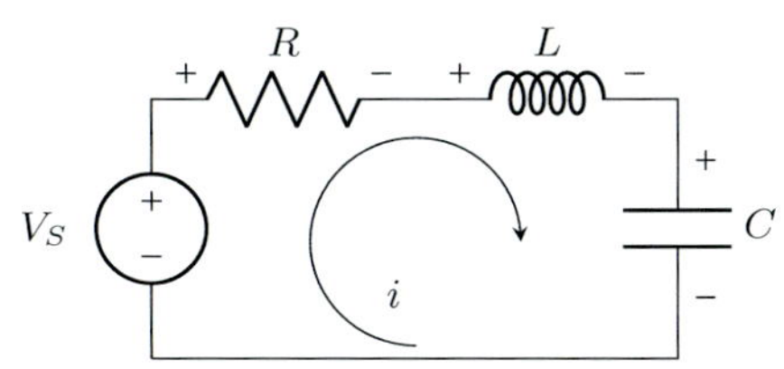

Solution: Ans. $\frac{d^2y}{dt^2} + \frac{R}{L}\frac{dy}{dt} + \frac{1}{LC}y = \frac{1}{LC}u.$

As before, V_S is a voltage source, R denotes a resistor, L an inductor, and C a capacitor. The loop denotes the reference direction for the current flowing through the circuit, $i(t)$. **Kirchoff's Voltage Law** states that the sum of the voltages around any loop must equal zero, yielding

$$-V_S + V_R + V_L + V_C = 0. \tag{10.3}$$

As before, the derivatives arise from two of the three **constitutive relations** for the circuit elements, namely

$$\begin{aligned} V_R &= R \cdot i, \\ V_L &= L\frac{di}{dt}, \\ i &= C\frac{dV_C}{dt}. \end{aligned} \tag{10.4}$$

From lines one and three of (10.4), we obtain

$$V_R = RC\frac{dV_C}{dt}, \tag{10.5}$$

and from lines two and three of (10.4), we obtain

$$\begin{aligned} V_L &= L\frac{d}{dt}\left(C\frac{dV_C}{dt}\right) \\ &= LC\frac{d^2V_C}{dt^2}. \end{aligned} \tag{10.6}$$

Substituting these into (10.3) yields

$$\begin{aligned} -V_S + RC\frac{dV_C}{dt} + LC\frac{d^2V_C}{dt^2} + V_C = 0 \\ \Updownarrow \text{ (rearranging terms)} \\ LC\frac{d^2V_C}{dt^2} + RC\frac{dV_C}{dt} + V_C = V_S. \end{aligned} \tag{10.7}$$

Upon defining $y := V_C$ and $u := V_S$, we have

$$\begin{aligned} LC\frac{d^2y}{dt^2} + RC\frac{dy}{dt} + y = u \\ \Updownarrow \text{ (dividing through by } LC) \\ \frac{d^2y}{dt^2} + \frac{R}{L}\frac{dy}{dt} + \frac{1}{LC}y = \frac{1}{LC}u. \end{aligned} \tag{10.8}$$

Note: Resistors heat up as the current increases, and this can change the "local resistance" of the element. Modeling this effect could lead to either a nonlinear or LTV model. ■

Example 10.4. *Redo the above example, but this time, assign* i*, the loop current, as the output. The input remains the applied voltage,* V_S*.*

Solution: **Ans.** $\frac{d^2y}{dt^2} + \frac{R}{L}\frac{dy}{dt} + \frac{1}{LC}y = \frac{1}{L}\frac{du}{dt}$, where we note that the input is differentiated.

By differentiating (10.3), we obtain

$$-\dot{V}_S + \dot{V}_R + \dot{V}_L + \dot{V}_C = 0. \tag{10.9}$$

This opening move makes the remaining computations easier. How would you know to do it? Experience, like that obtained in courses on electronic circuits.

After the opening move in (10.9), we next apply the constitutive relationships in (10.4). By differentiating its first and second lines, we obtain

$$\begin{aligned} \frac{dV_R}{dt} &= R\frac{di}{dt} \\ \frac{dV_L}{dt} &= L\frac{d^2i}{dt^2}, \end{aligned} \tag{10.10}$$

and from the third line, we obtain directly,

$$\frac{dV_C}{dt} = \frac{1}{C}i. \tag{10.11}$$

Substituting these into (10.9) yields

$$-\frac{dV_S}{dt} + R\frac{di}{dt} + L\frac{d^2i}{dt^2} + \frac{1}{C}i = 0. \tag{10.12}$$

Dividing through by $\frac{1}{L}$ and rearranging terms gives

$$\frac{d^2i}{dt^2} + \frac{R}{L}\frac{di}{dt} + \frac{1}{LC}i = \frac{1}{L}\frac{dV_S}{dt}. \tag{10.13}$$

Setting $y := i$ and $u := V_s$ then gives the answer. ■

(a)

(b)

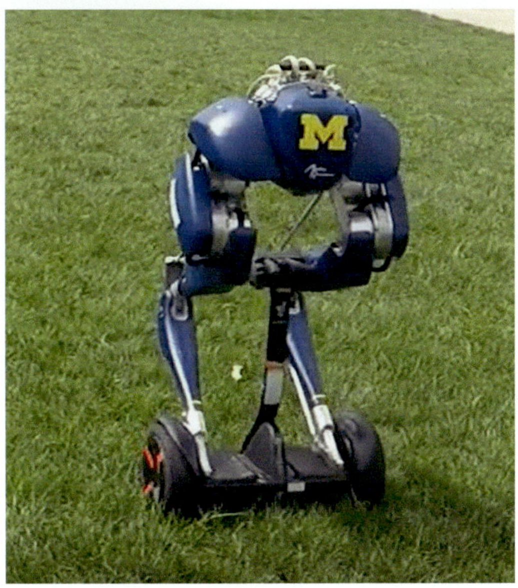

(c)

Figure 10.2: Segways take many forms. The one we are modeling is closer to image (a), which shows a pair of classic Segway transporters. Do you think the designer of the balance algorithm for the devices knew ahead of time the exact mass, moments of inertia, etc, for the riders? Or did they have to make the balance algorithm robust to a wide range of human and mechanical riders? To accomplish the latter takes more than our brief introduction to feedback control. You can read about the algorithm used on Cassie to achieve the very meta stunt of a robot riding another robot. **Credits:** (a) and (b) are Adobe Stock Photos, while (c) is from the Biped Lab.

10.2.3 A State-Variable Model of a (Planar) Segway Transporter

A Segway Transporter Is a Mobile Robot Ridden by a Human or Another Robot. In the following, we detail some of its key components.

- **Mechanical Components:**
  - *Wheels and Tires:* Two large wheels with durable tires designed for traction.
  - *Frame and Body:* Sturdy frame supporting the rider's platform, handlebar, and other components.
  - *Gear System:* Gears connect the electric motors to the wheels, enabling precise speed and movement control.
- **Electrical Components:**
  - *Batteries:* Rechargeable batteries provide power to the motors and control systems.
  - *Electric Motors:* Two independent brushless DC electric motors, one for each wheel, providing propulsion and control.
  - *Control Circuitry:* Electronic control units for managing power distribution, motor control, and balancing algorithms.
- **Sensing Components:**
  - *Gyroscopic Sensors:* Multiple gyroscopes provide data on the tilt and orientation of the transporter for balance control.
  - *Accelerometers:* Sensors measuring acceleration aid in maintaining balance and controlling motion.
  - *Encoders:* Encoders in the wheels measure rotation, which, when differentiated, aid in speed control and stability.

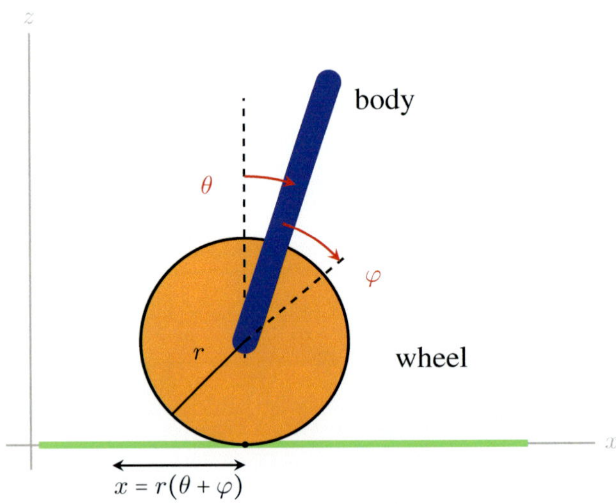

Figure 10.3: Coordinates used in the model of a planar Segway. θ is the lean angle of the body, measured CW from the vertical and φ is the rotation angle of the wheel with respect to the body. With these conventions, the distance of the wheel's axis from the origin is $r(\theta + \varphi)$, where r is the wheel's radius.

A planar set of **robot equations** (aka Lagrange's Equations of Motion) for the Segway Transporter is provided in Section 10.10. The model lumps the rider and the Segway platform into a single object with a combined mass, center of mass, and moment of inertia about its center of mass. The lean angle of the combined platform and rider is denoted by θ and is measured with respect to the vertical, with zero being upright. Each wheel of the Segway has a given mass, radius, and moment of inertia. Furthermore, each wheel is connected to a DC motor through gearing. The mass of the motors is lumped into the mass of the platform because they are physically bolted to it. The model includes the moment of inertia of each motor's rotor and the gear ratio. The control variable u in the model is the torque generated by the motor and applied to its rotor, typically called "motor torque." The torque applied between the platform and each wheel is multiplied by the gear ratio. The model does not include the electrical dynamics of the DC motors. Hence, we can build an even more realistic model that uses the voltage applied to the motor as the control input.

How does a Segway accelerate to a desired speed or brake to a stop? If you say it uses the motors, you are only half right. Why only half? Because if the lean angle is zero, that is, the platform+rider's center of mass is directly over the wheels, then the linear acceleration is zero. To accelerate forward, the rider must lean forward. Then, to prevent a face-plant, the motor plus control system must accelerate the wheels forward so as to drive the wheels under the center of mass of the platform+rider. Once the center of mass returns to being over the wheels, the linear acceleration becomes zero, and the Segway rolls along at a constant speed. To come to a stop, the rider leans opposite to the direction of travel. In a real 3D Segway, steering is accomplished by leaning the platform's handle to the left or right. In our planar (2D) Segway, there is no steering. In some sense, we are on train tracks.

Fact 10.5: Linearized (State-Variable) Model of the Segway Transporter

The linearization of the Segway Transporter about the upright equilibrium yields the state-variable control model

$$\frac{d}{dt}\begin{bmatrix}\theta\\ \varphi\\ \dot{\theta}\\ \dot{\varphi}\end{bmatrix}=\begin{bmatrix}0.000 & 0.000 & 1.000 & 0.000\\ 0.000 & 0.000 & 0.000 & 1.000\\ 50.937 & 0.000 & 0.000 & 0.000\\ -162.687 & 0.000 & 0.000 & 0.000\end{bmatrix}\begin{bmatrix}\theta\\ \varphi\\ \dot{\theta}\\ \dot{\varphi}\end{bmatrix}+\begin{bmatrix}0.000\\ 0.000\\ -21.468\\ 80.227\end{bmatrix}u, \tag{10.14}$$

where u is the motor torque and the angles θ and φ are defined in Fig. 10.3. The coefficients in the matrix are constants; thus, the system is LTI. The model can also be written as

$$\underbrace{\begin{bmatrix}\dot{x}_1\\ \dot{x}_2\\ \dot{x}_3\\ \dot{x}_4\end{bmatrix}}_{\dot{x}}=\underbrace{\begin{bmatrix}0.000 & 0.000 & 1.000 & 0.000\\ 0.000 & 0.000 & 0.000 & 1.000\\ 50.937 & 0.000 & 0.000 & 0.000\\ -162.687 & 0.000 & 0.000 & 0.000\end{bmatrix}}_{A}\underbrace{\begin{bmatrix}x_1\\ x_2\\ x_3\\ x_4\end{bmatrix}}_{x}+\underbrace{\begin{bmatrix}0.000\\ 0.000\\ -21.468\\ 80.227\end{bmatrix}}_{b}u, \tag{10.15}$$

where $x_1 := \theta$, $x_2 := \varphi$, $x_3 := \dot{\theta}$, and $x_4 := \dot{\varphi}$.

We next define interesting outputs for the model.

- **Lean Angle of the Segway Platform plus Rider:** $y_\theta = c_\theta \cdot x = \underbrace{\begin{bmatrix}1 & 0 & 0 & 0\end{bmatrix}}_{c_\theta}x$. We need this quantity to be near zero for the Segway to be upright.
- **Three different means to define speed:**
  - **Horizontal Speed Level 1:** $y_{v1} = c_{v1} \cdot x = \underbrace{\begin{bmatrix}0.000 & 0.000 & 0.000 & 0.250\end{bmatrix}}_{c_{v1}}x$ multiplies the radius of the wheel (0.25 m) by the relative angular velocity of the wheel with respect to the angle of the platform (in rad/s) to obtain a speed in m/s. The relative angular velocity is different from the wheel's angular velocity in the world frame when the platform also has angular velocity.
  - **Horizontal Speed Level 2:** $y_{v2} = c_{v2} \cdot x = \underbrace{\begin{bmatrix}0.000 & 0.000 & 0.250 & 0.250\end{bmatrix}}_{c_{v2}}x$, uses wheel radius and the absolute angular velocity of the wheel. The latter is the wheel's angular velocity in the world frame, even when the platform also has angular velocity.
  - **Horizontal Speed Level 3:** $y_{v3} = c_{v3} \cdot x = \underbrace{\begin{bmatrix}0.000 & 0.000 & 0.863 & 0.250\end{bmatrix}}_{c_{v3}}x$, is the horizontal velocity of the center of mass of the overall Segway. This seems like the correct quantity to control–**except**–the center of mass of the real system will vary with the mass and geometry of the rider. Engineering is hard!

For each of these outputs, $\dot{x} = Ax + bu, y = cx$ creates a SISO state-variable model for the Segway.

Note: When the Segway is upright and not wobbling wildly, $\dot{\theta}$, the angular velocity of the platform plus rider, will be very small compared to the angular velocity of the wheels. Hence, one may posit that the three horizontal speeds defined above are the same for all intents and purposes.

10.3 The Laplace Transform and an Algebraic Look at Linear ODEs

The Laplace transform is a powerful mathematical tool that plays a crucial role in the analysis and solution of linear time-invariant (LTI) ordinary differential equations (ODEs). Its ability to transform equations involving derivatives with respect to time into algebraic equations in the Laplace variable, s (aka the complex frequency domain), simplifies the operations of differentiation and integration in the time domain to mere multiplication and division in the frequency domain. Furthermore, the solution can be brought back from the frequency domain to the original time domain by taking the inverse Laplace transform. This technique is especially useful in engineering disciplines such as control systems, circuits, signal processing, and communications, where LTI systems are common.

The Laplace Transform

A function of time $f : \mathbb{R} \to \mathbb{R}$ is said to be **right-sided** if $f(t) = 0$ for all $t < 0$.

Definition 10.6. *For a right-sided piecewise continuous function, $f(t)$, its **Laplace transform**[a] is the improper integral,*

$$\begin{aligned} F(s) &:= \mathcal{L}\{f(t)\} := \int_{0^-}^{\infty} e^{-st} f(t)\, dt \\ &:= \lim_{a \to 0^-,\, b \to \infty} \int_a^b e^{-st} f(t)\, dt, \end{aligned} \tag{10.16}$$

*where $s = \sigma + \mathrm{i}\omega \in \mathbb{C}$ is called the **Laplace transform variable**.*

Note: *We use the notation "$\boldsymbol{f(t) \longleftrightarrow F(s)}$" to denote that $F(s)$ is the Laplace transform of $f(t)$ and $f(t)$ is the **inverse Laplace transform** of $F(s)$. The two functions $f(t)$ and $F(s)$ are also called **Laplace transform pairs**.*

We will see soon that the lower limit, 0^- (zero minus), is there to handle initial conditions for ODEs. Because the upper limit on the Laplace transform is infinite, the integral does not always exist. We address this issue next.

Definition 10.7. *The **region of convergence (ROC)** of $F(s)$ is*

$$\begin{aligned} ROC :=& \{s \in \mathbb{C} \mid e^{-st} f(t) \ \ \textit{is absolutely integrable}\} \\ =& \{s \in \mathbb{C} \mid \int_{0^-}^{\infty} |e^{-st} f(t)|\, dt < \infty\}. \end{aligned} \tag{10.17}$$

Note: *Because $s = \sigma + \mathrm{i}\omega$, and because $|e^{-(\sigma+\mathrm{i}\omega)t}| = |e^{-\sigma t} \cdot e^{-\mathrm{i}\omega t}| = e^{-\sigma t}$, the ROC can also be written as*

$$ROC := \{s \in \mathbb{C} \mid \int_{0^-}^{\infty} e^{-\sigma t} |f(t)|\, dt < \infty\}. \tag{10.18}$$

[a]Technically, this is the **unilateral Laplace transform**. There exists also a bilateral Laplace transform; because it is not used much in engineering, we do not introduce it.

Remark 10.8. ***(Optional Read)*** *The Laplace transform is a doubly improper integral due to the two limits. In theory, this implies that significant care is required when evaluating the integral. In practice, for all the functions that are important to engineers, we use Table 10.2 (or a larger version of it) to compute the Laplace transform and its inverse. As stated above, the use of 0^- (zero minus) as the lower limit is mainly due to the way that the Laplace transform interacts with the initial conditions of a linear ODE. In fact, the initial conditions are the limits from the left of $y(t)$ and its derivatives at $t = 0$. A second reason is the **unit impulse "function"**, the first function listed in Table 10.2. A course on signals and systems would delve more deeply into the motivation for using 0^-. Our goal is to provide more of a "Quick-Reference*

Guide" so that we can rapidly appreciate the power and insight unleashed by the Laplace transform before dedicating an entire course to its mastery.

Proposition 10.9: Two Key Properties of the Laplace Transform

(a) For all functions $f(t)$ of the form

$$t^n\, e^{at}\, u_{\rm stp}(t),$$
$$t^n\, e^{at}\, \sin(\omega t)\, u_{\rm stp}(t),$$

and

$$t^n\, e^{at}\, \cos(\omega t)\, u_{\rm stp}(t),$$

the region of convergence ROC of $F(s) = \mathcal{L}\{f(t)\}$ is

$$\text{ROC} = \{s \in \mathbb{C} \mid \text{real}(s - a) > 0\} = \{\sigma + \mathbf{i}\omega \in \mathbb{C} \mid \sigma > a\}.$$

(b) Suppose that $f(t)$ is continuously differentiable and both $f(t)$ and $f'(t)$ have Laplace transforms. Then,

$$\mathcal{L}\{f'(t)\} = sF(s) - f(0^-),$$

for $F(s) = \mathcal{L}\{f(t)\}$ and s in the intersection of their respective regions of convergence. In other words, differentiation in the Laplace domain corresponds to multiplication by the Laplace variable s.

The Laplace transforms of high-order derivatives are given in Table 10.1.

Note: For a SISO linear system, 0^- comes from the initial conditions on the output signal, $y(t)$. For the input signal, $u(t)$, its value at 0^- is ALWAYS zero. We have consistently multiplied input signals by the unit step function, $u_{\rm stp}(t)$, so that $\lim_{t\to 0^-} u^{(k)}(t) = 0$, for $k = 0, 1, 2, \ldots$

(Optional Read) Proof: (a) is immediate from Example 8.13.

To establish (b), we use integration by parts: $\int u\, dv = u \cdot v - \int v\, du$. Hence,

$$\begin{aligned}
\mathcal{L}\{f'(t)\} &= \int_{0^-}^{\infty} f'(t) e^{-st}\, dt \\
&= \int_{0^-}^{\infty} \underbrace{e^{-st}}_{u} \cdot \underbrace{f'(t)\, dt}_{dv} \\
&= \underbrace{e^{-st}}_{u} \cdot \underbrace{f(t)}_{v} \Big|_{t=0^-}^{t=\infty} - \int_{0^-}^{\infty} \underbrace{f(t)}_{v} \cdot \underbrace{(-s) \cdot e^{-st}\, dt}_{du} \\
&= \underbrace{e^{-st} \cdot f(t) \Big|_{t=0^-}^{t=\infty}}_{\text{must be careful here}} + \underbrace{s \int_{0^-}^{\infty} f(t) \cdot e^{-st}\, dt}_{s \cdot F(s)}.
\end{aligned}$$

The second term of the last row is $s \cdot F(s)$, the Laplace variable s times the Laplace transform of $f(t)$. As indicated, the first term takes a bit of care, but we covered the key technical part in Prop. 8.12-(b). For $s \in$ ROC of both Laplace transforms,

$$\lim_{t\to\infty} e^{-st} \cdot f(t) = 0.$$

Therefore, $e^{-st} \cdot f(t) \Big|_{t=0^-}^{t=\infty} = 0 - f(0^-)$, yielding

$$\mathcal{L}\{f'(t)\} = sF(s) - f(0^-).$$

Let us consider one of the higher derivatives. If we use the notation, $f''(t) = (f'(t))'$, then we have

$$\mathcal{L}\{f''(t)\} = \mathcal{L}\{(f'(t))'\} = s\mathcal{L}\{f'(t)\} - f'(0^-) = s\left(sF(s) - f(0^-)\right) - f'(0^-) = s^2 F(s) - sf(0^-) - f'(0^-),$$

assuming all indicated Laplace transforms exist. From here, a proof by induction gives the remainder of Table 10.1. ■

Table 10.1: Laplace Transforms of Derivatives with Nonzero Initial Conditions, assuming all indicated Laplace transforms exist.

Time Domain	Laplace Domain
$f'(t)$	$sF(s) - f(0^-)$
$f''(t)$	$s^2F(s) - sf(0^-) - f'(0^-)$
$f'''(t)$	$s^3F(s) - s^2f(0^-) - sf'(0^-) - f''(0^-)$
$\vdots$	$\vdots$
$f^{(n)}(t)$	$s^nF(s) - s^{n-1}f(0^-) - s^{n-2}f'(0^-) - \cdots - f^{(n-1)}(0^-)$
$f^{(n)}(t)$	$s^nF(s)$ when all initial conditions are zero.

Very Important: An input signal $u(t)$ never has (nonzero) initial conditions. Only output signals $y(t)$ can have nonzero initial conditions. In other symbols, for all input signals, $u(0^-) := \lim_{t\to 0^-} u(t) = 0$ and for all $k = 1, 2, \ldots$, $u^{(k)}(0^-) := \lim_{t\to 0^-} u^{(k)}(t) = 0$.

10.3.1 Common Laplace Transform Pairs

10.3.2 (Optional Read) How to Compute Laplace Transform Pairs by Hand

This section is an optional read because in practice we seldom compute Laplace transforms by hand, and in fact, even though they can be computed symbolically, that is rarely done either! Instead, we simply use a table.

Example 10.10. *Determine for which values of the complex variable $s = \sigma + \mathrm{i}\omega$ the following limits exist and are **zero**. Use the notation* real(s) *to denote σ, the real part of the complex number s, and* imag(s) *to denote ω, the imaginary part of the complex number s.*

(a) $\lim_{t\to\infty} e^{st}$

(b) $\lim_{t\to\infty} e^{-st}$

(c) $\lim_{t\to\infty} e^{-(s+a)t}$

(d) $\lim_{t\to\infty} t\,e^{-st}$

Solutions: For use in the following solutions, we note that

$$e^{st} = e^{(\sigma+\mathrm{i}\omega)t} = e^{\sigma t} \cdot e^{\mathrm{i}\omega t} = e^{\sigma t} \cdot \cos(\omega t) + \mathrm{i}\, e^{\sigma t} \cdot \sin(\omega t).$$

Hence, $\lim_{t\to\infty} e^{st}$ exists if, and only if, the two limits $\lim_{t\to\infty} e^{\sigma t} \cdot \cos(\omega t)$ and $\lim_{t\to\infty} e^{\sigma t} \cdot \sin(\omega t)$ both exist.

(a) **Ans.** $\lim_{t\to\infty} e^{st} = \lim_{t\to\infty} e^{\sigma t} \cdot e^{\mathrm{i}\omega t} = \begin{cases} 0 & \mathrm{real}(s) < 0 \\ 1 & s = 0 \\ \text{does not exist} & \mathrm{real}(s) \geq 0,\ \mathrm{imag}(s) \neq 0 \\ \infty & \mathrm{real}(s) > 0,\ \mathrm{imag}(s) = 0. \end{cases}$

The limit does not exist when real$(s) \geq 0$, imag$(s) \neq 0$ because $\lim_{t\to\infty} e^{\sigma t} \cdot \cos(\omega t)$ "oscillates" as $t \to \infty$; more precisely, for all $T > 0$, there exist $t_1, t_2 \geq T$ such that $e^{\sigma t_1} \cdot \cos(\omega t_1) = +1$ and $e^{\sigma t_2} \cdot \cos(\omega t_2) = -1$ as illustrated in the figures below.

Table 10.2: Common Laplace Transform Pairs. The unit step function $u_{\rm stp}(t)$ is included to emphasize that the Laplace transforms are only valid for signals that start at or after time $t = 0$; in other words, the signals must be zero for $t < 0$. In engineering, the unit impulse is typically defined as "$\delta(t) := \lim_{\epsilon \to 0+} \frac{1}{\epsilon} \text{rect}\left(\frac{t}{\epsilon}\right)$," whereas in mathematics, it is a "distribution." The unit impulse can be used to model impacts, such as a robot arm hitting a (rigid) obstacle.

Function $f(t), t \geq 0$	**Laplace Transform** $F(s)$	**ROC** (can be ignored)
$\delta(t)$	1	all s
$u_{\rm stp}(t)$	$\frac{1}{s}$	real(s) > 0
$t^n\, u_{\rm stp}(t)$	$\frac{n!}{s^{n+1}}$	real(s) > 0
$e^{at}\, u_{\rm stp}(t)$	$\frac{1}{s-a}$	real(s − a) > 0
$t^n e^{at}\, u_{\rm stp}(t)$	$\frac{n!}{(s-a)^{n+1}}$	real(s − a) > 0
$\sin(\omega t)\, u_{\rm stp}(t)$	$\frac{\omega}{s^2+\omega^2}$	real(s) > 0
$\cos(\omega t)\, u_{\rm stp}(t)$	$\frac{s}{s^2+\omega^2}$	real(s) > 0
$e^{at}\sin(\omega t)\, u_{\rm stp}(t)$	$\frac{\omega}{(s-a)^2+\omega^2}$	real(s − a) > 0
$e^{at}\cos(\omega t)\, u_{\rm stp}(t)$	$\frac{s-a}{(s-a)^2+\omega^2}$	real(s − a) > 0
$t\sin(\omega t)\, u_{\rm stp}(t)$	$\frac{2\omega s}{(s^2+\omega^2)^2}$	real(s) > 0
$t\cos(\omega t)\, u_{\rm stp}(t)$	$\frac{s^2-\omega^2}{(s^2+\omega^2)^2}$	real(s) > 0
$f(t-b)\cdot u_{\rm stp}(t-b), b \geq 0$	$e^{-bs}F(s)$	ROC of $F(s)$

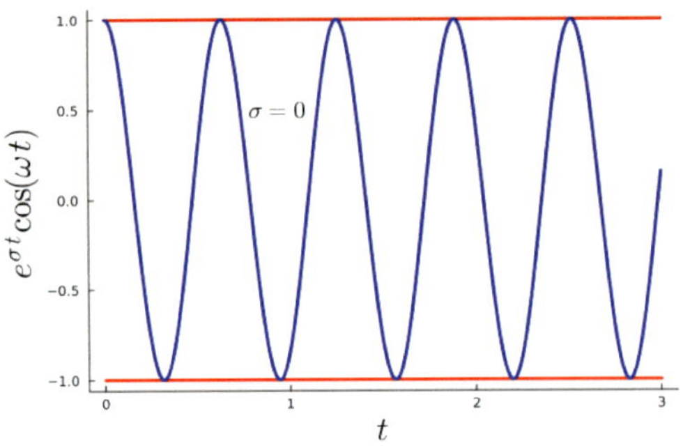

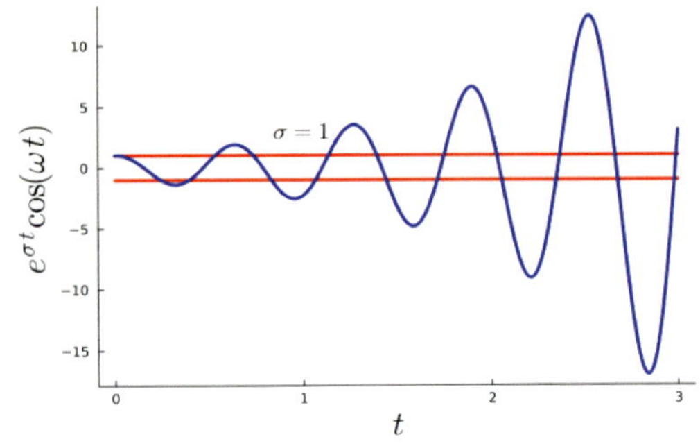

(b) **Ans.** $\lim_{t\to\infty} e^{-st} = \lim_{t\to\infty} e^{-\sigma t}\cdot e^{-\mathrm{i}\omega t} = \begin{cases} 0 & \text{real}(s) > 0 \\ 1 & s = 0 \\ \text{does not exist} & \text{real}(s) \leq 0,\ \text{imag}(s) \neq 0 \\ \infty & \text{real}(s) < 0,\ \text{imag}(s) = 0. \end{cases}$

(c) **Ans.**

$$\lim_{t\to\infty} e^{-(\sigma+a)t}\cdot e^{-\mathrm{i}\omega t} = \begin{cases} 0 & \text{real}(s) + a > 0 \\ 1 & s + a = 0 \\ \text{does not exist} & \text{real}(s) + a \leq 0,\ \text{imag}(s) \neq 0 \\ \infty & \text{real}(s) + a < 0,\ \text{imag}(s) = 0 \end{cases}$$
$$= \begin{cases} 0 & \text{real}(s) > -a \\ 1 & s = -a \\ \text{does not exist} & \text{real}(s) \leq -a,\ \text{imag}(s) \neq 0 \\ \infty & \text{real}(s) < -a,\ \text{imag}(s) = 0 \end{cases}$$

(d) **Ans.** $\lim_{t\to\infty} t\,e^{-st} = \lim_{t\to\infty} t\,e^{-\sigma t}\cdot e^{-\mathrm{i}\omega t} = \begin{cases} 0 & \text{real}(s) > 0 \\ \infty & s = 0 \\ \text{does not exist} & \text{real}(s) \leq 0,\ \text{imag}(s) \neq 0 \\ \infty & \text{real}(s) \leq 0,\ \text{imag}(s) = 0. \end{cases}$

Here, L'Hôpital's Rule is useful. Suppose that $\sigma < 0$. Then, $te^{\sigma t}$ presents us with infinity times zero in the limit, which we can rewrite as infinity over infinity, $\frac{t}{e^{-\sigma t}} = \frac{t}{e^{|\sigma| t}}$, because $\sigma < 0$. Hence, applying L'Hôpital's Rule gives:

$$\lim_{t\to\infty} te^{\sigma t} = \lim_{t\to\infty} \frac{t}{e^{|\sigma|t}} = \lim_{t\to\infty} \frac{1}{|\sigma|\,e^{|\sigma|t}} = 0.$$

■

Example 10.11. *Compute the Laplace transforms of the following signals, taking into account the limits computed in Example 10.10.*

(a) $u_{\text{stp}}(t)$

(b) $e^{-at}u_{\text{stp}}(t)$

(c) $tu_{\text{stp}}(t)$.

Solutions:

(a) **Ans.** $u_{\text{stp}}(t) \longleftrightarrow \frac{1}{s}$

Using antiderivatives,

$$\begin{aligned}\mathcal{L}\{u_{\rm stp}(t)\} &= \int_{0^-}^{\infty} e^{-st} u_{\rm stp}(t)\, dt \\ &= \int_{0}^{\infty} e^{-st}\, dt \quad \text{(applying definition of } u_{\rm stp}(t)\text{)} \\ &= \frac{-1}{s} e^{-st}\Big|_{t=0}^{t=\infty} \\ &= \frac{-1}{s}(0-1) \quad \mathrm{real}(s) > 0 \\ &= \frac{1}{s} \quad \mathrm{real}(s) > 0.\end{aligned}$$

The above calculation shows that $\mathcal{L}\{u_{\rm stp}(t)\}$ exists and is equal to $\frac{1}{s}$ for all $\{s \in \mathbb{C} \mid \mathrm{real}(s) > 0\}$. Recall that the set of all values in the complex plane for which the Laplace transform of a function exists is called its **region of convergence**. This set is very important for the bilateral transform, while it can be ignored completely for the unilateral Laplace transform. Why is that? It is because of Prop. 10.9-(a), which essentially states that the ROC can be computed by inspection for functions that commonly occur in engineering and STEM.

(b) **Ans.** $e^{-at} u_{\rm stp}(t) \longleftrightarrow \frac{1}{s+a}$

Using antiderivatives,

$$\begin{aligned}\mathcal{L}\{u_{\rm stp}(t)\} &= \int_{0^-}^{\infty} e^{-st} e^{-at} u_{\rm stp}(t)\, dt \\ &= \int_{0}^{\infty} e^{-(s+a)t}\, dt \\ &= \frac{-1}{s+a} e^{-(s+a)t}\Big|_{t=0}^{t=\infty} \\ &= \frac{-1}{s+a}(0-1) \quad \mathrm{real}(s) > -a \\ &= \frac{1}{s+a} \quad \mathrm{real}(s) > -a.\end{aligned}$$

(c) **Ans.** $t u_{\rm stp}(t) \longleftrightarrow \frac{1}{s^2}$

Using integration by parts, with $u = t$ and $dv = e^{-st}\, dt \implies v = -\frac{1}{s} e^{-st}$,

$$\begin{aligned}\mathcal{L}\{u_{\rm stp}(t)\} &= \int_{0^-}^{\infty} e^{-st} t u_{\rm stp}(t)\, dt \\ &= \underbrace{\int_{0}^{\infty} t e^{-st}\, dt}_{\int u\, dv} \\ &= \underbrace{-\frac{t}{s} e^{-st}}_{u \cdot v}\Big|_{t=0}^{t=\infty} - \underbrace{\int_{0}^{\infty} \frac{-1}{s} e^{-st}\, dt}_{\int v\, du} \\ &= -\frac{t}{s} e^{-st}\Big|_{t=0}^{t=\infty} + \frac{1}{s} \cdot \int_{0}^{\infty} e^{-st}\, dt \\ &= (0-0) + \frac{-1}{s^2} e^{-st}\Big|_{t=0}^{t=\infty} \\ &= \frac{-1}{s^2}(0-1) \quad \mathrm{real}(s) > 0 \\ &= \frac{1}{s^2} \quad \mathrm{real}(s) > 0.\end{aligned}$$

■

10.3.3 Computing Inverse Laplace Transforms by Hand

While we use tables to compute Laplace transforms, the task of computing inverse Laplace transforms requires more work. For example, the Laplace transform

$$F(s) = \frac{s+1}{(s+2)(s+3)} \tag{10.19}$$

is nowhere to be found in Table 10.2. However, it is easy to verify that

$$\frac{s+1}{(s+2)(s+3)} = \frac{2}{s+3} - \frac{1}{s+2}, \tag{10.20}$$

and hence, by Table 10.2, the inverse Laplace transform is:

$$\begin{aligned} f(t) &= \mathcal{L}^{-1}\{\frac{2}{s+3}\} - \mathcal{L}^{-1}\{\frac{1}{s+2}\} \\ &= 2e^{-3t}u_{\text{stp}}(t) - e^{-2t}u_{\text{stp}}(t). \end{aligned} \tag{10.21}$$

The process of going from (10.19) to (10.20) is called a **partial fraction expansion**, or PFE for short. We first encountered PFEs in Section 7.3.5.

Proposition 10.12: Partial Fraction Expansion

A proper rational function $F(s)$ with distinct roots $r_1, r_2, \ldots, r_n$ can be expanded as

$$F(s) = \sum_{j=1}^{n} \frac{k_j}{s - r_j},$$

where the coefficients k_j are called the **residues** of $F(s)$ at the **poles** r_j. The residues k_j can be computed using:

$$k_j = \lim_{s \to r_j} F(s)(s - r_j).$$

Remark: If r_j is complex, the associated residue k_j will generally be a complex number. Its complex conjugate k_j^* corresponds to the residue at the conjugate pole r_j^*.

Example 10.13. *Compute the inverse Laplace transforms of the following rational functions, using the method of partial fraction expansion, where necessary.*

(a) $\frac{1}{s^2+2s+2}$

(b) $\frac{s+3}{(s+1)(s-2)}$

(c) $\frac{s^2+4}{(s^2+1)(s+3)}$

Solutions:

(a) **Ans.** $\frac{1}{s^2+2s+2} \longleftrightarrow e^{-t}\sin(t)u_{\text{stp}}(t)$

Rewrite the denominator as a completed square:

$$s^2 + 2s + 2 = (s+1)^2 + 1.$$

The Laplace transform becomes:

$$\frac{1}{s^2+2s+2} = \frac{1}{(s+1)^2+1}.$$

Using the table of Laplace transforms:

$$\mathcal{L}^{-1}\left\{\frac{1}{(s+1)^2+1}\right\} = e^{-t}\sin(t)u_{\text{stp}}(t).$$

(b) **Ans.** $\frac{s+3}{(s+1)(s-2)} \longleftrightarrow \left(-\frac{2}{3}e^{-t} + \frac{5}{3}e^{2t}\right) u_{\rm stp}(t)$

Using partial fraction expansions,

$$\frac{s+3}{(s+1)(s-2)} = \frac{k_1}{s+1} + \frac{k_2}{s-2}.$$

Solving for k_1 and k_2,

$$k_1 = \lim_{s\to -1} \frac{(s+3)(s+1)}{(s+1)(s-2)} = \lim_{s\to -1} \frac{s+3}{s-2} = -\frac{2}{3}, \quad k_2 = \lim_{s\to 2} \frac{(s+3)(s-2)}{(s+1)(s-2)} = \lim_{s\to 2} \frac{s+3}{s+1} = \frac{5}{3}.$$

Substituting back,

$$\frac{s+3}{(s+1)(s-2)} = \frac{-\frac{2}{3}}{s+1} + \frac{\frac{5}{3}}{s-2}.$$

Using the inverse Laplace transform:

$$\mathcal{L}^{-1}\left\{\frac{-\frac{2}{3}}{s+1}\right\} = -\frac{2}{3}e^{-t}u_{\rm stp}(t), \quad \mathcal{L}^{-1}\left\{\frac{\frac{5}{3}}{s-2}\right\} = \frac{5}{3}e^{2t}u_{\rm stp}(t).$$

Therefore:

$$f(t) = \left(-\frac{2}{3}e^{-t} + \frac{5}{3}e^{2t}\right) u_{\rm stp}(t).$$

(c) **Ans.** $\frac{s^2+4}{(s^2+1)(s+3)} \longleftrightarrow \left(-\frac{3}{10}\cos(t) - \frac{9}{10}\sin(t) + \frac{13}{10}e^{-3t}\right) u_{\rm stp}(t)$

This problem is the most complicated, so we show all of the intermediate steps. **By the time we are done, you will be ready for software to compute inverse Laplace transforms.** We express

$$\frac{s^2+4}{(s^2+1)(s+3)} = \frac{k_1}{s-\mathbf{i}} + \frac{k_2}{s+\mathbf{i}} + \frac{k_3}{s+3}$$

as a partial fraction expansion.

Step 1: Solve for k_1, k_2, and k_3:

$$\begin{aligned} k_1 &= \lim_{s\to \mathbf{i}} \frac{(s^2+4)(s-\mathbf{i})}{(s^2+1)(s+3)} \\ &= \lim_{s\to \mathbf{i}} \frac{(s^2+4)(s-\mathbf{i})}{(s+\mathbf{i})(s-\mathbf{i})(s+3)} \\ &= \lim_{s\to \mathbf{i}} \frac{(s^2+4)}{(s+\mathbf{i})(s+3)} \\ &= \frac{\mathbf{i}^2+4}{(\mathbf{i}+\mathbf{i})(\mathbf{i}+3)} \\ &= \frac{-1+4}{2\mathbf{i}(\mathbf{i}+3)} \\ &= \frac{3}{2\mathbf{i}(\mathbf{i}+3)} \\ &= \frac{3}{-2+6\mathbf{i}} \\ &= \frac{3(-2-6\mathbf{i})}{(-2+6\mathbf{i})(-2-6\mathbf{i})} \\ &= \frac{-6-18\mathbf{i}}{40} \\ &= -\frac{3}{20} - \frac{9\mathbf{i}}{20}. \end{aligned}$$

$$
\begin{aligned}
k_2 &= \lim_{s\to-\mathbf{i}} \frac{(s^2+4)(s+\mathbf{i})}{(s^2+1)(s+3)} \\
&= \lim_{s\to-\mathbf{i}} \frac{(s^2+4)(s+\mathbf{i})}{(s+\mathbf{i})(s-\mathbf{i})(s+3)} \\
&= \lim_{s\to-\mathbf{i}} \frac{(s^2+4)}{(s-\mathbf{i})(s+3)} \\
&= \frac{(-\mathbf{i})^2+4}{(-\mathbf{i}-\mathbf{i})(-\mathbf{i}+3)} \\
&= \frac{-1+4}{-2\mathbf{i}(-\mathbf{i}+3)} \\
&= \frac{3}{-2\mathbf{i}(-\mathbf{i}+3)} \\
&= \frac{3}{-2-6\mathbf{i}} \\
&= \frac{3(-2+6\mathbf{i})}{(-2-6\mathbf{i})(-2+6\mathbf{i})} \\
&= \frac{-6+18\mathbf{i}}{40} \\
&= -\frac{3}{20}+\frac{9\mathbf{i}}{20} \\
&= k_1^*.
\end{aligned}
$$

$$
\begin{aligned}
k_3 &= \lim_{s\to-3} \frac{(s^2+4)(s+3)}{(s^2+1)(s+3)} \\
&= \lim_{s\to-3} \frac{s^2+4}{s^2+1} \\
&= \frac{((-3)^2+4)}{((-3)^2+1)} \\
&= \frac{(9+4)}{(9+1)} \\
&= \frac{13}{10}.
\end{aligned}
$$

Step 2: Substitute back: The partial fraction expansion becomes,

$$
\frac{s^2+4}{(s^2+1)(s+3)} = \frac{-\frac{3}{20}-\frac{9\mathbf{i}}{20}}{s-\mathbf{i}} + \frac{-\frac{3}{20}+\frac{9\mathbf{i}}{20}}{s+\mathbf{i}} + \frac{\frac{13}{10}}{s+3}.
$$

Step 3: Compute the inverse Laplace transform: Using the standard inverse Laplace transforms,

$$
\begin{aligned}
\mathcal{L}^{-1}\left\{\left(-\frac{3}{20}-\frac{9\mathbf{i}}{20}\right)\frac{1}{s-\mathbf{i}}\right\} &= \left(-\frac{3}{20}-\frac{9\mathbf{i}}{20}\right)e^{\mathbf{i}t}u_{\text{stp}}(t), \\
\mathcal{L}^{-1}\left\{\left(-\frac{3}{20}+\frac{9\mathbf{i}}{20}\right)\frac{1}{s+\mathbf{i}}\right\} &= \left(-\frac{3}{20}+\frac{9\mathbf{i}}{20}\right)e^{-\mathbf{i}t}u_{\text{stp}}(t), \\
\mathcal{L}^{-1}\left\{\frac{13}{10}\frac{1}{s+3}\right\} &= \frac{13}{10}e^{-3t}u_{\text{stp}}(t).
\end{aligned}
$$

In the following, we delete the unit step function to save space. We add it back at the end of the computations.

Step 4: Combine complex exponentials: Combine the terms with complex exponentials using Euler's formula:

$$e^{\mathbf{i}t} = \cos(t) + \mathbf{i}\sin(t), \quad e^{-\mathbf{i}t} = \cos(t) - \mathbf{i}\sin(t).$$

Consider

$$\left(-\frac{3}{20} - \frac{9\mathbf{i}}{20}\right)e^{\mathbf{i}t} + \left(-\frac{3}{20} + \frac{9\mathbf{i}}{20}\right)e^{-\mathbf{i}t}.$$

Using Euler's formula gives,

$$\left(-\frac{3}{20} - \frac{9\mathbf{i}}{20}\right)(\cos(t) + \mathbf{i}\sin(t)) + \left(-\frac{3}{20} + \frac{9\mathbf{i}}{20}\right)(\cos(t) - \mathbf{i}\sin(t)).$$

Expanding terms,

$$\left(-\frac{3}{20}\cos(t) - \frac{9\mathbf{i}}{20}\cos(t) - \frac{3\mathbf{i}}{20}\sin(t) - \frac{9}{20}\sin(t)\right) + \left(-\frac{3}{20}\cos(t) + \frac{9\mathbf{i}}{20}\cos(t) + \frac{3\mathbf{i}}{20}\sin(t) - \frac{9}{20}\sin(t)\right).$$

Simplifying the above yields

$$-\frac{6}{20}\cos(t) - \frac{18}{20}\sin(t),$$

and thus

$$-\frac{3}{10}\cos(t) - \frac{9}{10}\sin(t).$$

Step 5: Final result: After combining the inverse Laplace transforms for the complex and real roots, the time-domain function is

$$f(t) = -\frac{3}{10}\cos(t) - \frac{9}{10}\sin(t) + \frac{13}{10}e^{-3t},$$

yielding

$$f(t) = \left(-\frac{3}{10}\cos(t) - \frac{9}{10}\sin(t) + \frac{13}{10}e^{-3t}\right)u_{\text{stp}}(t)$$

once the unit step is added back in.

Note: The amount of algebra in the above solution is typical when performing PFEs by hand. The potential for errors at each step is significant and often frustrating. As we have emphasized throughout the course, however, modern computational tools can handle these algebraic manipulations quickly and reliably; see the solution to Example 10.14. ■

10.3.4 Algebraic Approach to Solving Linear ODEs

Our principal goal is to transform a linear time-invariant ODE into an algebraic equation and then underline the utility of this process. While in engineering practice we rarely use the Laplace transform to compute solutions to ODEs, doing so is the most direct way to observe derivatives being transformed into algebraic quantities. In that light, we give two examples that reduce the act of solving an ODE to dealing with rational functions of the Laplace variable s.

Example 10.14. *Use the Laplace transform to compute the solution to the linear time-invariant ODE*

$$\ddot{y}(t) + 5\dot{y}(t) + 6y(t) = u(t) + 7\dot{u}(t), \tag{10.22}$$

for $u(t) = e^{-t}u_{\text{stp}}(t)$ *and subject to the initial conditions* $y(0^-) = 2$ *and* $\dot{y}(0^-) = 3$.

Solution: **Ans.** $Y(s) = \frac{2(s^2+11s+7)}{s^3+6s^2+11s+6}$ and $y(t) = -3e^{-t}u_{\text{stp}}(t) + 22e^{-2t}u_{\text{stp}}(t) - 17e^{-3t}u_{\text{stp}}(t)$.

Using Table 10.1, we take the Laplace transform of both sides of (10.22), yielding

$$\left(s^2Y(s) - sy(0^-) - \dot{y}(0^-)\right) + 5\left(sY(s) - y(0^-)\right) + 6Y(s) = U(s) + 7sU(s). \tag{10.23}$$

At this point, we could use Table 10.2 to obtain $U(s) = \mathcal{L}\{e^{-t}u_{\text{stp}}(t)\} = \frac{1}{s+1}$, substitute in the given values for $y(0^-)$ and $\dot{y}(0^-)$, and do the algebra with one of the various symbolic manipulation packages (see below) to obtain a closed-form expression for the Laplace transform of the output, $Y(s)$. However, because this is our first example, we will also show how the calculations can be performed by hand because you will be expected to solve the problem that way in other engineering courses.

Continuing with a manual solution,

$$\left(s^2Y(s) - sy(0^-) - \dot{y}(0^-)\right) + 5\left(sY(s) - y(0^-)\right) + 6Y(s) = U(s) + 7sU(s)$$

$$\Updownarrow \text{ (move initial conditions to the right side)}$$

$$\left(s^2Y(s) + 5sY(s) + 6Y(s)\right) = \left(U(s) + 7sU(s)\right) + \left(sy(0^-) + \dot{y}(0^-) + 5y(0^-)\right)$$

$$\Updownarrow \text{ (group terms)}$$

$$\left(s^2 + 5s + 6\right)Y(s) = (1 + 7s)\,U(s) + (s + 5)\,y(0^-) + \dot{y}(0^-)$$

$$\Updownarrow \text{ (divide by } \left(s^2 + 5s + 6\right)\text{)}$$

$$Y(s) = \frac{1 + 7s}{s^2 + 5s + 6}U(s) + \frac{(s+5)\,y(0^-) + \dot{y}(0^-)}{s^2 + 5s + 6},$$

giving us the Laplace transform[1] of the output $Y(s)$ in terms of the input signal and the initial conditions

$$Y(s) = \underbrace{\frac{1 + 7s}{s^2 + 5s + 6}U(s)}_{\text{forced response}} + \underbrace{\frac{(s+5)\,y(0^-) + \dot{y}(0^-)}{s^2 + 5s + 6}}_{\text{initial condition response}} \tag{10.24}$$

As you may imagine, $\frac{1+7s}{s^2+5s+6}U(s)$ is called the **forced response** because it is the part of the output's response that depends on the input signal, $u(t)$. Similarly, $\frac{(s+5)y(0^-)+\dot{y}(0^-)}{s^2+5s+6}$ is called the **initial condition response** because it is the part of the output's response that depends on the initial conditions, $y(0^-)$ and $\dot{y}(0^-)$. The fact that the **total response** is the sum of the forced response and the initial condition response is called the **superposition principle**.

Soon, we will call $G(s) := \frac{7s+1}{s^2+5s+6}$ the **transfer function** of the input-output system. **Note that you can write it down without doing any computations whatsoever.** The denominator comes directly from the left side of the ODE, while the numerator comes from the right side. The name "transfer function" comes from $G(s)$ "transferring the effect of the input, $U(s)$, to the output, $Y(s)$", via

$$Y(s) = G(s)U(s).$$

The next step in the process is to use Table 10.2 to obtain $U(s) = \mathcal{L}\{e^{-t}u_{\text{stp}}(t)\} = \frac{1}{s+1}$ and to substitute in actual values for $y(0^-)$ and $\dot{y}(0^-)$. Because it's not that bad, we will write it out,

$$\begin{aligned}
Y(s) &= \frac{1 + 7s}{s^2 + 5s + 6}\boldsymbol{U(s)} + \frac{(s+5)\,\boldsymbol{y(0^-)} + \boldsymbol{\dot{y}(0^-)}}{s^2 + 5s + 6} \\
&= \frac{1 + 7s}{s^2 + 5s + 6}\,\frac{\mathbf{1}}{\boldsymbol{s+1}} + \frac{(s+5)\,\mathbf{2} + \mathbf{3}}{s^2 + 5s + 6} \\
&= \frac{2\left(s^2 + 11\,s + 7\right)}{s^3 + 6\,s^2 + 11\,s + 6},
\end{aligned}$$

after a bit of algebra.

To compute $y(t)$, we need to evaluate the inverse Laplace transform of $Y(s)$. A quick perusal of Table 10.2 shows that we are out of luck. Such a complicated Laplace transform is nowhere to be seen. As in Example 10.13, **the "trick" is to perform a partial fraction expansion (PFE)** of $Y(s)$.

The key step in computing a PFE is finding the denominator's roots. Here, we have

$$s^3 + 6\cdot s^2 + 11\cdot s + 6 = (s+3)\,(s+2)\,(s+1).$$

Hence, the PFE takes the form

$$\frac{2\cdot\left(s^2 + 11\cdot s + 7\right)}{s^3 + 6\cdot s^2 + 11\cdot s + 6} = \frac{k_1}{s+3} + \frac{k_2}{s+2} + \frac{k_3}{s+1}.$$

[1] Recall that all of the tedious algebra can be done with symbolic manipulation.

Instead of computing the coefficients by hand, we appeal to `SymPy`, yielding

$$Y(s) = \frac{-17}{s+3} + \frac{22}{s+2} - \frac{3}{s+1}.$$

Now, we can compute inverse Laplace transforms of each of these terms via Table 10.2, because

$$\frac{-17}{s+3} \longleftrightarrow -17e^{-3t}u_{\text{stp}}(t)$$

$$\frac{22}{s+2} \longleftrightarrow 22e^{-2t}u_{\text{stp}}(t)$$

$$\frac{-3}{s+1} \longleftrightarrow -3e^{-t}u_{\text{stp}}(t).$$

Therefore,

$$y(t) = -3e^{-t}u_{\text{stp}}(t) + 22e^{-2t}u_{\text{stp}}(t) - 17e^{-3t}u_{\text{stp}}(t).$$

Here is how the entire problem can be solved in code.

```
using SymPy

# Define the symbolic variable
@SymPy.syms s

# Define the input's Laplace transform and the initial conditions
U = 1/(s+1)
yzero = 2.0
dyzero = 3.0

# D(s) Y(s) = RHS

# Define the right-hand side (RHS) expression
RHS = (1+7s)*U + (s+5)*yzero + dyzero

# Define D(s)
D = s^2 + 5*s + 6

# Define Y as the RHS divided by D
Y = RHS / D

# Simplify Y
Y = SymPy.expand(Y) # Forces terms to be multiplied out
Y = SymPy.simplify(Y)
display(Y)

# Compute the partial fraction expansion
PFE = apart(Y, s)
```

Output $Y(s) = \frac{2\cdot\left(s^2+11\cdot s+7\right)}{s^3+6\cdot s^2+11\cdot s+6}$

$PFE = -\frac{17}{s+3} + \frac{22}{s+2} - \frac{3}{s+1}$

It is also possible to compute the inverse Laplace transform in code without first computing a partial fraction expansion. How sweet is that?

```
using PyCall

# Import Python's SymPy
sympy = pyimport("sympy")

# Define the symbols
s, t = sympy.symbols("s t")
```

```
# Define the expression Y(s)
Y = (2.0*s^2 + 22.0*s + 14.0)/(1.0*s^3 + 6.0*s^2 + 11.0*s + 6.0)

# Compute the inverse Laplace transform
y = sympy.inverse_laplace_transform(Y, s, t)

println(y)
```

Output

```
-17.0*exp(-3.0*t)*Heaviside(t) + 22.0*exp(-2.0*t)*Heaviside(t)
- 3.0*exp(-t)*Heaviside(t)
```

Note: `Heaviside(t)` is what we are denoting as $u_{\rm stp}(t)$. Sometimes `sympy` will express the answer in a funky way, with $e^{-3t}\, u_{\rm stp}$ factored out, such as

```
(-3*exp(2*t) + 22*exp(t) - 17)*exp(-3*t)*Heaviside(t)
```

When $e^{-3t}u_{\rm stp}(t)$ is multiplied through the expression in parentheses, the answer we gave above is recovered.

■

Amazing Observation: Suppose we did not really care about the exact form of the output $y(t)$, but only wanted to know that

$$\lim_{t\to\infty} y(t) = 0.$$

In that case, we do not need the exact values of the coefficients k_i. Once we know that the roots of the denominator polynomial have negative real parts, Table 10.2 implies that we have decaying exponentials, and the output converges to zero. Laplace must have been a genius!

Remark 10.15. *In your studies, you will often encounter complex mathematical problems that require symbolic computation, such as solving equations, performing integration, or calculating inverse Laplace transforms. While Julia offers powerful computational capabilities, its native symbolic mathematics packages,* `SymPy.jl` *and* `Symbolics.jl`*, sometimes lack certain advanced features or specific functions found in Python's SymPy library. To bridge this gap, you can use* `PyCall`*, a Julia package that allows one to call Python functions directly from Julia. This approach enables you to leverage the full power and extensive functionality of Python's SymPy, which is a mature library for symbolic mathematics. The integration of Python's capabilities within Julia's environment exemplifies the flexibility and interoperability of modern programming languages, allowing one to choose the best tool for each specific task and seamlessly combine them to enhance your computational experience.*

Example 10.16. *Use Laplace transforms to compute the solution to the SISO linear state-variable control system*

$$\begin{aligned}\underbrace{\begin{bmatrix}\dot{x}_1\\ \dot{x}_2\end{bmatrix}}_{\dot{x}} &= \underbrace{\begin{bmatrix}-5 & 1\\ -6 & 0\end{bmatrix}}_{A}\underbrace{\begin{bmatrix}x_1\\ x_2\end{bmatrix}}_{x} + \underbrace{\begin{bmatrix}7\\ 1\end{bmatrix}}_{b}u\\ y &= \underbrace{\begin{bmatrix}1 & 0\end{bmatrix}}_{c}x,\end{aligned} \tag{10.25}$$

for $u(t) = e^{-t}u_{\rm stp}(t)$ and subject to the initial conditions

$$x(0^-) := \begin{bmatrix}x_1(0^-)\\ x_2(0^-)\end{bmatrix} = \begin{bmatrix}2\\ 13\end{bmatrix}.$$

Solution: Ans. $Y(s) = \frac{2\left(s^2+11s+7\right)}{s^3+6s^2+11s+6}$ and $y(t) = -3e^{-t}u_{\rm stp}(t) + 22e^{-2t}u_{\rm stp}(t) - 17e^{-3t}u_{\rm stp}(t)$.

We perform the solution by hand to explain the **"tricky parts"**, such as, **what could it even mean to apply the Laplace transform to vectors and matrices? Answer: Because everything is linear, applying the Laplace transform row by row works great. We will observe that all the computations can be performed in matrix-vector form. Once we have everything in matrix-vector form, we can code it up and do it in a snap!**

To apply the Laplace transform row by row, we rewrite (10.25) as

$$\begin{aligned} \dot{x}_1 &= -5x_1 + x_2 + 7u \\ \dot{x}_2 &= -6x_1 + u \\ y &= x_1. \end{aligned} \tag{10.26}$$

Applying Laplace transforms line by line gives:

$$\begin{aligned} sX_1(s) - x_1(0^-) &= -5X_1(s) + X_2(s) + 7U(s) \\ sX_2(s) - x_2(0^-) &= -6X_1(s) + U(s) \\ Y(s) &= X_1(s). \end{aligned} \tag{10.27}$$

Moving the initial conditions to the right side yields,

$$\begin{aligned} sX_1(s) &= -5X_1(s) + X_2(s) + 7U(s) + x_1(0^-) \\ sX_2(s) &= -6X_1(s) + U(s) + x_2(0^-) \\ Y(s) &= X_1(s). \end{aligned} \tag{10.28}$$

We now rewrite the above in matrix-vector form. Define

$$X(s) := \begin{bmatrix} X_1(s) \\ X_2(s) \end{bmatrix}. \tag{10.29}$$

Then, we can write the first two rows of (10.28) as

$$\begin{aligned} sX(s) &= AX(s) + bU(s) + x(0^-) \\ &\Updownarrow \\ sX(s) - AX(s) &= bU(s) + x(0^-) \\ &\Updownarrow \\ sI_2X(s) - AX(s) &= bU(s) + x(0^-) \\ &\Updownarrow \\ (sI_2 - A)\, X(s) &= bU(s) + x(0^-), \end{aligned} \tag{10.30}$$

for A and b as defined in (10.25). Moreover, the last row becomes

$$Y(s) = cX(s), \tag{10.31}$$

with c as defined in (10.25). Solving for $X(s)$ in (10.30) gives

$$X(s) = (sI - A)^{-1}\left(bU(s) + x(0^-)\right). \tag{10.32}$$

Substituting (10.32) into (10.31) gives

$$Y(s) = c\,(sI - A)^{-1}\left(bU(s) + x(0^-)\right). \tag{10.33}$$

We observe that the initial conditions look like a modification of the input. The final answer can be rewritten as

$$\boxed{Y(s) = \underbrace{c\,(sI - A)^{-1}\, bU(s)}_{\text{forced response}} + \underbrace{c\,(sI - A)^{-1}\, x(0^-)}_{\text{initial condition response}}.} \tag{10.34}$$

The term multiplying the input, $G(s) := c\,(sI - A)^{-1}\,b$, will soon be called the **transfer function**. While $G(s)$ may not look like a rational function of s, it can be proven to be one using the adjoint method of computing the matrix inverse. We will compute the inverse symbolically to show the desired property for small examples.

Now that we have the general formula, we will do the computation of the above in code!

```
using SymPy, LinearAlgebra

# Define the symbolic variable
@SymPy.syms s

# Linear state variable model
A=[-5 1;-6 0.0]
b=[7.0; 1]
c=[1.0 0]

# Initial Conditions
x0=[2.0; 13]

# Laplace transform of the input
U = 1/(s+1)

# Compute Y(s)
Y = c*inv(s*I(2) - A)*b*U +  c*inv(s*I(2) - A)*x0

# Y is a 1 x 1 vector, so in Julia, we extract it
Y = Y[1]

# Simplify Y
Y = SymPy.expand(Y) # Forces terms to be multiplied out
Y = SymPy.simplify(Y)
display(Y)
println(Y)


# Compute the partial fraction expansion
PFE = apart(Y, s)
```

Output

$Y(s) = \frac{72 \cdot s^2 + 792 \cdot s + 504}{36 \cdot s^3 + 216 \cdot s^2 + 396 \cdot s + 216}$

```
(72.0*s^2 + 792.0*s + 504.0)/(36.0*s^3 + 216.0*s^2 + 396.0*s + 216.0)
```

$\text{PFE} = \frac{-3.000}{s + 1.000} + \frac{11.000}{0.500 \cdot s + 1.000} - \frac{5.667}{0.333 \cdot s + 1.000}$

If we look carefully, we can see that SymPy has not removed a factor of 36 from the numerator and denominator of $Y(s)$. Moreover, in the PFE, if we multiplied the second term by $\frac{2}{2}$ and the third term by $\frac{3}{3}$ we would obtain simpler answers, but we leave them as such because it does not matter.

Next, we obtain the inverse Laplace transform.

```
using PyCall

# Import Python's SymPy
sympy = pyimport("sympy")

# Define the symbols
s, t = sympy.symbols("s t")

# Define the expression Y(s)
Y = (72.0*s^2 + 792.0*s + 504.0)/(36.0*s^3 + 216.0*s^2 + 396.0*s + 216.0)

# Compute the inverse Laplace transform
y = sympy.inverse_laplace_transform(Y, s, t)

println(y)
```

Output

```
-17.0*exp(-3.0*t)*Heaviside(t) + 22.0*exp(-2.0*t)*Heaviside(t)
- 3.0*exp(-t)*Heaviside(t)
```

which is the same answer we obtained in the previous example. **Input-output and state-variable methods are really two sides of the same coin.**

10.3.5 Transfer Functions

A transfer function for an LTI (Linear Time-Invariant) input-output system is a Laplace (aka frequency) domain representation that describes the relationship between the input and output of a system **when the initial conditions are identically zero.** It is a ratio of two polynomials in terms of the Laplace variable s. The transfer function allows for the study of system stability without directly solving differential equations in the time domain. By examining the transfer function's "poles and zeros" (defined shortly), we can infer important properties about the system's response to a wide variety of inputs.

Transfer Function of a SISO Linear System

Definition 10.17. *The transfer function of the linear input output model*

$$y^{(n)} + a_{n-1}y^{(n-1)} + a_{n-2}y^{(n-2)} + \cdots + a_1\dot{y} + a_0 y = b_0 u + b_1 \dot{u} + \cdots + b_m u^{(m)}$$

*is the **rational function***

$$G(s) := \frac{Y(s)}{U(s)} = \frac{b_m s^m + \cdots + b_1 s + b_0}{s^n + a_{n-1}s^{n-1} + \cdots + a_1 s + a_0}. \tag{10.35}$$

*With this definition, $Y(s) = G(s) \cdot U(s)$ is precisely the Laplace transform of the linear system (10.1) when all **initial conditions are zero**.*

Note: When we need the transfer function of an input-output system, we simply write it down by inspection. No calculations are required.

Definition 10.18. *The transfer function of the SISO linear state-variable model $\dot{x} = Ax + bu$, $y = cx$ is*

$$G(s) := \frac{Y(s)}{U(s)} = c(sI - A)^{-1}b. \tag{10.36}$$

*With this definition, $Y(s) = G(s) \cdot U(s)$ is precisely the Laplace transform of the linear state-variable control system (10.2) when all **initial conditions are zero**.*

Note: Because the transfer function of a SISO linear state-variable model cannot be determined by inspection, for anything beyond 2×2 A-matrices, we use code to compute it.

10.3.6 The Segway Transporter à la Laplace

We derive a transfer function representation for the Segway from the state-variable model given in Fact 10.5.

Example 10.19. *For the Segway transporter, use one of the symbolic computation packages in Julia to compute the Laplace transform of the state-variable model $\dot{x} = Ax + bu$, assuming zero initial conditions,*

$$H(s) := \frac{X(s)}{U(s)} = (sI_4 - A)^{-1}b.$$

In addition, interpret the components of $H(s)$.

Solution: Ans. $H(s) = \begin{bmatrix} \frac{-21.468}{s^{2.000}-50.937} \\ \frac{80.227 \cdot s^2 - 594.021}{s^2 \cdot (s^2-50.937)} \\ \frac{-21.468 \cdot s}{s^2-50.937} \\ \frac{80.227 \cdot s^2 - 594.021}{s \cdot (s^2-50.937)} \end{bmatrix} = \begin{bmatrix} \frac{\theta(s)}{U(s)} \\ \frac{\varphi(s)}{U(s)} \\ \frac{\dot{\theta}(s)}{U(s)} \\ \frac{\dot{\varphi}(s)}{U(s)} \end{bmatrix}$.

The components of $H(s)$ are as labeled because the components of the state vector x are θ, φ, $\dot{\theta}$, and $\dot{\varphi}$. Moreover, because $\mathcal{L}\{\dot{\theta}\} = s \cdot \mathcal{L}\{\theta\}$ and $\mathcal{L}\{\dot{\varphi}\} = s \cdot \mathcal{L}\{\varphi\}$, we expect to see this reflected in the transfer functions that comprise $\frac{X(s)}{U(s)}$. Indeed, $\frac{\dot{\theta}(s)}{U(s)} = s \cdot \frac{\theta(s)}{U(s)}$ and $\frac{\dot{\varphi}(s)}{U(s)} = s \cdot \frac{\varphi(s)}{U(s)}$. Equivalently,

$$\frac{\theta(s)}{U(s)} = \frac{1}{s} \cdot \frac{\dot{\theta}(s)}{U(s)} \text{ and } \frac{\varphi(s)}{U(s)} = \frac{1}{s} \cdot \frac{\dot{\varphi}(s)}{U(s)}.$$

In other words, the Laplace variable s represents differentiation, and $\frac{1}{s}$ represents integration.

The following code computes the linearized state-variable model and the "transfer vector", $H(s) := \frac{X(s)}{U(s)} = (sI_4 - A)^{-1}b$. Here, we are using `SymPy`, but it could just as easily be done with `Symbolics` or `PyCall`. So many decisions....

```julia
using SymPy, LinearAlgebra

# SEGWAY PARAMETERS
function modelParameters(outSelect=0)
    g = 9.81 # m/s^2
    m_wh = 6 # kg mass of 2 wheels
    m_pend = 45+75 # kg mass of the person, platform, and the handlebar
    L = (45*.3 + 75*0.85)/(45+75) # m CoM for the person plus platform
        # measured from the axles of the wheels
    r_wh = 0.25 # m, Wheel radius
    J_pend = m_pend*L^2/6 # kg m^2 moment of inertia about
                # the center of mass
    J_wh = 0.1 # kg m^2 moment of inertia of each wheel
    J_rotor = 1e-4 # kg m^2 per rotor, one for each wheel
    N = 50 # gear ratio between each motor rotor and wheel
    if outSelect > 0
        # named tuple
        return (g=g, m_wh=m_wh, m_pend=m_pend, L=L, r_wh=r_wh, J_pend=J_pend,
                J_wh=J_wh, J_rotor=J_rotor, N=N)
    else # standard tuple
        return g, m_wh, m_pend, L, r_wh, J_pend, J_wh, J_rotor, N
    end
end


# bring the model into the workspace for later use
include("dyn_mod_planarSegway.jl")

# Set equilibrium Point
qe = [0.0;0.0]

F = dyn_mod_planarSegway(qe, 0*qe)
D = F.D
JacG = F.JacG
JacComx = F.JacComx
B = F.B

# build the linearized model about the given equilibrium point
A21 = -D\JacG
```

```
n, m = size(B)
A = [zeros(n,n) I(n); A21 zeros(n,n)]
b = [zeros(n,m); F.D \ B]
display(A)
display(b)


# Define the symbolic variable
@SymPy.syms s
H = inv(s*I(4) - A)*b

# Simplify H
H = SymPy.expand.(H) # Forces terms to be multiplied out
H = SymPy.simplify.(H)
#display(H)
println(H)
```

Output

```
4×4 Matrix{Float64}:
    0.0      0.0  1.0  0.0
    0.0      0.0  0.0  1.0
   50.9373   0.0  0.0  0.0
 -162.687    0.0  0.0  0.0

4×1 Matrix{Float64}:
   0.0
   0.0
 -21.467741053946913
  80.22714380349518

Sym[-1093.50791254434/(50.9372602267016*s^2 - 2594.60447940271);
(4086.55090116364*s^2 - 30257.7781854544)/(s^2*(50.9372602267016*s^2
 - 2594.60447940271));
-21.4677410539469*s/(s^2 - 50.9372602267016);
(4086.55090116364*s^2 - 30257.7781854544)/(s*(50.9372602267016*s^2
 - 2594.60447940271));;]
```

■

The primary control task for a Segway is to keep it upright. This means we need to keep the lean angle of the platform plus rider near zero. In fact, for a Segway, maintaining the Segway upright is the only task the feedback controller needs to perform, since it is assumed the rider will regulate speed through their body lean angle. Later in this chapter, we will also design a secondary controller that adjusts the overall lean angle so as to achieve a desired "speed".

Example 10.20. ***(Segway's Transfer Functions)*** *Compute the SISO transfer functions* $G_\theta(s) := \frac{Y_\theta(s)}{U(s)}$ *and* $G_{\rm v}(s) := \frac{Y_{\rm v}(s)}{U(s)}$*, where the latter uses the level-1 notion of speed defined in Fact 10.5.*

Solution: Ans. The code below carries out the computations required to obtain the results in (10.37); they are then presented pictorially in Figure 10.4.

$$\begin{aligned} G_\theta(s) &= \frac{-21.468}{s^2 - 50.937} \\ G_{\rm v}(s) &= \frac{-20.057(s^2 - 7.404)}{s\,(s^2 - 50.937)}. \end{aligned} \tag{10.37}$$

```
using LinearAlgebra, Symbolics

# Set equilibrium Point
qe = [0.0;0.0]
```

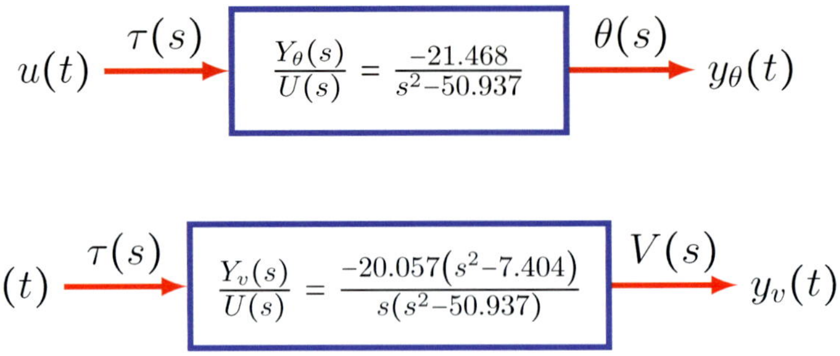

Figure 10.4: The Segway's transfer functions for the lean angle and cart speed are as given, where $u(t)$ is the motor torque. At the end of the chapter, we will seek to control **simultaneously** these two key quantities.

```

F = dyn_mod_planarSegway(qe, 0*qe)
D = F.D
JacG = F.JacG
B = F.B

# build the linearized model about the given equilibrium point
A21 = -D\JacG
# A21 = cleanUp(A21)
n, m = size(B)
A = [zeros(n,n) I(n); A21 zeros(n,n)]
b = [zeros(n,m); F.D \ B]
display(A)
display(b)

# Define the various outputs
c_th = [1.0 0 0 0]

# model params
g, m_wh, m_pend, L, r_wh, J_pend, J_wh, J_rotor, N = modelParameters()

c_v1 = [0 0 0 r_wh]


# Define the symbolic variables and transfer functions
@Symbolics.variables s
G_th = c_th*inv(s*I(4) - A)*b
G_v1 = c_v1*inv(s*I(4) - A)*b

# Simplify
G_th = Symbolics.simplify(G_th[1])
println(G_th)

println(" ")

G_v1 = Symbolics.simplify(G_v1[1])
println(G_v1)
```

Output

```
4×4 Matrix{Float64}:
    0.0        0.0  1.0  0.0
    0.0        0.0  0.0  1.0
   50.9373    -0.0  0.0  0.0
 -162.687     -0.0  0.0  0.0
```

```
4×1 Matrix{Float64}:
   0.0
   0.0
 -21.467741053946913
  80.22714380349518

-21.467741053946913 / (-50.93726022670157 + s^2)

(-148.50513185627278 + 20.056785950873795(s^2)) /
 (-50.93726022670157s + s^3)
```

■

Fact 10.21: Factoring the Segway's Transfer Function

Because transfer functions are rational functions of the Laplace variable s, the rules of ordinary algebra apply to them. While factoring an ODE seems far-fetched at best, expressing polynomials and rational functions in terms of factors is ordinary. Applying this idea to the Segway, its transfer function from motor torque to cart speed can be factored into two terms, as shown below. Indeed, $\frac{-20.057}{-21.468} = 0.934$ to three significant digits, while the other terms are "self-evident".

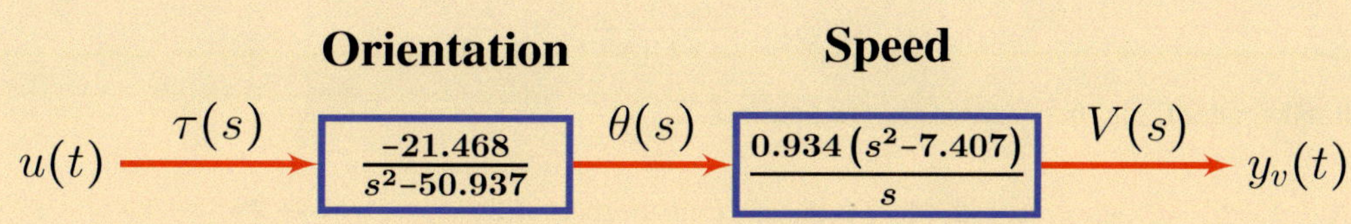

The above factorization shows mathematically how the orientation of the Segway's body directly affects its speed, something that is much harder to discern from the ODE itself. It suggests a two-pronged attack on how to control the Segway: first, stabilize the orientation and, second, use the orientation as a pseudo-actuator for the speed.

10.4 Poles, Zeros, and BIBO Stability

In Section 9.5.8, we studied linear ODEs of the form $\dot{x} = Ax,\ \ x(t_0) = x_0$, and learned that all solutions asymptotically converged to the origin if, and only if, the real parts of the eigenvalues of A have negative real parts. **The main message of this section is that the roots of the denominator of a transfer function work just like eigenvalues: if their real parts are negative, then bounded input signals give rise to bounded output signals.**

10.4.1 Poles and Zeros

A transfer function is **rational** if it can be expressed as the ratio of two polynomials.

Poles and Zeros

Definition 10.22. *For a rational transfer function $G(s) = \frac{N(s)}{D(s)}$, the polynomials $N(s)$ and $D(s)$ are said to be* ***coprime*** *if they have no common roots. In this case, the roots of $N(s)$ are the* ***zeros*** *of $G(s)$ and the roots of $D(s)$ are the* ***poles****.*

Note: If $N(s)$ and $D(s)$ have common factors, hence yielding common roots, the common factors must be canceled before determining the zeros and poles of the transfer function.

Example 10.23. *Determine the zeros and poles of the transfer functions*

(a) $G_1(s) = \frac{2}{s^2-1}$

(b) $G_2(s) = \frac{s+2}{s^2-1}$

(c) $G_3(s) = \frac{s+1}{s^2-1}$

(d) $G_4(s) = \frac{2s-4}{s^2+2s+2}$.

Solution:

(a) $N_1(s) = 2$ has no roots and $D_1(s) = s^2-1 = (s-1)(s+1)$ has roots ± 1. $N_1(s)$ and $D_1(s)$ are therefore coprime; it follows that $G_1(s)$ has no zeros and two poles at ± 1.

(b) $N_2(s) = s+2$ has one root, -2, which is distinct from the roots of $D_2(s) = s^2 - 1$, which are ± 1. Hence, $N_2(s)$ and $D_2(s)$ are coprime and it follows that $G_2(s)$ has one zero at -2 and two poles at ± 1.

(c) $N_3(s) = s+1$ has one root, -1, which is also a root of $D_3(s) = s^2-1$. Hence, $N_3(s)$ and $D_3(s)$ are *not coprime*. To determine the zeros and poles of $G_3(s)$, the common factor of $(s+1)$ is removed, yielding

$$G_3(s) = \frac{s+1}{(s+1)(s-1)} = \frac{1}{s-1}.$$

It follows that $G_3(s)$ has no zeros and one pole at 1.

(d) $N_4(s) = 2(s-2)$ has a root at 2, which is distinct from the roots of $D_4(s) = s^2+2s+2 = (s+1+\mathbf{i})(s+1-\mathbf{i})$, which are $-1 \pm \mathbf{i}$. $N_4(s)$ and $D_4(s)$ are therefore coprime; it follows that $G_4(s)$ has a zero at 2 and two poles at $-1 \pm \mathbf{i}$.

10.4.2 BIBO Stability

We treated bounded functions in Section 4.6. An **input or output signal** $s : [0,\infty) \to \mathbb{R}$ is **bounded** if there exists a constant M such that $|s(t)| \le M$ for all $t \ge 0$. Hence, signals such as constants, $\sin(\omega t)\, u_{\rm stp}(t)$, $\cos(\omega t)\, u_{\rm stp}(t)$, $t^k\, e^{at} u_{\rm stp}(t)$ for $a < 0$, and products of such functions are bounded. Functions such as $e^{at} u_{\rm stp}(t)$ and $t^k\, u_{\rm stp}(t)$ for $a > 0$ and $k > 0$ are unbounded.

BIBO Stability

Definition 10.24. *A system is* bounded-input bounded-output *stable, or* ***BIBO stable*** *for short, if the output is bounded whenever the input is bounded.*

If the system is ***not*** *BIBO stable, it is said to be* ***unstable****.*

Note: To be extra clear, the word "whenever" here means the output must remain bounded for "all possible" bounded inputs, $u : [0,\infty) \to \mathbb{R}$. That sounds like it could be hard to check!

Recall that a system is **causal** if present values of the output do not depend on future values of the input. If the system has a rational transfer function $G(s)$, then the system is causal if, and only if, the degree of the numerator is less than or equal to the degree of the denominator. **Transfer functions that satisfy this property have a special name; they are said to be proper.**

Proposition 10.25: Poles and BIBO Stability

A system with a proper rational transfer function is BIBO stable if, and only if, its poles have negative real parts. In other words, the stability of a rational transfer function

$$G(s) = \frac{N(s)}{D(s)},$$

is completely determined by its poles as long as the degree of the numerator polynomial is less than or equal to the degree of the denominator polynomial.

Note: The zeros do not play a role in stability.

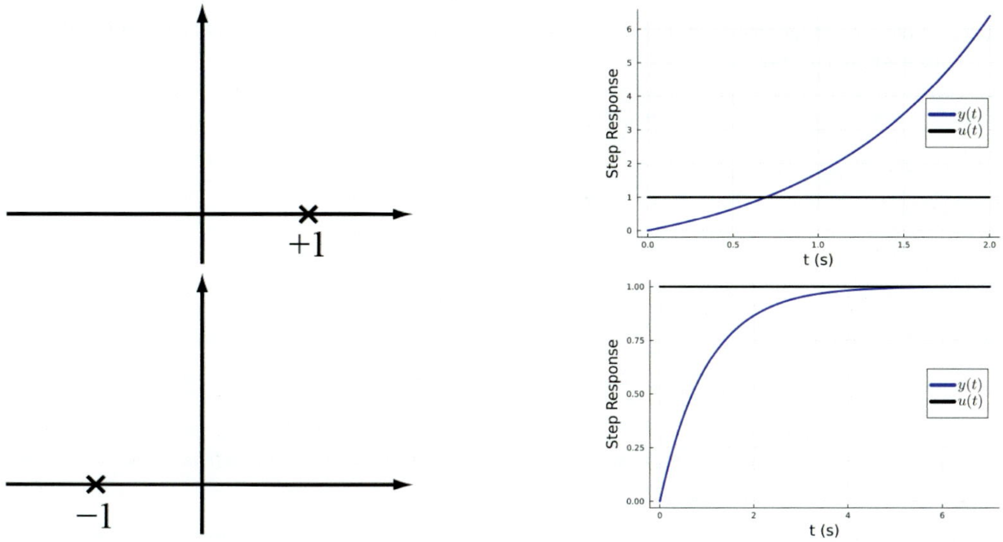

Figure 10.5: The positions of the poles are marked with a ×. For a positive real pole, bounded inputs can give rise to unbounded outputs, whereas for a negative real pole, bounded inputs always give rise to bounded outputs.

To illustrate the connections between time-domain models, transfer functions, poles, and BIBO stability, consider a system represented by a first-order differential equation with input $u(t)$ and output $y(t)$,

$$\dot{y}(t) - y(t) = u(t).$$

In the s-domain, the model becomes $sY(s) - Y(s) = U(s)$, yielding

$$Y(s) = G(s)U(s),$$

where the transfer function is

$$G(s) = \frac{1}{s-1}.$$

The pole is $s = +1$, and thus the system is unstable. Is it easy to give a bounded input that results in an unbounded output? Absolutely.

For definiteness, suppose the input is a unit step, $u_{\rm stp}(t)$. Then, from Table 10.2, $U(s) = \frac{1}{s}$, resulting in:

$$Y(s) = \frac{1}{s-1}\frac{1}{s}.$$

Applying the method of partial fraction expansion,

$$Y(s) = -\frac{1}{s} + \frac{1}{s-1}.$$

Taking the inverse Laplace transform yields

$$y(t) = \left(e^t - 1\right) u_{\rm stp}(t). \tag{10.38}$$

From this equation, the output $y(t)$ is unbounded, diverging to infinity as t goes to infinity, as shown in the upper half of **Fig. 10.5**.

On the other hand, for the model

$$\dot{y}(t) + y(t) = u(t),$$

the transfer function is

$$G(s) = \frac{1}{s+1}.$$

The pole is $s = -1$, and hence the system is BIBO stable. Computing the output in response to a unit-step input gives

$$y(t) = \left(1 - e^{-t}\right) u_{\rm stp}(t). \tag{10.39}$$

From this equation, the output converges to 1.0 as t goes to infinity, and, in particular, it is bounded for all $t \geq 0$, as illustrated in the lower half of Fig. 10.5.

Example 10.26. *Determine the stability of the transfer functions*

(a) $G_1(s) = \frac{2}{s^2-1}$

(b) $G_2(s) = \frac{s+2}{s^2-1}$

(c) $G_3(s) = \frac{s+1}{s^2-1}$

(d) $G_4(s) = \frac{2s-4}{s^2+2s+2}$

(e) $G_5(s) = \frac{s-1}{s}$,

where the first four transfer functions are taken from the previous example, and the last one is new.

Solutions: $G_1(s)$, $G_2(s)$ and $G_3(s)$ each has a pole at $+1$, and thus they are unstable. The poles of $G_4(s)$ are $-1 \pm \mathbf{i}$, which have negative real parts, and hence the transfer funciton is BIBO stable. $G_5(s)$ has a pole at the origin, and hence it is unstable. ■

At this point, the role of zeros seems somewhat mysterious. The zeros will come into the picture when closed-loop systems are analyzed.

Example 10.27. *(Optional Read) If a rational transfer function is not proper, do the poles still determine its stability? The short answer is no.*

Solution: If you are curious as to why not, consider $G(s) = s$, which corresponds to a differentiator, that is, the system

$$y(t) = \frac{du(t)}{dt},$$

where $u(t)$ is the input. The input $\sin(t^2)$ is bounded, since it assumes values between -1 and 1, but the corresponding output $y(t) = 2t\cos(t^2)$ is unbounded.

Some may think that the transfer function has no poles, and therefore, it is untrue that all of its poles have negative real parts. Instead of arguing semantics, consider the transfer function,

$$\begin{aligned} G(s) &= G_a(s) + G_b(s) \\ &= s + \frac{1}{s+2} \\ &= \frac{s(s+2)}{s+2} + \frac{1}{s+2} \\ &= \frac{s^2+2s+1}{s+2}, \end{aligned}$$

which is improper. A quick calculation shows that it has two zeros at -1 and a pole at -2. But the system is not BIBO stable because the bounded input $u(t) = \sin(t^2)$ produces the output

$$y(t) = y_a(t) + y_b(t),$$

where $y_a(t) = 2t\cos(t^2)$, which is unbounded, and the second term, being the output of a system with BIBO stable transfer function $G_b(s)$, must be bounded. The sum of an unbounded signal and a bounded signal is unbounded. Thus, a bounded input has produced an unbounded output, showing that the system is unstable.

■

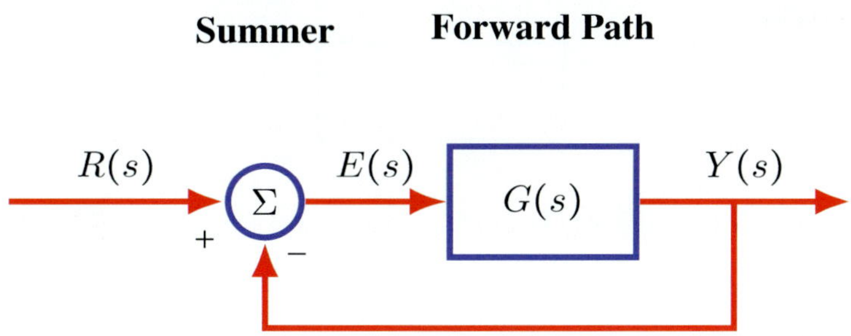

Figure 10.6: Unity negative feedback system.

10.5 Unity Feedback Systems

A common control system configuration, and the easiest one to study when first learning the subject of feedback control, is the **unity feedback system** shown in Fig. 10.6. It is also called a **negative feedback system** due to the minus sign on the feedback loop at the summer. In this figure, $Y(s)$ is the system output, $R(s)$ is the reference command, and $E(s)$ is the error between the output and the reference. The reference command $R(s)$ normally specifies the desired output of the system. When the reference command is a constant, the reference value is also called a **setpoint**. The transfer function in the **forward path** is $G(s)$, whereas the transfer function in the **feedback path** (the bottom loop) is 1, the source of the name **unity** feedback system.

10.5.1 Closed-Loop Transfer Functions

The block diagram in Fig. 10.6 is equivalent to the following set of linear equations:

$$E(s) = R(s) - Y(s) \tag{10.40}$$

$$Y(s) = G(s)E(s). \tag{10.41}$$

Solving these two equations for $Y(s)$ as a function of $R(s)$ results in:

$$\begin{aligned} Y(s) &= G(s)(R(s) - Y(s)) \\ &= G(s)R(s) - G(s)Y(s), \end{aligned}$$

and thus

$$Y(s) + G(s)Y(s) = G(s)R(s),$$

which is equivalent to

$$(1 + G(s))Y(s) = G(s)R(s).$$

Hence

$$\boxed{Y(s) = \frac{G(s)}{1 + G(s)} R(s),} \tag{10.42}$$

from which it follows that the transfer function from $R(s)$ to $Y(s)$ is

$$\boxed{\frac{Y(s)}{R(s)} = \frac{G(s)}{1 + G(s)}.} \tag{10.43}$$

The same steps can be repeated to determine the transfer function from $R(s)$ to $E(s)$,

$$E(s) = R(s) - G(s)E(s)$$

and

$$(1 + G(s))E(s) = R(s).$$

Thus,

$$\boxed{E(s) = \frac{1}{1 + G(s)} R(s),} \tag{10.44}$$

which yields the transfer function as

$$\boxed{\frac{E(s)}{R(s)} = \frac{1}{1+G(s)}.} \tag{10.45}$$

Example 10.28. *For the unity feedback system in Fig. 10.6, compute the poles of $\frac{Y(s)}{R(s)}$ and $\frac{E(s)}{R(s)}$ for each of the following forward-path transfer functions,*

(a) $G_1(s) = \frac{2}{s^2-1}$

(b) $G_2(s) = \frac{s+2}{s^2-1}$

(c) $G_3(s) = \frac{s+1}{s^2-1}$.

Solutions: Before starting the solutions, suppose you had the following ratio of rational numbers:

$$\frac{\frac{3}{4}}{1+\frac{3}{4}},$$

and you wanted to reduce it to a standard ratio of two integers. You would be unlikely to find the problem intimidating because you know how to clear the denominator as follows,

$$\begin{aligned}\frac{\frac{3}{4}}{1+\frac{3}{4}} &= \frac{3}{4+3} \quad \text{(multiply top and bottom by } 4) \\ &= \frac{3}{7}.\end{aligned}$$

Because there are no common factors, this is the final answer. If you somehow arrived at $\frac{6}{14}$, you would remove the common factor of two (aka 6 and 14 are not coprime). We are going to do the same for rational functions. **What is mind-bogglingly cool is that these rational functions are transfer functions and, THEREFORE, correspond to ODEs!**

(a) $G_1(s) = \frac{2}{s^2-1}$ **Ans.** Poles of $\frac{Y(s)}{R(s)}$ and $\frac{E(s)}{R(s)}$ are the same and equal to $\pm\mathbf{i}$. Hence, the closed-loop system is not BIBO stable.

$$\begin{aligned}\frac{Y(s)}{R(s)} &= \frac{G_1(s)}{1+G_1(s)} \\ &= \frac{\frac{2}{s^2-1}}{1+\frac{2}{s^2-1}} \quad \text{(multiply top and bottom by } (s^2-1)) \\ &= \frac{2}{(s^2-1)+2} \\ &= \frac{2}{s^2+1}.\end{aligned}$$

Hence, the poles are $\pm\mathbf{i}$. We did not have to be concerned about a common factor because the numerator is a constant and has no roots. In a similar fashion, we compute

$$\begin{aligned}\frac{E(s)}{R(s)} &= \frac{1}{1+G_1(s)} \\ &= \frac{1}{1+\frac{2}{s^2-1}} \quad \text{(multiply top and bottom by } (s^2-1)) \\ &= \frac{2(s^2-1)}{(s^2-1)+2} \\ &= \frac{2(s^2-1)}{s^2+1}.\end{aligned}$$

The numerator has roots ± 1, while the denominator has roots $\pm\mathbf{i}$. There are no common factors and hence the poles are $\pm\mathbf{i}$.

(b) $G_2(s) = \frac{s+2}{s^2-1}$ **Ans.** Poles of $\frac{Y(s)}{R(s)}$ and $\frac{E(s)}{R(s)}$ are the same and equal to $-\frac{1}{2} \pm \mathbf{i}\frac{\sqrt{3}}{2}$. Hence, the closed-loop system is BIBO stable.

$$\begin{aligned}
\frac{Y(s)}{R(s)} &= \frac{G_2(s)}{1+G_2(s)} \\
&= \frac{\frac{s+2}{s^2-1}}{1+\frac{s+2}{s^2-1}} \quad \text{(multiply top and bottom by } (s^2-1)) \\
&= \frac{s+2}{(s^2-1)+(s+2)} \\
&= \frac{s+2}{s^2+s+1}.
\end{aligned}$$

The root of the numerator is -2 and the roots of the denominator are $-\frac{1}{2} \pm \mathbf{i}\frac{\sqrt{3}}{2}$. Because there are no common factors, the poles are $-\frac{1}{2} \pm \mathbf{i}\frac{\sqrt{3}}{2}$. In a similar fashion, we compute

$$\begin{aligned}
\frac{E(s)}{R(s)} &= \frac{1}{1+G_2(s)} \\
&= \frac{1}{1+\frac{s+2}{s^2-1}} \quad \text{(multiply top and bottom by } (s^2-1)) \\
&= \frac{(s^2-1)}{(s^2-1)+(s+2)} \\
&= \frac{(s^2-1)}{s^2+s+1}.
\end{aligned}$$

The numerator has roots ± 1, while the denominator has roots $-\frac{1}{2} \pm \mathbf{i}\frac{\sqrt{3}}{2}$. There are no common factors and hence the poles are $-\frac{1}{2} \pm \mathbf{i}\frac{\sqrt{3}}{2}$.

(c) $G_3(s) = \frac{s+1}{s^2-1}$ **Ans.** The pole of $\frac{Y(s)}{R(s)}$ and $\frac{E(s)}{R(s)}$ is the same and equals $s = 0$. Hence, the closed-loop system is not BIBO stable.

$$\begin{aligned}
\frac{Y(s)}{R(s)} &= \frac{G_3(s)}{1+G_3(s)} \\
&= \frac{\frac{s+1}{s^2-1}}{1+\frac{s+1}{s^2-1}} \quad \text{(multiply top and bottom by } (s^2-1)) \\
&= \frac{s+1}{(s^2-1)+(s+1)} \\
&= \frac{s+1}{s^2+s} \\
&= \frac{s+1}{s(s+1)} \\
&= \frac{1}{s}.
\end{aligned}$$

The numerator has no roots (because we saw the cancellation and took care of it) and the root of the denominator is 0. Because there are no common factors, the pole is $s = 0$. In a similar fashion, we compute

$$\begin{aligned}
\frac{E(s)}{R(s)} &= \frac{1}{1+G_3(s)} \\
&= \frac{1}{1+\frac{s+1}{s^2-1}} \quad \text{(multiply top and bottom by } (s^2-1)) \\
&= \frac{(s^2-1)}{(s^2-1)+(s+1)} \\
&= \frac{(s^2-1)}{s^2+s} \\
&= \frac{(s+1)(s-1)}{s(s+1)} \\
&= \frac{s-1}{s}.
\end{aligned}$$

The numerator has a root at +1, while the denominator has a root at 0. There are no common factors and hence the pole is $s = 0$.

Note: The common factor in the numerator and denominator can be traced back to the common factor in the open-loop transfer function

$$G_3(s) = \frac{s+1}{s^2-1} = \frac{s+1}{(s-1)(s+1)}.$$

If the common factor had been removed BEFORE computing the closed-loop transfer functions, the resulting polynomials would not have had a common factor; that is, they, too, would have been coprime. As part of feedback design, we sometimes like to cancel a "STABLE zero" by a "STABLE pole". We will return to this point later. ■

We deliberately went through every step in each problem to underline the repetitive nature of the computations.

Proposition 10.29: Closed-Loop Transfer Functions

Suppose that $G(s) = \frac{N(s)}{D(s)}$ is the forward-path transfer function for the unity feedback system in Fig. 10.6. Then, the closed-loop transfer functions are:

$$\frac{Y(s)}{R(s)} = \frac{N(s)}{D(s)+N(s)}$$

and (10.46)

$$\frac{E(s)}{R(s)} = \frac{D(s)}{D(s)+N(s)}.$$

Moreover, if $N(s)$ and $D(s)$ are coprime, then so are the numerators and denominators of $\frac{Y(s)}{R(s)}$ and $\frac{E(s)}{R(s)}$.

Proof: Simple calculations show that the two transfer functions associated with a unity feedback system are as stated. Indeed,

$$\frac{Y(s)}{R(s)} = \frac{\frac{N(s)}{D(s)}}{1+\frac{N(s)}{D(s)}} = \frac{N(s)}{D(s)+N(s)},$$

and

$$\frac{E(s)}{R(s)} = \frac{1}{1+\frac{N(s)}{D(s)}} = \frac{D(s)}{D(s)+N(s)}.$$

In each case, the numerator and denominator have been multiplied by $D(s)$ to simplify the expressions to a ratio of polynomials.

Next, we prove that if $N(s)$ and $D(s)$ are coprime (no common roots), then the polynomials $N(s)$ and $D(s)+N(s)$ are coprime as well. To show this conclusion, suppose that s_1 is a root of $N(s)$, that is, $N(s_1) = 0$. Can it also be a root of $D(s)+N(s)$? The answer is no, because if $N(s_1) = 0$, then $D(s_1)+N(s_1) = D(s_1)$. Hence, if $D(s_1)+N(s_1) = 0$, then $D(s_1) = 0$, which shows that s_1 would then have to be a root of $D(s)$, which is not possible due to $N(s)$ and $D(s)$ being coprime. The identical argument applies to the second transfer function as well. ■

10.5.2 Closed-Loop Poles and Zeros

Proposition 10.29 allows us to define the closed-loop poles and zeros of a unity feedback system.

Closed-Loop Poles and Zeros

Definition 10.30. *For the unity feedback system in* Fig. *10.6, with forward-path transfer function* $G(s) = \frac{N(s)}{D(s)}$ *coprime, the* ***closed-loop poles*** *are the roots of* $D(s)+N(s)$ *and the* ***closed-loop zeros*** *are the roots of* $N(s)$. *It follows that the zeros of the forward-path transfer function are also the closed-loop zeros of the corresponding unity feedback loop*

Notes: (a) If $N(s)$ and $D(s)$ have common factors, hence yielding common roots, the common factors must be canceled before determining the zeros and poles of the closed-loop system. (b) Because we are normally interested in shaping the transient response of the transfer function from reference command to output, the zeros of interest to us are those of $\frac{Y(s)}{R(s)}$.

Even though it is a slight abuse of terminology, the zeros of $G(s)$ are often called the **open-loop zeros**, and the poles of $G(s)$ are often called the **open-loop poles**. This terminology makes sense because if one *opens the loop* by disconnecting the feedback path from the summer in Fig. 10.6, the transfer function from $R(s)$ to $Y(s)$ is simply $G(s)$.

Example 10.31. *Determine the open-loop (OL) and closed-loop (CL) poles and zeros when the forward-path (FP) transfer functions are as given in Example 10.28,*

(a) $G_1(s) = \frac{2}{s^2-1}$,

(b) $G_2(s) = \frac{s+2}{s^2-1}$,

(c) $G_3(s) = \frac{s+1}{s^2-1}$.

Solutions: Most of the work has been done in Examples 10.23, 10.26, and 10.28. The purpose here is to gather it all together and practice the new vocabulary.

FP $G(s)$	OL Zeros	OL Poles	CL Zeros	CL Poles
$\frac{2}{s^2-1}$	None	± 1	None	$\pm \mathbf{i}$
$\frac{s+2}{s^2-1}$	-2	± 1	-2	$-\frac{1}{2} \pm \mathbf{i}\frac{\sqrt{3}}{2}$
$\frac{s+1}{s^2-1}$	None	$+1$	None	0

When there are no open-loop zeros, there are no closed-loop zeros. Because the numerator and denominator of $G_3(s)$ are not coprime, we had to first remove their common factor of $(s+1)$ before completing the table.

■

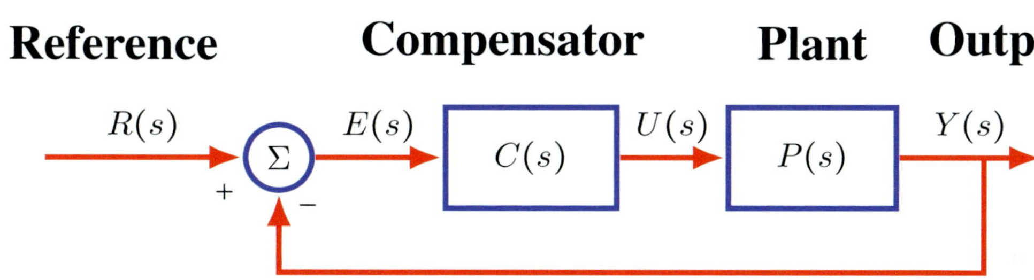

Figure 10.7: Cascade control system configuration, where $C(s)$ is the transfer function of the compensator and $P(s)$ is the transfer function of the plant. Their product, $G(s) = C(s)\,P(s)$, is the overall forward-path transfer function.

10.6 Cascade Control Architecture

Figure 10.7 is a refinement of the unity feedback system, where the forward path is written as a **cascade** connection of a **compensator** $C(s)$ and **plant** $P(s)$. Recall that "plant" is the generic term for the system that is to be controlled. The input to the plant is denoted $U(s)$. The plant input is the means by which the plant output is changed. For the Segway, the plant input was the torque at the wheels provided by the motors, while for the RLC circuit, it was the applied voltage.

The compensator is the algorithm that determines how to vary the input to the plant to regulate the plant output. In the cascade control configuration of Fig. 10.7, the compensator acts on $E(s) := R(s) - Y(s)$, the **error** between the reference command $R(s)$ and the measured output $Y(s)$, to produce the control input:

$$U(s) = C(s)E(s). \tag{10.47}$$

When the error is zero, it means that the plant output is exactly following or **tracking** the reference command. One way to think of the compensator is that it attempts to adjust the plant input so that the plant output tracks the reference command. Equivalently, the compensator seeks to drive the error between the reference and output signals to zero.

The closed-loop transfer functions, (10.43) and (10.45) in Section 10.5.1 are easily updated to this new situation by identifying $G(s) := C(s)P(s)$ as the forward-path transfer function. Doing so yields,

$$\boxed{\frac{Y(s)}{R(s)} = \frac{C(s)P(s)}{1+C(s)P(s)}} \tag{10.48}$$

and

$$\boxed{\frac{E(s)}{R(s)} = \frac{1}{1+C(s)P(s)}.} \tag{10.49}$$

Because $U(s) = C(s)E(s)$, it follows that

$$\boxed{\frac{U(s)}{R(s)} = \frac{C(s)}{1+C(s)P(s)}.} \tag{10.50}$$

All three of these transfer functions become important in most control system designs.

10.6.1 Two Common Compensators: Proportional (P) and Proportional-Derivative (PD)

We introduce here the two most commonly used compensators.

Proportional compensator

$$C(s) = K_{\mathrm{P}}, \tag{10.51}$$

where K_{P} is called a **proportional gain**. With a proportional compensator, the control signal is given by

$$u(t) = K_{\mathrm{P}}e(t) = K_{\mathrm{P}}(r(t) - y(t)).$$

Adjusting K_{P} allows the control signal to react more or less "vigorously" to the error between the commanded value of the output (aka reference signal), $r(t)$, and the measured value of the output, $y(t)$. It adds neither zeros nor poles to the forward path of the feedback loop. For a first-order plant, $P(s)$, gain adjustment alone can achieve stability, as we illustrate next.

Example 10.32. (Proportional feedback applied to an unstable first-order plant) *Suppose the plant has transfer function* $P(s) = \frac{1}{s-1}$ *and the compensator is a* proportional feedback $C(s) = K_{\mathrm{P}}$, *where* K_{P} *is a gain to be chosen. Determine the closed-loop transfer function* $\frac{Y(s)}{R(s)}$ *and evaluate its BIBO stability as a function of* K_{P}.

Solution:

$$\begin{aligned}\frac{Y(s)}{R(s)} &= \frac{C(s)\,P(s)}{1+C(s)\,P(s)}\\ &= \frac{K_{\mathrm{P}}\frac{1}{s-1}}{1+K_{\mathrm{P}}\frac{1}{s-1}} \quad \text{(multiply top and bottom by } (s-1))\\ &= \frac{K_{\mathrm{P}}}{s-1+K_{\mathrm{P}}}.\end{aligned}$$

The pole of the transfer function is the root of the denominator polynomial, $s-1+K_{\mathrm{P}} = 0$, yielding a single pole $1-K_{\mathrm{P}}$. The closed-loop system is, therefore, BIBO stable if, and only if, $K_{\mathrm{P}} > 1$. ■

Feedback control design would be way too easy if we could stabilize all plants with proportional feedback! The following result will make it easier to show a limitation of proportional feedback control.

Fact 10.33: Simplified Stability Test

For real numbers a_1 and a_0, the roots of the quadratic equation

$$s^2 + a_1 s + a_0 = 0 \tag{10.52}$$

have negative real parts if, and only if, $a_1 > 0$ and $a_0 > 0$.

Note: To apply the above result to $\bar{a}_2 s^2 + \bar{a}_1 s + \bar{a}_0 = 0$, with $\bar{a}_2 \neq 0$, we simply divide through by $\bar{a}_2$, yielding $a_1 = \frac{\bar{a}_1}{\bar{a}_2}$ and $a_0 = \frac{\bar{a}_0}{\bar{a}_2}$.

For a cubic equation, the result is, for real numbers a_2, a_1, and a_0, the roots of the equation

$$s^3 + a_2 s^2 + a_1 s + a_0 = 0$$

all have negative real parts if, and only if, $a_2 > 0$, $a_0 > 0$, and $a_1 - \frac{a_0}{a_2} > 0$. Similar results can be given for polynomials of arbitrary degree. The computations are often organized in a table called the Routh array, which is typically covered in an undergraduate feedback control course.

Example 10.34. (Proportional feedback applied to an unstable second-order plant) *Consider the plant $P(s) = \frac{1}{s(s-2)}$, with the proportional feedback, $C(s) = K_P$. Determine the transfer function $\frac{Y(s)}{R(s)}$ and evaluate its stability as a function of K_P.*

Solution: Evaluating the closed-loop transfer function gives

$$\frac{Y(s)}{R(s)} = \frac{K_P}{s^2 - 2s + K_P}.$$

The poles of the transfer function are the roots of the polynomial $s^2 - 2s + K_P = 0$. Applying the result of Fact 10.33, it is seen that the coefficient $a_1 = -2 < 0$ is independent of the value of the gain K_P, and hence at least one root will have a nonnegative real part, regardless of the value of K_P. The closed-loop system is, therefore, unstable for all values of the gain K_P, and **something more sophisticated than proportional feedback is required to stabilize the closed-loop system**. ■

PD or Proportional-Derivative Compensator

$$C(s) = K_P + K_D s = K_D\left(s + \frac{K_P}{K_D}\right) \tag{10.53}$$

has two gains, the proportional gain, K_P, which acts as before, and the **derivative gain**, K_D. With a proportional-derivative (PD) compensator, the control signal is given by

$$u(t) = K_P e(t) + K_D \dot{e}(t) = K_P(r(t) - y(t)) + K_D(\dot{r}(t) - \dot{y}(t)).$$

The derivative term allows the controller to react with "anticipation" to changes in the error signal. Indeed, if we approximate the derivative as

$$\dot{e}(t) \approx \frac{e(t + \delta t) - e(t)}{\delta t},$$

the control signal then depends on the error term at the future time, $t + \delta t$, giving it a head start on making a correction to the closed-loop system. Another way to look at it is that the linear approximation of the error signal at a point t_0 is

$$e(t) \approx e(t_0) + \underbrace{\frac{de(t_0)}{dt}}_{\dot{e}(t_0)}(t - t_0),$$

showing again that the derivative of e allows the prediction of future values of the error signal over small intervals of time. This discussion is meant to provide intuition, so if it seems hand-wavy or confusing to you, just ignore it. The Laplace transform will soon clear up lingering doubts.

A PD compensator adds a zero in the forward path at $-\frac{K_P}{K_D}$, but has no poles. The derivative action of a PD compensator will prove useful for stabilizing a system and for reducing oscillations.

Example 10.35. (PD compensation applied to an unstable second-order plant) *Suppose the plant is* $P(s) = \frac{1}{s(s-2)}$ *and* $C(s) = K_P + K_D s$. *Determine the transfer function* $\frac{Y(s)}{R(s)}$ *and evaluate its stability as a function of the controller gains.*

Solution: Evaluating the closed-loop transfer function gives

$$\frac{Y(s)}{R(s)} = \frac{K_P + K_D s}{(s^2 - 2s) + (K_P + K_D s)} = \frac{K_P + K_D s}{s^2 + (K_D - 2)s + K_P}.$$

The poles of the transfer function are the roots of the polynomial $s^2 + (K_D - 2)s + K_P = 0$. After identifying the coefficients $a_1 = (K_D - 2)$ and $a_0 = K_P$, Fact 10.33 yields that the closed-loop system's poles have negative real parts if, and only if, $K_P > 0$ and $K_D > 2$. ■

Is stability all we care about in a feedback loop? Hardly. We will next examine the steady-state error, that is, the long-term "accuracy" of the control system.

10.6.2 Steady-State Error

The **steady-state error**, or e_{SS}, is the asymptotic[2] value of the error signal defined in Figures 10.6 and 10.7. The steady-state error depends on the reference command and the forward-path transfer function. The expression for the error signal was derived in (10.44) and (10.49). When e_{SS} is small, the control system is accurately tracking the reference command as the system reaches a steady state. What happens between the closed-loop system being "excited" by an input signal and its reaching a steady state is called its **transient response**; this will be the subject of a separate section.

While the term "steady-state" can be defined for a range of reference signals, we will focus on the most common one, the unit-step function, that is, $R(s) = \frac{1}{s}$.

Proposition 10.36: Steady-State Error for Step Inputs Applied to BIBO Stable Closed-Loop Systems

Assume the forward path transfer function is $G(s) = \frac{N(s)}{D(s)}$ and the closed-loop system is BIBO stable. Then the steady-state error for a unit-step command is well-defined and is given by

$$e_{SS} = \lim_{s \to 0} \frac{D(s)}{D(s) + N(s)}. \tag{10.54}$$

Moreover, if $G(s)$ does not have a pole at the origin, then

$$e_{SS} = \frac{D(0)}{D(0) + N(0)} \neq 0,$$

while if $G(s)$ has one or more poles at the origin, then $D(0) = 0$ and the steady-state error is

$$e_{SS} = 0.$$

Note: To be clear, the result also holds for $G(s) = C(s)P(s)$ arising from a Cascade Controller Architecture.

Note: What happens to e_{SS} if the reference is $r(t) = r_0 \cdot u_{\text{stp}}(t)$, a scaled unit-step? Because $E(s) = \frac{1}{1+G(s)} R(s)$, it follows that if the reference is scaled by r_0, then so is the error. Hence, e_{SS} gives the *relative error*, or, said another way, the *percent error* in the output is $100\, e_{SS}\%$.

e_{SS}	% Error
0.10	10 %
0.25	25 %
1.00	100 %

Example 10.37. (Steady-state error) *Determine the steady-state error for a unit-step input when a unity feedback system is specified by*

[2]That is, its value as $t \to \infty$.

(a) $C_1(s) = 10$ *and* $P_1(s) = \frac{4}{s+4}$.

(b) $C_2(s) = 10$ *and* $P_2(s) = \frac{4}{s(s+4)}$.

(c) $C_3(s) = (10 + 2s)$ *and* $P_3(s) = \frac{1}{s(s-4)}$.

Solution: The step responses are shown in Fig. 10.8. The calculations for the time-domain expressions are given in Example 10.38 below. In engineering practice, if time-domain signals are required, they are usually computed numerically by a computer-aided control design program rather than by tedious hand calculations, but in other courses, you may need to do them by hand.

(a) $C_1(s) = 10$ and $P_1(s) = \frac{4}{s+4}$ **Ans.** $e_{SS} \approx 9.1\%$.

The forward-path transfer function is

$$C_1(s)P(s) = \frac{40}{s+4} =: \frac{N_1(s)}{D_1(s)}.$$

The pole of the closed-loop system satisfies

$$D_1(s) + N_1(s) = (s+4) + 40 = 0 \implies s = -44,$$

and thus, the closed-loop system is BIBO stable. For a step input, the steady-state error is

$$e_{SS} = \frac{D_1(0)}{D_1(0) + N_1(0)} = \frac{4}{40+4} = \frac{1}{11} \approx 0.091.$$

(b) $C_2(s) = 10$ and $P_2(s) = \frac{4}{s(s+4)}$. **Ans.** $e_{SS} = 0.0$.
The forward-path transfer function is

$$C_2(s)P_2(s) = \frac{40}{s(s+4)} =: \frac{N_2(s)}{D_2(s)}.$$

The poles of the closed-loop system satisfy

$$D_2(s) + N_2(s) = s(s+4) + 40 = s^2 + 4s + 40 = 0,$$

and thus by Fact 10.33, the closed-loop system is BIBO stable. For a step input, the steady-state error is

$$e_{SS} = \frac{D_2(0)}{D_2(0) + N_2(0)} = \frac{0}{0+40} = 0.0.$$

We emphasize that because the denominator of the forward-path transfer function has a pole at the origin, $D_2(0) = 0$.

(c) $C_3(s) = (10 + 2s)$ and $P_2(s) = \frac{1}{s(s-4)}$. **Ans.** e_{SS} is undefined because the closed-loop is unstable. In practice, we would see the error wildly oscillating with increasing amplitude because the closed-loop poles are complex, with a positive real part (which can be checked with the quadratic formula).
The forward-path transfer function is

$$C_3(s)P_3(s) = \frac{(10+2s)}{s(s-4)} =: \frac{N_3(s)}{D_3(s)}.$$

The pole of the closed-loop system satisfies

$$D_2(s) + N_2(s) = s(s-4) + (10+2s) = s^2 - 2s + 10 = 0,$$

and thus by Fact 10.33, the closed-loop system is unstable. Therefore, the steady-state error is undefined.

We show how the time domain signals in Fig. 10.8 are computed for part (a). The others are similar.

```
using ControlSystems, Plots, LaTeXStrings
# ControlSystems is slow to load

# Specify forward path transfer functions
C = 10.0
P = tf([4], [1.0, 4.0])
```

```
# build closed-loop tf
sys = feedback(C*P, 1)

# Compute and Plot the Step Response
t = 0:0.01:1.0

# Time trajectory
y, t = step(sys, t); y = y'

# Plot
p1 = plot(t, y, guidefont = 15, lw=3, color=:blue,  label=L"$y(t)$", legendfontsize=15,
     xlabel="t (s)", ylabel="Step Response", legend=:right)
Ref = ustep.(t)
Error = Ref - y
p1 = plot!(p1, t, Ref, lw=3, color=:black, legend=:right, label=L"$u(t)$" )
p1 = plot!(p1, t, Error, lw=3, color=:red, legend=:right, label=L"$e(t)$" )

png(p1, "C1P1StepResponse")
display(p1)
```

Output See Fig. 10.8. ■

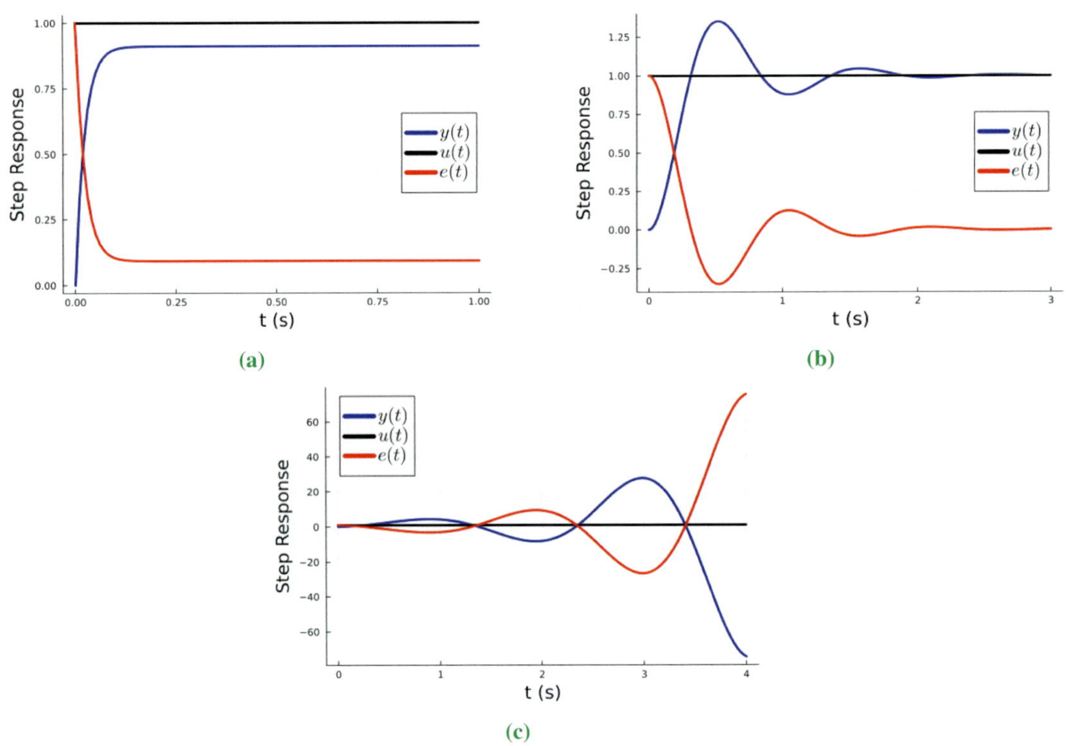

Figure 10.8: The step responses for Examples 10.37 and 10.38, with the error in red. In (a) and (b), a steady-state error is achieved, while in (c), the error does not have a well-defined limit due to the closed-loop system being unstable.

Example 10.38. *(Optional Read) For the transfer functions in Example 10.37, use code to determine time-domain expressions for $e(t)$ and then take limits to confirm the previous solutions.*

Solutions:

In each case, we use code to compute the inverse Laplace transform of $E(s) := \frac{1}{1+C(s)P(s)} \cdot \frac{1}{s}$, where $\frac{1}{s} = \mathcal{L}\{u_{\rm stp}(t)\}$.

(a) $C_1(s) = 10$ and $P_1(s) = \frac{4}{s+4}$ yield $e_1(t) = \frac{1}{11}u_{\rm stp}(t) + \frac{10}{11}e^{-44t}\, u_{\rm stp}(t)$ and hence $\lim\limits_{t\to\infty} e_1(t) = \dfrac{1}{11}$.

(b) $C_2(s) = 10$ and $P_2(s) = \frac{4}{s(s+4)}$ yield $e_2(t) = \frac{1}{3}e^{-2t}\, \sin(6t)\, u_{\rm stp}(t) + e^{-2t}\, \cos(6t)\, u_{\rm stp}(t)$ and hence $\lim\limits_{t\to\infty} e_2(t) = 0$.

(c) $C_3(s) = (10+2s)$ and $P_3(s) = \frac{1}{s(s-4)}$ yield $e_3(t) = e^t\, \cos(3t)\, u_{\rm stp}(t) - e^t\, \sin(3t)\, u_{\rm stp}(t)$ and hence $\lim\limits_{t\to\infty} e_3(t)$ is undefined. Indeed, $e_3(t)$ grows unbounded while oscillating between positive and negative values, never making up its mind between $+\infty$ and $-\infty$!

```julia
using PyCall

# Import Python's SymPy
sympy = pyimport("sympy")

# Define the symbols
s, t = sympy.symbols("s t")

Ustep = 1/s

# Define the expression E1(s)
C1 = 10 + 0*3
P1 = 4/(s+4)
E1 = (1/(1 + C1*P1))*Ustep

# Compute the inverse Laplace transform
e1 = sympy.inverse_laplace_transform(E1, s, t)

println("Part 1")
println(e1)
println(" ")

# Define the expression E2(s)
C2 = 10 + 0*3
P2 = 4/(s*(s+4))
E2 = (1/(1 + C2*P2))*Ustep

# Compute the inverse Laplace transform
e2 = sympy.inverse_laplace_transform(E2, s, t)

println("Part 2")
println(e2)
println(" ")

# Define the expression E3(s)
C3 = 10 + 2*s
P3 = 1/(s*(s-4))
E3 = (1/(1 + C3*P3))*Ustep

# Compute the inverse Laplace transform
e3 = sympy.inverse_laplace_transform(E3, s, t)

println("Part 3")
println(e3)
```

Output Recall that Heaviside(t) is the unit-step function, $u_{\rm stp}(t)$. Also note that $\lim\limits_{t\to\infty} u_{\rm stp}(t) = 1.0$

```
Part 1
Heaviside(t)/11 + 10*exp(-44*t)*Heaviside(t)/11
```

```
Part 2
(exp(-2*t)*sin(6*t)/3 + exp(-2*t)*cos(6*t))*Heaviside(t)

Part 3
(-exp(t)*sin(3*t) + exp(t)*cos(3*t))*Heaviside(t)
```

■

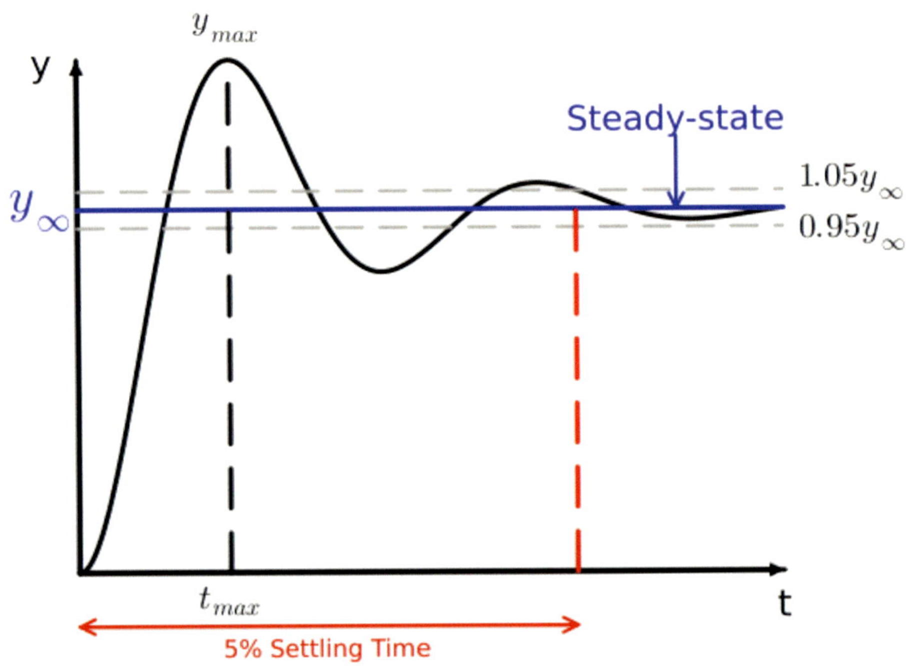

Figure 10.9: Typical step response of an underdamped system. The 5% settling time is the time it takes for the system to enter and then remain within the interval $[0.95y_\infty, 1.05y_\infty]$.

10.7 Transient Response of First- and Second-Order Systems

Thus far, we have learned how to compute closed-loop transfer functions, evaluate their stability, and analyze steady-state tracking performance for step inputs. This section examines the transient response of a system to an abrupt change in input, that is, the **step response**. The primary objective is to relate the qualitative features of the step response to the poles and zeros of the transfer function.

10.7.1 Common Performance Specifications

In this section, whether the system being considered is open-loop or closed-loop is irrelevant. Let the Laplace-transform representation of the system be

$$Y(s) = G(s)\frac{1}{s},$$

where $\frac{1}{s}$ is the Laplace transform of a unit-step input, $Y(s)$ is the output, and $G(s)$ is a BIBO stable transfer function. Let $y(t)$ be the resulting time response of the output, such as the one shown in Fig. 10.9. According to Section 10.6.2, because the transfer function is BIBO stable, $y(t)$ asymptotically converges to a steady-state value, denoted here by y_∞.

The key qualitative features of Fig. 10.9 are listed below:

1. **Peak value:** $y_{\max}$ is the peak value of the output.

In the case that the output monotonically approaches its steady-state value y_∞, as in **Fig. 10.10** for example, then $y_{\max}$ is defined to be y_∞.

2. **Percent overshoot:** the difference $y_{\max} - y_\infty$ is the amount the output exceeds its steady-state value, which could be zero, as indicated above. The percent overshoot is given by

$$\text{Percent overshoot} = \frac{y_{\max} - y_\infty}{y_\infty} \times 100\%$$

3. 5% **settling time** = $\min\{T \geq 0 : |y(t) - y_\infty| \leq 0.05 y_\infty, \forall t \geq T\}$. In other words, the 5% settling time T_s is the smallest $T > 0$ such that

$$0.95 y_\infty \leq y(t) \leq 1.05 y_\infty \tag{10.55}$$

for all $t \geq T$. Though not used here, the 2% settling time is fairly common as well.

4. **Peak Time:** Time to reach $y_{\max}$; it could be infinite.

5. **(Not shown) Rise time:** Time to increase from 10% to 90% of final value. Most commonly used for first-order systems or for over-damped second-order systems.

We will see in Section 10.8 that feedback design requires good rules of thumb for relating the overall system's performance to the zeros and poles of the transfer function. These rules of thumb are developed next.

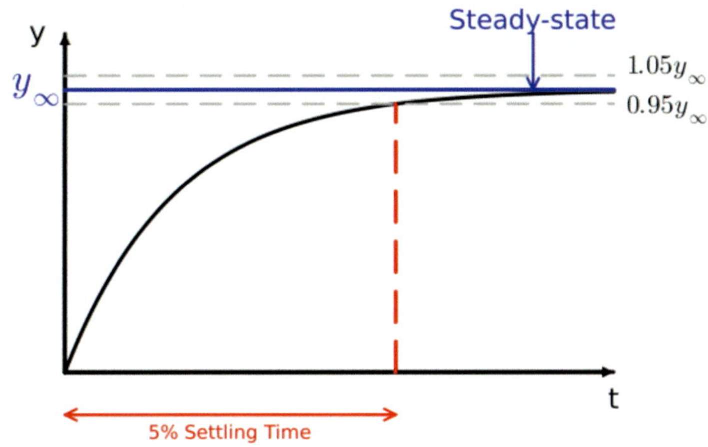

Figure 10.10: (First-Order Systems) The step response of a first-order system $\frac{k_0}{\tau s+1}$ has 10% – 90% rise time approximately equal to 2τ, no overshoot, and its 5% settling time is approximately 3τ. The peak value is the same as y_∞, meaning it is not achieved for any finite value of t.

10.7.2 First-Order System without a Zero

Consider a first-order system with unity DC gain, meaning that $G(0) = 1.0$, namely

$$G(s) = \frac{1}{\tau s + 1}. \tag{10.56}$$

When $\tau > 0$, it is called the **time constant** of the system. The pole is then given by $-1/\tau$. The step response can be computed to be

$$y(t) = \left(1 - e^{-\frac{t}{\tau}}\right) u_{\text{stp}}(t),$$

which is shown in **Fig. 10.10**. The time constant is the time it takes the step response to go from zero to 63.21% of its final value. The step response has no overshoot, and thus both the peak time and the 0% to 100% rise time are infinite.

Because it is easier to remember round numbers, it is common to estimate the time constant by the time it takes the step response to go from zero to 60% of its final value. In the spirit of the zero to 60% approximation for the time constant, the 10% – 90% rise time is approximately equal to 2τ and the 5% settling time is approximately equal to 3τ. These are summarized below for later use:

$$\boxed{\begin{aligned} \text{0\% to 60\% of final value} &= \tau, \\ \text{10\% to 90\% rise time} &= 2\tau, \\ \text{5\% settling time} &= 3\tau. \end{aligned}} \tag{10.57}$$

If the transfer function is instead $K\frac{1}{\tau s+1}$, so that the DC gain is K, the above approximations remain valid because they depend only on percentage changes in the output, and not on absolute values. This is an important advantage in expressing quantities in relative terms, or percentages.

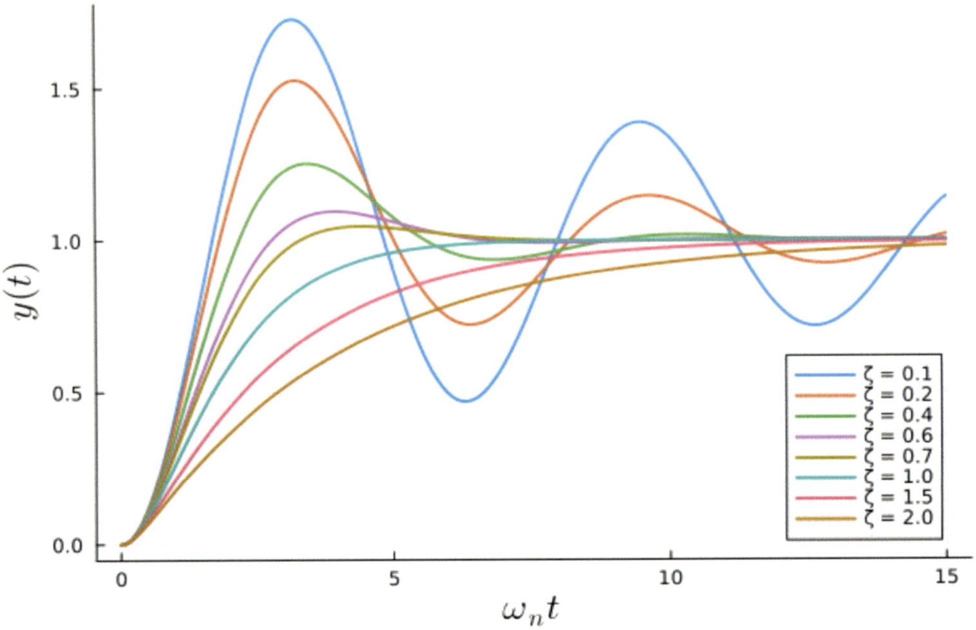

Figure 10.11: Step response of a standard second-order system, $G(s) = \frac{\omega_n^2}{s^2+2\zeta\omega_n s+\omega_n^2}$. There is overshoot for $0 < \zeta < 1$, while there is no overshoot for $\zeta \geq 1$. There is a trade-off between fast response and overshoot. **Note that the x-axis is scaled and equal to $\omega_n t$.**

10.7.3 Second-Order System without a Zero

Consider a second-order system with unity **DC gain**, that is, $G(0) = 1.0$,

$$G(s) = \frac{a_0}{s^2 + a_1 s + a_0}, \tag{10.58}$$

where $a_0 > 0$ and $a_1 > 0$ are assumed for BIBO stability. The step response of the system becomes straightforward to understand when the model is reparameterized in terms of the **damping ratio ζ** and **undamped natural frequency ω_n**, defined as

$$\omega_n^2 = a_0 \tag{10.59}$$

and

$$2\zeta\omega_n = a_1. \tag{10.60}$$

Using these definitions leads to:

$$G(s) = \frac{\omega_n^2}{s^2 + 2\zeta\omega_n s + \omega_n^2}, \tag{10.61}$$

where

$$\omega_n = \sqrt{a_0} \tag{10.62}$$

and

$$\zeta = \frac{a_1}{2\omega_n} = \frac{a_1}{2\sqrt{a_0}}. \tag{10.63}$$

The step response for a range of damping ratios is shown in **Fig. 10.11**. When the damping ratio ζ is between zero and one, there is overshoot and the output oscillates before settling to its final value. When the damping ratio is one or greater, there is no overshoot. The speed with which the step response converges to its steady-state value is determined by the undamped natural frequency. When ω_n is small, the response is sluggish, meaning that the settling time and rise time are long, and when ω_n is large, the response is rapid. For example, if $\omega_n = 1000$, the time scale in **Fig. 10.11** becomes milliseconds, whereas, if $\omega_n = 1/60$, the time scale is minutes.

The system in (10.61) is said to be in **standard form**. Note that if the model has instead the form

$$\widetilde{G}(s) = \frac{k}{s^2 + a_1 s + a_0}, \tag{10.64}$$

then it can be re-expressed as:

$$\widetilde{G}(s) = \frac{k}{a_0}\frac{a_0}{s^2 + a_1 s + a_0} = K\frac{\omega_n^2}{s^2 + 2\zeta\omega_n s + \omega_n^2}, \tag{10.65}$$

where $K = \frac{k}{a_0}$. The gain K introduces a change of scale in the system's output, but it does not change the percent overshoot, settling time, rise time, etc., because these are all relative quantities.

Poles

Applying the quadratic formula, the poles of the transfer function are computed to be:

$$s_{1,2} = -\zeta\omega_n \pm \omega_n\sqrt{\zeta^2 - 1}.$$

Depending on the sign of $\zeta^2 - 1$, three cases are possible:

- **underdamped** $0 < \zeta < 1$: a pair of complex conjugate poles;
- **critically damped** $\zeta = 1$: two repeated real poles; and
- **overdamped** $\zeta > 1$: two distinct real poles.

Step Response

The step responses for the three different types of standard second-order systems are now described in detail.

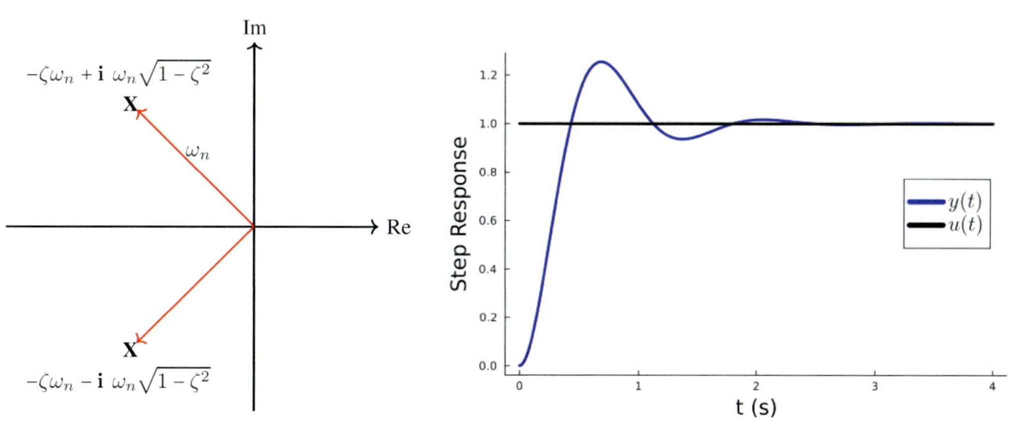

Figure 10.12: The complex poles of an underdamped system $(0 < \zeta < 1)$ give rise to overshoot and oscillations in the step response.

Underdamped ($0 < \zeta < 1$): The poles are complex and equal to

$$s_{1,2} = -\zeta\omega_n \pm \mathbf{i}\omega_n\sqrt{1-\zeta^2}. \tag{10.66}$$

Using the method of partial fraction expansion, the step response is computed to be:

$$y(t) = 1 - \frac{1}{\sqrt{1-\zeta^2}} e^{-\zeta\omega_n t} \sin\left(\omega_n\sqrt{1-\zeta^2}t + \theta\right) u_{\text{stp}}(t), \tag{10.67}$$

where

$$\cos(\theta) = \zeta.$$

The decaying, sinusoidal nature of the output is clearly seen.

The peak value is determined by computing the time derivative of the output, setting it to zero, and evaluating the relative maxima and minima. Doing so gives

$$y_{\max} = 1 + e^{\frac{-\zeta\pi}{\sqrt{1-\zeta^2}}}, \tag{10.68}$$

$$\text{percent overshoot} = 100e^{\frac{-\zeta\pi}{\sqrt{1-\zeta^2}}}\ \% \tag{10.69}$$

$$\text{peak time } t_{\max} = \frac{\pi}{\omega_n\sqrt{1-\zeta^2}}. \tag{10.70}$$

The $0 - 100\%$ rise time is given by:

$$t_r = \frac{\pi - \theta}{\omega_n\sqrt{1-\zeta^2}}. \tag{10.71}$$

Though the exact calculation of the settling time is difficult, for $0.3 \le \zeta \le 0.9$, the following approximation is usually sufficient,

$$5\% \text{ settling time} \approx \frac{3}{\zeta\omega_n}. \tag{10.72}$$

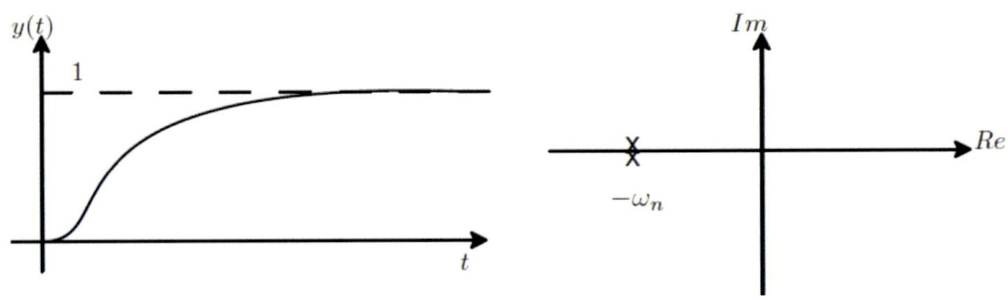

Figure 10.13: The step response of a critically damped ($\zeta = 1$) second-order system with no zeros is monotonic. The repeated real poles at $-\omega_n$ are denoted with a double X.

Critically damped ($\zeta = 1$): The transfer function has two, repeated real poles and can be written as

$$G(s) = \frac{\omega_n^2}{(s + \omega_n)^2}. \tag{10.73}$$

The step response is computed to be:

$$y(t) = \left(1 - e^{-\omega_n t} - \omega_n t e^{-\omega_n t}\right) u_{\text{stp}}(t). \tag{10.74}$$

By computing the time derivative of the output, it can be shown that $\dot{y}(t) > 0$ for all $t \ge 0$, and therefore, there are neither oscillations nor overshoot; moreover, the peak time is infinite. An approximation of the settling time is given by

$$5\% \text{ settling time} \approx \frac{4.8}{\omega_n}. \quad (10.75)$$

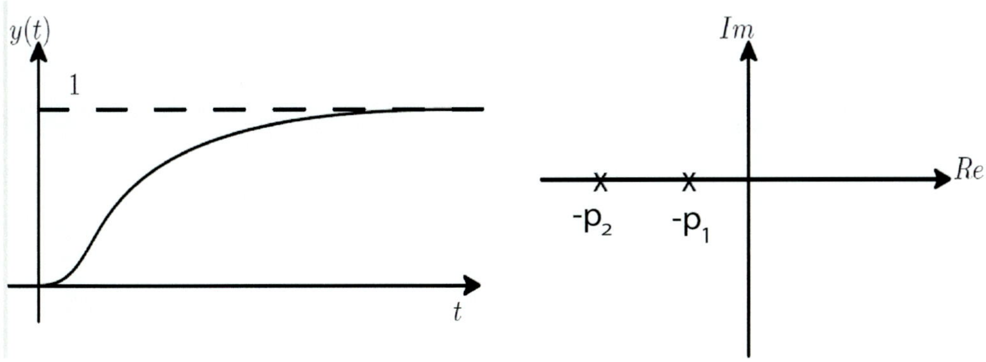

Figure 10.14: Step response of an overdamped, second-order system with no zeros.

Overdamped ($\zeta > 1$): The transfer function has two distinct real poles and can be written as

$$G(s) = \frac{p_1}{s+p_1} \cdot \frac{p_2}{s+p_2} = \frac{1}{\frac{1}{p_1}s+1} \cdot \frac{1}{\frac{1}{p_2}s+1},$$

where

$$\begin{aligned} p_1 &= \omega_n\left[\zeta - \sqrt{\zeta^2-1}\right] \\ p_2 &= \omega_n\left[\zeta + \sqrt{\zeta^2-1}\right]. \end{aligned} \quad (10.76)$$

Following the method of partial fraction expansion, the step response is computed to be

$$y(t) = \left(1 - \frac{p_2}{p_2-p_1}e^{-p_1 t} - \frac{p_1}{p_1-p_2}e^{-p_2 t}\right) u_{\mathrm{stp}}(t). \quad (10.77)$$

By computing the time derivative of the output, it can also be shown that $\dot{y}(t) > 0$ for all $t \geq 0$, which implies that the output monotonically approaches its final value, and thus there is no overshoot and the peak time is infinite, as shown in Fig. 10.14.

Because there is no overshoot, the primary qualitative feature of the output is its settling time.

1. For $1 < \zeta < 1.5$, the poles are close to one another and the terms in (10.77) decay at similar rates. An exact value for the settling time is difficult to compute, but a good approximation can be shown to be:

$$5\% \text{ settling time} \approx \frac{6.4}{\omega_n}(\zeta - 1) + \frac{4.8}{\omega_n}. \quad (10.78)$$

2. For $\zeta \geq 2$, the poles are far apart because $p_2 \geq 10p_1$. In this case, the last term in (10.77) decays much more rapidly than the second term. Moreover,

$$\frac{p_2}{p_2-p_1} \approx 1,$$

and thus,

$$y(t) \approx \left(1 - e^{-p_1 t}\right) u_{\mathrm{stp}}(t), \quad (10.79)$$

which is the step response of a first-order system with time constant $\tau = 1/p_1$, that is,

$$G(s) \approx \frac{1}{\frac{1}{p_1}s+1}. \quad (10.80)$$

Consequently, the 5% settling time can be estimated from the corresponding formula for a first-order system, namely

$$5\% \text{ settling time} \approx \frac{3}{p_1}.$$

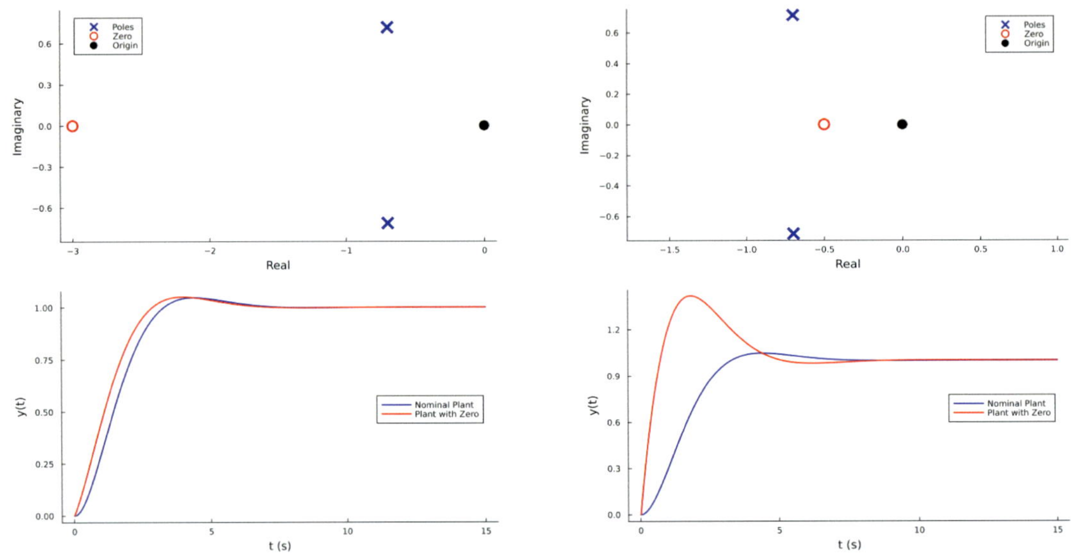

(a) (b)

Figure 10.15: (Stable Zeros Increase Overshoot) In (a), the zero is far to the left of the poles, and its effect if small, while in (b), the zero is near the poles and greatly increases the overshoot.

10.7.4 Effects of Zeros

Zeros in a transfer function can profoundly alter the transient response. The magnitude of their effect depends on where they are located with respect to the poles. This will be illustrated by considering a single zero in a second-order system,

$$\bar{G}(s) = \frac{(\frac{1}{z}s+1)\omega_n^2}{s^2+2\zeta\omega_n s+\omega_n^2}, \tag{10.81}$$

which has a zero at $-z$. For the purposes of our analysis, the zero has been introduced in such a way that it does not change the DC gain of the transfer function. While this excludes $z = 0$, it is fine from a practical point of view. The value of z can be either positive or negative, placing the zero in the left-half or right-plane plane, respectively.

As shown in **Figures 10.15** and **10.16**, the effects of a zero can be quite startling. It depends on the location of the zero.

Underdamped $0 < \zeta \le 1$

- $z < 0$, then the zero is in the right-half plane, which is officially called a **nonminimum phase zero**, though we will call it informally an **"unstable zero"**. A very special phenomenon happens: the step response exhibits **undershooting**, meaning that the output initially moves in a direction opposite to the commanded value. Zeros in the left-half plane do not cause undershooting.
- $|z| > 10\zeta\omega_n$, the zero can be neglected in the sense that the step response is essentially indistinguishable from that of a system without a zero.
- $|z| \approx 5\zeta\omega_n$ the zero can be neglected because the step response exhibits only a slight increase in overshoot and settling time from a system without a zero.
- $|z| \approx 2\zeta\omega_n$, the zero has a significant effect on the step response. For $z > 0$, the 0-100% rise time is reduced by 50%, and the overshoot is approximately doubled. For $z < 0$, the undershoot is roughly double the nominal overshoot.
- $|z| \approx \zeta\omega_n$, the zero has a profound effect on the step response. For $z > 0$, the 0-100% rise time is reduced by 75%, and the overshoot is quadrupled. For $z < 0$, the undershoot may be 100% or more.
- $|z| < 0.5\zeta\omega_n$, the step response is hardly recognizable as that of a second-order system!

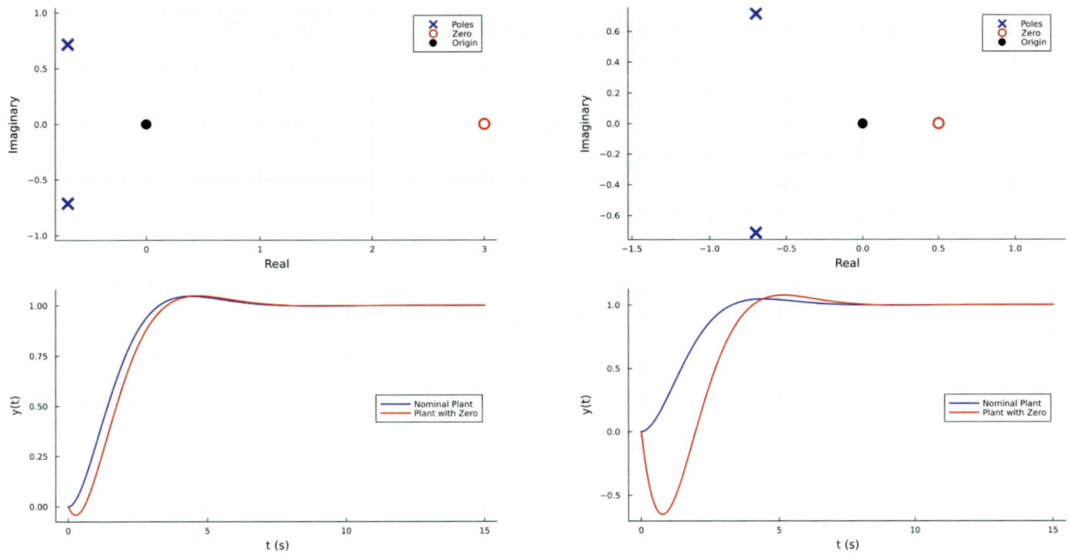

(a) (b)

Figure 10.16: (Unstable Zeros Cause Undershooting) In (a), the zero is far to the right of the origin, and its effect is small, while in (b), the zero is closer to the origin and causes a lot of undershooting.

Overdamped and Critically Damped, $\zeta \geq 1$

Zeros can also strongly affect the behavior of overdamped systems. When the zero is in the right-half plane, undershooting occurs. When the zero is in the left-half plane and is to the right of the two real poles, overshooting occurs. Even though there is overshooting, the response does not oscillate. The output reaches a peak value and then decays monotonically to its steady-state value. When the zero is between the two poles or to the left of both poles, no overshooting occurs.

10.7.5 (Optional Read) Why Zeros Alter the Transient Response

Let $\widetilde{Y}(s)$ be the output of the transfer function in (10.81), when the input is a unit step. $\widetilde{Y}(s)$ can be written as:

$$\widetilde{Y}(s) = Y(s) + \frac{1}{z} sY(s),$$

where $Y(s)$ is the step response of a standard second-order system without a zero. In the time domain, recalling that $\frac{d}{dt} \leftrightarrow s$, we have

$$\tilde{y}(t) = y(t) + \frac{1}{z}\frac{dy(t)}{dt}.$$

When z is large in magnitude, its reciprocal is small, and the added derivative term has little effect, meaning that $\tilde{y}(t) \approx y(t)$. When z is small in magnitude, its reciprocal is large, and the derivative term can dominate the sum, in which case, $\tilde{y}(t)$ is very different from $y(t)$. For zeros in the left-half plane, $z > 0$. It follows that when $y(t)$ is increasing, its derivative is positive, and therefore $\tilde{y}(t) = y(t) + \frac{1}{z}\frac{dy(t)}{dt} > y(t)$, leading to faster response and increased overshoot. When $z < 0$, it can be shown that:

$$\left.\frac{d\tilde{y}(t)}{dt}\right|_{t=0} < 0,$$

and thus $\tilde{y}(t)$ initially decreases, giving rise to undershooting.

10.8 Design of Cascade Compensators and Precompensators

Control system design is in some sense the opposite of control system analysis. In analysis, one seeks to predict the response of a system to inputs. In design, one seeks to select the components of a system so that its output exhibits "desired behavior". This section tackles the problem of designing closed-loop systems to achieve stability, rapid response, low oscillation, and low steady-state error. It will draw upon all of the results we have developed throughout the chapter.

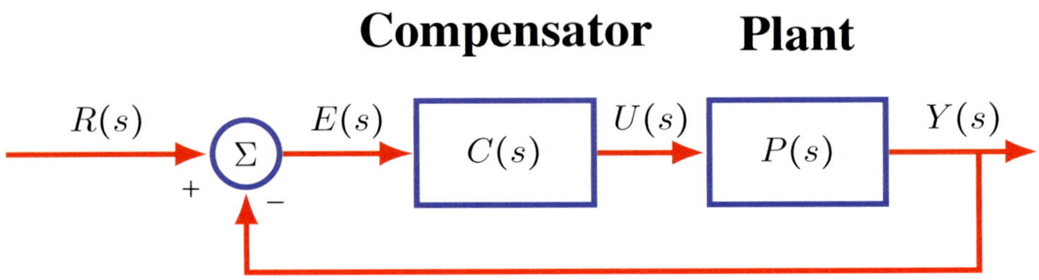

Figure 10.17: Reminder on a cascade compensator and plant in a unity feedback loop. This section seeks to design $C(s)$. It's easy to forget that we are manipulating ODEs, but thanks to Laplace, $s \longleftrightarrow \frac{d}{dt}$.

10.8.1 First-Order Systems

First-order systems provide an opportunity to explore feedback concepts without too many complications. Consider, therefore, a first-order plant, possibly unstable and without a zero,

$$P(s) = \frac{k_O}{s + p_O}. \tag{10.82}$$

The initial design objective is to create a stable, closed-loop system with a 5% settling time of T_s seconds for a step input. The design process begins by translating the specification for the transient response to the desired location of the closed-loop pole, or, equivalently for a first-order system, the desired time constant. From our analysis of first-order systems, we obtain

$$\boxed{\tau_{des} = \frac{T_s}{3}.} \tag{10.83}$$

Proportional Feedback for Stabilization of First-Order Plants

The next step in the design process is to select the compensator and solve for the unknown gain or gains. The simplest compensator is proportional feedback in a unity feedback configuration, as in **Fig. 10.17**; hence we start with

$$C(s) = K_P. \tag{10.84}$$

The closed-loop system is then

$$G_{cl}(s) = \frac{C(s)\,P(s)}{1 + C(s)\,P(s)} = \frac{K_P k_O}{s + p_O + K_P k_O},$$

which has time constant

$$\tau_{cl} = \frac{1}{p_O + K_P k_O}.$$

Setting τ_{cl} equal to τ_{des} and solving for the compensator gain yields

$$\boxed{K_P = \frac{\frac{1}{\tau_{des}} - p_O}{k_O}.} \tag{10.85}$$

The final step is to compute the step response and verify that the design meets the specifications.

Example 10.39. (Stable first-order system) *Suppose that $P(s) = \frac{2}{s+2}$, which has a time constant of 500 ms, and hence a 5% settling time of 1.5 s. It is desired that the closed-loop system be five times as fast, with a 5% settling time of 300 ms. Design an appropriate controller.*

Solution: A proportional compensator is selected. From (10.83), the desired time constant is 100 ms. Following the steps outlined above, or immediately applying (10.85), yields

$$K_P = 4.$$

The step response is shown in **Fig. 10.18**, along with the error between the reference command and the output. It is observed that the steady-state error is 0.2, in other words, the error is 20%.

A system that is open-loop unstable is illustrated next.

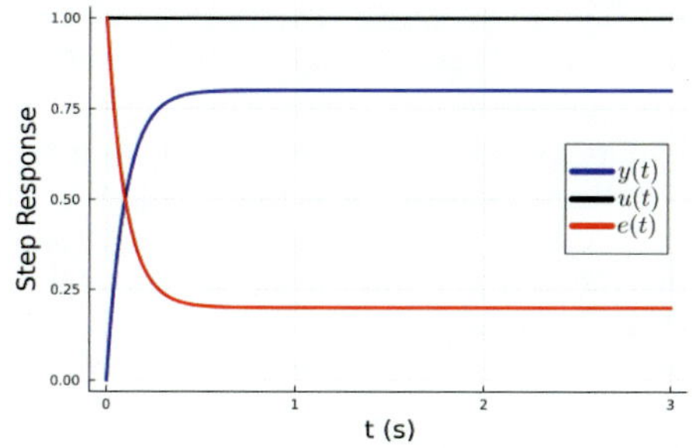

Figure 10.18: Step response and steady-state error for $P(s) = \frac{2}{s+2}$ and $C(s) = 4$. Black is the unit step input signal, blue is the closed-loop system's step response, and red is the tracking error of 20%.

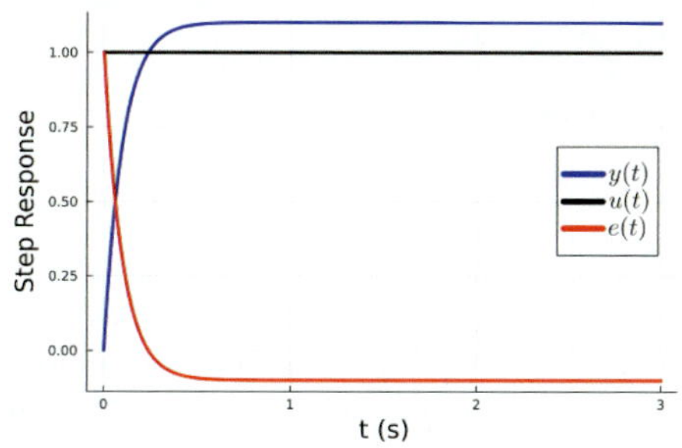

Figure 10.19: Step response and steady-state error for $P(s) = \frac{2}{s-1}$ and $C(s) = 5.5$. Black is the unit step input signal, blue is the closed-loop system's step response, and red is the tracking error of 10%.

Example 10.40. (Unstable first-order system) *Suppose that $P(s) = \frac{2}{s-1}$ and the objective is closed-loop stability with a 5% settling time of 300 ms. Design an appropriate controller.*

Solution: A proportional compensator is selected. From (10.83), the desired time constant is 100 ms and the required compensator gain is computed to be

$$K_P = 5.5$$

The step response is shown in **Fig. 10.19** along with the error between the reference command and the output. It is observed that the steady-state error is -0.1.

In most applications, the sign of the error is irrelevant, only its magnitude counts. Hence, even in the above case when $e_{SS} = -0.1$, we still say the error is 10%, meaning that its magnitude is 10%.

Fixing the Steady-State Error

In Examples 10.39 and 10.40, the steady-state error is 20% and 10%, respectively. Can it be reduced? There are two ways to accomplish this. Consider once again the general first-order plant given in (10.82) and the proportional compensator in (10.84). Applying Section 10.6.2 to the system yields the transfer function of the error,

$$\frac{E(s)}{R(s)} = \frac{s + p_0}{s + p_0 + K_P k_0}.$$

Assuming that K_P has been chosen to be consistent with BIBO stability (namely, $p_0 + K_P k_0 > 0$), then the steady-state error for a unit-step input is

$$e_{SS} = \frac{p_0}{p_0 + K_P k_0}. \tag{10.86}$$

It follows that

$$\lim_{|K_p|\to\infty} e_{SS} = 0.$$

The feedback gain can never be set to infinity, but the above analysis demonstrates that the magnitude of the steady-state error can be made as small as desired by selecting $|K_P|$ to be sufficiently large, assuming, once again, that closed-loop stability can be assured for large values of the feedback gain, and assuming we have a powerful enough actuator to handle the resulting control signal. Powerful motors on mobile devices require heavy batteries. Engineering is all about trade-offs.

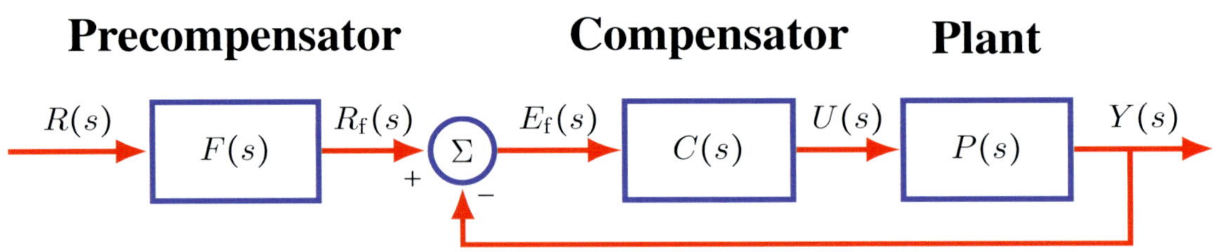

Figure 10.20: A unity feedback loop with a precompensator (aka prefilter) $F(s)$ added to the overall system. The prefilter can be used to ameliorate the effects of left-half plane zeros or to "fix" steady-state error. The overall structure of the diagram is called a **two-degree of freedom controller**: one degree of freedom is the forward-path compensator $C(s)$ that can now focus more on stabilization, while the second degree of freedom is the precompensator, $F(s)$, that can take care of some of the transient and steady-state tasks that were not handled by $C(s)$.

There is a second means of reducing the steady-state error that does not involve high controller gains. Figure 10.20 shows a precompensator inserted in the system. The true error remains as

$$E(s) := R(s) - Y(s),$$

which is different from the "filtered error" $E_f(s)$. After a bit of algebra, and assuming closed-loop stability, we compute for a unit-step input,

$$e_{SS} = \frac{1 + (1 - F(0))\, C(0)\, P(0)}{1 + C(0)\, P(0)} = \frac{1 + C(0)\, P(0) - F(0)\, C(0)\, P(0)}{1 + C(0)\, P(0)}. \tag{10.87}$$

Setting $e_{SS} = 0$ and solving for $F(0)$ yields

$$\boxed{F(0) = \frac{1 + C(0)\, P(0)}{C(0)\, P(0)}.} \tag{10.88}$$

For Example 10.39, we can let the precompensator be the constant gain,

$$F(s) := K_f = \frac{5}{4},$$

and for Example 10.40 we can let the precompensator be the constant gain,

$$F(s) := K_f = \frac{10}{9}.$$

All we are doing is exploiting linearity: by inserting a scale factor, a "dummy" reference signal, R_f, is created that achieves the desired steady-state error between the real reference signal and the measured output.

10.8.2 Second-Order Systems

We now consider a second-order system without a zero,

$$P(s) = \frac{k_0}{s^2 + a_1 s + a_0}. \tag{10.89}$$

As in the case of first-order systems, we will first address the transient response, which includes stability. When using a PD controller, it will insert a (stable) zero that will "mess up" our design considerations. In a second design phase, we will address the zero with a precompensator. We will also take up steady-state error.

The design process begins by translating the specifications for the transient response into the desired location of the closed-loop poles. We will continue to assume that one of the specifications is the 5% settling time. For simplicity and definiteness, we will take the second specification to be the percent overshoot. From Section 10.7.3, the poles are uniquely determined by the damping ratio ζ and the undamped natural frequency ω_n. If we further assume that $0 < \zeta < 1$, then the transient specifications relate to the desired closed-loop poles via

$$\begin{aligned} \zeta \quad &\leftrightarrow \quad \text{percent overshoot} \quad = \quad 100e^{\frac{-\zeta\pi}{\sqrt{1-\zeta^2}}}\ \% \\ \omega_n \quad &\leftrightarrow \quad 5\%\ \text{settling time} \quad \approx \quad \frac{3}{\zeta\omega_n}. \end{aligned} \tag{10.90}$$

In the above expression, the limit as $\zeta \to 1$ yields no overshoot. Common numbers to keep in mind for the damping ratio are given in the table below. Only novice control system designers use the exact formula for the damping ratio given by (10.90); experienced designers realize that some tweaking will be required in the end no matter what, because the model is never exact in practice.

ζ	$\approx$ Overshoot	True Overshoot
1.0	0 %	0.00 %
0.9	0 %	0.15 %
0.8	2 %	1.5 %
0.7	5 %	4.6 %
0.6	10 %	9.5 %
0.5	15 %	16.3 %

PD Control

The next step in the design process is to select the compensator and solve for the unknown gain or gains. A PD compensator will be assumed in a unity feedback configuration, as in Fig. 10.17. Hence we start with

$$C(s) = K_{\mathrm{P}} + K_{\mathrm{D}}s = K_{\mathrm{D}}\left(s + \frac{K_{\mathrm{P}}}{K_{\mathrm{D}}}\right).$$

The closed-loop system is then

$$\begin{aligned} G_{cl}(s) &= \frac{C(s)P(s)}{1 + C(s)P(s)} \\ &= k_0 \frac{K_{\mathrm{P}} + K_{\mathrm{D}}s}{s^2 + (a_1 + K_{\mathrm{D}}k_0)s + (a_0 + K_{\mathrm{P}}k_0)} \\ &= k_0 \frac{K_{\mathrm{D}}(s + K_{\mathrm{P}}/K_{\mathrm{D}})}{s^2 + (a_1 + K_{\mathrm{D}}k_0)s + (a_0 + K_{\mathrm{P}}k_0)}, \end{aligned}$$

which shows that the characteristic equation is

$$s^2 + \underbrace{(a_1 + K_{\mathrm{D}}k_0)}_{2\zeta\omega_n}\, s + \underbrace{(a_0 + K_{\mathrm{P}}k_0)}_{\omega_n^2} = 0.$$

Identifying the various terms and performing elementary algebra show the relationship between the compensator gains and the closed-loop poles to be

$$\begin{aligned} K_{\mathrm{P}} &= \frac{\omega_n^2 - a_0}{k_0} \\ K_{\mathrm{D}} &= \frac{2\zeta\omega_n - a_1}{k_0}. \end{aligned} \tag{10.91}$$

Due to the PD compensator, the closed-loop transfer function has a zero at $-K_P/K_D$. Recall that in Section 10.7.4 we showed that a zero can significantly alter the step response of a system, which will motivate a modification to our unity feedback configuration.

Before moving on, this video on a "poorly tuned" PD controller shows the importance of what we are studying: Control System Tuning by BPS.shorts.

Example 10.41. ***(Stabilizing an oscillator)*** *Given that $P(s) = \frac{1}{0.5s^2+2}$, which has poles $\pm\mathbf{i}\,2$, it is desired that the closed-loop system have less than 5% overshoot and a 5% settling time of 700 ms. Show that the step response of the open-loop system has sustained oscillations and then design a controller to meet the objectives for the closed-loop system.*

Solution: In the s-domain (aka Laplace domain), the response of the open-loop system to a step input is

$$\begin{aligned} Y(s) &= P(s)\frac{1}{s} \\ &= \frac{2}{(s+\mathbf{i}\,2)(s-\mathbf{i}\,2)}\frac{1}{s} \\ &= \frac{A_1}{s} + \frac{A_2}{s+\mathbf{i}\,2} + \frac{A_3}{s-\mathbf{i}\,2}, \end{aligned}$$

where $A_1 = \frac{1}{2}$, $A_2 = -\frac{1}{4}$, $A_3 = -\frac{1}{4}$. Taking the inverse Laplace transform term by term and applying Euler's formula leads to

$$\begin{aligned} y(t) &= \frac{1}{2}u(t) - \frac{1}{4}e^{-\mathbf{i}\,2t}u(t) - \frac{1}{4}e^{\mathbf{i}\,2t}u(t) \\ &= \frac{1}{2}\left(1-\cos(2t)\right)u(t). \end{aligned}$$

Hence, the step response of the open-loop system is a sustained oscillation at a frequency of 2 rad/s.

For the controller, a PD compensator is assumed. From the overshoot specification, the damping ratio is selected to be $\zeta = 0.7$. From the settling time specification and (10.90), the undamped natural frequency is computed to be $\omega_n = 6.1$. Following the control design steps outlined above, or immediately applying (10.91), yields:

$$K_P = 16.7 \text{ and } K_D = 4.3.$$

The step response is shown in **Fig. 10.21**, along with the error between the reference command and the output. The steady-state error is 11%. The 5% settling time is 1.2 s, almost twice the design value. The percent overshoot is evaluated to be

$$\frac{y_{\max} - y_\infty}{y_\infty} \times 100\% = \frac{1.13-0.89}{0.89} \times 100\% = 27\%,$$

more than five times the design value.

To understand why the transient performance differs form the design specifications, we have to recall that the PD compensator introduced a zero in the forward path of the feedback loop. Where did the zero of the PD compensator end up in relation to the closed-loop poles? The zero is at:

$$-\frac{K_P}{K_D} = -3.88,$$

which places it to the right of the poles, which have real part

$$-\zeta\omega_n = -4.27.$$

Based on **Fig. 10.15**, a large overshoot is expected. We can address this through the two-degree-of-freedom design in **Fig. 10.20**, where this time, $F(s)$ will be more than just a gain. The transfer function from reference to output is:

$$\frac{Y(s)}{R(s)} = F(s)\,\frac{C(s)\,P(s)}{1+C(s)\,P(s)},$$

which is simply the product of the transfer functions of the precompensator and the standard unity feedback loop.

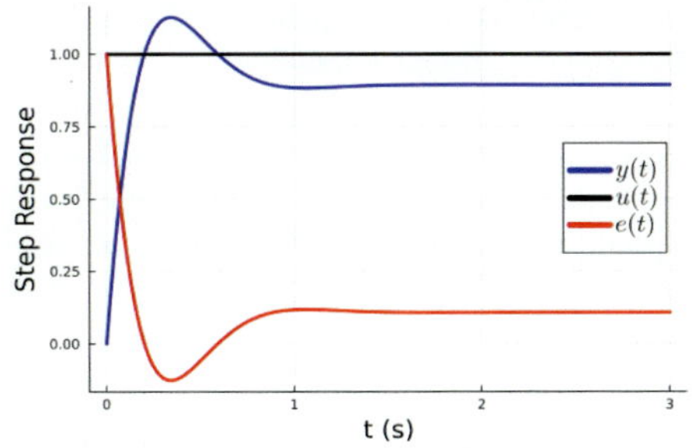

Figure 10.21: Step response and steady-state error for $P(s) = \frac{1}{0.5s^2+2}$ and $C(s) = 16.7 + 4.3s$. The zero in the forward path induced by the PD compensator (controller) causes more overshoot than is desired. This can be fixed with a precompensator.

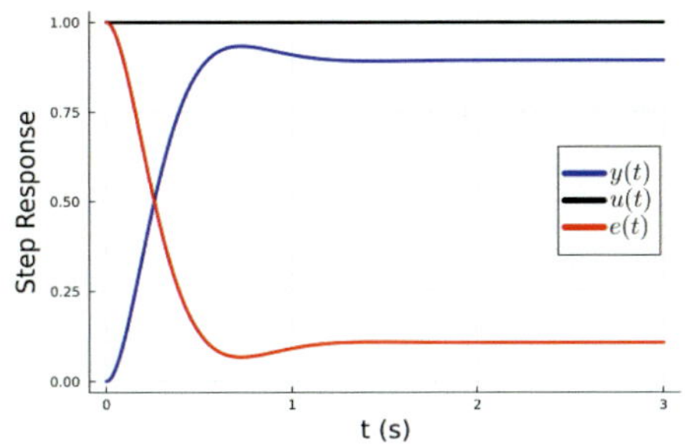

Figure 10.22: Step response when using a precompensator. The plant is $P(s) = \frac{1}{0.5s^2+2}$, $C(s) = 16.7+4.3s = 4.3(s+3.88)$ and $F(s) = \frac{3.88}{s+3.88}$.

The precompensator can be used to remove unwanted left-half plane zeros in the transfer function by the deliberate creation of **stable pole-zero cancellations**. If there are no unstable pole-zero cancellations, the control loop with the precompensator is BIBO stable if, and only if, both the precompensator and the control loop without the precompensator are BIBO stable. Moreover, if $F(0) = 1$, that is, the precompensator has unity DC gain, then the steady-state error for a unit-step input is the same, with or without the precompensator, namely

$$e_{\mathrm{SS}} = \frac{1}{1 + C(0)\,P(0)}\,.$$

Example 10.42. ***(Stabilizing an oscillator continued)*** *Consider again the system in Example 10.41, where it was desired that the closed-loop system have less than 5% overshoot and a 5% settling time of 700 ms. Use a two-degree-of-freedom controller to remove the troublesome zero introduced by the PD compensator.*

Solution: The PD compensator $C(s)$ is retained. The precompensator is selected as

$$F(s) = \frac{z}{s+z} = \frac{1}{\frac{s}{z}+1}, \tag{10.92}$$

where $z = \frac{K_{\mathrm{P}}}{K_{\mathrm{D}}} = 3.88$. This cancels the zero with a stable pole. The step response of the system is shown in Fig. 10.22. The overshoot and settling time specifications are now met.

Remark: We leave it to the reader to adjust the DC gain of the precompensator, $F(s)$, using (10.88) to zero the steady-state error.

10.8.3 Dealing with More Complicated Systems

The methods we have been following have allowed us to assign all of the poles of first- and second-order systems in order to meet a set of transient and steady-state design specifications. Assigning all the poles breaks down rapidly beyond the cases we have considered. To be sure, a few more tricks can be brought to bear, thereby slightly extending the range of systems that can be treated using the approach developed in this chapter, and these extensions will be outlined in the design examples of the next section. Nevertheless, faced with a plant with four poles and a zero, even these extensions do not work very often. So, what is an aspiring control designer to do?

At the junior or senior undergraduate level, most universities offer a course in feedback control. There, you may be exposed to the Root Locus Method of Evans and the Frequency Domain Methods of Nyquist and Bode. These methods will allow you to treat single-input, single-output, linear plants of considerable complexity. You will also be able to consider more comprehensive design specifications. At the MS level, many universities offer courses in multivariable linear systems, nonlinear systems, stochastic systems, adaptive systems, discrete-event systems, and much more. Control systems is a very active research area, rich with theoretical questions and exciting novel applications.

10.9 Feedback Design for a Linearized Model of a Planar Segway Transporter

The key transfer functions for the Segway are given in Example 10.20 and Fig. 10.4. We will stabilize the body lean angle before moving on to regulating speed.

10.9.1 Controlling Body Lean Angle

The relevant transfer function is

$$\frac{\theta(s)}{\tau(s)} := G_\theta(s) = \frac{-21.468}{s^2 - 50.937}. \tag{10.93}$$

The transfer function comes from the linearized model of the Segway developed in Section 10.3.6. It is a simple second-order transfer function with poles at $\pm\sqrt{50.937} = \pm 7.137$. Hence, the system is unstable and requires feedback control to remain upright.

Iteration # 1

We will use the time-domain specs from Example 10.42, namely less than 5% overshoot and a 5% settling time of 700 ms. We will also use a two-degree-of-freedom controller to remove the troublesome zero that we will introduce via a PD compensator.

We need to turn the specifications into gains for the PD compensator, using (10.91), where for this plant, $k_0 = -21.468$, $a_0 = -50.937$ and $a_1 = 0$. Turning the crank, we obtain,

$$\begin{aligned}
\zeta &= 0.7 \\
\omega_n &= \frac{3}{\zeta \cdot T_s} = 6.122 \\
K_P &= \frac{\omega_n^2 - a_0}{k_0} = -4.12 \\
K_D &= \frac{2\zeta\omega_n - a_1}{k_0} = -0.40,
\end{aligned}$$

hence the PD compensator is

$$C_{PD}(s) = K_D\left(s + \frac{K_P}{K_D}\right) = -0.28(s + 11.55).$$

With these parameters, we can compute the steady-state error,

$$e_{SS} = \frac{D(0)}{D(0) + N(0)} = \frac{a_0}{a_0 + k_0 K_P} = -1.36 \text{ or } 136\%,$$

which is unacceptably large. We choose to take a different approach to the performance specs.

```
# Define plant transfer function params
# P1 = k0/(s^2 + a1s + a0)
k0 = -21.468
a1 = 0
a0 = -50.937

# Set time domain specs for the closed-loop system
```

```
zeta = 0.7 # 5 % overshoot
Ts = 0.7

# Ts = 3.0/(zeta*wn)
wn = 3.0/(zeta*Ts)

# PD gains
Kp= (wn^2 - a0)/k0
Kd= (2*zeta*wn - a1)/k0

# Zero location for PD controller
# Kd*(s+z)
z = Kp/Kd

# Steady state error
# N(s)/D(S) = C(s)*P(s)
#           = (Kp + Kds)* k0/(s^2 + a1s + a0)
#
# eSS = D(0) / (D(0) + N(0))
eSS = a0 /(a0 + Kp*k0)

# Print relevant data
println("zeta = $zeta")
println("Ts = $Ts")
println("wn = $wn")
println("Kp = $Kp")
println("Kd = $Kd")
println("z  = $z")
println("eSS = $eSS")
```

Output

```
zeta = 0.7
Ts = 0.7
wn = 6.122448979591837
Kp = -4.118752632183022
Kd = -0.3992653517527749
z  = 10.315827842565596
eSS = -1.3588859666666666
```

Iteration #2

We now know that steady-state error is a big challenge for this system. We can flip the script and design for a given $e_{\rm SS}$ and overshoot (related to damping ratio, ζ), leaving the settling time (related to ω_n) as a free parameter.

For the design, we retain $\zeta = 0.7$ and specify $e_{\rm SS} = 20\%$. Both $e_{\rm SS} = 0.20$ and $e_{\rm SS} = -0.20$ satisfy the specification. Which one to choose? The one that gives real numbers in the calculations below! It turns out that we need $e_{\rm SS} = -0.20$ (try the other one, and you will agree).

Following this plan, from the expression for steady-state error, we obtain the proportional gain,

$$K_{\rm P} = \frac{a_0 - a_0\, e_{\rm SS}}{k_0\, e_{\rm SS}} = -14.236.$$

The undamped natural frequency is then:

$$\omega_n = \sqrt{K_{\rm P}\, k_0 + a_0} = 15.959,$$

and the derivative gain,

$$K_{\rm D} = \frac{2\zeta\, \omega_n - a_1}{k_0} = -1.041,$$

giving the PD compensator

$$\boxed{C(s) = -14.236 - 1.041\, s = -1.041(s + 13.679).} \tag{10.94}$$

The 5% settling time is:

$$T_s \approx \frac{3}{\zeta\,\omega_n} = 0.269.$$

In the code, we also design the precompensator, $\boldsymbol{F(s)}$, using (10.92) to cancel the stable zero introduced by the PD compensator and we adjust its DC gain using (10.88) to set the steady-state error to zero. The resulting transfer function from the commanded value for the lean angle, $\theta^{\rm des}(s)$, to $\theta(s)$ is then

$$\boxed{\frac{\theta(s)}{\theta^{\rm des}(s)} = F(s)\frac{C(s)\,P(s)}{1+C(s)\,P(s)} = \frac{254.685}{s^2+22.342\,s+254.685},} \tag{10.95}$$

after removing the stable pole-zero cancellation. We note that the transfer function has a DC gain of one.

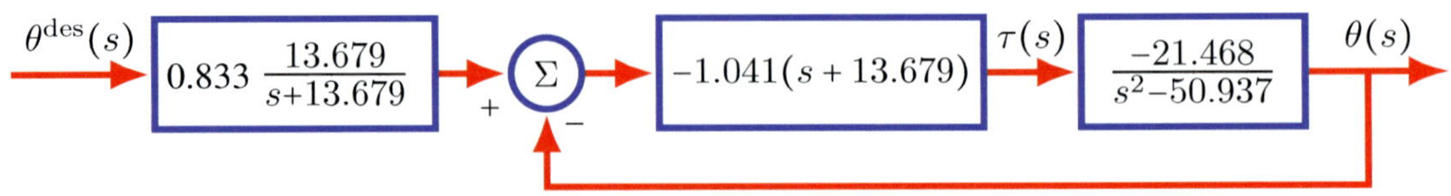

Figure 10.23: Closed-loop system to stabilize the lean angle or orientation of the Segway's body. The overall transfer function is given in (10.95), where the stable pole-zero cancellation has been removed. The result is a standard second-order system with a DC gain of one.

```
# Define plant transfer function params
# P = k0/(s^2 + a1s + a0)
k0= -21.468
a1 = 0
a0 = -50.937

# Specifications
zeta = 0.7 # 5% overshoot
eSS = -0.20

# Proportional Gain
Kp = (a0 - a0*eSS)/(k0*eSS)

# Undamped natural frequency
# becomes complex with eSS = +0.20
wn = sqrt(Kp*k0 + a0)

# Derivative gain
Kd= (2*zeta*wn - a1)/k0

# Zero location for PD controller
# Kd*(s+z)
z = Kp/Kd

# Settling time
Ts = 3/(zeta*wn)

# Steady state error to double check
# N(s)/D(S) = C(s)*P(s)
#           = (Kp + Kds)* k0/(s^2 + a1s + a0)
#
# eSS = D(0) / (D(0) + N(0))
eSS = a0 /(a0 + Kp*k0)

# precompensator DC gain
kF = (1 + Kp*k0/a0)/(Kp*k0/a0)

# Print relevant data
println("zeta = $zeta")
```

```
println("Ts = $Ts")
println("wn = $wn")
println("Kp = $Kp")
println("Kd = $Kd")
println("z  = $z")
println("eSS = $eSS")
println("kF = $kF")
```

Output

```
zeta = 0.7
Ts = 0.2685477580249118
wn = 15.958853342267417
Kp = -14.236165455561764
Kd = -1.0407301415676533
z  = 13.679017150514934
eSS = -0.20000000000000004
kF = 0.8333333333333333
```

Next, we check the step response to see that the specs are actually met.

```
using ControlSystems, Plots, LaTeXStrings

# Define the Plant, Controller, and Precompensator Transfer Functions
P = tf([k0], [1, a1, a0])  # Plant transfer function
C = tf([Kd, Kp], [1])  # Controller transfer function
F = tf([kF], [1/z, 1])  # Precompensator transfer function


# Combine the Transfer Functions
G = C * P

# Create a Unity Feedback Loop
CL = feedback(G, 1)

# Multiply by the precompensator
sys = F * CL

# Compute and Plot the Step Response
t = 0:0.01:1  # Time vector
y, t = step(sys, t)
p1 = plot(t, y', lw=3, label=false, guidefont = 15,
    xlabel="t (s)", ylabel=L"$\theta(t)$ Step Response (rad.)")



tfU = F*C/(1+C*P)
# Compute and Plot the Step Response
t = 0:0.01:1  # Time vector
u, t = step(tfU, t)

p2 = plot(t, u', lw=3, label=false, guidefont = 15,
    xlabel="t (s)", ylabel=L"$\tau(t)$ Step Response (N)")

display(p1)
display(p2)
```

Output See Fig. 10.24.

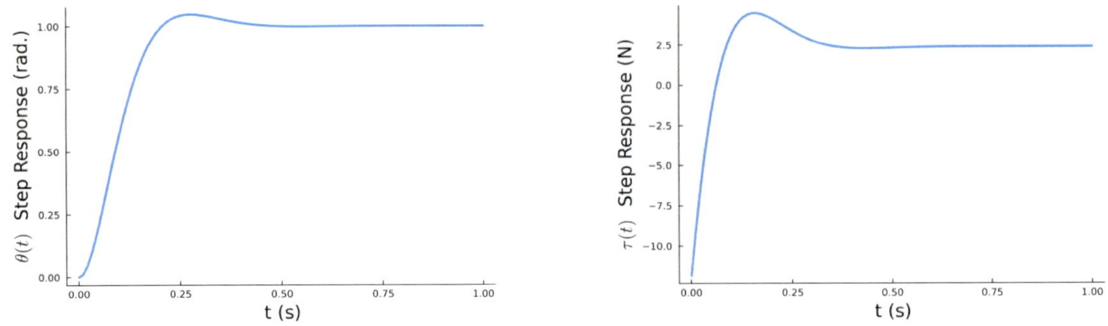

(a) (b)

Figure 10.24: **Step responses:** (a) Lean angle of the Segway and rider. The overshoot is 5% and the steady-state error is zero. (b) The torque required to achieve the step response. The peak of −10 N is a lot of torque, but you have to realize that a "unit step" in θ is one radian (approximately 60°). Because the system is linear, the torque required for 10° would be −1.67 N, which is realistic. The sign just specifies the direction the motor is spinning.

10.9.2 Controlling Speed and Lean Angle

Our objective now is to control speed, in addition to orientation. Of course, we cannot just add another actuator to the overall system. While we could perhaps switch back and forth from controlling one quantity for a few tens of milliseconds and then the other, that would make for a really interesting ride! We need a better idea. It is time to exploit the factorization in Fact 10.21, namely

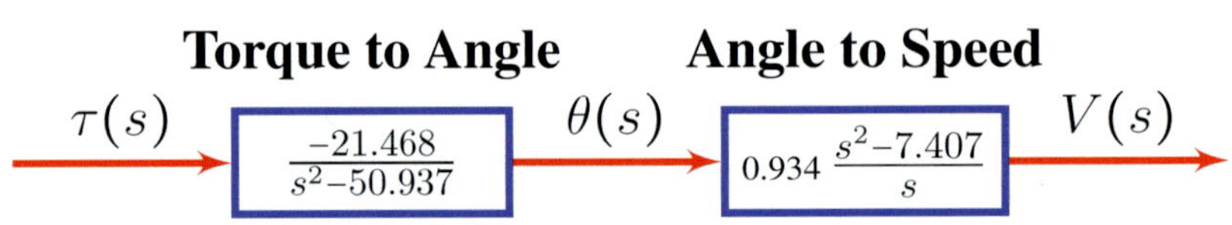

Recall that this comes from the two transfer functions for the Segway given in (10.37),

$$G_\theta(s) = \frac{\theta(s)}{\tau(s)} = -21.468\,\frac{1}{s^2 - 50.937}$$

and

$$G_v(s) = \frac{V(s)}{\tau(s)} = -20.057\,\frac{s^2 - 7.404}{s(s^2 - 50.937)},$$

which share a common term. Taking their ratio to eliminate the common factor yields

$$\frac{V(s)}{\theta(s)} = 0.934\,\frac{s^2 - 7.407}{s};$$

that is,

$$\boxed{V(s) = \left(0.934\,\frac{s^2 - 7.407}{s}\right)\theta(s).} \tag{10.96}$$

Hence, appending the transfer function $\left(0.934\,\frac{s^2-7.407}{s}\right)$ to the output of the feedback system in **Fig. 10.23** results in the new output being speed, and the input (which we treat as a pseudo actuator) being $\theta^{\rm des}$.

While **Fig. 10.25** may look complicated, from our work on regulating the body orientation, namely (10.95), we have that

$$\frac{\theta(s)}{\theta^{\rm des}(s)} = \frac{254.685}{s^2 + 22.342\,s + 254.685},$$

which allows us to replace the first three blocks of the feedback diagram with a standard second-order transfer function, yielding

$$V(s) = \left(0.934\,\frac{s^2 - 7.407}{s}\right)\left(\frac{254.685}{s^2 + 22.342\,s + 254.685}\right)\theta^{\rm des}(s),$$

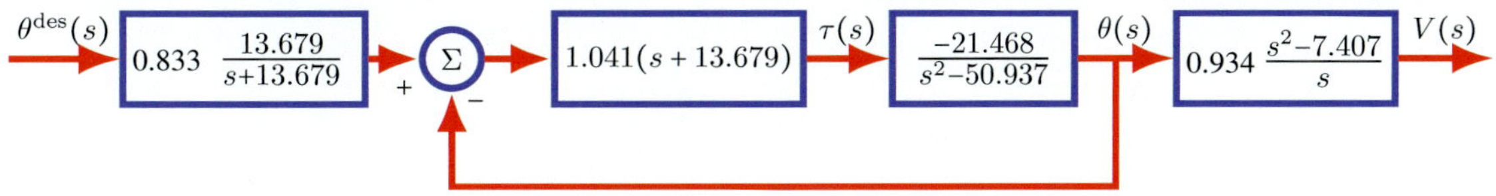

Figure 10.25: The speed Portion of the Segway's transfer function has been appended to the closed-loop system that stabilizes the orientation of the Segway's body.

or equivalently,

$$V(s) = \left(0.934\,\frac{(s-2.722)(s+2.722)}{s}\right)\left(\frac{254.685}{s^2+22.342\,s+254.685}\right)\theta^{\rm des}(s). \tag{10.97}$$

Hence, we can now focus on the transfer function,

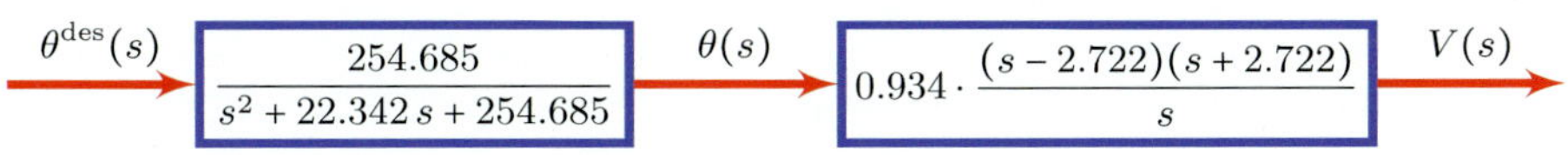

We need to stabilize the pole at the origin and mitigate the effects of the zeros. We cannot do anything about the unstable zero, but we can cancel the stable zero with the transfer function

$$\frac{2.722}{s+2.722}.$$

Because the poles of $\frac{254.685}{s^2+22.342\,s+254.685}$ are $-11.171 \pm 11.397\,\mathbf{i}$, which are sufficiently far into the left half plane, it turns out that, similar to a pole of a first-order transfer function, the pole at the origin can be stabilized with proportional feedback. Hence, for our second compensator, we propose the simple transfer function,

$$C_2(s) = K_2\frac{2.722}{s+2.722}, \tag{10.98}$$

where the gain K_2 will be tuned to achieve a "nice step response", that is, one that is sufficiently fast, but does not overshoot. We do not need to worry about the steady-state error because it will be zero due to the pole at the origin.

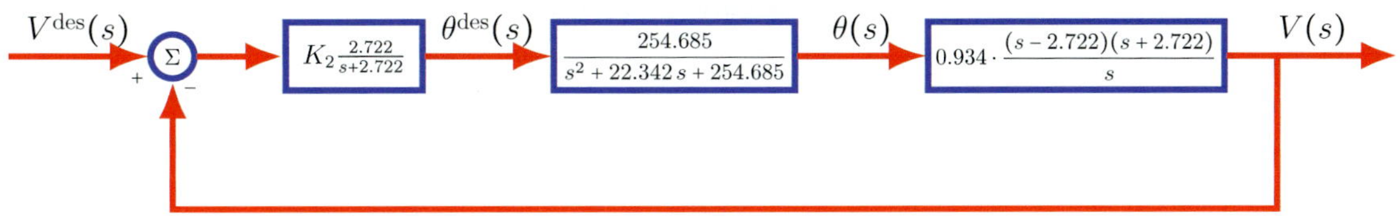

Figure 10.26: K_2 must be adjusted to obtain a nice step response. This can be done in simulation. Later, as an optional read, we show that the appropriate range for BIBO stability is $-0.227 < K_2 < 0$. The negative value for K_2 is due to the term $(s-2.722)$.

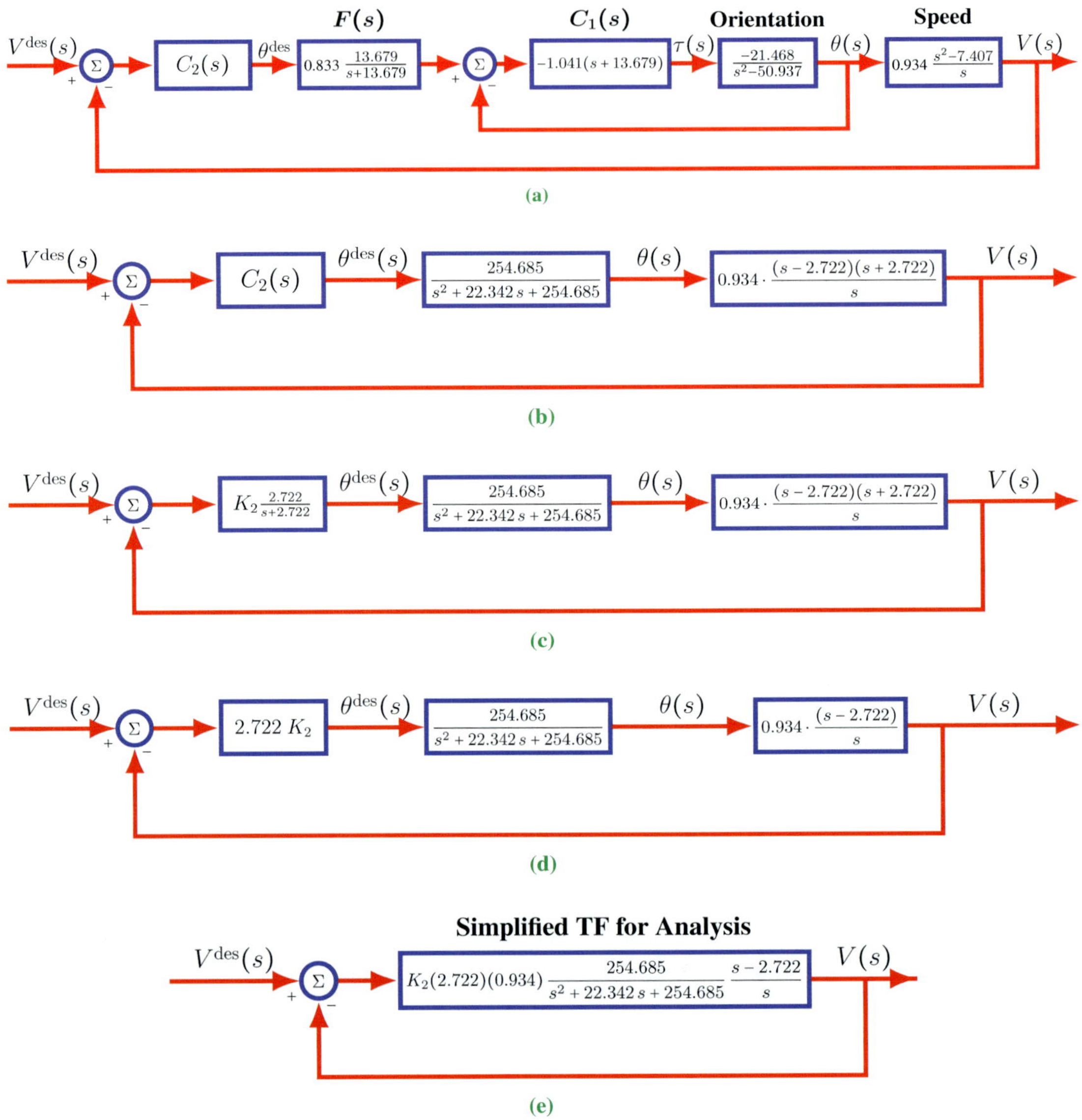

Figure 10.27: (Steps for Understanding the Design of the Speed Controller) This part of the design is harder because the transfer function involved is more complicated. However, because of how we designed $F(s)$ and $C_1(s)$, it simplifies greatly. The transfer function from θ^{des} to θ is a standard second-order transfer function; moreover, the numerator of the speed transfer function factors into two real zeros. This makes the design of $C_2(s)$ look much more approachable in (b) and (c). We hence propose that $C_2(s)$ as a proportional gain K and a "precompensator-like" term that cancels the "stable" zero. The overall transfer function simplifies in (e) to a third-order system with a zero. Its stability can be analyzed in closed-form.

(Optional Read) Analysis Required to Stabilize the Speed

To analyze how to regulate the Segway's forward speed using feedback control, we begin by constructing the full open-loop transfer function of the speed control system. The proposed control architecture consists of a compensator $C_2(s)$ in series with the system's dynamics, all placed in a unity feedback loop.

The three main components in the forward path are:

- The speed transfer function,
$$\frac{V(s)}{\theta(s)} = 0.934 \cdot \frac{(s-2.722)(s+2.722)}{s}.$$
- The closed-loop lean angle dynamics,
$$\frac{\theta(s)}{\theta^{\text{des}}(s)} = \frac{254.685}{s^2 + 22.342\,s + 254.685}.$$
- The proposed compensator for speed control that deliberately cancels a stable zero,
$$C_2(s) = K_2 \cdot \frac{2.722}{s+2.277}.$$

The full open-loop transfer function is the product of these three components,

$$G(s) = C_2(s) \cdot \left(\frac{\theta(s)}{\theta^{\text{des}}(s)}\right) \cdot \left(\frac{V(s)}{\theta(s)}\right) = C_2(s) \cdot \left(\frac{V(s)}{\theta(s)}\right) \cdot \left(\frac{\theta(s)}{\theta^{\text{des}}(s)}\right).$$

Substituting each part in, we obtain,

$$\begin{aligned} G(s) &= \left(K_2 \cdot \frac{2.722}{s+2.277}\right) \cdot \left(0.934 \cdot \frac{(s-2.722)(s+2.722)}{s}\right) \cdot \left(\frac{254.685}{s^2 + 22.342\,s + 254.685}\right) \\ &= (2.722K_2) \cdot \left(0.934 \cdot \frac{(s-2.722)}{s}\right) \cdot \left(\frac{254.685}{s^2 + 22.342\,s + 254.685}\right) \\ &= K_3 \cdot \left(\frac{s-2.722}{s}\right) \cdot \left(\frac{254.685}{s^2 + 22.342\,s + 254.685}\right), \end{aligned}$$

where we defined $K_3 := 2.542\,K_2$ and observed that the zero at $s = -2.722$ in the speed transfer function is canceled by the pole at $s = -2.722$ in the compensator, $C_2(s)$. The latter is an intentional cancellation of the stable zero by a stable pole, leaving only the unstable zero at $s = +2.722$.

The terms in the standard second-order system came from setting overshoot and 5%-settlng time. For a reason that will become clear shortly, we rewrite $G(s)$ as:

$$G(s) = K_3 \cdot \left(\frac{s - z_2}{s}\right) \cdot \left(\frac{\omega_n^2}{s^2 + 2\zeta\omega_n\, s + \omega_n^2}\right),$$

where $K_3 = 2.542\,K_2$, $z_2 = 2.722$, $\zeta = 0.7$, and $\omega_n = 15.959$. The closed-loop transfer function, with unity feedback, is then given by,

$$\frac{V(s)}{V^{\text{des}}(s)} = \frac{G(s)}{1+G(s)} = \frac{K_3\,\omega_n^2\,(s-z_2)}{s^3 + \underbrace{2\zeta\omega_n}_{a_2}\, s^2 + \underbrace{\omega_n^2(1+K_3)}_{a_1}\, s + \underbrace{(-K_3\,\omega_n^2\, z_2)}_{a_0}}.$$

This expression forms the basis for analyzing the stability and performance of the system from our simplified stability rules for third-order polynomials, namely,

$$\begin{aligned} a_0 > 0 &\iff K_3 < 0 \text{ because } z_2 > 0 \\ a_2 > 0 &\iff \text{automatic from the definitions of } \zeta \text{ and } \omega_n \\ a_1 - \frac{a_0}{a_2} > 0 &\iff 1 + K_3\left(1 + \frac{z_2}{2\,\zeta\,\omega_n}\right) > 0 \iff 1 + K_3\left(1 + \frac{z_2\,T_s}{6}\right) > 0 \iff K_3 > -\frac{6}{6 + z_2\,T_s}, \end{aligned}$$

where $0.269 = T_s \approx \frac{3}{\zeta\omega_n}$ is the value we used for the 5% settling time. The above computations yield:

$$-0.891 < K_3 < 0 \iff -0.351 < K_2 < 0$$

for the range of gains that will stabilize the speed loop, once we substitute in for T_s and $K_3 = 2.542\,K_2$.

Step Responses

The final feedback control diagram is given in **Fig. 10.28** and a family of step responses for various values of K is given in **Fig. 10.29**.

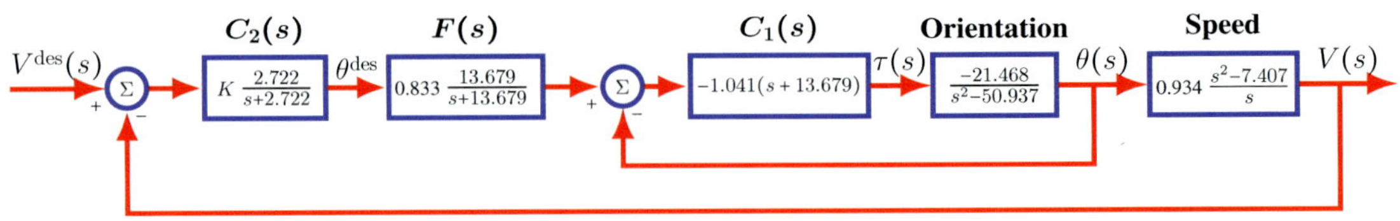

Figure 10.28: (Inner-Outer Feedback Loops for the Segway) There are two unity feedback loops. The inner loop is from **Fig. 10.23**, our controller for stabilizing the orientation of the Segway. It includes everything from $\theta^{\rm des}(s)$ (the input to the precompensator, $F(s)$) all the way to $\theta(s)$, the output of the orientation block. The other blocks and the left-most summing junction form the outer unity feedback loop for stabilizing the Segway's speed.

```
using ControlSystems, Plots, LaTeXStrings


# Plant transfer function for the inner feedback loop
# from theta desired to theta
Pa = tf([254.685], [1, 22.342, 254.685])

# remaining part of the Segway's transfer fuction from torque to speed
Pb = tf([0.934, 0, -0.934*7.407], [1.0, 0])

# design transfer function for the outer feedbck loop
P2 = Pa*Pb # Final Plant TF

# Search over gains to find a good one
# remainder of the controller is inside the for loop
K = -0.1*[1, 1.5, 2, 2.5]

# Create empty plot that we can add to in the loop
p1 = plot(guidefont = 15)

# For loop to compute multiple step responses
for i = 1:length(K)

# Compensator depending on K
# It cancels a stable zero
C2 = tf([K[i]*2.722], [1, 2.722])
G2 = C2*P2

# Outer Unity feedback loop
sys2 = feedback(G2, 1)

# Compute and Plot the Step Response
t = 0:0.01:5.0

# Time vector
v, t = step(sys2, t)

# Create a LaTeX string for the legend
K_label = string("K = ", round(K[i],digits=2))

# Add to plot
```

```
p1 = plot!(p1, t, v', lw=3, label=K_label)

end


# Make the plot
p1 = plot!(p1, xlabel="t (s)", ylabel=L"$v(t)$ Step Response (m/s)")
```

Output See **Fig. 10.29**.

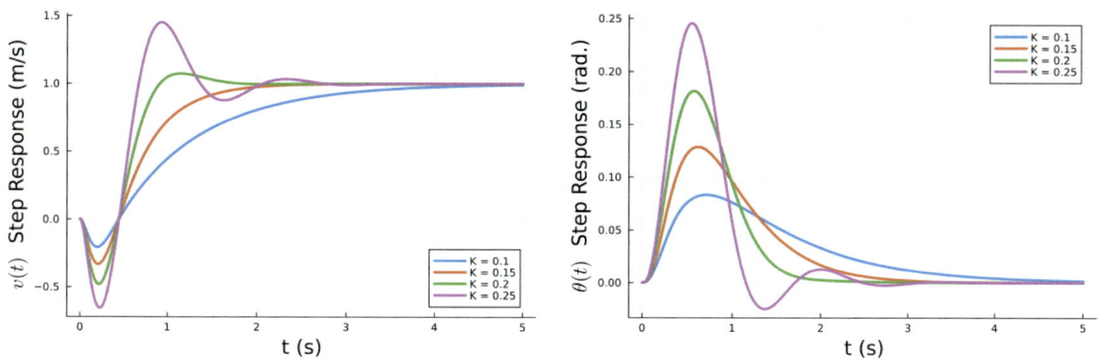

(a) (b)

Figure 10.29: Step responses: Part (a) shows the step response in speed for a range of gains $K := K_2$ in $C_2(s)$, while part (b) shows the corresponding lean angle trajectories. It is worth looking at both plots. Here, a unit step in speed means 1 m/s, a comfortable walking speed, hence somewhat slow in Segway terms. For $K = 0.2$, there is some overshoot in the speed response, but perhaps not too much. However, in the lean angle, we see approximately 0.17 radians, or $19.5°$ of undershoot, which is a lot. The safer bet is likely $K = 0.15$ or even 0.1.

10.9.3 The Real Deal: Implementing the Controller on the Nonlinear Model

It is now time for the rubber to meet the road, literally. Our goal is to implement the controller on the Segway's nonlinear model. We break the overall goal into simpler steps:

(a) Separate the controller terms in **Fig. 10.28** from the model terms.

(b) Transform the controller from the Laplace domain into ODEs in the time domain.

(c) Verify the controller on the linearized ODE model of the Segway before evaluating it on the nonlinear model. Because the controller worked well on the transfer function, it "has to work" on the ODE version of the linearized model, **unless our simulation code has a bug**.

(d) Replace the linear model with the nonlinear model. For "small" initial conditions, the response should be nearly identical to the linearized model. For "large" initial conditions, all bets are off.

Isolating the Controller

In **Fig. 10.28**, two of the transfer functions generate the Segway's orientation, $\theta(s)$, and cart speed, $V(s)$. These input signals to the controller will come from the nonlinear state-variable model of the Segway. On a real Segway or other mobile robot, they would come from sensors on the robots. When we remove model transfer functions from the block diagram, we are left with **Fig. 10.30**.

Where the feedback blocks required $\theta(s)$ and $V(s)$ from model blocks, we have moved them to the left of the diagram to show that they are now inputs to the controller along with the desired speed, $V^{\rm des}(s)$. The block diagram in **Fig. 10.30**

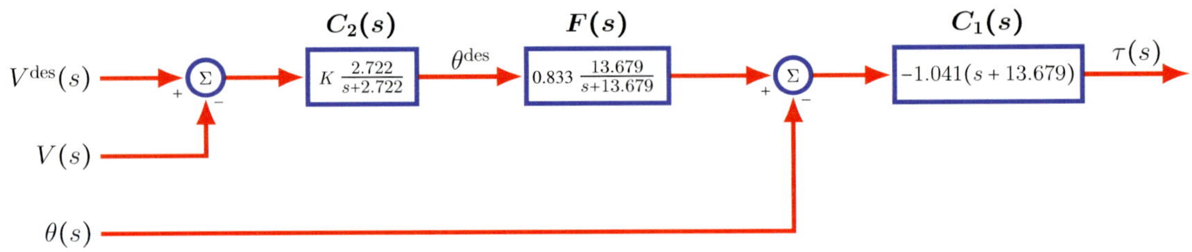

Figure 10.30: (Just the Controller for the Segway) Once the transfer function blocks for the Segway's model are removed from **Fig. 10.28**, we are left with just the controller blocks. The controller has three input signals: the desired speed, v^{des}, the measured speed, v, and the measured orientation or lean angle, θ. It has one output signal, the motor torque, τ.

is a representation of the following linear equation

$$
\begin{aligned}
\boldsymbol{\tau(s)} &= C_2(s)\,F(s)\,C_1(s)\,\Big(\boldsymbol{V^{\text{des}}(s) - V(s)}\Big) - C_1(s)\,\boldsymbol{\theta(s)} \\
&= \Big[\; C_2(s)\,F(s)\,C_1(s) \quad -C_1(s) \;\Big] \cdot \begin{bmatrix} \boldsymbol{V^{\text{des}}(s) - V(s)} \\ \boldsymbol{\theta(s)} \end{bmatrix},
\end{aligned}
\tag{10.99}
$$

where we have (temporarily) highlighted the input and output signals to make them easier to locate in the expressions.

Turning the Transfer Function Representation of the Controller into a State-Variable Model

To merge the control calculation in (10.99) with the ODEs of the Segway model (linear or nonlinear), we need to transform the transfer function computations back into ODEs. This is possible because each s in (10.99) corresponds to a $\frac{d}{dt}$, thanks to the Laplace transform. While the conversion is not so hard to accomplish by hand, it is better to do it in software. Here, we use the package `ControlSystems.jl`.

There is one sticky point: the PD controller $K_{\text{D}}\left(s + \frac{K_{\text{P}}}{K_{\text{D}}}\right)$ is not proper. The practical solution is to use a **"dirty derivative"** (yes, that is its name; you cannot make that up). The controller is rewritten as

$$K_{\text{D}}\left(\frac{s}{\tau_d s + 1} + \frac{K_{\text{P}}}{K_{\text{D}}}\right),$$

where τ_d is the time constant of the dirty derivative; in the code below, it is chosen to be 10 ms. With this modification, the PD controller is now proper and can be converted to an ODE.

```
using ControlSystems

# Build function to simplify process of going from Transfer Function
# Representation to State-variable model

function transferFunction2LinearODE(Sys_tf)
    # Convert to state-variable representation
    Sys_ss = ss(Sys_tf)
    # dx/dt = A x + B u
    #     y = C x + D U

    # Reduce to minimal and balanced state-space representation
    # Explained in Michigan EECS 565 Linear Feedback Control Systems
    #
    tol = 1e-3  # for stable pole-zero cancellations
    Sys_minimal = minreal(Sys_ss, tol)
    # Improve numerical charcteristics
    Sys_ODE, ~, ~ = balreal(Sys_minimal) # the tildes are important
```

```

    # Sys_ODE matrices
    #     A = Sys_ODE.A
    #     B = Sys_ODE.B
    #     C = Sys_ODE.C
    #     D = Sys_ODE.D

    return Sys_ODE
end
```

Output

```
transferFunction2LinearODE (generic function with 1 method)
```

```
using ControlSystems


# Create a causal (proper) PD controller that admits a state variable
# model
tau_d = 1e-2  # Dirty derivative time constant
s = tf("s")
s_dirty = s / (tau_d * s + 1) # dirty derivative

# Controllers and Filter
    # gains
K1 = -1.041  # Gain for C1(s)
K2 = 0.15    # Gain for C2(s)
K3 = 0.833   # gain for F(s)
    # TFs
C1 = K1 * (s_dirty + 13.679)
C2 = K2 * tf([2.722], [1, 2.722])
F = K3 * tf([13.679], [1, 13.679])

# Open-loop multi-input transfer function from [V^{\rm des}, V, theta]
# to tau
# Based on the diagram that isolates the feedback controller
contoller_TF = [C2 * F * C1  -C1]

# Convert to state-variable representation
contoller_ODE = transferFunction2LinearODE(contoller_TF)
Ac = contoller_ODE.A
Bc = contoller_ODE.B
Cc = contoller_ODE.C
Dc = contoller_ODE.D

println("State-Variable Representation of the Controller:")
display(Ac)
display(Bc)
display(Cc)
display(Dc)
```

Output

```
State-Variable Representation of the Controller:
3×3 Matrix{Float64}:
 -99.9969       -4.37006    1.09924
  -0.0695394    -2.91994    1.44874
  -0.00121461    1.44793  -13.4842

3×2 Matrix{Float64}:
  0.0517967  102.029
```

```
 -2.22881       0.073081
  0.560787      0.000209314

1×3 Matrix{Float64}:
 -102.029  -2.23001  0.560787

1×2 Matrix{Float64}:
 0.0  118.34
```

Function to implement the linearized Segway model

```
# Linear Segway Model

function SegwayLinearModel(dx, x, tau, t)
    A = [
        0.0        0.0  1.0  0.0
        0.0        0.0  0.0  1.0
       50.9373    -0.0  0.0  0.0
     -162.687     -0.0  0.0  0.0]
    b=[
       0.0
       0.0
     -21.467741053946913
       80.22714380349518]
    dx = A*x + b*tau
    return dx
end
```

Output

```
SegwayLinearModel (generic function with 1 method)
```

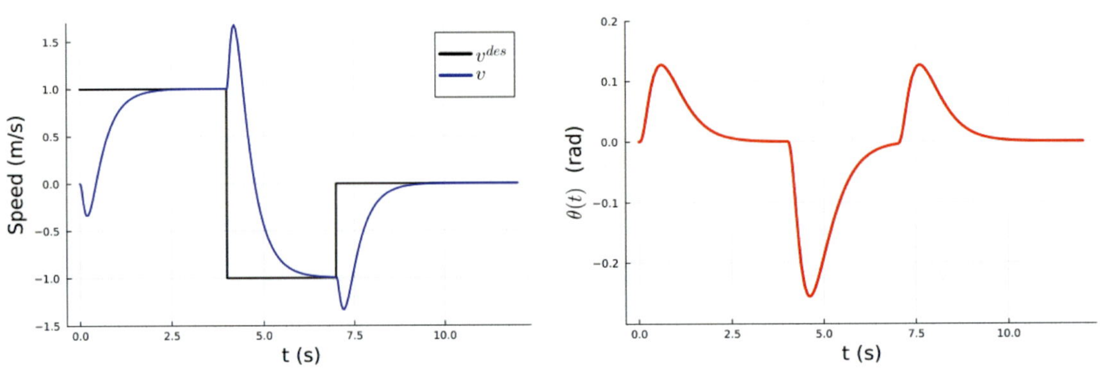

(a) (b)

Figure 10.31: Tracking a Speed Profile with the Linearized Segway Model: Part (a) shows a time-varying speed profile as the reference command and the closed-loop linear system's response, while part (b) shows the corresponding lean angle trajectory.

Put the Controller in Closed-Loop with the Linear Model

Custom function to make it easy to form a closed-loop system

```
# Custom function to make it easy to form a closed-loop system

function simulate_control_system(plant_model, contoller_ODE, K_IC,
                                 referenceCommand, xm0, xc0, tspan,p)
```

```
    # Combined state-space representation
    function combined_system(dx, x, p, t)
        # Extract model and controller states
        xm, xc = x[1:length(xm0)], x[(length(xm0)+1):end]    # Split the
        # state variables

        # Check if referenceCommand is a function or a constant
        if isa(referenceCommand, Function)
            # referenceCommand is a function
            Ref = referenceCommand(t)
        else
            # referenceCommand is a constant
            # Define behavior when referenceCommand is a constant
            Ref = referenceCommand
        end
        # Input to the controller
        # Calculate uc from K_IC, uRef, and xm
        uc = K_IC * [xm;Ref]
        # uc[1] = max(min(uc[1], 0.5), -0.5)
        # Output of the controller
        tau = contoller_ODE.C * xc + contoller_ODE.D * uc
        # tau is a 1 x 1 matrix. Must extract its value
        # Plant dynamics
        dxm = plant_model(dx[1:length(xm0)], xm, tau[1], t)
        dx[1 : length(xm0)] = dxm
        # Linear controller dynamics
        dx[(length(xm0)+1):end] = contoller_ODE.A * xc + contoller_ODE.B * uc
        # return for clarity
        return dx
    end
    # Form full set of initial conditions
    x0 = [xm0; xc0] # Combined initial state
    # Form and solve the differential equations
    prob = ODEProblem(combined_system, x0, tspan,p)
    sol = solve(prob,Tsit5())
    return sol
end
```

Output

```
simulate_control_system (generic function with 1 method)
```

Function to implement the linearized Segway model

```
# Linear Segway Model

function SegwayLinearModel(dx, x, tau, t)
    A = [
        0.0       0.0  1.0  0.0
        0.0       0.0  0.0  1.0
       50.9373   -0.0  0.0  0.0
     -162.687    -0.0  0.0  0.0]
    b=[
       0.0
       0.0
     -21.467741053946913
      80.22714380349518]
    dx = A*x + b*tau
    return dx
end
```

Output

```
SegwayLinearModel (generic function with 1 method)
```

Define a speed profile and simulate the closed-loop system

```julia
function vDes(t)
# Define a speed profile
    if t > 7
        vdes = 0.0
    elseif t > 4
        vdes = -1.0
    else
        vdes = 1.0
    end
    return vdes
end
using DifferentialEquations, Plots

function modelParameters(outSelect=0)
    g = 9.81 # m/s^2
    m_wh = 6 # kg mass of 2 wheels
    m_pend = 45+75 # kg mass of the person, platform, and the handlebar
    L = (45*.3 + 75*0.85)/(45+75) # m CoM for the person plus platform
    # measured from the axis of the wheels
    r_wh = 0.25 # m, Wheel radius
    J_pend = m_pend*L^2/6 # kg m^2 moment of inertia about the center of
    # mass
    J_wh = 0.1 # kg m^2 moment of intertia of each wheel
    J_rotor = 1e-4 # kg m^2 per rotor, one for each wheel
    N = 50 # gear ratio between each motor rotor and wheel
    if outSelect > 0
        # named tuple
        return (g=g, m_wh=m_wh, m_pend=m_pend, L=L, r_wh=r_wh, J_pend=J_pend,
            J_wh=J_wh, J_rotor=J_rotor, N=N)
    else # standard tuple
        return g, m_wh, m_pend, L, r_wh, J_pend, J_wh, J_rotor, N
    end
end

params = modelParameters(1)
xm0 = zeros(4,1)
xc0 = zeros(3,1)

# Interconnection matrix between model states, reference, and the inputs
# to the controller. Vdes assumed appended to the bottom of the
# vmoel's state vector
K_IC = zeros(2,length(xm0)+1)
# Row 1 forms Vdes - v
K_IC[1,5] = 1.0;
K_IC[1,4] = -params.r_wh # v =  dot_phi times wheel radius
# Row 2 picks off theta (orientation angle in radians)
K_IC[2,1] = 1.0

tspan = (0.0,12.0)
referenceCommand = 1.0
p=0.0
sol = simulate_control_system(SegwayLinearModel, contoller_ODE, K_IC,
                              vDes, xm0, xc0, tspan,p)

# Plot everything as a first check before making nice plots
pSol = plot(sol)
display(pSol)
```

```
# Reshape sol to be easier to work with
function niceFormat(sol)
    t = sol.t
    states = sol.u
    # Reshape each matrix to a row vector and stack them vertically
    x = vcat([reshape(mat, 1, :) for mat in states]...)
    return (t=t, x=x)
end

mySol=niceFormat(sol);

using Plots

t = mySol.t
x = mySol.x
theta = x[:,1]
phi = x[:,2]
dtheta = x[:,3]
dphi = x[:,4]

params = modelParameters(1)
# Level-1 Speed
v = params.r_wh*dphi

Ref = vDes.(t)

p1 = plot(t, Ref, lw=2, guidefont = 15, label=L"$v^{\rm des}$", legendfontsize=15,
     xlabel="t (s)", ylabel="Speed (m/s)", color=:black)
plot!(t, v, label=L"$v$", lw=2, color=:blue)

p2 = plot(t, theta, lw=3, label=false, guidefont = 15, legendfontsize=15,
    xlabel="t (s)", ylabel=L"$\theta(t)$  (rad)", color=:red)


display(p1)
display(p2)
```

Output The results are shown in Fig. 10.31.

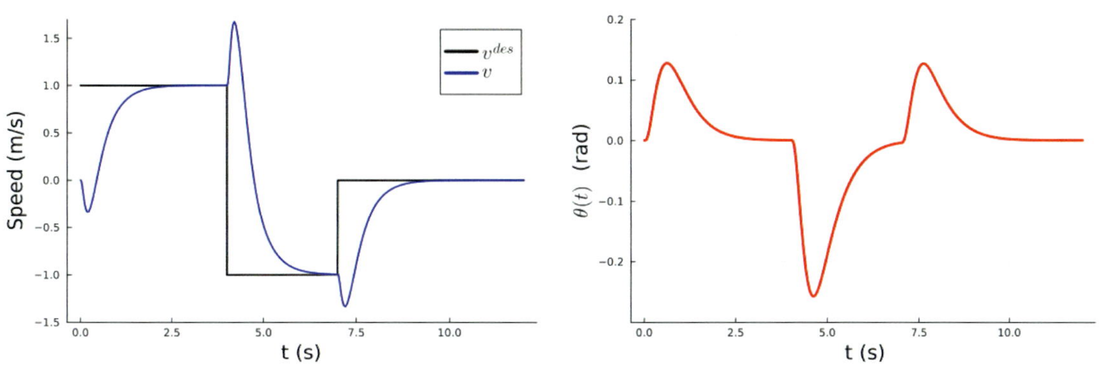

(a) (b)

Figure 10.32: **Tracking a Speed Profile with the Nonlinear Segway Model:** Part (a) shows a time-varying speed profile as the reference command and the closed-loop nonlinear system's response, while part (b) shows the corresponding lean angle trajectory. The results are indistinguishable from Fig. 10.31. Why is that? Because the controller maintains a small lean angle, the linear model is valid.

Put the Controller in Closed-Loop with the Nonlinear Model

We simply replace the linearized plant model with the nonlinear Lagrangian model of the Segway and re-run the previous closed-loop simulation code.

```
# bring the model into the workspace for later use
include("dyn_mod_planarSegway.jl")


function SegwayNonlinearModel(dx, x, tau, t)
     n = floor(Int, length(x) / 2) # In Julia n/2 is a Float64
     q = x[1:n]
     dq = x[n+1:end]
     model = dyn_mod_planarSegway(q, dq)
     dx1 = dq
     dx2 = (model.D) \ (-model.C*dq-model.G + model.B*tau) # note the use of
          # backslash
     dx=[dx1;dx2]
     return dx
end
```

Output The results are shown in Fig. 10.32

Our final trick of the trade: We add one line to the function that simulates the closed-loop system

```
# Original line
          uc = K_IC * [xm;Ref]
# Additional line
          uc[1] = max(min(uc[1], 0.5), -0.5)
```

which implements the following saturation on the error in the speed. Hence, if the speed change is huge, the controller does not over-respond. The downside is that the system is less responsive for both acceleration and deceleration. The upside is that the rider may do fewer face-plants. This idea can be very helpful for other feedback design problems.

$$\texttt{uc[1] = max(min(uc[1], 0.5), -0.5)} = \begin{cases} 0.5 & uc[1] > 0.5 \\ uc[1] & -0.5 \le uc[1] \le 0.5 \\ -0.5 & \text{otherwise.} \end{cases}$$

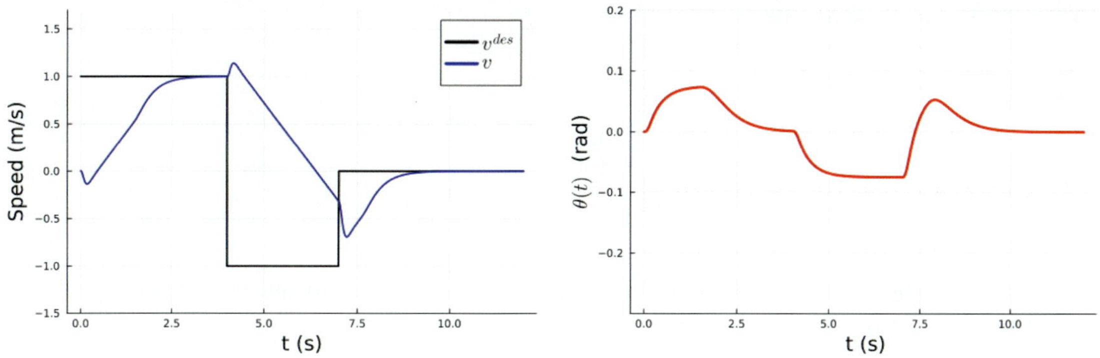

(a) (b)

Figure 10.33: Tracking a speed profile with the nonlinear Segway model after a saturation element was added to the feedback loop: Part (a) shows a time-varying speed profile as the reference command and the closed-loop nonlinear system's response, while part (b) shows the corresponding lean angle trajectory. The saturation element limits the controller's response to errors in tracking the reference speed to ± 0.5 m/s. This makes the system less responsive on the one hand, while drastically reducing lean angle on the other. Which is better? Depends on the planned use of the Segway.

10.10 (Optional Read) Robot Equations for a Planar Segway

The robot equations are based on the angles θ and φ as defined in Fig. 10.3.

```
function dyn_mod_planarSegway(q, dq)
# DYN_MOD_PLANARSEGWAY
# 2023-12-20 17:49:53
#
# Author: Grizzle
#
# Model NOTATION: D(q)ddq + C(q,dq)*dq + G(q) = B*tau
# The Robot Equations: From Lagrange's Equations of Motion
#
g, m_wh, m_pend, L, r_wh, J_pend, J_wh, J_rotor, N = modelParameters()
#
# Variable names for the model
theta, phi = q
dtheta, dphi = dq
#
D = zeros(2, 2)
  D[1, 1] = (J_pend*m_pend + J_pend*m_wh + (m_pend^2)*(r_wh^2) + (m_wh^2)*
            (r_wh^2) + 2.0J_wh*m_pend + 2.0J_wh*m_wh + (L^2)*(m_pend^2)*
            (cos(theta)^2) + (L^2)*(m_pend^2)*(sin(theta)^2) + 2.0m_pend*
            m_wh*(r_wh^2) + 2.0L*r_wh*(m_pend^2)*cos(theta) + 2.0L*m_pend*
            m_wh*r_wh*cos(theta)) / (m_pend + m_wh)
  D[1, 2] = 2.0J_wh + m_pend*(r_wh^2) + m_wh*(r_wh^2) + L*m_pend*r_wh*
            cos(theta)
  D[2, 1] = 2.0J_wh + m_pend*(r_wh^2) + m_wh*(r_wh^2) + L*m_pend*r_wh*
            cos(theta)
  D[2, 2] = 2.0J_wh + m_pend*(r_wh^2) + m_wh*(r_wh^2) + 2.0J_rotor*(N^2)
#
C = zeros(2, 2)
  C[1, 1] = (dtheta*(0.5(L^2)*(m_pend^2)*sin(2theta) - L*r_wh*(m_pend^2)*
            sin(theta) - (L^2)*(m_pend^2)*cos(theta)*sin(theta) - L*m_pend*
            m_wh*r_wh*sin(theta))) / (m_pend + m_wh)
  C[2, 1] = -L*dtheta*m_pend*r_wh*sin(theta)
#
G = zeros(2)
```

```
    G[1] = -L*g*m_pend*sin(theta)
    G[2] = 0
#
B = zeros(2, 1)
    B[2, 1] = 50
#
JacG = zeros(2, 2)
    JacG[1, 1] = -L*g*m_pend*cos(theta)
#
JacComx = zeros(2)
    JacComx[1] = (m_pend*(r_wh + L*cos(theta)) + m_wh*r_wh) / (m_pend + m_wh)
    JacComx[2] = (m_pend*r_wh + m_wh*r_wh) / (m_pend + m_wh)
#
    return (D=D, C=C, G=G, B=B, JacG=JacG, JacComx=JacComx)
end
```

10.11 (Optional Read) Proofs Associated with the Chapter

We start with a remark on Prop. 10.9: it is possible for a right-sided function $f(t)$ to have a Laplace transform and a derivative $f'(t)$ that is continuous on $[0,\infty)$, but yet $f'(t)$ does not have a Laplace transform. As an example, we use $f(t) = \sin\left(e^{t^2}\right) u_{\rm stp}(t)$; importantly, this function does not occur in engineering practice.

(a) **f has a Laplace transform:** For all $t \geq 0$ we have

$$|f(t)| = |\sin(e^{t^2})| \leq 1.$$

Therefore, for any $s \in \mathbb{C}$ with $\mathrm{real}(s) > 0$, the Laplace transform

$$F(s) = \int_0^\infty f(t) e^{-st}\, dt$$

satisfies

$$|F(s)| \leq \int_0^\infty \left|e^{-st}\right| dt \leq \int_0^\infty e^{-\mathrm{real}(s)t}\, dt = \frac{1}{\mathrm{real}(s)} < \infty.$$

Hence, f is Laplace transformable with region of convergence, $\mathrm{real}(s) > 0$.

(b) **f is continuously differentiable:** A straightforward application of the chain rule gives that

$$f'(t) = \cos\left(e^{t^2}\right) \cdot \frac{d}{dt}\left(e^{t^2}\right) = \cos\left(e^{t^2}\right) \cdot 2te^{t^2},$$

so that

$$f'(t) = 2te^{t^2} \cos\left(e^{t^2}\right)$$

exists and is continuous on $[0,\infty)$.

(c) **f' does not have a Laplace transform:** One has to show that the integral $\int_0^\infty f'(t) e^{-st}\, dt$ is divergent for all $s \in \mathbb{C}$. You can ask your favorite LLM to do this for you. The key issue is that the term e^{t^2} grows too rapidly and cannot be dominated by $e^{-\sigma t}$ for any $\sigma > 0$ finite. The oscillations due to $\cos\left(e^{t^2}\right)$ pose technical challenges in forming the bounds, but these can be overcome with methods developed in a course on real analysis.

■

Proof of Prop. 10.12 (Partial Fraction Expansion)

A proper rational function $F(s)$ with distinct roots $r_1, r_2, \ldots, r_n$ can be expanded as

$$F(s) = \sum_{j=1}^{n} \frac{k_j}{s - r_j},$$

where the coefficients k_j are called the residues of $F(s)$ at the poles r_j. The residues k_j can be computed using

$$k_j = \lim_{s \to r_j} F(s)(s - r_j).$$

Remark: If r_j is complex, the associated residue k_j will generally be a complex number. Its complex conjugate k_j^* corresponds to the residue at the conjugate pole r_j^*.

Proof: Start with the partial fraction form

$$F(s) = \sum_{j=1}^{n} \frac{k_j}{s - r_j}.$$

Multiply both sides by $(s - r_i)$, where r_i is a distinct root, which leads to

$$F(s)(s - r_i) = \sum_{j=1}^{n} k_j \frac{s - r_i}{s - r_j}.$$

At $s = r_i$, note that all terms in the summation vanish except the term where $j = i$, because for $j \neq i$, the numerator $(s - r_i)$ is zero, and hence

$$F(s)(s - r_i)\Big|_{s=r_i} = k_i.$$

It follows that

$$k_i = \lim_{s \to r_i} F(s)(s - r_i).$$

■

Proof of the Remark: Suppose $r_{j+1} = r_j^*$, where $r_j = a_j + \mathbf{i}\,\omega_j$ is a complex pole of $F(s)$, which means that $\omega_j \neq 0$. We start by writing the denominator associated with the poles r_j and r_j^* as the real polynomial,

$$(s - r_j)(s - r_j^*) = (s - a_j)^2 + \omega_j^2.$$

Define $G(s)$ by:

$$F(s) =: G(s) \cdot \frac{1}{(s - a_j)^2 + \omega_j^2} = G(s) \cdot \frac{1}{(s - r_j)(s - r_j^*)},$$

where $G(s)$ is a ratio of real polynomials because $F(s)$ is.

Step 1: Partial Fraction Expansion

Using a partial fraction expansion for the term $\frac{1}{(s-r_j)(s-r_j^*)}$, we can write:

$$F(s) = G(s) \cdot \left(\frac{k_j}{s - r_j} + \frac{k_{j+1}}{s - r_j^*} \right),$$

where k_j and k_{j+1} are the residues of $F(s)$ at r_j and r_j^*, respectively.

Step 2: Symmetry Property of $G(s)$

Because $F(s)$ has real coefficients, $G(s)$ also has real coefficients and satisfies the symmetry property:

$$G(s^*) = (G(s))^*.$$

Step 3: Relationship between k_j and k_{j+1}

From the residue formula, we know,

$$k_j = \lim_{s \to r_j} F(s)(s - r_j), \quad k_{j+1} = \lim_{s \to r_j^*} F(s)(s - r_j^*).$$

Substitute the expression for $F(s)$ into the formula for k_j,

$$k_j = \lim_{s \to r_j} G(s) \cdot \frac{1}{(s - r_j^*)} = G(r_j) \cdot \frac{1}{(r_j - r_j^*)} = G(r_j) \cdot \frac{1}{(2\,\mathbf{i}\,\omega_j)}.$$

Similarly, substitute the expression for $F(s)$ into the formula for k_{j+1},

$$k_{j+1} = \lim_{s \to r_j^*} G(s) \cdot \frac{1}{(s - r_j)} = G(r_j^*) \cdot \frac{1}{(r_j^* - r_j)} = G(r_j^*) \cdot \frac{1}{(-2\,\mathbf{i}\,\omega_j)}.$$

Step 3: Relationship between k_j and k_{j+1}

Using the symmetry property of $G(s)$, we know,

$$G(r_j^*) = G(r_j)^*.$$

Substitute this into the expression for k_{j+1},

$$k_{j+1} = G(r_j^*) \cdot \frac{1}{(-2\mathbf{i}\omega_j)} = G(r_j)^* \cdot \left(\frac{1}{(2\mathbf{i}\omega_j)}\right)^*.$$

Simplifying gives

$$k_{j+1} = \left(G(r_j) \cdot \frac{1}{(2\mathbf{i}\omega_j)}\right)^* = k_j^*.$$

Conclusion: The residue k_{j+1} at the pole r_j^* is the complex conjugate of the residue k_j at the pole r_j, as promised. ■

The purpose of partial fraction expansion is to express $F(s)$ as a sum of simple terms. This allows the inverse Laplace transform to be computed term by term, significantly simplifying the process of converting frequency-domain functions to their time-domain equivalents.

Proof of Prop. 10.25 (Poles and BIBO Stability)

A system with a proper rational transfer function is BIBO stable if, and only if, its poles have negative real parts. In other words, the stability of a rational transfer function

$$G(s) = \frac{N(s)}{D(s)},$$

is completely determined by its poles as long as the degree of the numerator polynomial is less than or equal to the degree of the denominator polynomial.

Proof: The proof assumes the poles are distinct, and the degree of the numerator polynomial is at least one less than the degree of the denominator polynomial. Moreover, we focus on showing that if there is a pole with a zero or positive real part, there is always a bounded input that results in an unbounded output.

Without loss of generality, we assume that the denominator polynomial is monic and write $D(s) = (s-r_1)(s-r_2)\cdots(s-r_n)$ as a product of its roots. Its partial fraction expansion becomes, therefore,

$$G(s) = \sum_{j=1}^{n} \frac{k_j}{s - r_j},$$

where r_j are the poles of the system, and k_j are the "residues", the official name of the coefficients of the partial fraction expansion! We make two observations:

- All of the terms k_j must be nonzero. This is because, otherwise, when the partial fraction expansion is recombined by placing the terms over a common denominator, the term corresponding to $(s - r_j)$ will not show up in the denominator, reducing its degree and eliminating r_j as a pole.
- If a pole is complex, its complex conjugate must also be a pole.

The response to a unit step input, $u_{\text{stp}}(t)$, is given by,

$$y(t) = \mathcal{L}^{-1}\left\{G(s) \cdot \frac{1}{s}\right\}.$$

We analyze two cases: real poles and complex poles.

Case 1: Real poles. If r_j is real, the corresponding term in $G(s)$ is $\frac{k_j}{s-r_j}$. For a unit step input,

$$Y_j(s) = \frac{k_j}{s - r_j}\frac{1}{s}.$$

- If $r_j \neq 0$, the Laplace transform becomes

$$\frac{k_j}{s-r_j}\frac{1}{s} = \alpha_0 \frac{1}{s} + \alpha_j \frac{1}{s-r_j} \quad \text{(by PFE)}$$
$$= -\frac{k_j}{r_j}\frac{1}{s} + \frac{k_j}{r_j}\frac{1}{s-r_j} \quad \text{(solving for the coefficients)},$$

and hence

$$y_j(t) = \frac{k_j}{r_j}\left(e^{r_j t} - 1\right) u_{\mathrm{stp}}(t).$$

If $r_j > 0$, the exponential term $e^{r_j t}$ grows unbounded as $t \to \infty$, causing $y(t)$ to blow up. If $r_j < 0$, the exponential term $e^{r_j t}$ decays to zero as $t \to \infty$.

- If $r_j = 0$, the term $\frac{1}{s-r_j}$ becomes $\frac{1}{s}$, resulting in

$$y_j(t) = k_j \mathcal{L}^{-1}\left\{\frac{1}{s^2}\right\} = k_j t\, u_{\mathrm{stp}}(t),$$

a ramp function, which also grows unbounded as $t \to \infty$.

Hence, for stability, all real poles must satisfy $r_j < 0$.

Case 2: Complex poles. Let $r_j = a_j + \mathbf{i}\omega_j$ with $\omega_j \neq 0$ (otherwise, it is not complex!), and assume r_j and its complex conjugate $r_{j+1} = r_j^* = a_j - \mathbf{i}\omega_j$ appear sequentially. The corresponding terms in $G(s)$ are $\frac{k_j}{s-r_j} + \frac{k_j^*}{s-r_j^*}$, and for a unit-step input,

$$Y_j(s) = \left(\frac{k_j}{s-r_j} + \frac{k_j^*}{s-r_j^*}\right)\frac{1}{s}.$$

Therefore,

$$\left(\frac{k_j}{s-r_j} + \frac{k_j^*}{s-r_j^*}\right)\frac{1}{s} = \alpha_0 \frac{1}{s} + \alpha_j \frac{1}{s-r_j} + \alpha_j^* \frac{1}{s-r_j^*} \quad \text{(by PFE)}$$
$$= -2\mathrm{real}\left(\frac{k_j}{r_j}\right)\frac{1}{s} + \frac{k_j}{r_j}\frac{1}{s-r_j} + \frac{k_j^*}{r_j^*}\frac{1}{s-r_j^*} \quad \text{(solving for the coefficients)}.$$

After some algebra, the inverse Laplace transform becomes

$$y_j(t) = 2\left[\mathrm{real}\left(\frac{k_j}{r_j}\right) e^{a_j t}\cos(\omega_j t) - \mathrm{imag}\left(\frac{k_j}{r_j}\right) e^{a_j t}\sin(\omega_j t) - \mathrm{real}\left(\frac{k_j}{r_j}\right)\right] u_{\mathrm{stp}}(t).$$

- If $a_j > 0$, the exponential term $e^{a_j t}$ grows unbounded as $t \to \infty$, causing $y(t)$ to blow up.
- If $a_j < 0$, the exponential term $e^{a_j t}$ decays to zero as $t \to \infty$.
- If $a_j = 0$, we replace the unit step input with another bounded input, $\sin(\omega_j t) u_{\mathrm{stp}}(t)$, where the frequency of the sinusoid, ω_j, is carefully chosen to match the natural frequency of the system, causing resonance.

  If $a_j = 0$, the poles are purely imaginary and the terms in the partial fraction expansion become

$$\frac{k_j}{s - \mathbf{i}\omega_j} + \frac{k_j^*}{s + \mathbf{i}\omega_j}.$$

  Combining these terms over the common denominator $(s^2 + \omega_j^2)$, the numerator becomes,

$$k_j(s + \mathbf{i}\omega_j) + k_j^*(s - \mathbf{i}\omega_j).$$

  Expanding the numerator,

$$k_j(s + \mathbf{i}\omega_j) + k_j^*(s - \mathbf{i}\omega_j) = (k_j + k_j^*)s + \mathbf{i}\omega_j(k_j - k_j^*).$$

$$k_j + k_j^* = 2\mathrm{real}\,(k_j)\,, \quad k_j - k_j^* = 2\,\mathbf{i}\,\mathrm{imag}\,(k_j)\,.$$

  Thus, the numerator simplifies to:

$$2\mathrm{real}\,(k_j)\, s + 2\mathrm{imag}\,(k_j)\, \omega_j.$$

The transfer function term becomes:

$$\frac{k_j}{s - \mathbf{i}\omega_j} + \frac{k_j^*}{s + \mathbf{i}\omega_j} = \frac{2\text{real}(k_j)\, s + 2\text{imag}(k_j)\,\omega_j}{s^2 + \omega_j^2}.$$

Now, consider the input $\sin(\omega_j t) u_{\text{stp}}(t)$, whose Laplace transform is:

$$\mathcal{L}\{\sin(\omega_j t) u_{\text{stp}}(t)\} = \frac{\omega_j}{s^2 + \omega_j^2}.$$

The output is:

$$y_j(t) = \mathcal{L}^{-1}\left\{\frac{(2\text{real}(k_j)\, s + 2\text{imag}(k_j)\,\omega_j)\omega_j}{(s^2 + \omega_j^2)^2}\right\}.$$

Split the numerator into two parts,

$$2\text{real}(k_j)\,\omega_j \frac{s}{(s^2 + \omega_j^2)^2} \quad \text{and} \quad 2\text{imag}(k_j)\,\omega_j \frac{\omega_j}{(s^2 + \omega_j^2)^2}.$$

Using standard inverse Laplace transforms,

- $\mathcal{L}^{-1}\left\{\frac{s}{(s^2+\omega_j^2)^2}\right\} = t\cos(\omega_j t)\, u_{\text{stp}}(t)$,
- $\mathcal{L}^{-1}\left\{\frac{\omega_j}{(s^2+\omega_j^2)^2}\right\} = t\sin(\omega_j t)\, u_{\text{stp}}(t)$.

The output becomes,

$$\begin{aligned} y_j(t) &= 2\,\text{real}(k_j)\,\omega_j\, t\cos(\omega_j t)\, u_{\text{stp}}(t) + 2\,\text{imag}(k_j)\,\omega_j\, t\sin(\omega_j t)\, u_{\text{stp}}(t) \\ &= \boldsymbol{t}\,(2\,\text{real}(k_j)\,\omega_j \cos(\omega_j t) + 2\,\text{imag}(k_j)\,\omega_j \sin(\omega_j t))\, u_{\text{stp}}(t). \end{aligned}$$

This clearly shows that the output grows linearly with time, indicating resonance and instability.

Thus, for BIBO Stability, all complex poles must have $\text{real}(r_j) = a_j < 0$.

Conclusion: For BIBO stability, the real parts of all poles must be strictly negative ($\text{real}(r_j) < 0$). Otherwise, it is always possible to excite the system with a bounded input that yields an unbounded output. ■

Proof of Prop. 10.36 (Steady-State Error for Step Inputs Applied to BIBO Stable Closed-Loop Systems)

Assume the forward path transfer function is $G(s) = \frac{N(s)}{D(s)}$ and the closed-loop system is BIBO stable. Then the steady-state error for a unit-step command is well-defined and is given by:

$$\mathbf{e_{SS}} = \lim_{s \to 0} \frac{\boldsymbol{D(s)}}{\boldsymbol{D(s) + N(s)}}. \tag{10.100}$$

Moreover, if $G(s)$ does not have a pole at the origin, then

$$e_{\text{SS}} = \frac{D(0)}{D(0) + N(0)} \neq 0,$$

while if $G(s)$ has one or more poles at the origin, then $D(0) = 0$ and the steady-state error is

$$e_{\text{SS}} = 0.$$

Proof: We give the proof for the case that the roots of $D(s) + N(s)$ are distinct. From the given information, $\frac{E(s)}{R(s)} = \frac{D(s)}{D(s)+N(s)}$. If we assume that $r(t) = u_{\text{stp}}(t)$, then $R(s) = \frac{1}{s}$ and $E(s) = \frac{D(s)}{D(s)+N(s)}\frac{1}{s}$. Let $r_1, r_2, \ldots, r_n$ be the roots of $D(s) + N(s)$. Then the partial fraction expansion of $E(s)$ becomes

$$E(s) = \frac{k_0}{s} + \sum_{i=1}^{n} \frac{k_i}{s - r_i},$$

and from Prop. 10.12,

$$k_0 = \lim_{s \to 0} \frac{s\, D(s)}{D(s) + N(s)} \frac{1}{s} = \frac{D(0)}{D(0) + N(0)}.$$

Taking the inverse Laplace transform term-by-term gives

$$e(t) = k_0 u_{\text{stp}}(t) + \sum_{i=1}^{n} k_i\, e^{r_i t} u_{\text{stp}}(t).$$

By BIBO stability of the closed-loop system, the real parts of the poles r_i are all negative. Hence,

$$\lim_{t\to\infty} e(t) = k_0 = \frac{D(0)}{D(0) + N(0)}$$

as claimed.

The poles of the open-loop transfer function, $\frac{Y(s)}{U(s)} = G(s) = \frac{N(s)}{D(s)}$, are the zeros of the closed-loop transfer function $\frac{E(s)}{R(s)} = \frac{D(s)}{D(s)+N(s)}$. Hence, $0 = k_0 = \frac{D(0)}{D(0)+N(0)}$ if, and only if, $D(0) = 0$. This completes the proof.

■

Index